Power Electronics in Renewable Energy Systems

Power Electronics in Renewable Energy Systems

Special Issue Editors

Teuvo Suntio
Tuomas Messo

MDPI • Basel • Beijing • Wuhan • Barcelona • Belgrade

MDPI

Special Issue Editors

Teuvo Suntio
Tampere University of Technology
Finland

Tuomas Messo
Tampere University of Technology
Finland

Editorial Office
MDPI
St. Alban-Anlage 66
4052 Basel, Switzerland

This is a reprint of articles from the Special Issue published online in the open access journal *Energies* (ISSN 1996-1073) from 2018 to 2019 (available at: https://www.mdpi.com/journal/energies/special_issues/power_electronics)

For citation purposes, cite each article independently as indicated on the article page online and as indicated below:

LastName, A.A.; LastName, B.B.; LastName, C.C. Article Title. *Journal Name* **Year**, *Article Number*, Page Range.

ISBN 978-3-03921-044-2 (Pbk)
ISBN 978-3-03921-045-9 (PDF)

Contents

About the Special Issue Editors

Teuvo Suntio received his PhD in electrical engineering from Helsinki University of Technology, Espoo, Finland, in 1992. He has worked in the power electronics-related industry for 22 years, and started his academic career in 1998. He is currently Professor in power electronics at Tampere University, Tampere, Finland. His research interests include the dynamic modeling and analysis of switched-mode converters in conventional and renewable energy applications.

Tuomas Messo received his PhD in electrical engineering from Tampere University of Technology, Tampere, Finland, in 2014. He is currently working at GE Grid Solutions, Tampere, Finland, and holds the position of Adjunct Professor in power electronics at Tampere University, Tampere, Finland. His research interests include the dynamic modeling, analysis, and control design of grid-connected three-phase power converters in renewable energy applications and microgrids.

Preface to "Power Electronics in Renewable Energy Systems"

The observed changes in weather conditions have accelerated the installation of renewable energy-based electricity systems around the world. Large-scale utilization of renewable energy sources in electricity production requires the use of power electronic converters to integrate the renewable energy systems into the power grids. This integration brings about certain challenges in terms of stability and robust performance of the power grids, which have to be solved before the wellbeing of the power grids can be guaranteed. This Special Issue of Energies aims to reveal the state-of-art in addressing interfacing problematics. According to the published papers, clear advancements have taken place, but the most critical issues remain unsolved. Direct power control with self-synchronizing synchronverters may be the most promising technique for solving the main stability problem, although many unsolved problems still persist. Another challenge in renewable energy production is the fluctuating nature of the available energy in renewable energy sources, which require utilization of stored energy to smooth the fluctuations. Different storage battery technologies are available, but their production may pose problems in the long term.

Teuvo Suntio, Tuomas Messo
Special Issue Editors

energies

MDPI

Editorial

Power Electronics in Renewable Energy Systems

Teuvo Suntio * and Tuomas Messo

Electrical Engineering Unit, Tampere University, Tampere 33720, Finland; tuomas.messo@tuni.fi
* Correspondence: teuvo.suntio@tuni.fi; Tel.: +358-400-828-431

Received: 13 May 2019; Accepted: 14 May 2019; Published: 15 May 2019

1. Introduction

Renewable energy-based generation of electrical energy is currently experiencing rapid growth in electrical grids. The dynamics of electrical grids are starting to change due to the large-scale integration of power electronic converters into the grid for facilitating the utilization of renewable energy. The main problem with the use of grid-connected power electronic converters is the negative incremental resistor behavior observed in their input and output terminal impedances, which makes the grid prone to harmonic stability problems that are observed nowadays more and more often. This problem is difficult or even impossible to remove because the converters have to be synchronized to the grid frequency, which actually creates the named output terminal-related problematic behavior. The input terminal-related problem is usually related to the output terminal feedback arrangement or to the grid synchronization actions. There are naturally many other problems, which can be related to the application of the renewable energy sources as an input source of the converters, and which can change their dynamic behavior profoundly.

The Special Issue of Energies "Power Electronics in Renewable Energy Systems" was intended to disseminate new promising methods to tackle the stability problems observed to take place in power grids, and to provide new information to support the understanding of the origin of those problems. The particular topics of interest in the original call for papers included, but were not limited to:

- Stability and modeling of large grid-connected PV and wind power plants;
- Dynamic modeling and control design of renewable energy converters in grid feeding, supporting, and forming modes;
- Impedance-based grid interaction studies;
- Issues related to control, stability, diagnostics and interfacing of energy storage in renewable energy systems;
- Voltage and frequency control of grids with high penetration of renewable distributed generation.

This special issue of Energies contains four invited submissions [1–4], three review articles [5–7], and twenty-three research articles [8–30] covering different topics in renewable energy systems. The authors' geographical distribution is:

China (14); Finland (4); Korea (2); The Netherlands (1); Japan (1); Israel (1); USA (1); Canada (1); India (1); Jordan (1); Malaysia (1); Italy (1); UK (1).

We thank the editorial staff and reviewers for their great efforts and help during the process.

2. Brief Overview of the Contributions to This Special Issue

The main contributions of the paper are briefly reviewed in the following subsections. The first subsection reviews the invited papers, the second subsection reviews the review papers, and the third subsection reviews the article-type papers that are not reviewed in first subsection. No topic-specific classification is applied.

2.1. Invited Papers

Suntio et al. [1] introduce the full picture of the source and load interaction formulations covering typical three-phase grid connected inverters, which are applied in renewable energy applications. The complexity of the analyses is obvious. The paper contains both simulated and experimental evidence to support the theoretical findings in the paper. It is clearly demonstrated that omitting the cross-coupling elements in the associated impedances will lead to very poor accuracy of the interaction analyses. Messo et al. [2] demonstrate that it is possible to implement an emulator for a grid-connected three-phase inverter by applying a high-bandwidth amplifier, which can be used effectively to study the dynamic behavior of AC microgrids. Sun et al. [3] study the behavior of the phase locked loop (PLL) of the grid-connected power converter during a phenomenon known as rate of change of frequency (ROCOF) and its influence on the stability of the grid. Such a phenomenon has been observed to take place in the Bonaire Island power grid as a consequence of a grid fault. According to the study, the PLL control bandwidth has a crucial role in the well-being of the power grid. Amer et al. [4] studied the maximum power point (MPP) tracking process, where the perturbation frequency was adaptive instead of the step size to ensure a fast and trouble-free tracking process. The demonstrations show that the proposed technique works well.

2.2. Review Papers

One of the review papers [1] was already reviewed in Section 2.1. Suntio [5] introduces the small-signal modeling and analysis of a peak-current-mode (PCM) controlled buck converter, which operates in discontinuous conduction mode (DCM). The modeling method is introduced already in 2001 but the given models are load-resistor affected and do not contain the effect of circuit parasitic elements. The paper shows that the load resistor hides the real location of the unstable pole, which is usually assumed to appear at the output–input voltage ratio of 2/3, but it can appear at the output–input voltage ratio of 1/2. It is also observed that the circuit elements have a significant contribution to the dynamic behavior as well. Tareen et al. [6] compare the use of static compensator (STATCOM) and active power filter (APF) in tackling the power quality issues in the case of high penetration of renewable energy sources in the power grid. The paper provides a comprehensive survey of the related topics and a wide list of key finds, which cannot be presented briefly in this overview.

2.3. Article Papers

Wang et al. [7] proposes a modified self-synchronized synchronverter, which works better in an unbalanced grid compared to the conventional self-synchronized synchronverters. The main benefit of the synchronverters is that there is no need for PLL function, which would eliminate the negative incremental resistor phenomenon in the output impedance of the grid-connected converter. Rizqiawan et al. [8] introduce the development of grid-connected inverter modules intended for teaching microgrid issues to electrical engineering students. Liu et al. [9] introduce issues related to the concept known as Energy Internet, where microgrid control is implemented based on the internet by utilizing the concept of an energy router. In this study, the primary energy sources are considered to be a photovoltaic array, an energy storage battery, and the battery of an electric vehicle. Opila et al. [10] introduce virtual oscillator control of voltage-sourced and current-controlled power converters. The paper is highly theoretical and lacks a direct practical connection to the real world. Anuradha et al. [11] introduce the design and analysis of a non-isolated three-port single ended primary inductance converter (SEPIC) for renewable energy source applications, where the input section of the converter is used as a modular section for interfacing different renewable energy sources. The problem in the paper is that the PV interfacing is not considered correctly, which is still quite a common issue in renewable energy interfacing studies. Park et al. [12] introduce the methods to control the speed of a turbine generator, where the energy of the system is extracted from different heat sources. The control algorithm indirectly estimates the required speed of the generator. The rest of the converter system is similar to the full-power converter systems used in wind power interfacing. Yan et al. [13] study the inertia and damping characteristics of a doubly-fed induction generator (DFIC) during a grid fault. The paper proposes methods to control the frequency in such a manner that the system stability is improved. Suntio [14] provides small-signal models for a peak current mode-controlled boost converter, which operates in discontinuous operation mode with all the power stage parasitic elements included. It is shown that the modeling technique developed in the early 2000s yields accurate models for the boost converter as well when the load resistor effect is removed. Yang et al. [15] study the oscillation phenomenon observed to take place between the parallel-connected grid-tied inverters in weak grid conditions. The damping method is based on adding two virtual impedances in series and in parallel with the original output impedance of the inverter. The practical experiments show that the proposed technique works. Xu et al. [16] introduce methods to provide grid-supporting functions in photovoltaic systems by utilizing battery energy storage to create a virtual synchronous generator. The power system is claimed to be able to operate without phase locked loop grid synchronizing. The concept is validated with simulations. Hu et al. [17] study the coordinated control of virtual synchronous generators. The developed strategy is validated by simulations. Liu et al. [18] study the ultra-short-term wind power prediction methods, which are based on multivariate phase–space reconstruction and linear regression. The developed method is shown to be more accurate than earlier developed methods. Li et al. [19] study the design of phase lock loop-based grid synchronizing aiming for fast transient behavior. The proposed concept is based on the application of adaptive notch filtering and moving average filtering as the inner loop of the phase locked loop. The experimental validation proves the improvement of the proposed technique. Salgado-Herrera et al. [20] study the total harmonic distortion (THD) in wind energy systems showing that the utilization of an active front-end converter to provide the DC-link voltage will reduce the overall THD. The concept is verified by simulation. Hu et al. [21] propose an improved droop control method based on the conventional droop control with a washout filter controller. The proposed method is validated by utilizing the hardware-in-the-loop (HIL) method. Merabet [22] studies the application of adaptive sliding mode speed control of a wind energy generator. The proposed technique is validated experimentally by utilizing a small-scale prototype system. Miceli et al. [23] study harmonic mitigation by means of computational methods in a single-phase five-level cascaded H-bridge multilevel inverter. The developed method is experimentally validated with a small-scale prototype. Tran et al. [24] study the control design of a grid-connected inverter under distorted grid conditions based on the linear-quadratic regulator technique. The proposed control technique is based on the internal model control principle. The experimental validation shows that the proposed technique works well. Li et al. [25] study the effect of adaptive resonant controllers on the stability of the grid-connected inverters under weak grid conditions. The validation of the studies is performed experimentally. The paper provides useful tips for implementing adaptive resonant controllers to avoid problems. Yan et al. [26] study the control methods of grid-connected inverters to make them mimic the characteristics of a synchronous generator. The proposed techniques are validated by simulations, which do not necessarily convince the readers. Yan et al. [27] study the adaptive maximum power point tracking-based control to create the properties of a synchronous generator. The reader may have problems understanding the proposed techniques because the authors have not explicitly specified in which operation mode the system is working. Dalala et al. [28] propose an algorithm for thermoelectric generators to track the maximum power point (MPP). The method is based on indirectly detecting the open-circuit voltage and estimating short circuit current, which are then used for tracking the MPP. The experimental waveforms show that the proposed technique tracks the MPP very quickly. Wang et al. [29] study the energy management in a micro-grid based on demand response. An optimization strategy is developed for minimizing the operating costs. The strategy is tested on a real case study. Liang et al. [30] study the balancing of the charges of embedded storage batteries in series-connected switching modules. The proposed control strategy is experimentally tested and shown to work well.

2.4. Discussions

The published papers represent the current main topics related to renewable energy. The lack of full understanding of the dynamics of the converters, which are applied in a renewable energy system, still dominates the discussions in this field even if hundreds of papers have already been published where the true nature has been explicitly presented. The lack of experimental validation will also usually reduce the acceptance of the information.

Conflicts of Interest: The authors declare no conflict of interest.

References

1. Suntio, T.; Messo, T.; Berg, M.; Alenius, H.; Reinikka, T.; Luhtala, R.; Zenger, K. Impedance-based interactions in grid-tied three-phase inverters in renewable energy applications. *Energies* **2019**, *17*, 464. [CrossRef]
2. Messo, T.; Luhtala, R.; Roinila, T.; de Jong, E.; Scharrenberg, R.; Calddognetto, T.; Mattavelli, P.; Sun, Y.; Fabian, A. Using high-bandwidth voltage amplifier to emulate grid-following inverter for ac microgrid dynamics studies. *Energies* **2019**, *12*, 379. [CrossRef]
3. Sun, Y.; de Jong, E.; Wang, X.; Yang, D.; Blaabjerg, F.; Cuk, V.; Cobben, J. The impact of PLL dynamics on the low inertia power grid: A case study of Bonaire Island power system. *Energies* **2019**, *12*, 1259. [CrossRef]
4. Amer, E.; Kuperman, A.; Suntio, T. Direct fixed-step power point tracking algorithms with adaptive perturbation frequency. *Energies* **2019**, *12*, 399. [CrossRef]
5. Suntio, T. Dynamic modeling and analysis of PCM-controlled DCM-operating buck converters—A reexamination. *Energies* **2018**, *11*, 1267. [CrossRef]
6. Tareen, W.; Aamir, M.; Mekhilef, S.; Nakaoka, M.; Seyedmahmoudian, M.; Horan, B.; Memon, M.; Baig, N. Mitigation of power quality issues due to high penetration of renewable energy sources in Electric gride systems using three-phase APF/STATCOM technologies: A review. *Energies* **2018**, *11*, 1491. [CrossRef]
7. Wang, X.; Chen, L.; Sun, D.; Zhang, L.; Nian, H. A modified self-synchronized synchronverter in unbalanced power grids with balanced currents and restrained power ripples. *Energies* **2019**, *12*, 923. [CrossRef]
8. Rizqiawan, A.; Hadi, P.; Fujita, G. Development of grid-connected inverter experiment modules for microgrid learning. *Energies* **2019**, *12*, 476. [CrossRef]
9. Liu, Y.; Li, Y.; Liang, H.; He, J.; Cui, H. Energy routing control strategy for integrated microgrids including photovoltaic, battery-energy storage and electric vehicles. *Energies* **2019**, *12*, 302. [CrossRef]
10. Opila, D.; Kintzley, K.; Shabshab, S.; Phillips, S. Virtual oscillator control of equivalent voltage-sourced and current-controlled power converters. *Energies* **2019**, *12*, 298. [CrossRef]
11. Anuradha, C.; Chellamma, N.; Maqsood, S.; Viljayalakshmi, S. Design and analysis of non-isolated three-port SEPIC converter for integrating renewable energy sources. *Energies* **2019**, *12*, 221. [CrossRef]
12. Park, H.-S.; Heo, H.-J.; Choi, B.-S.; Kim, K.-C.; Kim, J.-M. Speed control for turbine-generator of ORC power generation system and experimental implementation. *Energies* **2019**, *12*, 200. [CrossRef]
13. Yan, X.; Song, Z.; Xu, Y.; Sun, Y.; Wang, Z.; Sun, X. Study of inertia and damping characteristics of doubly fed induction generators and improved additional frequency control strategy. *Energies* **2019**, *12*, 38. [CrossRef]
14. Suntio, T. Modeling and analysis of a PCM-controlled boost converter designed to operate in DCM. *Energies* **2019**, *12*, 4. [CrossRef]
15. Yang, L.; Chen, Y.; Wang, H.; Luo, A.; Huai, K. Oscillation suppression method by two notch filters for parallel inverters under weak grid. *Energies* **2018**, *11*, 3441. [CrossRef]
16. Xu, H.; Su, J.; Liu, N.; Shi, Y. A grid-supporting photovoltaic system implemented by a VSG with energy storage. *Energies* **2018**, *11*, 3152. [CrossRef]
17. Hu, P.; Chen, H.; Cao, K.; Hu, Y.; Kai, D.; Chen, L.; Wang, Y. Coordinated control of multiple virtual synchronous generators in mitigating power oscillation. *Energies* **2018**, *11*, 2788. [CrossRef]
18. Liu, R.; Peng, M.; Xiao, X. Ultra-short-term wind power prediction based on multivariate phase space reconstruction and multivariate linear regression. *Energies* **2018**, *11*, 2763. [CrossRef]
19. Li, Y.; Yang, J.; Wang, H.; Ge, W.; Ma, Y. Leveraging hybrid filter for improving quasi-type-1 phase locked loop targeting fast transient response. *Energies* **2018**, *11*, 2472. [CrossRef]

20. Salgado-Herrera, N.; Campos-Gaona, D.; Anaya-Lara, O.; Medina-Rios, A.; Tapia-Sánchez, R.; Rodríguez-Rodríguez, J. THD reduction in wind energy system using type-4 wind turbine/PMSG applying the active front-end parallel operation. *Energies* **2018**, *11*, 2458. [CrossRef]

21. Hu, Y.; Wei, W. Improved droop control with washout filter. *Energies* **2018**, *11*, 2415. [CrossRef]

22. Merabet, A. Adaptive sliding mode speed control for wind energy experimental system. *Energies* **2018**, *11*, 2238. [CrossRef]

23. Miceli, R.; Schettino, G.; Viola, F. A novel computational approach for harmonic mitigation in PV systems with single-phase five-level CHBMI. *Energies* **2018**, *11*, 2100. [CrossRef]

24. Tran, T.; Yoon, S.-J.; Kim, K.-H. An LQR-based controller design for an LCL-filtered grid-connected inverter in discrete-time state-space under distorted grid environment. *Energies* **2018**, *11*, 2062. [CrossRef]

25. Li, X.; Lin, H. Stability analysis of grid-connected converters with different implementations of adaptive PR controllers under weak grid conditions. *Energies* **2018**, *11*, 2004. [CrossRef]

26. Yan, X.; Zhang, X.; Zhang, B.; Jia, Z.; Li, T.; Wu, M.; Jiang, J. A novel two-stage photovoltaic grid-connected inverter voltage-type control method with failure zone characteristics. *Energies* **2018**, *11*, 1865. [CrossRef]

27. Yan, X.; Li, J.; Wang, L.; Chao, S.; Lie, T.; Lv, Z.; Wu, M. Adaptive-MPPT-based control of improved photovoltaic virtual synchronous generators. *Energies* **2018**, *11*, 1834. [CrossRef]

28. Dalala, Z.; Saadeh, O.; Bdour, M.; Zahid, Z. A new maximum power point tracking (MPPT) algorithm for thermoelectric generators with reduced voltage sensors count control. *Energies* **2018**, *11*, 1826. [CrossRef]

29. Wang, Y.; Huang, Y.; Wang, Y.; Yu, H.; Li, R.; Song, S. Energy management for smart multi-energy complementary microgrid in presence of demand response. *Energies* **2018**, *11*, 974. [CrossRef]

30. Liang, H.; Guo, L.; Song, J.; Yang, Y.; Zhang, W.; Qi, H. State-of-charge balancing control of a modular multilevel converter with an integrated battery energy storage. *Energies* **2018**, *11*, 873. [CrossRef]

energies

MDPI

Review

Impedance-Based Interactions in Grid-Tied Three-Phase Inverters in Renewable Energy Applications

Teuvo Suntio [1,*], Tuomas Messo [1], Matias Berg [1], Henrik Alenius [1], Tommi Reinikka [1], Roni Luhtala [2] and Kai Zenger [3]

[1] Laboratory of Electrical Engineering, Tampere University, 33720 Tampere, Finland; tuomas.messo@tuni.fi (T.M.); matias.berg@tuni.fi (M.B.); henrik.alenius@tuni.fi (H.A.); tommi.reinikka@tuni.fi (T.R.)

[2] Laboratory of Automation and Hydraulics, Tampere University, 33720 Tampere, Finland; roni.luhtala@tuni.fi

[3] Department of Electrical Engineering and Automation, Aalto University, 02150 Espoo, Finland; kai.zenger@aalto.fi

* Correspondence: teuvo.suntio@tuni.fi; Tel.: +358-400-828-431

Received: 12 December 2018; Accepted: 28 January 2019; Published: 31 January 2019

Abstract: Impedance-ratio-based interaction analyses in terms of stability and performance of DC-DC converters is well established. Similar methods are applied to grid-connected three-phase converters as well, but the multivariable nature of the converters and the grid makes these analyses very complex. This paper surveys the state of the interaction analyses in the grid-connected three-phase converters, which are used in renewable-energy applications. The surveys show clearly that the impedance-ratio-based stability assessment are usually performed neglecting the cross-couplings between the impedance elements for reducing the complexity of the analyses. In addition, the interactions, which affect the transient performance, are not treated usually at all due to the missing of the corresponding analytic formulations. This paper introduces the missing formulations as well as explicitly showing that the cross-couplings of the impedance elements have to be taken into account for the stability assessment to be valid. In addition, this paper shows that the most accurate stability information can be obtained by means of the determinant related to the associated multivariable impedance ratio. The theoretical findings are also validated by extensive experimental measurements.

Keywords: source and load impedance; transient dynamics; stability; grid synchronization; power electronics; power grid

---

1. Introduction

The negative-incremental-resistor oscillations were observed to take place, in practice, already in the early 1970s when an LC-type input filter was connected at the input terminal of regulated converters as reported in References [1,2]. The development of the dynamic modeling method known as state-space averaging (SSA) in the early 1970s [3–5] enabled the theoretical studies of the origin of the input-filter interactions, which were published in the mid 1970s [6,7] by Middlebrook. He stated later that he applied the extra-element-theorem-based (EET) method [8,9] when developing the input-filter-design rules in References [6,7] for the cascaded input-filter-converter system. According to the EET method, the source or load-system-affected transfer function $G_{\mathrm{org}}^{\mathrm{S/L}}(s)$ of the converter can be given by

$$G_{\mathrm{org}}^{\mathrm{S/L}} = \frac{1 + Z_{\mathrm{n}-1}Y_{\mathrm{n}-2}}{1 + Z_{\mathrm{d}-1}Y_{\mathrm{d}-2}} \times G_{\mathrm{org}} \tag{1}$$

where $G_{org}(s)$ denotes the original or unterminated transfer function of the converter, $Z_{n-1}Y_{n-2}$ denotes the impedance-admittance product of the numerator polynomial, and $Z_{d-1}Y_{d-2}$ denotes the impedance-admittance product of the denominator polynomial, respectively. The formulation in Equation (1) is very useful, because it defines automatically the correct order of the impedance-like elements in the numerator and denominator impedance-admittance products, to perform the source/load analysis in a correct manner as instructed in References [10–12] and implied in Figure 1. The most crucial factor in the analysis of the cascaded systems is to recognize that the duality must be valid at the interface between the upstream and downstream subsystems. This means that the only valid source-sink pairs are Figure 1a,d as well as Figure 1b,c, respectively. The cascaded system will not be proper with the other source-sink combinations at the interface between the subsystems because of violating Kirchhoff's laws.

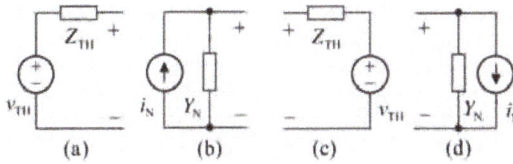

Figure 1. The source/sink equivalent circuits: (**a**) Thevenin's source, (**b**) Norton's source, (**c**) Thevenin's sink, and (**d**) Norton's sink.

As Equation (1) indicates, the theoretical formulation in Equation (1) does not contain impedance ratios, but it is sometimes easier to understand the behavior of the impedance-admittance products, when the product is considered as an impedance ratio as in References [6,7]. The minor-loop gain launched by Middlebrook in References [6,7] actually denotes the denominator product $Z_{d-1}Y_{d-2}$ as minor-loop gain, which can be given equally as Z_{d-1}/Z_{d-2}, because $Z_{d-2} = Y_{d-2}^{-1}$. The minor-loop gain actually equals $Z_{TH}Y_N$ according to Figure 1, which indicates explicitly that the numerator impedance of the impedance ratio (i.e., minor-loop gain) always equals the internal impedance of the voltage-type subsystem, and the denominator impedance equals the internal impedance of the current-type subsystem, respectively [10].

Stability of the cascaded system can be assessed based on $Z_{TH}Y_N$ by applying Nyquist stability criterion [13], because $(1 + Z_{TH}Y_N)^{-1}$ forms an impedance-based sensitivity function similarly as $(1 + L_x)^{-1}$ in control engineering [14,15], where L_x denotes the feedback-loop gain. If the phase or gain margin of the feedback loop is low then the sensitivity function will exhibit peaking, which affects the corresponding closed-loop transfer function, and it may cause deterioration in transient response or may make the converter more prone to instability [14,15]. The similar phenomena will take place also in case of $(1 + Z_{TH}Y_N)^{-1}$.

The impedance-admittance product $(Z_{n-1}Y_{n-2})$ of the numerator polynomial is not directly related to the system stability similarly as $Z_{TH}Y_N$ is. One of the elements in $Z_{n-1}Y_{n-2}$ equals either Z_{TH} or Y_N depending on the type of the source/load system (cf. Figure 1), and the other element (i.e., Z_{n-1} or Y_{n-2}) is a certain special impedance-like parameter, which will be introduced for the grid-tied three-phase inverters in Section 2. The special parameters of the DC-DC converters are defined in general and given also explicitly for a number of converters in Reference [12]. The impedance-based interactions via the numerator polynomial in Equation (1) may affect the control-related transfer functions or the internal input or output impedances that may deteriorate the transient behavior of the converter as demonstrated in Reference [11].

The input-filter-design rules, introduced in References [6,7], have been later extended to apply for stability and performance assessment in arbitrary systems as well, where the robust stability and performance are defined in the form of forbidden regions out of which the minor-loop gain $(Z_{TH}Y_N)$ should stay for the robust stability to exist [10,16–19] as illustrated in Figure 2. The forbidden region induced by the input-filter-design rules [6,7] is assumed to be outside of the circle, which has the radius of inverse of the gain margin and the center at the origin (cf. Figure 2, Middlebrook). This forbidden region is deemed to be excessively conservative for general usage, and therefore, reduced forbidden regions are proposed as discussed in Reference [17]. In regard to the input-filter design, Middlebrook's forbidden region is the only possible region for guaranteeing the stability of the cascaded input-filter-converter system (cf. Reference [10]). The smallest forbidden region is proposed in Reference [10] as a circle having the center at the critical point (−1,0) and the radius of inverse of the allowed maximum peaking in the corresponding sensitivity and complementary sensitivity functions as described in detail in Reference [10] (cf. Figure 2, MPC). In principle, the forbidden regions can be applied to the grid-connected three-phase converters and systems as well [20]. The multivariable nature of the grid-tied three-phase converters makes the performance and stability assessment more challenging [20–26], when the inverter-grid stability assessment has to be performed by applying generalized Nyquist stability criterion [27,28] instead of the simple Nyquist stability criterion [13].

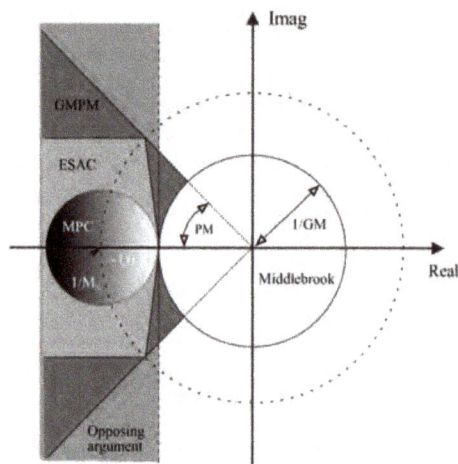

Figure 2. The evolution of forbidden regions according to Reference [11], where GMPM denotes gain and phase margin [16], ESAC Energy Systems Analysis Consortium [17], Opposing argument as the limited real part [18], and MPC maximum peak criteria [11].

The origin of the stability problems in grid-connected systems is usually the negative-incremental-resistor-like behavior at the input or output impedance of the three-phase converter [29–31] similarly as in DC-DC converters discussed in References [1,2]. In three-phase converters, the negative-incremental-resistor-like behavior is either the consequence of the grid synchronization [29–35] or the feedback control from the output-terminal variables [36–43]. In case of grid synchronization, the instability can be mitigated in some extends by lowering the control bandwidth of the phase locked loop (PLL) adaptively when the grid impedance is increasing as discussed in Reference [44]. The grid-connected converters synchronize themselves at the voltage of the connection point known usually as the point of common coupling (PCC). If the grid impedance is low, then the PCC voltage is in phase with the grid voltage. In case of weak grid [45–54], the grid impedance can be rather high, which changes the phase of the PCC voltage to deviate from the phase of the grid voltage depending on the level of the power supplied by the inverter [51–53]. In practice, this means that the inverter seems to supply reactive power into the grid even if the inverter output power is pure real power at PCC [53]. The total apparent power in grid naturally corresponds to the inverter output power. At the point, where the grid power is fully reactive power, the inverter becomes unstable due to trying to take power from the grid, but the switch control scheme of the converter bridge does not usually allow it. In case of grid-connected rectifiers, the same phenomenon will take place, but the direction of the power flow is reversed compared to the inverters [53]. The problem can be solved by modifying the PLL feedback controller in such a manner that its reference includes the information from the grid impedance seen from the PCC and the level of grid current [51]. In practice, this means that the grid impedance shall be measured online for modifying the PLL reference [53].

In grid-connected applications, the instability may take place, in principle, at any frequency depending on the behavior of the grid impedance, where the electrical resonances are of interest due to providing zero-degree phase behavior of the grid impedance at the resonant frequency [12,55,56]. The negative-incremental-resistance-like behavior of the grid-connected converter may reduce the damping of the grid and thus making it prone to electrical resonances, which may cause harmonic currents to appear [57–60]. The reason for the appearing of harmonic currents in PV applications may be also the inability of the inverter to maintain proper output-current waveforms under low-irradiance conditions in cloudy days, and especially, during the mornings and evenings [61,62].

The main objectives of this paper are to survey the state-of-art in the impedance-based interactions in grid-tied three-phase inverters and to introduce the implicit impedance-like parameters also for the grid-tied three phase converters for facilitating the better understanding of these phenomena, which are observed to take place in practical applications as well [63]. The main outcomes of the paper are the explicit proving that the cross-couplings of the impedance elements have to be taken into account for obtaining valid information on the state of stability as well as that the determinant of the impedance-ratio-based multivariable characteristic polynomial provides the most accurate information on the state of stability.

The rest of the paper is organized as follows: An introduction to the source and load-effect formulations for the grid-tied three-phase inverters are given in Section 2. Experimental and simulated evidence supporting the theoretical findings are given in Section 3, and the conclusions are finally drawn in Section 4.

2. Theoretical Formulation of Source and Load-Impedance Interactions

The grid-connected renewable energy systems have to be able to operate in grid-feeding, grid-supporting, and grid-forming modes [64,65] as well as performing smooth transfer between the grid-feeding and grid-forming modes of operation [66]. In grid-feeding and grid-supporting modes, the outmost feedback loops are taken from the input terminal of the converter (cf. Figure 3a, outer loop) [65,67]. In grid-forming mode, all the feedback loops are taken from the output terminal of the converter (cf. Figure 3, inner and outer loops) [64,65]. The photovoltaic (PVG) and wind energy (WEG) generators are known to be internally current-type input sources for power electronic converters [68–71]. Therefore, the input-terminal feedback is taken from the input-terminal voltage as illustrated in Figure 3a. Both of the renewable energy sources are known to be maximum-power-limited sources having one (i.e., WEG, PVG)) [72,73] or more (PVG) [74,75] maximum power points (MPP) at their power-voltage (PV) curves. In grid-feeding mode, when the feedback control is taken from the input-terminal variable [73], the input impedance of the converter usually stays passive, and therefore, the well-designed cascaded system composing of the energy source and the converter is stable. In grid-forming mode, when the outmost feedback is taken from the output-terminal variable [27,41–43,65,73], the input impedance of the converter will exhibit negative incremental-resistor-like characteristics [12]. In this case, the instability will take place, when the operating point of the converter enters into the MPP of the input energy source [69–77].

In order to understand the impedance-originated stability and performance-interaction phenomena, the analytical formulation of the corresponding source and load interactions will be derived in the subsequent sections based on the transfer functions of the associated converters and the interaction formulation proposed by Middlebrook in References [6,7]. In case of three-phase converters, the special impedance-like parameters are also of multivariable nature having direct (d) and quadrature (q) components as well as cross-coupling terms between the d and q components. In renewable energy applications, the input source of the converters is usually DC voltage, and therefore, the source output and converter input impedances are not of multivariable nature. Consequently, the source-interaction-related special impedance-like parameters can be solved quite easily in analytical forms as well. The three-phase grid connection means that all the load-interaction-related special impedance-like parameters will be of multivariable nature. The solving of the special impedance-like parameters in analytical forms will be very complicated due to the high complexity of the multivariable impedance-based sensitivity functions. Therefore, we will present those special parameters by neglecting the cross-coupling terms for giving the reader an idea of the nature of the load-interaction-related special parameters. The source and load-affected transfer functions can be solved easily in numerical forms by using, for example, Matlab^{TM} as demonstrated in [12].

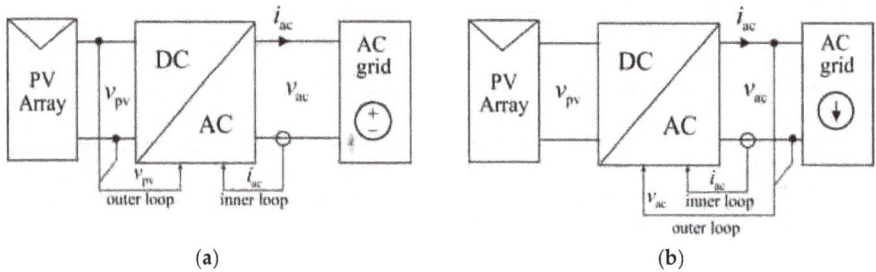

(a)   (b)

Figure 3. Operation modes of single-stage grid-connected PV energy system: (**a**) grid-feeding/ supporting mode, and (**b**) grid-forming mode.

2.1. Three-Phase VF-VO Inverter

The voltage-fed (VF) voltage-output (VO) inverter is used in the grid-forming-mode renewable energy systems (cf. Figure 3b), where the inverter takes care both the grid voltage and frequency. The loads connected to the grid will determine the power level and type of the inverter. In renewable energy applications, the input-terminal source is a voltage-type source, and the output terminal source is a three-phase current-type sink, respectively.

The transfer functions of the three-phase converters are given in synchronous reference frame or dq frame due to the multivariable nature of the converter [12,22] as

$$
\begin{bmatrix} \hat{i}_{in} \\ \hat{v}_{o-d} \\ \hat{v}_{o-q} \end{bmatrix} = \begin{bmatrix} Y_{in} & T_{oi-d} & T_{oi-q} & G_{ci-d} & G_{ci-q} \\ G_{io-d} & -Z_{o-d} & -Z_{o-qd} & G_{co-d} & G_{co-qd} \\ G_{io-q} & -Z_{o-dq} & -Z_{o-q} & G_{co-dq} & G_{co-q} \end{bmatrix} \begin{bmatrix} \hat{v}_{in} \\ \hat{i}_{o-d} \\ \hat{i}_{o-q} \\ \hat{d}_d \\ \hat{d}_q \end{bmatrix}
\tag{2}
$$

where the right-side vector contains the input variables, and the left-side vector the output variables, respectively. Equation (2) denotes actually a set of simultaneous equations, which define explicitly the transfer functions in the transfer function matrix. As an example, the element (1,1) of the transfer function known as the input admittance (Y_{in}) of the converter describes the relation $\hat{i}_{in}/\hat{v}_{in}$. The other transfer functions can be expressed in a similar manner following the rule dictated by $\hat{i}_{in}/\hat{v}_{in}$. The hat over the input and output variables denotes that the variables are small-signal variables.

Equation (2) can be formulated into a multivariable form according to Reference [12] as shown in Equations (3) and (4). The minus sign in front of the element (2,2) in Equation (4) is the consequence of the selected output-terminal-current direction (cf. Figure 3). The multivariable-form transfer functions in Equation (4) can be represented also in the form of multivariable linear circuit as given in Figure 4. The linear circuit can be utilized effectively to solve the source and load-affected transfer functions in multivariable form.

$$\begin{bmatrix} \begin{bmatrix} \hat{i}_{in} \\ 0 \end{bmatrix}^{\hat{i}_{in}} \\ \hat{v}_{o-d} \\ \hat{v}_{o-q} \end{bmatrix} = \begin{bmatrix} \begin{bmatrix} Y_{in} & 0 \\ 0 & 0 \end{bmatrix}^{Y_{in}} \\ \begin{bmatrix} G_{io-d} & 0 \\ G_{io-d} & 0 \end{bmatrix}^{G_{io}} \end{bmatrix} - \begin{bmatrix} \begin{bmatrix} T_{oi-d} & T_{oi-q} \\ 0 & 0 \end{bmatrix}^{T_{oi}} \\ \begin{bmatrix} Z_{o-d} & Z_{o-qd} \\ Z_{o-dq} & Z_{o-q} \end{bmatrix}^{Z_o} \end{bmatrix} \begin{bmatrix} \begin{bmatrix} G_{ci-d} & G_{ci-q} \\ 0 & 0 \end{bmatrix}^{G_{ci}} \\ \begin{bmatrix} G_{co-d} & G_{co-qd} \\ G_{co-dq} & G_{co-q} \end{bmatrix}^{G_{co}} \end{bmatrix} \begin{bmatrix} \begin{bmatrix} \hat{v}_{in} \\ 0 \end{bmatrix}^{\hat{v}_{in}} \\ \begin{bmatrix} \hat{i}_{o-d} \\ \hat{i}_{o-q} \end{bmatrix}^{\hat{i}_o} \\ \begin{bmatrix} \hat{d}_d \\ \hat{d}_q \end{bmatrix}^{\hat{d}} \end{bmatrix} \tag{3}$$

$$\begin{bmatrix} \hat{i}_{in} \\ \hat{v}_o \end{bmatrix} = \begin{bmatrix} Y_{in} & T_{oi} & G_{ci} \\ G_{io} & -Z_o & G_{co} \end{bmatrix} \begin{bmatrix} \hat{v}_{in} \\ \hat{i}_o \\ \hat{d} \end{bmatrix} \tag{4}$$

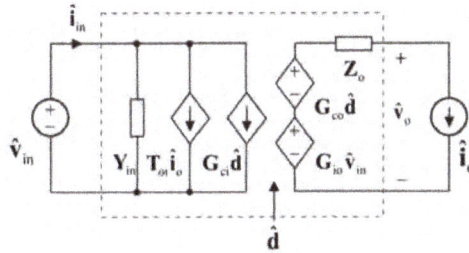

Figure 4. Linear multi-variable equivalent circuit of three-phase VF-VO DC-AC converter.

2.1.1. Source-Affected Transfer Functions

Figure 5 shows the corresponding equivalent circuit of Figure 4, when the converter is supplied by a non-ideal input source. According to Figure 5, the original input voltage $\hat{v}_{in}$ is changed as

$$\hat{v}_{in} = \hat{v}_{inS} - Z_{inS}\hat{i}_{in} \tag{5}$$

and

$$\hat{v}_{in} = [I + Z_{inS}Y_{in}]^{-1}\hat{v}_{inS} - [I + Z_{inS}Y_{in}]^{-1}Z_{inS}T_{oi}\hat{i}_o - [I + Z_{inS}Y_{in}]^{-1}Z_{inS}G_{ci}\hat{d} \tag{6}$$

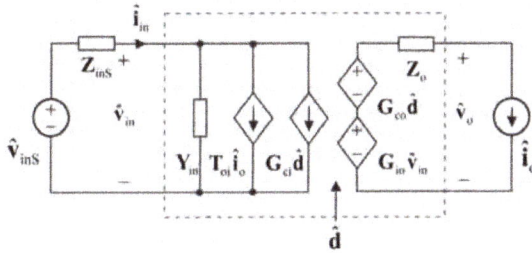

Figure 5. Linear multi-variable equivalent circuit of three-phase VF-VO DC-AC converter with a non-ideal source.

The source-affected set of multi-variable transfer functions can be obtained by replacing $\hat{v}_{in}$ in the upper row of Equation (4) by Equation (5) and in the bottom row of Equation (4) by Equation (6), respectively. These procedures yield the set of source-affected transfer functions as

$$\begin{bmatrix} \hat{i}_{in} \\ \hat{v}_o \end{bmatrix} = \begin{bmatrix} Y_{in}^S & T_{oi}^S & G_{ci}^S \\ G_{io}^S & -Z_o^S & G_{co}^S \end{bmatrix} \begin{bmatrix} \hat{v}_{inS} \\ \hat{i}_o \\ \hat{d} \end{bmatrix}$$

$$\begin{bmatrix} \hat{i}_{in} \\ \hat{v}_o \end{bmatrix} = \begin{bmatrix} [I + Y_{in}Z_{inS}]^{-1}Y_{in} & [I + Y_{in}Z_{inS}]^{-1}T_{oi} & [I + Y_{in}Z_{inS}]^{-1}G_{ci} \\ G_{io}[I + Z_{inS}Y_{in}]^{-1} & -(Z_o + G_{io}[I + Z_{inS}Y_{in}]^{-1}Z_{inS}T_{oi}) & G_{co} - G_{io}[I + Z_{inS}Y_{in}]^{-1}Z_{inS}G_{ci} \end{bmatrix} \begin{bmatrix} \hat{v}_{inS} \\ \hat{i}_o \\ \hat{d} \end{bmatrix} \tag{7}$$

Equation (7) shows that the input dynamics of the inverter (i.e., the elements (1,1), (1,2) and (1,3) in Equation (4)) and the input-to-output transfer function G_{io} (i.e., the element (2,1) in Equation (4)) are affected only via the impedance-based sensitivity function $[\mathbf{I} + \mathbf{Y}_{in}\mathbf{Z}_{inS}]^{-1}$ or $[\mathbf{I} + \mathbf{Z}_{inS}\mathbf{Y}_{in}]^{-1}$. It may be also obvious that the stability of the cascaded source-converter system can be assessed by means of $[\mathbf{I} + \mathbf{Y}_{in}\mathbf{Z}_{inS}]^{-1}$ and $[\mathbf{I} + \mathbf{Z}_{inS}\mathbf{Y}_{in}]^{-1}$ by applying the generalized Nyquist stability criterion [27,28]. The output impedance ($\mathbf{Z}_o$) (i.e., the element (2,2) in Equation (4)) and the control-to-output-voltage transfer function (G_{co}) (i.e., the element (2,3) in Equation (4)) are affected via the impedance-based sensitivity function $[\mathbf{I} + \mathbf{Z}_{inS}\mathbf{Y}_{in}]^{-1}$ as well as via a certain numerator polynomial, which contains, in general case, a multivariable impedance-admittance product. One of the multivariable elements, in this product, is the multivariable special parameter, which cannot be extracted directly from the matrix formulas in Equation (7), because the matrix equations cannot be manipulated into the EET form given in Equation (1) (Section 1). The special impedance parameters are known as input admittance at short-circuited output $\mathbf{Y}_{in-sco}$ and ideal input admittance $\mathbf{Y}_{in-\infty}$, respectively [12].

In this specific case, when the input source is a DC source, all the special parameters can be solved analytically as they are, because $[\mathbf{I} + \mathbf{Y}_{in}\mathbf{Z}_{inS}]^{-1}$ and $[\mathbf{I} + \mathbf{Z}_{inS}\mathbf{Y}_{in}]^{-1}$ can be given by

$$\begin{bmatrix} \frac{1}{1+Z_{inS}Y_{in}} & 0 \\ 0 & 1 \end{bmatrix} \tag{8}$$

According to Equations (3), (7), and (8), we can compute the source-affected transfer functions to be

$$\mathbf{Y}_{in}^S = \begin{bmatrix} \frac{Y_{in}}{1+Z_{inS}Y_{in}} & 0 \\ 0 & 0 \end{bmatrix} \mathbf{T}_{oi}^S = \begin{bmatrix} \frac{T_{oi-d}}{1+Z_{inS}Y_{in}} & \frac{T_{oi-q}}{1+Z_{inS}Y_{in}} \\ 0 & 0 \end{bmatrix} \mathbf{G}_{ci}^S = \begin{bmatrix} \frac{G_{ci-d}}{1+Z_{inS}Y_{in}} & \frac{G_{ci-q}}{1+Z_{inS}Y_{in}} \\ 0 & 0 \end{bmatrix} \mathbf{G}_{io}^S = \begin{bmatrix} \frac{G_{io-d}}{1+Z_{inS}Y_{in}} & 0 \\ \frac{G_{io-q}}{1+Z_{inS}Y_{in}} & 0 \end{bmatrix} \tag{9}$$

and

$$\mathbf{Z}_o^S = \begin{bmatrix} \frac{1+Z_{inS}Y_{in-sco-d}}{1+Z_{inS}Y_{in}}Z_{o-d} & \frac{1+Z_{inS}Y_{in-sco-qd}}{1+Z_{inS}Y_{in}}Z_{o-qd} \\ \frac{1+Z_{inS}Y_{in-sco-dq}}{1+Z_{inS}Y_{in}}Z_{o-dq} & \frac{1+Z_{inS}Y_{in-sco-q}}{1+Z_{inS}Y_{in}}Z_{o-q} \end{bmatrix} \tag{10}$$

where

$$\mathbf{Y}_{in-sco} = \begin{bmatrix} Y_{in-sco-d} & Y_{in-sco-qd} \\ Y_{in-sco-dq} & Y_{in-sco-q} \end{bmatrix} = \begin{bmatrix} Y_{in} + \frac{G_{io-d}T_{oi-d}}{Z_{o-d}} & Y_{in} + \frac{G_{io-d}T_{oi-q}}{Z_{o-qd}} \\ Y_{in} + \frac{G_{io-q}T_{oi-d}}{Z_{o-dq}} & Y_{in} + \frac{G_{io-q}T_{oi-q}}{Z_{o-q}} \end{bmatrix} \tag{11}$$

as well as

$$\mathbf{G}_{co}^S = \begin{bmatrix} \frac{1+Z_{inS}Y_{in-\infty-d}}{1+Z_{inS}Y_{in}}G_{co-d} & \frac{1+Z_{inS}Y_{in-\infty-qd}}{1+Z_{inS}Y_{in}}G_{co-qd} \\ \frac{1+Z_{inS}Y_{in-\infty-dq}}{1+Z_{inS}Y_{in}}G_{co-dq} & \frac{1+Z_{inS}Y_{in-\infty-q}}{1+Z_{inS}Y_{in}}G_{co-q} \end{bmatrix} \tag{12}$$

where

$$\mathbf{Y}_{in-\infty} = \begin{bmatrix} Y_{in-\infty-d} & Y_{in-\infty-qd} \\ Y_{in-\infty-dq} & Y_{in-\infty-q} \end{bmatrix} = \begin{bmatrix} Y_{in} - \frac{G_{io-d}G_{ci-d}}{G_{co-d}} & Y_{in} - \frac{G_{io-d}G_{ci-q}}{G_{co-qd}} \\ Y_{in} - \frac{G_{io-q}G_{ci-d}}{G_{co-dq}} & Y_{in} - \frac{G_{io-q}G_{ci-q}}{G_{co-q}} \end{bmatrix} \tag{13}$$

The expressions of the special parameters in Equations (11) and (13) equal the expressions defined for the corresponding DC-DC converters as explicitly given in Reference ([12], p. 143, Equation (3.32)).

2.1.2. Load-Affected Transfer Functions

Figure 6 shows the VF-VO inverter (cf. Figure 4) connected to the power grid via the PCC, where the power grid is represented by a generalized multivariable load system. It may be obvious that the interface at the output terminal of the load system (i.e., the terminal designated by $\hat{v}_{oL}$) is not accessible in general and therefore, the input admittance ($\mathbf{Y}_{inL}$) is the only measurable transfer function. As a consequence, we consider the equivalent circuit of the power grid to be composed of its input admittance denoted as $\mathbf{Y}_{\text{oL}}$ (i.e., $\mathbf{Y}_{\text{inL}}$ in Figure 6) and the constant-current sink denoted as $\hat{\mathbf{i}}_{\text{oL}}$ (i.e., $\mathbf{Z}_{\text{oL}} = 0$, $\mathbf{G}_{\text{ioL}} = \mathbf{T}_{\text{oiL}} = 1$).

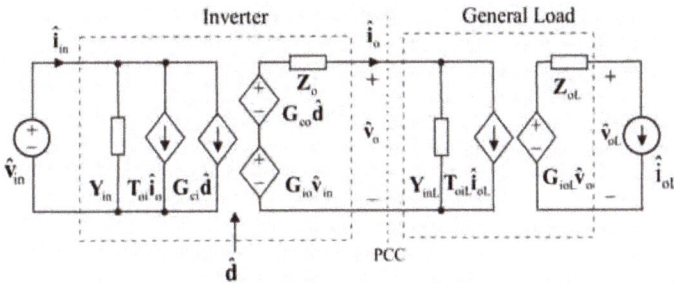

Figure 6. The VF-VO inverter connected to a generalized power grid.

Figure 7 shows the equivalent circuit of the converter in Figure 4, when it is terminated with a non-ideal load. According to Figure 7, the original output current $\hat{i}_o$ is changed as

$$\hat{i}_o = \hat{i}_{\text{oL}} + \mathbf{Y}_{\text{oL}} \hat{v}_o \tag{14}$$

and

$$\hat{i}_o = [\mathbf{I} + \mathbf{Y}_{\text{oL}} \mathbf{Z}_o]^{-1} \hat{i}_{\text{oL}} + [\mathbf{I} + \mathbf{Y}_{\text{oL}} \mathbf{Z}_o]^{-1} \mathbf{Y}_{\text{oL}} \mathbf{G}_{\text{io}} \hat{v}_{\text{in}} + [\mathbf{I} + \mathbf{Y}_{\text{oL}} \mathbf{Z}_o]^{-1} \mathbf{Y}_{\text{oL}} \mathbf{G}_{\text{co}} \hat{d} \tag{15}$$

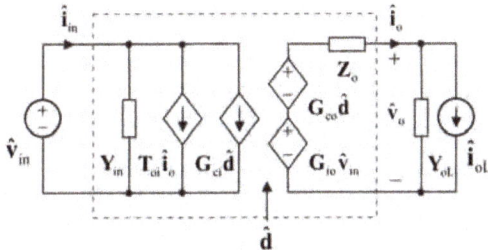

Figure 7. Linear multi-variable equivalent circuit of three-phase VF-VO DC-AC converter with a non-ideal load.

The load-affected set of transfer functions can be obtained by replacing $\hat{i}_o$ in the bottom row of Equation (4) by Equation (14) and in the upper row of Equation (4) by Equation (15), respectively. These procedures yield the set of load-affected transfer functions as

$$
\begin{bmatrix} \hat{i}_{\text{in}} \\ \hat{v}_o \end{bmatrix} = \begin{bmatrix} \mathbf{Y}_{\text{in}}^L & \mathbf{T}_{\text{oi}}^L & \mathbf{G}_{\text{ci}}^L \\ \mathbf{G}_{\text{io}}^L & -\mathbf{Z}_o^L & \mathbf{G}_{\text{co}}^L \end{bmatrix} \begin{bmatrix} \hat{v}_{\text{in}} \\ \hat{i}_{\text{oL}} \\ \hat{d} \end{bmatrix}
$$

$$
\begin{bmatrix} \hat{i}_{\text{in}} \\ \hat{v}_o \end{bmatrix} = \begin{bmatrix} \mathbf{Y}_{\text{in}} + \mathbf{T}_{\text{oi}}[\mathbf{I} + \mathbf{Y}_{\text{oL}} \mathbf{Z}_o]^{-1} \mathbf{Y}_{\text{oL}} \mathbf{G}_{\text{io}} & \mathbf{T}_{\text{oi}}[\mathbf{I} + \mathbf{Y}_{\text{oL}} \mathbf{Z}_o]^{-1} & \mathbf{G}_{\text{ci}} + \mathbf{T}_{\text{oi}}[\mathbf{I} + \mathbf{Y}_{\text{oL}} \mathbf{Z}_o]^{-1} \mathbf{Y}_{\text{oL}} \mathbf{G}_{\text{co}} \\ [\mathbf{I} + \mathbf{Z}_o \mathbf{Y}_{\text{oL}}]^{-1} \mathbf{G}_{\text{io}} & -[\mathbf{I} + \mathbf{Z}_o \mathbf{Y}_{\text{oL}}]^{-1} \mathbf{Z}_o & [\mathbf{I} + \mathbf{Z}_o \mathbf{Y}_{\text{oL}}]^{-1} \mathbf{G}_{\text{co}} \end{bmatrix} \begin{bmatrix} \hat{v}_{\text{in}} \\ \hat{i}_{\text{oL}} \\ \hat{d} \end{bmatrix} \tag{16}
$$

The load-affected multivariable transfer functions in Equation (16) can be solved easily with a proper software package such as Matlab^{TM} in numerical form. As shown in (3), both $\mathbf{Z}_o$ and $\mathbf{Y}_{\text{oL}}$ comprise of four distinct transfer functions, which makes $[\mathbf{I} + \mathbf{Z}_o \mathbf{Y}_{\text{oL}}]^{-1}$ and $[\mathbf{I} + \mathbf{Y}_{\text{oL}} \mathbf{Z}_o]^{-1}$ to be rather complicated in analytic form, and therefore, the further processing of the multivariable transfer functions in Equation (16) similarly as performed in Equations (9)–(13) is almost impossible although the elements (1,2), (2,1)–(2,3) in Equation (16) seem to be very simple. The elements (1,1) and (1,3) in Equation (16) include certain impedance-like special parameters, which are known as output impedances at open-circuit input terminal $\mathbf{Z}_{o-oci}(1,1)$ and ideal output impedance $\mathbf{Z}_{o-\infty}(1,3)$ [12], respectively. The elements (1,1) and (1,3) cannot be, however, put into the EET-method form according to Equation (1), and therefore, $\mathbf{Z}_{o-oci}$ and $\mathbf{Z}_{o-\infty}$ cannot be found as a full-order matrix equation. It shall be observed that all the transfer functions in Equation (16) related to $\hat{\mathbf{i}}_{oL}$ are not real in practice as discussed above.

We will give, in this paper, the load-affected transfer functions in Equation (16) by omitting the cross-coupling terms in $\mathbf{Z}_o$ and $\mathbf{Y}_{oL}$. Thus $[\mathbf{I} + \mathbf{Z}_o\mathbf{Y}_{oL}]^{-1}$ can be given by

$$\begin{bmatrix} \frac{1}{1+Z_{o-d}Y_{oL-d}} & 0 \\ 0 & \frac{1}{1+Z_{o-q}Y_{oL-q}} \end{bmatrix} \tag{17}$$

According to Equations (3), (16), and (17), we can compute the load-affected transfer functions to be

$$\mathbf{G}_{io}^{L} = \begin{bmatrix} \frac{G_{io-d}}{1+Z_{o-d}Y_{oL-d}} & 0 \\ \frac{G_{io-q}}{1+Z_{o-q}Y_{oL-q}} & 0 \end{bmatrix} \mathbf{Z}_{o}^{L} = \begin{bmatrix} \frac{Z_{o-d}}{1+Z_{o-d}Y_{oL-d}} & 0 \\ 0 & \frac{Z_{o-q}}{1+Z_{o-q}Y_{oL-q}} \end{bmatrix} \mathbf{G}_{co}^{L} = \begin{bmatrix} \frac{G_{co-d}}{1+Z_{o-d}Y_{oL-d}} & \frac{G_{co-qd}}{1+Z_{o-d}Y_{oL-d}} \\ \frac{G_{co-dq}}{1+Z_{o-q}Y_{oL-q}} & \frac{G_{co-q}}{1+Z_{o-q}Y_{oL-q}} \end{bmatrix}$$

$$\mathbf{T}_{oi}^{L} = \begin{bmatrix} \frac{T_{oi-d}}{1+Z_{o-d}Y_{oL-d}} & \frac{T_{oi-q}}{1+Z_{o-q}Y_{oL-q}} \\ 0 & 0 \end{bmatrix} \tag{18}$$

and

$$\mathbf{Y}_{in}^{L} = \begin{bmatrix} \left(\frac{1+Y_{oL-d}Z_{o-oci-d}}{1+Y_{oL-d}Z_{o-d}} + \frac{1+Y_{oL-q}Z_{o-oci-q}}{1+Y_{oL-q}Z_{o-q}} - 1 \right) Y_{in} & 0 \\ 0 & 0 \end{bmatrix} \tag{19}$$

where

$$\mathbf{Z}_{o-oci} = \begin{bmatrix} Z_{o-oci-d} & 0 \\ 0 & Z_{o-oci-q} \end{bmatrix} = \begin{bmatrix} Z_{o-d} + \frac{G_{io-d}T_{oi-d}}{Y_{in}} & 0 \\ 0 & Z_{o-q} + \frac{G_{io-q}T_{oi-q}}{Y_{in}} \end{bmatrix} \tag{20}$$

as well as

$$\mathbf{G}_{ci}^{L} = \begin{bmatrix} \left(\frac{1+Y_{oL-d}Z_{o-\infty-d}}{1+Y_{oL-d}Z_{o-d}} + \frac{1+Y_{oL-q}Z_{o-\infty-dq}}{1+Y_{oL-q}Z_{o-q}} - 1 \right) G_{ci-d} & \left(\frac{1+Y_{oL-q}Z_{o-\infty-q}}{1+Y_{oL-q}Z_{o-q}} + \frac{1+Y_{oL-d}Z_{o-\infty-qd}}{1+Y_{oL-d}Z_{o-d}} - 1 \right) G_{ci-q} \\ 0 & 0 \end{bmatrix} \tag{21}$$

where

$$\mathbf{Z}_{o-\infty} = \begin{bmatrix} Z_{o-\infty-d} & Z_{o-\infty-qd} \\ Z_{o-\infty-dq} & Z_{o-\infty-q} \end{bmatrix} = \begin{bmatrix} Z_{o-d} + \frac{T_{oi-d}G_{co-d}}{G_{ci-d}} & Z_{o-d} + \frac{T_{oi-q}G_{co-qd}}{G_{ci-d}} \\ Z_{o-q} + \frac{T_{oi-d}G_{co-dq}}{G_{ci-q}} & Z_{o-q} + \frac{T_{oi-q}G_{co-q}}{G_{ci-q}} \end{bmatrix} \tag{22}$$

The expressions of the special parameters in Equations (20) and (22) equal the expressions defined for the corresponding DC-DC converters as explicitly given in Reference ([12], p. 143, Equation (3.32)).

2.2. Three-Phase CF-CO Inverter

The current-fed (CF) current-output (CO) inverter is used in the grid-feeding-mode renewable energy systems (cf. Figure 3a), where the inverter synchronizes itself in the grid frequency and angle as well as supplies energy into the grid. In renewable energy applications, the input-terminal source is a voltage-type source, and the output terminal source is a three-phase grid, respectively.

The transfer functions of the three-phase converters are given in synchronous reference frame or dq frame due to the multivariable nature of the converter [12,22] as

$$
\begin{bmatrix} \hat{v}_{in} \\ \hat{i}_{o-d} \\ \hat{i}_{o-q} \end{bmatrix} = \begin{bmatrix} Z_{in} & T_{oi-d} & T_{oi-q} & G_{ci-d} & G_{ci-q} \\ G_{io-d} & -Y_{o-d} & -Y_{o-qd} & G_{co-d} & G_{co-qd} \\ G_{io-q} & -Y_{o-dq} & -Y_{o-q} & G_{co-dq} & G_{co-q} \end{bmatrix} \begin{bmatrix} \hat{i}_{in} \\ \hat{v}_{o-d} \\ \hat{v}_{o-q} \\ \hat{d}_d \\ \hat{d}_q \end{bmatrix} \tag{23}
$$

which can be given similarly as Equation (2) (Section 2.1) in multivariable mode as

$$
\begin{bmatrix} \hat{\mathbf{v}}_{in} \\ \hat{\mathbf{i}}_o \end{bmatrix} = \begin{bmatrix} \mathbf{Z}_{in} & \mathbf{T}_{oi} & \mathbf{G}_{ci} \\ \mathbf{G}_{io} & -\mathbf{Y}_o & \mathbf{G}_{co} \end{bmatrix} \begin{bmatrix} \hat{\mathbf{i}}_{in} \\ \hat{\mathbf{v}}_o \\ \hat{\mathbf{d}} \end{bmatrix} \tag{24}
$$

and represented by a linear multivariable equivalent circuit as given in Figure 8.

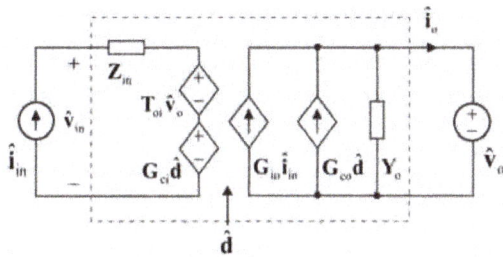

Figure 8. Linear multi-variable equivalent circuit of three-phase CF-CO DC-AC converter.

2.2.1. Source-Affected Transfer Functions

Figure 9 shows the corresponding equivalent circuit of Figure 8, when the converter is supplied by a non-ideal input source. According to Figure 9, the original input current $\hat{i}_{in}$ is changed as

$$
\hat{i}_{in} = \hat{i}_{inS} - Y_{inS}\hat{v}_{in} \tag{25}
$$

and

$$
\hat{i}_{in} = [\mathbf{I} + \mathbf{Y}_{inS}\mathbf{Z}_{in}]^{-1}\hat{i}_{inS} - [\mathbf{I} + \mathbf{Y}_{inS}\mathbf{Z}_{in}]^{-1}\mathbf{Y}_{inS}\mathbf{T}_{oi}\hat{v}_o - [\mathbf{I} + \mathbf{Y}_{inS}\mathbf{Z}_{in}]^{-1}\mathbf{Y}_{inS}\mathbf{G}_{ci}\hat{d} \tag{26}
$$

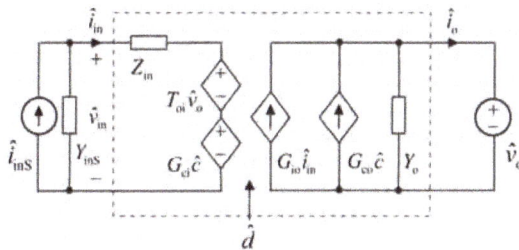

Figure 9. Linear multi-variable equivalent circuit of three-phase CF-CO DC-AC converter with a non-ideal source.

The source-affected set of multi-variable transfer functions can be obtained by replacing $\hat{i}_{in}$ in the upper row of Equation (24) by Equation (25) and in the bottom row of Equation (24) by Equation (46), respectively. These procedures yield the set of source-affected transfer functions as

$$\begin{bmatrix} \hat{v}_{in} \\ \hat{i}_o \end{bmatrix} = \begin{bmatrix} \mathbf{Z}_{in}^S & \mathbf{T}_{oi}^S & \mathbf{G}_{ci}^S \\ \mathbf{G}_{io}^S & -\mathbf{Y}_o^S & \mathbf{G}_{co}^S \end{bmatrix} \begin{bmatrix} \hat{i}_{inS} \\ \hat{v}_o \\ \hat{d} \end{bmatrix}$$

$$\begin{bmatrix} \hat{v}_{in} \\ \hat{i}_o \end{bmatrix} = \begin{bmatrix} [\mathbf{I} + \mathbf{Z}_{in}\mathbf{Y}_{inS}]^{-1}\mathbf{Z}_{in} & [\mathbf{I} + \mathbf{Z}_{in}\mathbf{Y}_{inS}]^{-1}\mathbf{T}_{oi} & [\mathbf{I} + \mathbf{Z}_{in}\mathbf{Y}_{inS}]^{-1}\mathbf{G}_{ci} \\ \mathbf{G}_{io}[\mathbf{I} + \mathbf{Y}_{inS}\mathbf{Z}_{in}]^{-1} & -(\mathbf{Y}_o + \mathbf{G}_{io}\mathbf{Y}_{inS}[\mathbf{I} + \mathbf{Z}_{in}\mathbf{Y}_{inS}]^{-1}\mathbf{T}_{oi}) & \mathbf{G}_{co} - \mathbf{G}_{io}\mathbf{Y}_{inS}[\mathbf{I} + \mathbf{Z}_{in}\mathbf{Y}_{inS}]^{-1}\mathbf{G}_{ci} \end{bmatrix} \begin{bmatrix} \hat{i}_{inS} \\ \hat{v}_o \\ \hat{d} \end{bmatrix}$$

(27)

Equation (27) shows that the input dynamics of the inverter (i.e., the elements (1,1), (1,2) and (1,3) in Equation (24)) and the input-to-output transfer function $\mathbf{G}_{io}$ (i.e., the element (2,1) in Equation (24)) are affected only via the impedance-based sensitivity function $[\mathbf{I} + \mathbf{Z}_{in}\mathbf{Y}_{inS}]^{-1}$ or $[\mathbf{I} + \mathbf{Y}_{inS}\mathbf{Z}_{in}]^{-1}$. It may be also obvious that the stability of the cascaded source-converter system can be assessed by means of the sensitivity functions by applying the generalized Nyquist stability criterion [27,28]. The output impedance ($\mathbf{Y}_o$) (i.e., the element (2,2) in Equation (24)) and the control-to-output-voltage transfer function ($\mathbf{G}_{co}$) (i.e., the element (2,3) in Equation (24)) are affected via the impedance-based sensitivity function $[\mathbf{I} + \mathbf{Z}_{in}\mathbf{Y}_{inS}]^{-1}$ as well as via a certain numerator polynomial, which contains, in general case, a multivariable impedance-admittance product. One of the multivariable elements, in this product, is the multivariable special parameter, which cannot be extracted directly from the matrix formulas in Equation (27), because they cannot be manipulated into the EET form given in Equation (1) (Section 1). The special impedance-like parameters are known as input impedance at open circuit output $\mathbf{Z}_{in-oco}$ and ideal input impedance $\mathbf{Z}_{in-\infty}$.

In this specific case, when the input source is a DC source, all the special parameters can be solved analytically as they are, because $[\mathbf{I} + \mathbf{Z}_{in}\mathbf{Y}_{inS}]^{-1}$ and $[\mathbf{I} + \mathbf{Y}_{inS}\mathbf{Z}_{in}]^{-1}$ can be given as

$$\begin{bmatrix} \frac{1}{1+Z_{in}Y_{inS}} & 0 \\ 0 & 1 \end{bmatrix}$$

(28)

According to Equations (23), (27), and (28), we can compute the source-affected transfer functions to be

$$\mathbf{Z}_{in}^S = \begin{bmatrix} \frac{Z_{in}}{1+Y_{inS}Z_{in}} & 0 \\ 0 & 0 \end{bmatrix} \mathbf{T}_{oi}^S = \begin{bmatrix} \frac{T_{oi-d}}{1+Y_{inS}Z_{in}} & \frac{T_{oi-q}}{1+Y_{inS}Z_{in}} \\ 0 & 0 \end{bmatrix} \mathbf{G}_{ci}^S = \begin{bmatrix} \frac{G_{ci-d}}{1+Y_{inS}Z_{in}} & \frac{G_{ci-q}}{1+Y_{inS}Z_{in}} \\ 0 & 0 \end{bmatrix} \mathbf{G}_{io}^S = \begin{bmatrix} \frac{G_{io-d}}{1+Y_{inS}Z_{in}} & 0 \\ \frac{G_{io-q}}{1+Y_{inS}Z_{in}} & 0 \end{bmatrix}$$

(29)

and

$$\mathbf{Y}_o^S = \begin{bmatrix} \frac{1+Y_{inS}Z_{in-oco-d}}{1+Y_{inS}Z_{in}}Y_{o-d} & \frac{1+Y_{inS}Z_{in-oco-qd}}{1+Y_{inS}Z_{in}}Y_{o-qd} \\ \frac{1+Y_{inS}Z_{in-oco-dq}}{1+Y_{inS}Z_{in}}Y_{o-dq} & \frac{1+Y_{inS}Z_{in-oco-q}}{1+Y_{inS}Z_{in}}Y_{o-q} \end{bmatrix}$$

(30)

where

$$\mathbf{Z}_{in-oco} = \begin{bmatrix} Z_{in-oco-d} & Z_{in-oco-qd} \\ Z_{in-oco-dq} & Z_{in-oco-q} \end{bmatrix} = \begin{bmatrix} Z_{in} + \frac{G_{io-d}T_{oi-d}}{Y_{o-d}} & Z_{in} + \frac{G_{io-d}T_{oi-q}}{Y_{o-qd}} \\ Z_{in} + \frac{G_{io-q}T_{oi-d}}{Y_{o-dq}} & Z_{in} + \frac{G_{io-q}T_{oi-q}}{Y_{o-q}} \end{bmatrix}$$

(31)

as well as

$$\mathbf{G}_{co}^S = \begin{bmatrix} \frac{1+Y_{inS}Z_{in-\infty-d}}{1+Y_{jnS}Z_{in}}G_{co-d} & \frac{1+Y_{inS}Z_{in-\infty-qd}}{1+Z_{ins}Y_{in}}G_{co-qd} \\ \frac{1+Y_{inS}Z_{in-\infty-dq}}{1+Y_{inS}Z_{in}}G_{co-dq} & \frac{1+Y_{inS}Z_{in-\infty-q}}{1+Z_{inS}Y_{in}}G_{co-q} \end{bmatrix}$$

(32)

where

$$\mathbf{Z}_{in-\infty} = \begin{bmatrix} Z_{in-\infty-d} & Z_{in-\infty-qd} \\ Z_{in-\infty-dq} & Z_{in-\infty-q} \end{bmatrix} = \begin{bmatrix} Z_{in} - \frac{G_{io-d}G_{ci-d}}{G_{co-d}} & Z_{in} - \frac{G_{io-d}G_{ci-q}}{G_{co-qd}} \\ Z_{in} - \frac{G_{io-q}G_{ci-d}}{G_{co-dq}} & Z_{in} - \frac{G_{io-q}G_{ci-q}}{G_{co-q}} \end{bmatrix}$$

(33)

The expressions of the special parameters in Equations (31) and (33) equal the expressions defined for the corresponding DC-DC converters as explicitly given in Reference ([12], p. 356, Equation (8.2))).

2.2.2. Load-Affected Transfer Functions

Figure 10 shows the CF-CO inverter (cf. Figure 8) connected to the power grid via the PCC, where the power grid is represented by a generalized multivariable load system. It may be obvious that the interface at the output terminal of the load system (i.e., the terminal designated by $\hat{i}_{oL}$) is not accessible in general and therefore, the input impedance (Z_{inL}) is the only measurable transfer function. As a consequence, we consider the equivalent circuit of the power grid to be composed of its input impedance denoted as Z_{oL} (i.e., Z_{inL}) and the constant-voltage sink denoted as $\hat{v}_{oL}$ (i.e., $Y_{oL} = 0$, $G_{ioL} = T_{oiL} = 1$).

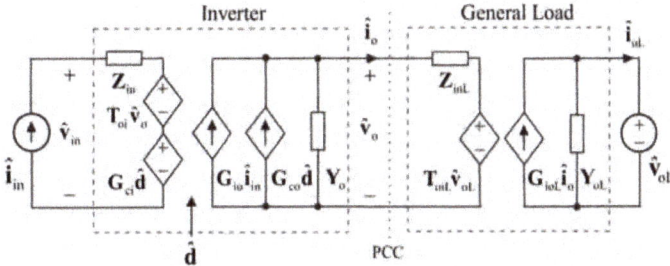

Figure 10. The CF-CO inverter connected to a generalized power grid.

Figure 11 shows the equivalent circuit of the converter in Figure 8, when it is terminated with a non-ideal load. According to Figure 11, the original output current $\hat{v}_o$ is changed as

$$\hat{v}_o = \hat{v}_{oL} + Z_{oL}\hat{i}_o \tag{34}$$

and

$$\hat{v}_o = [I + Z_{oL}Y_o]^{-1}\hat{v}_{oL} + [I + Z_{oL}Y_o]^{-1}Z_{oL}G_{io}\hat{i}_{in} + [I + Z_{oL}Y_o]^{-1}Z_{oL}G_{co}\hat{d} \tag{35}$$

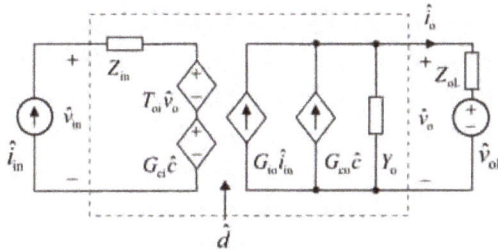

Figure 11. Linear multi-variable equivalent circuit of three-phase CF-CO DC-AC converter with a non-ideal load.

The load-affected set of transfer functions can be obtained by replacing $\hat{v}_o$ in the bottom row of Equation (24) by Equation (34), and in the upper row of Equation (24) by Equation (35), respectively. These procedures yield the set of load-affected transfer functions as

$$\begin{bmatrix} \hat{v}_{in} \\ \hat{i}_o \end{bmatrix} = \begin{bmatrix} Z_{in}^L & T_{oi}^L & G_{ci}^L \\ G_{io}^L & -Y_o^L & G_{co}^L \end{bmatrix} \begin{bmatrix} \hat{i}_{in} \\ \hat{v}_{oL} \\ \hat{d} \end{bmatrix}$$

$$\begin{bmatrix} \hat{v}_{in} \\ \hat{i}_o \end{bmatrix} = \begin{bmatrix} Z_{in} + T_{oi}[I+Z_{oL}Y_o]^{-1}Z_{oL}G_{io} & T_{oi}[I+Z_{oL}Y_o]^{-1} & G_{ci} + T_{oi}[I+Z_{oL}Y_o]^{-1}Z_{oL}G_{co} \\ [I+Y_oZ_{oL}]^{-1}G_{io} & -[I+Y_oZ_{oL}]^{-1}Y_o & [I+Y_oZ_{oL}]^{-1}G_{co} \end{bmatrix} \begin{bmatrix} \hat{i}_{in} \\ \hat{v}_{oL} \\ \hat{d} \end{bmatrix} \tag{36}$$

The load-affected multivariable transfer functions in Equation (36) can be solved easily with a proper software package such as Matlab^{TM} in numerical form. As shown in Equation (23), both Y_o and

$\mathbf{Z}_{oL}$ comprise of four distinct transfer functions, which makes $[\mathbf{I} + \mathbf{Y}_o\mathbf{Z}_{oL}]^{-1}$ and $[\mathbf{I} + \mathbf{Z}_{oL}\mathbf{Y}_o]^{-1}$ to be rather complicated. As a consequence, the further processing of the multivariable transfer functions in Equation (36) similarly as performed in Equations (30)–(33) is almost impossible although the elements (1,2), (2,1)–(2,3) in Equation (36) seem to be very simple. The elements (1,1) and (1,3) in Equation (36) include certain impedance-like special parameters, which are known as output admittance at short-circuited input terminal $\mathbf{Y}_{o-sci}(1,1)$ and ideal output admittance $\mathbf{Y}_{o-\infty}(1,3)$ [12], respectively.

Equation (36) shows that the output dynamics of the inverter (i.e., the elements (2,1), (2,2) and (2,3) in Equation (24)) and the reverse transfer function $\mathbf{T}_{io}$ (i.e., the element (1,2) in Equation (24)) are affected only via the impedance-based sensitivity function $[\mathbf{I} + \mathbf{Y}_o\mathbf{Z}_{oL}]^{-1}$ or $[\mathbf{I} + \mathbf{Z}_{oL}\mathbf{Y}_o]^{-1}$. It may be also obvious that the stability of the cascaded source-converter system can be assessed by means of the sensitivity functions by applying the generalized Nyquist stability criterion [27,28]. The input impedance ($\mathbf{Z}_{in}$) (i.e., the element (1,1) in Equation (24)) and the control-to-input-voltage transfer function ($\mathbf{G}_{ci}$) (i.e., the element (1,3) in Equation (24)) are affected via the impedance-based sensitivity function $[\mathbf{I} + \mathbf{Z}_{oL}\mathbf{Y}_o]^{-1}$ as well as via certain numerator polynomials, each of which contain a multivariable impedance-admittance product. One of the multivariable elements, in this product, is the multivariable special parameter, which cannot be extracted directly from the matrix formulas in Equation (36), because they cannot be manipulated into the EET form given in Equation (1) (Section 1). These special impedance-like parameters are known as output admittance at short-circuited input terminal ($\mathbf{Y}_{o-sci}$) and ideal output admittance ($\mathbf{Y}_{o-\infty}$) [12], respectively. It shall be observed that all the transfer functions in Equation (36) related to $\hat{v}_{oL}$ are not real in practice as discussed above.

We will give, in this paper, the load-affected transfer functions in Equation (36) by omitting the cross- coupling terms in $\mathbf{Y}_o$ and $\mathbf{Z}_{oL}$, which means that both of the sensitivity functions will be equal as well. Therefore $[\mathbf{I} + \mathbf{Y}_o\mathbf{Z}_{oL}]^{-1}$ can be given by

$$\begin{bmatrix} \frac{1}{1+Y_{o-d}Z_{oL-d}} & 0 \\ 0 & \frac{1}{1+Y_{o-q}Z_{oL-q}} \end{bmatrix} \tag{37}$$

According to Equations (23), (36), and (37), we can compute the load-affected transfer functions to be

$$\mathbf{G}_{io}^{L} = \begin{bmatrix} \frac{G_{io-d}}{1+Y_{o-d}Z_{oL-d}} & 0 \\ \frac{G_{io-q}}{1+Y_{o-q}Z_{oL-q}} & 0 \end{bmatrix} \quad \mathbf{Y}_{o}^{L} = \begin{bmatrix} \frac{Y_{o-d}}{1+Y_{o-d}Z_{oL-d}} & 0 \\ 0 & \frac{Y_{o-q}}{1+Y_{o-q}Z_{oL-q}} \end{bmatrix} \quad \mathbf{G}_{co}^{L} = \begin{bmatrix} \frac{G_{co-d}}{1+Y_{o-d}Z_{oL-d}} & \frac{G_{co-qd}}{1+Y_{o-d}Z_{oL-d}} \\ \frac{G_{co-dq}}{1+Y_{o-q}Z_{oL-q}} & \frac{G_{co-q}}{1+Y_{o-q}Z_{oL-q}} \end{bmatrix}$$
$$\mathbf{T}_{oi}^{L} = \begin{bmatrix} \frac{T_{oi-d}}{1+Y_{o-d}Z_{oL-d}} & \frac{T_{oi-q}}{1+Y_{o-q}Z_{oL-q}} \\ 0 & 0 \end{bmatrix} \tag{38}$$

and

$$\mathbf{Z}_{in}^{L} = \begin{bmatrix} \left(\frac{1+Z_{oL-d}Y_{o-sci-d}}{1+Y_{o-d}Z_{oL-d}} + \frac{1+Z_{oL-q}Y_{o-sci-q}}{1+Y_{o-q}Z_{oL-q}} - 1 \right) Z_{in} & 0 \\ 0 & 0 \end{bmatrix} \tag{39}$$

where

$$\mathbf{Y}_{o-sci} = \begin{bmatrix} Y_{o-sci-d} & 0 \\ 0 & Y_{o-sci-q} \end{bmatrix} = \begin{bmatrix} Y_{o-d} + \frac{G_{io-d}T_{oi-d}}{Z_{in}} & 0 \\ 0 & Y_{o-q} + \frac{G_{io-q}T_{oi-q}}{Z_{in}} \end{bmatrix} \tag{40}$$

as well as

$$\mathbf{G}_{ci}^{L} = \begin{bmatrix} \left(\frac{1+Z_{oL-d}Y_{o-\infty-d}}{1+Z_{oL-d}Y_{o-d}} + \frac{1+Z_{oL-q}Y_{o-\infty-dq}}{1+Z_{oL-q}Y_{o-q}} - 1 \right)G_{ci-d} & \left(\frac{1+Z_{oL-q}Y_{o-\infty-q}}{1+Z_{oL-q}Y_{o-q}} + \frac{1+Z_{oL-d}Y_{o-\infty-qd}}{1+Z_{oL-d}Y_{o-d}} - 1 \right)G_{ci-q} \\ 0 & 0 \end{bmatrix} \tag{41}$$

where

$$\mathbf{Y}_{o-\infty} = \begin{bmatrix} Y_{o-\infty-d} & Y_{o-\infty-qd} \\ Y_{o-\infty-dq} & Y_{o-\infty-q} \end{bmatrix} = \begin{bmatrix} Y_{o-d} + \frac{T_{oi-d}G_{co-d}}{G_{ci-d}} & Y_{o-d} + \frac{T_{oi-q}G_{co-qd}}{G_{ci-d}} \\ Y_{o-q} + \frac{T_{oi-d}G_{co-dq}}{G_{ci-q}} & Y_{o-q} + \frac{T_{oi-q}G_{co-q}}{G_{ci-q}} \end{bmatrix} \quad (42)$$

The expressions of the special parameters in Equations (40) and (42) equal the expressions defined for the corresponding DC-DC converters as explicitly given in Reference ([12], p. 356, Equation (8.2)).

2.3. Discussions

2.3.1. Special Impedance-like Parameters

The complexity to solve analytically and explicitly the load-affected transfer functions in Equation (16) (Section 2.1.2) and in Equation (36) (Section 2.2) may be understood according to the solved $[\mathbf{I} + \mathbf{Y}_o\mathbf{Z}_{oL}]^{-1}$ (cf. Equation (36)), which can be given by

$$\frac{\begin{bmatrix} 1 + Y_{o-q}Z_{oL-q} + Y_{o-dq}Z_{oL-qd} & Y_{o-d}Z_{oL-qd} + Y_{o-qd}Z_{oL-q} \\ Y_{o-q}Z_{oL-q} + Y_{o-dq}Z_{oL-d} & 1 + Y_{o-d}Z_{oL-d} + Y_{o-qd}Z_{oL-dq} \end{bmatrix}}{(1 + Y_{o-q}Z_{oL-q} + Y_{o-dq}Z_{oL-qd})(1 + Y_{o-d}Z_{oL-d} + Y_{o-qd}Z_{oL-dq}) - (Y_{o-d}Z_{oL-qd} + Y_{o-qd}Z_{oL-q})(Y_{o-q}Z_{oL-q} + Y_{o-dq}Z_{oL-d})} \quad (43)$$

It may be also obvious that the full-order special impedance parameters cannot be found in analytical form, because the required EET-method-like formulations (cf. Section 1, Equation 1) cannot be found.

All the ideal impedance-like special parameters given in Equations (13), (22), (33), and (42) are invariant to the feedback and feedforward schemes applied in the converter and to the type of load connected at the output terminal of the converter [12]. Consequently, the ideal parameters can be computed by means of the open-loop transfer functions. The other special parameters given in Equations (11), (20), (31), and (40) are invariant to those feedback arrangements, which are taken from the opposite terminal, where the parameters are assumed to be measured. They will change if the feedback is taken from the same terminal, where the parameters are assumed to be measured. This information dictates the way to compute the special parameters based on the corresponding open or closed-loop transfer functions.

2.3.2. Stability Assessment

The stability of the cascaded system comprising of the converter and its source or load subsystem can be assessed by the means of the multivariable impedance-based sensitivity functions $[\mathbf{I} + \mathbf{Z}_x\mathbf{Y}_y]^{-1}$ or $[\mathbf{I} + \mathbf{Y}_y\mathbf{Z}_x]^{-1}$, where the impedance and admittance matrixes comprise of the impedances and admittances related to the interface between the converter and its subsystem, by applying generalized Nyquist stability criterion [27,28]. The stability can be assessed by computing the determinant of $[\mathbf{I} + \mathbf{Z}_x\mathbf{Y}_y]$ or $[\mathbf{I} + \mathbf{Y}_y\mathbf{Z}_x]$ (i.e., the impedance-ratio-based multivariable characteristic polynomial) in a similar manner as in a single-input-single-output (SISO) system by applying the basic Nyquist stability criterion [13] but the critical point will be the origin (0,0) of the complex plane instead of $(-1,0)$. The form of the determinant is explicitly given in Equation (43) as its denominator. If the interface, at which the stability is to be assessed, is of SISO type as in the source-side interfaces in the above treated inverters in Sections 2.1.1 and 2.2.1 then the relevant impedance ratio comprises of the input and output impedances of the subsystems as shown in Equations (8) and (28). In this case, the basic Nyquist stability criterion can be applied to the corresponding impedance ratio in terms of the critical point $(-1,0)$, respectively.

If the grid is symmetrically loaded then $Z_d(Y_d)$ equals $Z_q(Y_q)$ and $Z_{qd}(Y_{qd})$ equals $-Z_{dq}(Y_{dq})$. Theoretically, the same equalities may be valid also in the inverter. As a consequence, $[\mathbf{I} + \mathbf{Z}_x\mathbf{Y}_y]^{-1}$ equals $[\mathbf{I} + \mathbf{Y}_y\mathbf{Z}_x]^{-1}$ and the stability can be assessed based on either of these sensitivity functions. In practice, the equalities may not be, however, valid.

As discussed in Section 1, the robust performance (i.e., sufficient phase (PM) and gain (GM) margins for not affecting significantly the original transfer functions via the peaking in the corresponding impedance-based sensitivity function [10]) can be given in terms of a specific forbidden region as given in Figure 1 (Section 1) in SISO-type interface. The least space requiring forbidden regions in Figure 1 is the MPC region, where the radius of circle equals the inverse of the maximum allowed peaking (M) in the sensitivity function (S). denoted by M. The maximum peaking can be given in terms of PM and GM according to ([14], pp. 92, 93) as

$$M_{PM} = \frac{1}{2\sin(PM/2)} M_{GM} = \frac{GM}{GM-1} \tag{44}$$

The MPC-forbidden region takes into account the combined peaking effect of PM and GM. In case of the DC-interfaced inverters, the DC-interface-related robust performance can be obtained by requiring the contour of the impedance-based minor-loop gain to stay out of the MPC-defined forbidden region as stated in Reference [10]. In this case, the impedance-based minor-loop gain is the same for all the source-affected transfer functions, and the maximum peaking can be explicitly defined by means of the PM and GM associated to the minor-loop gain.

In the general case, when the sensitivity function is also a multivariable function as in Equation (43), the load/source-affected transfer functions are composed of two or more elements having each a different impedance-based sensitivity function. Equation (43) is presented in Equation (45) in a more convenient form, and the load-affected $\mathbf{G}_{co}^{L}$ in Equation (36) applying Equation (45) is presented in its explicit form in Equation (46), respectively. Equation (46) shows explicitly the subcomponents of each load-affected elements of $\mathbf{G}_{co}^{L}$. If the same procedures are applied for $\mathbf{Z}_{in}^{L}$ and $\mathbf{G}_{ci}^{L}$ in Equation (36) then the number of subcomponents is higher than two, which will increase the complexity of the corresponding transfer functions significantly. Equation (45) indicates that we can find four different impedance-based sensitivity functions (S_i) as given in Equation (47), which may exhibit different peaking behavior, because the numerator of the sensitivity functions contains one common factor with the denominator.

$$\frac{\begin{bmatrix} 1+a & b \\ c & 1+d \end{bmatrix}}{(1+a)(1+b)-bc} \tag{45}$$

$$\mathbf{G}_{co}^{L} = \begin{bmatrix} G_{co-d}^{L} & G_{co-qd}^{L} \\ G_{co-dq}^{L} & G_{co-q}^{L} \end{bmatrix} = \frac{\begin{bmatrix} (1+a)G_{co-d}+cG_{co-dq} & (1+a)G_{co-qd}+cG_{co-q} \\ (1+d)G_{co-dq}+bG_{co-d} & (1+d)G_{co-q}+bG_{co-qd} \end{bmatrix}}{(1+a)(1+d)-bc} \tag{46}$$

$$S_1 = \frac{1+a}{(1+a)(1+d)-bc} \quad S_2 = \frac{b}{(1+a)(1+d)-bc} \quad S_3 = \frac{c}{(1+a)(1+d)-bc} \quad S_4 = \frac{1+d}{(1+a)(1+d)-bc} \tag{47}$$

If the cross-coupling terms are small as assumed in Section 2.2.2 then we have two different sensitivity functions as given in Equation (37). In this specific case, the stability of the system can be inferred by requiring that both of the impedance-ratio-based minor-loop gains satisfy Nyquist stability criterion [13]. In addition, the robust stability can be ensured by applying the MPC-based criteria to both of the minor-loop gains [10]. In general, it is not recommended to neglect the cross-coupling terms, because they may have significant impact on the validity of the stability information as discussed in References [78–82]. The complexity of the analysis with all the cross-coupling terms included is usually the reason for neglecting the cross-couplings [83–85].

In general case, the formulation for the requirements for the robust stability (RS) has to be performed based on the maximum singular value ($\bar{\sigma}$) of the multivariable sensitivity function $\mathbf{S}$, which equals $[\mathbf{I}+\mathbf{Y}_y\mathbf{Z}_x]^{-1}$ or $[\mathbf{I}+\mathbf{Z}_x\mathbf{Y}_y]^{-1}$ [14,15], and which can be easily computed by applying Matlab™ command svd ($\mathbf{S}$) (i.e., singular value decomposition). In case of full-matrix uncertainty as in this case, RS is guaranteed when $\bar{\sigma} < 1$ [14].

2.3.3. General Load

Figure 12 shows the multivariable equivalent circuits of the inverter output in grid-forming mode, and the power grid in multivariable mode. The source effect caused by the inverter (i.e., $\mathbf{Z}_{inv}$) in the general-load case can be computed in a similar manner as performed in Section 2.1.1, which yields

$$
\begin{bmatrix} \hat{i}_{pcc} \\ \hat{v}_{oL} \end{bmatrix} = \begin{bmatrix} \mathbf{Y}^S_{inL} & \mathbf{T}^S_{oiL} \\ \mathbf{G}^S_{ioL} & -\mathbf{Z}^S_{oL} \end{bmatrix} \begin{bmatrix} \hat{v}_{inv} \\ \hat{i}_{oL} \end{bmatrix}
$$

$$
\begin{bmatrix} \hat{i}_{pcc} \\ \hat{v}_{oL} \end{bmatrix} = \begin{bmatrix} [\mathbf{I}+\mathbf{Y}_{inL}\mathbf{Z}_{inv}]^{-1}\mathbf{Y}_{inL} & [\mathbf{I}+\mathbf{Y}_{inL}\mathbf{Z}_{inv}]^{-1}\mathbf{T}_{oiL} \\ \mathbf{G}_{ioL}[\mathbf{I}+\mathbf{Z}_{inv}\mathbf{Y}_{inL}]^{-1} & -(\mathbf{Z}_{oL}+\mathbf{G}_{ioL}[\mathbf{I}+\mathbf{Z}_{inv}\mathbf{Y}_{inL}]^{-1}\mathbf{Z}_{inL}\mathbf{T}_{oiL}) \end{bmatrix} \begin{bmatrix} \hat{v}_{inv} \\ \hat{i}_{oL} \end{bmatrix}
\tag{48}
$$

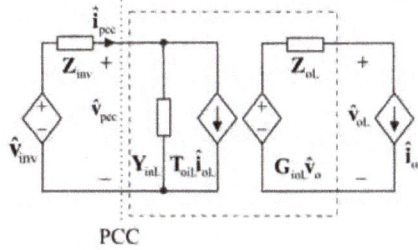

Figure 12. The equivalent circuits of the inverter output and the generalized power grid in grid-forming mode of operation.

Equation (48) shows that the stability of the interface can be assessed based on the output impedance of the inverter ($\mathbf{Z}_{inv}$) and the input admittance of the grid ($\mathbf{Y}_{inL}$) by applying General Nyquist stability criterion to $[\mathbf{I}+\mathbf{Y}_{inL}\mathbf{Z}_{inv}]^{-1}$ and $[\mathbf{I}+\mathbf{Y}_o\mathbf{Z}_{inL}]^{-1}$ as stated also in Section 2.1.2. The other properties of the cascaded system cannot be computed even if the voltage (cf. Equation (49)) and current (cf. Equation (48)) of PCC can be explicitly measured, because $\mathbf{T}_{oiL}$ cannot be measured in practice.

$$
\hat{v}_{pcc} = [\mathbf{I}+\mathbf{Z}_{inv}\mathbf{Y}_{inL}]^{-1}\hat{v}_{inv} - [\mathbf{I}+\mathbf{Z}_{inv}\mathbf{Y}_{inL}]^{-1}\mathbf{Z}_{inv}\mathbf{T}_{oiL}\hat{i}_{oL}
\tag{49}
$$

Figure 13 shows the multivariable equivalent circuits of the inverter output in grid-feeding mode, and the power grid in multivariable mode. The source effect caused by the inverter ($\mathbf{Y}_{inv}$) in the general-load case can be computed in a similar manner as performed in Section 2.2.1, which yields

$$
\begin{bmatrix} \hat{v}_{pcc} \\ \hat{i}_{oL} \end{bmatrix} = \begin{bmatrix} \mathbf{Z}^S_{inL} & \mathbf{T}^S_{oiL} \\ \mathbf{G}^S_{ioL} & -\mathbf{Y}^S_{oL} \end{bmatrix} \begin{bmatrix} \hat{i}_{inv} \\ \hat{v}_{oL} \end{bmatrix}
$$

$$
\begin{bmatrix} \hat{v}_{pcc} \\ \hat{i}_{oL} \end{bmatrix} = \begin{bmatrix} [\mathbf{I}+\mathbf{Z}_{inL}\mathbf{Y}_{inv}]^{-1}\mathbf{Z}_{inL} & [\mathbf{I}+\mathbf{Z}_{inL}\mathbf{Y}_{inv}]^{-1}\mathbf{T}_{oiL} \\ \mathbf{G}_{ioL}[\mathbf{I}+\mathbf{Y}_{inv}\mathbf{Z}_{inL}]^{-1} & -(\mathbf{Y}_{oL}+\mathbf{G}_{ioL}[\mathbf{I}+\mathbf{Y}_{inv}\mathbf{Z}_{inL}]^{-1}\mathbf{Y}_{inv}\mathbf{T}_{oiL}) \end{bmatrix} \begin{bmatrix} \hat{i}_{inv} \\ \hat{v}_{oL} \end{bmatrix}
\tag{50}
$$

Equation (50) shows that the stability of the interface can be assessed based on the output admittance of the inverter ($\mathbf{Y}_{inv}$) and the input impedance of the grid ($\mathbf{Z}_{inL}$) by applying Nyquist Generalized stability criterion to $[\mathbf{I}+\mathbf{Z}_{inL}\mathbf{Y}_{inv}]^{-1}$ and $[\mathbf{I}+\mathbf{Y}_{inv}\mathbf{Z}_{inL}]^{-1}$ as stated also in Section 2.2.2. In grid-feeding mode, the inverter synchronizes the voltage (cf. Equation (51) and current (cf. Equation (50) of the PCC interface. The load system can be assumed to be such that the elements of the system are basically invariant to each other. Therefore, $\mathbf{G}_{ioL}$ and $\mathbf{T}_{oiL}$ are equal as the circuit-theoretical reciprocity theorem implies as well [86]. The voltage (cf. Equation (50)) and current (cf. Equation (51)) of the PCC interface (cf. Figure 13)

$$
\hat{i}_{pcc} = [\mathbf{I}+\mathbf{Y}_{inv}\mathbf{Z}_{inL}]^{-1}\hat{i}_{inv} - [\mathbf{I}+\mathbf{Y}_{inv}\mathbf{Z}_{inL}]^{-1}\mathbf{Y}_{inv}\mathbf{T}_{oiL}\hat{v}_{oL}
\tag{51}
$$

Similarly, as in case of grid-forming-mode operation, the phase of i_{oL} cannot be determined based on the available information, because $\mathbf{T}_{oiL}$ cannot be measured in practice. This means that the synchronizing angle cannot be corrected to produce real power at certain point in the grid system without external knowledge on the phase behavior from the grid coordinated control facilities.

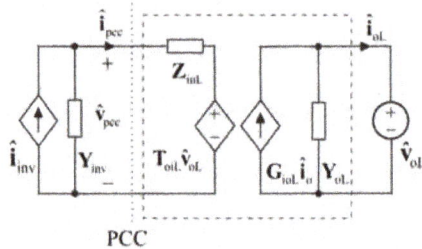

Figure 13. The equivalent circuits of the inverter output and the generalized power grid in grid-feeding mode of operation.

3. Experimental Evidence

The grid-forming-mode inverter, which is reported more in detail in Reference [26], is evaluated by applying Typhoon HIL real-time simulation setup, Boombox control platform, and Venable frequency response analyzer as shown in Figure 14.

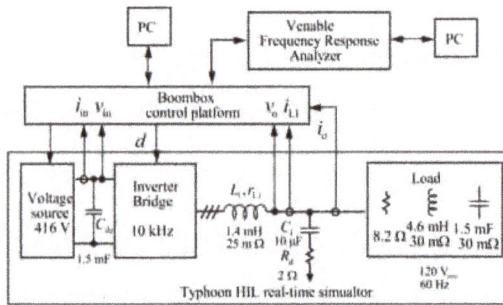

Figure 14. Grid-forming-mode inverter test setup.

The grid-feeding-mode inverter, which is reported more in detail Reference [87], is evaluated by using the real hardware setup as shown in Figure 15, where the PV (PVS7000) and grid (PAS15000) emulators are manufactured by Spitzenberger & Spies as well as the inverter is based on MyWay platform (MWINV-1044-SIC). The given frequency responses are measured applying the MIMO measurement technique reported in Reference [80], and the perturbation signals are injected via the grid emulator in dq domain.

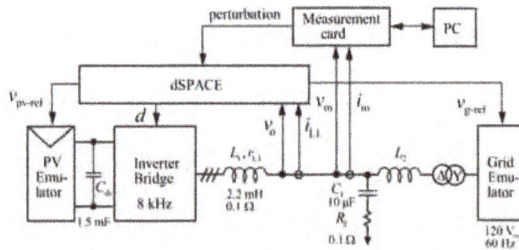

Figure 15. Grid-feeding-mode inverter test setup.

3.1. Grid-Forming-Mode Inverter

The inductor-current and output-voltage loop gains are given in Figures 16 and 17, where the effects of the different loads are visible: The unterminated (i.e., loaded by constant-current sink) loop gain is denoted by red color, the resistor-loaded loop gain by blue color, and the parallel-connected resistor-inductor-capacitor (RLC) loaded loop gain by black color, respectively. The unterminated loop (red) gain indicates unstable operation having the crossover frequency at 20 Hz. The resistor and RLC loading increases the crossover frequency to 300 Hz and they stabilize the converter as well. Figure 17 shows that the system is stable, when the output-voltage loop is properly tuned: Stabilization of the inverter requires that the voltage-loop crossover frequency is placed at higher frequency than the unstable pole of the system (cf. Figure 16 (red) vs. Figure 17 (red)).

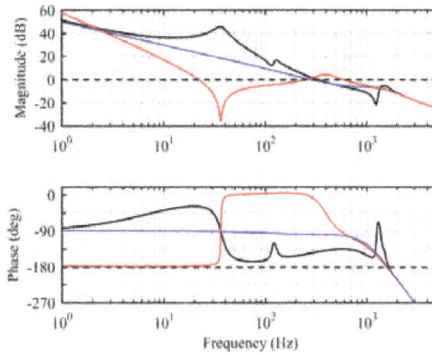

Figure 16. The inductor-current loop gain at unterminated mode (red), as resistor loaded (blue), and as parallel RLC loaded (black).

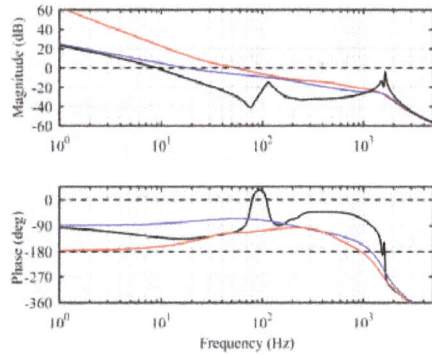

Figure 17. The output-voltage loop gain at unterminated mode (red), as resistor loaded (blue), and as parallel RLC loaded (black).

It may be obvious that the load impedance has significant effect on the converter dynamic behavior, which is actually the consequence of the rather high magnitudes of the output impedance as shown in Figure 18. The high-output impedance may be reduced by applying output current feedforward.

Figure 18 shows that the output impedance of the grid-forming inverter exhibits such symmetry that $Z_{o-d} = Z_{o-q}$ and $Z_{o-dq} = -Z_{o-qd}$ as discussed also in case of the grid impedance in Section 2.3.2. If the grid is asymmetrically loaded, then its impedance (admittance) may not have this property. It this property exists in inverter and grid then the two multivariable sensitivity functions will be identical.

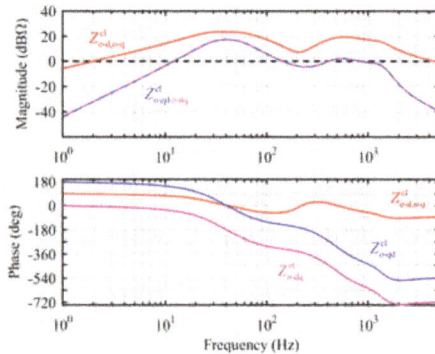

Figure 18. The measured closed-loop output impedances of the grid-forming inverter, where the red color denotes the closed-loop Z_{o-d} and Z_{o-q}, the blue color denotes Z_{o-qd}, and the magenta color Z_{o-dq}, respectively.

Figure 19 shows the measured (black) and predicted (red and blue) closed-loop input impedance of the grid-forming inverter, when the inverter is loaded with the parallel-connected RLC circuit (cf. Figure 14). The blue-colored input impedance denotes the case, where the analytical model includes all the output-terminal-side cross-couplings. The red-colored input impedance denotes the case, where the cross-couplings are omitted (cf. Equation (19), Section 2.1.2). According to Figure 19, the cross-couplings do not have significant contribution on the behavior of the load-affected input impedance. This phenomenon cannot be generalized, because the grid impedance may not exhibit similar symmetry in practice as in this special case.

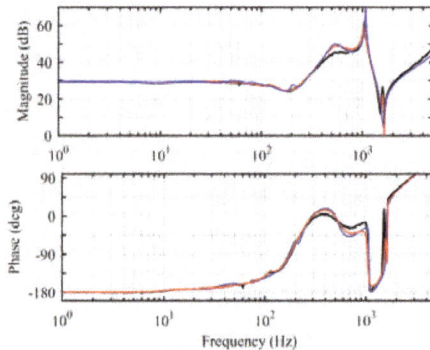

Figure 19. The RLC-load-affected input impedance of the grid-forming inverter, where the black color denotes the measured input impedance, the red color denotes the predicted input impedance with all the cross-couplings, and the blue color denotes the predicted input impedance omitting all the cross-couplings.

Figure 20 shows the frequency responses of the analytically predicted set of $Y_{in-\infty}$ given in Equation (13) (cf. Section 2.1.1). According to Reference [12], the low-frequency behavior of the closed-loop input admittance is dominated by the ideal input admittance, which in this case equals the d-component of $Y_{in-\infty}$ (cf. black line in Figure 20) having the value of approximately $-30 \text{ dB}\Omega^{-1}$ at the low frequencies. As an impedance, this value equals 30 dBΩ, respectively. Figure 19 shows the closed-loop input impedance of the inverter, where the low-frequency value equals approximately $Y_{in-\infty-d}^{-1}$ as stated in [12]. Similarly, as in DC-DC converters, the d-component of $Y_{in-\infty}$ exhibits negative-incremental admittance behavior as well.

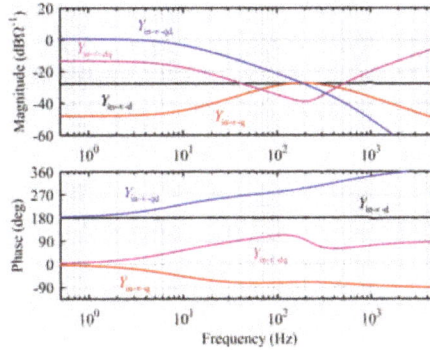

Figure 20. The set of predicted frequency responses of the multi-variable ideal input admittance $\mathbf{Y}_{\text{in}-\infty}$ computed according to Equation (13) in Section 2.1.1.

3.2. Grid-Feeding-Mode Inverter

The stability of the grid-feeding-mode inverter is evaluated by computing the eigenvalues of the multivariable impedance-based minor-loop gain (i.e., $\mathbf{Y}_{\text{inv}-\text{o}}\mathbf{Z}_{\text{grid}-\text{in}}$) and the Nyqyist plot of $\det\left[\mathbf{I} + \mathbf{Y}_{\text{inv}-\text{o}}\mathbf{X}_{\text{grid}-\text{in}}\right]$ as well as by computing the maximum singular value of the sensitivity function $\mathbf{S} = \left[\mathbf{I} + \mathbf{Y}_{\text{inv}-\text{o}}\mathbf{X}_{\text{grid}-\text{in}}\right]^{-1}$. The information given by the above-named impedance-based elements is validated by measuring the frequency response of the inductor-current loop as well as the time-domain behavior of the q component of the output current. The value of the series inductor L_2 (cf. Figure 15) is varied from 0 mH to 12 mH, respectively.

3.2.1. Impedance-Ratio-Based Analysis

Figure 21 shows the plots of the maximum singular value of $\mathbf{S} = \left[\mathbf{I} + \mathbf{Y}_{\text{inv}-\text{o}}\mathbf{X}_{\text{grid}-\text{in}}\right]^{-1}$ in respect to the frequency, when L_2 equals 0 and 12 mH, respectively. The red line shows that the robustness of stability is poor (i.e., $\bar{\sigma} > 1$) at the low frequencies up to 60 Hz and when the frequency exceeds 1 kHz.

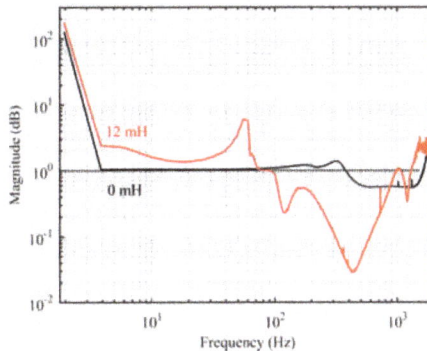

Figure 21. The maximum singular value of the multivariable impedance-based sensitivity function, where the black line denotes the singular value computed when L_2 equals 0 mH and the red line denotes the case when L_2 equals 12 mH.

Figure 22 shows the Bode plots of the eigenvalues of the multivariable impedance-based minor-loop gain (i.e., $\mathbf{Y}_{\text{inv}-\text{o}}\mathbf{X}_{\text{grid}-\text{in}}$) when L_2 equals 12 mH. The dashed lines denote the eigenvalues, when the cross-couplings are omitted. The dashed-line circles are placed at the frequencies, which indicate poor robustness of stability or unstable operation. The frequency response of λ_1

(red color) indicates that the phase margin is very low at the low frequencies, and the system is unstable approximately at 800 Hz (i.e., the derivative of the phase is negative, when crossing the ±180-degree line) [88]. The frequency response of λ_2 implies that the system is unstable approximately at 100 Hz and 1 kHz [89].

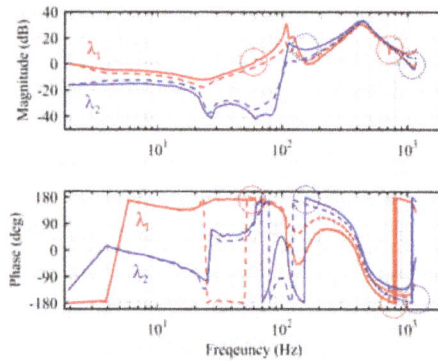

Figure 22. The Bode plots of the eigenvalues λ_1 (red) and λ_2 (blue), where the dashed lines denote the case having the cross-couplings omitted. The dashed-line circles denote the frequencies, where the stability of the system is questionable ($L_2 = 12$ mH).

Figure 23a shows the full Nyquist plot of λ_1, and Figure 23b shows the extended view of the plot in the vicinity of the critical point (−1,0). Figure 23b shows that the system is stable but the phase margin is very small.

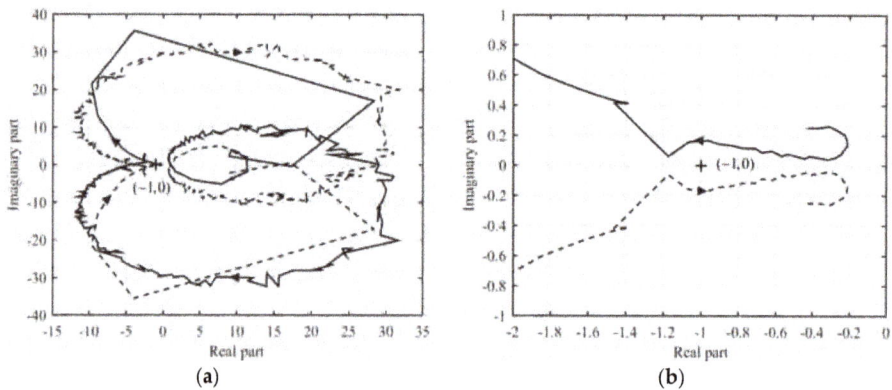

(a)

(b)

Figure 23. The Nyquist plot of λ_1: (**a**) the full plot, and (**b**) the extended plot in the vicinity of the point (−1,0) ($L_2 = 12$ mH). The solid line denotes contour for positive frequencies and the dashed line denotes the contour for negative frequencies. The arrowhead of the solid line shows the direction of increasing frequencies, respectively.

Figure 24a shows the full Nyquist plot of λ_2, and Figure 24b shows the extended view of the plot in the vicinity of the critical point (−1,0). Figure 24b implies that the system is unstable.

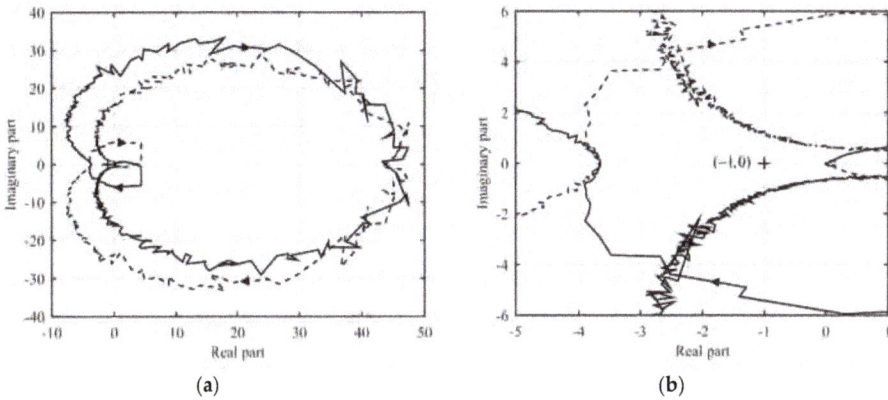

(a)   (b)

Figure 24. The Nyquist plot of λ_2: (**a**) the full plot, and (**b**) the extended plot in the vicinity of the critical point $(-1,0)$ ($L_2 = 12$ mH) (cf. the caption of Figure 23 for the explicit definitions of the other contents in the figure).

Figure 25 shows the Bode plot of $\det\left[\mathbf{I} + \mathbf{Y}_{\mathrm{inv}-o}\mathbf{X}_{\mathrm{grid}}\right]$, when L_2 equals 0, 7, and 12 mH, where the dashed lines indicate the frequency responses, from which the cross-couplings are omitted. The full-order and reduced-order responses are quite equal at the frequencies higher than 300 Hz but deviate from each other significantly at the frequencies in the range from 50 Hz to 300 Hz, respectively. Extracting stability information from Figure 25 is not similarly straightforward as from Figure 22.

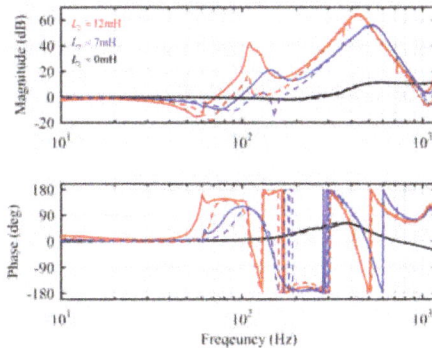

Figure 25. The frequency responses of $\det\left[\mathbf{I} + \mathbf{Y}_{\mathrm{inv}-o}\mathbf{X}_{\mathrm{grid}}\right]$, when L_2 equals 0, 7, and 12 mH. The solid lines denote the full-order responses, and the dashed lines denote the reduced-order responses, respectively.

Figure 26a shows the full-order Nyquist plot of $\det\left[\mathbf{I} + \mathbf{Y}_{\mathrm{inv}-o}\mathbf{X}_{\mathrm{grid}}\right]$, and Figure 26b shows the extended plot in the vicinity of the critical point $(0,0)$, when L_2 equals 12 mH. The solid line denotes the Nyquist contour for positive frequencies and the dashed line denotes the contour for negative frequencies, respectively. The arrowhead of the solid line denotes the direction of increasing frequency. Figure 26b indicates that the system is stable although the margins are extremely low.

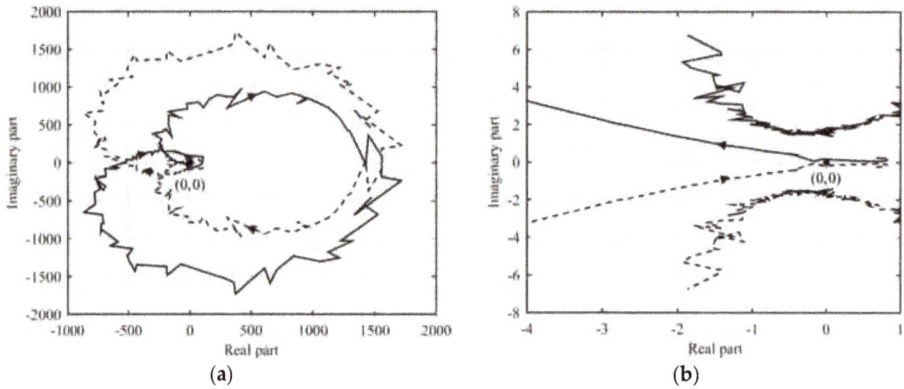

(a)   (b)

Figure 26. The Nyquist plot of $\det\left[\mathbf{I}+\mathbf{Y}_{\text{inv}-\text{o}}\mathbf{X}_{\text{grid}}\right]$: (a) the full plot, and (b) the extended plot in the vicinity of the critical point (0,0) (L_2 = 12 mH) (cf. the caption of Figure 23 for the explicit definitions of the other contents in the figure).

Figure 27 shows the extended Nyquist plots of the full (red) (cf. Figure 26b) and reduced-order (blue) contours in the vicinity of the critical point (0,0), when L_2 equals 12 mH. The reduced-order plot implies that the system is unstable, when encircling clockwise the critical point (0,0).

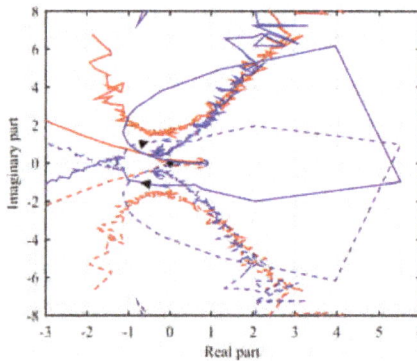

Figure 27. The extended Nyquist plots of the full (red) and reduced-order (blue) plots of $\det\left[\mathbf{I}+\mathbf{Y}_{\text{inv}-\text{o}}\mathbf{X}_{\text{grid}}\right]$, when L_2 equals 12 mH (cf. the caption of Figure 23 for the explicit definitions of the other contents in the figure).

The maximum singular-value plot in Figure 21 indicates that the robustness of stability is poor at the low frequencies up to 60 Hz, and at the frequencies exceeding 1 kHz (i.e., $\bar{\sigma} > 1$). The eigenvalue λ_2 (cf. Figure 24) implied that the system is unstable. The full-order Nyquist plot of $\det\left[\mathbf{I}+\mathbf{Y}_{\text{inv}-\text{o}}\mathbf{X}_{\text{grid}}\right]$ (cf. Figure 26) implied that the system is stable but the stability margins would be extremely low. As discussed in Reference [14], the Nyquist plot of $\det\left[\mathbf{I}+\mathbf{Y}_{\text{inv}-\text{o}}\mathbf{X}_{\text{grid}}\right]$ would give the most accurate prediction on the stability. It may be also obvious that the singular value does not directly indicate that the system is unstable.

3.2.2. Inductor-Current-Loop-Based Analysis

Figure 28 shows the measured (solid lines) inductor-current-loop frequency responses (Figure 28a, d-component, and Figure 28b, q-component), when L_2 equals 0 mH (blue lines) and 12 mH (red lines). Figure 28a shows that the crossover frequency (f_c) and phase margin of the designed d-component of current-loop gain (blue) equal 300 Hz and 30 degrees, respectively. In case of weak grid (red), the grid impedance has modified the d-component of current-loop gain to have three distinct gain crossovers approximately at 90 Hz, 122 Hz, and 143 Hz. The corresponding phase margins vary from 28 degrees to 63 degrees, respectively, which indicates that the inverter is stable. The figure shows also that the fourth gain crossover is very close to appear approximately at 1 kHz, where the margin is only 1 dB. The phase crossover frequency equals 1.3 kHz with a gain margin of 4 dB. The robustness of the stability is very poor especially in the frequency range from 20 Hz to 50 Hz, where the phase margin is close to zero.

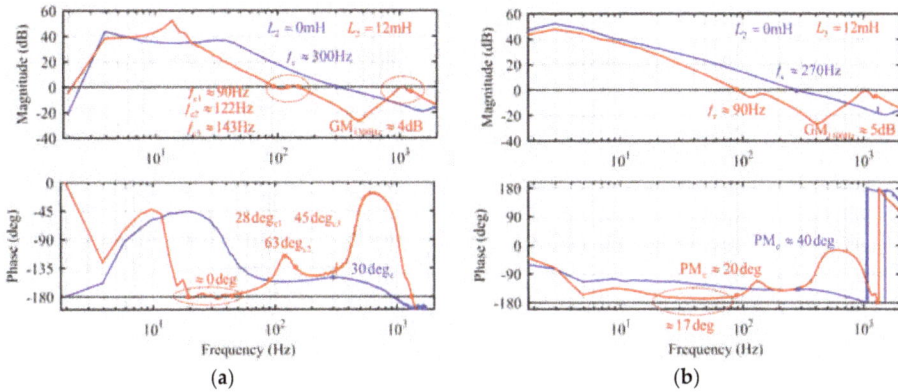

Figure 28. The measured (solid lines) current-loop gains at $L_2 = 0$ mH (blue) and 12 mH (red): (a) d-component, and (b) q-component.

Figure 28b shows that the crossover frequency (f_c) and phase margin of the designed q-component current loop (blue) equal 270 Hz and 40 degrees, respectively. In case of weak grid (red), the grid impedance has modified the q-component of the current-loop gain to have the crossover frequency approximately at 90 Hz with the phase margin of 20 degrees. As the figure shows, three other crossover frequencies are very close to appear similarly as in the d-component of the current-loop gain. The phase crossover frequency equals approximately 1.3 kHz with the gain margin of 5 dB. The corresponding phase margins vary from 28 degrees to 90 degrees, respectively, which indicates that the inverter is stable. The robustness of the stability is somewhat better than that of the d-component in the frequency range from 20 Hz to 50 Hz, where the phase margin is close to 17 degrees instead of close to zero. The stability information given by Figure 28 may not be absolutely true, because the cross-coupling terms will contribute also to the stability information.

Figure 29 shows the measured Nyquist plot of $\det[\mathbf{I} + \mathbf{L_C}]$ with cross-couplings (blue) and without cross-couplings (red), where $\mathbf{L_C}$ denotes the multivariable inductor-current feedback loop. The blue-colored Nyquist plot shows that the inverter is stable (i.e., no clockwise encirclement around the critical point (0,0)) but the margins would be low. The red-colored Nyquist plot implies that the inverter is unstable. Figures 26b and 27 (Section 3.2.1) give quite the same information on the state of stability.

Figure 30 shows the Nyquist plots of the eigenvalues λ_1 (red) and λ_2 (blue), where Figure 30a shows the whole plots, and Figure 30b, the extended part of the plots in the vicinity of the critical point $(-1,0)$. The plot of λ_2 implies that the system is unstable.

Figure 31 shows the plot of the maximum singular value of the sensitivity function $\mathbf{S_c} = [\mathbf{I} + \mathbf{L_c}]^{-1}$ at $L_2 = 0$ (black), and $L_2 = 12$ mH (red), respectively. The black-line response indicates that the phase margins of the original design are rather low (cf. Figure 28) and therefore, the original design is not robustly stable. The red-line responses indicate that the robustness of stability is lost at the frequencies from 50 Hz to 500 Hz as well as at the frequencies higher than 1 kHz.

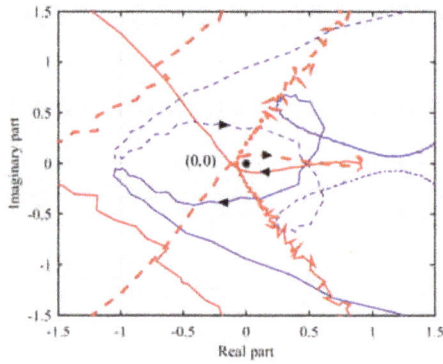

Figure 29. Nyquist plots of det$[\mathbf{I} + \mathbf{L}_C]$ with cross-couplings (blue line) and without cross-couplings (red line).

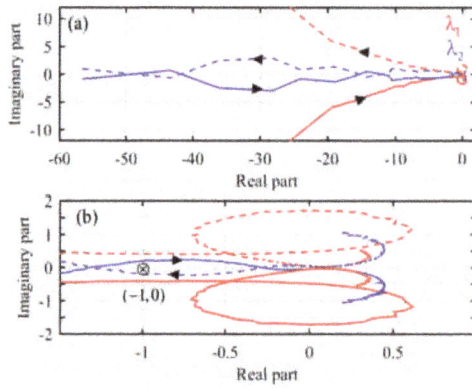

Figure 30. Nyquist plots of the current-loop-gain eigenvalues λ_1 (red) and λ_2: (**a**) full plots, and (**b**) extended plots in the vicinity of the critical point $(-1,0)$.

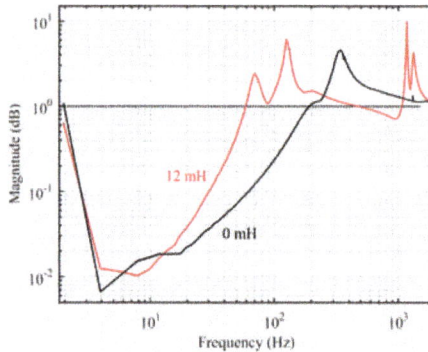

Figure 31. The maximum singular values of $\mathbf{S}_c = [\mathbf{I} + \mathbf{L}_c]^{-1}$ at $L_2 = 0$ (black line) and $L_2 = 12$mH.

The measured inductor-current-loop d and q components in Figure 28, the Nyquist plot of det$[\mathbf{I} + \mathbf{L}_C]$ in Figure 30, and the maximum singular values in Figure 31 imply that the stability of the converter is not robust and the converter may be very close to instability. Figure 32 shows the response of the inductor-current q component to a 5-A step change in its reference. The response indicates decaying oscillation approximately at 90 Hz, which equals the crossover frequency of the inductor-current-loop q component with the phase margin of 20 degrees. The response indicate definitively that the converter is stable similarly as the Nyquist plot of $\det\left[\mathbf{I} + \mathbf{Y}_{inv-o}\mathbf{X}_{grid}\right]$ in Figure 26b.

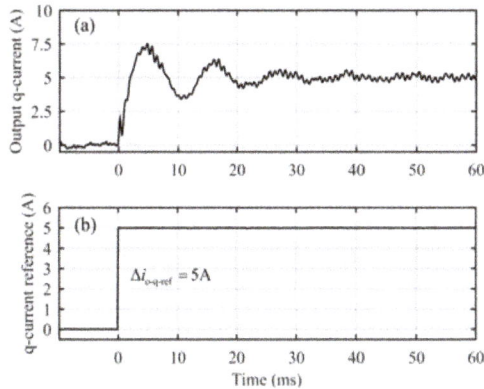

Figure 32. Response of inductor-current q component (**a**) to a 5A step change in its reference (**b**).

3.2.3. Discussions

The stability of the grid-feeding-mode inverter was assessed by means the measured output- terminal-side multivariable impedance ratio and the measured inductor-current-loop gains, when the weak-grid condition was emulated by adding a 12-mH series inductor in the grid side. The time-domain operation of the inverter implied stable operation. The Nyquist plots of $\det\left[\mathbf{I} + \mathbf{Y}_{inv-o}\mathbf{X}_{grid}\right]$ (cf. Figure 26b) and $\det[\mathbf{I} + \mathbf{L}_C]$ (cf. Figure 29) indicate explicitly that the inverter is stable but the stability margins are poor, which the time-domain step response of the inductor-current q component also confirmed explicitly (cf. Figure 32). The eigenvalue plots implied that the inverter is unstable. The impedance-ratio-based maximum-singular-value plot shows explicitly that the stability is not robust at the low frequencies (up 60 Hz) and at the high frequencies (>1 kHz). It was also shown that the obtainable stability information is very poor when the cross-couplings are neglected.

As discussed in Reference [89], the multivariable impedance measurements are extremely difficult to be performed due to the grid-impedance effect on the synchronization [90] as well as the highly varying nature of the grid impedance [91,92], and therefore, the accuracy of the measurements may be questionable. The same applies also to the measurements of the feedback-loop gains. The observed variance in the information provided by the different stability-assessment methods may be the consequence of the inaccurate measurement results as well.

4. Conclusions

The dynamic behavior and the factors affecting the sensitivity of the power electronic converters to the external impedance interactions are fully solved for the DC-DC converters as described in Reference ([12], Parts 2 and 3). The dynamic analysis of the three-phase power electronics converters is still in its infancy. The reason for this is obviously the high complexity of the elements affecting their dynamic behavior. The usual method to relax the complexity is to remove the cross-coupling terms of the associated multivariable transfer functions, which deteriorates also the obtainable stability information.

This paper has shown that the three-phase grid-tied converters have similar internal impedance-like parameters as the DC-DC converters have, which affect the converter sensitivity to the source and load-impedance-induced interactions. The source and load interactions of the three-phase grid-tied converter can be solved with ease by using proper software packages such as Matlab^{TM}. This paper provides the explicit formulations for the implicit internal parameters, which governs the interactions through the DC interface. The full-order formulations for the three-phase interface cannot be obtained, because the relevant source/load-affected transfer functions cannot be put into the form stipulated by the EET method. It is obvious that further studies are needed for analyzing the interactions and their relations to the implicit internal parameters.

Intensive research on the impedance-based stability analysis has been going on for several years already. The complexity of the analyses has led to simplifying the impedance-based minor-loop gains by omitting the cross-couplings between the different impedances constituting the minor-loop gain. The experimental measurements, in this paper, show clearly that the omitting of the cross-couplings will easily lead to inaccurate information on the robustness and state of the stability. We have shown, in this paper, that the stability assessment based on $\det\left[\mathbf{I} + \mathbf{Y}_{inv-o}\mathbf{X}_{grid}\right]$ may give the most accurate information on the state of stability, which was confirmed by assessing the stability by means of $\det[\mathbf{I} + \mathbf{L}_C]$ (i.e., the measured inductor-current feedback loop).

The content of the paper is intended to initialize extensive studies in the dynamic behavior of the three-phase grid-tied converters as well as in the measurement methods of dq-domain frequency responses to improve their accuracy.

Author Contributions: T.S. wrote the paper, T.M. has been responsible for the basic theories related to the inverters, M.B., H.A., T.R., and R.L. performed the practical measurements and provided the measurement data, K.Z. has been responsible for the theories related to the multivariable stability assessment.

Funding: This research received no external funding.

Conflicts of Interest: The authors declare no conflicts of interest.

References

1. Yuan, Y.; Biess, J.J. Some design aspects concerning input filters for DC-DC converters. In Proceedings of the IEEE Power Electronics Specialist Conference (IEEE PESC), Pasadena, CA, USA, 19–20 April 1971; pp. 66–76.
2. Sokal, N.O. System oscillations from negative input resistance at the power input of port switching regulator, amplifier, DC/DC converter, or DC/AC inverter. In Proceedings of the IEEE Power Electronics Specialist Conference (IEEE PESC), Pasadena, CA, USA, 11–13 June 1973; pp. 138–140.
3. Wester, G.W.; Middlebrook, R.D. Low-frequency characterization of switched dc-dc converters. *IEEE Trans. Aerosp. Electron. Syst.* **1973**, *AES-9*, 376–385. [CrossRef]
4. Middlebrook, R.D.; Ćuk, S. A general unified approach to modelling switching-converter power stages. In Proceedings of the IEEE Power Electronics Specialist Conference (IEEE PESC), Cleveland, OH, USA, 8–10 June 1976; pp. 18–34.
5. Middlebrook, R.D.; Ćuk, S. A general unified approach to modelling switching-converter power stages. *Int. J. Electron.* **1977**, *42*, 521–550. [CrossRef]
6. Middlebrook, R.D. Input filter considerations in design and application of switching regulators. In Proceedings of the Industry Application Society Annual Meeting (IEEE IAS), Chicago, IL, USA, 1–14 October 1976; pp. 91–107.
7. Middlebrook, R.D. Design techniques for preventing input-filter oscillations in switched-mode regulators. In Proceedings of the The 5th National Solid-State Power Conversion Conference (Powercon 5), San Francisco, CA, USA, 4–6 May 1978; pp. A3.1–A3.16.
8. Middlebrook, R.D. The two extra element theorem. In Proceedings of the Frontiers in Education Conference, West Lafayette, IN, USA, 21–24 September 1991; pp. 702–708.
9. Middlebrook, R.D. Null double injection and the extra element theorem. *IEEE Trans. Educ.* **1989**, *32*, 167–180. [CrossRef]
10. Vesti, S.; Suntio, T.; Oliver, J.A.; Prieto, R.; Cobos, J.A. Impedance-based stability and transient-performance assessment applying maximum peak criteria. *IEEE Trans. Power Electron.* **2013**, *28*, 2099–2104. [CrossRef]
11. Vesti, S.; Suntio, T.; Oliver, J.A.; Prieto, R.; Cobos, J.A. Effect of control method on impedance-based interactions in a buck converter. *IEEE Trans. Power Electron.* **2013**, *28*, 5311–5322. [CrossRef]

12. Suntio, T.; Messo, T.; Puukko, J. *Power Electronics Converters—Dynamics and Control in Conventional and Renewable Energy Applications*; Wiley-VCH: Weinheim, Germany, 2018.

13. Nyquist, H. Regeneration theory. *Bell. Syst. Tech. J.* **1932**, *11*, 126–147. [CrossRef]

14. Skogestad, S.; Postlethwaite, I. *Multivariable Feedback Control*; Wiley: Chichester, UK, 1996.

15. Maciejowski, J.M. *Multivariable Feedback Design*; Addison-Wesley: Wokingham, UK, 1989.

16. Wildrick, C.M.; Lee, F.C.; Cho, B.H.; Choi, B. A method of defining the load impedance specifications for a stable distributed power system. *IEEE Trans. Power Electron.* **1995**, *10*, 280–285. [CrossRef]

17. Sudhoff, S.D.; Glover, S.F.; Lamm, P.T.; Schmucker, D.H.; Delisle, D.E. Admittance space stability analysis of power electronic system. *IEEE Trans. Aerosp. Electron. Syst.* **2000**, *36*, 965–973. [CrossRef]

18. Feng, X.; Liu, J.; Lee, F.C. Impedance specifications for stable dc distributed power system. *IEEE Trans. Power Electron.* **2002**, *17*, 157–162. [CrossRef]

19. Riccobono, A.; Santi, E. Comprehensive review of stability criteria for DC power distribution systems. *IEEE Trans. Ind. Appl.* **2014**, *50*, 3525–3535. [CrossRef]

20. Wen, B.; Burgos, R.; Boroyevich, D.; Mattavelli, P.; Shen, Z. AC stability analysis and dq frame impedance specifications in power-electronics-based distributed power systems. *IEEE J. Emerg. Sel. Top. Power Electron.* **2017**, *5*, 1455–1465. [CrossRef]

21. Harnefors, L. Modeling of three-phase dynamic systems using complex transfer functions and transfer matrices. *IEEE Ind. Electron.* **2007**, *54*, 2239–2248. [CrossRef]

22. Messo, T.; Aapro, A.; Suntio, T. Generalized multivariable small-signal model of three-phase grid-connected inverter in DQ domain. In Proceedings of the IEEE 16th Workshop Control. Model. Power Electron. (COMPEL), Vancouver, BC, Canada, 12–15 June 2015; pp. 1–8.

23. Wang, Z.; Harnefors, L.; Blaabjerg, F. Unified impedance model of grid-connected voltage-source converters. *IEEE Trans. Power Electron.* **2018**, *33*, 1775–1787. [CrossRef]

24. Lu, D.; Wang, Z.; Blaabjerg, F. Impedance-based analysis of DC-link voltage dynamics in voltage-source converters. *IEEE Trans. Power Electron.* **2018**, in press. [CrossRef]

25. Aapro, A.; Messo, T.; Suntio, T. An accurate small-signal model of a three-phase VSI-based photovoltaic inverter with LCL filter to predict inverter output impedance. In Proceedings of the International Conference Power Electron. (ICPE ECCE Asia), Seoul, Korea, 1–5 June 2015; pp. 2267–2274.

26. Berg, M.; Messo, T.; Suntio, T. Frequency response analysis of load effect on dynamics of grid-forming inverter. In Proceedings of the Int. Power Elect. Conference (IPEC-Niigata ECCE Asia), Niigata, Japan, 20–24 May 2018; pp. 963–970.

27. MacFarlane, A.G.J.; Postlethwaite, I. The generalized Nyquist stability criterion. *Int. J. Control* **1977**, *25*, 81–127. [CrossRef]

28. Desoer, C.A.; Wang, Y.-T. On the generalized Nyquist stability criterion. *IEEE Trans. Autom. Control* **1980**, *25*, 187–196. [CrossRef]

29. Harnefors, L.; Buongiorno, M.; Lundberg, S. Input-admittance calculation and shaping for controlled voltage-source converters. *IEEE Trans. Ind. Electron.* **2007**, *54*, 3323–3334. [CrossRef]

30. Puukko, J.; Messo, T.; Suntio, T. Negative output impedance in three-phase grid-connected renewable energy source inverters based on reduced-order models. In Proceedings of the IET Conference on Renewable Power Generation (RPG 2011), Edinburgh, UK, 6–8 September 2011; pp. 1–6.

31. Harnefors, L.; Yepes, A.G.; Vidal, A.; Doval-Gandoy, J. Passivity-based controller design of grid-connected VSCs for prevention of electrical resonance instability. *IEEE Trans. Ind. Electron.* **2015**, *62*, 702–710. [CrossRef]

32. Zhang, C.; Cai, X.; Li, Z.; Rygg, A.; Molinas, M. Properties and physical interpretations of the dynamic interactions between voltage source converters and grid: Electrical oscillation and its stability control. *IET Power Electron.* **2017**, *10*, 894–902. [CrossRef]

33. Nousiainen, L.; Suntio, T. Simple VSI-based single-phase inverter: Dynamical effect of photovoltaic generator and multiplier-based grid synchronization. In Proceedings of the IET Conference on Renewable Power Generation (RPG 2011), Edinburgh, UK, 6–8 September 2011; pp. 1–6.

34. Khazaei, J.; Miao, Z.; Piyasinghe, L. Impedance-model-based MIMO analysis of power synchronization control. *Electr. Power Syst. Res.* **2018**, *154*, 341–351. [CrossRef]

35. Cho, Y.; Hur, K.; Kang, Y.C.; Muljadi, E. Impedance-based stability analysis in grid interconnection impact study owing to the increased adoption of converter-interfaced generators. *Energies* **2017**, *10*, 1355. [CrossRef]

36. Krismanto, A.U.; Mithulananthan, N.; Krause, O. Stability of renewable energy based microgrid in autonomous operation, Sustainable. *Energy Grids Netw.* **2018**, *13*, 134–147.

37. Krismanto, A.U.; Mithulananthan, N.; Kamwa, I. Oscillatory stability assessment of microgrid in autonomous operation with uncertainties. *IET Renew. Power Gener.* **2016**, *31*, 494–504. [CrossRef]

38. Krismanto, A.U.; Mithulananthan, N. Identification of modal interaction and small signal stability in autonomous microgrid operation. *IET Gener. Transm. Distrib.* **2018**, *12*, 247–257. [CrossRef]

39. Liu, H.; Sun, J. Voltage stability and control of offshore wind farms with AC collection and HVDC transmission. *IEEE J. Emerg. Sel. Top. Power Electron.* **2014**, *2*, 1181–1189.

40. Amin, M.; Molinas, M. Understanding the origin of oscillatory phenomena observed between wind farms and HVDC systems. *IEEE J. Emerg. Sel. Top. Power Electron.* **2017**, *5*, 378–392. [CrossRef]

41. Liu, Z.; Su, M.; Sun, Y.; Han, H.; Hou, X.; Guerrero, J.M. Stability analysis of DC microgrids with constant power load under distributed control methods. *Automatica* **2018**, *90*, 62–72. [CrossRef]

42. AL-Nussairi, M.; Bayindir, R.; Padmanaban, S.; Mohet-Popa, L.; Siano, P. Constant power loads (CPL) with microgrids: Problem definition, stability analysis and compensation techniques. *Energies* **2017**, *10*, 1656. [CrossRef]

43. Du, W.; Jiang, G.; Erickson, M.; Lasseter, R.H. Voltage-source control of PV inverter in a CERTS microgrid. *IEEE Trans. Power Deliv.* **2014**, *29*, 1726–1734. [CrossRef]

44. Cespedes, M.; Sun, J. Adaptive control of grid-connected inverters based on online grid impedance measurement. *IEEE Trans. Sustain. Energy* **2014**, *5*, 516–523. [CrossRef]

45. Krishnayya, P.C.S.; Piwko, R.J.; Weaver, T.L.; Bahrman, M.P.; Hammad, A.E. DC-transmission terminating at low short circuit ratio locations. *IEEE Trans. Power Deliv.* **1986**, *1*, 308–318. [CrossRef]

46. Diedrichs, V.; Beekmann, A.; Busker, K.; Nikolai, S.; Adloff, S. Control of wind power plants utilizing voltage source converter in high-impedance grids. In Proceedings of the 2012 IEEE Power and Energy Society General Meeting, San Diego, CA, USA, 22–26 July 2012; pp. 1–8.

47. Zhou, J.Z.; Ding, H.; Fan, S.; Zhang, Y.; Gole, A.M. Impact of short-circuit ratio and phase-locked-loop parameters on the small-signal behavior of a VSC-HVDC converter. *IEEE Trans. Power Deliv.* **2014**, *29*, 2287–2296. [CrossRef]

48. Etxegarai, P.; Egui, P.; Torres, E.; Iturregi, A.; Valverde, V. Review of grid connection requirements for generation assets in weak power grids. *Renew. Sustain. Energy Rev.* **2015**, *41*, 1501–1514. [CrossRef]

49. Schwanka Trevisan, A.; Mendonca, A.; Fisher, M.; Adloff, S.; Nikolai, S.; El-Deib, A. Process and tools for optimizing wind power projects connected to weak grids. *IET Renew. Power Gener.* **2018**, *12*, 539–546. [CrossRef]

50. Krishayya, P.C.; Adapa, R.; Holm, M. *IEEE Guide for Planning DC Links Terminating at AC Locations Having Low Short-Circuit Capacities*; IEEE Std 1204-1997 (R2003); IEEE: Piscataway, NJ, USA, 1997.

51. Wen, B.; Dong, D.; Boroyevich, D.; Burgos, R.; Mattavelli, P.; Shen, Z. Impedance-based analysis of grid-synchronization stability for three-phase parallel converters. *IEEE Trans. Power Electron.* **2016**, *31*, 26–38. [CrossRef]

52. Amin, M.; Rygg, A.; Molinas, M. Self-synchronization of wind farm in an MMC-based HVDC system: A stability investigation. *IEEE Trans. Energy Convers.* **2017**, *32*, 458–470. [CrossRef]

53. Suul, J.A.; D'Arco, S.; Rodriquez, P.; Molinas, M. Impedance-compensated grid synchronisation for extending the stability range of weak grids with voltage source converters. *IET Gener. Transm. Distrib.* **2016**, *10*, 1315–1326. [CrossRef]

54. Li, X.; Lin, H. Stability analysis of grid-connected converters with different implementations of adaptive PR controllers under weak grid conditions. *Energies* **2018**, *11*, 2004. [CrossRef]

55. Blanchard, J. History of electrical resonance. *Bell Syst. Tech. J.* **1941**, *20*, 415–433. [CrossRef]

56. Agbemuko, A.J.; Dominguez-Garcia, J.L.; Prieto-Araujo, E.; Gomis-Bellmunt, O. Impedance modelling and parametric sensitivity of a VSC-HVDC system: New insights on resonances and interactions. *Energies* **2018**, *11*, 845. [CrossRef]

57. Enslin, J.H.R.; Heskes, P.J.M. Harmonic interaction between a large number of distributed power inverters and the distribution network. *IEEE Trans. Power Electron.* **2004**, *19*, 1586–1593. [CrossRef]

58. Wang, X.; Blaabjerg, F.; Wu, W. Modeling and analysis of harmonic stability in an AC power-electronics-based power system. *IEEE Trans. Power Electron.* **2014**, *29*, 6421–6432. [CrossRef]

59. Pereira, H.A.; Freijedo, F.D.; Silva, M.M.; Mendes, V.F.; Teodorescu, R. Harmonic current prediction by impedance modeling of grid-tied inverters: A 1.4 MW PV plant case study. *Electr. Power Energy Syst.* **2017**, *93*, 30–38. [CrossRef]

60. Du, Y.; Lu, D.D.-C.; James, G.; Cornforth, D. Modeling and analysis of current harmonic distortion from grid connected PV inverters under different operating conditions. *Sol. Energy* **2017**, *94*, 182–194. [CrossRef]

61. Chicco, G.; Schlabbach, J.; Spertino, F. Experimental assessment of the waveform distortion in grid-connected photovoltaic installations. *Sol. Energy* **2009**, *83*, 1026–1039. [CrossRef]

62. Pakonen, P.; Hilden, A.; Suntio, T.; Verho, P. Grid-connected PV-power-plant-induced power quality problems—Experimental evidence. In Proceedings of the 2016 18th European Conference on Power Electronics and Applications (EPE'16 ECCE Europe), Karlsruhe, Germany, 5–9 September 2016; pp. 1–10.

63. Li, C. Unstable operation of photovoltaic inverter from filed experiences. *Power Eng. Lett.* **2018**, *33*, 1013–1015.

64. Rocabert, J.; Luna, A.; Blaabjerg, F.; Rodriguez, P. Control of power converters in AC microgrids. *IEEE Trans. Power Electron.* **2012**, *27*, 4734–4749. [CrossRef]

65. Viinamäki, J.; Suntio, T.; Kuperman, A. Grid-forming-mode operation of boost-power-stage converter in PV-interfacing applications. *Energies* **2017**, *10*, 1033. [CrossRef]

66. Wang, J.; Chang, N.; Feng, X.; Monti, A. Design of a generalized algorithm for parallel inverters for smooth microgrid transition operation. *IEEE Trans. Ind. Electron.* **2015**, *62*, 4900–4914. [CrossRef]

67. Blaabjerg, F.; Teodorescu, R.; Liserre, M.; Timbus, A.V. Overview of control and grid synchronization for distributed power generation. *IEEE Trans. Ind. Electron.* **2006**, *53*, 1398–1409. [CrossRef]

68. Kolesnik, S.; Sitbon, M.; Agranovich, G.; Kuperman, A.; Suntio, T. Comparison of photovoltaic and wind generators as dynamic input sources to power processing interfaces. In Proceedings of the 2nd International Conference Intelligent Energy & Power Syst. (IEPS), Kiev, Ukraine, 7–11 June 2016; pp. 1–5.

69. Kolesnik, S.; Kuperman, A. On the similarity between low-frequency equivalent circuits of photovoltaic and wind generators. *IEEE Trans. Energy Convers.* **2015**, *30*, 407–409. [CrossRef]

70. Kolesnik, S.; Kuperman, A. Analytical derivation of electric-side maximum power line for wind generators. *Energies* **2017**, *10*, 10. [CrossRef]

71. Kolesnik, S.; Sitbon, M.; Batzelis, E.; Suntio, T.; Kuperman, A. Solar irradiation independent expression for photovoltaic generator maximum power line. *IEEE J. Photovolt.* **2017**, *7*, 1416–1420. [CrossRef]

72. Xia, Y.; Ahmed, K.H.; Williams, B. Wind turbine power coefficient analysis of a new maximum power point tracking technique. *IEEE Trans. Ind. Electron.* **2013**, *60*, 1122–1132. [CrossRef]

73. Suntio, T.; Leppäaho, J.; Huusari, J.; Nousiainen, L. Issues on solar-generator interfacing with current-fed MPP-tracking converters. *IEEE Trans. Power Electron.* **2010**, *25*, 2409–2419. [CrossRef]

74. Mäki, A.; Valkealahti, S.; Suntio, T. Dynamic Terminal Characteristics of a Solar Generator, it7680. Available online: https://www.google.com.tw/url?sa=t&rct=j&q=&esrc=s&source=web&cd=1&ved=2ahUKEwiLmfHHm5XgAhVb7mEKHYNzAtwQFjAAegQIBBAB&url=https%3A%2F%2Fieeexplore.ieee.org%2Fdocument%2F5606786%2F&usg=AOvVaw3fqyKWELX8TcRTwHvS4mp3 (accessed on 25 May 2018).

75. Mäki, A.; Valkealahti, S. Effect of photovoltaic generator components on the number of MPPs under partial shading conditions. *IEEE Trans. Energy Convers.* **2013**, *28*, 1008–1017. [CrossRef]

76. Wyatt, J.; Chua, L. Nonlinear resistive maximum power theorem with solar cell application. *IEEE Trans. Circuits Syst.* **1983**, *30*, 824–828. [CrossRef]

77. Suntio, T. *Dynamic Profile of Switched-Mode Converter—Modeling, Analysis and Control*; Wiley-VCH: Weinheim, Germany, 2009.

78. Bakhshzadeh, M.K.; Wang, Z.; Blaabjerg, F.; Hjerrild, J.; Kocewiak, L.; Leth Bak, K.; Hesselbaek, B. Couplings in phase domain impedance modeling of grid-connected converters. *IEEE Trans. Power Electron.* **2016**, *31*, 6792–6796.

79. Liu, Z.; Liu, J.; Bao, W.; Zhao, Y. Infinity-norm of impedance-based stability criterion for three-phase AC distribution power systems with constant power loads. *IEEE Trans. Power Electron.* **2015**, *30*, 3030–3043. [CrossRef]

80. Roinila, T.; Messo, T.; Santi, E. MIMO-identification techniques for rapid impedance based stability assessment of three-phase systems in dq-domain. *IEEE Trans. Power Electron.* **2018**, *33*, 4015–4022. [CrossRef]

81. Wen, B.; Boroyevich, D.; Burgos, R.; Mattavelli, P.; Shen, Z. Small-signal stability analysis of three-phase AC systems in the presence of constant power loads based on measured d-q frame impedances. *IEEE Trans. Power Electron.* **2015**, *30*, 5952–5963. [CrossRef]

82. Wen, B.; Boroyevich, D.; Burgos, R.; Mattavelli, P.; Shen, Z. Analysis of d-q small-signal impedance of grid-tied inverters. *IEEE Trans. Power Electron.* **2016**, *31*, 675–687. [CrossRef]
83. Sun, J. Impedance-based stability criterion for grid-connected inverters. *IEEE Trans. Power Electron.* **2011**, *26*, 3075–3078. [CrossRef]
84. Wen, B.; Boroyevich, D.; Burgos, R.; Mattavelli, P.; Shen, Z. Inverse Nyquist stability criterion for grid-tied inverters. *IEEE Trans. Power Electron.* **2017**, *32*, 1548–1556. [CrossRef]
85. Dong, D.; Wen, B.; Boroyevich, D.; Mattavelli, P.; Xue, Y. Analysis of phase-locked loop low-frequency stability in three-phase grid-connected power converters. *IEEE Trans. Ind. Electron.* **2015**, *62*, 310–321. [CrossRef]
86. Tse, C.K. *Linear Circuit Analysis*; Addison-Wesley Longman: Harlow, UK, 1998.
87. Aapro, A.; Messo, T.; Suntio, T.; Roinila, T. Effect of active damping on output impedance of three-phase grid-connected converter. *IEEE Trans. Ind. Electron.* **2017**, *64*, 7532–7541. [CrossRef]
88. Francis, G.; Burgos, R.; Boroyevich, D.; Waqng, F.; Karimini, K. An algorithm and implementation system for measuring impedance in the D-Q domain. In Proceedings of the IEEE Energy Convers. Cong. & Expo, Phoenix, AZ, USA, 17–22 September 2011; pp. 3221–3228.
89. Liao, Y.; Wang, X. General rules of using bode plots for impedance-based stability analysis. In Proceedings of the IEEE 19th Workshop Control. Model. Power Electron. (COMPEL), Padua, Italy, 25–28 June 2018; pp. 1–6.
90. Gong, H.; Yang, D.; Wang, X. Impact of synchronization phase dynamics on dq impedance measurement. In Proceedings of the IEEE 19th Workshop Control. Model. Power Electron (COMPEL), Padova, Italy, 25–28 June 2018; pp. 1–5.
91. Jessen, L.; Fuchs, F.W. Modeling of inverter output impedance for stability analysis in combination with measured grid impedance. In Proceedings of the IEEE 16th Workshop Control. Model. Power Electron. (COMPEL), Vancouver, BC, Canada, 12–15 June 2015; pp. 1–7.
92. Jessen, L.; Gunter, S.; Fuchs, F.W.; Gottschalk, M.; Hinrichs, H.-J. Measurement result and performance analysis of the grid impedance in different low voltage grids for wide frequency band to support grid integration of renewables. In Proceedings of the IEEE Energy Conversion Cong. & Expo (IEEE ECCE), Montreal, QC, Canada, 20–24 September 2015; pp. 1960–1967.

energies

MDPI

Article

Using High-Bandwidth Voltage Amplifier to Emulate Grid-Following Inverter for AC Microgrid Dynamics Studies

Tuomas Messo [1,*], Roni Luhtala [2], Tomi Roinila [2], Erik de Jong [3], Rick Scharrenberg [4], Tommaso Caldognetto [5], Paolo Mattavelli [5], Yin Sun [3] and Alejandra Fabian [3]

1   Department of Electrical Energy Engineering, Tampere University of Technology, 33720 Tampere, Finland
2   Department of Automation and Hydraulics, Tampere University of Technology, 33720 Tampere, Finland;
     roni.luhtala@tut.fi (R.L.); tomi.roinila@tut.fi (T.R.)
3   Department of Electrical Engineering, Eindhoven University of Technology, 5600 Eindhoven,
     The Netherlands; Erik.deJong@dnvgl.com (E.d.J.); Yin.Sun@dnvgl.com (Y.S.);
     Alejandra.Fabian@dnvgl.com (A.F.)
4   DNV-GL, 6812 Arnhem, The Netherlands; Rick.Scharrenberg@dnvgl.com
5   Department of Management and Engineering, University of Padova, 35121 Padova, Italy;
     tommaso.caldognetto@unipd.it (T.C.); paolo.mattavelli@unipd.it (P.M.)
*   Correspondence: tuomas.messo@tuni.fi

Received: 12 December 2018; Accepted: 22 January 2019; Published: 25 January 2019

Abstract: AC microgrid is an attractive way to energize local loads due to remotely located renewable generation. The AC microgrid can conceptually comprise several grid-forming and grid-following power converters, renewable energy sources, energy storage and local loads. To study the microgrid dynamics, power-hardware-in-the-loop (PHIL)-based test setups are commonly used since they provide high flexibility and enable testing the performance of real converters. In a standard PHIL setup, different components of the AC microgrid exist as real commercial devices or electrical emulators or, alternatively, can be simulated using real-time simulators. For accurate, reliable and repeatable results, the PHIL-setup should be able to capture the dynamics of the microgrid loads and sources as accurately as possible. Several studies have shown how electrical machines, dynamic RLC loads, battery storages and photovoltaic and wind generators can be emulated in a PHIL setup. However, there are no studies discussing how a three-phase grid-following power converter with its internal control functions should be emulated, regardless of the fact that grid-following converters (e.g., photovoltaic and battery storage inverters) are the basic building blocks of AC microgrids. One could naturally use a real converter to represent such dynamic load. However, practical implementation of a real three-phase converter is much more challenging and requires special knowledge. To simplify the practical implementation of microgrid PHIL-studies, this paper demonstrates the use of a commercial high-bandwidth voltage amplifier as a dynamic three-phase power converter emulator. The dynamic performance of the PHIL setup is evaluated by identifying the small-signal impedance of the emulator with various control parameters and by time-domain step tests. The emulator is shown to yield the same impedance behavior as real three-phase converters. Thus, dynamic phenomena such as harmonic resonance in the AC microgrid can be studied in the presence of grid-following converters.

Keywords: DC-AC power converters; impedance emulation; stability analysis; power-hardware-in-the-loop

1. Introduction

Three-phase power electronics converters are essential building blocks of AC microgrids. They are used for interfacing distributed resources, such as photovoltaic generators, fuel cells and wind generators, electrical energy storage and electrical loads of various types. A conceptual AC microgrid formed by renewable generation, conventional synchronous generator, local DC and AC loads, energy storage and possibility for grid connection, is illustrated in Figure 1. The complexity of the microgrid increases when more sources and loads are replaced by the power electronics interfaces. Microgrids based solely on power electronic converters can be already found on ships, airplanes and tractions systems [1–3]. To reach power system with 100 percent renewable energy generation, the power system should rely on grid-forming converters to regulate AC voltage and frequency [4]. The presence of many different converters, from various manufacturers and with different internal control parameters, complicates model-based stability and power quality studies of the AC microgrid. Power-hardware-in-the-loop (PHIL) has been proven to be an effective method to study systems that are difficult to analyze using analytical or numerical models [5].

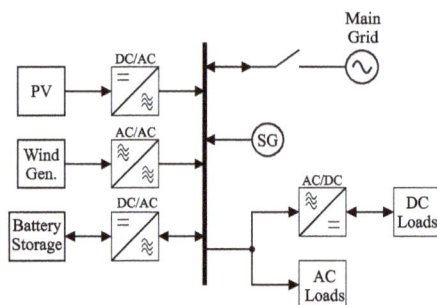

Figure 1. An AC microgrid in which dynamics are dominated by power electronics converters.

Power quality and stability problems caused by the interaction between the converters and the grid impedance is currently gaining significant amount of attention in industry and academia [6–8]. The converter–grid interface suffers from resonance if the ratio of converter and grid impedance does not satisfy the Nyquist stability criterion [9–11]. Moreover, the control performance and stability margin of a grid-forming converter can be significantly reduced by the load impedance seen at its output terminals [12]. The most challenging load impedances are RLC-resonant passive load and other power converters, such as grid-following inverters and active front-ends.

Dynamics of the AC microgrid can be studied using different simulation software, hardware-in-the-loop (HIL) or power-hardware-in-the-loop (PHIL) simulations and experimental tests on real equipment. HIL and controller-in-the-loop (CIL) simulations are effective methods to eliminate critical programming and hardware errors from the power converter control platform before actual prototype implementation [13]. PHIL tests have gained a lot of attention recently due to the fact that real converters can be tested in realistic conditions that are also well defined and repeatable. PHIL tests are no longer restricted to testing scaled-down prototypes since power amplifiers with a rating of several megawatts are available [14]. Moreover, the equipment in the PHIL test setup can be oversized to withstand large over-currents and over-voltages to enable reliable and fast protection and to avoid physical damage to converter under test.

Many papers have studied the use of the PHIL-concept to emulate the impedance of the power system in order to have an accurate electrical equivalent for the transmission line in order to test the performance of grid-following inverters [15]. The real grid impedance varies over time, which makes its emulation using a set of passive components impractical [16]. A three-phase back-to-back converter is used to emulate transient behavior and different load characteristics of an induction motor in [17], with a further extension covering saturation effects and experimental results in [18].

The emulation of an induction machine is considered in [19] using a combination of real-time digital simulator and back-to-back converter. In [20], a three-phase inverter is used to emulate passive RLC loads, including the possibility to emulate asymmetrical loading conditions. Furthermore, electrical emulation of AC loads has been demonstrated recently by several authors [21–24]. A PHIL test bench is presented in [25] to emulate an RLC-load impedance for unintentional islanding tests. Furthermore, the impedance behavior of the RLC load is considered and experimental impedance measurements up to 3 kHz are provided.

A vast amount of literature discusses impedance emulation of different components of the AC microgrid. However, a grid-following converter is an essential component in AC microgrid as it can be used as the DC-AC interface between various renewable source or energy storage and the AC microgrid. To allow reliable studies of dynamic phenomena of the AC microgrid, the PHIL-setup should be able to represent the small-signal impedance of the loads and sources as accurately as possible. Several studies have shown how electrical machines, dynamic RLC loads, battery storages and photovoltaic and wind generators can be emulated in a PHIL setup. However, there are no studies showing how a three-phase grid-following power converter with its control functions, such as AC current control, grid-voltage-feedforward and grid synchronization, should be emulated. One could naturally use a real inverter prototype to represent such dynamic load. However, practical implementation of a real three-phase converter is challenging and requires special knowledge on, e.g., the characteristics of the semiconductor switches, signal conditioning, over-current and over-voltage protection, auxiliary power circuitry design and electrical safety. Therefore, this paper shows how a standard laboratory voltage amplifier can be turned into a three-phase power converter, requiring only a passive AC filter to represent the passive parts of the converter impedance and a real-time simulator for implementing the control functions to represent the active parts of the converter impedance.

This paper discusses the control design and implementation of a power-hardware-in-the-loop (PHIL) setup based on a four-quadrant voltage amplifier to emulate the small-signal behavior of three-phase grid-following converter. The output impedance of the emulator is validated by frequency response measurements, which is shown to capture the small-signal dynamics caused by AC current control, grid-voltage feedforward and grid synchronization accurately. Therefore, the proposed PHIL setup can be applied to study stability and power quality problems caused by dynamic anomalies, such as harmonic resonance in a AC microgrid with a complex structure, as shown in Figure 1.

The main contributions of this paper are summarized as:

- It is shown how a standard voltage amplifier can be configured to represent a three-phase grid-following converter in an AC microgrid PHIL test setup.
- The output impedance of the emulator concept is verified by frequency response measurements.
- The emulator is able to represent the effects of current control, phase-locked-loop and grid voltage feedforward in its output impedance behavior.
- It is demonstrated how the internal dynamics of the voltage amplifier can be merged numerically to the dynamic model of the emulator, to necessitate accurate design of control loops and to develop the small-signal impedance model for validation of the frequency response measurements.

Section 2 describes the implementation and control design of the PHIL setup. Section 3 derives the small-signal impedance model, which is used for validation. Section 4 shows that the impedance emulator can replicate accurately the effects of phase-locked-loop, current control and grid voltage feedforward. Conclusions are derived in Section 5.

2. Emulating Converters Using a Laboratory Voltage Amplifier

Practical tests were carried out in the Flex Power Grid laboratory of DNV-GL, located in Arnhem, the Netherlands. The laboratory is equipped with a 16-channel voltage/current amplifier. The amplifier has several output stages, which were configured in groups of three, as illustrated in Figure 2. Each amplifier can operate in four quadrants, thus each can generate or absorb real and reactive power.

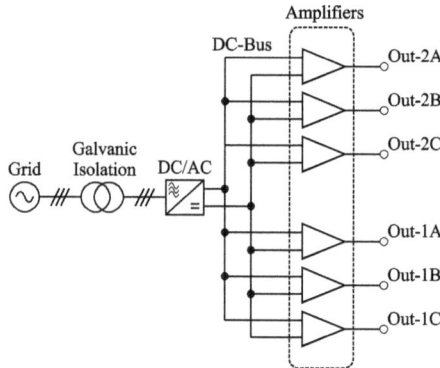

Figure 2. Six output channels of the power amplifier configured as two groups.

The exact parameters and internal control functions of the amplifiers were not known. Therefore, dynamic characterization had to be carried out before attempting to emulate the impedance of a grid-following converter. The output stages Out-1A to Out-1C were used to emulate the output terminals of a three-phase grid-following converter and output stages Out-2A to Out-2C were configured to emulate a three-phase current/voltage sink to generate the nominal loading conditions for the emulator. The rated power of a single amplifier was 35 kVA. Thus, the maximum power of the emulator was 105 kVA.

The closed-loop frequency response of amplifier group 1 was first identified, over which the control loops, mimicking inverter control dynamics, were later designed. A synchronous reference frame was realized using a real-time simulator from OPAL-RT. Transformation between the natural (abc) reference frame and the synchronous (dq0) reference frames was achieved using the transformation matrices defined in Equations (1) and (2). The principle of the measurement setup is shown in Figure 3. The frequency response of the lower amplifier group was measured by injecting a sine-sweep perturbation to its reference voltage vector $\mathbf{v}_{\mathrm{dq1\text{-}ref}}$ and by using a frequency response analyzer to extract the frequency response from the reference to the actual amplifier output voltage vector $\mathbf{v}_{\mathrm{dq1}}$. The amplifier was loaded by amplifier group 2, which was configured to act as a current sink. The phase-to-ground voltage was set to 230 V_{rms} and grid frequency to 50 Hz.

$$
\begin{bmatrix} v_d \\ v_q \\ v_0 \end{bmatrix} = \frac{2}{3} \begin{bmatrix} \cos(\theta_s) & \cos\left(\theta_s - \frac{2\pi}{3}\right) & \cos\left(\theta_s - \frac{4\pi}{3}\right) \\ -\sin(\theta_s) & -\sin\left(\theta_s - \frac{2\pi}{3}\right) & -\sin\left(\theta_s - \frac{4\pi}{3}\right) \\ \frac{1}{2} & \frac{1}{2} & \frac{1}{2} \end{bmatrix} \begin{bmatrix} v_a \\ v_b \\ v_c \end{bmatrix} \tag{1}
$$

$$
\begin{bmatrix} v_a \\ v_b \\ v_c \end{bmatrix} = \frac{2}{3} \begin{bmatrix} \cos(\theta_s) & \sin(\theta_s) & 1 \\ \cos\left(\theta_s - \frac{2\pi}{3}\right) & \sin\left(\theta_s - \frac{2\pi}{3}\right) & 1 \\ \cos\left(\theta_s - \frac{4\pi}{3}\right) & \sin\left(\theta_s - \frac{4\pi}{3}\right) & 1 \end{bmatrix} \begin{bmatrix} v_d \\ v_q \\ v_0 \end{bmatrix} \tag{2}
$$

Figure 4 shows the measured frequency response from the reference values to actual voltages produced by the amplifier as solid lines. According to the manufacturer, the amplification stays within ±1.5 dB up to 5 kHz and the gain experiences its largest magnitude (4 dB) at 17 kHz. However, under loaded condition, the +1.5 dB point was located roughly at 2.5 kHz and the maximum measured amplification was 9.3 dB. Thus, rather than rely on the declared characteristics, it is important to measure the amplifier frequency response in the actual test setup, since the operating point may alter the final response. Otherwise, the designed control loops of the inverter emulator will be erroneous. Moreover, the actual bandwidth may differ because the amplifiers were used in a three-phase configuration. The bandwidth of the amplifier was considered sufficient since grid-following converters usually employ current control with bandwidth of less than 1 kHz. As a comparison, the 2 kHz current control bandwidth of the induction motor emulator in [18] was considered sufficient.

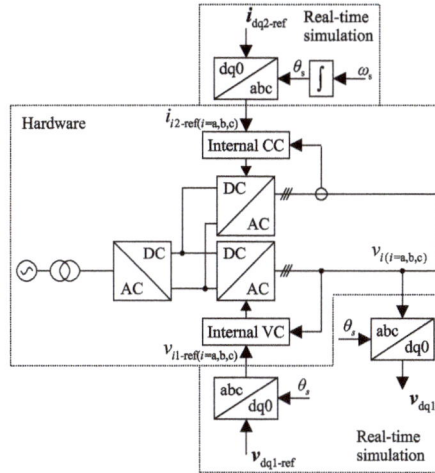

Figure 3. Test setup to measure closed-loop frequency response of the voltage amplifier under loaded condition.

Figure 4 shows the measured cross-coupling frequency responses as dashed lines, i.e., from reference of the d-component to produced q-component and reference of the q-component to produced d-component. The magnitude of both cross-coupling frequency responses was roughly −30 decibels. Thus, the amplifier itself can be considered not to cause significant cross-couplings between the d and q components. The measured frequency responses were later used to design control loops to necessitate accurate emulation of a grid-feeding converter. That is, the frequency responses in Figure 4 were transformed into numerical frequency response vectors and used in solving the dynamic model of the impedance emulator.

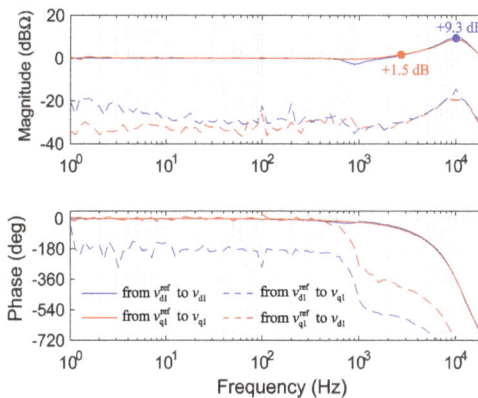

Figure 4. Measured frequency response of amplifier in internal voltage control mode when it is loaded by 50 A_{peak} AC current.

The main objective of this study was to use the voltage amplifier to emulate a grid-following three-phase converter. A 2 mH three-phase inductor was connected to the output of the amplifier, as illustrated in Figure 5. The inductor was needed to emulate the passive impedance properties caused by a real AC filter. Moreover, the inductor served as a decoupling impedance since both amplifier groups were internally controlled as voltage sources. The secondary purpose of the decoupling impedance was also to avoid over-currents during turn-on of the voltage amplifiers. The reference voltages for amplifier group 1 can be interpreted as three-phase switch control signals of the emulated inverter and reference voltages for amplifier group 2 as grid voltages.

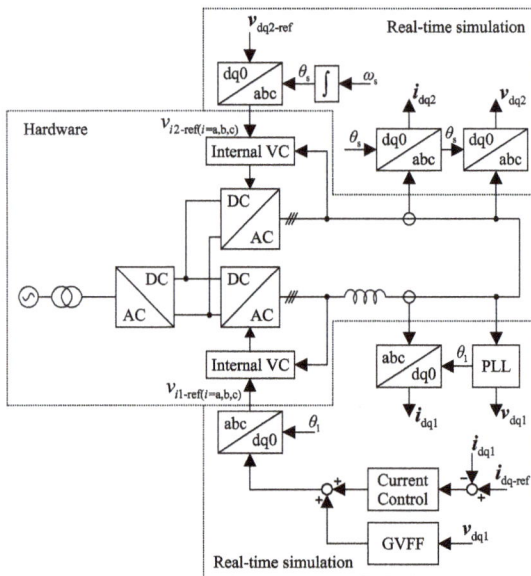

Figure 5. Test setup to verify the emulated impedance.

The output current of the emulator can be derived in the synchronous reference frame, assuming ideal voltage sources, and given as in Equation (3), where r_L is the resistance of the inductor and ω_s is the fundamental frequency of the AC microgrid.

$$\frac{d}{dt} i_L^{dq} = \frac{1}{L} \left(v_1^{dq} - r_L i_L^{dq} - j\omega_s i_L^{dq} - v_2^{dq} \right)$$ (3)

Dynamic model of the voltage amplifier can be represented in dq-domain using the measured frequency responses of Figure 4 and given as a transfer matrix according to Equation (4).

$$\begin{bmatrix} \hat{v}_{d1} \\ \hat{v}_{q1} \end{bmatrix} = \begin{bmatrix} G_{dd}(j\omega) & G_{qd}(j\omega) \\ G_{dq}(j\omega) & G_{qq}(j\omega) \end{bmatrix} \begin{bmatrix} \hat{v}_{d1}^{ref} \\ \hat{v}_{q1}^{ref} \end{bmatrix}$$ (4)

Since Equation (3) is linear, the dynamics of the voltage amplifier in Equation (4) can be substituted into Equation (3) to obtain the linearized state-space of the PHIL setup. Inductor currents $\hat{i}_{Ld1}$ and $\hat{i}_{L11}$ were selected as state variables, reference values of amplifier 1 $\hat{v}_d^{ref}$ and $\hat{v}_q^{ref}$ were selected as control variables, voltages produced by voltage amplifier $\hat{v}_d$ and $\hat{v}_q$ 2 were considered as grid voltages and inductor currents were selected as output variables. s denotes the derivative operator in the frequency domain. It should be noted that the amplifier dynamics are substituted into the state space model as numerical frd-type data vectors.

$$
s \begin{bmatrix} \hat{\imath}_{Ld1} \\ \hat{\imath}_{Lq1} \end{bmatrix} = \overbrace{\begin{bmatrix} -\dfrac{r_L}{L} & \omega_s \\ -\omega_s & -\dfrac{r_L}{L} \end{bmatrix}}^{A} \begin{bmatrix} \hat{\imath}_{Ld1} \\ \hat{\imath}_{Lq1} \end{bmatrix}
$$
$$
+ \underbrace{\begin{bmatrix} G_{dd}(j\omega)/L & G_{qd}(j\omega)/L & 0 & -1/L \\ G_{qd}(j\omega)/L & G_{qq}(j\omega)/L & -1/L & 0 \end{bmatrix}}_{B} \begin{bmatrix} \hat{e}_d^{ref} \\ \hat{e}_q^{ref} \\ \hat{v}_d \\ \hat{v}_q \end{bmatrix}
$$

(5)

$$
\begin{bmatrix} \hat{\imath}_d \\ \hat{\imath}_q \end{bmatrix} = \overbrace{\begin{bmatrix} 1 & 0 \\ 0 & 1 \end{bmatrix}}^{C} \begin{bmatrix} \hat{\imath}_{Ld1} \\ \hat{\imath}_{Lq1} \end{bmatrix}
$$

(6)

Transfer function matrix can be solved from Equations (5) and (6) according to the well known formula, $\mathbf{G} = \mathbf{C}\,(s\mathbf{I} - \mathbf{A})^{-1}\,\mathbf{B} + \mathbf{D}$, where the matrix D is zero. The result is a 2×4 matrix that contains the frequency responses of the emulator at open-loop according to Equation (7). It should be noted that the transfer functions are essentially in numerical form, since the actual measured frequency responses were embedded in the state-space in Equation (5).

$$
\begin{bmatrix} \hat{\imath}_{od} \\ \hat{\imath}_{oq} \end{bmatrix} = \overbrace{\begin{bmatrix} G_{codd}(j\omega) & G_{coqd}(j\omega) & -Y_{odd}(j\omega) & -Y_{oqd}(j\omega) \\ G_{codq}(j\omega) & G_{coqq}(j\omega) & -Y_{odq}(j\omega) & -Y_{oqq}(j\omega) \end{bmatrix}}^{G} \begin{bmatrix} \hat{e}_d^{ref} \\ \hat{e}_q^{ref} \\ \hat{v}_{gd} \\ \hat{v}_{gq} \end{bmatrix}
$$

(7)

Current control is implemented in the synchronous reference frame using two PI-controllers. The overall control block diagram is shown in Figure 6 where it is evident that the control dynamics of d- and q-component are coupled. Two loop gains can be solved from the block diagram and given as in Equations (8) and (9), where G_{cd} and G_{cq} are the controller transfer functions and G_{lp} is a first-order low-pass filter. Control design was done using loop-shaping method due to the fact that the transfer functions exist only in numerical form [26]. Control parameters are collected in Table 1.

$$
T_{dd} = G_{cd} G_{lp} G_{codd} - \frac{G_{coqd} G_{codq} G_{cd} G_{cq} G_{lp}^2}{1 + G_{cq} G_{lp} G_{coqq}}
$$

(8)

$$
T_{qq} = G_{cq} G_{lp} G_{coqq} - \frac{G_{coqd} G_{codq} G_{cd} G_{cq} G_{lp}^2}{1 + G_{cd} G_{lp} G_{codd}}
$$

(9)

The blue curve in Figure 7 shows the measured control loop gain of the current d-component T_{dd} compared with the numerical frequency response solved from Equation (8). The behavior of the q-component is essentially the same and not shown here. The control loop has crossover frequency of 500 Hz and phase margin of 59 degrees. During the measurement, the amplifier was connected to a second group of amplifiers, which were used to form stable grid voltages as in Figure 5. The measured loop gain matches with the frequency response derived in Equation (8). Thus, the emulator can be used to accurately capture the dynamics from AC current control of the grid-following inverter.

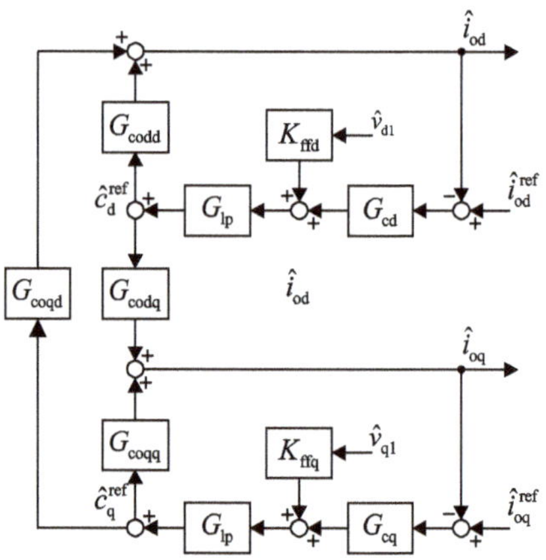

Figure 6. Current control dynamics.

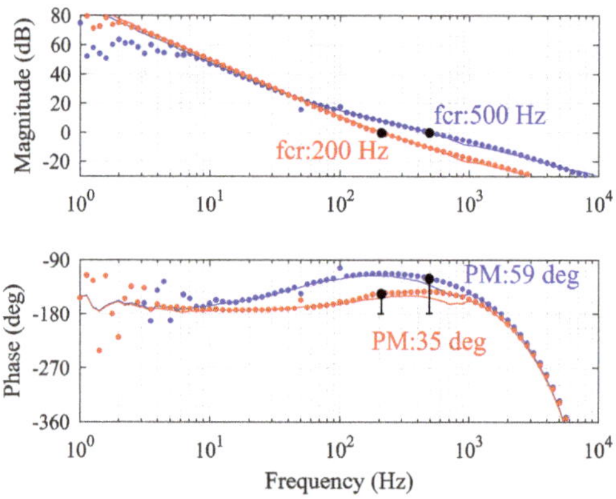

Figure 7. Modeled (solid) and measured (dotted) loop gain of d-current control for two different parameter settings.

Table 1. Parameters.

V_{ac}	230 V	$K_{P\text{-cc}}$	6.4	K_{ffd}	0 ... 1.0
P_{ac}	25 kW	$K_{I\text{-cc}}$	2018	K_{ffq}	0 ... 1.0
ω_s	$2\pi\cdot50$ Hz	$K_{P\text{-pll}}$	0.35	L	2 mH
f_{sample}	20 kHz	$K_{I\text{-pll}}$	21.9	r_L	200 mΩ

Three-phase converters synchronize their output currents with the grid voltages, thus a phase-locked-loop is implemented. Figure 8 shows the measured loop gain of the PLL and the numerical frequency response obtained from Equation (10), where G_{delay} is a third-order Pade-approximation of the control system delay $T_{delay} = 0.5/f_{sw}$. The designed crossover frequency was 20 Hz and phase margin 65 degrees. The measured PLL loop gain follows exactly the analytical model and, thus, the emulator can be used to emulate the effects of different grid synchronization algorithms.

$$T_{pll} = \frac{V_d}{s}\left(K_{P\text{-pll}} + \frac{K_{I\text{-pll}}}{s}\right)G_{delay} \tag{10}$$

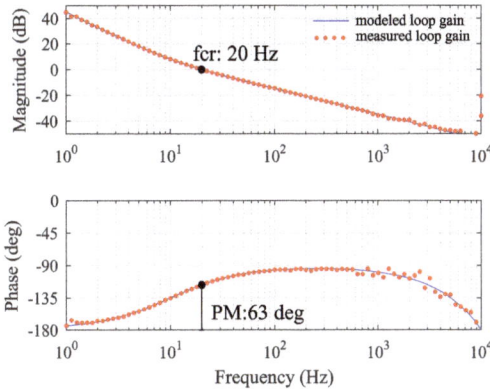

Figure 8. Modeled (solid) and measured (dotted) loop gain of the phase-locked-loop.

Figure 9 shows the response of AC currents to a symmetrical voltage dip of 25 percent. The currents are well regulated with a peak value of 50 A, which indicates that the system is not susceptible to transients and can be safely operated. A special note should be made regarding the small DC current flowing between the different amplifier groups. The DC current flow is caused by the fact that at DC there are two voltage sources in parallel with very small resistance in between. Thus, even small DC offset between the amplifier reference value may produce large DC current. In [19], an isolation transformer is used to cut the path of circulating DC current. However, in this work, the main goal was to find how accurately the grid-following converter impedance can be emulated, without the effect of transformers, cabling, etc. Moreover, for the used hardware, the DC component presents no problem, since each amplifier group can operate at nominal apparent power in all four quadrants at DC. For future application of these results, it is advised that an isolation transformer (or step-up transformer) is connected between the impedance emulator and the rest of the PHIL setup.

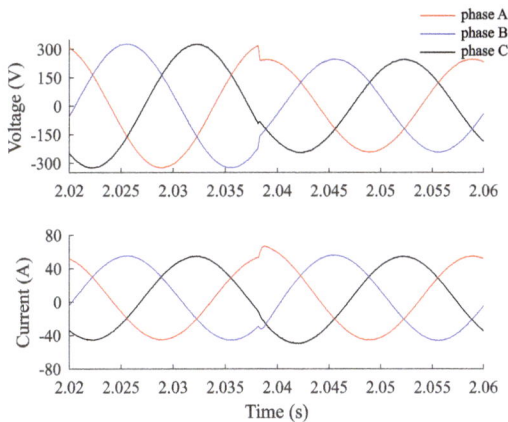

Figure 9. AC waveforms during a sudden symmetrical voltage dip of 25 percent with well-designed current control.

Performance of the current control was compared to a case where current control was intentionally deteriorated by reducing the crossover frequency and phase margin, corresponding to the loop gain illustrated in Figure 7 in red. Figure 10 shows the two cases where reference of the current d-component was suddenly changed from 10 to 50 A. It is evident that the voltage amplifier can reproduce the waveforms that are characteristic of sufficient and insufficient phase margin in the current control. Poorly damped oscillation in the step response is a symptom of low phase margin in the current control loop gain (35 degrees).

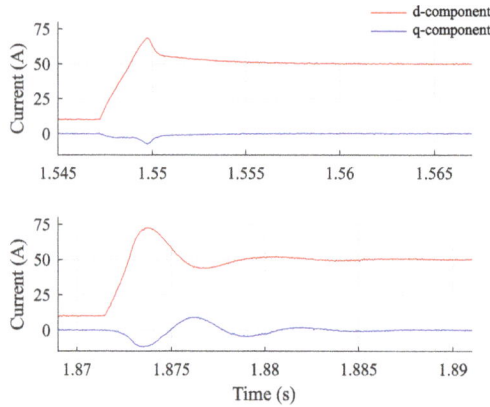

Figure 10. Step response of currents in dq-domain with well-designed (**upper**) and poorly designed (**lower**) current control.

3. Small-Signal Admittance Model and Verification of the Emulator

A grid-following converter is operated under AC-current control mode. From the point-of-view of circuit analysis, the converter should be modeled as a Norton-equivalent current source (or sink depending on the direction of power flow). Thus, in this section, the output admittance model is derived and the output admittance of the emulator is measured. However, the admittance can be easily understood as the inverse of impedance, i.e., for an ideal current source its output impedance is infinite and admittance is zero.

The current dynamics can be written using transfer matrices and vector-notation as in Equation (11) or, in short, as in Equation (12). Accordingly, a linear equivalent circuit can be drawn as in Figure 11. The control variable c' can be given in the control system reference frame by including the grid voltage feedforward and by taking into account the effect of the phase-locked-loop according to the small-signal relations in Equation (13) or by using transfer matrices as in Equation (14) [27].

$$
\begin{bmatrix} \hat{i}_{od} \\ \hat{i}_{oq} \end{bmatrix} = \begin{bmatrix} G_{codd} & G_{coqd} \\ G_{codq} & G_{coqq} \end{bmatrix} \begin{bmatrix} \hat{c}_d^{ref} \\ \hat{c}_q^{ref} \end{bmatrix} \\
- \begin{bmatrix} Y_{odd} & Y_{oqd} \\ Y_{odq} & Y_{oqq} \end{bmatrix} \begin{bmatrix} \hat{v}_{gd} \\ \hat{v}_{gq} \end{bmatrix}
$$

(11)

$$
\mathbf{i}_o = \mathbf{G}_{co}\mathbf{c} - \mathbf{Y}_o\mathbf{v}_o
$$

(12)

$$
\begin{bmatrix} \hat{x}_d' \\ \hat{x}_q' \end{bmatrix} = \begin{bmatrix} \hat{x}_d \\ \hat{x}_q \end{bmatrix} + \frac{T_{pll}}{V_d\left(T_{pll}\right)} \begin{bmatrix} 0 & X_q \\ 0 & -X_d \end{bmatrix} \begin{bmatrix} \hat{v}_{od} \\ \hat{v}_{oq} \end{bmatrix}
$$

(13)

$$
\mathbf{x}' = \mathbf{x} + \mathbf{G}_{pll}\mathbf{X}\mathbf{v}_o
$$

(14)

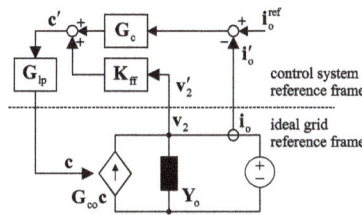

Figure 11. Equivalent small-signal circuit with control system.

The control variable can be given in the control system reference frame as in Equation (15), where the controller transfer matrix and feedforward gain matrix can be given as in Equations (16) and (17), respectively.

$$c' = \mathbf{G}_{\mathrm{lp}}\left(\mathbf{G}_{\mathrm{c}}\left(\mathbf{i}_{\mathrm{o}}^{\mathrm{ref}} - \mathbf{i}_{\mathrm{o}}'\right) + \mathbf{K}_{\mathrm{ff}}\mathbf{v}_{\mathrm{o}}'\right) \tag{15}$$

$$\mathbf{G}_{\mathrm{c}} = \begin{bmatrix} \left(K_{\mathrm{P\text{-}cc}} + \frac{K_{\mathrm{I\text{-}cc}}}{s}\right) & 0 \\ 0 & \left(K_{\mathrm{P\text{-}cc}} + \frac{K_{\mathrm{I\text{-}cc}}}{s}\right) \end{bmatrix} \tag{16}$$

$$\mathbf{K}_{\mathrm{ff}} = \begin{bmatrix} K_{\mathrm{ffd}} & 0 \\ 0 & K_{\mathrm{ffq}} \end{bmatrix} \tag{17}$$

By applying Equation (13) for each variable, the control variable c can be given in the ideal grid reference frame according to Equation (18).

$$\begin{aligned} c = & \mathbf{G}_{\mathrm{lp}}\left(\mathbf{G}_{\mathrm{c}}\mathbf{i}_{\mathrm{o}}^{\mathrm{ref}} - \mathbf{G}_{\mathrm{c}}\mathbf{i}_{\mathrm{o}} + \mathbf{K}_{\mathrm{ff}}\mathbf{V}_2\right) \\ & + G_{\mathrm{pll}}\left(\mathbf{G}_{\mathrm{lp}}\mathbf{K}_{\mathrm{ff}}\mathbf{V}_2 - \mathbf{G}_{\mathrm{lp}}\mathbf{G}_{\mathrm{c}}\mathbf{I}_{\mathrm{o}} - \mathbf{V}_1\right)\mathbf{v}_2 \end{aligned} \tag{18}$$

$$\mathbf{Y}_{\mathrm{o\text{-}c}} = \left(\mathbf{I} + \mathbf{G}_{\mathrm{lp}}\mathbf{G}_{\mathrm{co}}\mathbf{G}_{\mathrm{c}}\right)^{-1}\left[\mathbf{Y}_{\mathrm{o}} - \mathbf{G}_{\mathrm{lp}}\mathbf{K}_{\mathrm{ff}}\mathbf{G}_{\mathrm{co}} + G_{\mathrm{pll}}\mathbf{G}_{\mathrm{co}}\left(\mathbf{G}_{\mathrm{c}}\mathbf{G}_{\mathrm{lp}}\mathbf{I}_{\mathrm{o}} - \mathbf{G}_{\mathrm{lp}}\mathbf{K}_{\mathrm{ff}}\mathbf{V}_{\mathrm{o}} + \mathbf{C}\right)\right] \tag{19}$$

Substituting the control variable in Equation (12) and solving the admittance yields in Equation (19), where the matrices $\mathbf{I}_{\mathrm{o}}$, $\mathbf{V}_2$ and $\mathbf{V}_1$ include the steady-state operating point of AC current and voltage on both sides of the inductor. The matrices are required for transforming the variables between the ideal grid reference frame and the control system reference frame, and are given in the Appendix A. The emulated inverter was controlled at unity power factor, the reference value of the current d-component was 50 A and for the q-component was zero. However, it was observed that small amount of reactive power was flowing between the amplifiers and the actual steady-state value for current q-component was -2 A. The small reactive power exchange was presumably due to the internal capacitor of the voltage amplifier.

Figure 12 shows the admittance components given by the analytical model in blue solid line and the actual measured admittance in red dots. Moreover, the black line represents admittance identified from a switching model of a three-phase inverter with equivalent electrical and control parameters. The switching model was implemented using the SimScape software package in MATLAB Simulink. Thus, the black line represents the admittance that would be expected from a real converter. The cross-coupling admittance terms Y_{qd} and Y_{dq} are order of magnitude smaller than the direct and quadrature components Y_{dd} and Y_{qq}. Therefore, their effect on stability and power quality are minor and the deviation in the accuracy is irrelevant [28]. In the following, the accuracy of the emulator is determined by examining the d- and q-components.

The emulated admittance follows the expected value very precisely up to 500 Hz, which, not by coincidence, places about the crossover frequency of the current control, as shown in Figure 7. The PHIL setup can emulate all the admittance components precisely, including the cross-coupling terms. However, at frequencies higher than 1 kHz, there is a deviation between the expected admittance and the measured admittance. This is likely caused by the unknown output filter and internal active damping or feedforward control of the voltage amplifier. In fact, it is natural to expect the voltage amplifier to reduce the output impedance (increase admittance) outside the emulated control bandwidth. However, the PHIL setup can be concluded to replicate the admittance precisely up to the current control crossover, which is enough to characterize most of the impedance-based interactions, i.e., those caused by grid synchronization, grid voltage feedforward and current control.

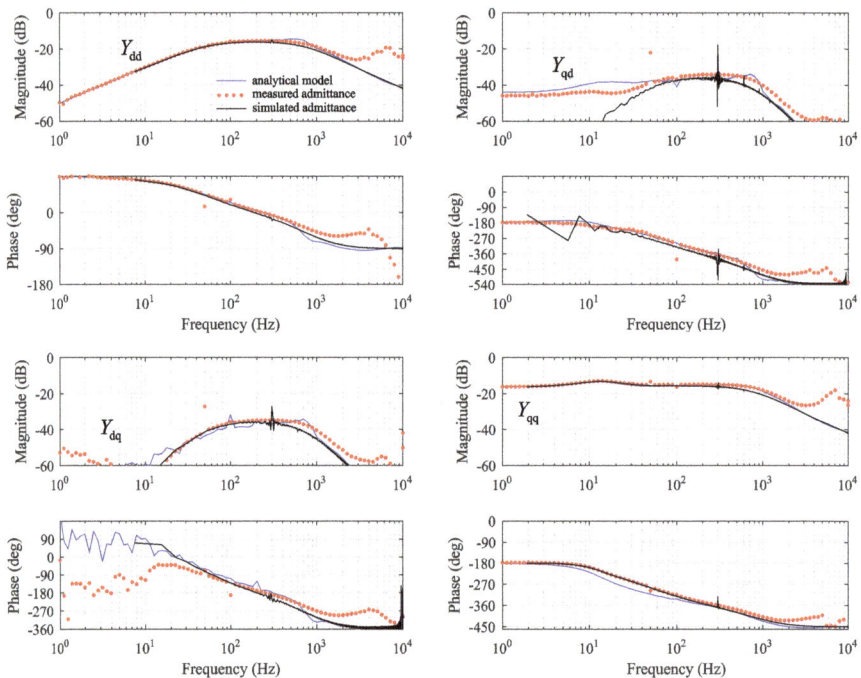

Figure 12. Comparison of expected and measured admittance.

4. Effect of Different Control Functions

The phase-locked-loop (PLL) has been shown to make the q-component of inverter impedance resemble a negative resistor below the PLL crossover frequency [27]. Figure 13 shows the measured admittance q-component with different PLL crossover frequencies. In each case, the PLL was tuned for a phase margin of approximately 65 degrees. The measured admittance in solid line gives precisely the same frequency response as the model, which is shown as the dotted line. Thus, the PHIL setup emulates the behavior of the PLL very precisely and the negative resistance region is extended with the increasing PLL crossover frequency. Thus, the PHIL is suitable for studying low-frequency instability caused by grid synchronization, such as the instability in weak grid condition or interaction between several parallel inverters.

To further demonstrate that the emulator represents the effect of the phase-locked-loop correctly, measurements from a low-power photovoltaic inverter prototype are provided in Figure 14. The prototype includes an LCL-type filter and is discussed further in [29]. The phase-locked-loop causes a negative conductance part to the admittance term, which can be seen at frequencies below the bandwidth of the PLL as a constant magnitude and 180-degree phase shift. Moreover, the magnitude of low-frequency admittance is given by the ratio of d-components of the current and voltage, which at the unity power factor is approximately I_{ac}/V_{ac}. Both the admittance measured from the proposed emulator setup and admittance measured from a real inverter have these properties, as summarized in the figure. The difference in magnitude at low frequencies can be explained by different power rating and at high frequencies the prototype differs in current control parameters and AC filter design. However, the effect of the PLL is identical in both cases, since the PLL crossover is set to 20 Hz.

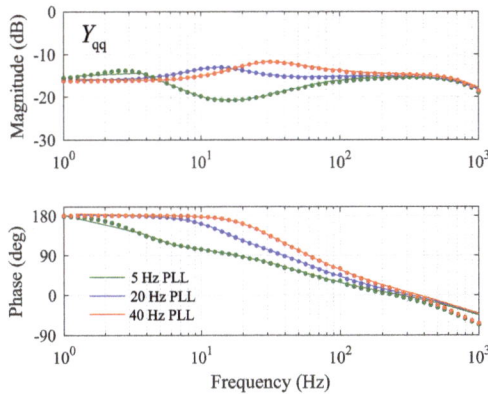

Figure 13. Admittance q-component with different phase-locked-loop crossover frequencies.

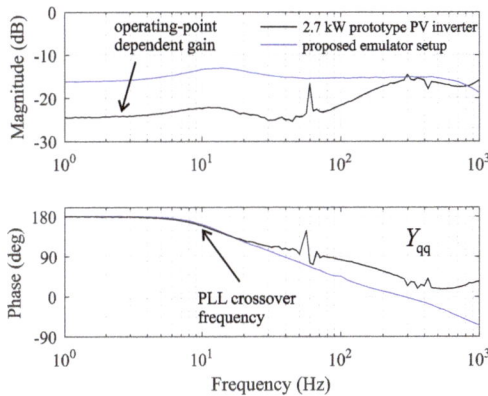

Figure 14. Admittance q-component measured from emulator and 2.7 kW PV inverter prototype.

The AC current control aims to make the inverter resemble more closely an ideal current source. In practice, the current control decreases the inverter output admittance below the crossover frequency, by introducing a parallel capacitive output admittance [26]. Figure 15 shows the measured admittance with three different current control crossover frequencies, with identical phase margin. It is evident that increasing the current control crossover frequency makes the admittance smaller, thus making the inverter to behave more closely as an ideal current source. The PHIL setup can, therefore, be used to characterize problems caused by low-bandwidth current control.

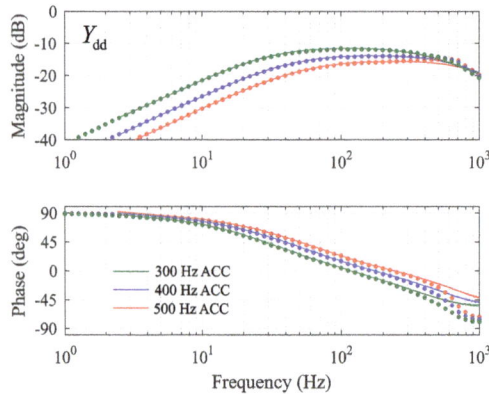

Figure 15. Admittance d-component with different current control crossover frequencies.

Grid voltage feedforward is often used to further decrease the inverter output admittance to mitigate the effect of background harmonics. Figures 16 and 17 show the measured admittance d- and q-component with different values of feedforward gain. Ideally the proportional feedforward gain of a three-phase two-level inverter would be selected as the inverse of DC side voltage, which in the figures is referred as the 100-percent gain. As can be seen in Figure 16, the admittance d-component decreases with increasing feedforward gain as in a real inverter. However, using high gain can make the admittance lose its passive characteristics, as indicated by the red curve, which is why the value of feedforward gain is usually limited. That is, the admittance is determined passive only at the frequencies where the phase-curve stays within ± 90 degrees.

As can be seen in Figure 17, the feedforward does not affect the admittance q-component at low frequencies, where the PLL dominates the admittance behavior. The effect of increasing the feedforward gain on the admittance q-component can be clearly seen in Figure 17 around few hundred hertz, where the admittance decreases when the feedforward gain approaches the nominal value. However, with the nominal value, i.e., 100 percent, the admittance loses its passive characteristics over a much wider bandwidth than what is caused by the PLL, since the phase curve rises above 90 degrees around 100 Hz. This happens regardless of the fact that PLL crossover is only 20 Hz.

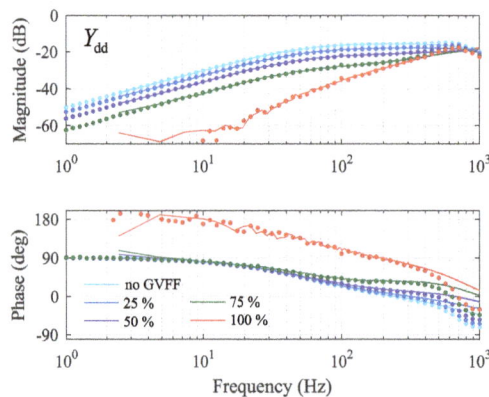

Figure 16. Admittance d-component with different gains for grid voltage feedforward.

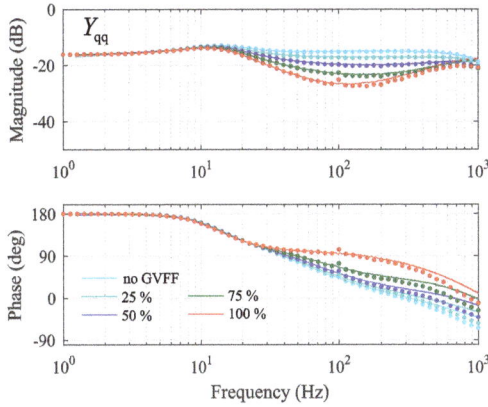

Figure 17. Admittance q-component with different gains for grid voltage feedforward.

To further validate the ability of the emulator setup to replicate the effect of grid voltage feedforward, measured admittance q-component is compared to measurements from the 2.7 kW prototype PV inverter. The admittances are shown in Figure 18 for the emulator setup on the left and for the PV inverter on the right. Even though the admittances cannot be directly compared due to different sizing and parameters of the PV inverter, it is clear that the grid voltage feedforward decreases the magnitude of the admittance at frequencies higher than the PLL crossover frequency. Thus, the emulator setup correctly replicates this effect.

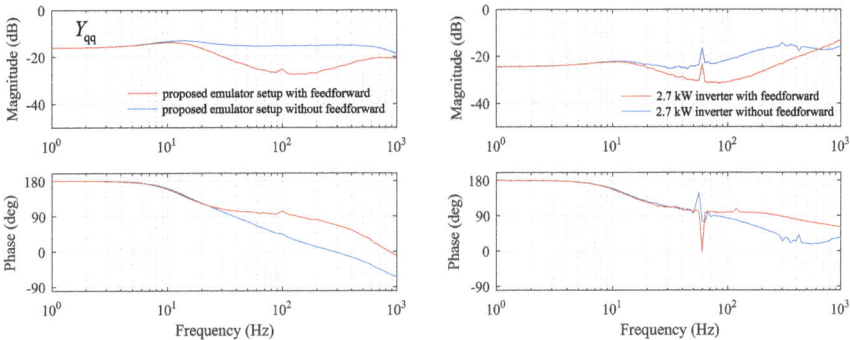

Figure 18. Admittance q-component with and without the grid voltage feedforward: from the proposed emulator (**left**); and from PV inverter prototype (**right**).

Based on the frequency response measurements, the power-hardware-in-the-loop can be effectively used to emulate the most important features of inverter admittance, including the effect of current control, phase-locked-loop and grid-voltage feedforward. Therefore, the PHIL setup is an attractive approach for stability and power quality analysis of future power systems with ever-increasing complexity, such as the AC microgrids. The steps to configure a voltage amplifier to emulate the small-signal impedance of a grid-connected converter can be summarized as:

- Characterize the internal small-signal dynamics of the amplifier, i.e., the transfer function from reference voltages to the actual voltages, and transform the measured frequency responses as frd-type data vectors as in Equation (4).

- Select and appropriate L or LCL-type AC filter to necessitate current control and to decouple the voltage amplifier from the rest of the AC system, i.e., to prevent hazardous over currents during turn-on.
- Include the measured frequency responses $G_{dd}(j\omega)$, $G_{qd}(j\omega)$, $G_{qq}(j\omega)$ and $G_{dq}(j\omega)$ in the state space representation in Equation (5) and solve numerically the open-loop transfer functions as in Equation (7).
- Tune the PI-controllers to regulate AC current d- and q-component to have the intended phase margin and crossover frequency using Equations (8) and (9).
- Tune the phase-locked-loop based on the loop gain transfer function in Equation (10).
- Verify the dynamic step response and output admittance of the emulator setup corresponds to the admittance given by the admittance model in Equation (19), as in Figures 12–17.

5. Conclusions

AC microgrids are becoming more common to energize local loads due to increased availability of distributed generation. AC microgrids conceptually comprise several grid-forming and grid-following power converters, renewable sources, energy storages and local loads. Power-hardware-in-the-loop (PHIL)-based studies have gained a lot of interest recently since they allow effortless dynamic tests of AC microgrids with full penetration of renewable energy generation. In a PHIL setup, different components of the microgrid exist as real devices or electrical emulators, or can be simulated using real-time simulators. An example of a PHIL-based simulator is an electrical photovoltaic emulator based on a linear amplifier or switching DC-DC converter. For accurate, reliable and repeatable results, the PHIL-setup should be able to represent the dynamics of the microgrid loads and sources as accurately as possible. Several studies have shown how electrical machines, dynamic RLC loads, battery storages and photovoltaic and wind generators should be emulated in a PHIL setup. However, there are no studies showing how a three-phase grid-following power converter with its internal control functions should be emulated. One could naturally use a real inverter prototype to represent such dynamic load. However, practical implementation of a real three-phase converter is much more challenging and requires special knowledge on, e.g., the characteristics of the semiconductor switches, signal conditioning, over-current and over-voltage protection and auxiliary power circuitry design. Therefore, this paper shows how a commercial laboratory voltage amplifier can be turned into a three-phase power converter, requiring only a passive AC filter and real-time simulator. Thus, researchers without special knowledge on design of inverters can study the dynamics caused by grid-following inverter-type loads in AC microgrids.

This paper discusses the control design and implementation of a power-hardware-in-the-loop (PHIL) setup based on a four-quadrant voltage amplifier to emulate the small-signal behavior of three-phase converter. It is shown by frequency domain measurements that the admittance of the PHIL-emulator follows precisely the expected impedance value within the current control bandwidth. Therefore, the PHIL setup can be applied to study stability and power quality problems in an AC microgrid. The effect of various control functions on the impedance, such as current control, grid synchronization and grid voltage feedforward, are successfully demonstrated.

Author Contributions: Conceptualization, T.M., T.R. and Y.S.; Methodology, T.M. and R.L.; Software, T.R., R.L., R.S. and A.F.; Validation, R.L., T.M. and R.S.; Formal Analysis, T.M. and R.L.; Resources, E.d.J.; Writing—Original Draft, T.M.; Writing—Review and Editing, P.M. and T.C.; Visualization, T.M.; Supervision, T.M. and E.d.J.; Project Administration, T.M. and E.d.J.; and Funding Acquisition, T.M., T.R. and E.d.J.

Funding: This research was funded by the European Research Infrastructure (ERIGrid, grant number 654113) and the Academy of Finland (Mitigation of harmonic resonances in three-phase renewable energy systems, grant number 297054). The implementation and measurements were carried out within a two-week research period hosted by DNV-GL in Arnhem, The Netherlands.

Acknowledgments: The authors would like to express their gratitude to rest of the staff of DNV-GL. Especially, for the friendly atmosphere and all the ideas and conversations, that helped to formulate the research topic and main outcomes. Moreover, the staff at the Padova University not mentioned in the text deserve warm thanks for lively discussions around the topic of this paper.

Energies **2019**, *12*, 379

Conflicts of Interest: The authors declare no conflict of interest.

Appendix A

$$\mathbf{V}_1 = \begin{bmatrix} 0 & \omega_s L I_d + r_L I_q + V_{q2} \\ 0 & -\omega_s L I_q - r_L I_d - V_{d2} \end{bmatrix} = \begin{bmatrix} 0 & 31 \\ 0 & -336 \end{bmatrix} \tag{A1}$$

$$\mathbf{V}_2 = \begin{bmatrix} 0 & V_{q2} \\ 0 & -V_{2d} \end{bmatrix} = \begin{bmatrix} 0 & 0 \\ 0 & -325 \end{bmatrix} \tag{A2}$$

$$\mathbf{I}_o = \begin{bmatrix} 0 & I_q \\ 0 & -I_d \end{bmatrix} = \begin{bmatrix} 0 & 2 \\ 0 & -50 \end{bmatrix} \tag{A3}$$

References

1. Khooban, M.H.; Dragicevic, T.; Blaabjerg, F.; Delimar, M. Shipboard microgrids: A novel approach to load frequency control. *IEEE Trans. Sustain. Energy* **2018**, *9*, 843–852. [CrossRef]
2. Roboam, X.; Sareni, B.; Andrade, A.D. More electricity in the air: Toward optimized electrical networks embedded in more-electrical aircraft. *IEEE Ind. Electron. Mag.* **2012**, *6*, 6–17. [CrossRef]
3. Pan, P.; Hu, H.; Yang, X.; Blaabjerg, F.; Wang, X.; He, Z. Impedance measurement of traction network and electric train for stability analysis in high-speed railways. *IEEE Trans. Power Electron.* **2018**, *33*, 10086–10100. [CrossRef]
4. Wang, X.; Blaabjerg, F. Harmonic stability in power electronic based power systems: Concept, modeling, and analysis. *IEEE Trans. Smart Grid* **2018**. [CrossRef]
5. Maniatopoulos, M.; Lagos, D.; Kotsampopoulos, P.; Hatziargyriou, N. Combined control and power hardware in-the-loop simulation for testing smart grid control algorithms. *IET Gener. Transm. Distrib.* **2017**, *11*, 3009–3018. [CrossRef]
6. Li, C. Unstable operation of photovoltaic inverter from field experiences. *IEEE Trans. Power Deliv.* **2018**, *33*, 1013–1015. [CrossRef]
7. Buchhagen, C.; Rauscher, C.; Menze, A.; Jung, J. BorWin1—First Experiences with harmonic interactions in converter dominated grids. In Proceedings of the International ETG Congress 2015: Die Energiewende—Blueprints for the New Energy Age, Bonn, Germany, 17–18 November 2015.
8. Luhtala, R.; Roinila, T.; Messo, T. Implementation of real-time impedance-based stability assessment of grid-connected systems using MIMO-identification techniques. *IEEE Trans. Ind. Appl.* **2018**, *54*, 5054–5063. [CrossRef]
9. Aapro, A.; Messo, T.; Roinila, T.; Suntio, T. Effect of active damping on output impedance of three-phase grid-connected converter. *IEEE Trans. Ind. Electron.* **2017**, *64*, 7532–7541. [CrossRef]
10. Wang, X.; Harnefors, L.; Blaabjerg, F. Unified impedance model of grid-connected voltage-source converters. *IEEE Trans. Power Electron.* **2018**, *33*, 1775–1787. [CrossRef]
11. Wen, B.; Boroyevich, D.; Burgos, R.; Mattavelli, P.; Shen, Z. Small-signal stability analysis of three-phase AC systems in the presence of constant power loads based on measured d-q frame impedances. *IEEE Trans. Power Electron.* **2015** *30*, 5952–5963. [CrossRef]
12. Berg, M.; Messo, T.; Suntio, T. Frequency response analysis of load effect on dynamics of grid-forming inverter. In Proceedings of the International Power Electronics Conference, IPEC-ECCE Asia, Niigata, Japan, 20–24 May 2018.
13. Kadam, A.H.; Menon, R.; Williamson, S.S. Traction inverter performance testing using mathematical and real-time controller-in-the-loop permanent magnet synchronous motor emulator. In Proceedings of the IECON 2016—42nd Annual Conference of the IEEE Industrial Electronics Society, Florence, Italy, 23–26 October 2016.

Energies **2019**, *12*, 379

14. McKinney, M.H.; Fox, J.C.; Collins, E.R.; Bulgakov, K.; Salem, T.E. Design, development, and commissioning of a multimegawatt test facility for renewable energy research. *IEEE Trans. Ind. Appl.* **2016**, *52*, 11–17. [CrossRef]
15. Kotsampopoulos, P.C.; Lehfuss, F.; Lauss, G.F.; Bletterie, B.; Hatziargyriou, N.D. The limitations of digital simulation and the advantages of PHIL testing in studying distributed generation provision of ancillary services. *IEEE Trans. Ind. Electron.* **2015**, *62*, 5502–5515. [CrossRef]
16. Jessen, L.; Günter, S.; Fuchs, F.W.; Gottschalk, M.; Hinrichs, H.J. Measurement results and performance analysis of the grid impedance in different low voltage grids for a wide frequency band to support grid integration of renewables. In Proceedings of the 2015 IEEE Energy Conversion Congress and Exposition (ECCE), Montreal, QC, Canada, 20–24 September 2015.
17. Masadeh, M.A.; Pillay, P. Power electronic converter-based three-phase induction motor emulator. In Proceedings of the 2016 IEEE International Conference on Power Electronics, Drives and Energy Systems (PEDES), Trivandrum, India, 14–17 December 2016.
18. Masadeh, M.A.; Amitkumar, K.S.; Pillay, P. Power electronic converter-based induction motor emulator including main and leakage flux saturation. *IEEE Trans. Transp. Electrif.* **2018**, *4*, 483–493. [CrossRef]
19. Vodyakho, O.; Steurer, M.; Edrington, C.S.; Fleming, F. An induction machine emulator for high-power applications utilizing advanced simulation tools with graphical user interfaces. *IEEE Trans. Energy Convers.* **2012**, *27*, 160–172. [CrossRef]
20. Meissner, M.; Gensior, A.; Merk, P.; Reincke-Collon, C. Load emulation in an inverter-dominated medium voltage island grid. In Proceedings of the 2014 IEEE 15th Workshop on Control and Modeling for Power Electronics (COMPEL), Santander, Spain, 22–25 June 2014.
21. Wang, J.; Yang, L.; Ma, Y.; Wang, J.; Tolbert, L.M.; Wang, F.F.; Tomsovic, K. Static and dynamic power system load emulation in a converter-based reconfigurable power grid emulator. *IEEE Trans. Power Electron.* **2016**, *31*, 3239–3251. [CrossRef]
22. Kanaan, H.Y.; Caron, M.; Al-Haddad, K. Design and implementation of a two-stage grid-connected high efficiency power load emulator. *IEEE Trans. Power Electron.* **2014**, *29*, 3997–4006. [CrossRef]
23. Rao, Y.S.; Chandorkar, M. Electrical load emulator for unbalanced loads and with power regeneration. In Proceedings of the 2012 IEEE International Symposium on Industrial Electronics, Hangzhou, China, 28–31 May 2012.
24. Hogan, D.J.; Gonzalez-Espin, F.; Hayes, J.G.; Foley, R.; Lightbody, G.; Egan, M.G. Load and source electronic emulation using resonant current control for testing in a microgrid laboratory. In Proceedings of the 2014 IEEE 5th International Symposium on Power Electronics for Distributed Generation Systems (PEDG), Galway, Ireland, 24–27 June 2014.
25. Caldognetto, T.; Dalla Santa, L.; Magnone, P.; Mattavelli, P. Power electronics based active load for unintentional islanding testbenches. *IEEE Trans. Ind. Appl.* **2017**, *53*, 3831–3839. [CrossRef]
26. Suntio, T.; Messo, T.; Puukko, J. *Power Electronic Converters: Dynamics and Control in Conventional and Renewable Energy Applications*; Wiley-VCH: Weinheim, Germany, 2017; pp. 633–661, ISBN 978-3-527-34022-4.
27. Messo, T.; Jokipii, J.; Makinen, A.; Suntio, T. Modeling the grid synchronization induced negative-resistor-like behavior in the output impedance of a three-phase photovoltaic inverter. In Proceedings of the 2013 4th IEEE International Symposium on Power Electronics for Distributed Generation Systems (PEDG), Rogers, AR, USA, 8–11 July 2013 .
28. Messo, T.; Aapro, A.; Suntio, T. Generalized multivariable small-signal model of three-phase grid-connected inverter in DQ-domain. In Proceedings of the 2015 IEEE 16th Workshop on Control and Modeling for Power Electronics (COMPEL), Vancouver, BC, Canada, 12–15 July 2015; pp. 1–8.
29. Messo, T.; Luhtala, R.; Aapro, A.; Roinila, T. Accurate impedance model of grid-connected inverter for small-signal stability assessment in high-impedance grids. In Proceedings of the 2018 International Power Electronics Conference (IPEC-Niigata 2018-ECCE Asia), Niigata, Japan, 20–24 May 2018; pp. 3156–3163.

energies

MDPI

Article

The Impact of PLL Dynamics on the Low Inertia Power Grid: A Case Study of Bonaire Island Power System

Yin Sun [1,2,*], E. C. W. (Erik) de Jong [1,2], Xiongfei Wang [3], Dongsheng Yang [3], Frede Blaabjerg [3], Vladimir Cuk [1] and J. F. G. (Sjef) Cobben [1,4]

[1] Electrical Energy System, Eindhoven University of Technology, 5612 AZ Eindhoven, The Netherlands; E.C.W.de.Jong@tue.nl (E.C.W.d.J.); v.cuk@tue.nl (V.C.); J.F.G.cobben@tue.nl (J.F.G.C.)
[2] Group Technology Research, DNV GL, 6812AR Arnhem, The Netherlands
[3] Energy Technology Department, Aalborg University, 9100 Aalborg, Denmark; xwa@et.aau.dk (X.W.); doy@et.aau.dk (D.Y.); fbl@et.aau.dk (F.B.)
[4] Strategy Consulting, Asset Management, Alliander, 6912RR Duiven, The Netherlands
* Correspondence: yin.sun@dnvgl.com

Received: 3 December 2018; Accepted: 27 March 2019; Published: 2 April 2019

Abstract: To prepare for the future high penetration level of renewable energy sources, the power grid's technical boundaries/constraints for the correct operation of powerelectronics interfaced devices need to be further examined and defined. This paper investigates the challenge of integrating Voltage Source Converters (VSC) into low inertia power grids, where the system frequency can vary rapidly due to the low kinetic energy buffer available, which used to be provided by the rotational inertia of synchronous generators. The impact of rate of change of frequency (ROCOF) on the PLL dynamics and its subsequent influence on the VSC power stage output is explained. The Bonaire island network is presented as case study. The performance of the VSC is analyzed under a fast ROCOF event, which is triggered by a short circuit fault. A down-scaled experiment is used to validate the Bonaire island network simulation results. It shows that the phase angle error measured by the synchronous-reference frame phase-locked loop (SRF-PLL) is proportional to the slope of the ROCOF and inversely proportional to its controller integral gain constant.

Keywords: ROCOF; PLL; error; low inertia; VSC

---

1. Introduction

A weak grid is characterized as an AC power system with a low short-circuit ratio (SCR) and/or inadequate mechanical inertia (IEEE standard 1204-1997 [1]). Some recent studies [2–10] on the voltage-sourced converter (VSC) integration into a weak grid address only a weak grid with a high grid impedance, whilst the challenges associated with low inertia is seldom discussed. With the increase in renewable penetration in the AC power system, the system frequency stability margin decreases as the system inertia decreases. This leads to rapid frequency variations in low inertia power grids. Typically, for a large inter-connected power system, the total kinetic energy buffer provided by all the synchronous generators in the system is large. In this case, local disturbances (e.g., generator trip, load rejection, short circuit fault etc.) cause only mild frequency variation thanks to the total system mechanical inertia. However this is not the case with low inertia power grids, such as the Bonaire island grid. In Figure 1, a fault occurs in the 12 kV network and it is cleared after roughly 400 ms. From Figure 1, the frequency (red) plummets from 50 Hz to 46 Hz within 400 ms (i.e., ROCOF = 10 Hz/s) whilst the active (blue) and reactive power (green) consumption in the network jumped during the fault.

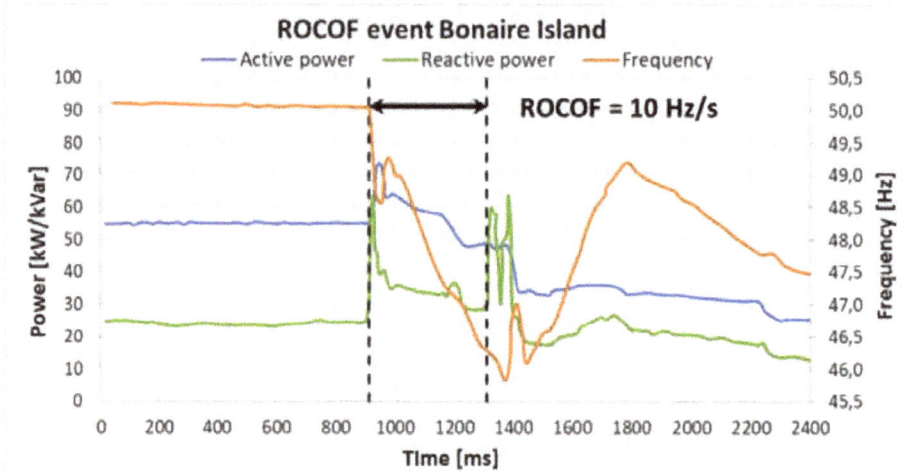

Figure 1. Fast rate of change of frequency (ROCOF) Event Triggered by a 12 kV short circuit fault on the Bonaire island grid—(red) system frequency y-axis on the right side, (blue) system active power consumption, (green) system reactive power consumption [11].

Inspired by this event and expected future challenges associated with battery storage, grid frequency support in a power grid with high penetration of renewable energy sources, this paper investigates the impact of fast rate of change of frequency (ROCOF) on the grid-connected VSC phase locked loop (PLL) dynamics. Although the ROCOF phenomena has already been mentioned in several papers concerning the design and analysis of the PLL alone for the anti-islanding detection [3,12,13] and the inertia emulation [14], the power grid mechanical inertia coupling is not considered. This paper explains the origin of the fault-induced fast ROCOF in the low inertia power grid. Thereafter, the mechanical inertia coupling is investigated using the case study of the Bonaire island, where the network model is validated and the mechanical inertia is represented by synchronous generator model in EMTDC/PSCAD. The case study results of the Bonaire island power grid is further verified by down-scaled experiments and the challenges associated with the fault-induced fast ROCOF on the grid-connected VSCs are discussed.

This paper is organized in four sections. Section 2 begins with the definition of the feedback control system error after which the PLL steady state error is derived. In Section 3, a detailed 850 kW VSC model created in EMTDC/PSCAD is introduced. Its stability is studied using a pole-zero diagram. In Section 4 the VSC dynamics under a fast ROCOF event is studied. First by considering the grid as a simple voltage source behind a given short circuit impedance. Then the VSC dynamic model is integrated into the validated Bonaire island power network model, where the mechanical inertia of diesel generators are also considered. With the coupling of the mechanical inertia, this case study investigates the VSC behavior under the fast ROCOF event triggered by a three-phase cable fault at the 12 kV level. A down-scaled experiment is performed to verify the EMTDC/PSCAD simulation results concerning the fault induced ROCOF in the Bonaire island power network. Finally, conclusions are drawn in Section 5.

2. PLL Modelling and Analysis

To understand the impact of fast ROCOF on the PLL dynamics, the definition of control system error is introduced first. Then the small signal dynamics of PLL is derived analytically with its steady state error expressed as a function of ROCOF frequency slew rate and PLL control integral gain constant (i.e., K_i). Despite the innovative concepts and implementations proposed in the literature, the basic PLL structure remains largely unaltered [15,16] but enhanced with input signal filtering (e.g., bandpass filter realized by Second Order Generalized Integrator (SOGI)) and adaptive frequency tracking capability. In this section the PLL is implemented with the synchronous-reference frame (SRF) commonly used for the majority of three phase grid-connected applications.

2.1. Feedback Control System Error

For a typical three-phase SRF-PLL, its small signal transfer function is shown in Figure 2.

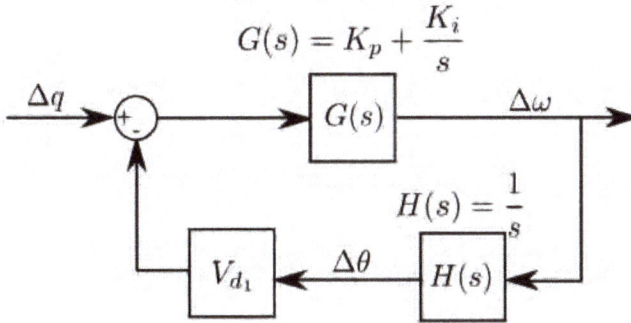

Figure 2. Phase locked loop (PLL) small signal closed loop transfer function diagram.

In Figure 2, Δq is the grid voltage space vector small perturbation projected on the q-axis of the SRF with respect to the steady-state operating point (i.e., q equals to zero), $\Delta \omega$ represents the PI controller output in rad/s, $\Delta \theta$ is the phase angle output, and V_{d_1} is the grid voltage space vector projected on the d-axis of the SRF when the perturbation Δq is small. In the per unit system, V_{d_1} can be normalized to 1, and Figure 2 can be simplified as shown in Figure 3.

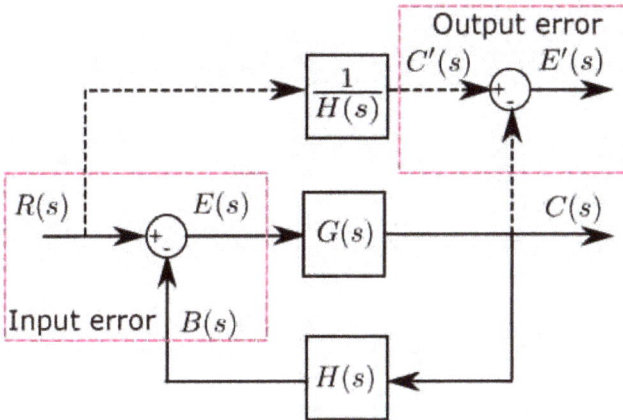

Figure 3. Feedback control system with input and output error definition.

To assess the small signal dynamic performance of a given feedback transfer function, the input error $E(s)$ is defined as the difference between input $R(s)$ and its closed loop feedback $B(s)$ i.e.,

$$E(s) = R(s) - B(s) = R(s) - C(s)H(s) \tag{1}$$

The relationship between input error and output error can be written as:

$$E'(s) = \frac{E(s)}{H(s)} \tag{2}$$

2.2. PLL Steady State Error

For the PLL feedback transfer function with $H(s) = \frac{1}{s}$, (2) provides the theoretical basis to explain the PLL steady-state frequency output error and phase angle output error. As the VSC power stage output depends on the accurate phase angle output from the PLL, we are thus interested in the phase angle output under the fast ROCOF event. When the system frequency is changing rapidly due to power system disturbances (e.g., generator trip, load change, faults etc.), the system frequency $f(t)$ and phase angle $\theta(t)$ deviation can be written as a function of time as follows:

$$f(t) = K_{ramp}t \tag{3}$$

$$\theta(t) = \frac{1}{2}K_{ramp}t^2 \tag{4}$$

where the $f(t)$ and $\theta(t)$ are the frequency and phase angle as function of time, t is the time, and K_{ramp} defines the ramp rate of frequency deviation. Following the input error definition (1), the steady-state phase angle error can be calculated by applying the final theorem and having $s \to 0$:

$$\lim_{s \to 0} sE(s) = \lim_{s \to 0} sR(s)\frac{1}{1 + G_{open}(s)H(s)}$$
$$= \lim_{s \to 0} \frac{K_{ramp}}{s^2}\frac{1}{1 + \frac{1}{s}(K_p + \frac{K_i}{s})} \tag{5}$$
$$= \lim_{s \to 0} \frac{K_{ramp}}{K_i}$$

Similarly the output frequency steady-state error can be calculated by inserting (5) into (2):

$$\lim_{s \to 0} sE'(s) = \lim_{s \to 0} s\frac{E(s)}{H(s)} = \lim_{s \to 0} \frac{sK_{ramp}}{K_i} = 0 \tag{6}$$

From (5), it can be seen that the phase angle will have a steady state error, which is a function of the ROCOF ramp rate K_{ramp} and PLL controller integral gain K_i. For a given ROCOF event, the K_{ramp} is largely fixed by the system inertia and the shortage/surplus of power caused by the transient event, hence the PLL output steady state phase angle error is determined by the integral gain K_i. To minimize the steady-state phase angle error during the fast ROCOF event, it is therefore beneficial to keep the PLL integral gain constant K_i high.

2.3. System Stability and PLL Controller Bandwidth

To allow for the rapid fundamental grid frequency tracking and phase angle determination, the PLL can be designed with a high control bandwidth. Yet, it is common practice to design the PLL with slow dynamics for stable operation. This is especially true under the high impedance power grid condition, where slower response of the PLL can filter out the terminal voltage variation caused by the active/reactive power injection [4] and limit the spurious harmonic current injection as a result of background voltage harmonics. Additionally, Wang et al. [17] points out that a high PLL control bandwidth could trigger harmonic instability of the VSC power stage output when the negative resistance region caused by the PLL impedance shaping effect intersects with the grid resonance point. Hence it is vital to design the PLL with slow dynamics in the high impedance grid for the overall VSC stable operation. Revisiting the conclusion from Section 2.2, one should opt to design the PLL with a high integral gain constant (K_i) yet low control bandwidth. From the control engineering textbook [18], the controller bandwidth is defined as the frequency, where the close loop gain equals to −3 db. The closed loop transfer function of $\Delta\theta$ can be expressed as:

$$G_{PLLcl} = \frac{\Delta\theta}{\Delta q} = \frac{GH}{1+GH}$$

$$= \frac{K_ps + Ki}{s^2 + K_ps + K_i} \tag{7}$$

If the closed loop second order system (7) is represented in terms of its closed-loop roots natural damping frequency (ω_n) and damping factor (ζ) [19]:

$$G_{PLLcl} = \frac{K_ps + Ki}{s^2 + K_ps + K_i}$$

$$= \frac{2\zeta\omega_n s + \omega_n^2}{s^2 + 2\zeta\omega_n s + \omega_n^2} \tag{8}$$

where $\omega_n = \sqrt{K_i}$ and $\zeta = \frac{K_p}{2\sqrt{K_i}}$. According to [18], when $\zeta = 0.707$ (optimal damping) the closed-loop bandwidth (ω_{bw}) of the second order system depicted in (8) can be approximated by its closed-loop roots natural damping frequency (ω_n). Since $\omega_n = \sqrt{K_i}$, it is thus inevitable to have high steady-state phase angle output error under the fast ROCOF event when the PLL control bandwidth is kept low. Two sets of PLL parameters with 45 degree phase stability margin are proposed in Table 1. The PLL_{low} parameter set operates a PLL with low control bandwidth (i.e., $\omega_{bw} \approx 2.8$ Hz). PLL_{high} parameter set operates a PLL with high control bandwidth (i.e., $\omega_{bw} \approx 28$ Hz).

Table 1. PLL Parameters Selected for the Study.

Bandwidth (ω_{bw})	K_p	K_i
PLL_{low}	8.4	100
PLL_{high}	84	10,000

In the Section 4, the PLL parameters proposed in Table 1 and its influence on the VSC power stage output will be studies under a fast ROCOF event.

3. VSC Modelling and Analysis

To allow a holistic analysis of the impact of the ROCOF on the VSC power stage output considering the PLL dynamics, a generic switching VSC model of a 850 kW wind turbine is introduced in this section.

3.1. VSC Simulation Model

The main control system and the electrical parameters chosen for the 850 kW VSC are shown in Table 2. A typical cascaded control scheme is assigned to the 850 kW VSC simulation model as shown in Figure 4, where the outer loops are realized by two parallel PI controllers regulating the DC bus voltage and the reactive power output to a constant and inner-loop is realized by using proportional resonance (PR) controllers regulating the inverter side current dynamics. The reference value is indicated with ∗ in their superscript. In Figure 4, v_c is the filter capacitor instantaneous phase to neutral voltage in the abc frame, i_1 is the inverter side instantaneous current in the abc frame, i_d^* and i_q^* are the current control loop references in-phase and quadrature with the grid voltage (i.e., v_c) in dq frame, P_1 and Q_1 are active and reactive power calculated at the filter capacitor side, v_{dc} and v_{dc}^* are the DC voltage and its reference setpoint respectively. K_p is the proportional gain of the PR controller and ω_0 is the frequency at which the PR controller resonant.

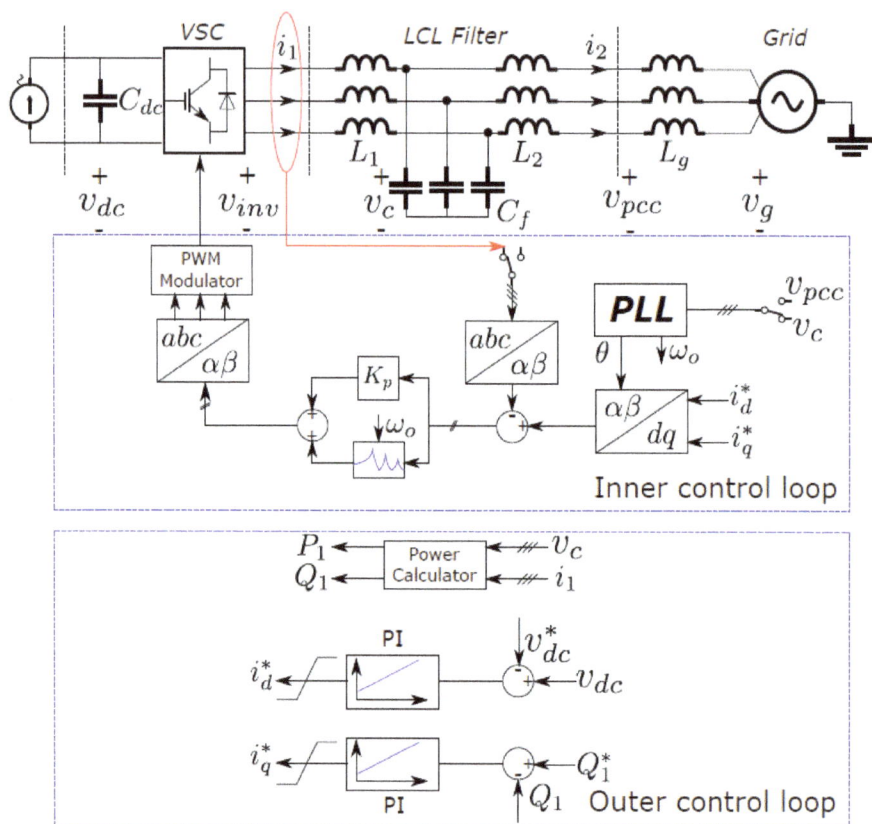

Figure 4. Voltage Source Converter (VSC) inner and outer control loop diagram for the study [20].

Table 2. VSC Main Parameters.

Parameter	Value	Unit
Rated Power	850	kW
DC Link Voltage V_{dc}	800	Volts
DC Link Capacitor C_{dc}	20	mF
AC Voltage V_{rms}	400	Volts
L_1 Inverter Side Inductor	80	μH
R_1 Resistance of L_1	0.001	Ohm
L_2 Grid Side Inductor	80	μH
R_2 Resistance of L_2	0.001	Ohm
C_f Filter Capacitor	425	μF
R_f ESR of C_f	0.01	Ohm
Sampling Time T_s	100	μs
Switching Frequency f_{sw}	5000	Hz
PR Proportional Gain K_p	1	p.u.
PR Integral Gain K_i	250	p.u.
PR Bandwidth ω_c	2	p.u.
DC Proportional Gain $K_{pv_{dc}}$	1	p.u.
DC Integral Gain $K_{iv_{dc}}$	100	p.u.
Q Proportional Gain K_{pQ}	1	p.u.
Q Integral Gain K_{iQ}	100	p.u.

3.2. VSC Stability Analysis

When the bandwidth of the inner and outer control loops are selected properly, then the inner loop and outer loop of the VSC control system can be considered as decoupled. From the VSC system stability point of view, the inner current control is directly interacting with the output filter circuit with a fixed operating point (i.e., reference signal received from outer control loop), hence the small-signal stability of VSC can be analyzed by deriving its closed-loop transfer function of current controller (Figure 5) and plotting it in a pole zero map (Figure 6). Figure 5, $G_{ic}(z)$ depicts the proportional resonance controller in Z domain with 10 kHz sampling time:

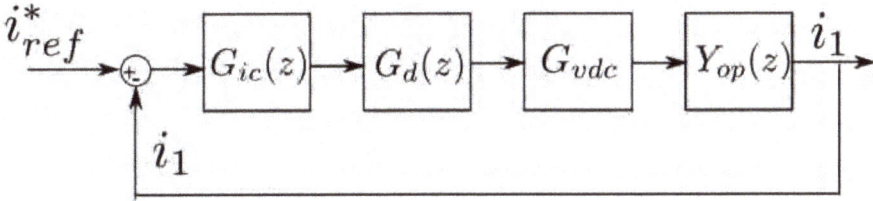

Figure 5. VSC Z Domain Current Control Loop Transfer Function Diagram.

$$G_{ic}(z) = \mathcal{Z}\{K_p + \frac{2K_i\omega_c}{s^2 + 2\omega_c s + \omega_0^2}\}\Big|_{Tustin}^{T_s=100\ \mu s} \tag{9}$$

where ω_o is the fundamental frequency output from the PLL, and ω_c is the bandwidth of the proportional resonance controller. $G_{v_{dc}}(z)$ is half of the v_{dc} voltage (i.e., 800/2 = 400 Volts) for a two-level VSC with bipolar switching:

$$G_{v_{dc}(z)} = 400 \tag{10}$$

$G_d(z)$ indicates one sampling cycle digital computation delay:

$$G_d(z) = z^{-1} \tag{11}$$

The LCL filter transfer function block $Y_{op}(z)$ can be written as:

$$Y_{op}(z) = \mathcal{Z}\{\frac{1}{L_f(s) + \frac{C_f(s)L_g(s)}{C_f(s)+L_g(s)}}\}\Big|_{ZOH}^{T_s=100\mu s} \tag{12}$$

$$L_f(s) = L_1 s + R_1$$
$$C_f(s) = \frac{1}{C_f s + R_f} \tag{13}$$
$$L_g(s) = L_2 s + R_2$$

The parameters used in (9), (10), (12), (13) can be found in Table 2. In Figure 6, the proportional resonance controller parameters are calculated in proportion to the gain constant K ($K_p = 1 \cdot K$; $K_i = 250 \cdot K$), where K varies from $0 \rightarrow 2$ in step of 0.1. From the pole zero map, it indicates that when $K = 2$ the current controller is critically stable. When $K = 1$ is chosen, the dominant pole is approaching the optimal damping ($\zeta = 0.707$) [18] and found the balance between the transient response speed and the overall system stability.

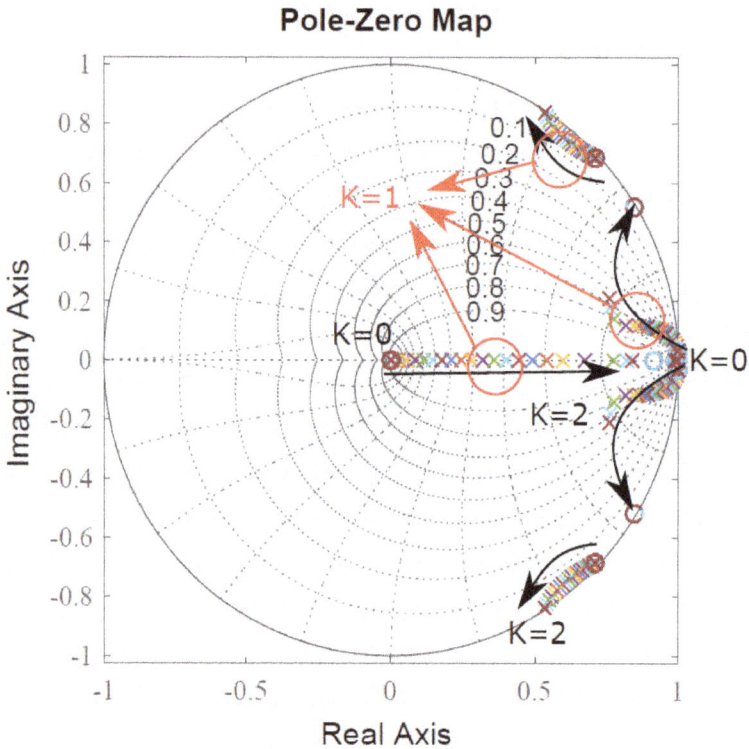

Figure 6. VSC Z-domain pole zero map of the inner current control loop.

4. Simulation and Experimental Results

This section presents the case study results of PLL dynamics under fast ROCOF. The detailed VSC model is firstly connected to a simple test network represented by an ideal voltage behind a short-circuit impedance, where the emulated fast ROCOF is triggered as a frequency ramp-down event. Next the VSC model is integrated with the validated Bonaire network model, and a fault-induced fast ROCOF event is triggered by applying a 400 ms three-phase fault on the 12 kV cable. Time domain simulation results under the fast ROCOF events are presented and a vector diagram based analysis is carried out to explain the phenomena. A down-scaled experiment is performed to verify the simulation results with Bonaire island power network.

4.1. Simulation Results—Simple Test Network

In this section, the detailed VSC electromagnetic transient model developed in Section 3.1 is connected to a simple test network model, where an ideal voltage source is connected in series behind a short-circuit impedance (Figure 7). To study the impact of a fast ROCOF on the PLL dynamics and the VSC power stage output, the ideal voltage source is triggered by a frequency ramp-down event (ROCOF = 10 Hz/s), where the mechanical inertia coupling is not considered.

Figure 8 presents the simulation results with a simple test network. The emulated frequency ramp-down event is shown by the green curve in Figure 8b, where system frequency starts to decline at 2 s and it settles at 46 Hz in 0.4 s (ROCOF = 10 Hz/s). Two sets of PLL parameters (see Table 1), namely the PLL with the high and low control bandwidth, are calculated on the same event. Looking at Figure 8b,c, the PLL with a high controller bandwidth (blue) tightly follow the frequency variation

(Figure 8b) with negligible phase error (Figure 8c) whilst the PLL with low controller bandwidth (red curves in Figure 8b,c respectively) exhibit inferior dynamic performance during the fast ROCOF event. Surprisingly the VSC power stage output (dashed line in Figure 8a) with low PLL bandwidth does not deviate significantly from its power set-point despite a significant phase angle error (red curve in Figure 8c).

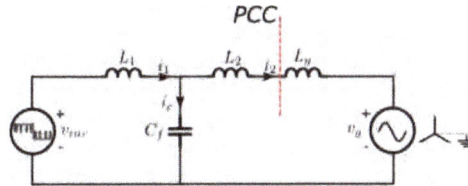

Figure 7. VSC dynamics under fast ROCOF connected to a simplified network.

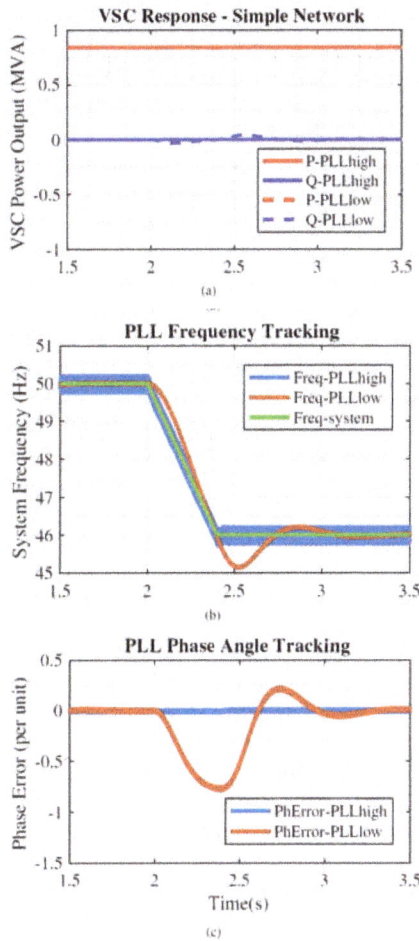

Figure 8. Simulation results with a simplified network and different PLL parameters (see Table 1)—(**a**) VSC output (**b**) PLL frequency tracking (**c**) PLL phase angle tracking.

In Figure 9, the direct (I^*_{dref}) and quadrature (I^*_{qref}) current reference is shown. For the case with a high PLL controller bandwidth, its current controller reference point is maintained the same during the fast ROCOF event (solid line in Figure 9). However, this is not the case when the PLL controller bandwidth is low. In order to maintain the VSC output power during the fast ROCOF event, the phase angle error introduced by the PLL will be counteracted by the outer controller loop which constantly regulates the direct (I^*_{dref}) and quadrature (I^*_{qref}) current reference (dotted and dashdotted line in Figure 9).

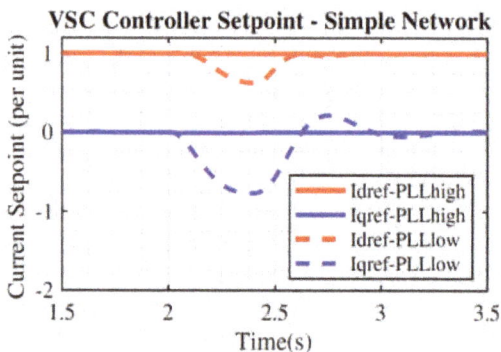

Figure 9. Current controller reference signal from dual outer loop controllers—PLL with low bandwidth (dash line), PLL with high bandwidth (solid line).

A detailed explanation of the fast ROCOF impact on the PLL dynamics and the subsequent VSC power stage output can be made by the vector diagram shown in Figure 10. Take the PLL with low controller bandwidth for example, when the fast ROCOF event initiates, a phase angle error (Δq) occurs between the actual grid voltage vector (U_{s1}) and the d-axis of rotating frame. Both i_d and i_q will project in phase and quadrature component on the actual grid voltage vector (U_{s1}). Effectively, this indicates the coupling of the active and reactive power in the control, and this can be compensated by the outer loop power flow controllers. In fact, the PI controller embedded in the outer loop controller increase the PLL controller from type II to type III making it capable of maintaining VSC power stage output despite the large phase angle error.

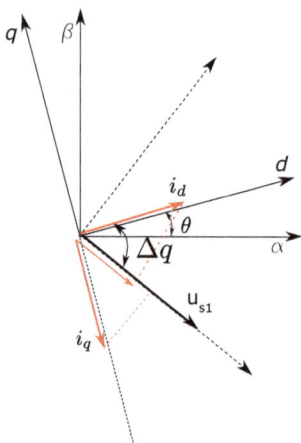

Figure 10. Vector diagram illustrating PLL dynamics under fast ROCOF.

4.2. Simulation Results—Bonaire Island Power System

In this section the simple test network is replaced by a simplified Bonaire island network (see Figure 11) with 5 diesel generators in service supplying a total system load of 12 MW/6 MVAr lumped at the main 12 kV distribution substation, technical details related to the network structure, the dynamic diesel generator controllers, and the validation results are included in [21]. For the simplified Bonaire island network model, the inertia aspect is included in the PSCAD generator model. Similar to the actual fault record in Figure 1, where the fast ROCOF is induced by the 12 kV system fault, here for simplicity a balanced three-phase fault on the 12 kV system (400 ms fault clearing time, 20% voltage dip) and the dynamic behavior of a generic 850 kW wind turbines model is observed under a fault induced fast ROCOF event.

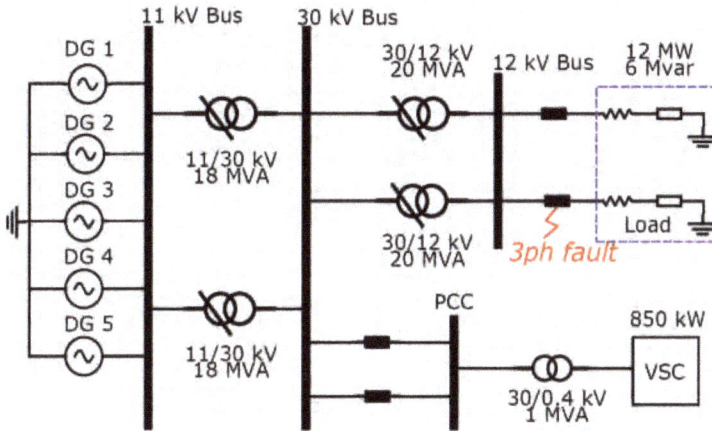

Figure 11. Single line diagram of Bonaire Island power system.

Figure 12 demonstrates the simulation results when the VSC is coupled to the simplified Bonaire island network. A simulated three phase 12 kV cable fault causes the total system consumption to increase (Figure 12b) and the system frequency decreases sharply from approximately 50 Hz to 47.5 Hz in 400 ms (Figure 12c). When the fault occurs in the network (for 20% voltage dip), the grid-connected VSC will run into the low voltage ride through (LVRT) mode and inject active/reactive power per grid code requirement. With reference to [7,22,23], anti-windup will freeze the outer loop controller integral input (i.e., set to 0) and the inner current controller current reference ($I^*_{dcode} \& I^*_{qcode}$) is calculated according to the grid code requirement (see Figure 13). For this study, the LVRT strategy sets the $I^*_{dcode} = 1.0\,pu$ and $I^*_{qcode} = 0.0\,pu$ for the maximum active power delivery.

In the case of a fault induced fast ROCOF, the LVRT strategy will fix the current controller reference given by the grid code requirement. For the PLL with a low control bandwidth, the large phase angle error will effectively cause the coupling of active and reactive power control as explained in Section 4.1 with the vector diagram (Figure 10). For the LVRT strategy with maximized active power delivery, the results from Figure 12a indicates that the VSC output with a low PLL control bandwidth (dashed line in Figure 12a) delivers less active power and consumes additional reactive power from the grid during the fault induced fast ROCOF event.

4.3. Experimental Results

A down-scaled experimental setup is built to verify the analytical/simulation results concerning the fast ROCOF effect on the PLL dynamics. Chroma 61845 has been used to emulate the low inertia grid condition of Bonaire island power system. The VSC is implemented by a Danfoss FC103P11KT 11 converter and the control algorithms are programmed in dSPACE1007. The parameters of the inverter are summarized in Table 3 and the experimental setup is shown in Figure 14.

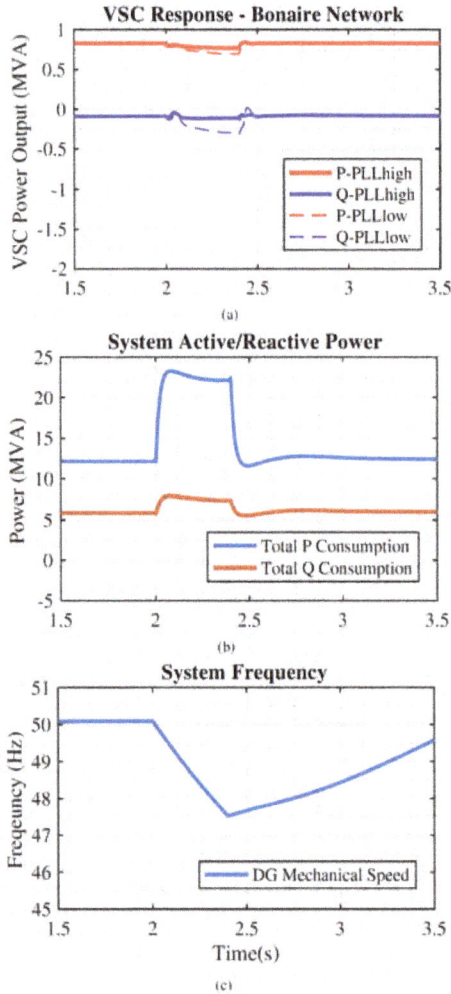

Figure 12. Simulation results with Bonaire island power network using different PLL control bandwidths—(**a**) VSC power output (**b**) Total active/reactive power consumption (**c**) System frequency measured by the diesel generator mechanical speed.

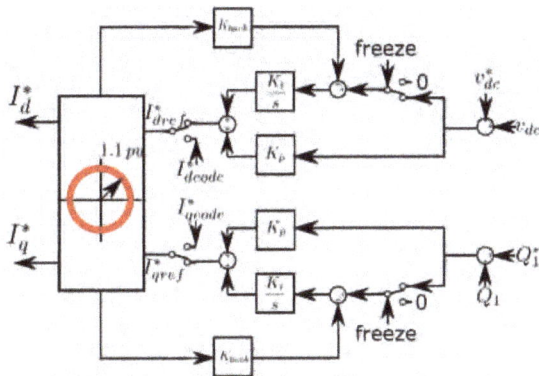

Figure 13. LVRT strategy for 850 kW voltage source converter.

Table 3. Paramters of inverter for experiment verification.

Parameter	Value	Unit
Rated Power	2	kW
Grid fundamental frequency	50	Hz
DC Link Voltage V_{dc}	730	Volts
DC Link Capacitor C_{dc}	1500	μF
AC Voltage V_{rms}	400	Volts
L_1 Inverter side inductor	1500	μH
L_t Equivalent grid-side inductor ($L_2 + L_g$)	1500	μH
C_f Filter capacitor	5	μF
Inverter control sampling frequency f_s	10	kHz
Inverter switching frequency f_{sw}	10	kHz

In this down-scaled experiment, a three-phase fault (20% voltage dip and 10 Hz/s ROCOF) is emulated by the regenerative grid simulator Chroma 61845 and cleared after 400 ms. Figure 15a demonstrates the experimental results when the inverter is operated with the low bandwidth PLL parameters (see Table 1). The inverter is firstly operated in normal operating condition with 2 kW active power output and 0 kVA reactive power output. Then the three-phase fault is initiated and cleared after 400 ms. During the fault, the frequency decreased from 50 Hz to 46 Hz (i.e., 10 Hz/s ROCOF), and the q-axis voltage deviated from zero indicating a large θ angle error.

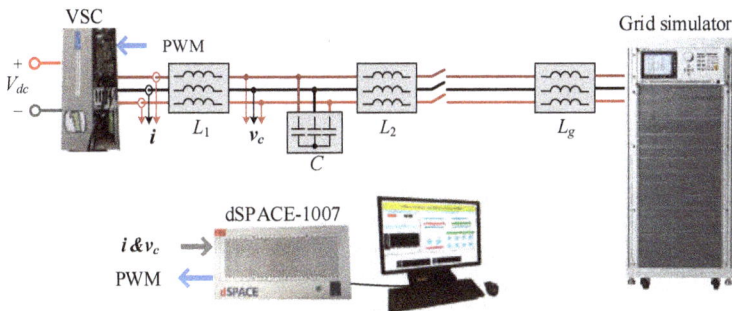

Figure 14. Single line diagram of bonaire island power system.

Figure 15b shows the experimental results when the inverter is operated with the high-bandwidth PLL parameters (see Table 1). For the LVRT strategy with maximized active power delivery, experimental results confirm that the VSC output with a low PLL control bandwidth (Figure 15a) will deliver less active power and consumes additional reactive power from the grid. With increased renewable energy penetration, less active power delivery in the low inertia grid will not only threaten the frequency stability but also worsen the transient frequency nadir. Additionally, extra consumption of reactive power from the grid during the fault condition will hinder the voltage recovery following the fault clearance.

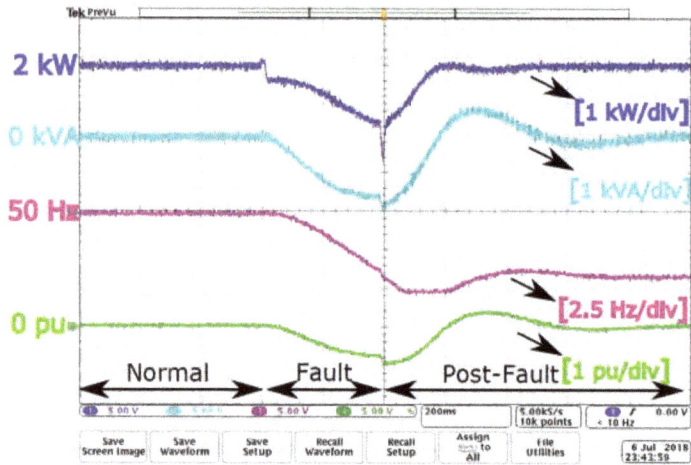

(a)

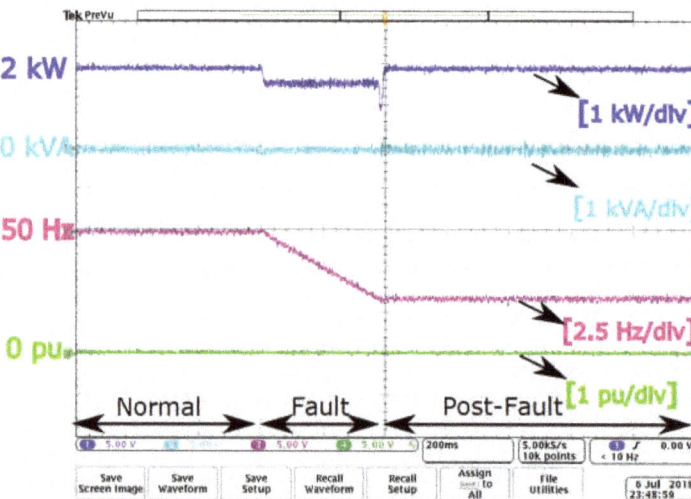

(b)

Figure 15. Experimental results with fault induced ROCOF of 10 Hz/s—(**a**) Inverter response during fault induced fast ROCOF with low bandwidth PLL parameters, (**b**) Inverter rersponse during fault induced fast ROCOF with high bandwidth PLL parameters.

5. Conclusions

Based on the fault record in the Bonaire island grid, this paper investigates a low inertia power grid scenario, where the system frequency varies rapidly due to the low kinetic energy buffer provided by the synchronous generator's mechanical inertia. For a PLL with low controller bandwidth, the phase angle error can effectively be counteracted by the outer loop PI controller regulation. However, for the fault induced fast ROCOF as recorded in the Bonaire island power grid, the existing LVRT strategy shown in the literature will freeze the outer loop PI controller and calculate the current controller reference directly according to the grid code requirement. Simulation results in a simplified Bonaire island network indicate that the VSC with a low PLL dynamics delivers less active power to the grid whilst it consumes additional reactive power during the fault induced fast ROCOF. A down-scaled experimental setup was used to verify the analysis/simulation results of the simplified Bonaire network.

Author Contributions: Writing—Review & Editing, Y.S., E.C.W.d.J., X.W., F.B., V.C.; Formal Analysis, Y.S.; Investigation, Y.S.; Validation, D.Y.; Supervision, J.F.G.C.

Funding: This research was funded by DNV GL Group Technology Research and 3E Stichting.

Acknowledgments: The author would like to acknowledge the financial support from DNV GL, Group Technology Research and the Stichting 3E. Furthermore, the author would like to extend his gratitude to ir. Wim Kuijpers, and WEB-Bonaire for facilitating the validation of the dynamic model and the provision of network data.

Conflicts of Interest: The authors declare no conflict of interest.

References

1. IEEE Standard. *IEEE Guide for Planning DC Links Terminating at AC Locations Having Low Short-Circuit Capacities*; IEEE: Piscataway, NJ, USA, 1997.
2. Durrant, M.; Werner, H.; Abbott, K.; Farag, A.; Durrant, M.; Wemer, H.; Abbottt, K. Model of a VSC HVDC Terminal Attached to a Weak AC System. In Proceedings of the 2003 IEEE Conference on Control Applications, Istanbul, Turkey, 25 June 2003; Volume 1, pp. 173–177.
3. Kim, B.H.; Sul, S.K. Stability-Oriented Design of Frequency Drift Anti-Islanding and Phase-Locked Loop under Weak Grid. *IEEE J. Emerg. Sel. Top. Power Electron.* **2017**, *5*, 760–774. [CrossRef]
4. Zhou, P.; Yuan, X.; Hu, J.; Huang, Y. Stability of DC-link voltage as affected by phase locked loop in VSC when attached to weak grid. In Proceedings of the IEEE Power and Energy Society General Meeting, National Harbor, MD, USA, 27–31 July 2014; pp. 1–5.
5. Huang, Y.; Yuan, X.; Hu, J.; Zhou, P.; Wang, D. DC-Bus Voltage Control Stability Affected by AC-Bus Voltage Control in VSCs Connected to Weak AC Grids. *IEEE J. Emerg. Sel. Top. Power Electron.* **2016**, *4*, 445–457. [CrossRef]
6. Lu, S.; Xu, Z.; Xiao, L.; Jiang, W.; Bie, X. Evaluation and Enhancement of Control Strategies for VSC Stations under Weak Grid Strengths. *IEEE Trans. Power Syst.* **2017**, *PP*, 1. [CrossRef]
7. Huang, Y.; Wang, D.; Shang, L.; Zhu, G.; Tang, H.; Li, Y. Modeling and Stability Analysis of DC-Link Voltage Control in Multi-VSCs with Integrated to Weak Grid. *IEEE Trans. Energy Convers.* **2017**, *32*, 1127–1138. [CrossRef]
8. Wu, G.; Liang, J.; Zhou, X.; Li, Y.; Egea-Alvarez, A.; Li, G.; Peng, H.; Zhang, X. Analysis and design of vector control for VSC-HVDC connected to weak grids. *CSEE J. Power Energy Syst.* **2017**, *3*, 115–124. [CrossRef]
9. Givaki, K.; Xu, L. Stability analysis of large wind farms connected to weak AC networks incorporating PLL dynamics. In Proceedings of the International Conference on Renewable Power Generation (RPG 2015), Beijing, China, 17–18 October 2015; pp. 1–6.
10. Fan, L.; Miao, Z. An Explanation of Oscillations Due to Wind Power Plants Weak Grid Interconnection. *IEEE Trans. Sustain. Energy* **2017**, *PP*, 1–2. [CrossRef]
11. Sun, Y. The Impact of Voltage-Source-Converters' Control on the Power System: The Stability Analysis of a Power Electronics Dominant Grid. Ph.D. Thesis, Department of Electrical Engineering, Eindhoven, The Netherlands, 2018.
12. Bifaretti, S.; Lidozzi, A.; Solero, L.; Crescimbini, F. Anti-islanding detector based on a robust PLL. *IEEE Trans. Ind. Appl.* **2015**, *51*, 398–405. [CrossRef]
13. Freitas, W.; Xu, W.; Affonso, C.M.; Huang, Z. Comparative analysis between ROCOF and vector surge relays for distributed generation applications. *IEEE Trans. Power Deliv.* **2005**, *20*, 1315–1324. [CrossRef]

14. Duckwitz, D.; Fischer, B. Modeling and Design of df/dt-based Inertia Control for Power Converters. *IEEE J. Emerg. Sel. Top. Power Electron.* **2017**, *5*, 1553–1564. [CrossRef]

15. Blaabjerg, F.; Teodorescu, R.; Liserre, M.; Timbus, A.V. Overview of control and grid synchronization for distributed power generation systems. *IEEE Trans. Ind. Electron.* **2006**, *53*, 1398–1409. [CrossRef]

16. Josep, M.; Vasquez, J.C.; Golestan, S.; Member, S.; Guerrero, J.M. Three-Phase PLLs: A Review of Recent Advances. *IEEE Trans. Power Electron.* **2017**, *32*, 1894–1907.

17. Wang, X.; Harnefors, L.; Blaabjerg, F. A Unified Impedance Model of Grid-Connected Voltage-Source Converters. *IEEE Trans. Power Electron.* **2017**, *33*, 1775–1787. [CrossRef]

18. Franklin, G.F.; Powell, J.D.; Emami-Naeini, A. *Feedback Control of Dynamic Systems*, 6th ed.; Addison-Wesley: Reading, MA, USA, 1994.

19. Chung, S.K. A phase tracking system for three phase utility interface inverters. *IEEE Trans. Power Electron.* **2000**, *15*, 431–438. [CrossRef]

20. Sun, Y.; DeJong, E.C.W.; Kuijpers, W.G.; Wang, X.; Blaabjerg, F.; Cuk, V.; Cobben, J.F.G. PLL Dynamics in Low Inertia Weak Grid. In Proceedings of the 16th International Workshop on Large-Scale Integration of Wind Power into Power System as well as on Transmission Networks for Offshore Wind Power Plants (WIW 17), Berlin, Germany, 25–27 October 2017; p. 369.

21. Sun, Y.; De Jong, E.; Cuk, V.; Cobben, S. 6MW solar plant integration feasibility study: Bonaire island case study. In Proceedings of the 2016 IEEE 17th Workshop on Control and Modeling for Power Electronics, COMPEL 2016, Trondheim, Norway, 27–30 June 2016.

22. Alepuz, S.; Busquets-Monge, S.; Bordonau, J.; Martinez-Velasco, J.A.; Silva, C.A.; Pontt, J.; Rodriguez, J. Control strategies based on symmetrical components for grid-connected converters under voltage dips. *IEEE Trans. Ind. Electron.* **2009**, *56*, 2162–2173. [CrossRef]

23. Ma, K.; Liserre, M.; Blaabjerg, F. Operating and loading conditions of a three-level neutral-point-clamped wind power converter under various grid faults. *IEEE Trans. Ind. Appl.* **2014**, *50*, 520–529.

energies

MDPI

Article

Direct Fixed-Step Maximum Power Point Tracking Algorithms with Adaptive Perturbation Frequency

Eyal Amer [1], Alon Kuperman [1,*] and Teuvo Suntio [2]

[1] Applied Energy Laboratory, Department of Electrical and Computer Engineering, Ben-Gurion University of the Negev, Beer-Sheva 8410501, Israel; eyalamer@gmail.com

[2] Laboratory of Electrical Energy Engineering, Tampere University of Technology, Tampere 33720, Finland; teuvo.suntio@tut.fi

*   Correspondence: alonk@bgu.ac.il; Tel.: +972-8-6461-599

Received: 20 December 2018; Accepted: 24 January 2019; Published: 27 January 2019

Abstract: Owing to the good trade-off between implementation and performance, fixed-step direct maximum power point tracking techniques (e.g., perturb and observe and incremental conductance algorithms) have gained popularity over the years. In order to optimize their performance, perturbation frequency and perturbation step size are usually determined a priori. While the first mentioned design parameter is typically dictated by the worst-case settling time of the combined energy conversion system, the latter must be high enough to both differentiate the system response from that caused by irradiation variation and match the finite resolution of the analog-to-digital converter in case of digital implementation. Well-established design guidelines, however, aim to optimize steady-state algorithm performance while leaving transients nearly untreated. To improve transient behavior while keeping the steady-state operation unaltered, variable step direct maximum power point tracking algorithms based on adaptive perturbation step size were proposed. This paper proposes a concept of utilizing adaptive perturbation frequency rather than variable step size, based on recently revised guidelines for designing fixed-step direct maximum power point tracking techniques. Preliminary results demonstrate the superiority of the proposed method over adaptive perturbation step size operation during transients, without compromising the steady state performance.

Keywords: photovoltaic generators; maximum power point tracking; step size; perturbation frequency

---

1. Introduction

Energy produced by a photovoltaic generator (PVG) is mainly dependent on a single (referred to as the "energy generating") parameter: Solar irradiation. In addition, PVG power is load dependent and affected by temperature. As a result, the PVG power curve is characterized by a single maximum power point (MPP) on a single unit level for a specific set of environmental variables. Consequently, generalized electrical characteristics of a PVG are represented by a family of power curves for a range of solar irradiations and temperatures [1]. Upon variation of one or both environmental variables, locations of MPP current and voltage—and hence power—change. Such a behavior calls for instantaneous maximum power point tracking (MPPT) in order to optimize PVG economical utilization [2].

Comparison of different MPPT algorithms presented in the literature so far may be found [3–11]. Owing to their inherently generic nature and relatively simple implementation, direct non-model-based techniques, such as perturb and observe (PO) [12,13], incremental conductance [14,15] and extremum seeking or ripple correlation control [16–20], are probably the most widely applied MPPT methods. While well-established, fixed step versions of MPPT algorithms suffer from the well-known trade-off between transient and steady-state operation [21]. In order to tackle this drawback, adjustable (or adaptive) step size versions of non-model-based algorithms have been proposed [22–38], in which the step size is adjusted, typically in proportion to the derivative of PVG power with respect to control variable. Indeed, such algorithms demonstrate superiority over the fixed step versions, achieved at the expense of more complex implementation. It is interesting to note that the performance similarity of fixed step direct non-model-based algorithms versions have been revealed [39,40], while the equivalence of major variable step size MPPT algorithms has been demonstrated [41].

Consider, without loss of generality, a photovoltaic energy conversion operating under a single-loop direct non-model based MPPT. For example, the interfacing power converter (IPC) duty cycle is the perturbed variable (see Figure 1a). A small perturbation Δd is injected into the system every T_p seconds. Following a T_Δ-lasting transient, the corresponding change in generated power is observed either at the PVG or load side (see Figure 1b), and the next perturbation polarity (in fixed step versions) and size (in variable step versions) is determined so that the operation point is driven towards the MPP. It was shown in References [13,21] that the maximum perturbation frequency (reciprocal of T_p) is limited by settling time of the generated power transient induced by the perturbation while the minimum perturbation step size depends on the maximum expected irradiation variation rate and sensing resolution. Recently, the authors of References [42,43] have refined the above design guidelines, proving that the maximum perturbation frequency should be designed at the short-circuit operating point while the minimum perturbation step size should be designed at the maximum power point, both corresponding to standard test conditions.

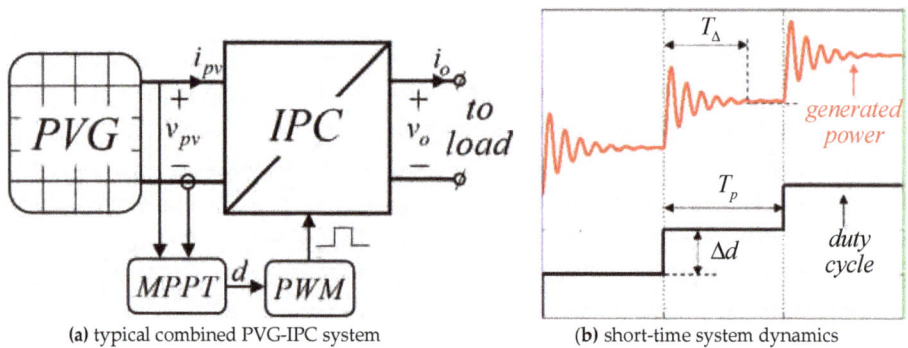

Figure 1. Photovoltaic energy conversion under direct non-model based maximum power point tracking (MPPT) technique

Once the design guidelines briefly reviewed above are respected, stable three-point behavior is ensured in steady-state [21]. It is therefore commonly assumed that if the selected T_p is too small, the MPPT algorithm can be confused and the operating point may become unstable, bringing disordered or chaotic behaviors into the system. On the other hand, selecting a T_p that is too long penalizes MPPT convergence speed and efficiency. Therefore, in order to improve the performance of the fixed-step based MPPT algorithm, variable step size methods should utilize adaptive step size while keeping the perturbation frequency constant. Increasing the perturbation step size when the operation point is far from the MPP and decreasing it in the MPP vicinity is, therefore, the main concept of such algorithms. Unfortunately, step size increase yields correspond to an increase of transients. In addition, because MPP voltage is nearly independent of irradiation, variable step MPPT algorithms tend to be confused upon irradiation changes [21]. On the other hand, applications of direct non-model based MPPT algorithms with a perturbation period much lower than the settling time of the system response were reported in References [44,45]. It was shown that faster responses to irradiance changes were achieved, yet steady-state oscillations were larger than those for the perturbation frequency dictated by design guidelines. Nevertheless, experimental investigation alone was conducted without a solid analytical background. The only attempt made to vary the perturbation frequency followed by a theoretical framework was made in Reference [46]. However, the effect of the PVG on the system dynamics was completely disregarded and, therefore, the presented outcomes are not consistent with the design guidelines above. The effect of PVG on generated power dynamics is extremely significant [47,48] and must be considered during MPPT algorithm design, since the boundary value of the perturbation frequency is dependent on PVG parameters.

Inspired by the promising results demonstrated in References [44,45], this paper proposes to combine the advantages of operating with high perturbation frequency values during transients and the design-guideline-imposed values of the perturbation frequency in steady state, while maintaining the step size at minimum value, as dictated by the fixed-step algorithm design guidelines. Such a concept yields a fixed-step variable-perturbation-frequency MPPT algorithm. The proposed technique has the ability of accelerating the transients caused by either system initialization away from MPP or sudden irradiation changes while maintaining accuracy during a steady-state regime. Preliminary results demonstrate the superiority of the proposed approach over fixed-step and fixed-frequency methods, as well as over variable-step approaches.

The rest of the paper is organized as follows: A review of PVG properties is briefly given in Section 2. Combined PVG-IPC-load dynamics are derived in Section 3. The principle of adaptive step MPPT algorithms is given in Section 4, followed by the proposed concept of adaptive perturbation frequency algorithms in Section 5. An example comparing the performance of different approaches is discussed in Section 6. The conclusions are drawn in Section 7.

2. The Photovoltaic Generator

A generalized PVG equivalent circuit is shown in Figure 2a [49]. It consists of a photo current source i_P; a current source i_D, representing the current of k parallel-connected semiconductor diodes, is given by

$$i_D = \sum_k I_{0k} \left(\exp\left\{ \frac{v_{PV} - r_S i_{PV}}{\alpha_k V_T} \right\} - 1 \right)$$

(1)

where I_{0k} and α_k symbolize the reverse saturation current and ideality factor of the k-th diode, and V_T denotes thermal voltage; equivalent capacitance is c_{PV}, and the equivalent shunt and series resistances are r_{SH} and r_S, respectively. The equivalent capacitance c_{PV} is small and its value may be neglected since it is typically offset by the value of IPC input capacitance. Equivalent circuit components are environmental variables dependent as follows: the photocurrent i_P depends on both irradiation and temperature; the diode current i_D is temperature dependent and irradiation independent. The resistances are typically considered independent environmental variables, even though r_{SH} possesses some irradiation dependence.

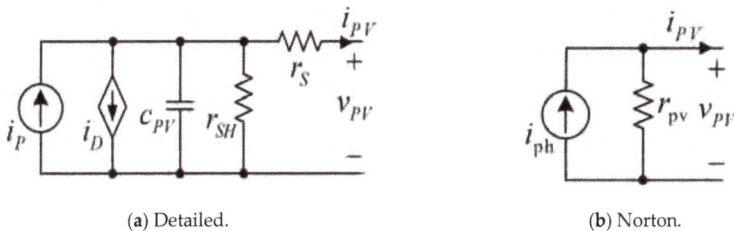

(a) Detailed.   (b) Norton.

Figure 2. Photovoltaic generator (PVG) equivalent circuit.

All the equivalent circuit parameters may be estimated either from experimental measurements or extracted from the manufacturer's datasheets [50,51]. Consequently, the detailed PVG equivalent circuit may be rearranged into a simplified dynamic Norton representation, as shown in Figure 2b. Norton and detailed equivalent circuit parameters are related as

$$r_{pv} = r_S + r_{SH}||r_D, \quad i_{ph} = i_P \frac{r_{SH}||r_D}{r_{pv}} \tag{2}$$

with

$$r_D = \frac{\partial v_{pv}}{\partial i_D} \tag{3}$$

representing the equivalent dynamic resistance of i_D. It may be concluded that the value of PVG dynamic resistance r_{pv} depends on both environmental variables and is also influenced by the operating point. In cases where a typical single-diode equivalent circuit is considered (i.e., for $k = 1$ in Equation (1)), PVG dynamic resistance may be reformulated into

$$r_{pv} = r_S + \frac{r_{SH}}{1 + W\left(\frac{I_0 r_{SH}}{\alpha V_T} \exp\left(\frac{r_{SH}(i_P - I_0 - i_{PV})}{\alpha V_T}\right)\right)}, \tag{4}$$

where $W(\cdot)$ stands for the Lambert-W function. According to the analysis in Reference [31], the practical PVG dynamic resistance at short circuit (SC) and open circuit (OC) conditions may be approximated as

$$r_{pv}|_{SC} \approx r_{SH} \tag{5}$$

and

$$r_{pv}|_{OC} \approx r_S, \tag{6}$$

respectively, since $W(x) << 1$ for $x << 1$, $W(x) >> 1$ for $x >> 1$ and $r_{SH} >> r_S$. It is then concluded that in PVG, dynamic resistance generally resides within the $[r_S, r_{SH}]$ region of values throughout the whole operation range. Therefore, for any MPP, the following holds;

$$r_{pv}|_{OC} < r_{pv}|_{MPP} < r_{pv}|_{SC}. \tag{7}$$

3. Photovoltaic Generator (PVG) Power Dynamics in a Combined PVG-IPC-Load System

Referring to the systems in Figures 1a and 2b while defining PVG dynamic conductance as $Y_{pv} = 1/r_{pv}$, combined small-signal system dynamics are given by [52,53]

$$\begin{aligned}
\hat{v}_{pv} &= \frac{Z_{in}}{1 + Z_{in}Y_{pv}}\hat{i}_{ph} + \frac{T_{oi}}{1 + Z_{in}Y_{pv}}\hat{v}_o + \frac{G_{ci}}{1 + Z_{in}Y_{pv}}\hat{d} \\
\hat{i}_{pv} &= \frac{1}{1 + Z_{in}Y_{pv}}\hat{i}_{ph} - \frac{Y_{pv}T_{oi}}{1 + Z_{in}Y_{pv}}\hat{v}_o - \frac{Y_{pv}G_{ci}}{1 + Z_{in}Y_{pv}}\hat{d},
\end{aligned} \tag{8}$$

where $\hat{d}$ denotes the IPC duty cycle; G_{ci}, T_{oi}, and Z_{in} symbolize the IPC control-to-input-voltage transfer function, output-to-input voltage transfer function, and input impedance, respectively. The temperature effect is disregarded in Equation (8) due to its relatively slow dynamics. The resulting small signal PVG power dynamics are given by [42]

$$\hat{p}_{pv} = I_{pv}\hat{v}_{pv} + V_{pv}\hat{i}_{pv} + \hat{i}_{pv}\hat{v}_{pv} \tag{9}$$

with (I_{pv}, V_{pv}) representing the PVG operating point, further rearranged as

$$\hat{p}_{pv} \approx V_{pv}\left(\frac{1}{R_{pv}} - \frac{1}{r_{pv}}\right)\hat{v}_{pv} - \frac{1}{r_{pv}}\hat{v}_{pv}^2 \tag{10}$$

with

$$R_{pv} = \frac{V_{pv}}{I_{pv}} \tag{11}$$

representing PVG static resistance. Taking into account the following static and dynamic resistances relation

$$
\begin{aligned}
r_{pv} &\gg R_{pv}, & v_{pv} &\ll v_{pv}\big|_{MPP} \\
r_{pv} &\approx R_{pv}, & v_{pv} &\approx v_{pv}\big|_{MPP} \\
r_{pv} &\ll R_{pv}, & v_{pv} &\gg v_{pv}\big|_{MPP}
\end{aligned}
\tag{12}
$$

The small-signal dynamics of PVG power may be approximated as

$$
\hat{p}_{pv} \approx
\begin{cases}
I_{pv}\hat{v}_{pv}, & v_{pv} < v_{pv}\big|_{MPP} \\
-\frac{1}{R_{pv}}\hat{v}_{pv}^2, & v_{pv} \approx v_{pv}\big|_{MPP} \\
-\frac{V_{pv}}{r_{pv}}\hat{v}_{pv}, & v_{pv} > v_{pv}\big|_{MPP}
\end{cases}
\tag{13}
$$

The generalized control-to-input-voltage transfer function of the combined PVG-IPC-Load system may be obtained from Equation (8) as

$$
G_{ci}^{pv}(s) = \frac{G_{ci}}{1 + Z_{in}Y_{pv}} = -V_O \frac{\omega_n^2(1 + s/\omega_{z\text{-}esr})}{s^2 + 2\zeta_{pv}\omega_n s + \omega_n^2}.
\tag{14}
$$

While the parameters values in Equation (14) depend on IPC topology, the structure of Equation (14) is IPC topology independent. Therefore, in case of a small-signal duty cycle perturbation given by

$$
\hat{d}(s) = \frac{\Delta d}{s},
\tag{15}
$$

corresponding generalized PVG voltage response is given in Laplace and time domains by

$$
\hat{v}_{pv}(s) = \hat{d}\cdot G_{ci}^{pv} = -V_O\Delta d\left(\frac{1}{s} - \frac{s + 2\zeta_{pv}\omega_n - \frac{\omega_n^2}{\omega_{z\text{-}esr}}}{s^2 + s2\zeta_{pv}\omega_n + \omega_n^2}\right)
\tag{16}
$$

and

$$
\hat{v}_{pv}(t) = V_O\Delta d\left(1 - \frac{\sqrt{1 + \frac{\omega_n}{\omega_{z\text{-}esr}}\left[\frac{\omega_n}{\omega_{z\text{-}esr}} - 2\zeta_{pv}\right]}}{\sqrt{1 - \zeta_{pv}^2}}\exp(-\zeta_{pv}\omega_n t)\sin\left[\omega_d t + \tan^{-1}\left\{\frac{\sqrt{1 - \zeta_{pv}^2}}{\zeta_{pv} - \frac{\omega_n}{\omega_{z\text{-}esr}}}\right\}\right]\right),
\tag{17}
$$

respectively, with

$$
\omega_d = \omega_n\sqrt{1 - \zeta_{pv}^2}
\tag{18}
$$

and

$$
0 < \zeta_{pv} < 1.
\tag{19}
$$

The practical assumption $\omega_{z\text{-}esr} \gg \omega_n$ further simplifies Equation (17) as

$$
\hat{v}_{pv}(t) \approx V_O\Delta d\left(1 - \frac{1}{\sqrt{1 - \zeta_{pv}^2}}\exp(-\zeta_{pv}\omega_n t)\sin\theta(t)\right),
\tag{20}
$$

where

$$
\theta(t) = \omega_d t + \tan^{-1}\left\{\frac{\sqrt{1 - \zeta_{pv}^2}}{\zeta_{pv}}\right\}.
\tag{21}
$$

Combining Equation (20) with Equation (13) yields

$$
\hat{p}_{pv}(t) \approx
\begin{cases}
-I_{pv}V_O\Delta d\left(1 \pm \dfrac{1}{\sqrt{1-\zeta_{pv}^2}}\exp(-\zeta_{pv}\omega_n t)\cdot \sin\theta(t)\right), & v_{pv} \ll v_{pv}|_{MPP} \\[3mm]
\dfrac{(V_{DC}\Delta d)^2}{R_{pv}}\left(1 - \dfrac{1}{\sqrt{1-\zeta_{pv}^2}}\exp(-\zeta_{pv}\omega_n t)\cdot \sin\theta(t)\right)^2, & v_{pv} \approx v_{pv}|_{MPP} \\[3mm]
\dfrac{V_{pv}V_O\Delta d}{r_{pv}}\left(1 \pm \dfrac{1}{\sqrt{1-\zeta_{pv}^2}}\exp(-\zeta_{pv}\omega_n t)\cdot \sin\theta(t)\right), & v_{pv} \gg v_{pv}|_{MPP}
\end{cases}
\quad . \tag{22}
$$

Settling the time of the PVG power transient, imposed by duty cycle perturbation, is then dictated by the corresponding envelope behavior, given by

$$
env\left(\hat{p}_{pv}(t)\right) \approx
\begin{cases}
-I_{pv}V_O\Delta d\left(1 \pm \dfrac{1}{\sqrt{1-\zeta_{pv}^2}}\exp(-\zeta_{pv}\omega_n t)\right), & v_{pv} \ll v_{pv}|_{MPP} \\[3mm]
\dfrac{(V_{DC}\Delta d)^2}{R_{pv}}\left(1 \pm 2\dfrac{1}{\sqrt{1-\zeta_{pv}^2}}\exp(-\zeta_{pv}\omega_n t) + \dfrac{1}{1-\zeta_{pv}^2}\exp(-2\zeta_{pv}\omega_n t)\right), & v_{pv} \approx v_{pv}|_{MPP} \\[3mm]
\dfrac{V_{pv}V_O\Delta d}{r_{pv}}\left(1 \pm \dfrac{1}{\sqrt{1-\zeta_{pv}^2}}\exp(-\zeta_{pv}\omega_n t)\right), & v_{pv} \gg v_{pv}|_{MPP}
\end{cases}
\tag{23}
$$

The corresponding settling times are then obtained as

$$
T_\Delta \approx -
\begin{cases}
\dfrac{1}{\zeta_{pv}\omega_n}\ln\left(\Delta\sqrt{1-\zeta_{pv}^2}\right), & v_{pv} \ll v_{pv}|_{MPP} \\[3mm]
\dfrac{1}{\zeta_{pv}\omega_n}\ln\left(\dfrac{\Delta}{2}\sqrt{1-\zeta_{pv}^2}\right), & v_{pv} \approx v_{pv}|_{MPP} \\[3mm]
\dfrac{1}{\zeta_{pv}\omega_n}\ln\left(\Delta\sqrt{1-\zeta_{pv}^2}\right), & v_{pv} \gg v_{pv}|_{MPP}
\end{cases}
\quad . \tag{24}
$$

It should be noted that since ζ_{pv} is dependent on r_{pv} (the two are inversely proportional), settling times must be evaluated considering Equations (5)–(7). Typically, settling time increases monotonically with the decrease of ζ_{pv} (i.e., with the increase of PVG dynamic resistance). Therefore, $T_\Delta|_{OC} < T_\Delta|_{MPP} < T_\Delta|_{SC}$, and the longest settling time is expected at an SC condition, establishing the operating point for perturbation frequency design in MPPT algorithms with a fixed perturbation frequency. The value of a reciprocal of perturbation frequency should then obey [42]

$$
T_p \geq T_\Delta|_{SC}. \tag{25}
$$

4. Maximum Power Point Tracking (MPPT) with Adaptive Step Size

MPPT methods with adaptive step are based on directly or indirectly defining an objective function given by [22,34]

$$
y[k] = \frac{p_{pv}[k] - p_{pv}[k-1]}{d[k] - d[k-1]} = \frac{\Delta p_{pv}}{\Delta d}, \tag{26}
$$

where k is the sampling instant. Since

$$
y[k]
\begin{cases}
> 0, & v_{pv} \ll v_{pv}|_{MPP} \\
\approx 0, & v_{pv} \approx v_{pv}|_{MPP} \\
< 0, & v_{pv} \gg v_{pv}|_{MPP}
\end{cases}
\quad , \tag{27}
$$

The way of adjusting the duty cycle is as follows:

$$
\begin{cases}
increase\, d, & y[k] > 0 \\
decrease\, d, & y[k] < 0 \\
maintain\, d, & y[k] \approx 0
\end{cases}
\quad . \tag{28}
$$

Thus, the typical strategy of adapting the duty cycle perturbation at the $(k+1)$th sampling instant is given by

$$d[k+1] = d[k] + N \cdot y[k]. \tag{29}$$

with N referred to as the scaling factor. The size of

$$\Delta d[k] = N \cdot y[k] \tag{30}$$

must be kept between two bounds,

$$\Delta d_{\min} < \Delta d[k] < \Delta d_{\max}, \tag{31}$$

selected as follows. The value of Δd_{min} should be such that the corresponding imposed steady-state PVG power difference, Δp_{pv}, is higher than the steady-state PVG power difference caused by irradiation change during MPPT algorithm perturbation interval to satisfy [21]:

$$\Delta d_{\min} = \frac{1}{G_0} \sqrt{\frac{V_{pv}|_{MPP} \cdot K_{ph} \cdot |\dot{G}_s| \cdot T_p}{H \cdot V_{pv}|_{MPP} + Y_{pv}|_{MPP}}}, \tag{32}$$

where the value of the MPP voltage corresponds to standard test conditions, $K_{ph} = \frac{\partial i_{pv}}{\partial G_s}$ denotes the PVG material constant, G_0 signifies the DC gain of the duty cycle to PVG voltage transfer function in Equation (14), $\dot{G}_s$ represents the solar irradiation change rate, and $H = -\frac{1}{2} \frac{\partial^2 i_{pv}}{\partial^2 v_{pv}}\Big|_{MPP}$. It is important to highlight that since Equation (32) depends on environmental conditions, the combination of parameters leading to the highest value of $\Delta d_{\min}$ must be utilized, taking into account the worst case of irradiation change rate. It should be emphasized that in addition to irradiation variations, finite resolution of the utilized analog-digital converter (ADC) should also be considered upon selection of Δd_{min} [21]. On the other hand, the selected value of $\Delta d_{\max}$ must keep the PVG voltage within a feasible operation range for all expected operation conditions, satisfying [21]:

$$\Delta d_{\max} = N \frac{\Delta P_{pvmax}}{\Delta V_{pvmax}}, \tag{33}$$

In fixed-step MPPT algorithms, the high value of Δd leads to accelerated convergence, traded off for steady-state accuracy (see Equation (22)) since higher values of Δd impose higher power deviations around MPP in steady state. It was shown in [41] that Equation (29) is the discrete-time version of integral-based adjustment

$$d(t) = d(t - T_p) + \underbrace{\frac{N}{T_p} \int_{t-T_p}^{t} y(\tau) d\tau}_{\Delta d}, \tag{34}$$

i.e., the size of Δd is adjusted according to the objective function value. Unfortunately, while the method of choosing Δd_{min} is relatively simple, the selection of Δd_{max} is often based on a trial-and-error approach, and the value resulting from this process is suitable only for a given system operating under specific operating conditions. Moreover, it is well-known that the Perturb-and-Observe algorithm is confused by sudden irradiation changes, causing significant transients. In some cases this confusion is partially cured by Incremental Conductance which does not solve the issue completely [15] and is additionally prone to noise due to its inherent differentiation operation. Lastly, note that the steady-state value of power perturbation is given by (cf. Equation (22))

$$\Delta \hat{p}_{pv}(t) \approx \begin{cases} -I_{pv} V_O \Delta d, & v_{pv} << v_{pv}|_{MPP} \\ \frac{(V_{DC} \Delta d)^2}{R_{pv}}, & v_{pv} \approx v_{pv}|_{MPP} \\ \frac{V_{pv} V_O \Delta d}{r_{pv}}, & v_{pv} >> v_{pv}|_{MPP} \end{cases} \quad (35)$$

i.e., $\Delta \hat{p}_{pv} \sim \Delta d$ when the operating point is away from the MPP and $\Delta \hat{p}_{pv} \sim (\Delta d)^2$ when the operating point is in the vicinity of MPP in addition to being operating region dependent. When Δd is increased, corresponding overshoots also rise, aggravating unwanted transients. Therefore, it would be desirable to keep the duty cycle at Δd_{min} or slightly above at all times while increasing the algorithm convergence during transients by other means.

5. Maximum Power Point Tracking (MPPT) with Adaptive Perturbation Frequency

Consider a time-domain generalization according to Equation (15)

$$\hat{d}(t) = \Delta d \cdot u(t - t_0) \quad (36)$$

with $\Delta d < 0$ (for demonstration purposes only, without loss of generality) denoting a step-like duty cycle perturbation at arbitrary $t = t_0$. Observing the corresponding time-domain PVG power response $\hat{p}_{pv}(t - t_0)$, it may be concluded from Equation (22) that

$$\hat{p}_{pv}(t) \begin{cases} < 0, & v_{pv} << v_{pv}|_{MPP} \\ > 0, & v_{pv} \approx v_{pv}|_{MPP} \\ > 0, & v_{pv} >> v_{pv}|_{MPP} \end{cases} \quad (37)$$

for any $t > t_0$. This means that in case the PVG operating point is not in the vicinity of MPP, the sign of power perturbation induced by duty cycle perturbation does not change and may be theoretically detected for any $t > t_0$. On the other hand, the steady-state value of the PVG power would not be accurately estimated if sampled before $t = t_0 + T_p$. Therefore, it is proposed to keep the duty cycle perturbation constant at Δd_{min} to obtain maximum accuracy in steady state while adjusting the reciprocal of perturbation frequency (instead of Equation (28)) as

$$\begin{cases} \text{decrease } T_p \text{ below } T_\Delta|_{SC}, & y[k] > 0 \\ \text{decrease } T_p \text{ below } T_\Delta|_{SC}, & y[k] < 0 \\ \text{maintain } T_p = T_\Delta|_{SC}, & y[k] \approx 0 \end{cases} \quad (38)$$

A possible strategy of adapting the reciprocal of perturbation frequency at $(k + 1)$th sampling instant is given by

$$T_p[k + 1] = T_p[k] + M \cdot y[k] \quad (39)$$

with M referred to as a scaling factor. The size of T_p should be kept between two bounds,

$$T_{pmin} < T_p[k] < T_\Delta|_{SC} \quad (40)$$

with the lower boundary selected using a similar line of thinking as the selection in Δd_{min} above. In general, the value of T_{pmin} will also be dictated by irradiation-induced variation and system resolution. In-depth investigation is left for future work, as only the concept of variable perturbation frequency is introduced here. The proposed algorithm is expected to be less confused by sudden irradiation changes than the variable step one, since the duty cycle remains unchanged. The increase of algorithm convergence time during transients is achieved by means of enlarged perturbation frequency. Moreover, the steady-state value of power perturbation is given by (cf. Equations (22) and (35))

$$\Delta \hat{p}_{pv}(t) \approx \begin{cases} -I_{pv}V_O\Delta d_{min}, & v_{pv} << v_{pv}|_{MPP} \\ \frac{(V_{DC}\Delta d_{min})^2}{R_{pv}}, & v_{pv} \approx v_{pv}|_{MPP} \\ \frac{V_{pv}V_O d_{min}}{r_{pv}}, & v_{pv} >> v_{pv}|_{MPP} \end{cases} \tag{41}$$

i.e., the resulting $\Delta \hat{p}_{pv}$ is now operating region dependent only, since Δd does not change.

6. Example

Consider the system utilized for perturbation frequency design guideline verification in References [42,43], consisting of:

- A 14.6 W PVG with a maximum power point current of $I_{pv}|_{MPP}= 0.9$ A, a maximum power point voltage $V_{pv}|_{MPP} = 16.2$ V, a short-circuit current of 1 A and an open-circuit voltage of 19.2 V for values of environmental variables given by $G_s = 500$ W, $T = 45\,°C$. Under these conditions, shunt and series PVG resistances are estimated as $r_{SH} \approx 1000\,\Omega$ and $r_S \approx 0.91\,\Omega$;
- A 100 KHz pulse width modulated boost power stage operating as IPC, terminated by a 26 V voltage source.

The system is shown in Figure 3 with the rest of the relevant parameters values indicated. PVG voltage v_{pv} and current i_{pv} are the measured variables. Output per-unit PVG characteristics are depicted in Figure 4. Note that $r_{pv}|_{MPP}= 18\Omega$ so that $r_{pv}|_{OC} \ll r_{pv}|_{MPP} \ll r_{pv}|_{SC}$, as predicted by Equation (7).

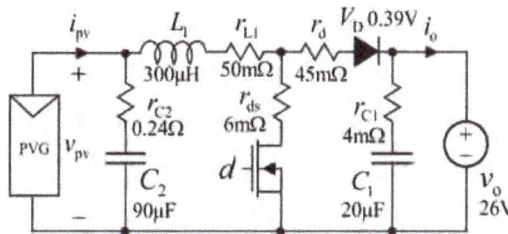

Figure 3. Boost-interfacing power converter (IPC)-based solar energy conversion system.

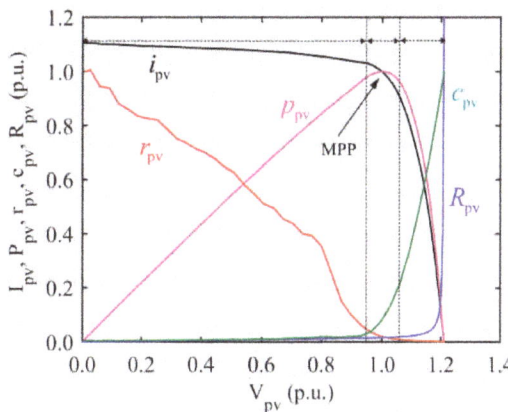

Figure 4. Output per-unit photovoltaic generator (PVG) characteristics.

The following settling times were revealed in Reference [42] for $\Delta = 0.05$:

$$T_\Delta \approx \begin{cases} 5.6ms, & v_{pv} \ll v_{pv}|_{MPP} \\ 3.3ms, & v_{pv} \approx v_{pv}|_{MPP} \\ 1.5ms, & v_{pv} \gg v_{pv}|_{MPP} \end{cases}, \tag{42}$$

Thus, $T_p > 5.6$ ms should be selected. In order to estimate Δd_{min}, the following parameters were used: material constant $K_{ph} = 1.9$ mA, saturation current $I_s = 1.097 \cdot 10^{-10}$A, and ideality factor $\eta = 1.0$, with an irradiation change rate $\dot{G}_s$ of 100 W/m^2/s and a 12-bit ADC with 3-V full-scale voltage span. The resulting minimum duty cycle perturbation step size was obtained as $\Delta d_{min} = 0.021$. Selecting, for demonstration purposes, $T_p = 5.7$ ms and $\Delta d = 0.025$, Figure 5 demonstrates the results of sweeping the converter duty cycle. It may be concluded that the PVG power curve is sampled with settling times matching Equation (40). Moreover, the claim that the direction of PVG power induced by duty cycle perturbation at $t = t_0$ does not change for any $t > t_0$ is well evident. For $v_{pv} < v_{pv}|_{MPP}$, negative Δd yields positive Δp_{pv} while for $v_{pv} > v_{pv}|_{MPP}$, negative Δd yields negative Δp_{pv}, as predicted by Equation (23).

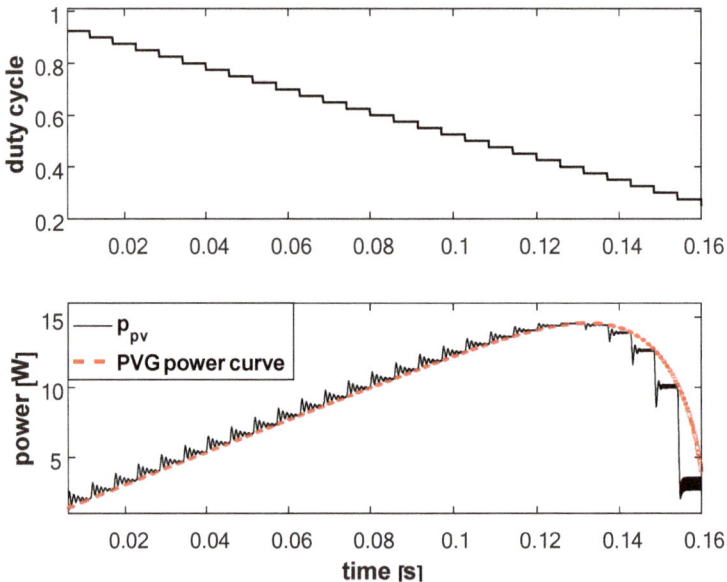

Figure 5. Duty cycle sweeping results.

In order to perform a demonstrative comparison of the system performance under different values of duty cycle and perturbation frequencies, the system was initialized away from MPP under 300 W/m^2 irradiation to observe initial convergence to the MPP. Next, the irradiation was increased to 500 W/m^2 at $t = 0.17$s to examine the corresponding response of the MPPT algorithm.

In the first case, the constant duty cycle constant perturbation frequency MPPT algorithm with $\Delta d = \Delta d_{min} = 0.021$ and $T_p = T_\Delta|_{SC} = 5.6$ ms was applied. The results are shown in Figure 6. The system converges to the MPP corresponding to 300 W/m^2 irradiation after ~120 ms and then oscillates around the MPP in three discrete steps, as predicted in [21]. Upon irradiation change, the system converges to the new MPP almost instantaneously due to the fact that the MPP voltage is nearly insensitive to irradiation [54]. Therefore, the converter duty cycle should not change significantly upon irradiation variation. This is evident in Figure 6.

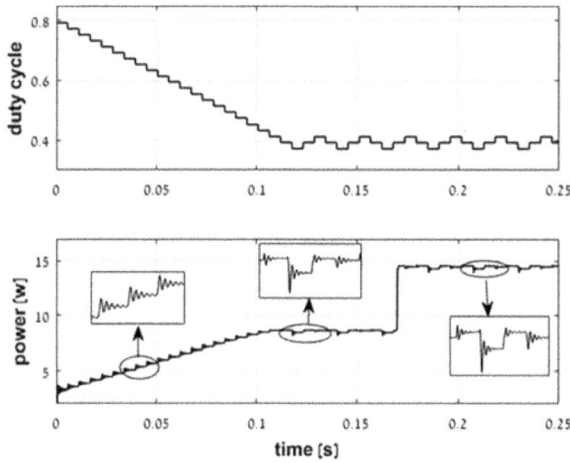

Figure 6. Performance under the constant duty cycle and constant perturbation frequency values of Δd = 0.021, T_p = 5.6 ms.

In the second case, the constant duty cycle constant perturbation frequency MPPT algorithm with $\Delta d = 5 \cdot \Delta d_{min} = 0.1$ and $T_p = T_\Delta|_{SC} = 5.6$ ms was applied, and the duty cycle was significantly increased. The results are shown in Figure 7. The system converges to the MPP corresponding to 300 W/m² irradiation after ~25 ms—five times faster than in the previous case, as expected—and then oscillates around the MPP in three discrete steps. Nevertheless, the differences between corresponding levels of PVG power and MPP power are much higher than in the previous case, resulting in significant steady state power losses. Moreover, upon irradiation change, the system does not instantaneously converge to the new MPP, since the duty cycle moves away from its optimum region due to confusion and then returns in about 200 ms.

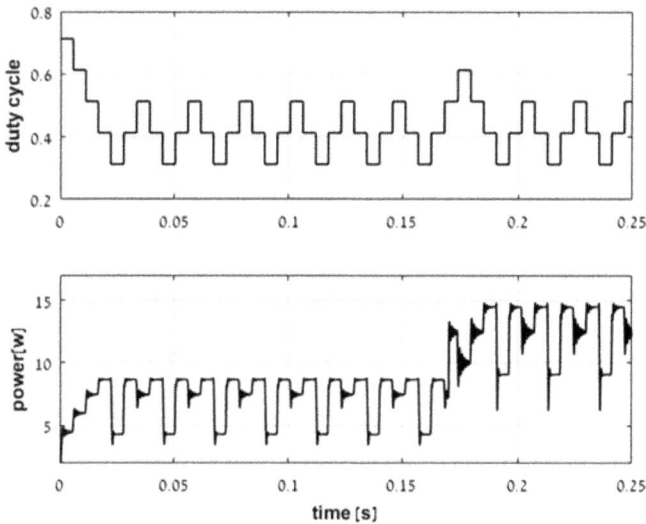

Figure 7. Performance under the constant duty cycle and constant perturbation frequency values of Δd = 0.1, T_p = 5.6 ms.

In the third case, the constant duty cycle constant perturbation frequency MPPT algorithm with $\Delta d = \Delta d_{min} = 0.021$ and $T_p = 0.1 \cdot T_\Delta|_{SC} = 0.56$ ms was applied, and the perturbation frequency was significantly increased. The results are shown in Figure 8. The system converges to the MPP corresponding to 300 W/m^2 irradiation after ~12.5 ms—ten times faster than in the first case, as expected—and then oscillates around the MPP in five, rather than three, discrete steps, entering chaotic mode [21]. Consequently, the differences between corresponding levels of PVG power and MPP power are higher than in the first case, resulting in higher steady state power losses. Upon irradiation change, the system instantaneously converges to the new MPP and oscillates around the MPP in three discrete steps, as desired. It may be observed that the power response never settles following a duty cycle perturbation, as predicted.

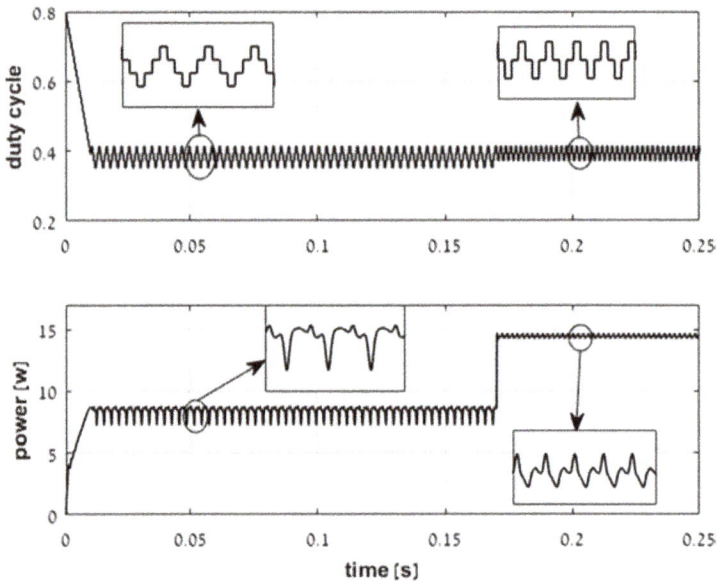

Figure 8. Performance under constant duty cycle and constant perturbation frequency values of $\Delta d = 0.021$, $T_p = 0.56$ ms.

In the fourth case, the variable duty cycle constant perturbation frequency MPPT algorithm with (cf. Equation (31)) $\Delta d_{min} = 0.021 < \Delta d < 0.1 = 5 \cdot \Delta d_{min}$ and $T_p = T_\Delta|_{SC} = 5.6$ ms was applied. The duty cycle was adapted according to Equation (29). The results are shown in Figure 9. The system converges to the MPP corresponding to 300 W/m^2 irradiation after ~60 ms (two times faster than in the first case yet two times slower than in the second) and then oscillates around the MPP in three discrete steps, similar to the first case. However, upon irradiation change, the system undergoes a significant transient and settles in the new MPP only after ~300 ms. This transient is probably the main drawback of perturbative MPPT algorithms with variable step size.

In the last case, the constant duty cycle variable perturbation frequency MPPT algorithm with (cf. Equation (40)) $\Delta d = \Delta d_{min} = 0.021$ and $0.1 \cdot T_\Delta|_{SC} = 0.56$ ms $< T_p < T_\Delta|_{SC} = 5.6$ ms was applied. The perturbation frequency was adapted according to Equation (39). The results are shown in Figure 10. The system converges to the MPP corresponding to 300W/m^2 irradiation after ~40ms (three times faster than in the first case yet three times slower than in the third) and then oscillates around the MPP in three discrete steps, similar to the first case. Upon irradiation change, the system converges to the new MPP after a transient lasing a single perturbation period (5.6 ms) and then oscillates around the MPP in three discrete steps, as desired.

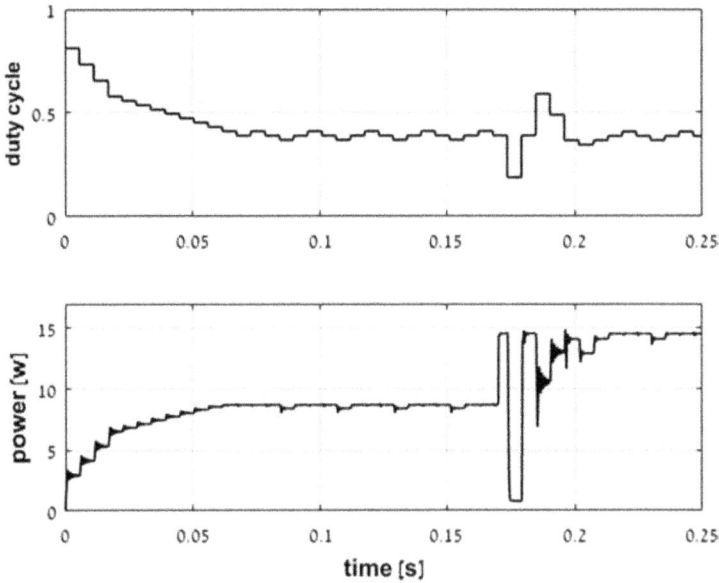

Figure 9. Performance under the variable duty cycle and constant perturbation frequency values of $0.021 < \Delta d < 0.1$, $T_p = 5.6$ ms.

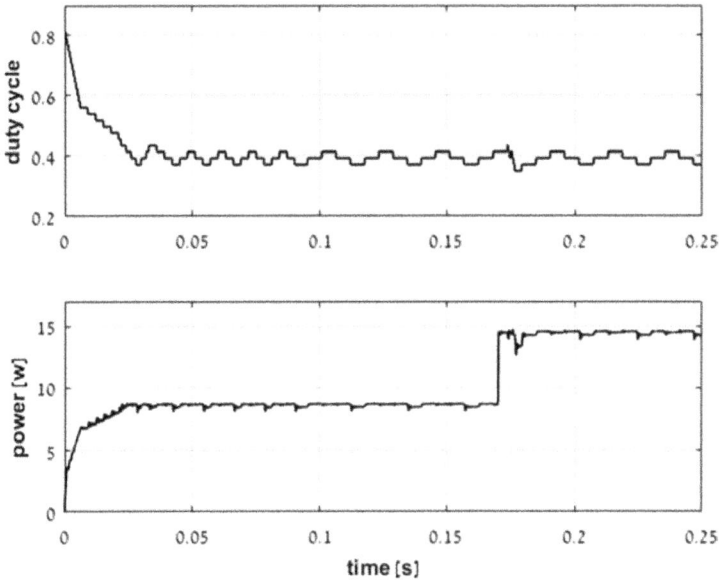

Figure 10. Performance under the constant duty cycle and variable perturbation frequency values of $\Delta d = 0.021$, 0.56 ms $< T_p < 5.6$ ms.

It may then be concluded that the constant duty cycle variable perturbation frequency MPPT algorithm seems to present a better trade-off between steady-state and transient performance than the fixed step fixed perturbation frequency and variable step fixed perturbation frequency MPPT algorithms.

7. Conclusions

In this paper, the concept of using an MPPT algorithm with an adaptive perturbation frequency instead of the commonly used variable step size was proposed. The analytical background was presented, based on recently revised design guidelines for designing fixed-step direct maximum power point tracking methods. Where possible, it is proposed to keep the step size at a minimum at all times while increasing perturbation frequency during transients and keeping it unchanged (equal to the maximum allowed by design guidelines) in a steady state. Such an approach helps to eliminate algorithm confusion during sudden irradiation changes and prevent excess transients caused by operation with an increased perturbation step. Preliminary results demonstrate the superiority of the proposed method over adaptive perturbation step size operation during transients, gained without compromising steady state performance. Since only the concept of variable perturbation frequency is introduced here, in-depth investigation is left for future work, to yield comprehensive design guidelines by deriving the lower bound of perturbation step time analytically, based on both system parameters and the behavior of environmental variables.

Author Contributions: E.A. conducted the research referred to in the paper during his M.Sc. studies. A.K. wrote the paper and supervised the research related to the paper with T.S.

Funding: This research received no external funding.

Conflicts of Interest: The authors declare no conflict of interest.

References

1. Gadelovits, S.; Kuperman, A.; Sitbon, M.; Aharon, I.; Singer, S. Interfacing renewable energy sources for maximum power transfer—Part I: Statics. *Renew. Sustain. Energy Rev.* **2014**, *31*, 501–508. [CrossRef]
2. Kolesnik, S.; Sitbon, M.; Gadelovits, S.; Suntio, T.; Kuperman, A. Interfacing renewable energy sources for maximum power transfer—Part II: Dynamics. *Renew. Sustain. Energy Rev.* **2015**, *51*, 1771–1783. [CrossRef]
3. de Brito, M.A.G.; Galotto, L., Jr.; Sampaio, L.P.; Melo, G.D.E.; Canesin, C.A. Evaluation of the main MPPT techniques for photovoltaic applications. *IEEE Trans. Ind. Electron.* **2013**, *60*, 1156–1167. [CrossRef]
4. Esram, T.; Chapman, P.L. Comparison of photovoltaic array maximum power point tracking techniques. *IEEE Trans. Energy Convers.* **2007**, *22*, 439–449. [CrossRef]
5. Faranda, R.; Leva, S. Energy comparison of MPPT techniques for PV systems. *WSEAS Trans. Power Syst.* **2008**, *3*, 446–455.
6. Lyden, S.; Haque, M.E. Maximum power point tracking techniques for photovoltaic systems: A comprehensive review and comparative analysis. *Renew. Sustain. Energy Rev.* **2015**, *52*, 1504–1518. [CrossRef]
7. Xiao, W.; Zeineldin, H.H.; Zhang, P. Statistic and parallel testing procedure for evaluating maximum power point tracking algorithms of photovoltaic power systems. *IEEE J. Photovolt.* **2013**, *3*, 1062–1069. [CrossRef]
8. Subudhi, B.; Pradhan, R. A comparative study on maximum power point tracking techniques for photovoltaic power systems. *IEEE Trans. Sustain. Energy* **2013**, *4*, 89–98. [CrossRef]
9. Latham, A.M.; Pilawa-Podgurski, R.; Odame, K.M.; Sullivan, C.R. Analysis and optimization of maximum power point tracking algorithms in the presence of noise. *IEEE Trans. Power Electron.* **2013**, *28*, 3479–3494. [CrossRef]
10. Andrean, V.; Chang, P.C.; Lian, K.L. A review and new problems discovery of four simple decentralized maximum power point tracking algorithms—Perturb and observe, incremental conductance, golden section search, and Newton's quadratic interpolation. *Energies* **2018**, *11*, 2966. [CrossRef]
11. Danandeh, M.A.; Mousavi, S.M. Comparative and comprehensive review of maximum power point tracking methods for PV cells. *Renew. Sustain. Energy Rev.* **2018**, *82*, 2743–2767. [CrossRef]
12. Koutroulis, E.; Kalaitzakis, K.; Voulgaris, N.C. Development of a microcontroller-based, photovoltaic maximum power point tracking control system. *IEEE Trans. Power Electron.* **2001**, *16*, 46–54. [CrossRef]
13. Femia, N.; Petrone, G.; Spagnuolo, G.; Vitelli, M. Optimization of perturb and observe maximum power point tracking method. *IEEE Trans. Power Electron.* **2005**, *20*, 963–973. [CrossRef]
14. Hussein, K.H.; Muta, I.; Hoshino, T.; Osakada, M. Maximum photovoltaic power tracking: An algorithm for rapidly changing atmospheric conditions. *IEE Proc. Gener. Transm. Distrib.* **1995**, *142*, 59–64. [CrossRef]

15. Elgendy, M.A.; Zahawi, B.; Atkinson, D.J. Assessment of the incremental conductance maximum power point tracking algorithm. *IEEE Trans. Sustain. Energy* **2013**, *4*, 108–117. [CrossRef]

16. Leyva, R.; Alonso, C.; Queinnec, I.; Cid-Pastor, A.; Lagrange, D.; Martinez-Salamero, L. MPPT of photovoltaic systems using extremum seeking control. *IEEE Trans. Aerosp. Electron. Syst.* **2006**, *42*, 249–258. [CrossRef]

17. Ghaffari, A.; Krstic, M.; Seshagiri, S. Power optimization and control in wind energy conversion systems using extremum seeking. *IEEE Trans. Contr. Syst. Technol.* **2014**, *22*, 1684–1695. [CrossRef]

18. Brunton, S.L.; Rowley, C.W.; Kulkarni, S.R.; Clarkson, C. Maximum power point tracking for photovoltaic optimization using ripple-based extremum seeking control. *IEEE Trans. Power Electron.* **2010**, *25*, 2531–2540. [CrossRef]

19. Esram, T.; Kimball, J.W.; Krein, P.T.; Chapman, P.L.; Midya, P. Dynamic maximum power point tracking of photovoltaic arrays using ripple correlation control. *IEEE Trans. Power Electron.* **2006**, *21*, 1282–1291. [CrossRef]

20. Bazzi, A.M.; Krein, P.T. Ripple correlation control: An extremum seeking control perspective for real time optimization. *IEEE Trans. Power Electron.* **2014**, *29*, 988–995. [CrossRef]

21. Femia, N.; Petrone, G.; Spagnuolo, G.; Vitelli, M. *Power Electronics and Control Techniques for Maximum Energy Harvesting in Photovoltaic Systems*; CRC Press: Boca Raton, FL, USA, 2013.

22. Liu, F.; Duan, S.; Liu, F.; Liu, B.; Kang, Y. A variable step size INC MPPT method for PV systems. *IEEE Trans. Ind. Electron.* **2008**, *55*, 2622–2628.

23. Libo, W.; Zhengming, Z.; Jianzheng, L. A single stage three phase grid connected photovoltaic system with modified MPPT method and reactive power compensation. *IEEE Trans. Energy Convers.* **2007**, *22*, 881–886. [CrossRef]

24. Kim, R.-Y.; Lai, J.-S.; York, B.; Koran, A. Analysis and design of maximum power point tracking scheme for thermoelectric battery energy storage system. *IEEE Trans. Ind. Electron.* **2009**, *56*, 3709–3716.

25. Xiao, W.; Dunford, W.G. A modified adaptive hill climbing MPPT method for photovoltaic power systems. In Proceedings of the 35th IEEE Annual Power Electronics Specialists Conference (PESC 04), Aachen, Germany, 20–25 June 2004; pp. 1957–1963.

26. Pandey, A.; Dasgupta, N.; Mukerjee, A.K. Design issues in implementing MPPT for improved tracking and dynamic performance. In Proceedings of the 32nd IEEE Annual Industrial Electronics Conference (IECON 2006), Paris, France, 7–10 November 2006; pp. 4387–4391.

27. Mei, Q.; Shan, M.; Liu, L.; Guerrero, J.M. A novel improved variable step-size incremental-resistance MPPT method for PV systems. *IEEE Trans. Ind. Electron.* **2011**, *58*, 2427–2434. [CrossRef]

28. Lee, K.-J.; Kim, R.-Y. An adaptive maximum power point tracking scheme based on a variable scaling factor for photovoltaic systems. *IEEE Trans. Energy Convers.* **2012**, *27*, 1002–1008. [CrossRef]

29. Khanna, R.; Zhang, Q.; Stanchina, W.E.; Reed, G.F.; Mao, Z.-H. Maximum power point tracking using model reference adaptive control. *IEEE Trans. Power Electron.* **2014**, *29*, 1490–1499. [CrossRef]

30. Manganiello, P.; Ricco, M.; Petrone, G.; Monmasson, E.; Spagnuolo, G. Optimization of perturbative PV MPPT methods through online system identification. *IEEE Trans. Ind. Electron.* **2014**, *61*, 6812–6821. [CrossRef]

31. Sitbon, M.; Schacham, S.; Kuperman, A. Disturbance observer based voltage regulation of current mode boost converter interfaced photovoltaic generator. *IEEE Trans. Ind. Electron.* **2015**, *62*, 5776–5785. [CrossRef]

32. Manganiello, P.; Ricco, M.; Petrone, G.; Monmasson, E.; Spagnuolo, G. Dual Kalman filter based identification and real-time optimization of PV systems. *IEEE Trans. Ind. Electron.* **2015**, *62*, 7266–7275. [CrossRef]

33. Li, C.; Chen, Y.; Zhou, D.; Liu, J.; Zeng, J. A high-performance adaptive incremental conductance MPPT algorithm for photovoltaic systems. *Energies* **2016**, *9*, 288. [CrossRef]

34. Kou, Y.; Xia, Y.; Ye, Y. Fast variable step maximum power point tracking method for photovoltaic systems. *J. Renew. Sustain. Energy* **2015**, *7*, 043126. [CrossRef]

35. Amir, A.; Amir, A.; Selvaraj, J.; Rahim, N.A.; Abusorrah, A.M. Conventional and modified MPPT techniques with direct control and dual scaled adaptive step size. *Sol. Energy* **2017**, *157*, 1017–1031. [CrossRef]

36. Tan, C.Y.; Rahim, N.A.; Selvaraj, J. Employing dual scaling mode for adaptive hill climbing method n buck converter. *IET Renew. Power Gen.* **2015**, *9*, 1010–1018. [CrossRef]

37. Ahmed, E.M.; Shoyama, M. Variable step size maximum power point tracker using a single variable for stand alone battery storage PV systems. *J. Power Electron.* **2011**, *11*, 218–227. [CrossRef]

38. Ahmed, E.M.; Shoyama, M. Scaling factor design based variable step size incremental resistance maximum power point tracking for PV systems. *J. Power Electron.* **2012**, *12*, 164–171. [CrossRef]

39. Kjaer, S.B. Evaluation of the 'Hill Climbing' and the 'Incremental Conductance' maximum power point trackers for photovoltaic power systems. *IEEE Trans. Energy Convers.* **2012**, *27*, 922–929. [CrossRef]
40. Sera, D.; Mathe, L.; Kerekes, T.; Spataru, S.V.; Teodorescu, R. On the perturb-and-observe and incremental conductance MPPT methods for PV systems. *IEEE J. Photovolt.* **2013**, *3*, 1070–1078. [CrossRef]
41. Kolesnik, S.; Kuperman, A. On the equivalence of major variable step size MPPT algorithms. *IEEE J. Photovolt.* **2016**, *6*, 590–594. [CrossRef]
42. Kivimäki, J.; Kolesnik, S.; Sitbon, M.; Suntio, T.; Kuperman, A. Revisited perturbation frequency design guideline for direct fixed-step maximum power point tracking algorithms. *IEEE Trans. Ind. Electron.* **2017**, *64*, 4601–4609. [CrossRef]
43. Kivimäki, J.; Kolesnik, S.; Sitbon, M.; Suntio, T.; Kuperman, A. Design guidelines for multi-loop perturbative maximum power point tracking algorithms. *IEEE Trans. Power Electron.* **2018**, *33*, 1284–1293. [CrossRef]
44. Elgendy, M.A.; Atkinson, D.J.; Zahavi, B. Experimental investigation of the incremental conductance maximum power point tracking algorithm at high perturbation rates. *IET Renew. Power Gener.* **2016**, *10*, 133–139. [CrossRef]
45. Elgendy, M.A.; Zahavi, B.; Atkinson, D.J. Operating characteristics of the P&O algorithm at high perturbation frequencies for standalone PV systems. *IEEE Trans. Energy Convers.* **2015**, *30*, 189–198.
46. Jiang, Y.; Qahouq, J.A.A.; Haskew, T.A. Adaptive step size with adaptive perturbation frequency digital MPPT controller for a single-sensor photovoltaic solar system. *IEEE Trans. Power Electron.* **2013**, *28*, 3195–3205. [CrossRef]
47. Nousiainen, L.; Puukko, J.; Mäki, A. Photovoltaic generator as an input source for power electronic converters. *IEEE Trans. Power Electron.* **2013**, *28*, 3028–3037. [CrossRef]
48. Suntio, T.; Messo, T.; Aapro, A.; Kivimäki, J.; Kuperman, A. Review of PV generator as an input source for power electronic converters. *Energies* **2017**, *10*, 1076. [CrossRef]
49. Lineykin, S.; Averbukh, M.; Kuperman, A. An improved approach to extracting the single-diode equivalent circuit parameters of a photovoltaic cell/panel. *Renew. Sustain. Energy Rev.* **2014**, *30*, 282–289. [CrossRef]
50. Averbukh, M.; Lineykin, S.; Kuperman, A. Obtaining PV panel operational curves for arbitrary cell temperatures and solar irradiation densities from standard conditions data. *Prog. Photovolt. Res. Appl.* **2013**, *21*, 1016–1024.
51. Lineykin, S.; Averbukh, M.; Kuperman, A. Issues in modeling amorphous silicon photovoltaic modules by single-diode equivalent circuit. *IEEE Trans. Ind. Electron.* **2014**, *61*, 6785–6793. [CrossRef]
52. Suntio, T.; Viinamaki, J.; Jokipii, J.; Messo, T.; Kuperman, A. Dynamic characterization of power electronics interfaces. *IEEE J. Emerg. Sel. Top. Power Electron.* **2014**, *2*, 949–961. [CrossRef]
53. Sitbon, M.; Leppaaho, J.; Suntio, T.; Kuperman, A. Dynamics of photovoltaic-generator-interfacing voltage-controlled buck power stage. *IEEE J. Photovolt.* **2015**, *5*, 633–670. [CrossRef]
54. Kuperman, A.; Averbukh, M.; Lineykin, S. Maximum power point matching versus maximum power point tracking for solar generators. *Renew. Sustain. Energy Rev.* **2013**, *19*, 11–17. [CrossRef]

energies

MDPI

Review

Dynamic Modeling and Analysis of PCM-Controlled DCM-Operating Buck Converters—A Reexamination

Teuvo Suntio

Laboratory of Electrical Energy Engineering, Tampere University of Technology, 33720 Tampere, Finland; teuvo.suntio@tut.fi; Tel.: +358-400-828-431

Received: 25 April 2018; Accepted: 14 May 2018; Published: 15 May 2018

Abstract: Peak-current-mode (PCM) control was proposed in 1978. The observed peculiar behavior caused by the application of PCM-control in the behavior of a switched-mode converter, which operates in continuous conduction mode (CCM), has led to a multitude of attempts to capture the dynamics associated to it. Only a few similar models have been published for a PCM-controlled converter, which operates in discontinuous conduction mode (DCM). PCM modeling is actually an extension of the modeling of direct-duty-ratio (DDR) or voltage-mode (VM) control, where the perturbed duty ratio is replaced by proper duty-ratio constraints. The modeling technique, which produces accurate PCM models in DCM, is developed in early 2000s. The given small-signal models are, however, load-resistor affected, which hides the real dynamic behavior of the associated converter. The objectives of this paper are as follows: (i) proving the accuracy of the modeling method published in 2001, (ii) performing a comprehensive dynamic analysis in order to reveal the real dynamics of the buck converter under PCM control in DCM, (iii) providing a method to improve the high-frequency accuracy of the small-signal models, and (iv) developing control-engineering-type block diagrams to facilitate the development of generalized transfer functions, which are applicable for PCM-controlled DCM-operated buck, boost, and buck-boost converters.

Keywords: peak-current-mode control; dynamic modeling; duty-ratio constraints; discontinuous conduction mode

---

1. Introduction

Peak-current-mode (PCM) control of switched-mode converters was publically introduced in 1978 [1,2], and it quickly became a very popular control method because of the beneficial features that it provides in converter operation and protection, as discussed in [3]. A huge number of modeling approaches has been proposed since the introduction of the PCM control in continuous conduction mode (CCM), as discussed and referenced in [4]. It has been recently shown in [4] that the accurate dynamic models of PCM-controlled converters, which operates in CCM, can be obtained by developing such duty-ratio constraints that include the duty-ratio gain (F_{m}), which becomes infinite at the mode-limit duty ratio (i.e., the maximum duty ratio after which the converter enters into harmonic mode of operation). Such models can be found, for example, from [4–8]. Only a few similar analytical models applicable for discontinuous conduction (DCM) are published [9–14].

The models in [9] are derived assuming that the inductor current does not exist as a state variable (i.e., no feedback from the inductor current is considered), because all the energy in the inductor is dissipated within the cycle, as assumed in [15]. This assumption is not valid, because the time-averaged inductor current does exist and it is a continuous state variable as discussed explicitly in [16]. A modeling technique is introduced in [10], which produces duty-ratio constraints with infinite duty-ratio gain (F_{m}) at the mode limit between the DCM and CCM operation but the models are load-resistor affected, which modifies the dynamics of the converter significantly, as discussed in [4,17]. The models presented in [11] are given in an implicit form including the load-resistor effect, which makes the model validation a challenging task, because the unterminated small-signal transfer functions cannot be recovered based on the given equivalent circuits. A discrete-time-modeling approach is presented in [12] for a buck converter including also PCM control in DCM, but it does not give any explicit transfer functions for comparison. The average and small-signal behavior of PCM-controlled boost converter based on numerical analysis methods is provided in [18] but no explicit analytic transfer functions are presented either. The unterminated PCM state spaces are given in [13] (pp. 139–144) and in [14] (pp. 222–224) for a buck converter based on the method introduced in [10], but the transfer functions are not solved for comparison.

As discussed in [4], the experimental verifications do not usually prove the validity of the models especially at the high frequencies due to the existence of un-modeled circuit elements at the input and/or output terminals, which affect the measurements either through the source or load-effect phenomena [14] (pp. 38–40). A good example of such a phenomenon can be found, for example, from [19] (cf. Figure 16 in [19]). Figure 1 shows the experimentally measured (red line), the analytically predicted (black line), and the simulation-based (blue line) frequency responses of the control-to-output-voltage transfer functions of a direct-duty-ratio (DDR) controlled buck converter analyzed in this paper [20]. The effect of the external circuit elements is explicitly visible, especially, in the behavior of the phase at the higher frequencies (cf. red line, >10 kHz). The predicted (black line) and simulation-based (blue line) frequency responses have very good match with each other. The ability to predict correctly the high frequency phase behavior (i.e., $\geq 1/10$th of switching frequency) is a quite important feature of the small-signal models from the control-design point of view. Therefore, it is well justified to perform the model validation by using Matlab^{TM} Simulink environment, where all the circuit elements are perfectly known.

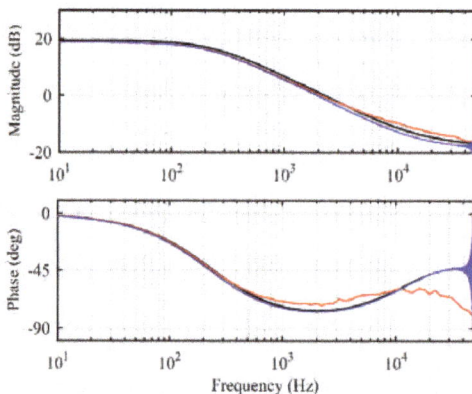

Figure 1. Measured (red line), analytically predicted (black line), and simulation-based (blue line) control-to-output-voltage frequency responses of direct-duty-ratio (DDR)-controlled buck converter operating in discontinuous conduction mode (DCM) at the input voltage of 20 V.

The investigations of this paper show that the modeling method in [10] will produce highly accurate unterminated small-signal models, when the parasitic circuit elements are taken into account. If the parasitic elements are omitted then the accuracy of the simplified models is not acceptable, especially, at the low input-voltage levels. The investigations show clearly also that an unterminated buck converter will become unstable in open loop at $M \approx 1/2$, where M denotes V_0/V_{in}. The load-resistor-affected instability will take place at $M \approx 2/3$ as predicted earlier in [10,11] as well as demonstrated explicitly in [21]. The existence of the right-half-plane (RHP) pole requires to designing the output-voltage feedback-control loop to have the crossover frequency higher than the RHP pole for stability to exist. The RHP pole will move into higher frequencies along the increase in M and D

(i.e., duty ratio) requiring careful selection of the feedback-loop crossover frequency for ensuring the stability of the converter. Therefore, the new findings concerning the location of the RHP pole have real scientific as well as practical values.

The rest of the paper is organized as follows: Section 2 introduces briefly the PCM-modeling method developed in [10] in a generalized unterminated form, and provides the corresponding explicit transfer functions for a buck converter. Section 3 presents the validation of the developed small-signal models by utilizing Matlab^{TM} Simulink-based switching models, and the pseudo-random binary-sequence-based frequency-response measurement technique introduced in [22,23]. The conclusions are provided finally in Section 4.

2. PCM-Control Modeling

The accurate small-signal modeling method of DCM-operated direct-duty-ratio (DDR) or voltage-mode (VM) controlled converters was establish in the late 1990s in [24,25] and later elaborated in a more convenient form in [16]. It is claimed in [26] that the small-signal models in [25] are not accurate enough, but the paper does not explicitly provide the required correction elements to improve the model accuracy. The small-signal models in [24,25] are load-resistor affected, which will hide the unterminated dynamic behavior of the converter, especially, at the low frequencies as well as which affects also the location of the low-frequency system poles [17]. Therefore, we will apply the methods presented in [16] for obtaining the small-signal DDR state space with the parasitic circuit elements included, which is utilized also in the corresponding PCM modeling. Figure 1, in Section 1, proves explicitly that the method described in [16] produces highly accurate unterminated small-signal models, when all the parasitic elements are included in the model. The duty ratio is generated in a DRR-controlled converter by means of a fixed pulse-width-modulator (PWM) ramp signal. In a PCM-controlled converter, the duty ratio is generated by means of the up slope of the instantaneous inductor current. As a consequence of this, the small-signal state space of a PCM-controlled converter can be found by developing proper duty-ratio constraints of the form [4,10,14]:

$$\hat{d} = F_m \left(\hat{x}_c - \sum_{i=1}^{n} q_i \hat{x}_i \right) \tag{1}$$

where F_m denotes duty-ratio gain, x_c the control variable (i.e., control current (i_{co})), and q_i the feedback or feedforward gain related to variable x_i, which can be either a state, or input variable of the converter as well as n the number of input and state variables [4]. The modeling is finalized by replacing the perturbed duty ratio ($\hat{d}$) by (1) in the linearized state space of the corresponding DDR-controlled converter [4,10,14]. The hat over the variables in (1) indicates that the corresponding variables are small-signal variables. This notation method is applied in rest of the paper.

2.1. Generalized Duty-Ratio Constrains in DCM

Figure 2 shows the inductor-current waveforms, the control current (i_{co}), and the inductor-current compensating ramp (m_c) during one switching cycle in DCM under dynamic conditions. According to [16], the real state variables, which will produce the dynamic behavior of the converter up to half the switching frequency, are the time-averaged values of the instantaneous variables (i.e., inductor currents and capacitor voltages), where the averaging is performed within one cycle. The time-averaged variables are denoted in this paper by $\langle x_i \rangle$. Figure 2 shows that at the time instant $t = (k + d)T_s$ the variables in Figure 2 are linked together by:

$$i_{co} - m_c d T_s = \sum_{i=1}^{n} \langle i_L \rangle + \Delta i_L \tag{2}$$

which is known as the comparator equation, because the associated PWM comparator will change state, when the condition determined by (2) is valid [13]. The only unspecified variable in (2) is Δi_L

(cf. Figure 2), which can be solved by applying the definition of the time-averaged inductor current (i.e., $\langle i_\text{L} \rangle$) at $t = (k + d)T_\text{s}$:

$$\langle i_\text{L} \rangle = \frac{m_1 d(d + d_1)T_\text{s}}{2} \tag{3}$$

where d_1 can be solved based on the inductor-current waveforms in Figure 2 as $d_1 = m_1 d / m_2$ [10]. Thus Δi_L can be given by:

$$\Delta i_\text{L} = m_1 dT_\text{s} - \langle i_\text{L} \rangle = m_1 dT_\text{s} - \frac{m_1(m_1 + m_2)d^2 T_\text{s}}{2m_2} \tag{4}$$

and the corresponding comparator equation in (2) by:

$$i_\text{co} - m_\text{c} dT_\text{s} = \langle i_\text{L} \rangle + m_1 dT_\text{s}(1 - \frac{d(m_1 + m_2)}{2m_2}) \tag{5}$$

where m_1 and m_2 denote the absolute values of the up and down slopes of the inductor current as denoted in Figure 2.

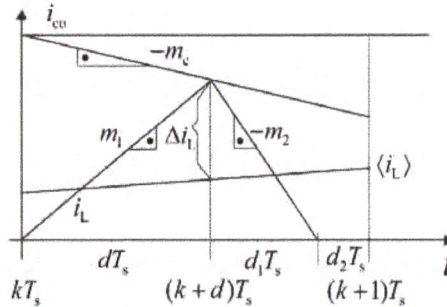

Figure 2. Inductor-current waveforms in DCM including the control current (i_co) and compensation ramp (m_c).

The coefficients in the small-signal duty ratio constraints in (1) can be found by substituting the up and down slope of the inductor current with their topology-based values in (5) as well as linearizing (5) at a certain operating point. The linearization requires to applying the partial-derivative-based method due to the highly nonlinear nature of the comparator equation in (5) for obtaining the required coefficients in (1) (cf. pp. 60, 61, [14]). Thus the duty-ratio gain (F_m) can be given in a generalized form (Note: In this case, only the duty ratio (d) is a variable, and all the other variables are constant) by:

$$F_\text{m} = \frac{1}{T_\text{s}\left(M_\text{c} + \frac{M_1(M_2 - D(M_1 + M_2))}{M_2}\right)} \tag{6}$$

which indicates that F_m becomes infinite, when the duty ratio (D) equals:

$$D = \frac{M_2}{M_1 + M_2} + \frac{M_2}{M_1(M_1 + M_2)} \cdot M_\text{c} \tag{7}$$

Equation (7) defines the mode limit for the converter operation at the switching frequency, where the first term denotes actually the mode limit between the DCM and CCM operation [14], because the only operational condition in DCM, where $DM_1 - D'M_2$ equals zero, is the boundary between DCM and CCM (i.e., the boundary conduction mode (BCM) [11]). It equals symbolically the same value, which defines the mode limit at $D = 0.5$ in the PCM-controlled converter in CCM, when

$M_c = 0$ [4]. Figure 3 shows the inductor-current waveforms, when the converter is driven into the harmonic modes of operation. The figure shows definitively that the buck converter can adopt both even and odd harmonic modes as discussed also in [10]. The reason for the existence of the odd harmonics is actually the existence of an RHP pole in the converter open-loop dynamics. In case of CCM operation, only the even harmonic modes are possible as explained in detail in [4]. In DCM operation, the mode-limit duty ratio does not either equal the average duty ratio of the harmonic operation as it does in CCM operation [4].

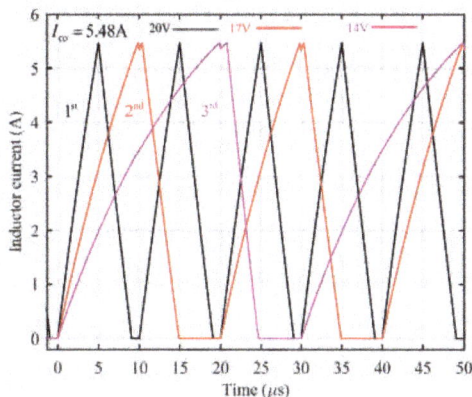

Figure 3. Simulation-based inductor-current behavior in harmonic modes in a peak-current-mode (PCM)-controlled buck converter.

Equation (5) can be developed in terms of duty ratio (D) for a second-order converter with $M_c = 0$ as [13]:

$$\frac{M_1(M_1 + M_2)T_s}{2M_2} \cdot D^2 - M_1 T_s \cdot D - \Delta I_L = 0 \tag{8}$$

where $\Delta I_L = I_{co} - I_L$. According to (8), we can compute that ΔI_{L-min} will be limited to:

$$\Delta I_{L-min} = \frac{M_1 M_2 T_s}{2(M_1 + M_2)} \tag{9}$$

at the duty ratio of:

$$D_{max} = \frac{M_2}{M_1 + M_2} \tag{10}$$

which equals D in (7) (i.e., the first term). At the higher duty ratios, Equation (8) does not have any more real-valued solutions indicating that the converter enters into harmonic operation mode as shown in Figure 3.

The steady-state comparator equation in (8) can be developed further in terms of the input-to-output gain (M) (i.e., $M = V_o/V_{in}$), and K (i.e., $K = 2L/T_s/R_{eq}$ and $R_{eq} = V_o/I_o$). It should be observed that M_1 and M_2 denote the inductor current up and down slopes as absolute values, and they have to be expressed as a function of M according to the behavior of the corresponding converter. In addition, the duty ratio (D) has to be replaced by the converter specific formula, which can be given for the buck converter as $D = M\sqrt{K/(1-M)}$ [11]. These procedures yield for the buck converter as:

$$M^3 - M^2 + K\left(\frac{I_{co}R_L}{2V_{in}}\right)^2 = 0 \tag{11}$$

Equation (11) can be further developed as:

$$\left(M - \frac{2}{3}\right)^2 \left(M - \left(\frac{3I_{co}R_L}{4V_{in}}\right)^2\right) = 0 \tag{12}$$

which indicates that there exists a double root at $M = 2/3$ in (11), which means that there are no real-valued solutions for $M > 2/3$ in open loop, as discussed also in [11,12] as well as explicitly demonstrated in [21]. It is explicitly proved in [10] that the mode limit at $M = 2/3$ does not exist, when the output-voltage feedback loop is closed. The dynamic analysis will reveal that the mode limit at $M = 2/3$ produces an RHP pole (i.e., the converter is unstable), which does not take place in the CCM converter. Therefore, the output-voltage feedback can remove the RHP pole when its control bandwidth is higher than the RHP pole. The boost and buck-boost converters do not have similar anomalies as the buck converter has in the open-loop behavior as discussed also in [11]. Equation (11) is actually load-resistor affected, and therefore, it does not correctly predict the location of the actual RHP pole in a buck converter as will be shown later in Section 2.3.

2.2. Small-Signal Model of DDR-Controlled Buck Converter in DCM

The averaged complete (i.e., including all parasitic elements) state space of a DDR-controlled buck converter shown in Figure 4, which operates in DCM, can be given according to [14,16] by:

$$\frac{d\langle i_L \rangle}{dt} = \frac{d(\langle v_{in} \rangle + (R_1 - R_2)\langle i_L \rangle + V_D)}{L} - \frac{2\langle i_L \rangle}{dT_s} \cdot \frac{R_1\langle i_L \rangle + \langle v_C \rangle - r_C\langle i_o \rangle + V_D}{\langle v_{in} \rangle - R_2\langle i_L \rangle - \langle v_C \rangle + r_C\langle i_o \rangle}$$

$$\frac{d\langle v_C \rangle}{dt} = \frac{\langle i_L \rangle}{C} - \frac{\langle i_o \rangle}{C}$$

$$\langle i_{in} \rangle = \frac{d^2 T_s}{2L}(\langle v_{in} \rangle - R_2\langle i_L \rangle - \langle v_C \rangle + r_C\langle i_o \rangle) \tag{13}$$

$$\langle v_o \rangle = \langle v_C \rangle + r_C C\frac{d\langle v_C \rangle}{dt}$$

$$R_1 = r_L + r_d + r_C \qquad R_2 = r_L + r_{ds} + r_C$$

from which the operating point can be derived by setting the derivatives to zero and denoting the circuit variables (i.e., voltages and currents) by capital letters yielding:

$$I_L = I_o I_{in} = \frac{((r_L + r_d)I_o + V_o + V_D)}{(V_{in} + (r_d - r_{ds})I_o + V_D)} \cdot I_o$$

$$V_o = V_C \tag{14}$$

$$D = \sqrt{\frac{2LI_o}{T_s} \cdot \frac{V_o + V_D + (r_L + r_d)I_o}{(V_{in} - V_o - (r_L + r_{ds})I_o)(V_{in} + V_D + (r_d - r_{ds1})I_o)}}$$

The small-signal state space can be derived by linearizing the averaged state space in (13) by applying the partial-derivatives-based method (cf. pp. 60, 61, [14]) at a certain operating point (14) yielding (15):

$$\frac{d\hat{i}_L}{dt} = -\frac{A_1}{L}\hat{i}_L - \frac{A_2}{L}\hat{v}_C + \frac{A_3}{L}\hat{v}_{in} + \frac{A_4}{L}\hat{i}_o + \frac{V_e}{L}\hat{d}$$

$$\frac{d\hat{v}_C}{dt} = \frac{\hat{i}_L}{C} - \frac{\hat{i}_o}{C} \tag{15}$$

$$\hat{i}_{in} = -B_1 R_2\hat{i}_L - B_1\hat{v}_C + B_1\hat{v}_{in} + B_1 r_C\hat{i}_o + I_e\hat{d}$$

$$\hat{v}_o = \hat{v}_C + r_C C\frac{d\hat{v}_C}{dt}$$

where:

$$A_1 = D(R_2 - R_1) + \frac{2L}{DT_s}\left(\frac{V_o+V_D+(2R_1-r_C)I_o}{V_{in}-V_o-(R_2-r_C)I_o} - \frac{R_2 I_o(V_o+V_D+(R_1-r_C)I_o)}{(V_{in}-V_o-(R_2-r_C)I_o)^2}\right)$$

$$A_2 = \frac{2LI_o}{DT_s}\frac{V_{in}+V_D+(R_1-R_2)I_o}{(V_{in}-V_o-(R_2-r_C)I_o)^2}$$

$$A_3 = D + \frac{2LI_o}{DT_s}\left(\frac{V_o+V_D+(R_1-r_C)I_o}{(V_{in}-V_o-(R_2-r_C)I_o)^2}\right) \quad (16)$$

$$A_4 = \frac{2LI_o r_C}{DT_s}\frac{V_{in}+V_D+(R_1-R_2)I_o}{(V_{in}-V_o-(R_2-r_C)I_o)^2}$$

$$B_1 = \frac{D^2 T_s}{2L}$$

and:

$$V_e = V_{in} + V_D + (R_1 - R_2)I_o + \frac{2LI_o}{D^2 T_s}\left(\frac{V_o+V_D+(R_1-r_C)I_o}{V_{in}-V_o-(R_2-r_C)I_o}\right)$$

$$I_e = \frac{DT_s}{L}(V_{in} - V_o - (R_2 - r_C)I_o)$$

$$R_1 = r_L + r_d + r_C \quad (17)$$

$$R_2 = r_L + r_{ds} + r_C$$

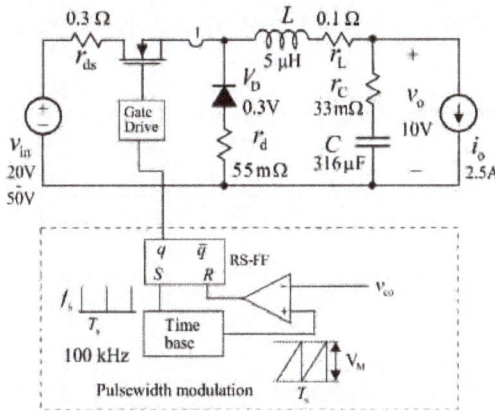

Figure 4. The power stage of the DDR-controlled buck converter in DCM.

2.3. Small-Signal Models of PCM-Controlled Buck Converter in DCM

The power stage of the PCM-controlled buck converter including the resistive load and the values of components are given in Figure 5, where the power stage equals the power stage in Figure 4. The generalized comparator equation for the second-order converters has been given earlier in (5), and the inductor-current up and down slopes, which are valid for a buck converter, are given explicitly in (18). According to (5) and (18), the corresponding unterminated duty-ratio constraints can be computed to be as given in (19) and in (20), respectively:

$$m_1 = \frac{\langle v_{in}\rangle - R_2\langle i_L\rangle - \langle v_C\rangle + r_C\langle i_o\rangle}{L}$$

$$m_2 = \frac{R_1\langle i_L\rangle + \langle v_C\rangle - r_C\langle i_o\rangle + V_D}{L} \quad (18)$$

$$\hat{d} = F_m(\hat{i}_{co} - q_L\hat{i}_L - q_C\hat{v}_C - q_{in}\hat{v}_{in} - q_o\hat{i}_o) \quad (19)$$

where:

$$F_m = \cfrac{1}{T_s\left(M_c + \frac{(V_{in}-V_o-(R_2-r_C)I_o)(V_o+V_D-DV_{in}+(D'R_1+DR_2-r_C)I_o)}{L(V_o+V_D+(R_1-r_C)I_o)}\right)}$$

$$q_L = 1 - \frac{DT_s}{L}R_2 + \frac{D^2 T_s}{2L(V_o+V_D+(R_1-r_C)I_o)}\left(\begin{array}{c}(R_1-R_2)V_o-(R_1-R_2)(2R_2-r_C)I_o \\ +\frac{R_1(V_{in}+(R_1-R_2)I_o)(V_{in}-V_o-(R_2-r_C)I_o)}{V_o+V_D+(R_1-r_C)I_o}\end{array}\right)$$

$$q_C = -\frac{DT_s}{L}\left(1 - \frac{D}{2}\frac{(V_{in}+V_D+(R_1-R_2)I_o)(V_{in}+(R_1-R_2)I_o)}{(V_o+V_D+(R_1-r_C)I_o)^2}\right) \qquad (20)$$

$$q_{in} = \frac{DT_s}{L}\left(1 - \frac{D}{2}\cdot\frac{2V_{in}-V_o+(R_1-2R_2+r_C)I_o}{V_o+V_D+(R_1-r_C)I_o}\right)$$

$$q_o = \frac{DT_s r_C}{L}\left(1 - \frac{D}{2}\cdot\frac{(V_{in}+(R_1-R_2)I_o)(V_{in}+V_D+(R_1-R_2)I_o)}{(V_o+V_D+(R_1-r_C)I_o)^2}\right)$$

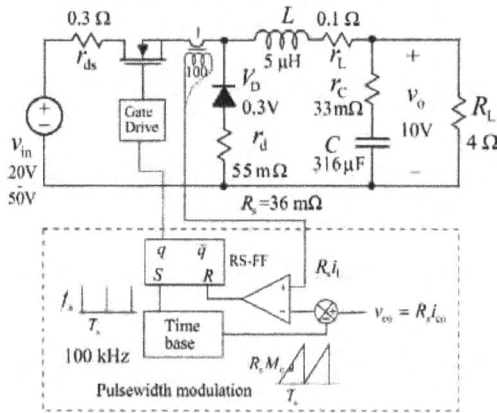

Figure 5. The power stage of the PCM-controlled buck converter in DCM.

In Equation (20), R_1 and R_2 are defined in (17), and M_c denotes the inductor-current compensation ramp in A/s. As discussed in the beginning of this section, the PCM state space can be obtained from the DDR state space in (15) by replacing the perturbed duty ratio by (19). As an outcome of this process, the small-signal state space valid for a PCM-controlled buck converter operating in DCM can be given by:

$$\frac{di_L}{dt} = -\frac{A_1+F_m q_L V_e}{L}\hat{i}_L - \frac{A_2+F_m q_C V_e}{L}\hat{v}_C + \frac{A_3-F_m q_{in}V_e}{L}\hat{v}_{in} + \frac{A_4-F_m q_o V_e}{L}\hat{i}_o + \frac{F_m V_e}{L}\hat{i}_{co}$$

$$\frac{dv_C}{dt} = \frac{i_L}{C} - \frac{i_o}{C} \qquad (21)$$

$$\hat{i}_{in} = -(B_1 R_2 + F_m q_L I_e)\hat{i}_L - (B_1 + F_m q_C I_e)\hat{v}_C + (B_1 - F_m q_{in}I_e)\hat{v}_{in} + (B_1 r_C - F_m q_o I_e)\hat{i}_o + F_m I_e \hat{i}_{co}$$

$$\hat{v}_o = \hat{v}_C + r_C C \frac{dv_C}{dt}$$

where A_{1-4} and B_1 are defined explicitly in (16) and V_e, and I_e in (17) as well as F_m, q_1, q_c, q_{in} and q_0 in (20), respectively.

The transfer functions representing the dynamics of the converter can be solved by applying proper software packages such as, for example, Matlab^{TM} Symbolic Toolbox. The symbolic-form transfer functions representing the input-side dynamics (i.e., the control-to-input-current transfer function, the output-current-to-input-current transfer function, and the input impedance) are very long, and thus only the transfer functions representing the output-side dynamics (i.e., the control-to-output-voltage transfer function ($G_{co-o} = \hat{v}_o/\hat{i}_{co}$), audiosusceptibility ($G_{io-o} = \hat{v}_o/\hat{v}_{in}$), and output impedance ($Z_{o-o} = \hat{v}_o/\hat{i}_o$)) are given explicitly in this paper in (22).

Usually, the DCM state spaces are given omitting all the parasitic circuit elements as a function of M and K (cf. pp. 222–224, [14]). We will show later in Section 3 that the simplified transfer functions do not represent correctly the dynamic behavior of the buck converter analyzed in this paper, and therefore, they are not given here. We will use the simplified transfer functions, however, in certain cases for providing better physical insight into the converter dynamics, when performing approximate analyses.

The three transfer functions comprising the output dynamics of the PCM-controlled buck converter in DCM are given in (22). The unterminated denominator (Δ) of the transfer functions is given in (23). The load-resistor-affected denominator is given in (24):

$$\Delta Z_{o-o}^{PCM} = \frac{sL + A_1 - A_4 + F_m V_e (q_L + q_o)}{LC}(1 + sr_C C)$$

$$\Delta G_{io-o}^{PCM} = \frac{A_3 - F_m q_{in} V_e}{LC}(1 + sr_C C) \tag{22}$$

$$\Delta G_{co-o}^{PCM} = \frac{F_m V_e}{LC}(1 + sr_C C)$$

where the unterminated denominator Δ equals:

$$s^2 + s\frac{A_1 + F_m q_L V_e}{L} + \frac{A_2 + F_m q_C V_e}{LC} \tag{23}$$

As discussed in [11], the PCM-controlled converters are highly damped converters, which means that the poles of the system are highly separated (i.e., the low-frequency pole ($\omega_{p\text{-LF}}$) lies close to origin, and the high-frequency pole ($\omega_{p\text{-HF}}$) lies close to infinity). Thus the poles can be approximated from (22) with quite high accuracy by utilizing the properties of a second-order polynomial, and the high separation of the poles, which yield (Note: the last simplified terms in (24) are computed assuming $M_c = 0$):

$$\omega_{p\text{-LF}} \approx -\frac{A_2 + F_m q_C V_e}{(A_1 + F_m q_L V_e)C} \approx -\frac{1 - 2M}{1 - M} \cdot \frac{1}{R_{eq}C}$$

$$\omega_{p\text{-HF}} \approx -\frac{A_1 + F_m q_L V_e}{L} \approx -\frac{D}{M - D} \cdot \frac{R_{eq}}{L} \tag{24}$$

where $M = V_o/V_{in}$ and $R_{eq} = V_o/I_o$ as well as $D = M\sqrt{K/(1-M)}$ and $K = 2L/T_s/R_{eq}$ (cf. p. 164, [14]).

Equation (24) shows explicitly that $\omega_{p\text{-LF}}$ becomes an RHP pole, when $M > 0.5$ (i.e., the minus sign becomes a plus sign), and it moves into higher frequencies in RHP, when M and D increases. $\omega_{p\text{-HF}}$ stays always as a left-half-plane (LHP) pole, because $M \geq D$, and it moves towards infinity, when M and D increases.

The full-order load-resistor-affected denominator can be given according to (25) but it does not give enough information to understand the effect of the load resistor on the system poles:

$$s^2(1 + \frac{r_C}{R_L}) + s(\frac{1}{R_L C} + \frac{A_1 + F_m q_L V_e}{L} + \frac{(A_1 - A_4 + F_m V_e (q_L + q_o))r_C}{R_L L}) + \frac{A_2 + F_m q_C V_e}{LC} + \frac{A_1 - A_4 + F_m V_e (q_L + q_o)}{LCR_L} \tag{25}$$

Equation (26) is derived from (25) by omitting the parasitic circuit elements as well as by transforming it into a more customary form according to [10]:

$$s^2 + s\left(\frac{1}{R_L C} + \frac{R_{eq}\sqrt{\frac{K}{1-M}} + 2F_m V_{in}}{L}\right) + \frac{1}{LC}(\frac{R_{eq}}{R_L} + \frac{1}{1-M})\sqrt{\frac{K}{1-M}} + 2F_m V_{in}(\frac{1}{R_L} + q_C)) \tag{26}$$

from which the simplified system poles can be solved at fully resistive load (i.e., $R_L = R_{eq}$) as:

$$\omega_{\text{p-LF}} \approx -\frac{\frac{1}{LC}\left(\frac{2-M}{1-M}\right)\sqrt{\frac{K}{1-M}}+2F_m V_{in}\left(\frac{1}{R_{eq}}+q_C\right)}{\frac{1}{R_{eq}C}+\frac{R_{eq}\sqrt{\frac{K}{1-M}}+2F_m V_{in}}{L}} \approx -\frac{\frac{(2-3M)D}{(M-D)(1-M)}}{\frac{L}{R_{eq}}+\frac{D}{M-D}\cdot R_{eq}C} \approx -\frac{(2-3M)}{R_{eq}C(1-M)}$$

$$\omega_{\text{p-HF}} \approx -\left(\frac{1}{R_{eq}C}+\frac{R_{eq}\sqrt{\frac{K}{1-M}}+2F_m V_{in}}{L}\right) \approx -\left(\frac{1}{R_{eq}C}+\frac{D}{M-D}\cdot\frac{R_{eq}}{L}\right) \approx -\frac{D}{M-D}\cdot\frac{R_{eq}}{L}$$

(27)

Equation (27) shows that the load-resistor-affected low-frequency pole moves into RHP, when $M > 2/3$, which complies with the instability condition predicted by Equation (12) in Section 2.1. The high-frequency pole equals the high-frequency pole in (24) and stays an LHP pole.

The load-resistor-affected denominator of transfer functions derived from the equivalent circuit representing the dynamics of the buck converter in [11] is explicitly given in [10] as:

$$s^2 + s\left(\frac{1}{R_{eq}C} + \frac{R_{eq}(1-M)}{L(1-2M)}\right) + \frac{1}{LC}\cdot\frac{2-3M}{1-2M}$$

(28)

from which the system poles can be approximated to be as:

$$\omega_{\text{p-LF}} \approx -\frac{2-3M}{R_{eq}C(1-M)}$$
$$\omega_{\text{p-HF}} \approx -\left(\frac{1}{R_{eq}C}+\frac{R_{eq}(1-M)}{L(1-2M)}\right) \approx -\frac{R_{eq}(1-M)}{L(1-2M)}$$

(29)

When studying carefully the system poles in (29) then it is obvious that the high-frequency pole becomes an RHP pole when $M > 0.5$, and the low-frequency pole becomes an RHP pole when $M > 2/3$. In practice, this means that the converter should be unstable under resistive load already when $M > 0.5$ but the converter has not been observed to behave like that. This means that the modeling method introduced in [11] does not provide correct second-order transfer functions.

The behavior of the system poles is presented in Table 1 in case of the converter in Figure 5, where the high separation of the poles is clearly visible. In addition, the table shows that the instability will take place already at the input voltage of 21.6 V, where $M < 0.5$ due to the contribution of the power-stage losses. The determining factor in the appearing of the open-loop instability is that the zeroth-order coefficients in (23) and (25) become negative, which indicates that one of the roots of (23) and (25) lies in RHP. This happens, because the feedback gain q_C (i.e., the output-capacitor-voltage feedback gain) (cf. Equation (20)) is negative. Actually, the missing of the negative sign of the first-order term indicates that the low-frequency pole is an RHP pole. It is obvious that the appearance of the instability can be controlled by the inductor-loop compensation (M_c), which will reduce F_m (cf. Equation (20)). This form of instability has not been reported earlier even if comprehensive analyses have been performed, for example, in [12]. The reason for this is that the load resistor affects the location of the poles as is visible in the load-resistor-affected poles given in (27) as well as in Table 1 (i.e., two right most columns) [17]. The investigations of this paper show that the load-resistor-affected RHP pole appears in vicinity of $M = 2/3$ as discussed in [10–14] and derived explicitly in Section 2.1 (Equation (12)) and in (27). The power-stage losses will shift the appearance of the instability into an operating point, where $M < 2/3$ as clearly visible in Table 1. The instability in vicinity of $V_{in} \approx 17$ V is also clearly visible in Figure 3 as the second-harmonic mode of operation.

The entries in Table 1 are computed by using the complete models. The coarsely approximated load-resistor-affected system poles in (27) (i.e., the last terms) yield $\omega_{\text{p-LF}} = -126$ Hz and $\omega_{\text{p-HF}} = -307$ kHz at the input voltage of 20 V. The coarsely approximated unterminated system poles in (24) (i.e., the last terms) yield $\omega_{\text{p-LF}} \approx 0$ Hz and $\omega_{\text{p-HF}} = -307$ kHz. These figures indicate that the simplified models will not predict accurately the location of the system poles.

The input-to-output transfer function (G_{io-o}^{PCM}) (known also as audiosusceptibility in [11]) in (22) can be nullified by providing M_c such that $A_3 - F_m q_{in} V_e = 0$. The approximate value of M_c can be computed to be:

$$M_c = \frac{M(1-M)}{2-M} \frac{V_{in}}{L} \qquad (30)$$

which complies with the value given in [12] (note: the definition of M_c in [12] differs from the definition of M_c in this paper; when the difference is taken into account, the values are equal).

Table 1. Behavior of the system poles (ω_{p-LF} & ω_{p-HF}) as a function of M and D with $M_c = 0$, $V_o = 10$ V, and $I_o = 2.5$ A.

V_{in}	M	D	ω_{p-LF}	ω_{p-HF}	$\omega_{p-LH}^{R_L}$	$\omega_{p-HF}^{R_L}$
50 V	0.2	0.117	−87 Hz	−160 kHz	−210 Hz	−160 kHz
30 V	0.333	0.218	−56 Hz	−188 kHz	−180 Hz	−188 kHz
21.6 V	0.463	0.344	1.2 Hz	−233 kHz	−123 Hz	−233 kHz
20 V	0.5	0.388	10 Hz	−253 kHz	−67 Hz	−253 kHz
17.5 V	0.57	0.489	114 Hz	−311 kHz	−12 Hz	−311 kHz
17.2 V	0.58	0.505	133 Hz	−323 kHz	5.7 Hz	−323 kHz

2.4. Generalized Small-Signal Transfer Functions Applicable for Buck, Boost, and Buck-Boost Conveters

The set of PCM transfer functions, which are valid for the buck, boost, and buck-boost converters, can be developed in a generalized form from the block diagrams presented in Figure 6 based on the generalized duty-ratio constraints in (31). In Figure 5, the transfer functions denoted by the superscript 'DDR' corresponds to the set of transfer functions of the corresponding DDR-controlled converter. The coefficients of (31) do not equal exactly the coefficients given in (20), because the output voltage is used as such in computing the inductor-current slopes as Figure 6 explicitly implies. The coefficient H_{sr} denotes a series resonant circuit, which is placed in the inductor-current feedback loop, for correcting the high-frequency phase behavior of the transfer functions (cf. [4]). H_{sr} is given in (32), where $\omega_{sr} = 2\pi f_s$ and $\zeta = 0.5$ in case of buck converter. The series-resonant circuit in (32) differs significantly from the series-resonant circuit utilized in [4], where $\omega_{sr} = \pi f_s$ and $\zeta = 0$. The validity of the proposed H_{sr} in case of PCM-controlled DCM buck converter is discussed in more detail in Section 3.

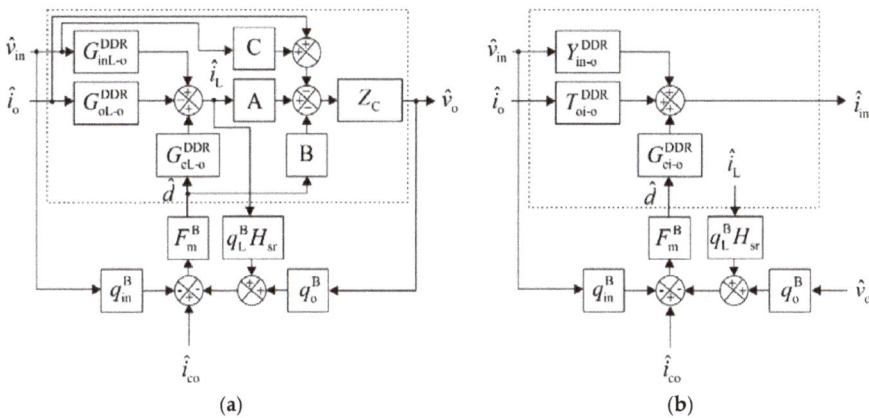

Figure 6. The control-engineering-type block diagrams for computing the general transfer functions representing (**a**) the output dynamics, and (**b**) the input dynamics of the basic second-order converters in DCM.

$$\hat{d} = F_m^B(\hat{i}_{co} - q_L^B H_{sr}\hat{i}_L - q_{in}^B\hat{v}_{in} - q_o^B\hat{v}_o) \tag{31}$$

$$H_{sr} = 1 + s\frac{2\zeta}{\omega_{sr}} + \frac{s^2}{\omega_{sr}^2} \tag{32}$$

The general set of transfer functions, which is valid for the buck, boost, and buck-boost converters, can be given as shown in (33) (the output dynamics) and (34) (the input dynamics), respectively:

$$G_{io-o}^{PCM} = \frac{\hat{v}_o}{\hat{v}_{in}} = \frac{(1+\frac{F_m q_L^B H_{sr}B}{A})G_{io-o}^{DDR} - F_m(q_{in}^B + \frac{q_L^B H_{sr}C}{A})G_{co-o}^{DDR}}{1+L_c+L_v}$$

$$Z_{o-o}^{PCM} = \frac{\hat{v}_o}{\hat{i}_o} = \frac{(1+\frac{F_m q_L^B H_{sr}B}{A})Z_{o-o}^{DDR} + \frac{F_m q_L^B H_{sr}}{A}G_{co-o}^{DDR}}{1+L_c+L_v} \tag{33}$$

$$G_{co-o}^{PCM} = \frac{\hat{v}_o}{\hat{i}_{co}} = \frac{F_m^B G_{co-o}^{DDR}}{1+L_c+L_v}$$

$$Y_{in-o}^{PCM} = \frac{\hat{i}_{in}}{\hat{v}_{in}} = Y_{in-o}^{DDR} - \frac{F_m^B G_{ci-o}^{DDR}}{1+L_c+L_v}(q_{in}^B + \frac{q_L^B H_{sr}C}{A} + (q_o^B + \frac{q_L^B H_{sr}}{AZ_C})G_{io-o}^{DDR})$$

$$T_{oi-o}^{PCM} = \frac{\hat{i}_{in}}{\hat{i}_o} = T_{oi-o}^{DDR} + \frac{F_m G_{ci-o}^{DDR}}{1+L_c+L_v}((q_o^B + \frac{q_L^B H_{sr}}{AZ_C})Z_{o-o}^{DDR} - \frac{q_L^B H_{sr}}{A}) \tag{34}$$

$$G_{ci-o}^{PCM} = \frac{\hat{i}_{in}}{\hat{i}_{co}} = \frac{F_m G_{ci-o}^{DDR}}{1+L_c+L_v}$$

where L_c and L_v denote the inductor-current and output-voltage loop gains (i.e., G_{cL-o}^{DDR} and G_{co-o}^{DDR} denote the control-to-inductor-current and control-to-output-voltage transfer functions of the DDR-controlled converter) given in (35):

$$L_c = F_m q_L^B H_{sr} G_{cL-o}^{DDR}$$
$$L_v = F_m q_o^B G_{co-o}^{DDR} \tag{35}$$

and Z_C denotes the impedance of the output capacitor, as well as A = 1, B = 0, and C = 0 for a buck converter, and A, B, and C are defined in (36) for boost and buck-boost converters, respectively:

$$A = 1 - \frac{D^2 T_s}{2L}(r_L + r_{ds}) \quad B = \frac{DT_s}{L}V_{in} \quad C = \frac{D^2 T_s}{2L} \tag{36}$$

The duty-ratio constrains applicable for (32) and (33) can be obtained from (19) by setting $r_C = 0$ as:

$$F_m^B \approx F_m$$
$$q_L^B \approx q_L$$
$$q_{in}^B \approx q_{in} \tag{37}$$
$$q_o^B \approx q_C$$

as well as G_{cL-o}^{DDR} and G_{co-o}^{DDR} for computing G_{co-o}^{PCM} in (33) can be given for a buck converter as:

$$G_{cL-o}^{DDR} = \frac{\hat{i}_L}{\hat{i}_{co}} = \frac{V_e s}{L(s^2 + s\frac{A_1}{L} + \frac{A_2}{LC})} \quad G_{co-o}^{DDR} = \frac{\hat{v}_o}{\hat{i}_{co}} = \frac{V_e(1 + sr_C C)}{LC(s^2 + s\frac{A_1}{L} + \frac{A_2}{LC})} \tag{38}$$

where the coefficients $A_{1,2}$ are given in (16) and V_e in (17), respectively.

3. Simulink-Based Model Validation

As stated earlier, the applied Simulink models for a buck converter are explicitly given in [27]. The principles of the frequency-response-measuring method is described in [22,23]. The method is realized as a Simulink m-file for measuring the frequency responses from the switching models based on Simulink. The pseudo-random binary-sequence technique [22,23], which is utilized in measuring the frequency responses, will produce increased distribution in the data points, when the injection frequency approaches half the switching frequency [4]. The used component values are defined in Figure 5 (Section 2.3). In the presented frequency responses, the inductor-current sensing resistor (R_s) is assumed to be 1 Ω instead of 36 mΩ (cf. Figure 5) (note: R_s affects the gain of the control-related transfer functions by multiplying the corresponding transfer functions with $1/R_s$ [4]), and the inductor-current compensation $M_c = 0$. As stated in [10,12], the inductor-current-loop compensation is not required in a DCM-operated converter similarly as it is a necessity in CCM-operated converter, because the absolute mode limit will take place in the boundary between the DCM and CCM operation, and the DCM converter has to be designed accordingly. The converter in Figures 4 and 5 is designed to operate in DCM up to $D = 0.75$ (i.e., $K_{crit} = D' = 2Lf_s/R_{eq} = 0.25$; cf. [11]).

3.1. Open-Loop Validation

Figure 7 shows the set of control-to-output-voltage transfer functions, where the simulated transfer functions are denoted by dashed lines (i.e., unterminated: red and load-resistor affected: blue). The solid black lines denote the predicted transfer functions, respectively. The operating point of the converter corresponds to M = 0.5. The red responses in Figure 7 indicate that the unterminated converter would be unstable in open loop due to the existence of an RHP pole (i.e., the zeroth-order term in the denominator has become negative, which has caused a change of 180 degrees in the phase behavior at the low frequencies). The dashed blue line (i.e., the load-resistor-affected response) indicate that the load resistor has shifted the appearance of the RHP pole to the higher values of M compared to the unterminated case. Figure 7 shows also that the simulated phase behavior deviates from the averaged-model-based prediction as it does in PCM-controlled CCM converter as well [4]. The phase deviation starts already in vicinity of 10 kHz (i.e., at 1/10th of switching frequency). In CCM, the phase deviation is observed to start at 1/5th of the switching frequency [4]. This kind of phase deviation has not been reported earlier to take place in a PCM-controlled DCM-operated converter. As Figure 7 indicates, the predicted and simulated responses match very well. The unterminated G_{co-o} is solved computationally according to $G_{co-o} = (1 + Z_{o-o}/R_L) \cdot G_{co-o}^{R_L}$ by means of the corresponding simulated responses (i.e., Z_{o-o} and $G_{co-o}^{R_L}$) [4].

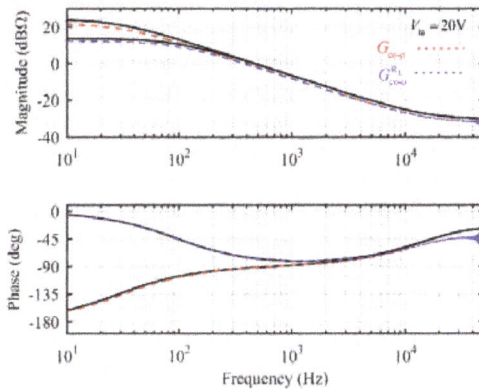

Figure 7. The control-to-output-voltage transfer function as unterminated (G_{co-o}) (red dashed line) and as resistor-load-affected ($G_{co-o}^{R_L}$) (blue dashed line) at the input voltage of 20 V. The predicted responses are denoted by solid black line and the simulated by dashed line, respectively.

Figure 8 shows the simulated (dashed red line) and predicted (solid black line) control-to-output-voltage transfer functions, when the high-frequency extension (H_{sr}) in Equation (32) with

$\omega_{sr} = 2\pi f_s$ (i.e., f_s denotes the switching frequency) and $\zeta = 0.5$ (i.e., ζ denotes the damping factor) have been taken into account in the prediction.

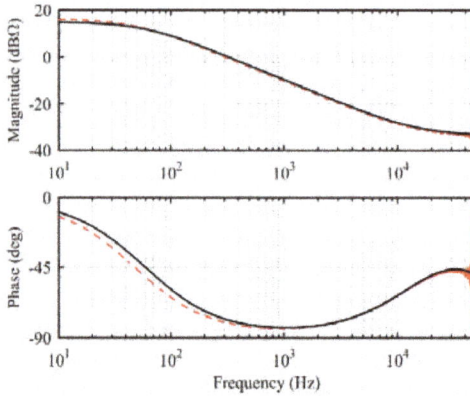

Figure 8. The unterminated control-to-output-voltage transfer function (G_{co-o}) (measured: dashed read line, predicted: solid black line) at the input voltage of 30 V. The high-frequency extension (H_{sr}) (Equation (32)) is added into the prediction.

The match between the simulated and predicted responses is very good. In case of PCM control in CCM, the $\omega_{sr} = \pi f_s$ and $\zeta = 0$, respectively, for obtaining similar match between the simulated and predicted responses as in [4].

Figure 9 shows the simulated (dashed lines) and predicted (solid black lines) output impedances of the converter at the input voltages of 20 V (read color) and 30 V (blue color). The figure shows that the average-model-based prediction matches well also at the high frequencies. The same phenomenon takes place also in PCM-controlled CCM converter as shown in [4]. The absence of the excess phase shift is caused by the output-terminal capacitor, which removes the internal high-frequency behavior from the output impedance (cf. [4]). The open-loop instability at $M = 0.5$ is also clearly visible in the output impedance. The output impedance can be measured directly even if the converter is loaded with a resistor or whatever load impedance.

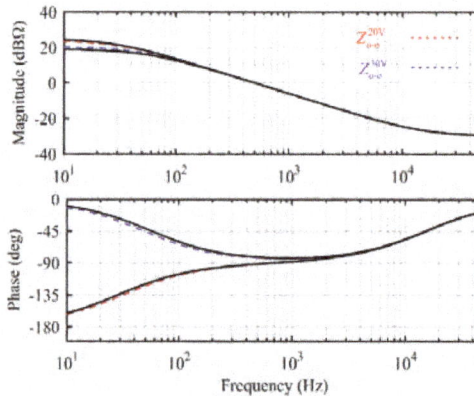

Figure 9. The output impedance (Z_{o-o}) (measured: dashed lines, predicted: solid black line) at the input voltage of 20 V (red color) and 30 V (blue color). The high-frequency extension (H_{sr}) (Equation (32)) is added into the predictions.

Figure 10 shows the comparison of the predicted frequency responses of the control-to-output-voltage transfer functions when the predictions are based on the simplified models (solid lines: 20 V (red) and 50 V (blue)) and on the complete models (dashed lines: 20 V (red) and 50 V (blue)). The responses show that both the models predict the responses with equal accuracy at high input voltage but the prediction accuracy of the simplified model is quite poor at the low input voltages.

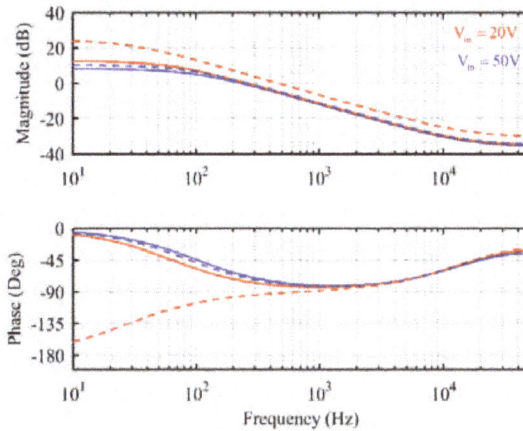

Figure 10. The predicted frequency responses of the control-to-output transfer functions at the input voltage of 20 V (red lines) and 50 V (blue lines). The solid lines denote the responses predicted by means of the simplified models and the dashed lines the responses predicted by the complete models.

3.2. Closed-Loop Validation

The output-voltage loop was designed to have the crossover frequency of 10 kHz and the phase margin (PM) of 60 degrees by using Type-2 controller shown in (39). The controller design was performed at the input voltage of 50 V. The controller zero (ω_z) was placed at $1/\sqrt{LC}$, and the high-frequency pole (ω_p) at $\pi f_s/4$, which require to use the controller gain (K_{cc}) of 851,138 to obtain the crossover frequency of 10 kHz, respectively:

$$G_{cc} = \frac{K_{cc}(1+s/\omega_z)}{s(1+s/\omega_p)} \tag{39}$$

Figure 11 shows the simulated (dashed lines) and predicted (solid black lines) output-voltage loop gains at the input voltage of 20 V (magenta color), 30 V (blue color), and 50 V (red color). The figure shows also that the converter is stable at 20 V, because the crossover frequency is much higher than the frequency of the RHP pole (cf. Table 1 for the low-frequency-pole locations). The figure shows clearly that the crossover frequency and PM does not change along the changes in the input voltage. This behavior is characteristic to the PCM control as discussed in [4]. Figure 7 indicates that the control design can be performed by using the load-resistor-affected $G_{co-o}^{R_L}$, when the output-voltage feedback-loop crossover frequency is placed at the frequencies of 1/10th of switching frequency or higher. In case of output-current-feedback control, it is required to use unterminated models, because the load resistor affects, especially, the high-frequency part of the transfer function as discussed and demonstrated in [4,17].

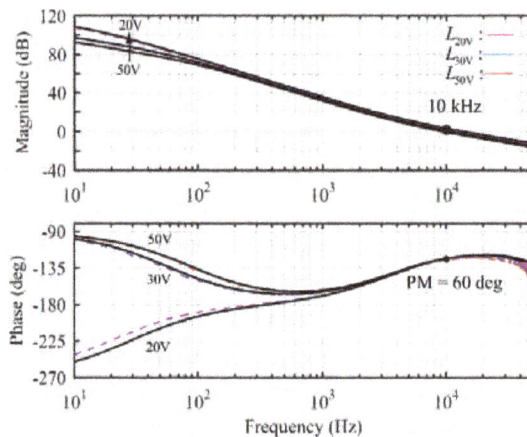

Figure 11. The simulated (dashed lines) and predicted (solid black lines) output-voltage loop gains at the input voltage of 20 V (magenta color), 30 V (blue color), and 50 V (red color), respectively. The high-frequency extension (H_{sr}) (Equation (32)) is added into the prediction.

4. Conclusions

The investigations of this paper show evidently that the method to model the dynamic behavior of a PCM-controlled converter in DCM, which was published in 2001 [10], yields very accurate predictions, when the parasitic circuit elements and the high-frequency extension proposed in this paper are included. In addition, the investigations show that the phase behavior of the DCM-operated PCM-controlled converter exhibits also excess phase shift at the high frequencies but the phase-shift behavior differs from that of the CCM-operated converter [4]. In addition, the buck converter will be unstable in open loop (i.e., an RHP pole appears), when $M > 0.5$, which differs from the earlier predicted instability condition at $M > 2/3$. The stability can be ensured in closed loop by designing the crossover frequency of the feedback control loop to be higher than the RHP pole. The PCM-controlled DCM-operated converter does not need to be compensated either, because the mode limit between the basic switching frequency and the second-harmonic modes of operation will take place at the boundary between the DCM and CCM mode of operation. The mode limit will take place, because the duty-ratio gain becomes infinite at the DCM-CCM-mode boundary similarly as the CCM converter at $D = 0.5$ without the inductor-current-loop compensation. This paper proposes also useful control-engineering-type block diagrams for solving the generalized transfer functions, which are applicable for buck, boost, and buck-boost converters in DCM similarly as presented in [4,13,14] for the PCM-controlled converters operating in CCM.

Conflicts of Interest: The author declares no conflict of interest.

References

1. Deisch, C.W. Simple switching control method changes power converter into a current source. In Proceedings of the IEEE Power Electronics Specialists Conference (IEEE PESC), Syracuse, NY, USA, 13–15 June 1978; pp. 300–306.
2. Capel, A.; Ferrante, G.; O'Sullivan, D.; Weinberg, A. Application of injected current model for the dynamic analysis of switching regulators with the new concept of LC^3 modulator. In Proceedings of the IEEE Power Electronics Specialists Conference (IEEE PESC), Syracuse, NY, USA, 13–15 June 1978; pp. 135–147.
3. Redl, R.; Sokal, N.O. Current-mode control, five different types, used with the three basic classes of power converters: Small-signal ac and large-signal DC characterization, stability requirements, and implementation

of practical circuits. In Proceedings of the IEEE Power Electronics Specialists Conference (IEEE PESC), Toulouse, France, 24–28 June 1985; pp. 771–785.

4. Suntio, T. On dynamic modeling of PCM-controlled converters—Buck converter as an example. *IEEE Trans. Power Electron.* **2018**, *33*, 5502–5518. [CrossRef]

5. Ridley, R.B. A new continuous-time model for current-mode control. *IEEE Trans. Power Electron.* **1991**, *6*, 271–280. [CrossRef]

6. Tan, F.D.; Middlebrook, R.D. A unified model for current-programmed converters. *IEEE Trans. Power Electron.* **1995**, *10*, 397–408. [CrossRef]

7. Sun, J.; Bass, R.M. A new approach to averaged modeling of PWM converters with current-mode control. In Proceedings of the IEEE International Conference Industrial Electronics Control Instrumentation (IEEE IECON), New Orleans, LA, USA, 9–14 November 1997; pp. 599–604.

8. Suntio, T.; Hankaniemi, M.; Roinila, T. Dynamical modeling of peak-current-mode-controlled converter in continuous conduction mode. *J. Simul. Model. Pract. Theory* **2007**, *15*, 1320–1337. [CrossRef]

9. Ridley, R.B. A new continuous-time model for current-mode control with constant frequency, constant on-time and constant off-time in CCM and DCM. In Proceedings of the IEEE Power Electronics Specialists Conference (IEEE PESC), San Antonio, TX, USA, 11–14 June1990; pp. 382–389.

10. Suntio, T. Analysis and modeling of peak-current-mode controlled buck converter in DICM. *IEEE Trans. Ind. Electron.* **2001**, *48*, 127–135. [CrossRef]

11. Erickson, R.W.; Maksimović, D. *Fundamentals of Power Electronics*, 2nd Ed. ed; Kluwer Academic Publishers: Norwell, MA, USA, 2001.

12. Fang, C.-C. Unified discrete-time modeling of buck converter in discontinuous mode. *IEEE Trans. Power Electron.* **2011**, *26*, 2335–2342. [CrossRef]

13. Suntio, T. *Dynamic Profile of Switched-Mode Converter—Modeling, Analysis and Control*; Wiley-VCH: Weinheim, Germany, 2009.

14. Suntio, T.; Messo, T.; Puukko, J. *Power Electronic Converters—Dynamics and Control in Conventional and Renewable Energy Applications*; Wiley-VCH: Weinheim, Germany, 2017.

15. Ćuk, S.; Middlebrook, R.D. A general unified approach to modeling switching DC-to-DC converters in discontinuous conduction mode. In Proceedings of the IEEE Power Electronics Specialists Conference (IEEE PESC), Palo Alto, CA, USA, 14–16 June 1977; pp. 36–57.

16. Suntio, T. Unified average and small-signal modeling of direct-on-time control. *IEEE Trans. Ind. Electron.* **2006**, *53*, 287–295. [CrossRef]

17. Suntio, T. Load-resistor-affected dynamic models in control design of switched-mode converters. *EPE J.* **2018**. [CrossRef]

18. Davoudi, A.; Jatskevich, J.; Chapman, P.L. Numerical dynamic characterization of peak current-mode–controlled DC-DC converters. *IEEE Trans. Circuits Syst. II Exp. Briefs* **2009**, *56*, 906–910. [CrossRef]

19. Smithson, S.C.; Williamson, S.S. A unified state-space model of constant-frequency current-mode-controlled power converters in continuous conduction mode. *IEEE Trans. Ind. Electron.* **2015**, *62*, 4514–4524. [CrossRef]

20. Hankaniemi, M.; Karppanen, M.; Suntio, T. Dynamical characterization of voltage-mode controlled buck converter operating in CCM and DCM. In Proceedings of the Power Electronics & Motion Control Conference (EPE-PEMC), Portoroz, Slovenia, 30 August–1 September 2006; pp. 54–59.

21. Sugimoto, Y.; Sai, T.; Watanabe, K.; Abe, M. Feedback loop analysis and optimized compensation slope of the current-mode buck DC-DC converter in DCM. *IEEE Trans. Circuits Syst. I Reg. Pap.* **2015**, *62*, 311–319. [CrossRef]

22. Roinila, T.; Helin, T.; Vilkko, M.; Suntio, T.; Koivisto, H. Circular correlation based identification of switching power converter with uncertainty analysis using fuzzy density approach. *J. Simul. Model. Pract. Theory* **2009**, *17*, 1043–1058. [CrossRef]

23. Roinila, T.; Vilkko, M.; Suntio, T. Fast loop gain measurement of switched-mode converter using binary signal with specified Fourier amplitude spectrum. *IEEE Trans. Power Electron.* **2009**, *24*, 2746–2755. [CrossRef]

24. Sun, J.; Mitchell, D.M.; Greuel, M.F.; Krein, P.T.; Bass, R.M. Modeling of PWM converters in discontinuous conduction mode—A reexamination. In Proceedings of the IEEE Power Electron. Specialists Conference (IEEE PESC), Fukuoka, Japan, 17–22 May 1998; pp. 615–622.

25. Sun, J.; Mitchell, D.M.; Greuel, M.F.; Krein, P.T.; Bass, R.M. Averaged modeling of PWM converters operating in discontinuous mode. *IEEE Trans. Power Electron.* **2001**, *16*, 482–492.

26. Davoudi, A.; Jatskevich, J.; De Rybel, T. Numerical state-space average-value modeling of PWM DC-DC converters operating in DCM and CCM. *IEEE Trans. Power Electron.* **2006**, *21*, 1003–1012. [CrossRef]

27. Suntio, T.; Kivimäki, J. Physical insight into the factors affecting the load-transient response of a buck converter. In Proceedings of the 16th European Conference Power Electronics Applications (EPE ECCE EUROPE), Lappeenranta, Finland, 26–28 August 2014; pp. 1–10.

energies

MDPI

Review

Mitigation of Power Quality Issues Due to High Penetration of Renewable Energy Sources in Electric Grid Systems Using Three-Phase APF/STATCOM Technologies: A Review

Wajahat Ullah Khan Tareen [1,2,*], Muhammad Aamir [3], Saad Mekhilef [1], Mutsuo Nakaoka [1], Mehdi Seyedmahmoudian [4], Ben Horan [5], Mudasir Ahmed Memon [1] and Nauman Anwar Baig [2]

[1]   Power Electronics and Renewable Energy Research Laboratory (PEARL), Department of Electrical Engineering, University of Malaya, Kuala Lumpur 50603, Malaysia; saad@um.edu.my (S.M.); nakaoka@pe-news1.eee.yamaguchi-u.ac.jp (M.N.); memon.mudasir@usindh.edu.pk (M.A.M.)
[2]   Department of Electrical Engineering, International Islamic University, Islamabad 44000, Pakistan; nauman.anwar@iiu.edu.pk
[3]   Department of Electrical Engineering, Bahria University, Islamabad 44000, Pakistan; muhammadaamir.buic@bahria.edu.pk
[4]   School of Software and Electrical Engineering, Swinburne University of Technology, VIC 3122, Australia; mseyedmahmoudian@swin.edu.au
[5]   School of Engineering, Deakin University, Waurn Ponds, VIC 3216, Australia; ben.horan@deakin.edu.au
*   Correspondence: wajahat.tareen@iiu.edu.pk or wajahattareen@gmail.com; Tel.: +0092-332-574-4848

Received: 20 April 2018; Accepted: 23 May 2018; Published: 7 June 2018

Abstract: This study summarizes an analytical review on the comparison of three-phase static compensator (STATCOM) and active power filter (APF) inverter topologies and their control schemes using industrial standards and advanced high-power configurations. Transformerless and reduced switch count topologies are the leading technologies in power electronics that aim to reduce system cost and offer the additional benefits of small volumetric size, lightweight and compact structure, and high reliability. A detailed comparison of the topologies, control strategies and implementation structures of grid-connected high-power converters is presented. However, reducing the number of power semiconductor devices, sensors, and control circuits requires complex control strategies. This study focuses on different topological devices, namely, passive filters, shunt and hybrid filters, and STATCOMs, which are typically used for power quality improvement. Additionally, appropriate control schemes, such as sinusoidal pulse width modulation (SPWM) and space vector PWM techniques, are selected. According to recent developments in shunt APF/STATCOM inverters, simulation and experimental results prove the effectiveness of APF/STATCOM systems for harmonic mitigation based on the defined limit in IEEE-519.

Keywords: FACTS devices; active power filter; static compensator; control strategies; grid-connected converter; SPWM; SVM

---

1. Introduction

Electricity is an indication of comfortable life, and the demand for this energy source is increasing. Development in power industries increases the number of linear and nonlinear loads in each system. In nonlinear load conditions, many solid-state switching converters draw reactive power and current harmonics from the AC grid. These nonlinear loads generate harmonics, which produce disturbance and directly influence human life. Nowadays, each piece of equipment, power system, and service, such as furnaces, computer power supplies, communication systems, renewable energy systems, electrical power generations, and high-voltage systems, requires a continuous power supply. Researchers and different power companies are continuously exploring solutions to power quality problems [1], such as harmonics, system imbalance, load balancing, excess neutral current, and power system grid intrusions. Evidently, the increasing demand for nonlinear loads produces harmonics in the power system, thereby resulting in poor power quality. Flexible AC transmission system (FACTS) devices, such as static compensators (STATCOMs) and active power filters (APFs), are the most dominant technologies available for industrial and commercial purposes and are deemed the solution to this power quality problem [2–4].

In the past, harmonic grid problems were solved using passive filter (PF) devices [5]. These filters, together with low-cost solutions for power quality issues, are considered the initial stage of development in mitigating current harmonics [6,7]. APFs are considered the second stage of development and an effective solution to overcome the limitation of PFs. However, the size, cost, and rating of APFs are considerably increased by the increasing demand for power system capacity [8,9]. To overcome the issues of shunt APFs, a third stage of development consists of hybrid power filter (HPF) devices, which comprise hybrid combinations of PFs with shunt APFs [10]. Furthermore, HPF technology is evaluated in the fourth stage of development as a unified power quality conditioner (UPQC) [11]. Previous research [12,13] on power quality improvement has led to important developments in FACTS devices, namely, the static volt-ampere reactive (VAR) compensator (SVC), dynamic voltage restorer (DVR), and distribution STATCOM (DSTATCOM) [14–16]. Such improvements compensate for the mitigation problems related to power quality, including poor load power factor, load harmonics, imbalanced load conditions, and DC offset in loads.

Many shunt APFs are impractical for use for high-power-rating components. Therefore, transformerless and reduced switch count topologies are the leading technologies in power electronics that aim to reduce system cost and concurrently provide such benefits as small volumetric size, light weight, remarkably compact structure, and high reliability. Modern APF topologies are rapidly replacing the standard ones because of their effective performance, efficient response, and favorable cost and size attributes. Nevertheless, reducing the number of power semiconductor devices, power conversion circuits, sensors, and control circuits requires complicated control strategies [17,18].

Despite the importance of decreasing the number of power components, no study focuses on reduced switch count topologies and complex control strategies in shunt APFs and STATCOMs [19]. The present study focuses on the comprehensive review of advanced reduced switch count topologies, specifically SAPFs and STATCOMs, control schemes for reconfigurable voltage-fed-type inverters (VSIs). The control structures and possibilities of implementing grid-connected power converters in different reference frames are discussed and compared. Furthermore, the overview of the complete control schemes and enhanced control strategies, including the most appropriate control strategies, such as sinusoidal pulse width modulation (PWM) and space vector PWM techniques, are presented. Comparisons and characteristics of operating reduced switch and leg count VSIs are concluded according to a well-surveyed literature summary. Finally, a summary and recommendations for future research are highlighted.

2. Harmonics and International Standards

The amount or penetration of harmonics had previously been largely increasing, thereby affecting the performance and efficiency of the system [20]. Harmonics are generated in the system because of non-linear or critical loads. Harmonics exist in a power grid or power distribution network in the form of series and parallel resonances generated by harmonic current loads, which increase the source voltage and current total harmonic distortion. However, modern technologies in power quality improvement mitigate current, voltage, active and reactive power, voltage zero-crossing, and other issues, such as harmonics, voltage sags and swells, notches and flickers, spikes and glitches, and voltage imbalance [21].

Harmonic sources are generally classified into two types on the basis of system impedance, namely, current and voltage types [22]. A diode rectifier with smoothing inductor is categorized as a current harmonic source because it produces the current harmonics at the source side, which is the input of the rectifier. A diode rectifier with smoothing capacitor is a voltage harmonic source, which is affected by the AC-side impedance and generates harmonics at the output of the inverter. Two international standards of IEEE-519, namely, IEEE Recommended Practices and Requirements for Harmonic Control in Electrical Power Systems and IEC 61000-3-2 [23,24], provide limitations and guidelines for manufacturers and power utility companies that are connected to the power grid. These standards address harmonic issues, such as current harmonics, power quality, grounding, and voltage harmonics.

3. Methods for Mitigating Harmonics

Different methods are adopted to mitigate harmonic contents in power grid-connected APF systems [25]. PFs, APFs, and STATCOMs are improved technologies that solve harmonic power quality problems [26].

3.1. Shunt PFs

PFs were introduced to mitigate harmonic compensation as an initial solution to power quality issues in the power distribution network. This technology presents a simple economical solution that consists of different series and parallel combinations of inductors, capacitors, and damping resistors [27]. Figure 1 shows the commonly used PF configuration [28].

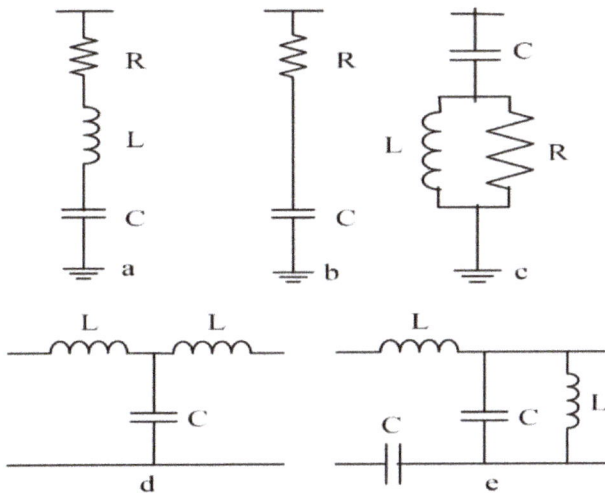

Figure 1. PF structures: (**a**) single-tuned; (**b**) first-order high-pass; (**c**) second-order high-pass; (**d**) LCL; and (**e**) LLCC filters.

Each operation is influenced by the fundamental source impedance. The filtering characteristics depend on the values of the inductance and capacitance sets. The operation is tuned according to the required harmonic order, such as first, second, and third order, to track the requisite harmonics. Studies have presented many PF design techniques, such as series, shunt, single-tuned, double-tuned, low-pass, high-pass, bandpass, LCL and LCC filters. The PF configurations are installed in series with the power distribution system to provide high impedance and cancel the flow of harmonic current. PFs are generally coupled with a thyristor-controlled reactor to improve harmonic mitigation and reactive current component compensation. Some of the key practical limitations are as follows:

- PFs require a separate filter for each harmonic current, and their filtering range is limited.
- PFs allow only one component (either a harmonic or a fundamental current component) to pass at a time.
- Large amounts of harmonic current saturate or overload the filter and cause series resonance with the AC source, thereby resulting in excessive harmonic flow into the PFs.
- PFs amplify source-side harmonic contents because of the impedance in the source of parallel and series negative resonances between the grid and the filter [29].
- The design parameters of PFs in an AC system depend on the system operating frequency, which changes around its nominal value according to variable load conditions.
- PFs only eliminate frequencies to which they are tuned, thus resulting in limited compensation, large size, and tuning issues.

3.2. Shunt APFs

APFs were introduced and investigated as a solution to the limitations of PFs. An APF consists of an active switching device and passive-energy storage devices, such as inductors and capacitors, which provide superior compensation characteristics, including voltage and current harmonics, voltage imbalance compensation for utilities, and current imbalance compensation for consumers. APFs mitigate reactive power, neutral current, changing line impedance, frequency variation, voltage notch, sudden voltage distortion, transient disturbance, and voltage balance and improve the power factor of voltage and current in medium-power systems [30].

Different APF topologies and control methodologies have been proposed and progressively investigated as a perceived solution to critical issues in high-power load applications [31,32]. APFs are classified into many categories in accordance to subsequent measures. The circuit structure of an APF commonly includes a voltage-source PWM inverter with a DC-link capacitor. Evidently, current-source APFs are superior in terms of compensating current dynamics. However, voltage-source APFs perform better than current-source APFs in terms of filter losses and PWM carrier harmonic reduction. Table 1 shows the survey results for shunt and series APFs [11].

Table 1. Comparison of active power filters (APFs) in power quality improvement techniques.

Category	Shunt APF	Series APF
Connection with system	Parallel with distribution system	Connected in series with distribution system
Action	Current source	Voltage source
Filter rating	Voltage rated at full load rating Current rating comprises partially harmonic and partly reactive current components	Current rated at full load rating Voltage rating is partially compensated voltage component
Functioning	Harmonic load current filtering Compensation for reactive current Mitigation of current unbalance	Mitigation of voltage harmonics, sag, and swells Mitigation of current harmonics Compensation of reactive current Mitigation of current unbalances
Characteristics of compensation	Source impedance exerts no effect on compensation for current source loads.	Source and load impedance exert no effect on compensation for voltage source loads.
Application	Injected current may cause excess current when applied to a voltage source load.	A low-impedance parallel branch (for improvement of power factor) when working with current source load
Load considered	Nonlinear/inductive current source loads or harmonics containing current source loads	Nonlinear/capacitive voltage source loads or harmonics containing voltage source loads

The basic compensation principle of shunt APFs is eliminating the current harmonics generated by critical loads. Generally, APFs are installed in a shunt position near the nonlinear load to compensate for the effect of harmonic nonlinearity [33]. APFs eliminate harmonics by injecting reactive or compensating current at the point of common coupling (PCC) into the power network. Each APF generates inverse harmonics as a mirror image to the nonlinear load harmonics, thereby canceling the harmonics and leaving the fundamental component to make the source current purely sinusoidal, as depicted in Figure 2 [28].

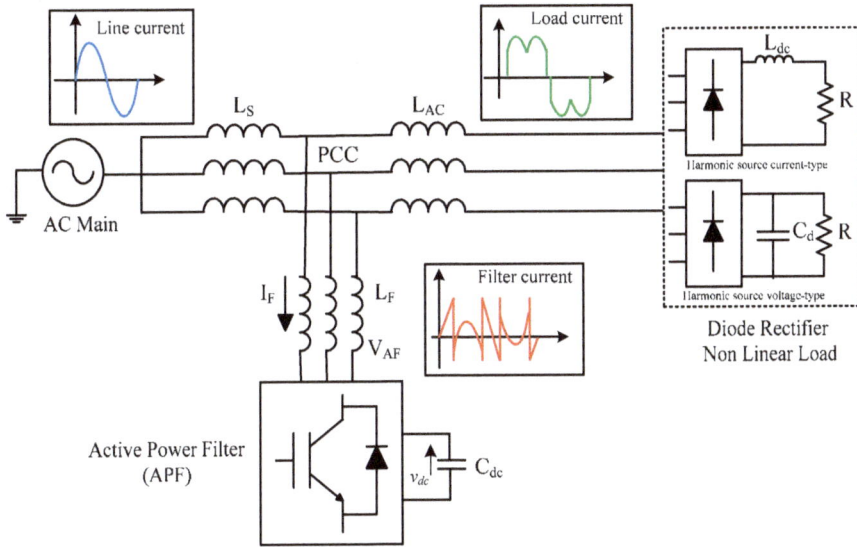

Figure 2. Basic compensation principle of the SAPF.

3.3. STATCOM

FACTS devices are increasingly required in modern transmission systems for high-power transfers [34]. Two types of FACTS devices, namely, SVCs and STATCOMs [35], can restore active power current and dynamic reactive power compensation [36]. These devices consist of the modular unit of the voltage source converter (VSC) equipped with insulated-gate bipolar transistors (IGBTs) for rapid switching and controlling using the PWM scheme. A STATCOM is a shunt device that is connected with PCC to provide voltage support and active and reactive power, increase transient stability and improve damping. A brief detailed classification of the FACTS devices is shown in Figure 3 [37,38].

At the distribution level, STATCOM devices are operated as an active filter to achieve harmonic filtering, power factor correction, neutral current compensation, and load balancing [39]. APFs are an advanced technology that offers faster dynamic response, smaller size, lower system cost, and higher performance under low-voltage oscillations than SVC devices. Energy storage devices improve power quality and provide rapid controllable transient response to the bus voltage by injecting or absorbing the corresponding amount of reactive power into or from the system. Such devices maintain the minimum values of amplitude, phase, and frequency to control voltage flickers during fault events. In addition, the inconsistent flow of reactive power between the supply power grid and the loads is prevented [40]. Table 2 presents the best available solution for the specific compensation challenges [41–43].

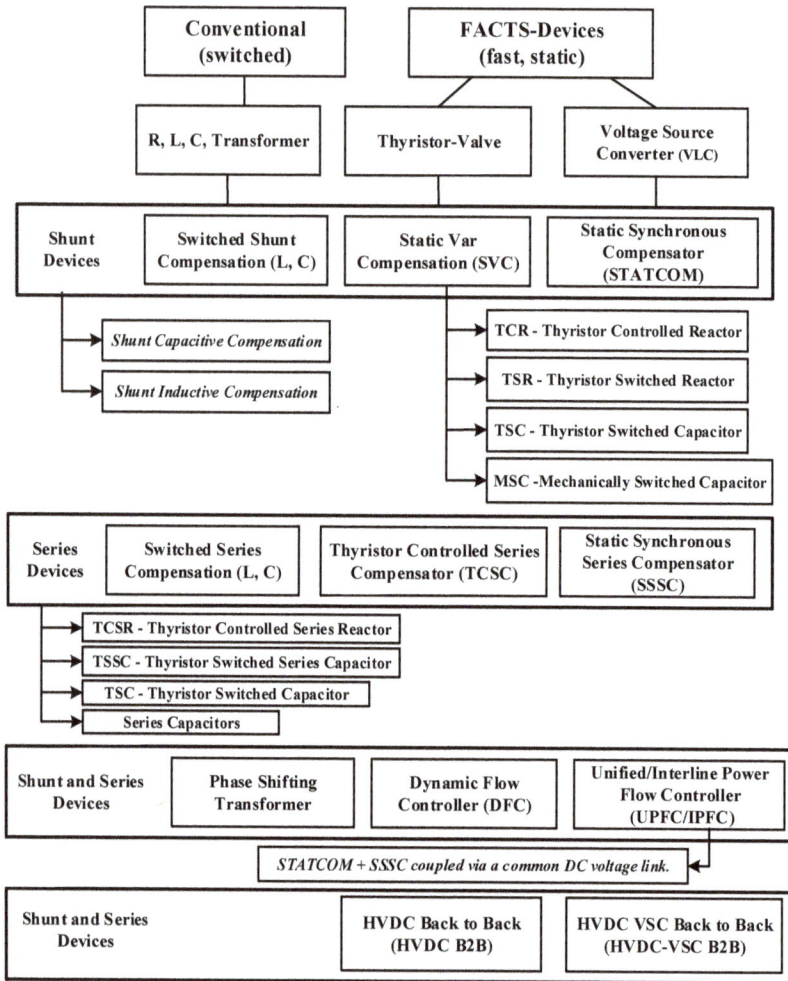

Figure 3. Overview of the major FACTS devices.

The following sections discuss the high penetration of renewable energy sources in the electrical power grid system for reactive power compensation and harmonic mitigation problems.

3.3.1. Multilevel PV-STATCOM Applications in Grid-Connected Systems

The conventional two- and three-level inverter configurations are not suitable for photovoltaic (PV) applications in high-power-rating systems [44,45]. Many configurations are inefficient in harvesting maximum power, have a low utilization factor, and require a large AC filter at the output for high power quality. A grid-connected solar PV power conditioning inverter conventionally requires a transformer for galvanic isolation to match the grid voltages with the inverter output. The concept of PV-STATCOM systems is required to regulate real and reactive power through the converter and mitigate the limitations of conventional PV inverters [46]. The cascaded H-bridge (CHB) inverter in multilevel PV-STATCOMs is an advanced configuration for high-solar-power applications and has isolated input DC sources and no additional DC power source requirement for STATCOM devices [47,48].

Table 2. Comparison of the advanced technology solution for the reactive power compensation challenges.

Technology	MSC(DN)/MSR	SVC	SVC PLUS (STATCOM)	Hybrid STATCOM	Synchronous Condenser
Application	Compensation of predictable load	Fast dynamic compensation and voltage recovery during faults	Fast dynamic compensation and voltage recovery during faults	Fast dynamic compensation and voltage recovery during faults	Provision of short-circuit power, inertia, dynamic compensation, and voltage recovery during faults
Switching	Limited switching only	Unlimited switching	Unlimited switching	Unlimited switching	Continuous operation
V/I characteristic	No response	Good overvoltage performance	Superior under voltage performance	Superior under voltage performance	Good overload capability
Control range	Adjustable by MSC and MSR ranges	Adjustable by branch ranges	Symmetrical output: Adjustable range	Unsymmetrical output: Adjustable range	Adjustable by generator size
Redundancy	No inbuilt redundancy	Inbuilt redundancy in thyristor valves	Inbuilt redundancy in power modules	Inbuilt redundancy in power modules and in thyristor valves	Depending on solution
Harmonics	Susceptible to harmonics	TCR is source of harmonics—AC filters required	Harmonically self-compensated—no filters required	Harmonically self-compensated—no filters required	Not susceptible to harmonics
Response time	2–5 cycles, depending on breaker	2–3 cycles	1.5–2 cycles	1.5–2 cycles	seconds
Operation and maintenance	Very low, depending on breaker	Low, primarily visual inspection	Very low, primarily visual inspection	Very low, primarily visual inspection	Low, inspection every 3–4 years
Losses at 0 MVAR output power	0%	0.3% of the rated output power	0.15% of the rated output power	0.15% of the rated output power	~1% of the rated output power
Availability	>99%	>99%	>99%	>99%	>98%

In the study of [49], a PV inverter operated as a STATCOM for the optimal use of the system, provided active power during daytime, when solar energy is available, and provided reactive power when irradiation was low. Similarly, in several studies, PV inverters operated as STATCOMs during daytime to provide the reactive power compensation in a distribution utility network [50,51]. However, the performance of the system depends on the rated capacity of the inverter, the constant DC-link voltage, and the maximum power point (MPP) instant time tracking in the distribution utility network [52]. In the study of [53], a PV-STATCOM was adopted with active power filtering against power quality issues in the PV system, such as transient voltages, voltage flickering, and harmonics. Therefore, a PV-active filter-STATCOM is designed for harmonic mitigation and continuous regulation of reactive power for grid applications.

In the study of [54], a SHAPF based on a cascaded H-bridge multilevel inverter (CHB-MLI) was tested for high-power PV applications with reduced filter size, maximized PV cell power extraction, and improved utilization factor through reactive power compensation. With advanced level controls, the PV-STATCOM system can be adopted for active filter applications. Furthermore, PV-STATCOM inverters improve the stability of voltage-sensitive loads during grid fault conditions [55]. In the study of [56,57], a PV-STATCOM without a DC–DC converter was tested; reduced switch count configurations controlled the DC-link voltage through the operation of STATCOM, and the PV-STATCOM system improved the utilization factor and power quality by compensating the reactive power reference in the range of 0–1 P.U. with reduced system size and cost. In study [58] emphasized the necessity of multifunction PV inverters and different active and passive controls in resolving power quality issues. Furthermore, three different controls, namely, *abc*, *d–q*, and alpha–beta controls, were discussed with a comparison of suitable topologies on the basis of consumer applications [59]. A brief review of anti-islanding techniques and power quality issues were discussed in the study of [60].

A comprehensive review of power quality, stability, and protection and the negative impacts of PV systems due to voltage parameter control and static compensation techniques in the power distribution grid were discussed in the studies of [61,62]. Meanwhile, [63] investigated a CHB-based inverter and modular multilevel converters for STATCOM applications and compared reactive power controls for large PV applications [64].

Independent DC links make CHB-based inverters a suitable candidate for PV and STATCOM applications because of the following advantages: (a) a small filter produces low THD; (b) an output transformer is unnecessary because of the increased number of modules, which extend the output voltage level with low rating; (c) the cost and size of the power components are reduced by the low-voltage-level operation; (d) the cost and size of the power components and auxiliary equipment are reduced because the voltage levels in each module are low. A modified selective harmonic elimination PWM (SHE-PWM) technique was discussed in the study of [65] for a transformerless STATCOM that was tested using the CHB configuration for grid applications [66,67]. The independent DC-link sources in PV-STATCOM applications reduce system cost. Therefore, the CHB-MLI multilevel configuration with phase-shifted modulation techniques is suitable for large-scale systems [68]. An advanced selective swapping scheme was adopted in a 154-kV 21-level CHB inverter-based STATCOM system to reduce the issue of DC ripple voltage [69].

In the study of [70,71], a sinusoidal voltage waveform was achieved using SHE-PWM. Furthermore, in the study of [72], the sensors at the DC-links were eliminated by adopting the control due to the nonlinear parameter uncertainties presented for the CHB-MLI in the study. Zero-sequence voltage and negative sequence current schemes were tested under one- and two-line faults for a CHB-based STATCOM to reduce switching losses [73]. Another active power balance control scheme was tested to regulate the reactive power for a CHB inverter; the positive- and negative-sequence control components were monitored for DC-link voltage balance under the balanced or imbalanced conditions of the grid voltage [74]. Therefore, a PV-STATCOM inverter operates on two AC-source controllers at the same frequency level. The sources with high voltage amplitude provide reactive power to the other sources. In this study, the magnitude of the inverter output depends on the reference reactive power signal. Similarly, the sources with leading phase angle provide active power to the other sources. In conclusion, the STATCOM can be used to eliminate harmonics and compensate for lagging and leading VAR.

STATCOM devices aim to provide rapid load voltage regulation along the controllable exchange of reactive power in the system, as well as achieve power quality improvement, reactive power control, voltage regulation, power swing, and improvement of transmission line capacity during power system faults [75]. However, many other power converters, such as DSTATCOMs [76], UPQCs [77], and DVRs [78], also eliminate harmonics and imbalance issues in the generation and utilization of the system [79,80]. A DSTATCOM is a synchronous voltage generator or a static synchronous compensator with a coupling transformer, an inverter, and an energy storage device similar to that of the configuration of a STATCOM. The DSTATCOM injects or absorbs uninterrupted capacitive and inductive reactive power into or from the distribution system [81].

The AC-controllable VSC can be used as a multifunction STATCOM with modified control structure to provide harmonic elimination, reactive power, and load imbalance compensation under nonlinear loads and non-ideal main voltage conditions. The potential applications of STATCOM devices such as APFs, are tested for active power filtering, and reactive power supplies are considerably attractive to distribution engineers [82]. Table 3 demonstrates reactive power-compensating devices against various control parameters that are related to the said devices [83,84]. These parameters are automatic voltage control, STATCOM, SVC and thyristor-controlled series capacitor.

Table 3. Comparison of control parameters affected by reactive power-compensating devices.

Parameters	Static Capacitors	Capacitor & Reactor Bank	AVC	STATCOM	SVC	TCSC	UPFC
Reactive power	**	***	**	****	***	**	****
Active power		**	**	*	*	**	
Voltage stability	**	**	**	****	***	***	****
Voltage control	**	**	**	****	***	**	****
Flicker control		*		****	***		****
Harmonic reduction		*					****
Power flow control						***	****
Oscillation damping		*		***	**	***	****

High number of "*" is preferred.

3.3.2. Wind Turbine STATCOM (WT-STATCOM) Applications in Grid-Connected Systems

The evolution of power electronic technology has created two generation of FACTS devices in wind farms. These devices are classified into two generations, namely, the old thyristor-type and the advanced VSC-type [38,85], and are used to mitigate voltage instability, reactive power problems [86], and harmonic distortion to improve the power quality of the network [87].

The thyristor-based FACTS devices in wind farms are SVCs. The main structure of the SVC consists of capacitors and inductances that are controlled by thyristors. The STATCOM consists of a main component of modular VSC, which is equipped with IGBTs that are controlled by PWM. Figure 4 shows the structure of an induction generator-based wind turbine and a STATCOM that is connected to a grid and a wind turbine terminal [88]. A coupling transformer is installed to provide galvanic isolation and to control the flow of reactive power during steady- and transient-state conditions [88]. With these modified control strategies, the STATCOMs are superior to other mitigation methods, such as SVCs and series-saturated reactors during normal and transient conditions [89].

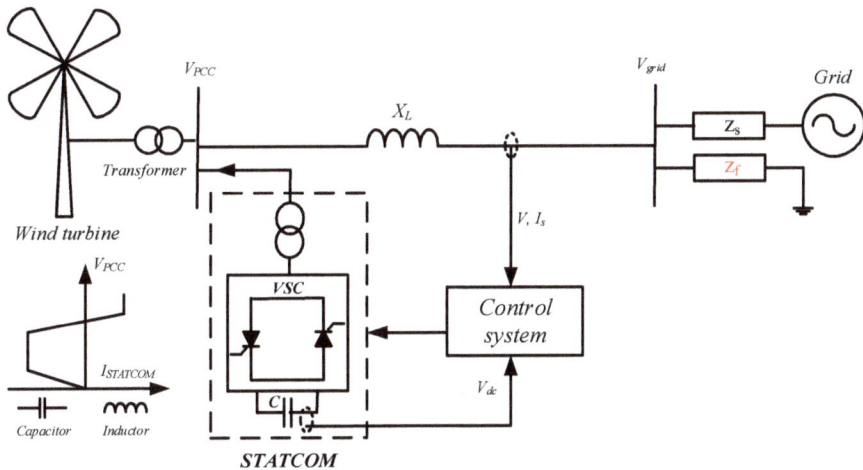

Figure 4. The structure of the WT based FSIG connected to STATCOM.

An energy storage system (ESS) with STATCOM provides a new robust decentralized control operation for high-wind-power applications [90]. The controller operates on the linear quadratic output–feedback method and provides a promising solution for high-wind-power systems [91].

Another robust STATCOM control was presented [92], and it controls the active and reactive power and pitch angle of wind turbine induction generators for improved low-voltage ride through (LVRT) capability. The limitation of this control is the enhanced torque produced inside the induction machine, which produces a high maximum torque and stresses the drivetrain during fault recovery. Therefore, an advanced indirect torque control (ITC)-based control is used to limit the maximum torque of the STATCOM controller [93].

In another study [94], the STATCOM and SVC were compared in terms of LVRT development. The STATCOM, which is 15% cheaper than the SVC, is the more economical solution. In conclusion, SVC devices have limited capabilities at low voltage levels. However, advanced STATCOMs can maintain reactive power output over a wide range of voltages with a smaller structure and faster control response than SVC [94].

Generally, a STATCOM FACTS device is installed for an entire wind farm [86,95], multilevel full H-bridge whereas an SVC-type thyristor-controlled resistor (TCR) device is installed for a single wind turbine. Besides reactive power regulation, these FACTS devices provide isolation and firewall protection to wind turbines against the effects of grid disturbances and connect the wind turbines with low X/R ratio or weak grids [96]. A study [97] therefore used a hybrid electrolyser and fuel cell system to control the ramp rate of wind turbine power and enhance the voltage quality at the PCC. The system stopped the inrush power flows in the electrolyser and the fuel cell path by controlling the frequent start-up and shutdown of the fuel cell stacks, which are effects associated with the grid-connected wind farms. STATCOM-based control methods [98] are adopted to regulate grid reactive power and maintain a suitable level of voltage sag on the grid and prevent the wind turbine from disconnecting from the grid during faults. The unified power flow controller is one of the most advanced FACTS devices used in wind applications and consists of two power converters. One converter controls as a STATCOM and the other operates as a static synchronous series compensator to control power flow with the limitation of high cost [99,100]. During grid faults, the reactive power demand can be compensated by an external dynamic STATCOM and an ESS [101], such as hybrid battery–super capacitor energy storage [102]. The ESS system with STATCOM provides a promising solution for wind power system applications [91].

4. Standard Classification of Shunt APFs

Generally, APFs are classified into two categories, namely, DC and AC power filters. DC–APFs are designed with a thyristor configuration for high-power drives [55] and high-voltage DC systems. An AC–APF configuration consists of active solutions, such as active power quality and line conditioners and instantaneous reactive power compensators for current and voltage harmonics. Shunt APFs are classified under three categories, namely, topology-, converter-, and phases-type configurations. Topology-type filters can be delegated as shunt, series, and hybrid APFs (HAPFs), as shown in Figure 5.

Converter-type filters are classified as VSI and current-fed-type inverters (CSI). Figure 6a illustrates the configuration of the CSI–APF, a bridge structure topology that consists of a diode that is connected in series with a semiconductor self-commutating switching device (IGBT) [28]. The diode is used for reverse voltage blocking with the interfacing DC energy storage inductor in the system. This topology shows high value losses and requires high-value installed AC power capacitors. Furthermore, the dynamic response time is slow, and additional complex control is required to regulate the harmonic current.

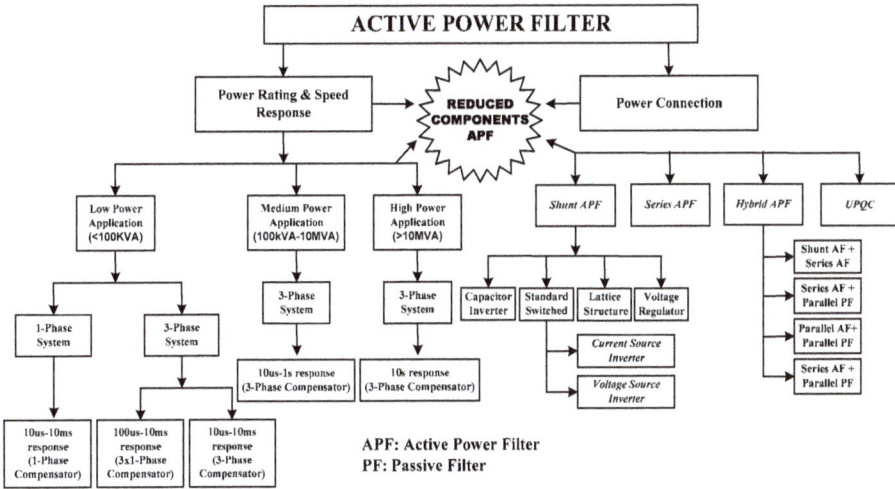

Figure 5. Hierarchical structure of APF classifications.

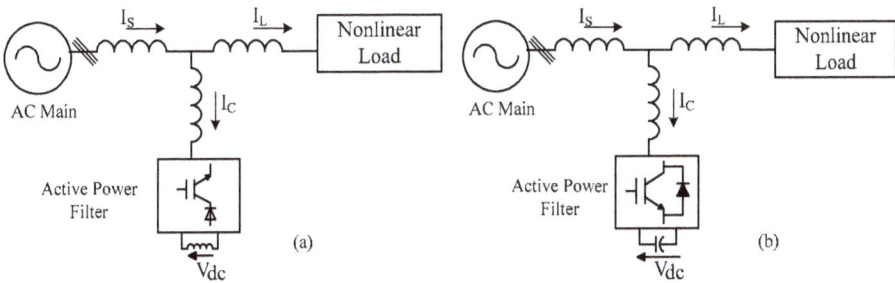

Figure 6. (a) Current-fed- and (b) voltage-fed-type inverters.

The VSI–APF is illustrated in Figure 6b. This inverter is installed at the load tail in the power network and fed from the energy storage capacitors. This category is more widely used than CSI–APFs and present advantages on PWM–CSI, such as easy installation, simple circuit, low cost, and convertibility to multistep versions with low switching frequencies. These inverters are generally used to eliminate current harmonics, compensate reactive power, and improve imbalanced current. The load phase-type connection is contingents upon three configurations, namely, two- (single phase), three- (three-phase without neutral), and four-wire (three-phase with neutral) configurations [103]. Furthermore, in terms of connections with main power systems, VSIs are divided into three configurations, namely, series, shunt, and arrangements of series and parallel as UPQC [58].

Figure 7a shows series APF topology [104]. The PWM waveform is injected in the system on an instantaneous basis to maintain a pure sinusoidal voltage waveform across the load. A matching transformer is used near the load end to eliminate voltage harmonics. This transformer balances the load terminal voltage and reduces the negative sequence voltage helping in controlling the voltage regulation and harmonic propagation caused by resonance in three-phase electric utilities. This configuration is less used in the industry than other configurations to address the limitation of handling high load currents, high system current ratings, losses, and filter sizes. Table 4 shows a critical technical and economic comparison of the power quality improvement techniques found in cited publications regarding shunt APFs [105].

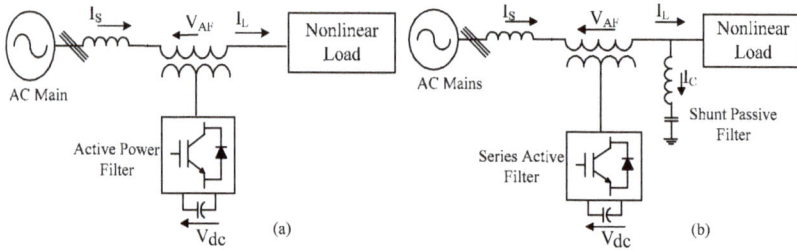

Figure 7. (**a**) Series and (**b**) hybrid APFs.

Table 4. Comparison of APFs in terms of power quality improvement techniques.

Attributes	Types of Filter				
	Passive Filter	Active Filter	Hybrid Filter	DSTATCOM	UPQC
Reactive power compensation	Poor	Good	Good	Excellent	Excellent
Harmonic suppression	Fixed	Adjustable	Fixed	Adjustable	Adjustable
Resonance	May exist	No	No	No	No
Load compensation	Not provided	Not provided	Not provided	Excellent	Good
Power rating of power converter	-	High	small	Highest	small
Power converter switches	-	6	4, 6	4, 6, 12	4, 6, 8, 12, 18, 24
Total cost	Lowest	High	Moderate	High	Highest

Cost depends upon the amount of switches.

Shunt Hybrid APFs

HAPFs are low-cost systems that compensate harmonic and voltage regulations. Such a system is a combination of passive and active filter devices, as depicted in Figure 5b. HAPFs are arranged into three topology configurations, namely, the combinations of parallel APF with shunt PF, series APF with shunt PF, and APF in series with shunt PF, as shown Figure 8a–c. In the series APF with shunt PF, the active filter is connected across the series to provide high impedance and isolation path for the harmonics to follow in the shunt PF. This configuration is commonly tested in medium-voltage level systems to provide reactive power, voltage harmonic compensation, and three-phase voltage balancing [106].

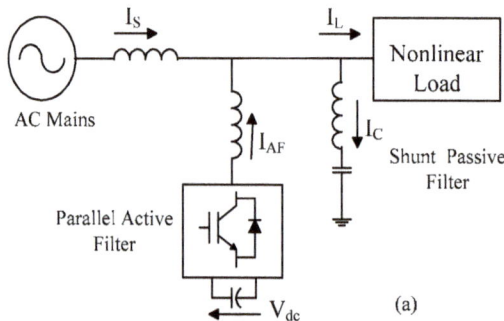

Figure 8. *Cont.*

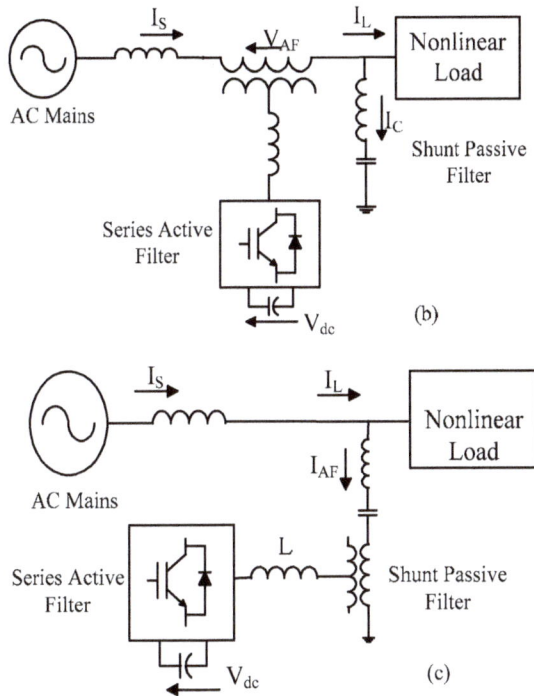

Figure 8. (**a**) Shunt APF with shunt PF; (**b**) Series APF with shunt PF; and (**c**) APF connected in series with shunt PF.

Table 5 summarizes the utilization of shunt APF and HAPF configuration for specific applications [11]. In the shunt APF with shunt PF configuration, both components are connected in shunt position to the main power supply. The shunt APF cancels the low-order current harmonics left by the shunt PF. Both systems are used to eliminate the fundamental reactive power and high-order load current harmonics in high-power systems. Nevertheless, such a system uses a large amount of passive components. Thus, this system is limited to single-load system applications. The series-connected APF with shunt PF is used in medium- and high-voltage applications, in which the DC-link voltage is maintained constant by the APF circuit. Moreover, PFs maintain the fundamental voltage component of the grid to a minimum value to reduce system size, cost, and voltage stress on active switches during filtering [107].

Table 5. Selection of shunt and hybrid APFs for specific applications.

Number	Application	Types of Filter			
		Series Active Filter	Shunt Active Filter	Hybrid Filter (Active Series and Passive Shunt)	Hybrid Filter (Active Series and Active Shunt)
1	Voltage harmonic compensation	*		*	*
2	Voltage flicker reduction	*	*		*
3	Removing voltage sags	*	*	*	*
4	Improving voltage regulation	*	*	*	*
5	Reactive power compensation		*	*	*
6	Current harmonic compensation		*	*	*
7	Neutral current compensation		*	*	

Table 5. *Cont.*

Number	Application	Types of Filter			
8	Improving load balancing		*		
9	(1 + 4)	*			*
10	(1 + 2 + 3 + 4)	*			*
11	(1 + 4 + 5 + 6)			*	*
12	(1 + 5)			*	*
13	(5 + 6)		*	*	*
14	(5 + 6 + 7 + 8)		*		
15	(5 + 6 + 8)		*		*
16	(6 + 8)		*		
17	(5 + 7 + 8)		*		

"*" indicates the filter system can preform the mention operation.

Circuits are further classified into three additional categories on the basis of the number of phase configurations, namely, single-phase two-wire (1P2W), three-phase three-wire (3P3W), and three-phase four-wire (3P4W) systems [108]. These circuit categories are applied at high frequency in low-, medium-, and high-power systems, respectively [68,109]. 1P2W filters are further classified as passive–passive [69], passive–active [70], and active–active systems [71,110,111]. Similarly, 3P3W filters are divided into passive–passive [72], passive–active [73], and active–active systems [67]. Finally, 3P4W filters are further classified as passive–passive, passive–active, and active–active systems [67,112]. Operating a three-phase power supply in a single-phase load system results in imbalanced neutral current and reactive power load problems. Therefore, many limitations are addressed by the four-wire hybrid configuration. However, the capacitor midpoint switch type is adopted for low-rating applications and the four-leg switch type is for controlling the neutral in the HAPF applications. Some of the practical limitations of shunt APFs are as follows:

- The initial installation cost is high.
- The control structure and design are considerably complex. Moreover, the increased harmonics and losses complicate filter control.
- With rapid dynamic current response and high-power rating system demand, the APF presents a design trade-off.

5. Advanced Classification of APF/STATCOM

Cost reduction is an important aspect in designing power converters, particularly in low-power-range industrial applications, such as power filters, variable-speed motor drives, uninterruptible power supplies, and static frequency changers [113]. Table 6 shows the results of testing a reduced number of power converters or inverters for minimizing losses and increasing system feasibility [114–117]. Moreover, the attributes and parameters of different DC–AC inverters are compared. On the basis of the total number of switches or reduced switch count configuration, modified APF inverters are classified as AC–AC power converter, parallel inverter, and split DC-leg inverter topologies.

Table 6. Comparison of different DC–AC power inverter topologies.

Topologies	Electrical Isolation	Efficiency (%)	Advantages	Disadvantages
Single bus inverter with two paralleled half bridge	No	-	-Minimum component count	-Large dc filter components
Dual bus inverter with two split half bridge single	No	-	-Reliability and flexibility	-High component count
phase 3 wire inverter	Yes	-	-Small passive component	-Complex control; for non-isolated circuit
Dual phase inverter with transformer	Yes	-	-Boosting capability	-Higher cost and size
Three-phase PWM inverter	Yes	~98%	-Simple design and control	-
High frequency link inverter	Yes	~96%	-Boosting capability	-Highly complex; higher cost and size
Z source inverter	No	~98%	-Boosting capability; save cost, no need for extra dc/dc converter	-Complex control; current stress is high
LLCC resonant inverter	No	~95%	-Lower current ripples; soft switching techniques	-Low power density; needs large volume and weight of resonant filter magnetic components

5.1. AC–AC Power Converter Topology

Figure 9a illustrates the system configuration of 3P3W AC–AC APF power converter topology that is connected with the distribution network and three-phase nonlinear harmonic-producing loads. The H-bridge converter circuit, represented by "H" is connected with a single DC-link capacitor (C_{dc}) [118,119]. This circuit consumes different numbers of switches installed in series and shunt and hybrid combinations of SAPF systems. The compensating current is injected into the AC distribution network at the PCC by coupling AC inductors or transformers [120]. The 3P3W AC–AC APF–VSI configuration consists of six and nine semiconductor devices, compared with the twelve switches in the conventional circuit [121–123]. The H-type APF topology, together with the DC-link capacitor, successfully reduces heavy switching power loss and minimizes voltage stress on active switches. However, this topology exhibits drawbacks in the form of DC-link voltage imbalance and minimum adept DC-link voltage [124].

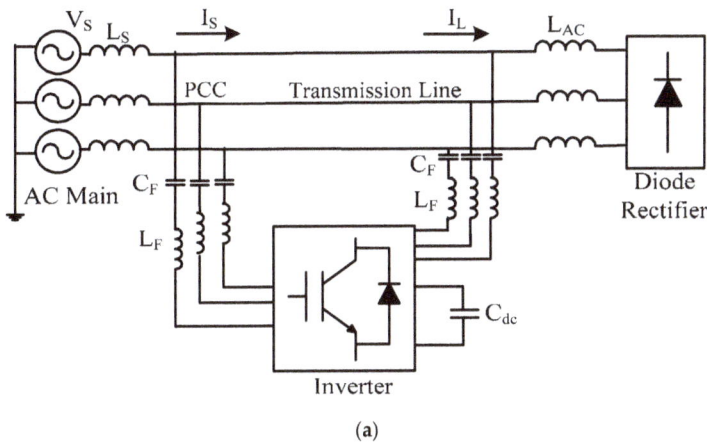

(a)

Figure 9. *Cont.*

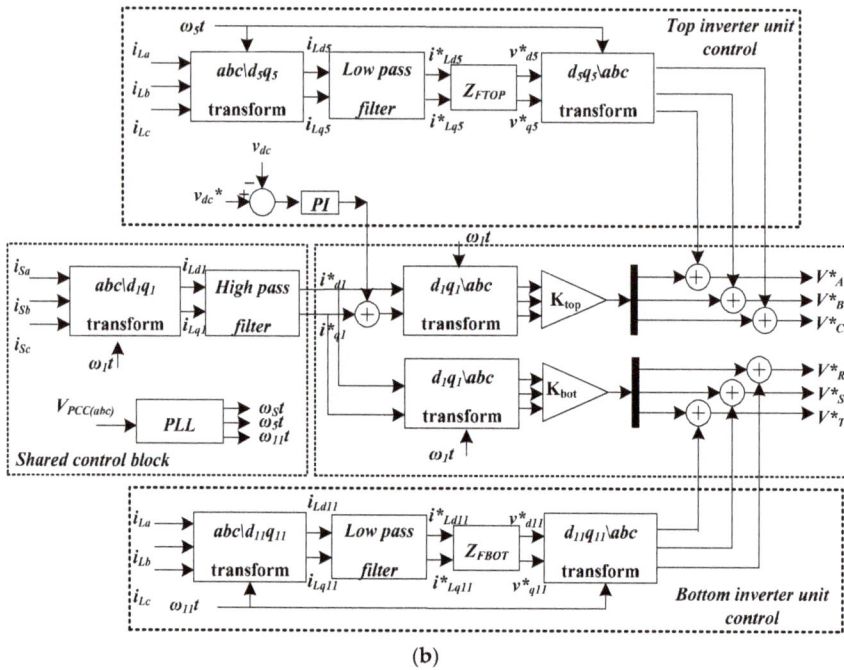

(b)

Figure 9. (a) AC–AC inverter topology and (b) control block of the proposed hybrid power filter based on SSTL inverter.

A six-switch inverter circuit, which consists of a two-leg inverter coupled with a single DC-link capacitor, is presented [125]. The number of semiconductor devices is reduced from a two-leg nine-switch inverter to a low-cost two-leg six-switch inverter circuit. The six- and nine-switch APF–VSI is connected in series with two passive LC filters in shunt position to the AC power transmission line without installing any matching transformer. The improved compact circuit design displays higher compensation results than the conventional 3P3W APF–VSI. A combination of feedback and feed-forward control loops is adopted as an advanced control scheme [126,127]. Figure 9b illustrates the control block diagram, which controls each unit and consists of three subsystems [128], which are a top and a bottom inverter unit and a shared control unit. Each PF is tuned at different harmonic frequencies to mitigate the 5th and 7th harmonics at the top inverter and the 11th and 13th harmonics at the bottom inverter. This PF also maintains the balance of DC-link voltage compensation. To reduce the DC-link current through the DC side of the system, a capacitor is installed between the DC-link poles and the PCC [129]. However, the shared control unit operates with PLL scheme, which consists of a simple and robust voltage pre-filter. This unit generates a quasi-square wave, thereby proving its advanced compensation and harmonic contents compared with those of conventional APF systems. Additionally, the H-bridge modular structure type reduces manufacturing cost and presents rapid production rate and high reliability. The controller should track a single capacitor voltage, thereby reducing the complexity of the voltage regulation, and requires additional voltage sensors in terms of multilevel inverter configurations [130,131].

5.2. Parallel-Inverter APF Topology

A dual-hybrid converter stage or parallel-inverter topology is depicted in Figure 10 [132]. In this topology, the rectifier and inverter are coupled with a parallel DC-link storage device in the APF topology but maintains a large number of switching devices. The conventional coupled back-to-back H-bridge inverter is a well-established, low-voltage configuration used in many industrial applications [133,134]. This inverter usually consists of twelve switches. On the contrary, the improved configuration eliminates four switches from each inverter, thereby reducing the total number of switches to eight; the scheme is adopted to eliminate a single leg in each power converter by connecting the third phase to the negative terminal of the individual VSI [135]. The new configuration increases system reliability and decreases voltage stress across the active switches.

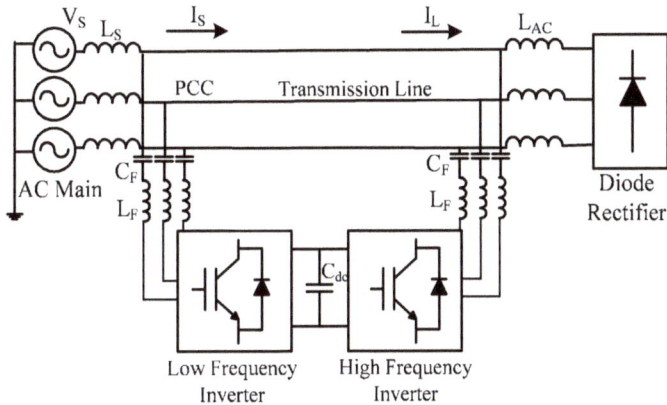

Figure 10. Parallel inverter topology.

A large-value DC capacitor is needed for stabilizing voltage balance across the DC-link capacitors and producing reactive power in high-power dynamic load systems. However, the DC-side energy storage component is the main limitation contributing to circuit failure. The DC energy storage components shorten the lifespan of the converter [136]. In addition, these components are expensive and largely fail in a power electronic circuit. Many factors, such as dissipation of heat, degradation of energy parameter, and high value of voltage capability, contribute to the high failure rate. The DC energy storage component is arranged in the aluminum electrolytic capacitors, which act as filters and energy buffers to the AC voltage ripples in the APF system. The aging of the aluminum electrolytic capacitors increases its internal resistance and contributes the most to frequent damage in operation [137]. In addition, these capacitors are expensive, large, and heavy and primarily cause power converter failure.

Switch reduction leads to a complex control and structure design for stopping the flow of zero-sequence current that circulate between the two power converters. Hence, the most suitable solution is installing a transformer or isolating the DC capacitors [135]. The parallel inverter topology presents limitations in the form of an oversized DC-link capacitor, restricted amplitude sharing, and limited phase shift at the output terminals. To overcome the issue of switching losses and electromagnetic interferences due to high switching frequency, soft switching techniques, such as zero-voltage transition and zero-current transition schemes, are adopted [138]. A control technique is implemented on the basis of feed-forward and feedback loops, as shown in Figure 9b. The low-frequency inverter (LFI), tuned at 550 Hz, mitigates low-order harmonics, and maintains the DC-link voltage and reactive power demand on the system constant. In the arrangement, the high-frequency inverter (HFI), tuned at 750 Hz, eliminates the remaining high-level harmonics. The control block functions in two loop modes, namely, feed-forward loop, which controls the LFI and improves the dynamic response from the system, and feedback loop, which operates the HFI and obtains high steady-state and harmonic compensation results. The Clarke and Park transformation technique is adopted to determine the cosine and sine components. The synchronous reference frame (SRF) method is used to generate the reference current for the parallel APF inverter topology [139]. Table 7 compares mainstream power converter topologies according to different topology parameters [140,141].

Table 7. Comparison of mainstream power converter topologies.

Series	Converter Topology Features	Diode Rectifier	2L-B2B VSC	ZSI	Multi-Level Converter	Matrix Converter	Nine Switch AC-AC Converter
1	Need controlled switches	None	Less	Less	Large	Large	Least
2	Circuit configuration	Simple	Simple	Simple	Complex	Complex	Simple
3	Cost	Very low	Moderate	High	Very high	high	Low
4	DC-link capacitor	Yes	Yes	Yes	Yes	No	Yes
5	Operational stages	Two	Two	Two	Two	One	One
6	Waveform quality	Good	Better	Better	Best	Better	Depends
7	Harmonic distortion	High	Moderate	Low	Least	Low	Depends
8	Switches losses	None	High	High	Low	Low	High
9	Conduction losses	Low	Low	Low	Highest	High	Low
10	Reliability	High	Low	High	Low	High	Low
11	Bi-directional power flow	No	Yes	Yes	Yes	Yes	Yes
12	Control complexity	Easy	Moderate	Moderate	Most complex	More complex	complex

5.3. Split DC-Leg Inverter Topology

Split DC-link topology provides a neutral common point for three-phase VSI, 3P3W, and 3P4W systems [142,143], and uses two connector pairs to split the single leg, thus providing a neutral path or midpoint connection [144,145], as depicted in Figure 11. The three-leg split capacitor and four-leg VSI-based topologies are the most demanding configurations for the 3P3W APF system. However, the two-level VSI configuration is inappropriate for filtering and harmonic compensation in high-power applications. The four-switch (B4) inverter uses four switches and four diodes, unlike the practical six-switch inverter (B6) [146]. Split DC-link topology uses few semiconductor devices, a feature that helps the neutral current, which consists of a small fundamental value of AC components. Table 8 compares the performance of the split DC-leg APF with the conventional APF topologies [147].

Table 8. Performance comparison of split DC-leg APF with conventional APF topologies.

Split DC-Link topology		Conventional Topology	
Advantages	Disadvantages	Advantages	Disadvantages
Simple design	Unequal voltage sharing in between the split capacitors legs	Handle unbalanced and nonlinear conditions	Need two or many extra switches
Fewer converter switches	Need an expensive capacitors	Low DC-bus voltage	Complicated control strategy
Simple and fast current tracking control	Unbalanced and nonlinear loads reason a split voltages perturbation	AC output voltage can be greater (about %15) than the output of split DC-link topology	-
-	Need a neutral point balancing technique	Lower ripple in the DC-link voltage	-

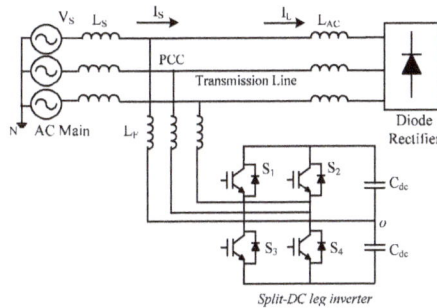

Figure 11. Four-switch DC-split voltage source inverter topology.

The drawback of split DC-link topology is its need for an expensive and large capacitor value to achieve equal voltage sharing between split capacitors [148]. Under severe imbalanced and nonlinear conditions, a large amount of neutral current flows through the neutral path, thus causing perturbation in the control scheme. However, due to its circuitry, split DC-link topology utilizes less expensive capacitors to provide a maximum available line–line peak voltage ($V_{dc}/2$) and maintain a low-ripple DC-link voltage. The dual-bridge inverter practices the B4 technique and eliminates variations in the current and voltage of DC-link capacitors. In 3P4W inverters, the AC voltage is 15% higher than in split DC-link inverters [149,150]. Hence, the three phase four-leg (3P4L) inverter shows superior performance under imbalanced and nonlinear conditions at the cost of a complicated control scheme. PWM or space vector modulation (SVM) techniques are adopted to generate reference signals for the PWM inverter. An overview of advanced core technologies for the power quality systems of high-power electronics is illustrated in Figure 12 [88,139].

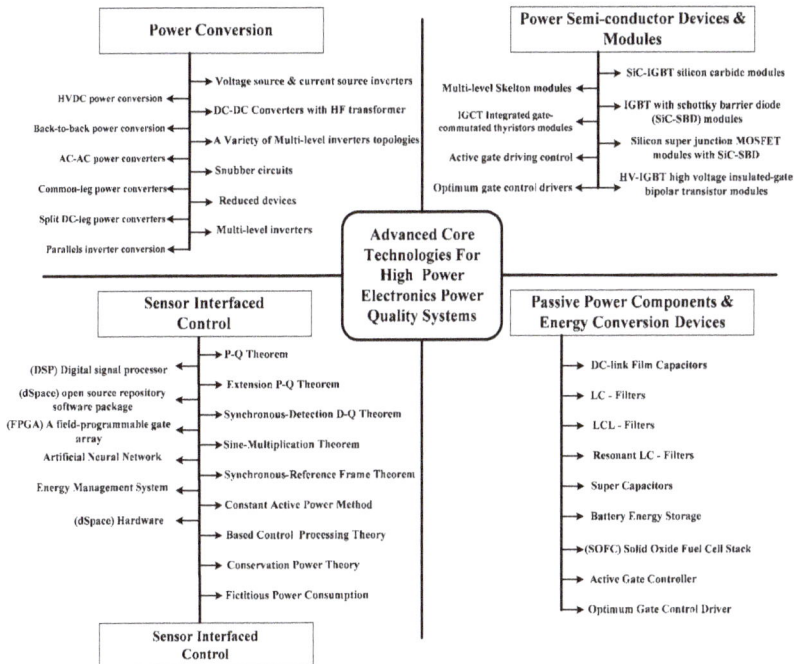

Figure 12. Overview of advanced core technologies for power quality systems of high-power electronics.

6. APFs/STATCOM Control Techniques

The primary operation of power converters depends on the control modulation strategies for controlling parameters such as switching losses, THD of the output current or voltage, amplitude, frequency and phase synchronization, sudden dynamic response, and power factor correction [151]. Generally, APF control is divided into two sections [152], specifically, reference signal estimation technique. As depicted in Figure 13, reference signal estimation is further classified into two sub-categories: current or voltage reference synthesis and current or voltage reference calculation [153].

Control signal techniques are sub-categorized further into two domains, namely, frequency-domain and time-domain as shown in Figure 13. However, depending upon open- and closed-loop concepts, closed-loop control is sub-classified into constant inductor current, constant capacitor voltage, linear voltage control, and optimization techniques [154]. These techniques are further arranged into modern control techniques for output voltage control, current regulation, harmonic filtering, and compensation in all critical circumstances [155,156].

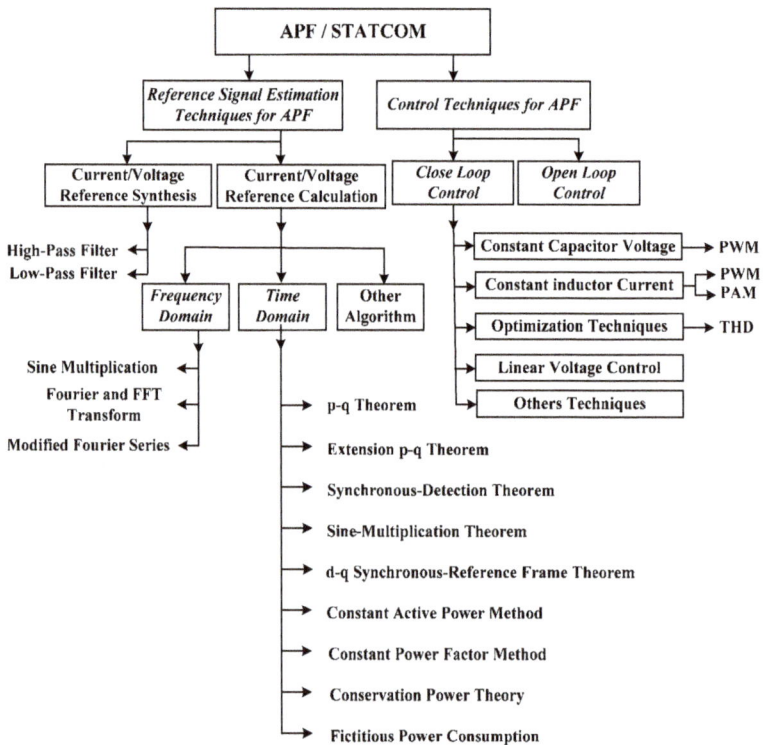

Figure 13. Different control techniques.

Frequency-domain techniques are more accurate and efficient than time-domain approaches for generating certain reference signals. Techniques such as Fourier transform analysis [157], sine multiplication technique [158], and modified Fourier series technique [159] estimate the existing harmonics in the system and generate reference signal in response to the present harmonics. However, compared with time-domain techniques, frequency-domain techniques are limited by their need for complex control circuitry. Moreover, these techniques require more time (one cycle) than time–domain methods to calculate Fourier coefficients, reference currents, and voltage signals for the sampling of variable coefficients. Meanwhile, time-domain techniques use instantaneous *p–q* theory [160], synchronous *d–q* reference frame theorem [161], synchronous detection theorem [120,162], constant power factor algorithm [163], PI controller [156], fuzzy controller [164], hysteresis controller [165], and sliding-mode controller [166]. Methods such as dividing frequency control [167] and extension *p–q* theorem [168] are implemented to generate the reference signals, which implement the *p–q* technique. Similarly, the synchronous *d–q* reference, known as SRF, practices the notch filter method [169] for calculating the voltage and current reference for the synchronously rotating reference frame, as in the case of the DC bus feedback voltage-fed APFs. Besides these control methods, other controllers are widely practiced in the operation of APF systems, including indirect current and adaptive DC-link control [47], PWM control [170,171], sine wave triangle PWM [172], deadbeat control [173], adaptive fuzzy dividing frequency controller [174], synchronous flux detection algorithm [175], neural network [176] and reactive current extraction [177].

7. Advanced Control Techniques for APFs/STATCOM

The reduction of switching components in any system leads to certain limitations. To overcome these, two advanced control techniques are adopted for operating the reduced switch count inverter, namely, SPWM and SVPWM, which operate in open- and closed-loop strategies, respectively. SVM is more complex than SPWM in optimizing control; switching schemes are selected according to the appropriate designs of the inverter and the topology [178]. This section provides a detailed review of the two advanced modulation methods and compares their advantages and limitations in matching the reduced switch count inverter topologies [171].

7.1. Sinusoidal Pulse Width Modulation

SPWM is one of the standard modulation techniques in switching power converters [179]. To control the gate switching signal for inverter operations, a low-frequency sinusoidal reference signal as comparator is compared with a high-frequency triangular carrier signal. The comparator output defines the operating range of switching orders. In addition, the key factor to consider in modulator design is amplitude distortion, which is caused by variations in the DC-voltage source. To control the desired line voltage frequency, the frequency of the modulating sinusoidal signal defines the inverter output. The amplitude distortion of the PWM waveforms stops the amplitude of the fundamental component and produces low-order harmonic contents [180]. However, THD and power dissipation are two of the key issues in high-power converter applications. Thus, a fundamental frequency SPWM control method is adopted to minimize switching losses and optimize harmonic contents. The SPWM modulation scheme can easily be implemented for single- and multiple-carrier applications. Multicarrier SPWM control techniques increase the performance of high-level inverters.

Sinusoidal SPWM is the most enhanced technique in PWM [181] and offers the key benefits of easy execution, low THD and harmonic output, and low switching losses. Similar to the six-switch converter, the SPWM is implemented in a reduced four-switch configuration. The comparators and the carrier signals are similar to those in the conventional SPWM; the only difference lies in the command to control the reference signal pattern for controlling the four-switch power converter. Reduced switch count inverters are not symmetrical; thus, the phase shift between the reference signals changes to compensate for the DC-bus voltage fluctuations [182]. Moreover, minimal single-phase current passes through DC-link capacitors but at the cost of complex control strategies and additional hardware. Phase voltages are preferred for the power converter presented in Figure 12, as discussed below:

$$v_a = V_m \sin(\omega t) \tag{1}$$

$$v_b = V_m \sin\left(\omega t - \frac{2\pi}{3}\right) \tag{2}$$

$$v_c = V_m \sin\left(\omega t + \frac{2\pi}{3}\right) \tag{3}$$

When the third-leg phase of the VSI has no control, the middle connection point of the DC-link (point O) is taken as the reference (as demonstrated in Figure 11).Therefore:

$$v_{an} = V_a - V_c = \sqrt{3}V_m \sin\left(\omega t - \frac{2\pi}{6}\right)$$ (4)

$$v_{bn} = V_b - V_c = \sqrt{3}V_m \sin\left(\omega t - \frac{2\pi}{2}\right)$$ (5)

$$v_{cn} = V_c - V_c = 0$$ (6)

The infinite value of the modulation bandwidth of periodic signal produces harmonics. Theoretically, these harmonics are neglected in SPWM control; however, the harmonics in the carrier and data modulation signal are filtered out by digital filters, such as Butterworth, Chebyshev, Bessel, Elliptic, and ITEA, which minimize the time integral and enhance the filtering functions to eliminate carrier signal harmonics [183].

7.2. Space Vector Pulse Width Modulation

SVPWM is an enhanced control method for reduced switch count inverters and directly uses the system parameters to determine the correct switching states and identify each switching vector [184,185]. SVPVM determines the duty cycle according to the modulation scheme. The switching vectors are divided into six sectors in the complex space plane of (a, b). All the sectors are separated by combining the turn-on and turn-off switching states of the power inverter. However, the lookup table and sector identification make SVM a complex scheme for determining the switching intervals for all vectors. The two adjacent switching state vectors are identified by the reference vectors. These vectors compute the turn-on and turn-off states of each switch; the sectors or subsectors find the switching sequences and increase the n-level of the power inverter, which requires a microprocessor and complex algorithms. The limitation of SVM is that it needs a considerable amount of time to carry out fundamental calculations, thereby causing delays in the process [186]. To overcome this problem, a large value of system reactive components and advanced deadbeat control are used [187]. The fundamental voltage ratios and harmonic compensation show better results in SVM schemes than in SPWM. Furthermore, SPWM also has a peak output voltage that is 15% higher than triangular carrier signal-type modulation techniques [188].

Generally, the SVM scheme is used in multilevel power inverter systems and employs a maximum number of different-level carrier waves to compare with the reference voltage signals, thus generating the n-level space vector that consists of per-sector and n-switching states for positive, zero, and negative switching sequences [189,190]. These vectors are divided among three groups, namely, small, middle, and large value vectors. Furthermore, the reference voltage depends on the voltage vectors and its dwelling times during the sampling period (T_s) for the output voltage. The switching status and the pole voltages are determined by the switching status and can be calculated as follows:

$$V_{NZ} = \frac{1}{3}(V_{AZ} + V_{BZ} + V_{CZ})$$ (7)

$$v_a = V_{AN} = V_{AZ} - V_{NZ}$$ (8)

$$v_a = V_{BN} = V_{BZ} - V_{NZ}$$ (9)

$$v_a = V_{CN} = V_{CZ} - V_{NZ}$$ (10)

The combinations of the two-leg four-switch (S_1–S_4) converters result in four space vectors. Expression (11) is used to calculate the vector representation of the three phase variables:

$$\begin{bmatrix} x_\alpha \\ x_\beta \end{bmatrix} = \frac{2}{3}\begin{bmatrix} 1 & -\frac{1}{2} & -\frac{1}{2} \\ 0 & \frac{\sqrt{3}}{2} & -\frac{\sqrt{3}}{2} \end{bmatrix}\begin{bmatrix} x_a \\ x_b \\ x_c \end{bmatrix}$$ (11)

Note that the B4 converter does not have zero vectors, unlike the B6 converter. Based on the calculation of the remaining active vectors, the three pole voltages do not have the same potential during the operation. To calculate the sectors, the volt–second integral is expressed in Expression (12):

$$\vec{V}T_{sw} = \vec{V}_1 t_1 + \vec{V}_2 t_2 + \vec{V}_3 t_3 + \vec{V}_4 t_4 \tag{12}$$

The vector decomposition of the above equation along with the time weights restriction can be written as:

$$\begin{aligned}
\vec{V}T_{sw}\cos\varphi &= -V_1 t_1 \cos\tfrac{\pi}{3} + V_2 t_2 \cos\tfrac{\pi}{6} + V_3 t_3 \cos\tfrac{\pi}{3} \\
\vec{V}T_{sw}\sin\varphi &= -V_1 t_1 \sin\tfrac{\pi}{3} - V_2 t_2 \sin\tfrac{\pi}{6} - V_3 t_3 \sin\tfrac{\pi}{3} \\
T_{sw} &= t_1 + t_2 + t_3
\end{aligned} \tag{13}$$

The modulation time is calculated by solving Expression (13) as follows:

$$t_2 = \frac{\sqrt{2}VT_{SW}}{2}\cos\left(\varphi - \frac{\pi}{6}\right) \tag{14}$$

$$t_1 = -t_2 + T_{SW}\left(\frac{1}{2} - \frac{\sqrt{2}V\sin(\varphi)}{E}\right) \tag{15}$$

$$t_3 = T_{SW} - t_1 - t_2 \tag{16}$$

The calculated vectors choose the sequence for SVM modulation with high symmetry and low switching frequency, as explained in study of [191,192]. Table 9 summarizes the evaluation of harmonic detection methods for APF systems and power grid-tied practical applications [193].

8. Performance Evaluation of APF/STATCOM System

A filtering system is needed to analyze the performance of APF/STATCOM applications under nonlinear load conditions. In the evaluation process, several features are important to calculate the performance of the APF system, such as THD of output current and voltage, transient and steady-state response, reactive power compensation and in some cases unity input power factor. The model specification and performance of the filter are evaluated in accordance with standards such as IEEE 519. The system consists of six active IGBT module inverters with a DC-link capacitor installed in the shunt position with the power grid system. All the experimental parameters and results are explain in the study [194], where a 2 kVA laboratory prototype test-rig is tested with the 5 kW three-phase diode rectifier load. The utility voltage, load current, and utility current waveforms of the system are seriously distorted and non-sinusoidal, thereby showing the nonlinear diode rectifier load as having a THD of 39.48%. According to the IEEE standard 519, the minimum THD value in utility current harmonics should be less than 5% and the preferred input unity power factor should not be less than 0.8. As explain in the study, all the waveforms are nearly sinusoidal after compensation and the THD of the utility current is reduced to 4.2%, with an output voltage THD of 3%. However, the rated DC-link capacitor is selected to keep the voltage ripple under 1% and maintain a DC voltage of 300 V, which is sufficient for proper and effective APF/STATCOM operation. To demonstrate the response time against the sudden changes in nonlinear load, the system is tested during the step load changes (step-on and step-off operations). The response time of the prototype is fast, achieving excellent performance during the load change (0 to 100%), and vice versa. Table 10 summarizes the market review of the available STATCOM, SVC and APF products for low and high power applications [195–201]. Traditionally, different manufacture sizes of STATCOM, SVC and APF devices are used to fulfil the power grid requirement [202–205].

Table 9. Evaluation of harmonic detection methods.

Parameters	Fast Fourier Transform FFT	Discrete Fourier Transform DFT	Recursive Discrete Fourier Transform RDFT	Synchronous Fundamental DQ Frame	Synchronous Individual Harmonic DQ Frame	Instantaneous Power PQ Theory	Generalized Integrators
Number of Sensors (For a Case of Three-Phase Application)	Three currents	Three currents	Three currents	Three currents, two/three voltages	Three currents, two/three voltages	Three currents, three voltages	Three currents
Number of Numerical Filters Required by the Harmonic Detection Algorithm	0	0	0	$2 \times$ HPFs	$2 \times$ LPFs $\times N^*$	$2 \times$ HPFs	$2 \times N^*$
Additional Tasks Required by the Harmonic Detection Algorithm	Windowing, synchronization	Windowing, synchronization		PLL	PLL	Voltage Preprocessing	·
Calculation Burden (Excluding the Numerical Filters)	-	-	+	+	-	+	-
Numerical Implementation Issues	Calculation Burden,	Calculation Burden,	Instability for low precision	Filtering	Filtering, Tuning	Filtering	Tuning control
Related Algorithms or Implementations	Similar FFT algorithms	·	Rotating frame	Filter type	Filter type	Filter type; other theories *pq*, *p-pq0*	Resonant filters type
Applications in Single- or Three-Phase Systems	Both1-ph/3-ph	Both1-ph/3-ph	Both1-ph/3-ph	Inherently 3-ph	Inherently 3-ph	Inherently 3-ph	Both 1-ph/3-ph
Usage of the Voltage Information in the Algorithm	No	No	No	Yes	Yes	Yes	No
Method's Performance for Unbalanced and Pre distorted Line Voltages	++	++	++	+	+	-	++
Method's Performance for Unbalanced Load Currents	++	++	++	+	++	++	+
Applied for Selective Harmonic Compensation	No	Yes	Yes	No	Yes	No	Yes
Transient Response Time	-	-	+	++	+	++	+
Steady-State Accuracy	+	+	+	-	+	+	-

"+" indicates an increase in performance; "—" indicates a decrease in performance.

Table 10. STATCOMs, SVCs and APFs technology products for high power applications available in the market.

Available Sizes of STATCOM			
Company	Product Name/Types	Voltage Level	Single Unit Capacity
ABB	PCS 6000 STATCOM	Several-Typical (11, 20, 21, 33, 138) KV	(6 . . . .16) MVAR
HITACHI	STATCOM	(66) KV	(20) MVAR
DONGFANG HITACHI (CD) ELECTRIC CONTROL EQUIPMENTS CO., LTD.	DHSTATCOM	(6) KV	(600 . . . .46000) KVAR
CONDENSATOR DOMINIT GMBH	KLARA-S, KLARA-M, KLARA-I	Several-Typical (400/525/690) V	(5/10, 6/12, 12/25) KVAR
GAMESA ELECTRIC	STATCOM	(11, 8 . . . ..34, 5) KV	(1, 5) MVAR
STATCOM SOLUTIONS PTY LTD	d105/d315	Several-Typical (200 . . . ..265) V	(5 . . . .15) KVA
ADF POWER TUNING	ADF P700 STATCOM	(6–36) KV	(1 . . . . . . 10) MVA
ADDNEW	STATCOM/SVG	(6, 10, 35) KV	(3) MVAR
AMSC	D-VAR	Up to (46) KV	(±2 . . . .100 s) MVAR
GAMESA	STATCOM-EN	(11.8 . . . .34.5) KV (Step-up Transformer)	(1.5) MVAR
MERUSPOWER	M-STATCOM (Merus M8000)	All voltages via Transformer	(1.3) MVAR
PONOVO	AccuVar ASVC	(3, 6, 10, 20, 35) KV	(±1 . . . ±18) MVAR ASVC-100 type (±10 . . . ±50) MVAR ASVC-200 type
S AND C ELECTRIC	The Purewave DSTATCOM	(0.48 . . . ..35) KV	(±1.23) MVAR/3.3 MVAR
SIEMENS	SVC Plus	Up to (36) KV (Transformer less)	(±25 . . . ..±50) MVAR
Available sizes of SVC			
ABB	SVC	(69) KV	(+50/−40) MVAR
GE Power	SVC	(33 . . . . 380) KV	(0 . . . .300) MVAR
ADDNEW	FC-TCR	(6, 10, 35) KV	(0 . . . .200) MVAR
ADDNEW	TSC	(6 . . . ..10) KV	(0.15 . . . .3) MVAR
ADDNEW	TCR	(6 . . . ..35) KV	(1 . . . .150) MVAR
PONOVO	SVC (FC-TCR)	(6 . . . ..66) KV	(0 . . . .400) MVAR
RXPE	TCR	(6, 10, 27.5, 35, 66) KV	(6 . . . .300) MVAR
SIEMENS	SVC classic (TSC-TCR)	(6 . . . . . . 800) KV	(40 . . . .800) MVAR
Available sizes of APF			
CONDENSATOR DOMINIT GMBH	NQ2501/NQ2502	Several-Typical (200–480, ±10%) V	(41.5 . . . . . . .41.5) KVA
ADF POWER TUNING	ADF P100-70/480, ADF P100-100/480, ADF P100-130/480, ADF P100-90/690	Several-Typical (208–480, 480–690) V	(49 . . . . . . .108) KVA
DELTA ELECTRONICS, INC	APF2000	(200–480) V	(22) KVA
SCHNEIDER	AccuSine PCS+ (LV active filters)	(380 . . . 690) V	(50 . . . .250) KVA
SCHAFFNER	FN3420 ECO sine active	(500–600) V	
SIEMENS	4RF1010-3PB0	(380–480) V	

9. Key Analysis on Configuration and Control Structure

Existing research on shunt APFs and STATCOM devices that focus on the reduction of components and their effects on the control strategies was reviewed. A typical analysis of the configurations and control structure techniques is presented below.

9.1. Limitations in Configuration Structure

The main limitations in the split DC-link power converter topology are as follows:

-   In B4 inverters, the third phase is connected clearly to the middle point or neutral point of the DC-link capacitors. The DC-bus current directly charges one of the capacitors and discharges the other. These dynamics unbalance current and voltage loading between the capacitors that discharge at a faster rate than the other, thus causing high current ripple in the imbalanced output waveform [206].
-   To compensate for the DC-bus voltage fluctuation issues [207], the removed single-leg terminal is connected to the negative terminal of the DC-bus PWM-VSI inverter and stops the imbalanced charging of the DC-link capacitors. Furthermore, the AC film capacitor stores the power ripples connected to the AC terminals to stop the flow of decoupling power ripples and provide balanced output currents and voltages [208].
-   A large DC-link voltage variation is shown in B8 split DC-leg converter applications. Both systems are operated at the same frequency and synchronized; thus, no fundamental current flows through the shared DC link. This outcome is a limitation in addition to the low AC voltage of the individual B4 power converter coupled with the shared DC-link capacitor.
-   In the three-phase system, a phase circulating current [209] flows through the DC-link capacitors. Thus, the capacitors are exposed to low-frequency harmonics, thereby limiting the use of high DC-link capacitor values. The AC–AC power converter configuration presents superior overall performance than the DC-bus midpoint configuration in terms of low THD and harmonic compensation capability because of the balanced current and voltage, as well as the minimum current ripple in the imbalanced output waveform.

9.2. Limitations in Control Structure Techniques

-   The need for voltage feed-forward and cross-coupling in SRF is the main limitation of the control structure. The phase angle of the grid voltage is required to start the control operation.
-   In the stationary reference frame, the PR controller reduces the complexity of the control structure in terms of current regulation as it has no need of the phase angle, unlike the dq-frame.
-   The adaptive band hysteresis controller increases the complexity of the control structure in the natural reference frame. However, the deadbeat controller simplifies the control scheme. Therefore, an individual control is required in each phase in case of individual phase PLLs and grid voltage to generate the current reference.
-   The hysteresis and deadbeat controllers do not consider low-order harmonics in the implementation process in harmonic compensators due to their fast dynamics.
-   In practical structures, both controllers require a sampling capability's hardware to compensate the positive sequence and need two filters, two transformation modules, and one controller, thus limiting its practical application in the dq-frame.
-   Table 11 illustrates the strengths and weaknesses of the APF control techniques [115,210,211].

Table 11. Strengths and weaknesses of APF/STATCOM control techniques.

METHOD	STRENGTH	WEAKNESS
PI-Controller	- Simple design and implementation - No complex circuit - Fast response time - Minimum Fluctuation	- Limitation of the control bandwidth with harmonic currents at high-frequency signals - Show limitation in the feedback system with constant parameters
Hysteresis Control	- light system of reckoning - Simple, robust, cost effective and easy implementation - Fastest control in transient - Fast speed of control loop - No need of oscillator or error amplifier	- High switching losses for small hysteresis band - Switching frequency variation and large frequency variations - Problem in filter design - Phases interferences - Resonance problems
Dead-Beat Control	- Fast control response - Wireless transmission - Fast transient response with low THD in the lower sampling frequency	- Control operation is depended on the data of the APF parameters - Requires a precise model of the filter to reach the desired performance. - Sensitive to the parametric variations of the controlled system and high THD for nonlinear loads
Reference Prediction	- Problem with static loads	- Filtering effect with load changing
Multi-rate Sampling	- Fast response - fast discretization in control variables	- Need fast control devices like FPGAs, AD converters - Slow-sampling rate
Phase-angle Correction	- Delay effect	- Take more time in calculation
One Cycle Control	- Simple design with flip-flops, comparators and clock - Better tracing transient waveform - Good time and dynamic response	- Difficult control implementation (Hardware) and modification - Complex hardware structure - Need an integrator at high speed
Adaptive Neural Network	- Does not require statistical training - No need underlying input data distribution - Can used a wide variety of functions and capture different patterns - They are the actual human operating system	- Difficult calculation and time consuming - Difficult to design and model analytically - Need a large sample size data packet - The system have a black box nature - They over-fit data outside of data training range (unpredictable)
Neural-Network Predicting Reference	- Active compensation among loads - Simple design and satisfactory accuracy - Mutual methods and faster control parameters adaptation	- Needed DC-link load measurement - Not implemented in unknown load - Needed large number of parameters - Performance evaluation, stuck in local minima
Selective Harmonics Compensation	- Parameters are frequency depended	- Power increase with the harmonic compensation

Table 11. *Cont.*

METHOD	STRENGTH	WEAKNESS
Master-Slave Control	- Discover and solve the problem in slave deadlock conflict affectively - Simple design and circuit - Better efficiency among the master-slave systems	- Master should be built robust and powerful - All system is depended upon the master - Complex and difficult master system - Need a high cost in communication - Reduction in real time abilities
Predictive Control	- Possibility to include nonlinearities of the system [212]. - Used to minimize switching frequency for high-power inverters and maintaining the current error within a specified bound [213]. - Allows achieving more precise current control with minimum THD and harmonic noise.	- Requires a precise model of the filter to reach the desired performance. - This method needs a lot of calculations.
Sliding Mode Control (SMC)	Exhibits reliable performance during transients. Shows an acceptable THD if it is designed well.	- The problem of the Chattering Phenomenon in discrete implementation. - The difficulty of designing a controller for both a good transient and zero steady state performance.
Fuzzy Control Methods	- Insensitive to parametric variations and operation points. - Sophisticated technique, easy to design and implement a large-scale nonlinear system.	Slow control method.
Repetitive Controller (RC)	These controllers are implemented as harmonic compensators and current controllers. They show robust performance for periodic disturbances and ensure a zero steady-state error at all the harmonic frequencies.	Is not easy to stabilize for all unknown load disturbances and cannot obtain very fast response for fluctuating load.

9.3. Key Findings

The following are the shortcomings of the current research as revealed by the review:

(1) In parallel inverter topology, the output voltage per phase at different frequencies generates transitions, which block the forbidden states. This voltage effectively limits the range of reference amplitudes and phase shifts.

(2) Generally, in reduced switch count power converters, the modulation strategy adopted is SPWM to switch and compensate for the DC-bus voltage fluctuation issues [214]. By contrast, in reduced switch count converters, the phase shift does not track the three-phase balance reference signal in the symmetry order.

(3) Switch reduction generally leads to interdependencies between AC input and output frequencies, unlike full-bridge converters. This restriction limits the references for modulation in operating the power converters at the same frequency. Voltage doubling and semiconductor stress are not issues in the B4 converter, unlike in the nine-switch H6 converter, because of the favorable maximum modulation ratio of unity [215].

(4) Reduced switch count (four-switch) topologies face more limitations in their switching states than conventional six-switch converters. Findings indicate that the removed leg terminal that is connected to either the upper positive DC-link terminal or the lower negative DC-link terminal is not achievable.

(5)   In the B6 converter, two switching states, (0, 0) and (1, 1), are stated as zero vectors, which stop the flow of the current toward the load. In the B4 converter, the current flows even in zero-vector states. Therefore, in two other switching states, (0, 1) and (1, 0), the resulting uncontrolled current flows through the common phase because of the direct connection between the DC-link capacitor and the AC terminal.

(6)   The PLL synchronizes the power inverter modulation to the power grid and provides freedom in designing the modulation index caused by phasing the angle in between the grids and by modulating waves to adjust the maximum magnitude for unity output.

(7)   Eliminating the active switches creates an unequal thermal distribution among the remaining switches at the expense of reduced structure, conduction losses, switching losses, and low system cost.

(8)   In the split converter, the third-phase current flows directly through DC-link capacitors, thus exposing the converter to low-frequency harmonics, which need a high-value-rated capacitor.

(9)   In the two-leg rectifier (multiply by 2/pi = 0.6), the output power gain is lower than that of the three-leg rectifier (multiply by 1.6), thereby increasing the current rating of the active switching components.

10. Upcoming Trends

The new trend in the field of power electronics aims to minimize the number of power semiconductor components, such as IGBT switches, to reduce the overall price of power converter devices. Designing cost-effective topologies on the basis of the reduced number of semiconductor devices in the range of 10 kilowatts and above has always been attractive to researchers. From these facts, the transformerless system is showing important developments toward a mature level. With advancements in microprocessors, controllers, and fast switching devices, long-lasting and proficient APF/STATCOM systems have been proposed with highly rated megawatt ranges, improved performance, enhanced efficiency, and most importantly, low costs for varying applications. Moreover, reduced switch count inverter topologies limit the cost, size, switching losses, and complexity of the control structure, as well as the algorithm and interface circuits. Similarly, new growth in efficient modulation techniques can help guarantee high reliability, fast transient and dynamic response, low THD, excellent harmonic and reactive power compensation, and current and voltage regulation of power electronics.

New trends in hybrid topologies aim to develop advanced APF and STATCOM systems that have low-rated power component systems and added dual functionality for improved performance. Similarly, a technology for achieving new growth in multilevel power inverters on the basis of the reduced number of components is becoming a popular research direction. The excessive penetration of renewable devices in the power transmission network creates various power quality challenges for engineers and researchers. However, FACTS devices, such as STATCOMs, SVCs, and DSTATCOMs, have been successful in mitigating the issues of power quality, including voltage sags, transients, harmonics, and damping oscillations. Further research and improvements in APF technology have recently been performed in terms of dual-terminal inverters, shared legs between inverters, and rectifiers to substitute for split-capacitor configurations. Consequently, next-generation power semiconductor devices and packaging of silicon carbide (*SiC*) IGBT power modules with Schottky barrier diode modules, integrated gate-commutated thyristor modules, and SiC MOSFET can be effectively used.

11. Conclusions

This paper discussed the topological and control schemes of APF and STATCOM devices. A transformerless and reduced switch count structure for power converters and control strategies were reviewed. A discussion about development stages in configuration with respect to low cost, low volumetric size and weight, compact structure, and high reliability was given. Different grid-connected control schemes, and their implementation structures, as well as the extraction of harmonic reference signals, were presented. This review discussed different devices, namely, PFs, shunt APFs, shunt hybrid filters, STATCOMs, SVCs, UPQCs, and DSTATCOMs, all of which are typically used to enhance power quality and mitigation of load harmonics. Dual-terminal inverters, transformerless, multilevel inverters, shared legs between inverters, shared legs between rectifiers, split capacitor configurations, and new-generation semiconductor devices deserve future research attention. Finally, simulation and experimental results verified the effectiveness of APF/STATCOM topologies for harmonic mitigation in accordance with IEEE-519.

Author Contributions: All the authors contributed in completing the Paper. The main idea of the paper was presented by W.U.K.T. while M.A., M.S. and B.H. helped in accumulating the research data in this review paper. M.A.M. and N.B. prepared the format of the paper. S.M. and M.N. is the head of research team.

Acknowledgments: The authors would like to acknowledge the financial support received from the University of Malaya, Malaysia, through Frontier Research Grant No. FG007-17AFR and Innovative Technology Grant No. RP043B-17AET.

Conflicts of Interest: The authors declare no conflict of interest.

Abbreviations

APF	Active power filters
B4	Four-switch inverter
CSI	Current-fed-type inverters
CHB	Cascaded H-bridge
DFT	Discrete Fourier transform
DSP	Digital signal processor
dSpace	Digital Signal Processor for Applied and Control Engineering
DSTATCOM	Distribution STATCOM
DQ	Synchronous Fundamental Frame
DVR	Dynamic voltage restorer
ESS	Energy storage system
FACTS	Flexible AC transmission system
FFT	Fast Fourier transform
GPGA	Field programmable gate array
HAPFs	Hybrid APF
HF	High frequency
HPF	High pass filter
HV	High voltage
IEEE	Institute of Electrical and Electronics Engineers
IEC	International Electro-technical Commission
IGBT	Insulated-gate bipolar transistors
ITC	Indirect torque control
Ki	Integral gain
Kp	Proportional gain
LPF	Low pass filter
LVRT	Low-voltage ride through
MLI	Multilevel inverter
MOSFET	Metal-oxide-semiconductor field-effect transistor
MPP	Maximum power point
PCC	Point of common coupling
PF	Passive filter
PI	Proportional integral controller
PLL	Phase locked loop
PQ	Instantaneous power theory
PV	Photovoltaic

PWM	Pulse width modulation
RC	Repetitive Controller
RDFT	Recursive discrete Fourier Transform
SAPF	Shunt active power filter
SBD	Schottky barrier diode
SHE	Selective harmonic elimination
SiC	Silicon carbide
SMC	Sliding Mode Control
SOFC	Solid oxide fuel call
SPWM	Sinusoidal Pulse Width Modulation
SRF	Synchronous-reference-frame
STATCOM	static compensator
SVC	Static volt-ampere reactive VAR compensator
SVM	Space vector modulation
SVPWM	Space vector Pulse Width Modulation
TCR	Thyristor-controlled resistor
THD	Total Harmonic Distortion
UPQC	Unified power quality conditioner (UPQC)
VSC	Voltage Source Converter
VSI	Voltage-fed-type inverters
WT	Wind turbine
1P2W	Single-phase two-wire
3P3W	Three-phase three-wire
3P4W	Three-phase four-wire
3P4L	Three phase four-leg

References

1. Singh, B.; Al-Haddad, K.; Chandra, A. A review of active filters for power quality improvement. *IEEE Trans. Ind. Electron.* **1999**, *46*, 960–971. [CrossRef]
2. Hamadi, A.; Rahmani, S.; Al-Haddad, K. A Hybrid Passive Filter Configuration for VAR Control and Harmonic Compensation. *IEEE Trans. Ind. Electron.* **2010**, *57*, 2419–2434. [CrossRef]
3. Bhattacharya, A.; Chakraborty, C.; Bhattacharya, S. Shunt compensation. *IEEE Ind. Electron. Mag.* **2009**, *3*, 38–49. [CrossRef]
4. Sirjani, R.; Rezaee Jordehi, A. Optimal placement and sizing of distribution static compensator (D-STATCOM) in electric distribution networks: A review. *Renew. Sustain. Energy Rev.* **2017**, *77*, 688–694. [CrossRef]
5. Beres, R.N.; Wang, X.; Liserre, M.; Blaabjerg, F.; Bak, C.L. A review of passive power filters for three-phase grid-connected voltage-source converters. *IEEE J. Emerg. Sel. Top. Power Electron.* **2016**, *4*, 54–69. [CrossRef]
6. Ringwood, J.V.; Simani, S. Overview of modelling and control strategies for wind turbines and wave energy devices: Comparisons and contrasts. *Annu. Rev. Control* **2015**, *40*, 27–49. [CrossRef]
7. Ahmed, K.H.; Finney, S.J.; Williams, B.W. Passive filter design for three-phase inverter interfacing in distributed generation. In Proceedings of the Compatibility in Power Electronics, Gdansk, Poland, 29 May–1 June 2007; pp. 1–9.
8. Lam, C.S.; Choi, W.H.; Wong, M.C.; Han, Y.D. Adaptive DC-Link Voltage-Controlled Hybrid Active Power Filters for Reactive Power Compensation. *IEEE Trans. Power Electron.* **2012**, *27*, 1758–1772. [CrossRef]
9. Espi, J.; Garcia-Gil, R.; Castello, J. Capacitive Emulation for LCL-Filtered Grid-Connected Converters. *Energies* **2017**, *10*, 930. [CrossRef]
10. Litran, S.P.; Salmeron, P. Analysis and design of different control strategies of hybrid active power filter based on the state model. *IET Power Electron.* **2012**, *5*, 1341–1350. [CrossRef]
11. Prakash Mahela, O.; Gafoor Shaik, A. Topological aspects of power quality improvement techniques: A comprehensive overview. *Renew. Sustain. Energy Rev.* **2016**, *58*, 1129–1142. [CrossRef]
12. Mithulananthan, N.; Canizares, C.A.; Reeve, J.; Rogers, G.J. Comparison of PSS, SVC, and STATCOM controllers for damping power system oscillations. *IEEE Trans. Power Syst.* **2003**, *18*, 786–792. [CrossRef]

13. Colak, I.; Kabalci, E.; Fulli, G.; Lazarou, S. A survey on the contributions of power electronics to smart grid systems. *Renew. Sustain. Energy Rev.* **2015**, *47*, 562–579. [CrossRef]

14. Singh, B.; Solanki, J. A Comparison of Control Algorithms for DSTATCOM. *IEEE Trans. Ind. Electron.* **2009**, *56*, 2738–2745. [CrossRef]

15. Singh, B.; Mukherjee, V.; Tiwari, P. A survey on impact assessment of DG and FACTS controllers in power systems. *Renew. Sustain. Energy Rev.* **2015**, *42*, 846–882. [CrossRef]

16. Tokiwa, A.; Yamada, H.; Tanaka, T.; Watanabe, M.; Shirai, M.; Teranishi, Y. New Hybrid Static VAR Compensator with Series Active Filter. *Energies* **2017**, *10*, 1617. [CrossRef]

17. Qazi, S.H.; Mustafa, M.W. Review on active filters and its performance with grid connected fixed and variable speed wind turbine generator. *Renew. Sustain. Energy Rev.* **2016**, *57*, 420–438. [CrossRef]

18. Tareen, W.U.K.; Mekhilef, S.; Nakaoka, M. A transformerless reduced switch counts three-phase APF-assisted smart EV charger. In Proceedings of the 2017 IEEE Applied Power Electronics Conference and Exposition (APEC), Tampa, FL, USA, 26–30 March 2017; pp. 3307–3312.

19. Jordehi, A.R. Particle swarm optimisation (PSO) for allocation of FACTS devices in electric transmission systems: A review. *Renew. Sustain. Energy Rev.* **2015**, *52*, 1260–1267. [CrossRef]

20. Shmilovitz, D. On the definition of total harmonic distortion and its effect on measurement interpretation. *IEEE Trans. Power Deliv.* **2005**, *20*, 526–528.

21. Barros, J.; Diego, R.I. A review of measurement and analysis of electric power quality on shipboard power system networks. *Renew. Sustain. Energy Rev.* **2016**, *62*, 665–672. [CrossRef]

22. Lascu, C.; Asiminoaei, L.; Boldea, I.; Blaabjerg, F. High Performance Current Controller for Selective Harmonic Compensation in Active Power Filters. *IEEE Trans. Power Electron.* **2007**, *22*, 1826–1835. [CrossRef]

23. Kanjiya, P.; Khadkikar, V.; Zeineldin, H.H. Optimal Control of Shunt Active Power Filter to Meet IEEE Std. 519 Current Harmonic Constraints Under Nonideal Supply Condition. *IEEE Trans. Ind. Electron.* **2015**, *62*, 724–734. [CrossRef]

24. Bollen, M.H. What is power quality? *Electr. Power Syst. Res.* **2003**, *66*, 5–14. [CrossRef]

25. Mahela, O.P.; Shaik, A.G.; Gupta, N. A critical review of detection and classification of power quality events. *Renew. Sustain. Energy Rev.* **2015**, *41*, 495–505. [CrossRef]

26. Khadem, S.K.; Basu, M.; Conlon, M.F. Parallel operation of inverters and active power filters in distributed generation system—A review. *Renew. Sustain. Energy Rev.* **2011**, *15*, 5155–5168. [CrossRef]

27. Wang, Y.; Kuckelkorn, J.; Zhao, F.-Y.; Spliethoff, H.; Lang, W. A state of art of review on interactions between energy performance and indoor environment quality in Passive House buildings. *Renew. Sustain. Energy Rev.* **2017**, *72*, 1303–1319. [CrossRef]

28. Büyük, M.; Tan, A.; Tümay, M.; Bayındır, K.Ç. Topologies, generalized designs, passive and active damping methods of switching ripple filters for voltage source inverter: A comprehensive review. *Renew. Sustain. Energy Rev.* **2016**, *62*, 46–69. [CrossRef]

29. Wu, J.-C.; Jou, H.-L.; Wu, K.-D.; Hsiao, H.-H. Three-phase four-wire hybrid power filter using a smaller power converter. *Electr. Power Syst. Res.* **2012**, *87*, 13–21. [CrossRef]

30. Bouzelata, Y.; Kurt, E.; Altın, N.; Chenni, R. Design and simulation of a solar supplied multifunctional active power filter and a comparative study on the current-detection algorithms. *Renew. Sustain. Energy Rev.* **2015**, *43*, 1114–1126. [CrossRef]

31. Mehrasa, M.; Pouresmaeil, E.; Zabihi, S.; Rodrigues, E.M.G.; Catalão, J.P.S. A control strategy for the stable operation of shunt active power filters in power grids. *Energy* **2016**, *96*, 325–334. [CrossRef]

32. Mohd Zainuri, M.; Mohd Radzi, M.; Che Soh, A.; Mariun, N.; Abd Rahim, N.; Teh, J.; Lai, C.-M. Photovoltaic Integrated Shunt Active Power Filter with Simpler ADALINE Algorithm for Current Harmonic Extraction. *Energies* **2018**, *11*, 1152. [CrossRef]

33. Qiao, W.; Harley, R.G.; Venayagamoorthy, G.K. Coordinated Reactive Power Control of a Large Wind Farm and a STATCOM Using Heuristic Dynamic Programming. *IEEE Trans. Energy Convers.* **2009**, *24*, 493–503. [CrossRef]

34. De la Villa Jaen, A.; Acha, E.; Exposito, A.G. Voltage Source Converter Modeling for Power System State Estimation: STATCOM and VSC-HVDC. *IEEE Trans. Power Syst.* **2008**, *23*, 1552–1559. [CrossRef]

35. Arsoy, A.B.; Liu, Y.; Ribeiro, P.F.; Wang, F. StatCom-SMES. *IEEE Ind. Appl. Mag.* **2003**, *9*, 21–28. [CrossRef]

36. Pathak, A.K.; Sharma, M.P.; Bundele, M. A critical review of voltage and reactive power management of wind farms. *Renew. Sustain. Energy Rev.* **2015**, *51*, 460–471. [CrossRef]

37. Singh, B.; Saha, R.; Chandra, A.; Al-Haddad, K. Static synchronous compensators (STATCOM): A review. *IET Power Electron.* **2009**, *2*, 297–324. [CrossRef]

38. Adamczyk, A.G.; Teodorescu, R.; Rodriguez, P.; Mukerjee, R.N. FACTS devices for large wind power plants. In Proceedings of the EPE Wind Energy Chapter Symposium, Stafford, UK, 15–16 April 2010.

39. Belouda, M.; Jaafar, A.; Sareni, B.; Roboam, X.; Belhadj, J. Integrated optimal design and sensitivity analysis of a stand alone wind turbine system with storage for rural electrification. *Renew. Sustain. Energy Rev.* **2013**, *28*, 616–624. [CrossRef]

40. Mahela, O.P.; Shaik, A.G. A review of distribution static compensator. *Renew. Sustain. Energy Rev.* **2015**, *50*, 531–546. [CrossRef]

41. Siemens. *Flexible AC Transmission Systems (FACTS), Parallel Compensation, Comprehensive Solutions for Safe and Reliable Grid Operation*; Siemens: Munich, Germany, 2016.

42. Hingorani, N.G. Flexible AC transmission. *IEEE Spectr.* **1993**, *30*, 40–45. [CrossRef]

43. Habur, K.; O'Leary, D. *FACTS—Flexible Alternating Current Transmission Systems: For Cost Effective and Reliable Transmission of Electrical Energy*; Siemens-World Bank Document-Final Draft Report; Siemens: Erlangen, Germany, 2004.

44. Norambuena, M.; Rodriguez, J.; Kouro, S.; Rathore, A. A novel multilevel converter with reduced switch count for low and medium voltage applications. In Proceedings of the 2017 IEEE Energy Conversion Congress and Exposition (ECCE), Cincinnati, OH, USA, 1–5 October 2017; pp. 5267–5272.

45. Vijayaraja, L.; Kumar, S.G.; Rivera, M. A review on multilevel inverter with reduced switch count. In Proceedings of the 2016 IEEE International Conference on Automatica (ICA-ACCA), Curico, Chile, 19–21 October 2016; pp. 1–5.

46. Vavilapalli, S.; Padmanaban, S.; Subramaniam, U.; Mihet-Popa, L. Power Balancing Control for Grid Energy Storage System in Photovoltaic Applications—Real Time Digital Simulation Implementation. *Energies* **2017**, *10*, 928. [CrossRef]

47. Sridhar, V.; Umashankar, S. A comprehensive review on CHB MLI based PV inverter and feasibility study of CHB MLI based PV-STATCOM. *Renew. Sustain. Energy Rev.* **2017**, *78*, 138–156. [CrossRef]

48. Chang, W.-N.; Liao, C.-H. Design and Implementation of a STATCOM Based on a Multilevel FHB Converter with Delta-Connected Configuration for Unbalanced Load Compensation. *Energies* **2017**, *10*, 921. [CrossRef]

49. Varma, R.K.; Khadkikar, V.; Seethapathy, R. Nighttime Application of PV Solar Farm as STATCOM to Regulate Grid Voltage. *IEEE Trans. Energy Convers.* **2009**, *24*, 983–985. [CrossRef]

50. Varma, R.K.; Rahman, S.A.; Vanderheide, T. New Control of PV Solar Farm as STATCOM (PV-STATCOM) for Increasing Grid Power Transmission Limits During Night and Day. *IEEE Trans. Power Deliv.* **2015**, *30*, 755–763. [CrossRef]

51. Varma, R.K.; Das, B.; Axente, I.; Vanderheide, T. Optimal 24-hr utilization of a PV solar system as STATCOM (PV-STATCOM) in a distribution network. In Proceedings of the 2011 IEEE Power and Energy Society General Meeting, San Diego, CA, USA, 24–29 July 2011; pp. 1–8.

52. Junbiao, H.; Solanki, S.K.; Solanki, J.; Schoene, J. Study of unified control of STATCOM to resolve the Power quality issues of a grid-connected three phase PV system. In Proceedings of the 2012 IEEE PES Innovative Smart Grid Technologies (ISGT), Washington, DC, USA, 16–20 January 2012; pp. 1–7.

53. Seo, H.R.; Kim, G.H.; Jang, S.J.; Kim, S.Y.; Park, S.; Park, M.; Yu, I.K. Harmonics and reactive power compensation method by grid-connected Photovoltaic generation system. In Proceedings of the 2009 International Conference on Electrical Machines and Systems, Tokyo, Japan, 15–18 November 2009; pp. 1–5.

54. Demirdelen, T.; Kayaalp, R.İ.; Tumay, M. Simulation modelling and analysis of modular cascaded multilevel converter based shunt hybrid active power filter for large scale photovoltaic system interconnection. *Simul. Model. Pract. Theory* **2017**, *71*, 27–44. [CrossRef]

55. Varma, R.K.; Rahman, S.A.; Sharma, V.; Vanderheide, T. Novel control of a PV solar system as STATCOM (PV-STATCOM) for preventing instability of induction motor load. In Proceedings of the 2012 25th IEEE Canadian Conference on Electrical and Computer Engineering (CCECE), Montreal, QC, Canada, 29 April–2 May 2012; pp. 1–5.

56. Toodeji, H.; Farokhnia, N.; Riahy, G.H. Integration of PV module and STATCOM to extract maximum power from PV. In Proceedings of the 2009 International Conference on Electric Power and Energy Conversion Systems, Sharjah, UAE, 10–12 November 2009; pp. 1–6.

57. Luo, L.; Gu, W.; Zhang, X.-P.; Cao, G.; Wang, W.; Zhu, G.; You, D.; Wu, Z. Optimal siting and sizing of distributed generation in distribution systems with PV solar farm utilized as STATCOM (PV-STATCOM). *Appl. Energy* **2018**, *210*, 1092–1100. [CrossRef]

58. Zeng, Z.; Yang, H.; Zhao, R.; Cheng, C. Topologies and control strategies of multi-functional grid-connected inverters for power quality enhancement: A comprehensive review. *Renew. Sustain. Energy Rev.* **2013**, *24*, 223–270. [CrossRef]

59. Hassaine, L.; Olias, E.; Quintero, J.; Salas, V. Overview of power inverter topologies and control structures for grid connected photovoltaic systems. *Renew. Sustain. Energy Rev.* **2014**, *30*, 796–807. [CrossRef]

60. Karimi, M.; Mokhlis, H.; Naidu, K.; Uddin, S.; Bakar, A.H.A. Photovoltaic penetration issues and impacts in distribution network—A review. *Renew. Sustain. Energy Rev.* **2016**, *53*, 594–605. [CrossRef]

61. Mahmud, N.; Zahedi, A. Review of control strategies for voltage regulation of the smart distribution network with high penetration of renewable distributed generation. *Renew. Sustain. Energy Rev.* **2016**, *64*, 582–595. [CrossRef]

62. Kow, K.W.; Wong, Y.W.; Rajkumar, R.K.; Rajkumar, R.K. A review on performance of artificial intelligence and conventional method in mitigating PV grid-tied related power quality events. *Renew. Sustain. Energy Rev.* **2016**, *56*, 334–346. [CrossRef]

63. Vivas, J.H.; Bergna, G.; Boyra, M. Comparison of multilevel converter-based STATCOMs. In Proceedings of the 2011 14th European Conference on Power Electronics and Applications, Birmingham, UK, 30 August–1 September 2011; pp. 1–10.

64. Agrawal, R.; Jain, S. Comparison of reduced part count multilevel inverters (RPC-MLIs) for integration to the grid. *Int. J. Electr. Power Energy Syst.* **2017**, *84*, 214–224. [CrossRef]

65. Najjar, M.; Moeini, A.; Bakhshizadeh, M.K.; Blaabjerg, F.; Farhangi, S. Optimal Selective Harmonic Mitigation Technique on Variable DC Link Cascaded H-Bridge Converter to Meet Power Quality Standards. *IEEE J. Emerg. Sel. Top. Power Electron.* **2016**, *4*, 1107–1116. [CrossRef]

66. Haw, L.K.; Dahidah, M.S.A.; Almurib, H.A.F. SHE-PWM Cascaded Multilevel Inverter With Adjustable DC Voltage Levels Control for STATCOM Applications. *IEEE Trans. Power Electron.* **2014**, *29*, 6433–6444. [CrossRef]

67. Song, W.; Huang, A.Q. Fault-Tolerant Design and Control Strategy for Cascaded H-Bridge Multilevel Converter-Based STATCOM. *IEEE Trans. Ind. Electron.* **2010**, *57*, 2700–2708. [CrossRef]

68. Yiqiao, L.; Nwankpa, C.O. A new type of STATCOM based on cascading voltage-source inverters with phase-shifted unipolar SPWM. *IEEE Trans. Ind. Appl.* **1999**, *35*, 1118–1123. [CrossRef]

69. Gultekin, B.; Gercek, C.O.; Atalik, T.; Deniz, M.; Bicer, N.; Ermis, M.; Kose, K.N.; Ermis, C.; Koc, E.; Cadirci, I.; et al. Design and Implementation of a 154-kV ± 50-Mvar Transmission STATCOM Based on 21-Level Cascaded Multilevel Converter. *IEEE Trans. Ind. Appl.* **2012**, *48*, 1030–1045. [CrossRef]

70. Gultekin, B.; Ermis, M. Cascaded Multilevel Converter-Based Transmission STATCOM: System Design Methodology and Development of a 12 kV ± 12 MVAr Power Stage. *IEEE Trans. Power Electron.* **2013**, *28*, 4930–4950. [CrossRef]

71. Nunes, W.; Encarnação, L.; Aredes, M. An Improved Asymmetric Cascaded Multilevel D–STATCOM with Enhanced Hybrid Modulation. *Electronics* **2015**, *4*, 311–328. [CrossRef]

72. De León Morales, J.; Mata-Jiménez, M.T.; Escalante, M.F. Adaptive scheme for DC voltages estimation in a cascaded H-bridge multilevel converter. *Electr. Power Syst. Res.* **2011**, *81*, 1943–1951. [CrossRef]

73. Hatano, N.; Ise, T. Control Scheme of Cascaded H-Bridge STATCOM Using Zero-Sequence Voltage and Negative-Sequence Current. *IEEE Trans. Power Deliv.* **2010**, *25*, 543–550. [CrossRef]

74. Lee, C.T.; Wang, B.S.; Chen, S.W.; Chou, S.F.; Huang, J.L.; Cheng, P.T.; Akagi, H.; Barbosa, P. Average Power Balancing Control of a STATCOM Based on the Cascaded H-Bridge PWM Converter with Star Configuration. *IEEE Trans. Ind. Appl.* **2014**, *50*, 3893–3901. [CrossRef]

75. Divan, D.; Moghe, R.; Prasai, A. Power Electronics at the Grid Edge : The key to unlocking value from the smart grid. *IEEE Power Electron. Mag.* **2014**, *1*, 16–22. [CrossRef]

76. Ertao, L.; Yin, X.; Zhang, Z.; Chen, Y. An Improved Transformer Winding Tap Injection DSTATCOM Topology for Medium-Voltage Reactive Power Compensation. *IEEE Trans. Power Electron.* **2018**, *33*, 2113–2126.

77. Devassy, S.; Singh, B. Modified p-q Theory Based Control of Solar PV Integrated UPQC-S. In Proceedings of the 2016 IEEE Industry Applications Society Annual Meeting, Portland, OR, USA, 2–6 October 2016.

78. Swain, S.; Ray, P.K. Short circuit fault analysis in a grid connected DFIG based wind energy system with active crowbar protection circuit for ridethrough capability and power quality improvement. *Int. J. Electr. Power Energy Syst.* **2017**, *84*, 64–75. [CrossRef]

79. Bayindir, R.; Colak, I.; Fulli, G.; Demirtas, K. Smart grid technologies and applications. *Renew. Sustain. Energy Rev.* **2016**, *66*, 499–516. [CrossRef]

80. Mansoor, M.; Mariun, N.; Toudeshki, A.; Abdul Wahab, N.I.; Mian, A.U.; Hojabri, M. Innovating problem solving in power quality devices: A survey based on Dynamic Voltage Restorer case (DVR). *Renew. Sustain. Energy Rev.* **2017**, *70*, 1207–1216. [CrossRef]

81. Jaalam, N.; Rahim, N.A.; Bakar, A.H.A.; Tan, C.; Haidar, A.M.A. A comprehensive review of synchronization methods for grid-connected converters of renewable energy source. *Renew. Sustain. Energy Rev.* **2016**, *59*, 1471–1481. [CrossRef]

82. Crosier, R.; Wang, S.; Jamshidi, M. A 4800-V grid-connected electric vehicle charging station that provides STACOM-APF functions with a bi-directional, multi-level, cascaded converter. In Proceedings of the 2012 Twenty-Seventh Annual IEEE Applied Power Electronics Conference and Exposition (APEC), Orlando, FL, USA, 5–9 February 2012; pp. 1508–1515.

83. Saqib, M.A.; Saleem, A.Z. Power-quality issues and the need for reactive-power compensation in the grid integration of wind power. *Renew. Sustain. Energy Rev.* **2015**, *43*, 51–64. [CrossRef]

84. Patrao, I.; Figueres, E.; González-Espín, F.; Garcerá, G. Transformerless topologies for grid-connected single-phase photovoltaic inverters. *Renew. Sustain. Energy Rev.* **2011**, *15*, 3423–3431. [CrossRef]

85. Llorente Iglesias, R.; Lacal Arantegui, R.; Aguado Alonso, M. Power electronics evolution in wind turbines—A market-based analysis. *Renew. Sustain. Energy Rev.* **2011**, *15*, 4982–4993. [CrossRef]

86. Rubio, J.L.O. Aplicaciones de los dispositivos FACTS en generadores eólicos. *Técnica Ind.* **2008**, *276*, 36.

87. Shafiullah, G.M.; Oo, A.M.T.; Shawkat Ali, A.B.M.; Wolfs, P. Potential challenges of integrating large-scale wind energy into the power grid—A review. *Renew. Sustain. Energy Rev.* **2013**, *20*, 306–321. [CrossRef]

88. Chen, Z.; Guerrero, J.M.; Blaabjerg, F. A Review of the State of the Art of Power Electronics for Wind Turbines. *IEEE Trans. Power Electron.* **2009**, *24*, 1859–1875. [CrossRef]

89. Woei-Luen, C.; Yuan-Yih, H. Controller design for an induction generator driven by a variable-speed wind turbine. *IEEE Trans. Energy Convers.* **2006**, *21*, 625–635.

90. Hossain, M.J.; Pota, H.R.; Ramos, R.A. Improved low-voltage-ride-through capability of fixedspeed wind turbines using decentralised control of STATCOM with energy storage system. *IET Gener. Transm. Distrib.* **2012**, *6*, 719–730. [CrossRef]

91. Muyeen, S.M.; Takahashi, R.; Murata, T.; Tamura, J.; Ali, M.H. Application of STATCOM/BESS for wind power smoothening and hydrogen generation. *Electr. Power Syst. Res.* **2009**, *79*, 365–373. [CrossRef]

92. Hossain, M.J.; Pota, H.R.; Ugrinovskii, V.A.; Ramos, R.A. Simultaneous STATCOM and Pitch Angle Control for Improved LVRT Capability of Fixed-Speed Wind Turbines. *IEEE Trans. Sustain. Energy* **2010**, *1*, 142–151. [CrossRef]

93. Suul, J.A.; Molinas, M.; Undeland, T. STATCOM-Based Indirect Torque Control of Induction Machines During Voltage Recovery After Grid Faults. *IEEE Trans. Power Electron.* **2010**, *25*, 1240–1250. [CrossRef]

94. Molinas, M.; Suul, J.A.; Undeland, T. Low Voltage Ride Through of Wind Farms With Cage Generators: STATCOM Versus SVC. *IEEE Trans. Power Electron.* **2008**, *23*, 1104–1117. [CrossRef]

95. Popavath, L.; Kaliannan, P. Photovoltaic-STATCOM with Low Voltage Ride through Strategy and Power Quality Enhancement in a Grid Integrated Wind-PV System. *Electronics* **2018**, *7*, 51. [CrossRef]

96. Sannino, A.; Svensson, J.; Larsson, T. Power-electronic solutions to power quality problems. *Electr. Power Syst. Res.* **2003**, *66*, 71–82. [CrossRef]

97. Kasem, A.H.; El-Saadany, E.F.; El-Tamaly, H.H.; Wahab, M.A.A. Power ramp rate control and flicker mitigation for directly grid connected wind turbines. *IET Renew. Power Gener.* **2010**, *4*, 261–271. [CrossRef]

98. Yuvaraj, V.; Deepa, S.N.; Rozario, A.P.R.; Kumar, M. Improving Grid Power Quality with FACTS Device on Integration of Wind Energy System. In Proceedings of the 2011 Fifth Asia Modelling Symposium, Kuala Lumpur, Malaysia, 24–26 May 2011; pp. 157–162.

99. Howlader, A.M.; Senjyu, T. A comprehensive review of low voltage ride through capability strategies for the wind energy conversion systems. *Renew. Sustain. Energy Rev.* **2016**, *56*, 643–658. [CrossRef]

100. Leandro, G.C.; Soares, E.L.; Rocha, N. Single-phase to three-phase reduced-switch-count converters applied to wind energy conversion systems using doubly-fed induction generator. In Proceedings of the 2017 Brazilian Power Electronics Conference (COBEP), Juiz de Fora, Brazil, 19–22 November 2017; pp. 1–6.

101. Kook, K.S.; Liu, Y.; Atcitty, S. Mitigation of the wind generation integration related power quality issues by energy storage. *Electr. Power Qual. Util. J.* **2006**, *12*, 77–82.

102. Chowdhury, M.M.; Haque, M.E.; Aktarujjaman, M.; Negnevitsky, M.; Gargoom, A. Grid integration impacts and energy storage systems for wind energy applications—A review. In Proceedings of the 2011 IEEE Power and Energy Society General Meeting, San Diego, CA, USA, 24–29 July 2011; pp. 1–8.

103. Miveh, M.R.; Rahmat, M.F.; Ghadimi, A.A.; Mustafa, M.W. Control techniques for three-phase four-leg voltage source inverters in autonomous microgrids: A review. *Renew. Sustain. Energy Rev.* **2016**, *54*, 1592–1610. [CrossRef]

104. Zhaoan, W.; Qun, W.; Weizheng, Y.; Jinjun, L. A series active power filter adopting hybrid control approach. *IEEE Trans. Power Electron.* **2001**, *16*, 301–310. [CrossRef]

105. Khadkikar, V. Enhancing Electric Power Quality Using UPQC: A Comprehensive Overview. *IEEE Trans. Power Electron.* **2012**, *27*, 2284–2297. [CrossRef]

106. Mulla, M.A.; Rajagopalan, C.; Chowdhury, A. Hardware implementation of series hybrid active power filter using a novel control strategy based on generalised instantaneous power theory. *IET Power Electron.* **2013**, *6*, 592–600. [CrossRef]

107. Salmeron, P.; Litran, S.P. Improvement of the Electric Power Quality Using Series Active and Shunt Passive Filters. *IEEE Trans. Power Deliv.* **2010**, *25*, 1058–1067. [CrossRef]

108. Shivashankar, S.; Mekhilef, S.; Mokhlis, H.; Karimi, M. Mitigating methods of power fluctuation of photovoltaic (PV) sources—A review. *Renew. Sustain. Energy Rev.* **2016**, *59*, 1170–1184. [CrossRef]

109. Rastogi, M.; Mohan, N.; Edris, A.A. Hybrid-active filtering of harmonic currents in power systems. *IEEE Trans. Power Deliv.* **1995**, *10*, 1994–2000. [CrossRef]

110. Planas, E.; Andreu, J.; Gárate, J.I.; Martínez de Alegría, I.; Ibarra, E. AC and DC technology in microgrids: A review. *Renew. Sustain. Energy Rev.* **2015**, *43*, 726–749. [CrossRef]

111. Singh, S.; Gautam, A.R.; Fulwani, D. Constant power loads and their effects in DC distributed power systems: A review. *Renew. Sustain. Energy Rev.* **2017**, *72*, 407–421. [CrossRef]

112. Ghosh, A.; Ledwich, G. A unified power quality conditioner (UPQC) for simultaneous voltage and current compensation. *Electr. Power Syst. Res.* **2001**, *59*, 55–63. [CrossRef]

113. Taher, S.A.; Afsari, S.A. Optimal location and sizing of DSTATCOM in distribution systems by immune algorithm. *Int. J. Electr. Power Energy Syst.* **2014**, *60*, 34–44. [CrossRef]

114. Kirubakaran, A.; Jain, S.; Nema, R.K. A review on fuel cell technologies and power electronic interface. *Renew. Sustain. Energy Rev.* **2009**, *13*, 2430–2440. [CrossRef]

115. Baroudi, J.A.; Dinavahi, V.; Knight, A.M. A review of power converter topologies for wind generators. *Renew. Energy* **2007**, *32*, 2369–2385. [CrossRef]

116. Balikci, A.; Akpinar, E. A multilevel converter with reduced number of switches in STATCOM for load balancing. *Electr. Power Syst. Res.* **2015**, *123*, 164–173. [CrossRef]

117. Tareen, W.U.; Mekhilef, S.; Seyedmahmoudian, M.; Horan, B. Active power filter (APF) for mitigation of power quality issues in grid integration of wind and photovoltaic energy conversion system. *Renew. Sustain. Energy Rev.* **2017**, *70*, 635–655. [CrossRef]

118. Patnaik, S.S.; Panda, A.K. Three-level H-bridge and three H-bridges-based three-phase four-wire shunt active power filter topologies for high voltage applications. *Int. J. Electr. Power Energy Syst.* **2013**, *51*, 298–306. [CrossRef]

119. Junior, R.L.S.; Lazzarin, T.B.; Barbi, I. Reduced Switch Count Step-up/Step-down Switched-Capacitor Three-Phase AC-AC Converter. *IEEE Trans. Ind. Electron.* **2018**. [CrossRef]

120. Heydari, M.; Fatemi, A.; Varjani, A.Y. A Reduced Switch Count Three-Phase AC/AC Converter with Six Power Switches: Modeling, Analysis, and Control. *IEEE J. Emerg. Sel. Top. Power Electron.* **2017**, *5*, 1720–1738. [CrossRef]

121. Limongi, L.R.; Bradaschia, F.; Azevedo, G.M.S.; Genu, L.G.B.; Filho, L.R.S. Dual hybrid power filter based on a nine-switch inverter. *Electr. Power Syst. Res.* **2014**, *117*, 154–162. [CrossRef]

122. Bradaschia, F.; Limongi, L.R.; Cavalcanti, M.C.; Neves, F.A.S. A generalized scalar pulse-width modulation for nine-switch inverters: An approach for non-sinusoidal modulating waveforms. *Electr. Power Syst. Res.* **2015**, *121*, 302–312. [CrossRef]

123. Heydari, M.; Varjani, A.Y.; Mohamadian, M.; Fatemi, A. Three-phase dual-output six-switch inverter. *IET Power Electron.* **2012**, *5*, 1634–1650. [CrossRef]

124. Kolar, J.W.; Friedli, T.; Rodriguez, J.; Wheeler, P.W. Review of Three-Phase PWM AC–AC Converter Topologies. *IEEE Trans. Ind. Electron.* **2011**, *58*, 4988–5006. [CrossRef]

125. Limongi, L.R.; da Silva Filho, L.R.; Genu, L.G.B.; Bradaschia, F.; Cavalcanti, M.C. Transformerless Hybrid Power Filter Based on a Six-Switch Two-Leg Inverter for Improved Harmonic Compensation Performance. *IEEE Trans. Ind. Electron.* **2015**, *62*, 40–51. [CrossRef]

126. Hyosung, K.; Seung-Ki, S. Compensation voltage control in dynamic voltage restorers by use of feed forward and state feedback scheme. *IEEE Trans. Power Electron.* **2005**, *20*, 1169–1177.

127. Luo, A.; Xu, X.; Fang, L.; Fang, H.; Wu, J.; Wu, C. Feedback-Feedforward PI-Type Iterative Learning Control Strategy for Hybrid Active Power Filter With Injection Circuit. *IEEE Trans. Ind. Electron.* **2010**, *57*, 3767–3779. [CrossRef]

128. Bhattacharya, A.; Chakraborty, C.; Bhattacharya, S. Parallel-Connected Shunt Hybrid Active Power Filters Operating at Different Switching Frequencies for Improved Performance. *IEEE Trans. Ind. Electron.* **2012**, *59*, 4007–4019. [CrossRef]

129. Lee, T.L.; Wang, Y.C.; Li, J.C.; Guerrero, J.M. Hybrid Active Filter With Variable Conductance for Harmonic Resonance Suppression in Industrial Power Systems. *IEEE Trans. Ind. Electron.* **2015**, *62*, 746–756. [CrossRef]

130. Memon, M.; Saad, M.; Mubin, M. Selective Harmonic Elimination in Multilevel Inverter using Hybrid APSO Algorithm. *IET Power Electron.* **2018**. [CrossRef]

131. Vemuganti, H.P.; Sreenivasarao, D.; Kumar, G.S.; Spandana, A.S. Reduced carrier PWM scheme with unified logical expressions for reduced switch count multilevel inverters. *IET Power Electron.* **2018**, *11*, 912–921. [CrossRef]

132. Qin, J.; Saeedifard, M. Predictive Control of a Modular Multilevel Converter for a Back-to-Back HVDC System. *IEEE Trans. Power Deliv.* **2012**, *27*, 1538–1547.

133. Wang, H.; Liserre, M.; Blaabjerg, F. Toward Reliable Power Electronics: Challenges, Design Tools, and Opportunities. *IEEE Ind. Electron. Mag.* **2013**, *7*, 17–26. [CrossRef]

134. Gui, Y.; Lee, Y.O.; Han, Y.; Chung, C.C. Novel passivity-based controller design for Back-to-back STATCOM with asymmetrically structured converters. In Proceedings of the 2012 IEEE Power and Energy Society General Meeting, San Diego, CA, USA, 22–26 July 2012; pp. 1–6.

135. Bhattacharya, A.; Chakraborty, C.; Bhattacharya, S. A reduced switch transformer-less dual hybrid active power filter. In Proceedings of the 2009 35th Annual Conference of IEEE Industrial Electronics, Porto, Portugal, 3–5 November 2009; pp. 88–93.

136. Venet, P.; Perisse, F.; El-Husseini, M.H.; Rojat, G. Realization of a smart electrolytic capacitor circuit. *IEEE Ind. Appl. Mag.* **2002**, *8*, 16–20. [CrossRef]

137. Liu, X.; Loh, P.C.; Wang, P.; Blaabjerg, F. A Direct Power Conversion Topology for Grid Integration of Hybrid AC/DC Energy Resources. *IEEE Trans. Ind. Electron.* **2013**, *60*, 5696–5707. [CrossRef]

138. Lin, B.R.; Shih, K.L. Analysis and implementation of a softswitching converter with reduced switch count. *IET Power Electron.* **2010**, *3*, 559–570. [CrossRef]

139. Donghua, C.; Shaojun, X. Review of the control strategies applied to active power filters. In Proceedings of the 2004 IEEE International Conference on Electric Utility Deregulation, Restructuring and Power Technologies, Hong Kong, China, 5–8 April 2004; Volume 2, pp. 666–670.

140. Tripathi, S.M.; Tiwari, A.N.; Singh, D. Grid-integrated permanent magnet synchronous generator based wind energy conversion systems: A technology review. *Renew. Sustain. Energy Rev.* **2015**, *51*, 1288–1305. [CrossRef]

141. Ahmed, H.F.; Cha, H.; Khan, A.A. A Single-Phase Buck Matrix Converter With High-Frequency Transformer Isolation and Reduced Switch Count. *IEEE Trans. Ind. Electron.* **2017**, *64*, 6979–6988. [CrossRef]

142. Singh, A.R.; Patne, N.R.; Kale, V.S. Adaptive distance protection setting in presence of mid-point STATCOM using synchronized measurement. *Int. J. Electr. Power Energy Syst.* **2015**, *67*, 252–260. [CrossRef]

143. Jacobina, C.B.; de Freitas, I.S.; Lima, A.M.N. DC-Link Three-Phase-to-Three-Phase Four-Leg Converters. *IEEE Trans. Ind. Electron.* **2007**, *54*, 1953–1961. [CrossRef]

144. Liang, J.; Green, T.C.; Feng, C.; Weiss, G. Increasing Voltage Utilization in Split-Link, Four-Wire Inverters. *IEEE Trans. Power Electron.* **2009**, *24*, 1562–1569. [CrossRef]

145. Liu, H.-B.; Mao, C.-X.; Lu, J.-M.; Wang, D. Three-phase four-wire shunt APF-STATCOM using a four-leg converter. *Power Syst. Prot. Control* **2010**, *38*, 11–17.

146. Broeck, H.W.V.D.; Wyk, J.D.V. A Comparative Investigation of a Three-Phase Induction Machine Drive with a Component Minimized Voltage-Fed Inverter under Different Control Options. *IEEE Trans. Ind. Appl.* **1984**, *IA-20*, 309–320. [CrossRef]

147. Rodriguez, P.; Pindado, R.; Bergas, J. Alternative topology for three-phase four-wire PWM converters applied to a shunt active power filter. In Proceedings of the IEEE 2002 28th Annual Conference of the Industrial Electronics Society, Sevilla, Spain, 5–8 November 2002; Volume 4, pp. 2939–2944.

148. Dos Santos, E.C.; Jacobina, C.B.; Dias, J.A.A.; Rocha, N. Single-Phase to Three-Phase Universal Active Power Filter. *IEEE Trans. Power Deliv.* **2011**, *26*, 1361–1371. [CrossRef]

149. Lohia, P.; Mishra, M.K.; Karthikeyan, K.; Vasudevan, K. A Minimally Switched Control Algorithm forThree-Phase Four-Leg VSI Topology toCompensate Unbalanced and Nonlinear Load. *IEEE Trans. Power Electron.* **2008**, *23*, 1935–1944. [CrossRef]

150. Munjewar, S.S.; Thombre, S.B.; Mallick, R.K. Approaches to overcome the barrier issues of passive direct methanol fuel cell—Review. *Renew. Sustain. Energy Rev.* **2017**, *67*, 1087–1104. [CrossRef]

151. Hoon, Y.; Mohd Radzi, M.; Hassan, M.; Mailah, N. Control Algorithms of Shunt Active Power Filter for Harmonics Mitigation: A Review. *Energies* **2017**, *10*, 2038. [CrossRef]

152. Zaveri, T.; Bhalja, B.; Zaveri, N. Comparison of control strategies for DSTATCOM in three-phase, four-wire distribution system for power quality improvement under various source voltage and load conditions. *Int. J. Electr. Power Energy Syst.* **2012**, *43*, 582–594. [CrossRef]

153. Badihi, H.; Zhang, Y.; Hong, H. Active power control design for supporting grid frequency regulation in wind farms. *Annu. Rev. Control* **2015**, *40*, 70–81. [CrossRef]

154. Scali, C.; Farnesi, M. Implementation, parameters calibration and field validation of a Closed Loop Performance Monitoring system. *Annu. Rev. Control* **2010**, *34*, 263–276. [CrossRef]

155. Gonzalez, S.A.; Garcia-Retegui, R.; Benedetti, M. Harmonic Computation Technique Suitable for Active Power Filters. *IEEE Trans. Ind. Electron.* **2007**, *54*, 2791–2796. [CrossRef]

156. Tavana, M.R.; Khooban, M.-H.; Niknam, T. Adaptive PI controller to voltage regulation in power systems: STATCOM as a case study. *ISA Trans.* **2017**, *66*, 325–334. [CrossRef] [PubMed]

157. Monadi, M.; Amin Zamani, M.; Ignacio Candela, J.; Luna, A.; Rodriguez, P. Protection of AC and DC distribution systems Embedding distributed energy resources: A comparative review and analysis. *Renew. Sustain. Energy Rev.* **2015**, *51*, 1578–1593. [CrossRef]

158. Wang, B.; Cathey, J.J. DSP-controlled, space-vector PWM, current source converter for STATCOM application. *Electr. Power Syst. Res.* **2003**, *67*, 123–131. [CrossRef]

159. Bina, M.T.; Bhat, A.K.S. Averaging Technique for the Modeling of STATCOM and Active Filters. *IEEE Trans. Power Electron.* **2008**, *23*, 723–734. [CrossRef]

160. Moghadasi, A.; Sarwat, A.; Guerrero, J.M. A comprehensive review of low-voltage-ride-through methods for fixed-speed wind power generators. *Renew. Sustain. Energy Rev.* **2016**, *55*, 823–839. [CrossRef]

161. İnci, M.; Bayındır, K.Ç.; Tümay, M. Improved Synchronous Reference Frame based controller method for multifunctional compensation. *Electr. Power Syst. Res.* **2016**, *141*, 500–509. [CrossRef]

162. Cañizares, C.A.; Pozzi, M.; Corsi, S.; Uznunovic, E. STATCOM modeling for voltage and angle stability studies. *Int. J. Electr. Power Energy Syst.* **2003**, *25*, 431–441. [CrossRef]

163. De Araujo Ribeiro, R.L.; de Azevedo, C.C.; de Sousa, R.M. A Robust Adaptive Control Strategy of Active Power Filters for Power-Factor Correction, Harmonic Compensation, and Balancing of Nonlinear Loads. *IEEE Trans. Power Electron.* **2012**, *27*, 718–730. [CrossRef]

164. Amoozegar, D. DSTATCOM modelling for voltage stability with fuzzy logic PI current controller. *Int. J. Electr. Power Energy Syst.* **2016**, *76*, 129–135. [CrossRef]

165. Moghbel, M.; Masoum, M.A.S.; Fereidouni, A.; Deilami, S. Optimal Sizing, Siting and Operation of Custom Power Devices with STATCOM and APLC Functions for Real-Time Reactive Power and Network Voltage Quality Control of Smart Grid. *IEEE Trans. Smart Grid* **2017**, *PP*. [CrossRef]

166. Zribi, M.; Alrifai, M.; Rayan, M. Sliding Mode Control of a Variable-Speed Wind Energy Conversion System Using a Squirrel Cage Induction Generator. *Energies* **2017**, *10*, 604. [CrossRef]

167. Varma, R.K.; Salehi, R. SSR Mitigation with a New Control of PV Solar Farm as STATCOM (PV-STATCOM). *IEEE Trans. Sustain. Energy* **2017**, *8*, 1473–1483. [CrossRef]

168. Božiček, A.; Blažič, B.; Papič, I. Time–optimal current control with constant switching frequency for STATCOM. *Electr. Power Syst. Res.* **2010**, *80*, 925–934. [CrossRef]

169. Wang, L.; Lam, C.S.; Wong, M.C. Selective Compensation of Distortion, Unbalanced and Reactive Power of a Thyristor-Controlled LC-Coupling Hybrid Active Power Filter (TCLC-HAPF). *IEEE Trans. Power Electron.* **2017**, *32*, 9065–9077. [CrossRef]

170. Behrouzian, E.; Bongiorno, M.; Teodorescu, R. Impact of Switching Harmonics on Capacitor Cells Balancing in Phase-Shifted PWM-Based Cascaded H-Bridge STATCOM. *IEEE Trans. Power Electron.* **2017**, *32*, 815–824. [CrossRef]

171. Tareen, W.U.K.; Mekhilef, S. Three-phase Transformerless Shunt Active Power Filter with Reduced Switch Count for Harmonic Compensation in Grid-Connected Applications. *IEEE Trans. Power Electron.* **2018**, *33*, 4868–4881. [CrossRef]

172. Liu, X.; Lv, J.; Gao, C.; Chen, Z.; Chen, S. A Novel STATCOM Based on Diode-Clamped Modular Multilevel Converters. *IEEE Trans. Power Electron.* **2017**, *32*, 5964–5977. [CrossRef]

173. Mishra, S.S.; Mohapatra, A.; Satpathy, P.K. Grid Integration of Small Hydro Power Plants Based on PWM Converter and D-STATCOM. In *Artificial Intelligence and Evolutionary Computations in Engineering Systems: Proceedings of ICAIECES 2016*; Dash, S.S., Vijayakumar, K., Panigrahi, B.K., Das, S., Eds.; Springer: Singapore, 2017; pp. 617–631.

174. Suganthi, L.; Iniyan, S.; Samuel, A.A. Applications of fuzzy logic in renewable energy systems—A review. *Renew. Sustain. Energy Rev.* **2015**, *48*, 585–607. [CrossRef]

175. Hasani, A.; Haghjoo, F. A Secure and Setting-Free Technique to Detect Loss of Field in Synchronous Generators. *IEEE Trans. Energy Convers.* **2017**, *32*, 1512–1522. [CrossRef]

176. Athari, H.; Niroomand, M.; Ataei, M. Review and Classification of Control Systems in Grid-tied Inverters. *Renew. Sustain. Energy Rev.* **2017**, *72*, 1167–1176. [CrossRef]

177. Lu, D.; Wang, J.; Yao, J.; Wang, S.; Zhu, J.; Hu, H.; Zhang, L. Clustered Voltage Balancing Mechanism and its Control Strategy for Star-Connected Cascaded H-Bridge STATCOM. *IEEE Trans. Ind. Electron.* **2017**, *64*, 7623–7633. [CrossRef]

178. Goodwin, G.C.; Mayne, D.Q.; Chen, K.-Y.; Coates, C.; Mirzaeva, G.; Quevedo, D.E. An introduction to the control of switching electronic systems. *Annu. Rev. Control* **2010**, *34*, 209–220. [CrossRef]

179. De Rossiter Correa, M.B.; Jacobina, C.B.; da Silva, E.R.C.; Lima, A.M.N. A General PWM Strategy for Four-Switch Three-Phase Inverters. *IEEE Trans. Power Electron.* **2006**, *21*, 1618–1627. [CrossRef]

180. Moeed Amjad, A.; Salam, Z. A review of soft computing methods for harmonics elimination PWM for inverters in renewable energy conversion systems. *Renew. Sustain. Energy Rev.* **2014**, *33*, 141–153. [CrossRef]

181. Sahoo, S.; Bhattacharya, T. Phase Shifted Carrier Based Synchronized Sinusoidal PWM Techniques for Cascaded H-Bridge Multilevel Inverters. *IEEE Trans. Power Electron.* **2018**, *33*, 513–524. [CrossRef]

182. Liu, C.; Wu, B.; Zargari, N.R.; Xu, D.; Wang, J. A Novel Three-Phase Three-Leg AC/AC Converter Using Nine IGBTs. *IEEE Trans. Power Electron.* **2009**, *24*, 1151–1160.

183. Colak, I.; Kabalci, E. Developing a novel sinusoidal pulse width modulation (SPWM) technique to eliminate side band harmonics. *Int. J. Electr. Power Energy Syst.* **2013**, *44*, 861–871. [CrossRef]

184. Zhang, Y.; Wu, X.; Yuan, X. A Simplified Branch and Bound Approach for Model Predictive Control of Multilevel Cascaded H-Bridge STATCOM. *IEEE Trans. Ind. Electron.* **2017**, *64*, 7634–7644. [CrossRef]

185. Monroy-Morales, J.; Campos-Gaona, D.; Hernández-Ángeles, M.; Peña-Alzola, R.; Guardado-Zavala, J. An Active Power Filter Based on a Three-Level Inverter and 3D-SVPWM for Selective Harmonic and Reactive Compensation. *Energies* **2017**, *10*, 297. [CrossRef]

186. Dehnavi, S.M.D.; Mohamadian, M.; Yazdian, A.; Ashrafzadeh, F. Space Vectors Modulation for Nine-Switch Converters. *IEEE Trans. Power Electron.* **2010**, *25*, 1488–1496. [CrossRef]

187. Barghi Latran, M.; Teke, A. Investigation of multilevel multifunctional grid connected inverter topologies and control strategies used in photovoltaic systems. *Renew. Sustain. Energy Rev.* **2015**, *42*, 361–376. [CrossRef]

188. Wang, W.; Luo, A.; Xu, X.; Fang, L.; Chau, T.M.; Li, Z. Space vector pulse-width modulation algorithm and DC-side voltage control strategy of three-phase four-switch active power filters. *IET Power Electron.* **2013**, *6*, 125–135. [CrossRef]

189. Camacho, A.; Castilla, M.; Miret, J.; Vicuña, L.G.d.; Guzman, R. Positive and Negative Sequence Control Strategies to Maximize the Voltage Support in Resistive-Inductive Grids During Grid Faults. *IEEE Trans. Power Electron.* **2018**, *33*, 5362–5373. [CrossRef]

190. Jayabalan, M.; Jeevarathinam, B.; Sandirasegarane, T. Reduced switch count pulse width modulated multilevel inverter. *IET Power Electron.* **2017**, *10*, 10–17. [CrossRef]

191. Prabaharan, N.; Palanisamy, K. A comprehensive review on reduced switch multilevel inverter topologies, modulation techniques and applications. *Renew. Sustain. Energy Rev.* **2017**, *76*, 1248–1282. [CrossRef]

192. Monfared, M.; Rastegar, H.; Kojabadi, H.M. Overview of modulation techniques for the four-switch converter topology. In Proceedings of the 2008 IEEE 2nd International Power and Energy Conference, Johor Bahru, Malaysia, 1–3 December 2008; pp. 803–807.

193. Asiminoael, L.; Blaabjerg, F.; Hansen, S. Detection is key—Harmonic detection methods for active power filter applications. *IEEE Ind. Appl. Mag.* **2007**, *13*, 22–33. [CrossRef]

194. Tareen, W.U.; Mekhilef, S. Transformer-less 3P3W SAPF (three-phase three-wire shunt active power filter) with line-interactive UPS (uninterruptible power supply) and battery energy storage stage. *Energy* **2016**, *109*, 525–536. [CrossRef]

195. ADF Power Tuning. Products adf-p700-statcom. Available online: https://adfpowertuning.com/products/adf-p700-statcom (accessed on 1 July 2017).

196. AMSC (NASDAQ: AMSC). Dynamic Volt-Amp Reactive (D-VAR) Compensation Solution. Available online: http://www.amsc.com/gridtec/utility_reactive_power_solutions (accessed on 15 July 2017).

197. Kato, T.; Ito, T.; Aihara, T.; Namatame, S. Development of a 20-MVA STATCOM for Flicker Suppression. *Hitachi Rev.* **2007**, *56*, 133.

198. Lyons, J.P.; Vlatkovic, V.; Espelage, P.M.; Esser, A.A.M. High Power Motor Drive Converter System and Modulation Control. U.S. Paten US5910892A, 8 July 1999.

199. Bousseau, P.; Fesquet, F.; Belhomme, R.; Nguefeu, S.; Thai, T.C. Solutions for the grid integration of wind farms—A survey. *Wind Energy* **2006**, *9*, 13–25. [CrossRef]

200. Yu, Q.; Li, P.; Liu, W.; Xie, X. Overview of STATCOM technologies. In Proceedings of the 2004 IEEE International Conference on Electric Utility Deregulation, Restructuring and Power Technologies, Hong Kong, China, 5–8 April 2004; pp. 647–652.

201. Bagnall, T.; Ritter, C.; Ronner, B.; Maibach, P.; Butcher, N.; Thurnherr, T. PCS6000 STATCOM ancillary functions: Wind park resonance damping. In Proceedings of the European Wind Energy Conference and Exhibition 2009, Marseille, France, 16–19 March 2009.

202. Xu, X.; Edmonds, M.J.; Bishop, M.; Sember, J. Application of distributed static compensators in wind farms to meet grid codes. In Proceedings of the Asia-Pacific Power and Energy Engineering Conference (APPEEC), Shanghai, China, 27–29 March 2012; pp. 1–5.

203. Jiao, Z.; Xingyan, N.; Mingjun, D.; Zefeng, Q.; Qirui, L. Research on control strategy of cascade STATCOM under unbalanced system voltage. In Proceedings of the 2012 China International Conference on Electricity Distribution (CICED), Shanghai, China, 10–14 September 2012; pp. 1–4.

204. Bryantsev, A.; Bazylev, B.; Lur'e, A.; Raichenko, M.; Smolovik, S. Compensators of reactive power for controlling and stabilizing the voltage of a high-voltage electrical network. *Russ. Electr. Eng.* **2013**, *84*, 57–64. [CrossRef]

205. Qiao, C.; Jin, T.; Smedley, K.M. One-cycle control of three-phase active power filter with vector operation. *IEEE Trans. Ind. Electron.* **2004**, *51*, 455–463. [CrossRef]

206. Kim, J.; Hong, J.; Nam, K. A Current Distortion Compensation Scheme for Four-Switch Inverters. *IEEE Trans. Power Electron.* **2009**, *24*, 1032–1040.

207. Gi-Taek, K.; Lipo, T.A. VSI-PWM rectifier/inverter system with a reduced switch count. *IEEE Trans. Ind. Appl.* **1996**, *32*, 1331–1337. [CrossRef]

208. Byoung-Kuk, L.; Tae-Hyung, K.; Ehsani, M. On the feasibility of four-switch three-phase BLDC motor drives for low cost commercial applications: Topology and control. *IEEE Trans. Power Electron.* **2003**, *18*, 164–172. [CrossRef]

209. Blaabjerg, F.; Freysson, S.; Hansen, H.H.; Hansen, S. A new optimized space-vector modulation strategy for a component-minimized voltage source inverter. *IEEE Trans. Power Electron.* **1997**, *12*, 704–714. [CrossRef]

210. Kesler, M.; Ozdemir, E. Synchronous-Reference-Frame-Based Control Method for UPQC Under Unbalanced and Distorted Load Conditions. *IEEE Trans. Ind. Electron.* **2011**, *58*, 3967–3975. [CrossRef]

211. Angulo, M.; Ruiz-Caballero, D.A.; Lago, J.; Heldwein, M.L.; Mussa, S.A. Active Power Filter Control Strategy With Implicit Closed-Loop Current Control and Resonant Controller. *IEEE Trans. Ind. Electron.* **2013**, *60*, 2721–2730. [CrossRef]

212. Bouzid, A.M.; Guerrero, J.M.; Cheriti, A.; Bouhamida, M.; Sicard, P.; Benghanem, M. A survey on control of electric power distributed generation systems for microgrid applications. *Renew. Sustain. Energy Rev.* **2015**, *44*, 751–766. [CrossRef]

213. Rodriguez, J.; Kazmierkowski, M.P.; Espinoza, J.R.; Zanchetta, P.; Abu-Rub, H.; Young, H.A.; Rojas, C.A. State of the art of finite control set model predictive control in power electronics. *IEEE Trans. Ind. Inform.* **2013**, *9*, 1003–1016. [CrossRef]

214. Zhang, L.; Loh, P.C.; Gao, F. An Integrated Nine-Switch Power Conditioner for Power Quality Enhancement and Voltage Sag Mitigation. *IEEE Trans. Power Electron.* **2012**, *27*, 1177–1190. [CrossRef]

215. Fatemi, A.; Azizi, M.; Mohamadian, M.; Varjani, A.Y.; Shahparasti, M. Single-Phase Dual-Output Inverters With Three-Switch Legs. *IEEE Trans. Ind. Electron.* **2013**, *60*, 1769–1779. [CrossRef]

![energies logo] *energies*

MDPI

Article

A Modified Self-Synchronized Synchronverter in Unbalanced Power Grids with Balanced Currents and Restrained Power Ripples

Xiaohe Wang [1], Liang Chen [1], Dan Sun [1,*], Li Zhang [2] and Heng Nian [1]

1   College of Electrical Engineering, Zhejiang University, Hangzhou 310027, China;
    21410053@zju.edu.cn (X.W.); 21410077@zju.edu.cn (L.C.); nianheng@zju.edu.cn (H.N.)
2   State Key Laboratory of Operation and Control of Renewable Energy & Storage Systems, China Electric
    Power Research Institute, Beijing 100192, China; zhangli82@epri.sgcc.com.cn
*   Correspondence: sundan@zju.edu.cn; Tel.: +86-1368-578-8025

Received: 31 January 2019; Accepted: 6 March 2019; Published: 10 March 2019

Abstract: This paper proposes a modified self-synchronized synchronverter for unbalanced power grids. Small signal analysis of the conventional synchronverter shows that its stability margin around 50 Hz is very limited. Thus, power ripples will be caused at the frequency of 50 Hz. Filter- based current feeding loops are adopted in the conventional synchronverter in order to enhance its stability and eliminate power ripples. In addition, the characteristics of the conventional synchronverter in unbalanced power grids are analyzed, and an improved strategy using a resonant controller is proposed to restrain the current harmonics and power ripples. The parameter design is also studied for the proposed synchronverter. Experimental studies prove that the proposed strategy can achieve precise self-synchronization when the grid voltage is unbalanced, and the power-control performance is also improved significantly.

Keywords: synchronverter; power ripple elimination; resonant controller; unbalanced power grid

1. Introduction

With the rapid development of renewable power generation technologies such as wind power and photovoltaic, distributed generation system has become an effective way to meet load growth, reduce environmental pollution, and improve energy efficiency. DC/AC converters are one of the most common interfaces between the distributed generation system and the power grid. Traditional vector control strategies of DC/AC converters mainly aim to achieve a fast dynamic response [1]. Moreover, the power control of renewable energy is often set as maximum power point tracking (MPPT) in order to harvest as much energy as possible [2,3]. However, compared with the synchronous generators (SG), the DC/AC converters controlled by these strategies have no inertia, and they cannot provide frequency support for the power grid [4,5].

In order to achieve grid-friendly control performance similar as the SG, virtual synchronous generator (VSG) strategies are proposed [6]. The active power control loop of VSG is designed based on the swing equations, and the reactive power control loop is similar as the droop control strategy. In addition, the renewable power generation system should work with enough margins to support the grid, or the energy storage devices are needed to provide the extra active power. Therefore, the inertia property of SG can be inherited by the VSG [7,8]. In most of the VSG strategies, the phase-locked loop (PLL) is needed to detect the phase angle of grid voltages in order to synchronize with the grid [9], or to provide grid frequency information [10]. The PLL is a nonlinear control block which may complicate the system analysis. Therefore, some PLL excluded VSG strategies are proposed [11,12], where the converter can be synchronized with grid voltage by power balance of swing equations instead of tracking the phase angle of grid voltage directly by PLL. Among these VSG strategies, the synchronverter proposed in [13,14] can emulate the model of SG more accurately. Its reactive power control loop is designed based on the excitation control of SG rather than the droop control. Therefore, the synchronverter makes the renewable power generation system interfaced with the grid like conventional SG and have been paid much attention [15].

The works analyzed above all focus on synchronverters in balanced three-phase systems. However, since the distributed renewable power generation system is usually installed in remote or offshore areas, the grid faults cannot be avoided. Therefore, the performance of synchronverter needs to be investigated under both normal and non-ideal grid conditions. Single-phase voltage drop is one kind of common grid faults which will result in unbalanced power grids [16]. When the grid voltage becomes unbalanced, the negative sequence grid voltage will cause negative sequence grid currents and double frequency power ripples, which severely deteriorate the performance of the converter [17]. However, there is no literature focused on the modification of synchronverter in unbalanced power grids. Resonant controllers have been widely adopted in the improved control strategies under unbalanced power grid conditions to deal with the negative sequence components and proven to be an effective way to deal with this problem. Different kinds of resonant controllers are combined with the vector control (VC) and direct power control (DPC), i.e., second-order generalized integrator (SOGI) [18] reduced-order generalized integrator (ROGI) [19] and reduced-order vector integrator (ROVI) [20]. Theoretical analyses and experimental studies have verified the effectiveness of the resonant based strategies in unbalanced power grid. However, no research has been made to investigate the validity of resonant controller in synchronverter in unbalanced power grid. This is one of the motivations of this study.

Though being a grid-friendly converter, the stability of the synchronverter should be re-evaluated because there is a considerable parameter inconsistence between the synchronverter and physical conventional SG [21]. The stability analysis and control parameter tuning of the synchronverter are not easy due to its characteristics of more complicated nonlinearity. Global stability of the synchronverter was investigated in [22]. However, the stability analysis of the synchronverter in an unbalanced power grid, as well as the impact of the modified controller to the stability of the synchronverter, will be more complicated, and has not been examined yet. The linearization method has been widely adopted to analyze synchronverter systems. In [23], a small-signal sequence impedance model was established to compare the characteristics between VSG and the traditional strategy for grid-connected inverters. In [24], small signal model of synchronverter is established to analyze the stability of the battery system. The small signal models are also established in [25,26] to design the parameters of synchronverter in order to achieve better dynamic performance. It can be seen that the small signal model can reflect the stability of the system at specific steady operation point in a wide frequency range. Therefore, it is meaningful to establish the small signal model for synchronverter in both balanced and unbalanced power grid to study its stability performance and design the system parameters.

This paper proposes a modified self-synchronized synchronverter for unbalanced power grids and analyzes the system performance using a small signal model. The contributions of this paper are summarized as the following three aspects. (1) The small signal model for a synchronverter is established and it is observed that the stability margin around 50 Hz is very limited, especially when the control gain of reactive power increases. (2) Lowpass filters are applied to the current feeding loop to enhance the stability of the synchronverter and eliminate the power ripples at the frequency of 50 Hz. The effectiveness of the power ripple elimination method is verified by small signal analyses. (3) A resonant-based strategy is proposed to restrain the current harmonics and power ripples in unbalanced power grid. The parameters are designed according to the small signal analyses, and the effectiveness is also validated by the small signal analyses. In a word, the goal of this paper is to improve the quality of the powers and currents flowed into the grid, and make the self-synchronized process more precise under unbalanced grid condition.

The rest of this paper is organized as follows: in Section 2, the conventional self-synchronized synchronverter in a balanced power grid is introduced, and a power ripple elimination method is proposed. Small signal model analyses prove that the proposed method can enhance the system stability and reduce power ripples. In Section 3, the influence of unbalanced grid voltage on the conventional synchronverter is analyzed and a resonant-based strategy is proposed to restrain the current harmonics and power ripples in unbalanced power grid. The parameter design method of the proposed strategy is analyzed. Then in Section 4, comparative experimental studies are conducted to confirm the effectiveness of the proposed strategy in both the self-synchronized mode and power-control mode. Finally, the conclusions are presented in Section 5.

2. Self-Synchronized Synchronverter in a Balanced Power Grid and Its Power Ripple Elimination

2.1. Self-Synchronized Synchronverter

The topology of a DC/AC converter is illustrated in Figure 1. The reference point of v_a, v_b, v_c is the neutral of the grid, and the reference point of e_a, e_b, e_c is the half of DC link voltage. To mimic the characteristics of the SG, the synchronverter is proposed according to the mechanical and excitation equations of the SG. The control diagram of the conventional synchronverter in [13] is shown in Figure 2, where the physical circuit is simplified as the mathematical model. In order to simplify the small signal analyses, the control scheme of the synchronverter is based on the reference frame where the d-axis is aligned to the rotor of synchronverter, which is named as control dq frame in this paper.

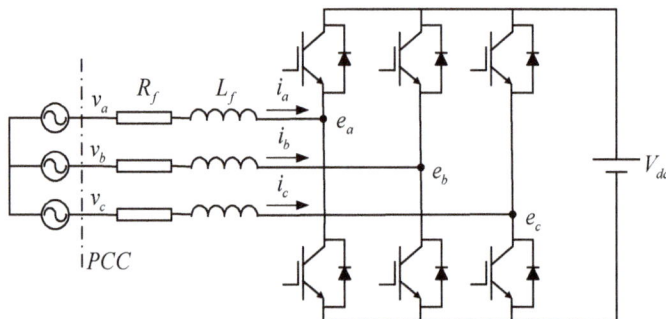

Figure 1. Topology of DC/AC converter.

The superscript r indicates that the variables are in the control dq frame. J is the moment of inertia, D is the damping factor, K is the integral factor. The phase angle θ_r of the converter voltage, also known as the back electrical magnetic field (EMF) of VSG, is produced by the active power controller. The amplitude of the converter voltage E is produced by the reactive power controller. ω_n is the rated angular frequency which is 100π (50 Hz) in this paper. The i_{dq}^r can be expressed as:

$$
\begin{bmatrix} i_d^r \\ i_q^r \end{bmatrix} = \frac{2}{3} \begin{bmatrix} \cos\theta_r & \cos(\theta_r - 2\pi/3) & \cos(\theta_r + 2\pi/3) \\ -\sin\theta_r & -\sin(\theta_r - 2\pi/3) & -\sin(\theta_r + 2\pi/3) \end{bmatrix} \begin{bmatrix} i_a \\ i_b \\ i_c \end{bmatrix} \tag{1}
$$

The reference voltage e_{dq}^r in the control dq frame is calculated as:

$$
\begin{cases} e_d^r = 0 \\ e_q^r = -E = -\omega_r \psi_f \end{cases} \tag{2}
$$

The electromagnetic active and reactive powers are calculated as:

$$\begin{cases} P_e = e_q^r i_q^r = -\omega_r \psi_f i_q^r \\ Q_e = e_q^r i_d^r = -\omega_r \psi_f i_d^r \end{cases} \tag{3}$$

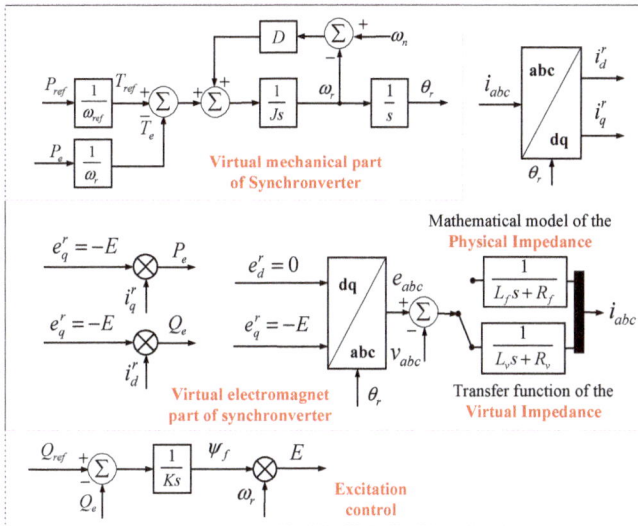

Figure 2. Control diagram of the conventional synchronverter.

To synchronize with the grid voltage without PLL, a self-synchronized method can be adopted [14]. As the grid currents are zero before the switch is closed, the control process is not effective. Therefore, virtual impedances are introduced to simulate the inductance and resistance between the converter and the grid. The virtual currents are calculated according to the grid voltages, converter voltages and virtual impedance as follows:

$$i_{vabc} = \frac{1}{L_v s + R_v}(e_{abc} - v_{abc}) \tag{4}$$

Therefore, when the virtual powers (calculated in the same way as (3)) are controlled to be zero, the virtual currents will be zero simultaneously, which means that the converter voltages have been synchronized with the grid voltages. Finally, when the connection process is finished, the real grid currents are used to replace the virtual currents to calculate the feedback powers.

2.2. Small Signal Analysis and Power Ripple Elimination

The small signal model of synchronverter is developed in synchronous reference frame aligned to the fundamental grid voltage vector [27], which is called system dq frame in this paper in order to distinguish from the control dq frame. In this paper, variables with superscripts s means they are in the system dq frame. Its phase angle θ_1 is calculated as:

$$\theta_1 = \omega_1 t \tag{5}$$

In the normal operation, ω_1 is equal to ω_n. The phase angle difference δ between the system dq frame and control dq frame is calculated as:

$$\delta = \theta_r - \theta_1 \tag{6}$$

The relationships between variables in the system dq frame and their counterparts in control frame are shown as:

$$x_{dq}^s = \begin{bmatrix} x_d^s \\ x_q^s \end{bmatrix} = \begin{bmatrix} \cos\delta & -\sin\delta \\ \sin\delta & \cos\delta \end{bmatrix} \begin{bmatrix} x_d^r \\ x_q^r \end{bmatrix}$$
$$x_{dq}^r = \begin{bmatrix} x_d^r \\ x_q^r \end{bmatrix} = \begin{bmatrix} \cos\delta & \sin\delta \\ -\sin\delta & \cos\delta \end{bmatrix} \begin{bmatrix} x_d^s \\ x_q^s \end{bmatrix} \tag{7}$$

For the synchronverter system studied in the paper, the x in (7) includes output current i_{dq}, grid voltage v_{dq}, and converter voltage e_{dq}.

The derivative of the currents in the system dq frame can be expressed as follows:

$$\begin{cases} di_d^s/dt = (\omega_1 L_f i_q^s - R_f i_d^s + e_d^s - v_d^s)/L_f \\ di_q^s/dt = (-\omega_1 L_f i_d^s - R_f i_q^s + e_q^s - v_q^s)/L_f \end{cases} \tag{8}$$

According to Figure 2, the swing equation of synchronverter is expressed as:

$$d\omega_r/dt = [T_m - T_e - D_p(\omega_r - \omega_1)]/J \tag{9}$$

where $T_m = P_{ref}/\omega_n$, is the nominal virtual mechanical torque input of synchronverter.

The excitation model of synchronverter can be expressed as:

$$d\psi_f/dt = K(Q_{ref} - Q_e) \tag{10}$$

Combining the equations above, the nonlinear model of the synchronverter can be obtained as follows:

$$\begin{cases} di_d^s/dt = (\omega_1 L_f i_q^s - R_f i_d^s + \omega_r \psi_f \sin\delta - v_d^s)/L_f \\ di_q^s/dt = (-\omega_1 L_f i_d^s - R_f i_q^s - \omega_r \psi_f \cos\delta - v_q^s)/L_f \\ d\omega_r/dt = [T_m - T_e - D_p(\omega_r - \omega_1)]/J \\ d\delta/dt = \omega_r - \omega_1 \\ d\psi_f/dt = K(Q_{ref} - Q_e) \end{cases} \tag{11}$$

Linearizing these equations around the steady state, the small signal model can be expressed in state space form as:

$$\begin{cases} \dot{x} = Ax + Bu \\ y = Cx + Du \end{cases} \tag{12}$$

where the state vector x includes the variables shown in (13):

$$x = \left[\Delta i_d^s, \Delta i_q^s, \Delta\omega, \Delta\delta, \Delta\psi_f\right]^T \tag{13}$$

input vector u includes variables shown in (14):

$$u = \left[\Delta v_d^s, \Delta v_q^s, \frac{\Delta P_{ref}}{\omega_n}, \Delta Q_{ref}\right]^T \tag{14}$$

and output vector y includes variables shown in (15):

$$y = \left[\Delta i_d^s, \Delta i_q^s, \Delta T_e, \Delta Q_e\right]^T \tag{15}$$

For simplicity, the 'Δ', denoting perturbation variables, can be omitted when it comes to the linearized small signal system.

The detailed expressions of *A*, *B*, *C* and *D* in (12) can be calculated by Matlab, which are shown as follows:

$$A = \begin{bmatrix} -R_f/L_f & \omega_1 & \psi_f \sin\delta/L_f & \omega_r\psi_f \cos\delta/L_f & \omega_r \sin\delta/L_f \\ -\omega_1 & -R_f/L_f & -\psi_f \cos\delta/L_f & \omega_r\psi_f \sin\delta/L_f & -\omega_r \cos\delta/L_f \\ -\psi_f \sin\delta/J & \psi_f \cos\delta/J & -D_p/J & -\psi_f(i_d^s \cos\delta + i_q^s \sin\delta)/J & (i_q^s \cos\delta + i_d^s \sin\delta)/J \\ 0 & 0 & 1 & 0 & 0 \\ K\omega_r\psi_f \cos\delta & K\omega_r\psi_f \sin\delta & K\psi_f(i_d^s \cos\delta + i_q^s \sin\delta) & K\omega_r\psi_f(i_q^s \cos\delta - i_d^s \sin\delta) & K\omega_r(i_d^s \cos\delta + i_q^s \sin\delta) \end{bmatrix}$$

$$B = \begin{bmatrix} -1/L_s & 0 & 0 & 0 \\ 0 & -1/L_s & 0 & 0 \\ 0 & 0 & 1/J & 0 \\ 0 & 0 & 0 & 0 \\ 0 & 0 & 0 & K \end{bmatrix} \quad D = \begin{bmatrix} 0 & 0 & 0 & 0 \\ 0 & 0 & 0 & 0 \\ 0 & 0 & 0 & 0 \\ 0 & 0 & 0 & 0 \end{bmatrix}$$

$$C = \begin{bmatrix} 1 & 0 & 0 & 0 & 0 \\ 0 & 1 & 0 & 0 & 0 \\ \psi_f \sin\delta & -\psi_f \cos\delta & 0 & \psi_f(i_d^s \cos\delta + i_q^s \sin\delta) & i_d^s \sin\delta - i_q^s \cos\delta \\ -\omega_r\psi_f \cos\delta & -\omega_r\psi_f \sin\delta & -\psi_f(i_d^s \cos\delta + i_q^s \sin\delta) & -\omega_r\psi_f(i_d^s \cos\delta - i_q^s \sin\delta) & -\omega_r(i_d^s \cos\delta + i_q^s \sin\delta) \end{bmatrix}$$

In this study, a small scale 1000 W converter system is used to test the synchronverter control scheme. Its parameters are listed in Table 1.

Table 1. Parameters of VSG system.

System Parameters	Value
Grid voltage	130 V
Rated frequency	50 Hz
Input resistance	0.1 Ω
Input inductance	6 mH
DC bus voltage	280 V

The damping coefficient *D* is a pre-defined parameter according to desired active power drooping profile, e.g., when the grid frequency decreased by 0.5 Hz, the converter shall generate an additional 100% rated power to support the grid. *J* and *K* are considered as tuning parameters. Their impacts on the dynamic of synchronverter are illustrated by pole-zero maps as shown in Figures 3 and 4 using the symbol of blue cross according to (12).

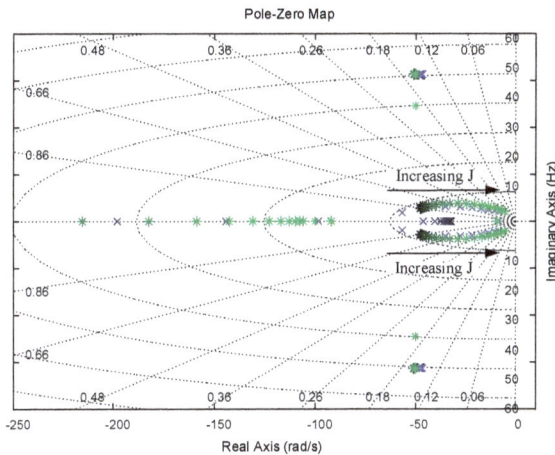

Figure 3. The impact of *J* (increased from 0.06 to 6) and the introduced current lowpass filters on the poles distribution of synchronverter system. Cross: poles without current lowpass filters. Asterisk: poles with current lowpass filters.

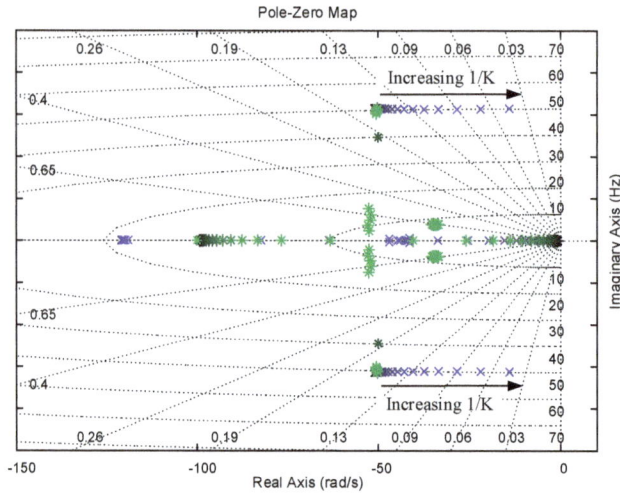

Figure 4. The impact of $1/K$ (increased from 0.075 to 7.5) and the introduced current lowpass filters on the poles distribution of synchronverter system. Cross: poles without current lowpass filters. Asterisk: poles with current lowpass filters.

It's observed in Figures 3 and 4 that with the increase of J and $1/K$, the poles are pushed to the right half of the plain. It is also observed that a pair of conjugated poles at 50 Hz are damped weakly and the situation becomes even worse when the control gain of reactive power ($1/K$) increases. Such a limited stability margin could jeopardize the stable operation of synchronverter due to insufficient noise rejection capability. Thus, there will be power ripples at the frequency of 50 Hz.

In order to eliminate the power ripples, two lowpass filters are applied to the current feeding loop in the synchronverter. The filters are designed in the control dq frame, so the frequency of filter is selected as 0 Hz. The filtered currents can be expressed as follows:

$$i_{fd}^r = \frac{1/\omega_c}{s + 1/\omega_c} i_d^r \quad i_{fq}^r = \frac{1/\omega_c}{s + 1/\omega_c} i_q^r \tag{16}$$

where ω_c is the cutoff frequency. The grid frequency is likely to drift in the situation where synchronverter is applied. Thus, in order to sufficiently reject noises higher than 50 Hz and leave enough margin for grid frequency drift, the cut-off frequency is chosen as 16 Hz.

Equation (16) can also be expressed as follows:

$$\begin{cases} di_{fd}^r/dt = (i_d^r - i_{fd}^r)/\omega_c \\ di_{fq}^r/dt = (i_q^r - i_{fq}^r)/\omega_c \end{cases} \tag{17}$$

Considering (17), the state space model in (12) is modified accordingly as:

$$\begin{cases} \dot{x}_f = A_f x_f + B_f u \\ y = C_f x_f + D_f u \end{cases} \tag{18}$$

where:

$$x_f = \left[i_d^s, i_q^s, i_{fd}^r, i_{fq}^r, \omega_r, \delta, \psi_f \right]^T \tag{19}$$

The detailed expressions of A_f, B_f, C_f and D_f in (18) can be calculated by Matlab, which are very complicated. Therefore, they are not presented for the conciseness of the paper.

The effectiveness of the proposed lowpass filter based current feeding loops is also demonstrated in Figures 3 and 4 using the symbol of green asterisk according to (18). It can be seen that the conjugated poles of modified synchronverter at 50 Hz move to the left and are less sensitive to the increase of $1/K$.

The effectiveness of the proposed filter based power ripple elimination method is illustrated by the bode diagram of transfer function from $[v_d, v_q]^T$ to $[P_e, Q_e]^T$ as shown in Figure 5. It's observed that compared with the conventional strategy, the peak gain from $[v_d, v_q]^T$ to the powers, which lies at 50 Hz, is considerably reduced by adding the lowpass filter. Therefore, the power ripples caused by the disturbance $[v_d, v_q]^T$ will be restrained effectively. In addition, the lowpass filter has limited impact on the transfer function from reference $[P_{ref}, Q_{ref}]^T$ to feedback $[P_e, Q_e]^T$, which indicates that the lowpass filter has little impact on the power reference tracking of the conventional synchronverter.

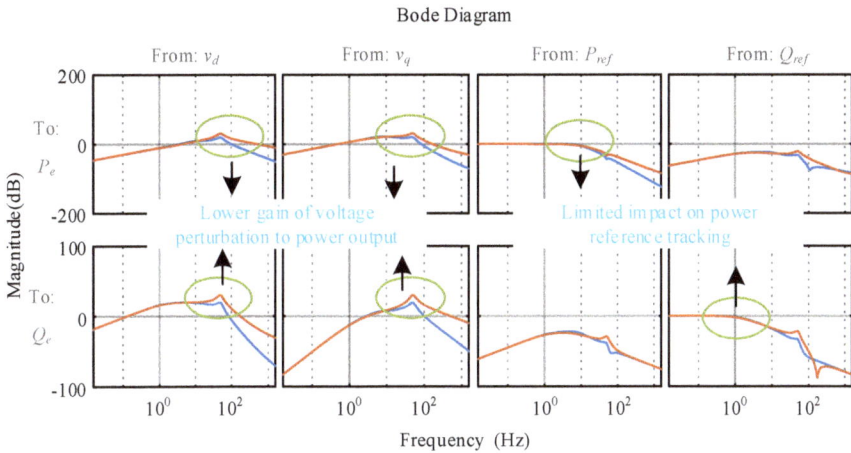

Figure 5. Bode diagram illustrating noise rejection enhancement by adding lowpass filters. Red: without lowpass filters. Blue: with lowpass filters.

3. Self-Synchronized Synchronverter in an Unbalanced Power Grid

3.1. Influence of Unbalanced Grid Voltage

The grid voltage and current have both the negative and positive sequence components when the grid voltage becomes unbalanced, which will cause double frequency ripple components in the reactive and active powers [20]. Power control of conventional synchronverter has limited bandwidth. Thus, power ripples cannot be suppressed in unbalanced power grid, which will seriously affect the control performance of the synchronverter.

In addition, the self-synchronization method proposed in [14] needs to generate a converter voltage very close to the grid voltage in order to avoid the current rushing at the moment when the converter is connected to the grid. However, when the grid voltage becomes unbalanced, the negative sequence current components (-100 Hz in synchronous reference frame) cannot be damped in the conventional synchronverter, which means that the converter voltage cannot precisely follow the grid voltage. Therefore, it's necessary to improve the conventional self-synchronized synchronverter in unbalanced power grid to restrain the negative sequence current and further guarantee the precision of the self-synchronization.

3.2. Modified Self-Synchronized Synchronverter

In the proposed modified self-synchronized synchronverter, the control scheme is expanded into a higher dimensional multiple input multiple output (MIMO) system, i.e., output currents reference $\left[i_{dref}^r, i_{qref}^r\right]^T$ are added into input vector $\boldsymbol{u}$ in (18). The output current references are set to zero.

Then, suppressing double frequency current oscillations in synchronous reference frame is equivalent to guaranteeing current reference tracking only at ±100 Hz. A SOGI based resonant controller is adopted in order to realize this purpose in this paper. The transfer function of the resonant controller is shown as:

$$H_R(s) = \frac{2k_r\omega_{cr}s}{s^2 + 2\omega_{cr}s + \omega_s{}^2} \tag{20}$$

where k_r is the gain parameter; ω_{cr} is the cutoff frequency; ω_s is the selected frequency (±100 Hz). Since the grid frequency is likely to drift in the situation where synchronverter is applied, the cutoff frequency ω_{cr} shall not be too small [27]. In this paper, ω_{cr} is chosen as 10 rad/s, so that the gain of resonant controller is reduced to 0.77 times the nominal gain when the grid frequency is drifted by 0.5 Hz from the nominal 50 Hz.

The modified control diagram is shown in Figure 6. The inputs of resonant controller are the errors between reference and feedback currents (virtual current in the self-synchronization mode, grid current in the power-control mode) and the outputs of resonant controller are added to converter voltage reference generated by power control part of synchronverter. The final converter voltage reference e^r_{fd} and e^r_{fq} can be expressed as follows:

$$\begin{cases} e^r_{fd} = e^r_{cd} + e^r_d = e^r_{cd} \\ e^r_{fq} = e^r_{cq} + e^r_q = e^r_{cq} - \omega_r\psi_f \end{cases} \tag{21}$$

where e^r_{cd} and e^r_{cq} are the outputs of the resonant controllers.

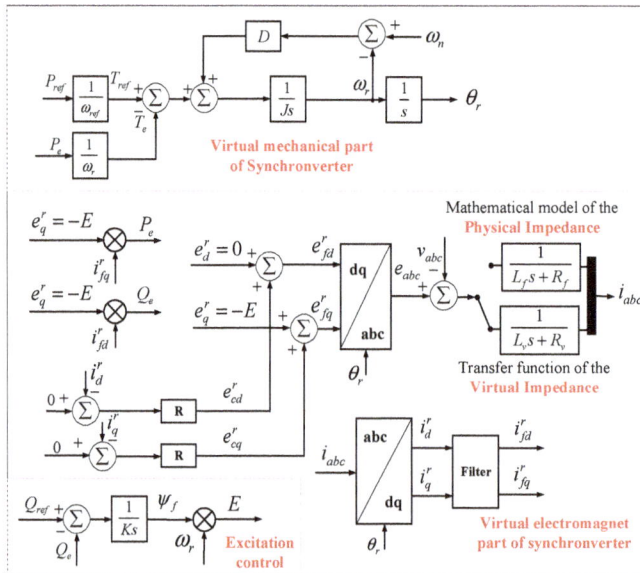

Figure 6. Control diagram of the proposed self-synchronized synchronverter in unbalanced power grid.

It should be noted that the sum of the three phase virtual currents calculated by (4) will not be zero in unbalanced power grid. Thus, the calculation process of the virtual currents is revised as (22):

$$i_{va} = \frac{e_a - v_a}{L_v s + R_v} \quad i_{vb} = \frac{e_b - v_b}{L_v s + R_v} \quad i_{vc} = -i_{va} - i_{vb} \tag{22}$$

To analyze the characteristics of the proposed strategy, the small signal model of synchronverter needs to be expanded considering two extra resonant controllers. According to (20), the following equations can be obtained:

$$\begin{cases} d^2e^r_{cd}/dt^2 + 2\omega_{cr}de^r_{cd}/dt + \omega_s^2 e^r_{cd} = -2k_r\omega_{cr}di^r_d/dt \\ d^2e^r_{cq}/dt^2 + 2\omega_{cr}de^r_{cq}/dt + \omega_s^2 e^r_{cq} = -2k_r\omega_{cr}di^r_q/dt \end{cases} \tag{23}$$

Equation (23) can also be expressed as follows:

$$\begin{cases} d\gamma_{d1}/dt = \gamma_{d2} \\ d\gamma_{d2}/dt = -2\omega_{cr}\gamma_{d2} - \omega_s^2\gamma_{d1} - 2k_r\omega_{cr}di^r_d/dt \\ d\gamma_{q1}/dt = \gamma_{q2} \\ d\gamma_{q2}/dt = -2\omega_{cr}\gamma_{q2} - \omega_s^2\gamma_{q1} - 2k_r\omega_{cr}di^r_q/dt \end{cases} \tag{24}$$

where:

$$\begin{cases} \gamma_{d1} = e^r_{cd} \\ \gamma_{d2} = de^r_{cd}/dt \\ \gamma_{q1} = e^r_{cq} \\ \gamma_{q2} = de^r_{cq}/dt \end{cases} \tag{25}$$

Considering (21) and (24), the new state space can be expressed as follows:

$$\begin{cases} \dot{x}_r = A_r x_r + B_r u \\ y = C_r x_r + D_r u \end{cases} \tag{26}$$

where x_r includes additional states introduced by resonant controllers, which can be shown as follows:

$$x_r = \left[i^s_d, i^s_q, i^r_{fd}, i^r_{fq}, \omega_r, \delta, \psi_f, \gamma_{d1}, \gamma_{d2}, \gamma_{q1}, \gamma_{q2} \right]^T \tag{27}$$

The detailed expressions of A_r, B_r, C_r and D_r in (26) can be calculated by Matlab, which are very complicated. Therefore, they are not presented for the conciseness of the paper.

Nevertheless, it needs to be clarified that because of the unbalanced power grid, there are additional periodic components in the steady operation point. However, since the overall system shows lowpass filter characteristic and limited unbalance degree, the modelling error caused by the periodic components in steady state is acceptable [28].

The comparison of bode diagrams of both the conventional and modified synchronverter are illustrated in Figure 7 according to (12) and (26). The red line in Figure 7 is the bode diagram of conventional synchronverter and it's observed that the magnitude of transfer function from $[v_d, v_q]^T$ to $\left[i^r_d, i^r_q \right]^T$ at 100 Hz is much higher than 0 dB. It means that the negative sequence component in unbalanced grid voltage will be amplified in the control system and finally results in large negative sequence output current. The resonant controllers enable the synchronverter to have current reference tracking capability at 100 Hz in synchronous reference frame because of the unity gain of the transfer function from $\left[i^r_{dref}, i^r_{qref} \right]^T$ to $\left[i^r_d, i^r_q \right]^T$ at 100 Hz. After adding resonant controllers, the magnitude of the transfer function from $[v_d, v_q]^T$ to $\left[i^r_d, i^r_q \right]^T$ at 100 Hz reduces significantly. This verifies the effectiveness of proposed negative sequence current suppression strategy using a resonant controller. In addition, the magnitude of transfer function from $[v_d, v_q]^T$ to $[P_e, Q_e]^T$ at 100 Hz also reduces significantly, which means that the double frequency power ripples can also be reduced by using resonant controller.

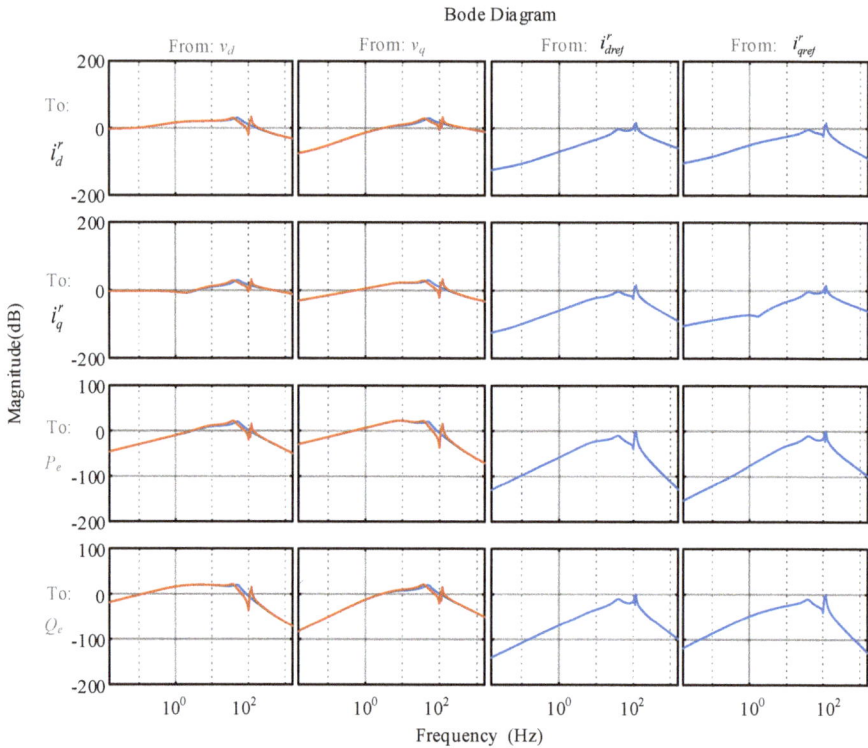

Figure 7. Bode diagram illustrating the effect of proposed resonant controller. Blue line: modified synchronverter with additional resonant controller. Red line: conventional synchronverter.

3.3. Parameter Design of The Resonant Controller

As the resonant controller has a wide resonance peak against frequency variation, the impact of resonant controller to the system stability needs to be considered, and the resonant controller gain k_r needs to be designed properly. According to the analyses of power control loop demonstrated in Section 2.2, the parameters $D, J, 1/K$ are chosen as 100, 0.6, 0.75, respectively, and the bandwidth of resonant controller is chosen as 10 rad/s. In order to investigate the influence of the parameter k_r to the stability of the control system, the pole-zero map of the modified synchronverter with different k_r is shown in Figure 8 according to Equation (26). In addition, to verify the effectiveness of the proposed filter based power ripple elimination method in unbalanced power grid, the results with and without the current lowpass filters are shown as blue asterisk and green cross, respectively.

It can be observed that two additional pairs of conjugate poles are added at around 100 Hz, which are caused by resonant controllers. It can be seen from Figure 8 that the increasing k_r will drive the poles to the right, and too large k_r may push the poles related to resonant controller to the right half of the plane so that the system will become unstable. Moreover, Figure 8 also shows that the current lowpass filter can make the poles at around 50 Hz move to the left, and it has very limited impact on poles related to the resonant controller. Based on the analyses above, the control parameters for the modified synchronverter can be selected as shown in Table 2, where all the parameters are normalized values.

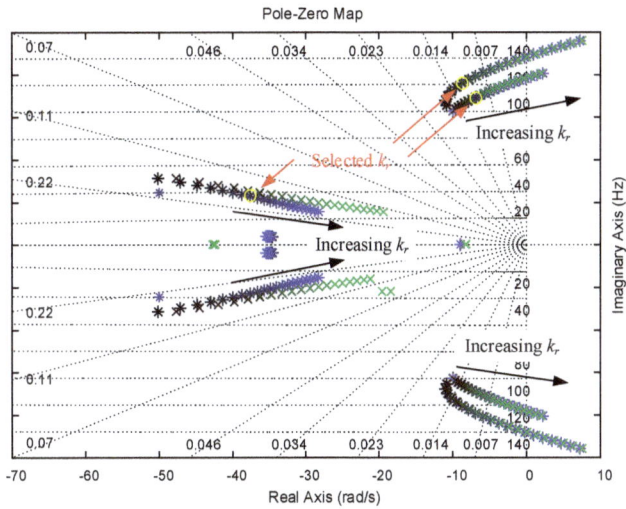

Figure 8. The impact of *kr* (increased from 0 to 6.3) and the introduced current lowpass filters on the poles distribution of synchronverter system. Cross: poles without current lowpass filters. Asterisk: poles with current lowpass filters.

Table 2. Parameters of VSG system.

System Parameters	Value
Virtual inertia J	0.6
Damping coefficient D	100
Excitation integral gain $1/K$	0.75
Gain of resonant controller k_r	1.86

4. Experimental Studies

4.1. Experimental Setup

Comparative experimental studies are conducted for the conventional and proposed synchronverters in an unbalanced power grid. The experimental platform is set up as shown in Figure 9. A TMS320F2812 Digital Signal Processor is used as the microprocessor. The unbalanced grid voltages are simulated using a transformer by making one phase of the voltages drop to 80%. The dc-bus voltage is assumed to be controlled as a constant value by the distributed generation system. Parameters of the synchronverter are shown in Tables 1 and 2. The switching frequency and sampling frequency are both set to 10 kHz. All the experimental results are shown in Figures 10–14.

4.2. Experimental Results

The dynamic responses of the conventional and proposed self-synchronized synchronverters during the grid connection process are shown in Figure 10.

Before the grid connection, the voltage errors between the grid and converter voltages are shown in α and β axes separately. And the voltage errors are changed to active and reactive powers when the switch is closed to show the dynamic response of the powers. The voltage errors contain ripples in the conventional synchronverter as shown in Figure 10a, which means that the synchronization is not precise. Thus, the currents are very large when the switch is closed. Compared with the conventional synchronverter, the voltage errors are effectively limited in the proposed synchronverter as shown in Figure 10b. In addition, the grid currents are smaller and damped much faster after the switch is closed. Therefore, the converter can be connected to the grid more smoothly when the proposed synchronverter is used in an unbalanced power grid.

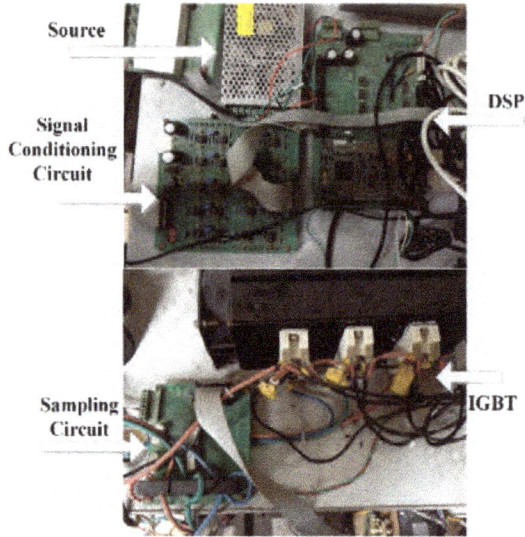

Figure 9. The experimental platform.

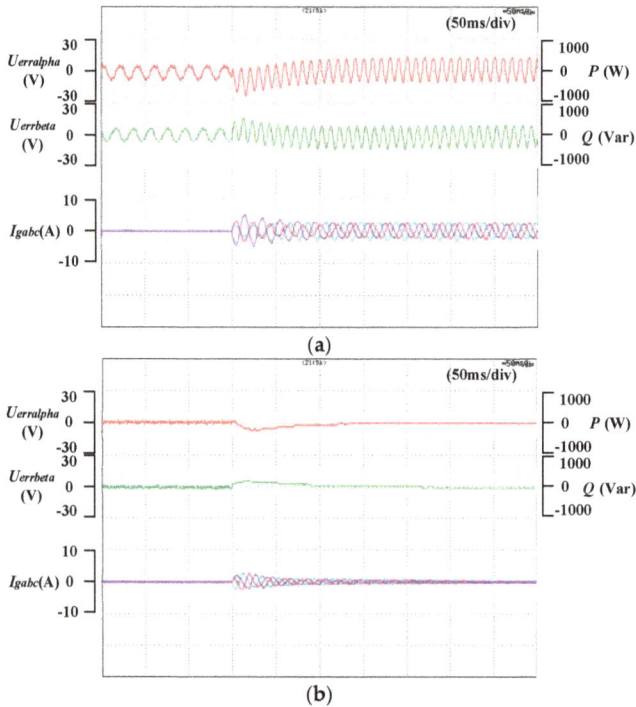

Figure 10. Dynamic responses during the grid connection process. (**a**) Conventional self-synchronized synchronverter. (**b**) Proposed self-synchronized synchronverter.

The steady-state power responses of the conventional and proposed synchronverters are shown in Figure 11. The Total Harmonic Distribution (THD) analyses of grid currents are shown in Figure 12, and the Fast Fourier Transform (FFT) analyses of active power are shown in Figure 13. In the conventional synchronverter, there are negative (-50 Hz) and third sequence (150 Hz) components in grid currents as shown in Figure 12a, and large ripples in both the reactive and active powers in Figure 11a. Current harmonics are eliminated in the proposed synchronverter as shown in Figure 12b. In addition, the power ripples are also effectively reduced in Figure 11b. Therefore, it can be concluded that the resonant based synchronverter can achieve much better control performance in unbalanced power grid compared with the conventional one.

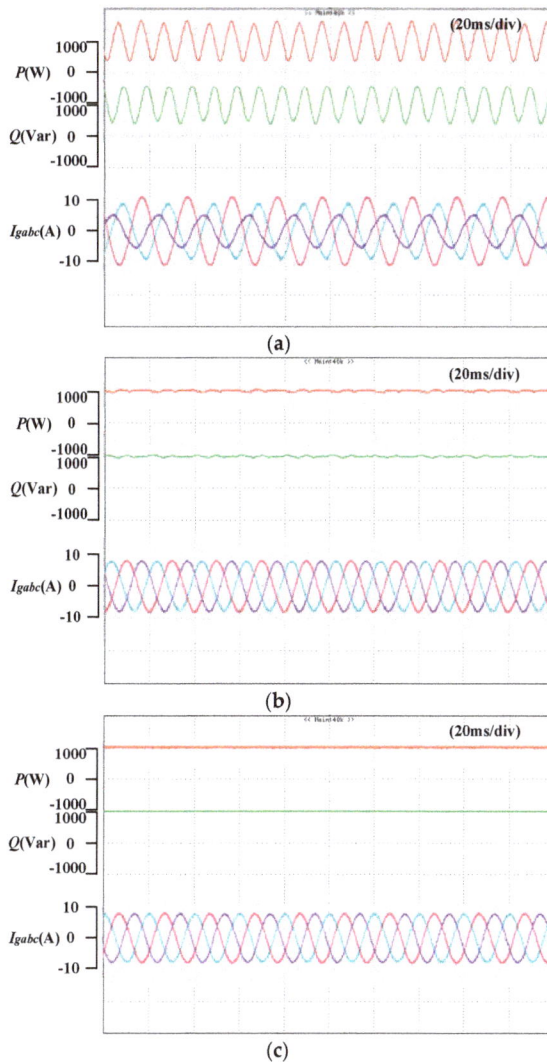

Figure 11. Steady-state responses. (**a**) Conventional synchronverter. (**b**) Proposed synchronverter without filter. (**c**) Proposed synchronverter with filter.

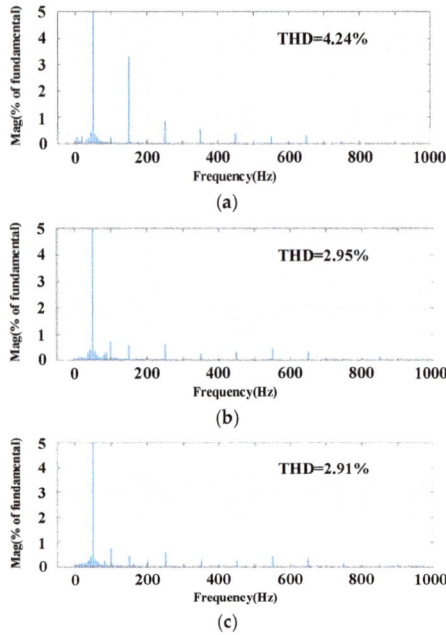

Figure 12. Grid currents THD analyses. (**a**) Conventional synchronverter. (**b**) Proposed synchronverter without filter. (**c**) Proposed synchronverter with filter.

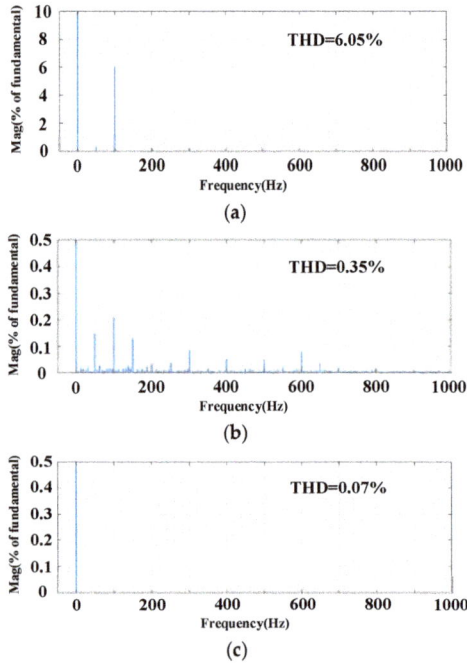

Figure 13. FFT analyses of active power. (**a**) Conventional synchronverter. (**b**) Proposed synchronverter without filter. (**c**) Proposed synchronverter with filter.

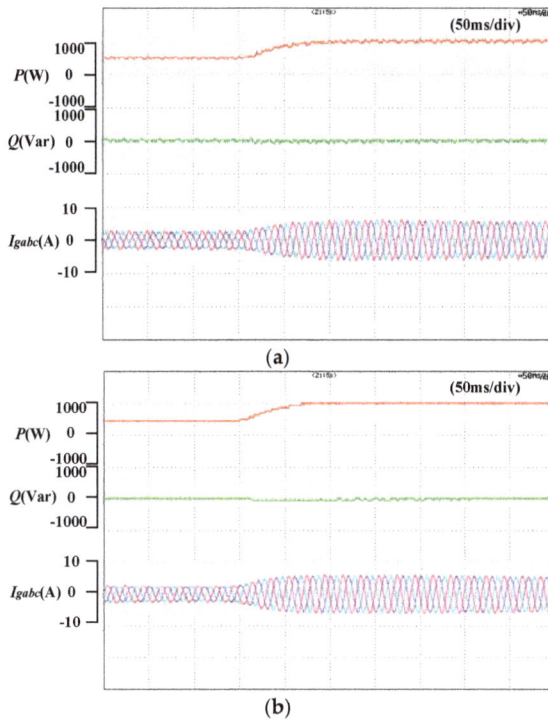

Figure 14. Active power step responses. (**a**) Proposed synchronverter without filter. (**b**) Proposed synchronverter with filter.

Compared with Figure 11b, the proposed synchronverter with filter in Figure 11c can further reduce the power ripples. It can be observed in Figure 13b that the power ripple components are mainly concentrated on the multiples of the 50 Hz, which coincides with the theoretical analyses. And the power ripples can be eliminated as shown in Figure 13c, which verifies the effectiveness of the proposed power ripple elimination method.

Besides, the impact of proposed filter on the dynamic of power control is shown in Figure 14. It's observed that the active power step responses of the proposed synchronverters without and with filter are almost identical, which verifies the analysis illustrated in Figure 5 that the filter doesn't affect the dynamic performance of the synchronverter.

5. Conclusions

A modified self-synchronized synchronverter has been proposed in this paper for unbalanced power grids. Based on the small signal analyses, it is observed that the stability margin of the conventional synchronverter around 50 Hz is very limited. Thus, a filter has been used to enhance the stability and eliminate the power ripples. In addition, a resonant-based strategy has been proposed to control the 100 Hz oscillations. Comparative experimental studies show that in the self-synchronized mode, the proposed strategy can make the synchronization more precise and the grid connection process smoother. In the power-control mode, the negative and third sequence harmonic components in grid currents can be eliminated, and the power ripples can be restrained. In addition, the power ripples can be further eliminated by the filter based method, and the dynamic control performance is not affected by the filter.

By employing the proposed strategy, the synchronverter can synchronize with the grid precisely without PLL in an unbalanced power grid, and the quality of the powers and currents flowing into the grid can be improved. Therefore, the stability of the grid can be supported by the renewable power generation systems under both balanced and unbalanced grid conditions and the efficiency of the generation systems can be increased because it does not need to disconnect from the grid under slight grid fault conditions. These merits lay the foundation for the better application of the renewable power generation in non-ideal power grid, which also lead to the great engineering practical and economic potential of the proposed strategy.

Author Contributions: Conceptualization, X.W. and D.S.; methodology, L.C. and X.W.; software, X.W. and L.C.; validation, X.W. and L.C.; formal analysis, X.W. and L.C.; data curation, D.S., L.Z. and H.N.; writing—original draft preparation, X.W. and L.C.; writing—review and editing, D.S. and H.N.; supervision, D.S. and H.N.; project administration, D.S. and L.Z.; funding acquisition, D.S. and L.Z.

Funding: This research was supported by Zhejiang Provincial Natural Science Foundation of China (No. LZ18E070001); National Natural Science Foundation of China, (No. 51622706); and Open Fund of Operation and Control of Renewable Energy & Storage Systems (China Electric Power Research Institute) (No. 1810-00893).

Conflicts of Interest: The authors declare no conflict of interest.

Nomenclature

v, e	Grid, converter voltage vectors.
i, i_v	Grid, virtual current vectors.
ψ_f	Exciting flux linkage.
P_e, Q_e	Active, reactive electromagnetic powers.
R_f, R_v	Grid, virtual resistances.
L_f, L_v	Grid, virtual inductances.
ω_1, ω_r	Synchronous, converter angular frequencies.
θ_1, θ_r	Synchronous, converter phase angles.

References

1. Huang, Y.; Zhai, X.; Hu, J.; Liu, D.; Lin, C. Modeling and Stability Analysis of VSC Internal Voltage in DC-Link Voltage Control Timescale. *IEEE J. Emerg. Sel. Top. Power Electron.* **2018**, *6*, 16–28. [CrossRef]
2. Subudhi, B.; Pradhan, R. A Comparative Study on Maximum Power Point Tracking Techniques for Photovoltaic Power Systems. *IEEE Trans. Sustain. Energy* **2013**, *4*, 89–98. [CrossRef]
3. Esram, T.; Chapman, P.L. Comparison of Photovoltaic Array Maximum Power Point Tracking Techniques. *IEEE Trans. Energy Convers.* **2007**, *22*, 439–449. [CrossRef]
4. Kundur, P.; Paserba, J.; Ajjarapu, V.; Andersson, G.; Bose, A.; Canizares, C.; Hatziargyriou, N.; Hill, D.; Stankovic, A.; Taylor, C.; et al. Definition and classification of power system stability IEEE/CIGRE joint task force on stability terms and definitions. *IEEE Trans. Power Syst.* **2004**, *19*, 1387–1401.
5. Duckwitz, D.; Fischer, B. Modeling and Design of df/dt-based Inertia Control for Power Converters. *IEEE J. Emerg. Sel. Top. Power Electron.* **2017**, *5*, 1553–1564. [CrossRef]
6. D'Arco, S.; Suul, J.A. Virtual synchronous machines-Classification of implementations and analysis of equivalence to droop controllers for microgrids. In Proceedings of the 2013 IEEE Grenoble Conference, Grenoble, France, 16–20 June 2013; pp. 1–7.
7. D'Arco, S.; Suul, J.A. Equivalence of Virtual Synchronous Machines and Frequency-Droops for Converter-Based MicroGrids. *IEEE Trans. Smart Grid* **2014**, *5*, 394–395. [CrossRef]
8. Sakimoto, K.; Miura, Y.; Ise, T. Stabilization of a power system with a distributed generator by a Virtual Synchronous Generator function. In Proceedings of the 8th International Conference on Power Electronics-ECCE Asia, Jeju, South Korea, 30 May–3 June 2011; pp. 1498–1505.
9. Wu, H.; Ruan, X.; Yang, D.; Chen, X.; Zhao, W.; Lv, Z.; Zhong, Q. Small-Signal Modeling and Parameters Design for Virtual Synchronous Generators. *IEEE Trans. Ind. Electron.* **2016**, *63*, 4292–4303. [CrossRef]

10. Liu, J.; Miura, Y.; Ise, T. Comparison of Dynamic Characteristics Between Virtual Synchronous Generator and Droop Control in Inverter-Based Distributed Generators. *IEEE Trans. Power Electron.* **2016**, *31*, 3600–3611. [CrossRef]

11. D'Arco, S.; Suul, J.A.; Fosso, O.B. Control system tuning and stability analysis of Virtual Synchronous Machines. In Proceedings of the 2013 IEEE Energy Conversion Congress and Exposition, 15–19 September 2013; pp. 2664–2671.

12. Amin, M.R.; Aizam Zulkifli, S. A framework for selection of grid-inverter synchronisation unit: Harmonics, phase-angle and frequency. *Renew. Sustain. Energy Rev.* **2017**, *78*, 210–219. [CrossRef]

13. Zhong, Q.C.; Weiss, G. Synchronverters: Inverters That Mimic Synchronous Generators. *IEEE Trans. Ind. Electron.* **2011**, *58*, 1259–1267. [CrossRef]

14. Zhong, Q.C.; Nguyen, P.L.; Ma, Z.; Sheng, W. Self-Synchronized Synchronverters: Inverters without a Dedicated Synchronization Unit. *IEEE Trans. Power Electron.* **2014**, *29*, 617–630. [CrossRef]

15. Natarajan, V.; Weiss, G. Almost global asymptotic stability of a grid-connected synchronous generator. *arXiv* **2016**; arXiv:161004858.

16. Yazdani, A.; Iravani, R. A unified dynamic model and control for the voltage-sourced converter under unbalanced grid conditions. *IEEE Trans. Power Deliv.* **2006**, *21*, 1620–1629. [CrossRef]

17. Hu, Y.; Zhu, Z.Q.; Odavic, M. Instantaneous Power Control for Suppressing the Second Harmonic DC Bus Voltage under Generic Unbalanced Operating Conditions. *IEEE Trans. Power Electron.* **2017**, *32*, 3998–4006. [CrossRef]

18. Nian, H.; Cheng, P.; Zhu, Z.Q. Independent Operation of DFIG-Based WECS Using Resonant Feedback Compensators under Unbalanced Grid Voltage Conditions. *IEEE Trans. Power Electron.* **2015**, *30*, 3650–3661. [CrossRef]

19. Cheng, P.; Nian, H. Collaborative Control of DFIG System during Network Unbalance Using Reduced-Order Generalized Integrators. *IEEE Trans. Energy Convers.* **2015**, *30*, 453–464. [CrossRef]

20. Cheng, P.; Nian, H. Direct power control of voltage source inverter in a virtual synchronous reference frame during frequency variation and network unbalance. *IET Power Electron.* **2016**, *9*, 502–511. [CrossRef]

21. Natarajan, V.; Weiss, G. Synchronverters with better stability due to virtual inductors, virtual capacitors and anti-windup. *IEEE Trans. Ind. Electron.* **2017**, *64*, 5994–6004. [CrossRef]

22. Zhong, Q.C.; Konstantopoulos, G.C.; Ren, B.; Krstic, M. Improved Synchronverters with Bounded Frequency and Voltage for Smart Grid Integration. *IEEE Trans. Smart Grid* **2018**, *9*, 786–796. [CrossRef]

23. Wu, W.; Zhou, L.; Chen, Y.; Luo, A.; Dong, Y.; Zhou, X.; Xu, Q.; Yang, L.; Guerrero, J.M. Sequence-Impedance-Based Stability Comparison between VSGs and Traditional Grid-Connected Inverters. *IEEE Trans. Power Electron.* **2019**, *34*, 46–52. [CrossRef]

24. Rodríguez-Cabero, A.; Roldan-Perez, J.; Prodanovic, M. Synchronverter small-signal modelling and eigenvalue analysis for battery systems integration. In Proceedings of the 2017 IEEE 6th International Conference on Renewable Energy Research and Applications, San Diego, CA, USA, 5–8 November 2017; pp. 780–784.

25. Dong, S.; Chen, Y.C. A Method to Directly Compute Synchronverter Parameters for Desired Dynamic Response. *IEEE Trans. Energy Convers.* **2018**, *33*, 814–825. [CrossRef]

26. Dong, S.; Chen, Y.C. Adjusting Synchronverter Dynamic Response Speed via Damping Correction Loop. *IEEE Trans. Energy Convers.* **2017**, *32*, 608–619. [CrossRef]

27. Wen, B.; Boroyevich, D.; Burgos, R.; Mattavelli, P.; Shen, Z. Analysis of D-Q Small-Signal Impedance of Grid-Tied Inverters. *IEEE Trans. Power Electron.* **2016**, *31*, 675–687. [CrossRef]

28. Gelb, A.; Vander Velde, W.E. *Multiple-Input Describing Functions and Nonlinear System Design*; McGraw-Hill: New York, NY, USA, 1968.

energies

MDPI

Article

Development of Grid-Connected Inverter Experiment Modules for Microgrid Learning

Arwindra Rizqiawan [1,*,†], **Pradita Hadi** [2,*,†] **and Goro Fujita** [2,*]

[1] School of Electrical Engineering and Informatics, Institut Teknologi Bandung, Bandung 40132, Indonesia

[2] College of Engineering, Shibaura Insitute of Technology, Tokyo 135-8548, Japan

* Correspondence: windra@stei.itb.ac.id (A.R.); na15503@shibaura-it.ac.jp (P.H.); gfujita@sic.shibaura-it.ac.jp (G.F.)

† These authors contributed equally to this work.

Received: 19 December 2018; Accepted: 30 January 2019; Published: 1 February 2019

Abstract: New paradigms in the modern power system should be introduced to student of electrical engineering, or engineer in training, as early as possible. Besides class-room study, experimental exercise may be introduced to help the student understand the concept of microgrid. One main challenge is the power electronics converter, which connects the distributed energy source to the existing power grid. This study modeled and developed a grid-connected inverter that is useful for providing a close to real application for a student or engineer in training. This development is important for microgrid learning to give practical perspective to the student. A grid-connected inverter for distributed generation was developed at laboratory scale. The grid-connected inverter was developed modularly to make it easier for the student to understand the basic concept of grid-connected inverter building blocks, as well as its function as a whole. The developed grid-connected inverter was intended to be able to operate on two different mode: grid-forming mode and grid-injecting mode. Experiments were conducted to verify the results.

Keywords: inverter; grid-connected; microgrid; experiment; modules

1. Introduction

The concern of climate change leads to urgent calls to reduce greenhouse gas emissions. Electrical energy sector offers easier and relatively faster implementation to reduce greenhouse gas emissions by proposing utilization of renewable energy sources that other sectors such as transportation. Distributed generation concept is sought to be the best way to incorporate renewable energy sources in our existing power grid network, while also improving the grid quality [1,2]. Traditional renewable energy sources for distributed generation power system include bio-mass, solar photovoltaic, fuel cell, and wind [3]. Besides those mentioned sources, distributed generation also proposes combined heat and power (CHP) concept. CHP concept aims to improve our total energy efficiency by reducing and utilizing thermal losses of energy production [4,5].

Distributed generation concept brings larger consequences that significant changes are inevitable in the power system. Traditional power grid network has unidirectional energy transfer from the centralized bulk power generating units to the customers. The consumers do not have any active contribution to the power system by generating local energy sources because all power comes from the utility company [6]. Contrarily, distributed generation requires active bidirectional power delivery, either from the utility to the consumer or the reverse in the case of the local generation producing excess energy. It needs adjustments on several aspects of power system, such as energy business, reverse power transfer, voltage stabilization,

Energies **2019**, *12*, 476; doi:10.3390/en12030476   www.mdpi.com/journal/energies

islanding scheme, etc. Distributed generation is an technological improvement, which should be able to work together with the current traditional centralized system to obtain a reliable and sustainable power system [5].

Microgrid is a special derivation of the distributed generation concept. Microgrid takes a system approach that views distributed generation and its associated load as a "cell" or "subsystem" of a larger power system network [7]. This approach allows local control of generation and load consumption, thus reducing the central dispatch control [8]. Since this subsystem has characteristics normally found in power systems but at a smaller size, the term *microgrid* is used [7]. Today, several microgrid testbeds are developed, such as microgrid testbed by Consortium for Electric Reliability Technology Solutions (CERTS) in U.S. [9]; Hachinohe, Aichi, and Kyoto Eco-Energy Project in Japan [10]; and many others.

Microgrid is an active distribution network. It is a distributed generation that utilizes renewable energy sources to supply energy to the load at distribution voltage level. The advantage of using distribution voltage level is the renewable energy sources do not need to have big capacity, rather distributed but in smaller size. Since small capacity renewable energy sources are sufficient, microgrid can utilize local energy potentials, which lead to energy independent communities. This point is a very interesting feature of microgrid for developed countries, which still have remote areas without electricity. Microgrid offers flexibility to be operated under two modes: interconnected and stand-alone from the main power grid. Interconnection with the main grid is very important because of the intermittent characteristics of renewable energy sources.

These new paradigms in the modern power system should be introduced to students as early as possible. Besides class-room study, experimental exercises may be introduced to help students to understand the concept of microgrid. One main challenge is the power electronics converter, which connects the distributed energy source to the existing power grid. Power electronics converter plays an important role to deliver power from the distributed energy source to the main grid. This research is intended as the preliminary work on developing grid-connected inverter for microgrid learning. The current target was to develop a grid-connected inverter that can control its power output, therefore can be operated in two different modes: grid-forming mode and grid-injecting mode.

2. Grid-Connected Inverter

Grid-connected inverter is a special type of AC-DC converter, the output voltage of which is connected to the available electrical grid. Contrary to applications such as Uninterruptible Power Supply (UPS) or Adjustable Speed Drive (ASD), where AC-DC converter output is connected directly to the load (or electric motor, in the case of ASD), grid-connected inverter is always connected to the electrical grid. Therefore, grid-connected inverter should always produce output that meet the synchronization condition with the electrical grid. Failing to meet the synchronization condition will make the AC-DC converter be in lost synchronism state and no longer in grid connection. Grid-connected inverter can have either single-phase or three-phase topology, however this section only discusses three-phase grid-connected inverter.

Figure 1 shows simple representation of grid-connected inverter concept. It is connected to the AC grid through filter inductance L. The control block processes the desired output variables through feedback from its terminal. It controls any kind of variables, such as output power, voltage, current, or DC voltage, depending on the grid-connected inverter's particular function. Synchronization block senses and processes the grid's variable into control block providing the necessary information to keep the DC-AC converter synchronized with AC grid. This representation was the basic block of the proposed experimental modules.

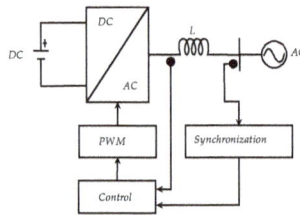

Figure 1. Simple representation of grid-connected inverter.

3. Proposed Experimental Modules

3.1. Grid-Forming Mode Control

Grid-connected inverter generates constant magnitude and frequency of three-phase voltage under grid-forming mode control [11]. This mode works when microgrid is not connected to the main grid, thus it supplies its own local load from the renewable energy source. Figure 2 shows the complete grid-forming mode control of grid-connected inverter. It consists of double-loop control of current and voltage control section, voltage reference section, and measurement section. Voltage reference can be either internal or external three-phase signal through *dq* transformation and PLL. This scheme generates three-phase voltage *v*, which follows the voltage reference.

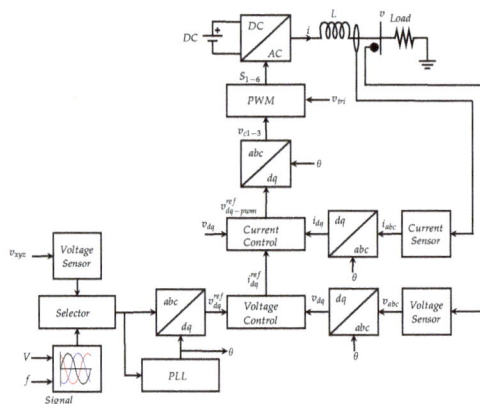

Figure 2. Grid-forming control of grid-connected inverter.

To construct the grid-forming mode control of grid-connected inverter, several blocks should be prepared.

3.1.1. *dq* Transformation

dq transformation, often called Park transformation, is a space vector transformation of three-phase time-domain signals from a stationary phase coordinate system (*abc*) to a rotating coordinate system (*dq*) [12]. The *dq* transform reduces three-phase AC quantities (f_a, f_b, and f_c) into two DC quantities (f_d and f_q). The DC quantities provide easier control or filtering process. Another advantage is that active and reactive power can be controlled independently by using *dq* quantities.

Power invariant formulation of *dq* transformation of any variable, f, of three-phase signals into *dq* quantities can be expressed by Equation (1).

$$
\begin{bmatrix} f_d \\ f_q \end{bmatrix} = \sqrt{\frac{2}{3}} \begin{bmatrix} \cos(\theta) & \cos(\theta - \frac{2\pi}{3}) & \cos(\theta + \frac{2\pi}{3}) \\ -\sin(\theta) & -\sin(\theta - \frac{2\pi}{3}) & -\sin(\theta + \frac{2\pi}{3}) \end{bmatrix} \begin{bmatrix} f_a \\ f_b \\ f_c \end{bmatrix}
\tag{1}
$$

where θ is phase displacement between the rotating and fixed coordinate system at each time t.

3.1.2. Phase-Locked Loop (PLL)

Phase-locked loop (PLL) is a popular method for synchronizing utility network and grid-connected power electronics converter [13,14]. In such kind of applications, PLL's role is essential to provide accurate and fast detection of utility phase angle for generating reference signal of grid-connected inverter.

The common method to realize PLL on three-phase system is by using synchronous frame PLL. Figure 3 shows the block diagram of synchronous frame PLL. Instantaneous phase angle θ is detected by synchronizing the reference frame with the vector of three-phase voltage (v_a, v_b, and v_c). To be locked to the phase angle of utility voltage vector, the Proportional–Integral (PI) controller drives v_q to be zero, as given by the reference v_q*, which gives the angular speed quantities. Phase angle θ is simply obtained by integrating the angular speed.

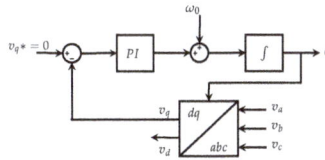

Figure 3. Block diagram of Phase-Locked Loop (PLL).

3.1.3. Current Control

Current control acts as inner loop section on the control of grid-connected inverter [15,16]. Figure 4 shows the block diagram of *dq* cross-decoupling current control scheme, where both i_d and i_q are controlled using independent PI controller to track reference values i_d^{ref} and i_q^{ref}, respectively. Output of PI controller, together with the terminal voltage in *dq* and cross drop voltage in the inductor filter L, generates the voltage references v_{d-pwm}^{ref} and v_{q-pwm}^{ref}. Those voltage references drive the Pulse Width Modulation (PWM) to produce the corresponding command for the switches of the grid-connected inverter.

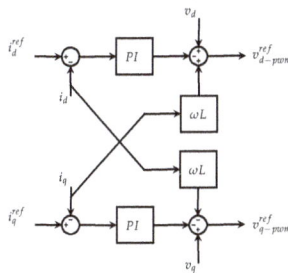

Figure 4. Current control block diagram.

3.1.4. Voltage Control

Voltage control acts as outer loop in the grid-forming mode of grid-connected inverter [17]. Figure 5a,b shows independent control of voltage in dq frame by using two different PI controllers. PI controller tracks the error between the reference voltage, v_d^{ref} (or v_q^{ref}), and actual output voltage, v_d (or v_q). It generates corresponding signal for the reference current controller (Section 3.1.3), which in turn determines the required current of the grid-connected inverter.

In the grid-forming mode, the reference voltage can be obtained from either signal generator or voltage sensor measurement through abc to dq transformation. Figure 6 shows the scheme of voltage reference generation blocks. Signal generator receives input of frequency, f, and magnitude, V, to generate three-phase signal reference. This block is mainly used when the grid-connected inverter is on stand-alone operation supplying power to the local loads, thus it should provide stable output voltage.

The other method to generate voltage reference is by using three-phase measurement on voltage sensor, and then transforming into the corresponding dq voltage reference. This method is suitable when the grid-connected inverter is synchronized with the main grid, for which the synchronized set of three-phase output needs to be produced by grid-connected inverter. Three-phase voltage of the main grid on the point of synchronization is measured through voltage sensor, and then the phase angle is detected by using PLL to obtain proper dq transformation angle. Voltage control tracks this set of reference voltage, therefore it is expected that the output voltage of grid-connected inverter is synchronized with the main grid.

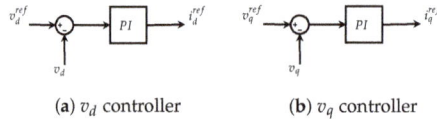

(a) v_d controller   (b) v_q controller

Figure 5. Voltage control block diagram.

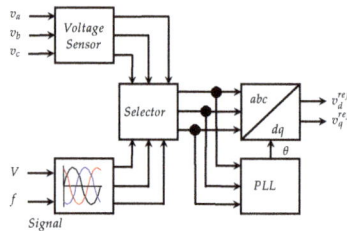

Figure 6. Voltage reference generation scheme.

3.1.5. PWM Generation

Voltage references for PWM generated by current control, v_{d-pwm}^{ref} and v_{q-pwm}^{ref}, are still in dq frame. However, the sinusoidal PWM method needs three modulating signal displaced 120° as its input values. Therefore, the voltage references v_{d-pwm}^{ref} and v_{q-pwm}^{ref} should be transformed back to abc reference frame by using inverse Park transformation. To ensure that the output of grid-connected inverter remains synchronized with the desired voltage, the phase angle θ obtained from PLL in Figure 6 is used for the inverse dq transformation. The results are three controlled reference signals, v_{c1}, v_{c2}, and v_{c3}, for the PWM block. Figure 7 shows the schematic of this PWM generation scheme.

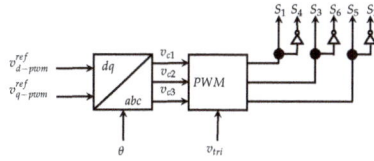

Figure 7. PWM generation scheme using *dq* transformation.

Other method is to use polar coordinate to transform PWM voltage references v_{d-pwm}^{ref} and v_{q-pwm}^{ref} into magnitude m and phase angle δ. Angle θ from PLL is used to synchronize with the voltage reference, therefore the signal generator block produces balanced three-phase reference voltage. For instance, the phase A reference voltage is shown in Equation (2), while the other phases are obtained from phase A by shifting it accordingly. Figure 8 shows this PWM generation scheme using polar transformation.

$$v_{c1} = m\sin(\theta) \tag{2}$$

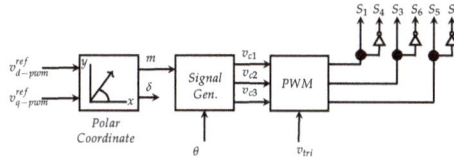

Figure 8. PWM generation scheme using polar transformation,

3.2. Grid-Injecting Mode Control

Grid-injecting mode control operates when grid-connected inverter is connected with the main grid, after successful synchronization process. During this mode, grid-connected inverter delivers the controlled-power output to the main grid, while the bus voltage and frequency are maintained by the main grid. Figure 9 shows the complete grid-injecting mode control of grid-connected inverter. Similar to the grid-forming mode control, it consists of double-loop control of current and power control section, power reference section, and measurement section. Grid-connected inverter operation can be either constant power operation or droop power operation. This scheme generates output power, which follows the desired power reference. For microgrid learning, this option allows students to more easily study the role of grid-connected inverter in the modern power system. Constant power control is suitable for renewable power system, which are passively connected to the system. The power delivered to the grid is held constant, possibly provided by external maximum power point tracking (MPPT) control. The droop control is also provided so the grid-connected inverter can actively play a role in supporting the grid. This kind of control is required if the renewable power system should provide ancillary services to support the voltage or frequency of the grid.

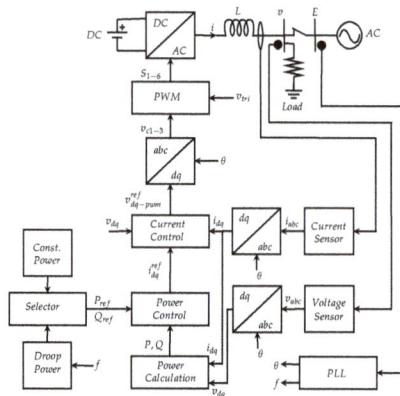

Figure 9. Grid-injecting control of grid-connected inverter.

3.2.1. Power Control

Similar to the voltage control on grid-forming mode, power control acts as outer loop in the grid-injecting mode of grid-connected inverter [16,17]. Figure 10a,b shows independent control of power in *dq* frame by using two different PI controllers. PI controller tracks the error between the reference power, P_{ref}, and actual output power, P. It generates corresponding signal for the reference of current controller, which in turn determines the required current of the grid-connected inverter. A similar concept is also applied for reactive power control through controlling Q_{ref} and Q.

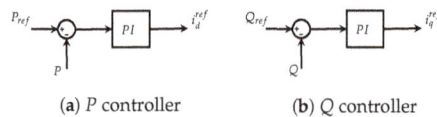

(**a**) P controller   (**b**) Q controller

Figure 10. Power control block diagram.

3.2.2. Droop Power Control

Droop power control is originally applied in parallel operation of synchronous generator to ensure equitable load sharing between generating units [18]. In the field of power electronics control, application of droop control was started in the UPS [19]. Afterwards, droop power control is also applied in the power control of renewable energy based-inverter for microgrid and distributed generation system [15,20,21].

The power flowing from sending side, *A*, to the receiving side, *B*, of Figure 11 can be expressed as follows [22,23].

$$S = P + jQ = V_s I^*$$

(3)

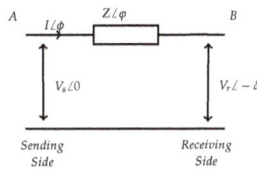

Figure 11. Power flow from sending to receiving side.

Since the current flowing can be represented as $I = \frac{V_s - V_r}{Z}$, Equation (3) can also be represented as

$$S = V_s \left(\frac{V_s - V_r}{Z} \right)^* \tag{4}$$

$$= V_s \left(\frac{V_s - V_r e^{j\delta}}{Z e^{-j\varphi}} \right) \tag{5}$$

$$= \frac{V_s^2}{Z} e^{j\varphi} - \frac{V_s V_r}{Z} e^{(j\varphi + \delta)} \tag{6}$$

Combining Equations (3) and (6), the active and reactive power can be obtained:

$$P = \frac{V_s^2}{Z} \cos(\varphi) - \frac{V_s V_r}{Z} \cos(\varphi + \delta) \tag{7}$$

$$Q = \frac{V_s^2}{Z} \sin(\varphi) - \frac{V_s V_r}{Z} \sin(\varphi + \delta) \tag{8}$$

Considering that $Z e^{j\varphi} = R + jX$, Equations (7) and (8) can also expressed as

$$P = \frac{V_s}{R^2 + X^2} \left(R(V_s - V_r \cos \delta) + X V_r \sin \delta \right) \tag{9}$$

$$Q = \frac{V_s}{R^2 + X^2} \left(-R V_r \sin \delta + X(V_s - V_r \cos \delta) \right) \tag{10}$$

Combining and rearranging Equations (9) and (10) gives

$$V_r \sin \delta = \frac{XP - RQ}{V_s} \tag{11}$$

$$V_s - V_r \cos \delta = \frac{RP + XQ}{V_s} \tag{12}$$

In most cases, the reactance is much bigger than resistance, $X \gg R$, thus the resistance can be neglected. If the power angle δ is also small enough, then by approximating, $\sin \delta \approx \delta$ and $\cos \delta \approx 1$. Therefore, Equations (11) and (12) become

$$\delta \approx \frac{XP}{V_s V_r} \tag{13}$$

$$(V_r - V_s) \approx \frac{XQ}{V_s} \tag{14}$$

Under previous considerations, Equations (13) and (14) show that power angle δ is proportional to the active power P, while the voltage drop $(V_r - V_s)$ is proportional to the reactive power Q. It means that the δ can be controlled through P and V_r through Q. Since by controlling the frequency the power angle is also controlled, by using relation on Equation (13) the real power flow has relation with the frequency.

Frequency and the voltage magnitude can be controlled independently through active and reactive power controlling. This concept is called droop power control, which can be represented as follows.

$$f - f_0 = -k_P(P - P_0) \tag{15}$$
$$V - V_0 = -k_Q(Q - Q_0) \tag{16}$$

where f_0, V_0, P_0, and Q_0 are rated frequency, voltage, active power, and reactive power, respectively. f, V, P, and Q are the corresponding values obtained through droop regulation. Figure 12 shows the voltage and frequency droop characteristics.

The droop power concept shown in Equations (15) and (16) can be applied in grid-injecting mode control of grid-connected inverter to obtain power reference. It can be applied either to both frequency and voltage droop or frequency droop only, thus grid-connected inverter is operated under unity power factor.

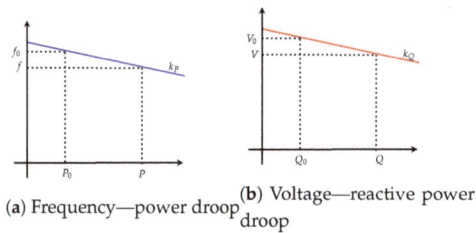

(a) Frequency—power droop (b) Voltage—reactive power droop

Figure 12. Droop characteristics.

4. Experimental Verification

4.1. Grid-forming Mode Experiments

The experiment conducted on grid-connected inverter connected to the three-phase source to represent infinite bus is shown in Figure 13. The details of developed control system is shown in Figure 14. The value of inductor and load are shown in Table 1. These values were to ensure that the proposed system has high X/R ratio (Approximately X/R = 22) so that Equations (15) and (16) were valid for the grid-injecting mode experiments. A 50 V voltage was generated in the PCC side of the three-phase source, as shown in Figure 13, and then grid-connected inverter was run under voltage control mode. Both voltages at PCC side and INV side were monitored. The complete experimental modules set-up is shown in Figure 15.

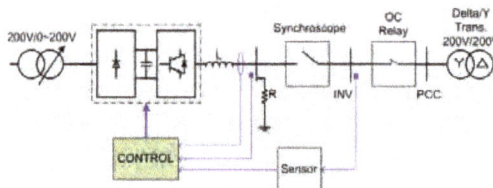

Figure 13. Experiment schematic for three-phase grid-connected operation.

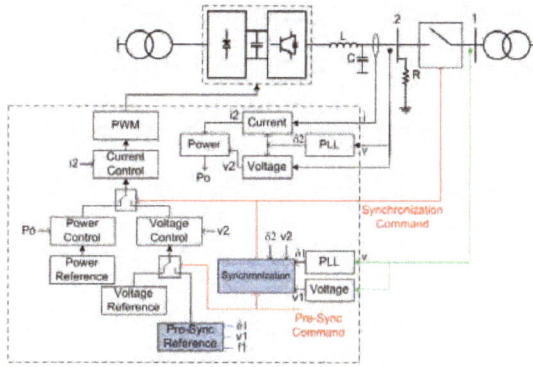

Figure 14. Control concept of grid-connected inverter.

Figure 15. Snapshot of the experimental set up.

Table 1. Experiment parameters.

Parameter	Value
Capacity	1 kVA
DC voltage	337 V
L	33 mH
R_L	1.4 Ω
R_{Load}	125 Ω

Figure 16 shows the experiment result of voltage control mode of grid-connected inverter. It can be seen that the inverter generated voltage that followed both magnitude and phase angle of the reference voltage at PCC bus. Voltage control mode was operated to build necessary voltage level and frequency on the output of grid-connected inverter based on the reference obtained from the voltage of PCC bus. Synchronization could be executed after voltages on INV and PCC bus had the same quantities in terms of voltage magnitude, phase sequence, and frequency.

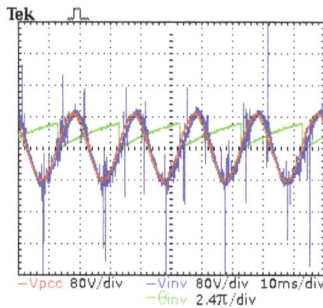

Figure 16. Experiment result of voltage control mode of grid-connected inverter.

Synchronization of grid-connected inverter and three-phase source at 200 V can be seen in Figure 17a, which shows the sequence during synchronization process. After the rms voltage reached the desired voltage, in this case 200 V, the phase adjustment was conducted at point 2, and then synchronization was executed at point 3. Instantaneous voltage at synchronization point 3 is shown at Figure 17b; it can be seen that synchronization was executed smoothly at t_sync. Figure 18 shows snapshot of both terminal voltage of grid-connected inverter and three-phase source after the synchronized operation. It can be seen that both voltages were synchronized in terms of magnitude, phase angle, and frequency.

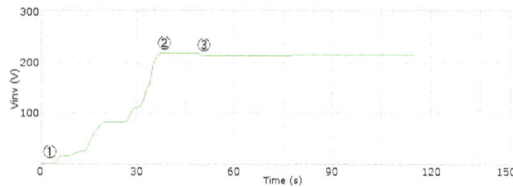

(**a**) Grid-connected inverter rms output voltage

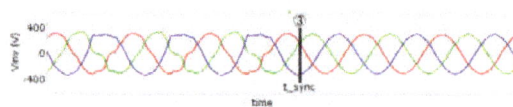

(**b**) Grid-connected inverter instantaneous output voltage

Figure 17. Synchronization sequence on grid-connected inverter: (1) grid-connected inverter is started; (2) phase-adjustment; and (3) synchronization.

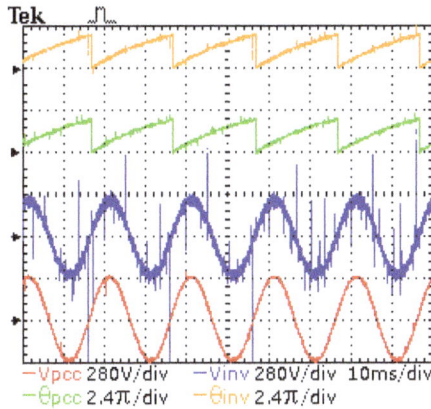

Figure 18. Output terminal voltage of grid-connected inverter and three-phase grid.

4.2. Grid-Injecting Mode Experiments

The same schematic (Figure 13) was used for this experiment. The grid-connected inverter was changed into power control mode after successful synchronization with three-phase source grid. The target was to control the output power of grid-connected inverter, which followed the given power reference values. This experiment was conducted by setting four different reference values: 0.05 pu (50 W), 0.1 pu (100 W), 0.15 pu (150 W), and 0.2 pu (200 W).

Figure 19 shows the results of constant power control on three-phase source connected experiment. It can be seen that the power control worked well by providing the output power according to the given reference values, as shown in Figure 19a. Since the grid-connected inverter was connected to strong three-phase grid, the voltage at the INV bus and the frequency were relatively constant during the power change conditions, as shown in Figure 19b,c. In contrast to the active power, the reactive power was intentionally kept constant at zero during changes in active power to ensure that all of the active power provided by the DC side was transferred to the grid side. Figure 20a,b shows the controlled reactive power during changes in active power. The total harmonic distortion (THD) for voltage and current were 1.5% and 11%, respectively, as shown in Figure 21a,b.

Another way to make grid-connected inverter actively involved in the power management of microgrid distributed generation system is by using droop power control. In this method, grid-connected inverter power control determines its output power by considering the frequency of the system and the pre-determined droop slope. Therefore, the balance of power and frequency is obtained from the power delivered by droop power control of grid connected inverter together with the droop characteristic of other source, if any, in the microgrid distributed generation system.

To verify the droop power control, an experiment on grid-connected inverter synchronized with synchronous generator was conducted. The synchronous generator was chosen because the frequency is easy to fluctuate compared to the stiff three-phase source grid connection. The grid-connected inverter was equipped with droop power control in response of changing on the load demand side. The observed case in this experiment had variable k_P of 5%. To provide easy comparison, in this experiment, the load demand was also changed from 150 W to 100 W.

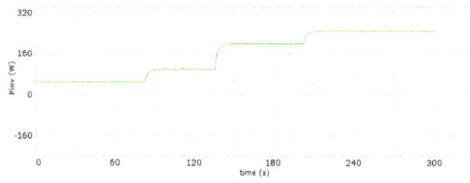

(a) Grid-connected inverter output power

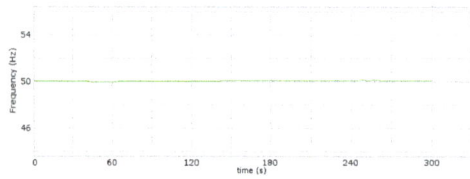

(b) Frequency of the system

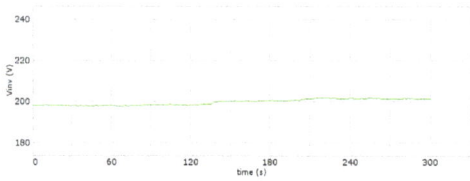

(c) Voltage at PCC bus

Figure 19. Experiment results of constant power control on three-phase grid-connected operation.

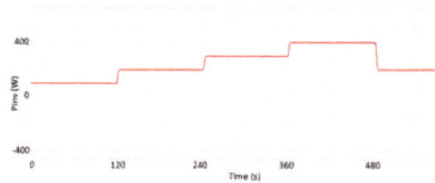

(a) Controlled active power

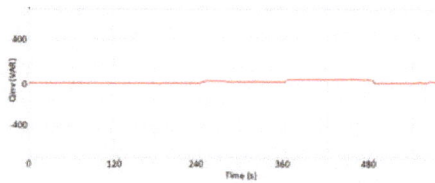

(b) Controlled reactive power

Figure 20. Active and reactive power control performance.

(a) Inverter output voltage THD

(b) Inverter output current THD

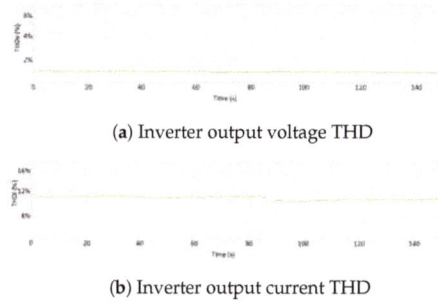

Figure 21. Voltage and current total harmonic distortion (THD) level.

Figure 22 shows the results of experiment of 5% droop slope. Following the load demand change, the grid-connected inverter reduced its output power while the synchronous generator kept its output power constant, the result being that the frequency could be kept constant, as shown in Figure 22c. Figure 22a shows that synchronous generator was constant during load demand change because the inherent droop characteristic of synchronous generator was less flat compared to the droop characteristic of grid-connected inverter.

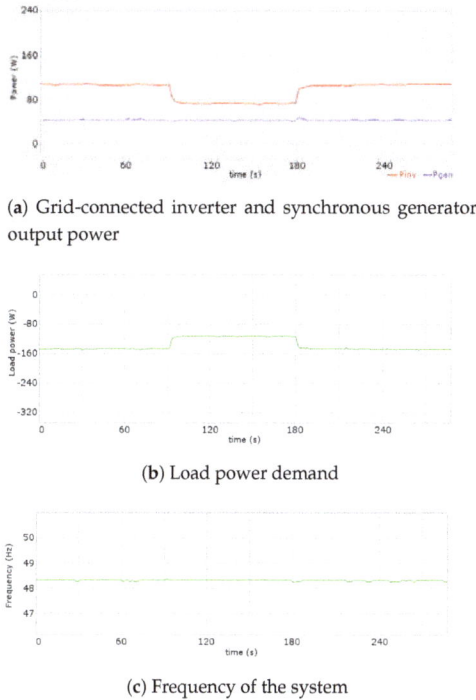

(a) Grid-connected inverter and synchronous generator output power

(b) Load power demand

(c) Frequency of the system

Figure 22. *Cont.*

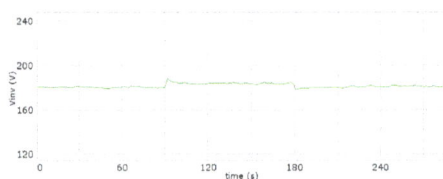

(**d**) Voltage at PCC bus

Figure 22. Experiment results of droop power control on synchronous generator connected operation.

5. Conclusions

A grid-connected inverter for distributed generation was developed at laboratory scale. This device was equipped with voltage control, power control, and synchronization control. Therefore, it can be used as grid-forming or grid-injecting inverter on microgrid power system. Voltage control under grid-forming mode could generate stable three-phase output voltage from internal or external voltage reference. Power control under grid-injecting mode could generate stable output power, under both constant power reference or droop power regulation. Experiments were conducted to verify the results.

Author Contributions: Conceptualization, A.R.; methodology, A.R.; software, A.R.; validation, A.R., and P.H.; formal analysis, A.R.; investigation, A.R., and P.H; resources, P.H.; data curation, P.H.; writing—original draft preparation, A.R.; writing—review and editing, P.H. and G.F.; visualization, A.R.; supervision, G.F.; project administration, G.F.; and funding acquisition, G.F.

Funding: This research received no external funding.

Conflicts of Interest: The authors declare no conflict of interest.

References

1. Ochoa, L.F.; Harrison, G.P. Minimizing energy losses: Optimal accomodation and smart operation of renewable distributed generation. *IEEE Trans. Power Syst.* **2011**, *26*, 198–205. [CrossRef]
2. Guttromson, R.T. Modeling distributed energy resource dynamics on the transmission system. *IEEE Trans. Power Syst.* **2002**, *17*, 1148–1153. [CrossRef]
3. Masters, G.M. *Renewable and Efficient Electric Power System*; Wiley-Interscience: Hoboken, NJ, USA, 2004.
4. Chowdhury, S.; Chowdhury, S.P.; Crossley, P. *Microgrids and Active Distribution Networks*; IET: London, UK, 2009.
5. Nissen, M.B. High performance development as distributed generation. *IEEE Potential* **2009**, *28*, 25–31. [CrossRef]
6. Jenkins, N.; Ekanayake, J.B.; Strbac, G. *Distributed Generation*; IET: London, UK, 2010.
7. Lasseter, B. Microgrid [distributed power generation]. In Proceedings of the 2001 IEEE Power Engineering Society Winter Meeting, Columbus, OH, USA, 28 January–1 February 2001. [CrossRef]
8. Lasseter, R.H.; Paigi, P. Microgrid: A conceptual solution. In Proceedings of the 35th Annual IEEE Power Electronics Specialist Conference, Aachen, Germany, 20–25 June 2004.
9. Nikkhajoei, H.; Lasseter, R. Distributed generation interface to the certs microgrid. *IEEE Trans. Power Deliv.* **2009**, *24*, 1598–1608. [CrossRef]
10. Kroposki, B.; Lasseter, R.; Ise, T.; Morozumi, S.; Papathanassiou, S.; Hatziargyriou, N. Making microgrids work. *IEEE Power Energy Mag.* **2008**, *6*, 40–53. [CrossRef]
11. Mohd, A.; Ortjohann, E.; Morton, D.; Omari, O. Review of control techniques for inverters parallel operation. *Electr. Power Syst. Res.* **2010**, *80*, 1477–1487. [CrossRef]
12. Krause, P.C.; Wasynczuk, O.; Sudhoff, S.D. *Analysis of Electric Machinery*; IEEE Press: New York, NY, USA, 1995.

13. Guo, X.Q.; Wu, W.Y.; Gu, H.R. Phase locked loop and synchronization methods for grid-interfaced converters: A review. *Przeglad Elektrotchniczny (Electr. Rev.)* **2011**, *87*, 182–187.

14. Hsieh, G.C.; Hung, J.C. Phase-locked loop techniques—A survey. *IEEE Trans. Ind. Electron.* **1996**, *43*, 609–615. [CrossRef]

15. Chung, I.Y.; Liu, W.; Cartes, D.A.; Collins, E.G.; Moon, S.I. Control method of inverter-interface distributed generators in a microgrid system. *IEEE Trans. Ind. Appl.* **2010**, *46*, 1078–1088. [CrossRef]

16. Delghavi, M.B.; Yazdani, A. A unified control strategy for electronically interfaced distributed energy resources. *IEEE Trans. Power Deliv.* **2012**, *27*, 803–812. [CrossRef]

17. Rocabert, J.; Luna, A.; Blaabjerg, F.; Rodriguez, P. Control of power converters in ac microgrid. *IEEE Trans. Power Electron.* **2012**, *27*, 4734–4748, . [CrossRef]

18. Kundur, P. *Power System Stability and Control*; McGraw-Hill: New York, NY, USA, 1994.

19. Chandorkar, M.C.; Divan, D.M.; Adapa, R. Control of parallel connected inverters in standalone ac supply system, *IEEE Trans. Ind. Appl.* **1993**, *29*, 136–143. [CrossRef]

20. Borup, U.; Blaabjerg, F.; Enjeti, P.N. Sharing of nonlinear load in parallel-connected three-phase converters. *IEEE Trans. Ind. Appl.* **2001**, *37*, 1817–1823. [CrossRef]

21. Majumder, R.; Ledwich, G.; Ghosh, A.; Chakrabarti, S.; Zare, F. Droop control of converter-interfaced microsources in rural distributed generation. *IEEE Trans. Power Deliv.* **2010**, *25*, 2768–2778. [CrossRef]

22. Brabandere, K.D.; Bolsens, B.; den Keybus, J.V.; Woyte, A.; Driesen, J.; Belmans, A. A voltage and frequency droop control method for parallel inverters. *IEEE Trans. Power Electron.* **2007**, *22*, 1107–1115. [CrossRef]

23. Weedy, B.M.; Cory, B.J. *Electric Power System*; Wiley: New York, NY, USA, 1998.

energies

MDPI

Article

Energy Routing Control Strategy for Integrated Microgrids Including Photovoltaic, Battery-Energy Storage and Electric Vehicles

Yingpei Liu *, Yan Li, Haiping Liang, Jia He and Hanyang Cui

School of Electrical and Electronic Engineering, North China Electric Power University, Baoding 071003, China; 2172213081@ncepu.edu.cn (Y.L.); lianghaiping@ncepu.edu.cn (H.L.); 2172213191@ncepu.edu.cn (J.H.); 2172213073@ncepu.edu.cn (H.C.)
* Correspondence: liuyingpei@ncepu.edu.cn

Received: 30 November 2018; Accepted: 15 January 2019; Published: 18 January 2019

Abstract: The Energy Internet is an inevitable trend of the development of electric power system in the future. With the development of microgrids and distributed generation (DG), the structure and operation mode of power systems are gradually changing. Energy routers are considered as key technology equipment for the development of the Energy Internet. This paper mainly studies the control of the LAN-level energy router, and discusses the structure and components of the energy router. For better control of the power transmission of an energy router, the energy routing control strategy for an integrated microgrid, including photovoltaic (PV) energy, battery-energy storage and electric vehicles (EVs) is studied. The front stage DC/DC converter of the PV system uses maximum power point tracking (MPPT) control. The constant current control is used by the bidirectional DC/DC converter of the battery-energy storage system and the EV system when they discharge. The DC/AC inverters adopt constant reactive power and constant DC voltage control. Constant current constant voltage control is adopted when an EV is charged. The control strategy model is simulated by Simulink, and the simulation results verify the feasibility and effectiveness of the proposed control strategy. The DG could generate reactive power according to the system instructions and ensure the stable output of the DC voltage of the energy router.

Keywords: Energy Internet; energy router; microgrid; electric vehicle; PV; battery-energy storage

---

1. Introduction

In the *third industrial revolution*, Rifkin, an American economist, proposes the concept of Energy Internet and mainly describes the application of the Energy Internet in the future [1,2], which has drawn worldwide attention and studies [3,4]. With the continuous development of Internet+, microgrid, renewable energy and DG, the structure and operation mode of the traditional power system are gradually changing [5–7].

The Energy Internet is a new generation of intelligent network, having electric power information as its core. The energy domain and information technologies interact with each other and then make efficient use of different energy sources [8–10]. The Energy Internet is characterized by networking and distribution, and various international research organizations have done a lot of research works based on different needs and conditions. Germany has proposed the concept of E-Energy and the Energy Internet plan, and studied the system design and management mode [11]. The Future Renewable Electric Energy Delivery and Management (FREEDM) has proposed the concept of energy router and designed energy routers based on solid state transformers (SSTs). The energy router is considered as the infrastructure for building the future energy Internet [12]. Japan's "Digital Power Grid Alliance" manages and dispatches the electric energy in the corresponding region based on a "Digital Power

Grid Router" [13]. In China's Energy Internet project, the large power grid is regarded as the backbone WAN and the microgrid as the local area network LAN. All components are connected through integrated information energy switching equipment [14]. Almost all of the above Energy Internet prototypes contain energy routers or power routers, indicating that energy routers are one of the key technologies of the Energy Internet [15,16]. Therefore, the Energy Internet is also defined as a multi-level distributed energy sharing complex network based on energy routers and energy LANs (units) with the microgrid as its basic structure [17]. The Energy Internet architecture based on energy routers is shown in Figure 1.

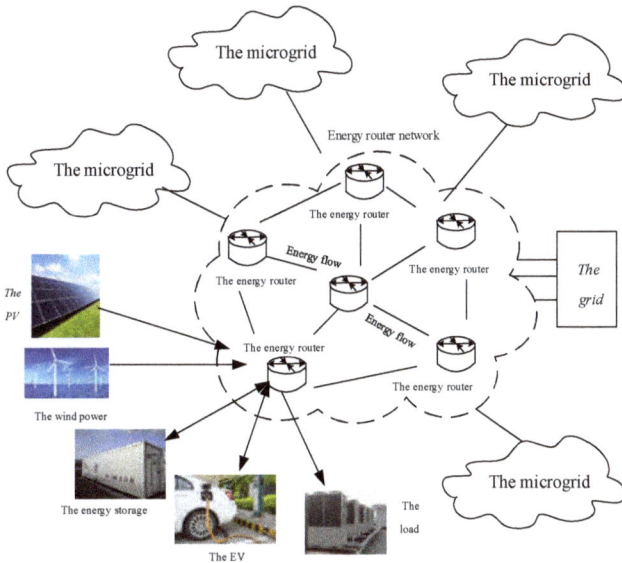

Figure 1. The Energy Internet architecture based on energy routers.

As the main component of the Energy Internet, the energy router has comprehensive functions such as energy-exchange, communication and energy management [18]. At present, energy router research is still in a primary stage. The aim of this paper is to study the control of a LAN-level energy router. This paper discusses the structure and compositions of the energy router, and focuses on the energy transmission control strategy of the energy router for different distributed generations, such as PV, battery-energy storage and EVs. The energy routing models of PV, energy storage and EVs are established and simulated by Simulink. The simulation results verify the feasibility and effectiveness of the proposed control strategy. The DG could generate reactive power according to the system instructions and ensure the stable output of the DC voltage of the energy router. The structure of the paper is as follows: in Section 2, related work is presented. The architecture of the energy router is described in Section 3. The energy routing control strategy is described in Section 4. The simulation and results are presented in Section 5. Finally, Section 6 gives the conclusions.

2. Related Work

At present, energy router research is still in a primary stage. In [19], a distributed power routing control strategy based on the Dijkstra algorithm method for the management and coordination control of power flow in energy routers was put forward. A three-layer tree architecture of an intelligent distributed energy network based on the multi-energy router concept is drawn in [20], where the typical characteristics of the energy router are summarized and analyzed. In [21], the authors presented the main circuit structure of a multi-interface energy router based on a power electronic converter and introduce the control strategy of each module of an energy router. A distributed energy router is designed in [22] and its control mode is designed to ensure the stability of AC and DC buses. However, different DGs have different characteristics and requirements. The common DGs are wind power, PV, energy storage and EVs [23,24]. The PV power generation has the characteristics of randomness, volatility and weak reactive power support [25]. Theories show that distributed energy storage is an effective approach to solve the problems of PV mentioned above [26,27]. EVs are controllable in time and mobile in space. They could be used as micro sources (V2G) to release electric energy. They could also be used as a load (charging), which is inserted into the grid to store electricity [28–30]. In fact, energy routers need to ensure the normal operation and quality of the power grid, but it is also important to ensure the reasonable flow of distributed energy, meet the load demand and control the correct and safe energy flow intelligently.

This paper mainly studies the control of a LAN-level energy router, discusses the structure and compositions of the energy router, and focuses on the energy transmission control strategy of the energy router for different distributed generations, such as PV, battery-energy storage and EVs.

3. The Architecture of the Energy Router

Since the development trend of energy routers is to combine the energy interface layer of the unified regional energy LAN with the distributed generation, the energy router needs various types of energy input interfaces and corresponding control loops. The architecture of an energy router is shown in Figure 2.

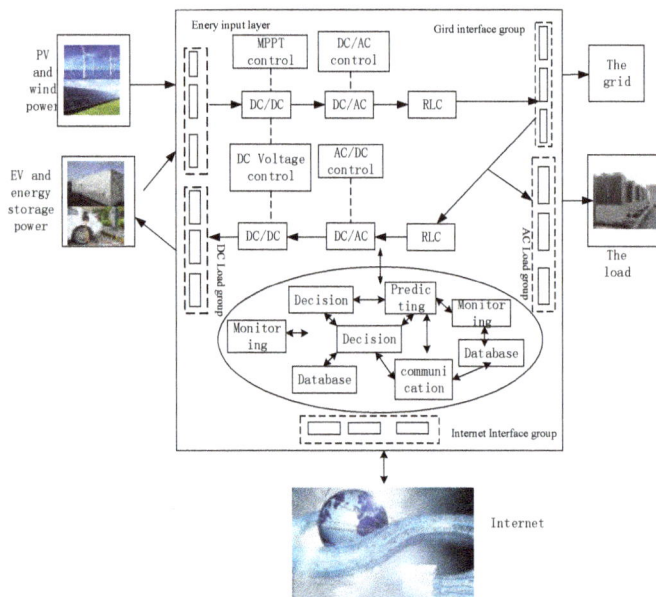

Figure 2. The architecture of an energy router.

The energy router can be divided into three parts: the control part, the energy transmission loop and the Internet communication unit. The energy router network information unit has the functions of receiving Internet information, collecting and processing data, data analysis, prediction, interaction and decision-making. The energy transmission layer and control part work according to different requirements. The three parts interact and ultimately achieve high efficiency and safety of energy utilization.

In terms of the energy routing control, different control methods are adopted for different DGs. The PV system uses the unidirectional DC/DC converter to achieve one-way flow of power, at the same time to realize MPPT. A bidirectional DC/DC converter is used for the battery-energy storage system and EV system when they discharge. The DC/AC inverter of the AC side adopts constant reactive power and constant DC voltage control, which enables the PV, battery-energy storage and EVs to generate reactive power according to the grid instructions and ensure stable output of DC voltage of energy router. The constant current constant voltage control is adopted when an EV is charged. The energy routing models of PV, energy storage and the EVs are established and simulated based on Simulink.

4. The Energy Routing Control Strategy

4.1. EV and PV Battery-Energy Storage Power System

In order to study the energy routing strategy for the integrated microgrid, including PV, battery-energy storage and EVs, their structures are designed in this paper as shown in Figure 3, which includes a PV system, battery-energy storage system, EV system, load and the inverter control systems of each system.

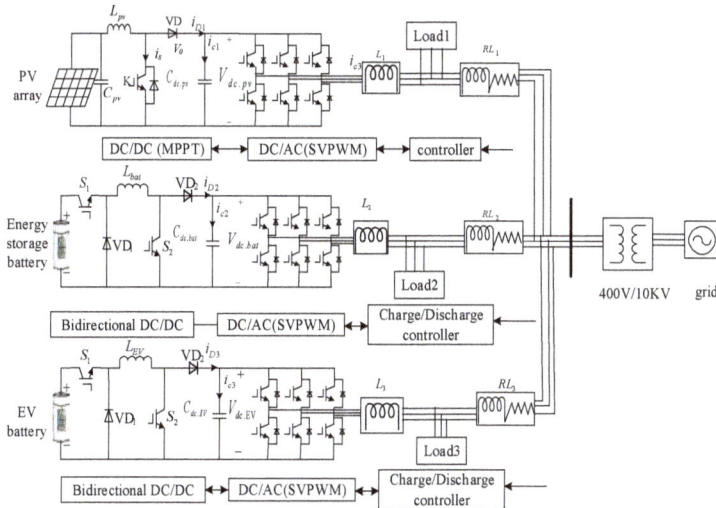

Figure 3. Structure of a photovoltaic (PV) battery-energy storage hybrid power system with EVs.

In Figure 3, S_1, S_2 and VD_1, VD_2 are the switches and diodes. L_{pv}, L_{EV}, L_{bat} are the boost/buck circuit inductances of the PV system, energy storage system and EV system, respectively. $C_{dc.pv}$, $C_{dc.EV}$, $C_{dc.bat}$ are their DC capacitances, respectively. $V_{dc.pv}$, $V_{dc.bat}$, $V_{dc.EV}$ are the DC voltages. The C_{pv} is the boost circuit capacitances of PV. L_1, L_2, L_3 are the converter reactors. Load1, Load2, Load3 are the loads of system. i_{c1}, i_{c2}, i_{c3}, i_{D3}, i_{D2} and i_{D1} are the currents of the diode and DC capacitances. RL_1, RL_2, RL_3 are the grid-connected capacitors and inductances.

4.2. Energy Routing Control Strategy of PV System

The power generation control system of PV is divided into two control subsystems, which are unidirectional DC/DC and DC/AC, respectively. The control goals of PV system are to achieve the MPPT operation of the PV array, stabilize the converter DC voltage and output reactive power according to the system instruction.

4.2.1. Unidirectional DC/DC Control Strategy

The front-stage converter of the PV power generation control system adopts unidirectional DC/DC control to realize one-way power flow and simultaneously realize MPPT control. The DC/DC converter realizes input and output voltages control by adjusting the duty cycle of insulated-gate bipolar transistor (IGBT). The principle of PV front-stage DC/DC control is shown in Figure 4. I_{PV} is the current of PV array, V_{PV} is the DC voltage of the PV array.

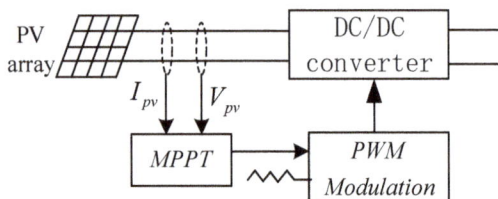

Figure 4. The principle of PV front-stage DC/DC control.

MPPT refers to the tracking and control of DGs, so that the maximum power output can be obtained in all circumstances. The common MPPT methods of PV arrays are the constant voltage method, disturbance observation method, and incremental conductance method. There is a maximum power point when a PV array operates under certain conditions. The main principle of the disturbance observation method is to apply a periodic constant step disturbance to the output voltage of PV array ΔV (or a constant step disturbance to the output current of the PV array ΔI). If the ΔP is positive, it means the working point voltage is less than the maximum power point voltage, and the disturbance in the original direction will continue to increase. Otherwise, the working point voltage will be larger than the maximum power point, and a negative direction will be added. The working point of the PV cell will be kept close to the MPPT. In this paper, the PV power generation system uses disturbance observation which is the most commonly used method.

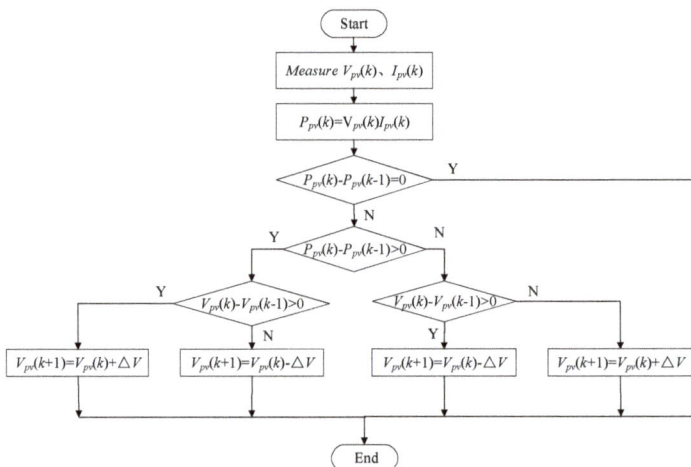

Figure 5. Maximum power point tracking (MPPT) algorithm program based on disturbance observation.

In this way, by referring to a self-optimizing search program, which is shown in Figure 5, the constant step length disturbance is added to the PV output voltage. The change of current and pre-change power is compared in real time, so that PV array can work at the maximum power point as much as possible.

4.2.2. DC/AC Control Strategy

In Figure 6, the PV power generation inverter adopts double loop control, the outer ring is constant DC voltage and constant reactive power control, the inner ring is current control. $V_{q.grid.ref}$ and $V_{d.grid.ref}$ are respectively the instruction values of the d and q axis components of the grid side voltage. $V_{dc.pv}$ is the output voltage actual value of the unidirectional DC/DC converter of PV, Q_{grid1} is the reactive power actual value of the grid. They are the input of outer ring value of the inverter control system, and are compared respectively with the DC voltage instruction value of PV $V_{dc.pv.ref}$ and network side reactive power instruction value $Q_{grid1.ref}$.

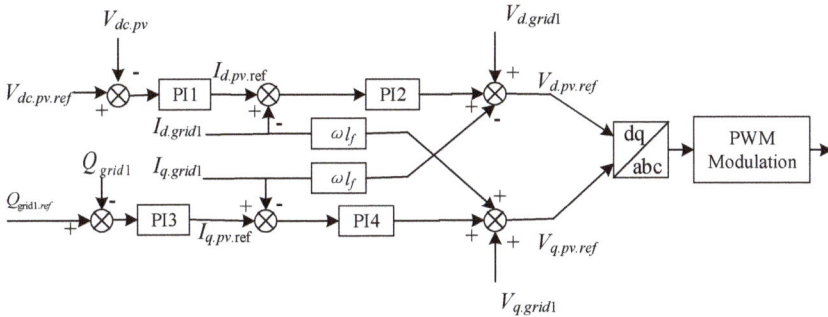

Figure 6. Control principle of PV power generation inverter.

The input current instruction value of PV $I_{d.pv.ref}$ and $I_{q.pv.ref}$ are obtained through the outer loop proportional integral (PI) controller for synchronous rotating coordinate system under the current inner loop controller. After coordinate transformation, the d, q components of the grid side DC current $I_{d.grid1}$ and $I_{q.grid1}$ are obtained, and they are compared with their instruction value of corresponding current d and q components $I_{d.grid1.ref}$ and $I_{q.grid1.ref}$. The six PWM trigger pulses are obtained through PI controller and PWM modulation of the inner loop. Finally, IGBT is controlled to realize non-static adjustment of constant DC voltage and constant reactive power. The space vector pulse width modulation (SVPWM) modulation is adopted in this paper, and inductance capacitance (LC) filtering method is used to suppress the current harmonic generated by it in order to make it work normally.

4.3. Energy Routing Control Strategy of Battery-Energy Storage System

The battery-energy storage power generation control system is divided into two control subsystems, which are bidirectional DC/DC converter and DC/AC inverter. Its goals are:

- achieving two-way flow of power in battery-energy storage system
- achieving constant reactive power control
- achieving constant DC voltage control

In Figure 7, the front stage converter of energy storage adopts directional DC/DC, which could increase the voltage or decrease the voltage. It is mainly composed of the inductance L_{bat}, capacitor C_{bat}, IGBT switches S_1, S_2, diode VD_1 and VD_2.

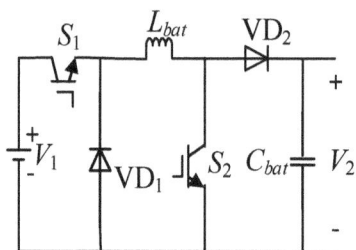

Figure 7. The circuit topology of Buck-Boost converter.

By controlling the conduction ratio of PWM to achieve its switches on and off. The output voltages, the duty ratio of the two switch tubes are set as V_1, V_2 and D_1, D_2 respectively, the relation of them as shown in Equation (1):

$$\frac{V_1}{V_2} = \frac{1 - D_2}{D_1} \tag{1}$$

When the charge-discharge current instruction value $I_{bat.ref}$ shown in Figure 6 is positive, the bidirectional DC/DC is in the boost state. Otherwise, the bidirectional DC/DC is in the buck state.

Figure 8 is the schematic diagram of the battery-energy storage control system. It consists of a bidirectional DC/DC converter control and inverter control. The principle of the bidirectional DC/DC control has been explained previously, the inverter is controlled by dual closed-loop control of voltage and current.

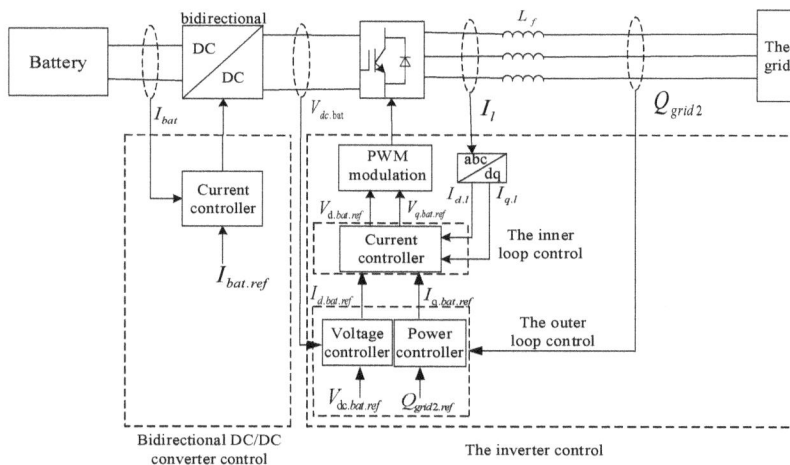

Figure 8. The schematic diagram of battery-energy storage control system.

$V_{dc.bat}$ is the actual DC voltage on the DC side of the inverter of the battery-energy storage system, Q_{grid2} is the actual reactive power on the grid side of the grid. They are respectively controlled by the voltage outer ring and the power outer ring, and then the instruction values of d and q axis of current control $I_{d.bat.ref}$ and $I_{q.bat.ref}$ are obtained. The current collected by the grid I_l is transformed through coordinate transformation to obtain DC/AC inverter of d and q components of AC side current $I_{d.l}$, $I_{q.l}$. The $I_{d.l}$ and $I_{q.l}$ are compared with $I_{d.bat.ref}$ and $I_{q.bat.ref}$, respectively, and the PWM trigger pulse is obtained through the current controller, which ensures the non-static adjustment of DC voltage/reactive power.

4.4. Energy Routing Control Strategy of EVs

When the EV is a micro source (V2G), it adopts the same control strategy as a battery-energy storage control system, with bidirectional DC/DC control at the front stage and constant DC voltage constant reactive power of DC/AC inverter control at the after stage.

At present, lithium ion batteries are widely used in EVs. When the EV in this paper is used as system load (charging), its charging control mode needs to consider the characteristics of lithium ion batteries. The control purposes of EV charging machines are to improve the charging efficiency of the battery, shorten the charging time and extend the battery service life. As we all know, shortening the charging time will lead to a sharp decline in battery life, therefore, the charging time and battery life of EV should be considered comprehensively. This control strategy, mainly including AC/DC inverter voltage stability control, DC/DC buck chopper constant current control and DC/DC buck chopper constant voltage control, is shown in Table 1.

Table 1. Control strategy of charging machine.

EV	AC/DC Part	DC/DC Part
Control mode	Constant DC voltage	Constant current constant voltage

The control principle of the AC/DC inverter is same as that of discharge. Two different control modules of constant current and constant pressure are established on the DC/DC control side respectively. The switching of constant current and constant voltage charging mode is realized according to the battery terminal voltage. The current loop PI control is adopted in the constant current stage. By changing the duty cycle ratio, the on-off of the switch tube is controlled and the output current is kept constant. The control structure of the inverter is shown in Figure 9.

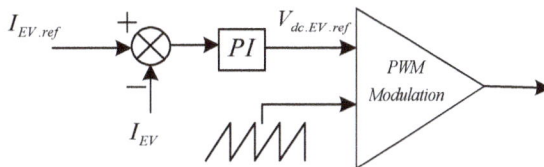

Figure 9. The structure of constant current control.

Constant voltage control adopts double closed loop control of the external power battery terminal voltage and internal inductance current. The current inner loop is to control according to the current instruction output by the voltage of the outer loop, and improve the dynamic response of the system. The outer voltage loop is used to maintain the stability of the power battery terminal voltage. The inverter control structure shown in Figure 10.

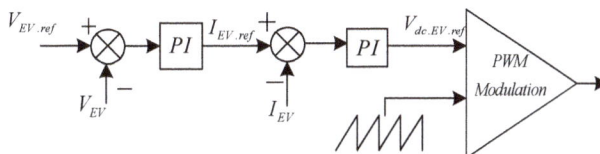

Figure 10. The structure of constant voltage control.

5. Simulation and Results

In order to verify the effectiveness of the control strategy proposed in this paper, a simulation study is carried out. In this paper, a complete simulation structure diagram of the energy routing control of the PV, battery-energy storage and EV system is built under Simulink. The simulation parameters are shown in Tables 2–4.

Table 2. Simulation parameters of the PV system.

Meaning and Unit	Number	Meaning and Unit	Number	PI Parameters	Number
Parallel number of the PV array	48	DC voltage of inverter $V_{dc.pv.ref}$/V	800	K_{P1}	2
Series number the PV array	10	Sampling frequency/kHZ	1000	K_{i1}	100
Short-circuit current/A	60.9	Boost circuit capacitance C_{pv}/μF	60×10^3	K_{P2}	1200
Open-circuit voltage/V	853.1	Boost circuit inductance L_{pv}/mH	50×10^{-3}	K_{i2}	1
Optimum operating voltage/V	729.1	DC container $C_{dc.pv}$/μF	750	K_{P3}	1
Optimum operating voltage/A	273.1	Converter reactor L_1/mH	1	K_{i3}	50
Rated power/kW	199.1	The cycle of PWM	10×10^{-3}	K_{P4}	20
Load1/MW + MVar	0.02 + 0.02	-	-	K_{i4}	20

Table 3. Simulation parameters of the battery-energy storage system.

Meaning and Unit	Number	PI Parameters	Number
DC voltage of inverter $V_{dc.bat.ref}$/V	800	K_{P1}	2
Rated voltage of energy storage battery/V	400	K_{i1}	100
Rated capacity of energy storage battery/Ah	200	K_{P2}	1000
Initial state of charge (SOC) of the energy storage battery/%	50	K_{i2}	1
Charge discharge current/A	±110	K_{P3}	1
Start time of bidirectional DC/DC converter/s	0.05	K_{i3}	50
Sampling frequency/kHZ	1000	K_{P4}	20
Boost/Buck circuit inductance L_{bat}/mH	5	K_{i4}	20
DC container $C_{dc.bat}$/μF	2200	-	-
Converter reactor L_2/mH	2	-	-
The cycle of PWM	10×10^{-3}	-	-
Load2/MW + MVar	0.02 + 0.02	-	-

Table 4. Simulation parameters of the EV system.

Meaning and Unit	Number	PI Parameters	Number
DC voltage of inverter $V_{dc.bat.ref}$/V	800	K_{P1}	2
Rated voltage of EV/V	400	K_{i1}	100
Rated capacity of EV/Ah	200	K_{P2}	1000
Initial state of charge (SOC) of the EV/%	50	K_{i2}	1
Charge discharge current/A	±110	K_{P3}	1
Termination voltage at constant current charging stage/V	440	K_{i3}	50
Start time of bidirectional DC/DC converter/s	0.05	K_{P4}	20
Sampling frequency/kHZ	1000	K_{i4}	20
Boost circuit inductance L_{bat}/mH	5	-	-
DC container $C_{dc.bat}$/μF	2200	-	-
Converter reactor L_2/mH	2	-	-
The cycle of PWM	10×10^{-3}	-	-
Load3/MW + MVar	0.02 + 0.02	-	-

5.1. Example 1: The Reactive Power Instruction Values of PV, Battery-Energy Storage and EV Are all 0 Var

Figure 11 shows the PV simulation results when the reactive power instruction values of PV, battery-energy storage and EV are all 0 Var. As seen in Figure 11a, the initial temperature of the PV array is stable at about 25 °C. As depicted in Figure 11b, the light intensity of the PV array decreases at around 1s and increases back to the original state at around 3 s.

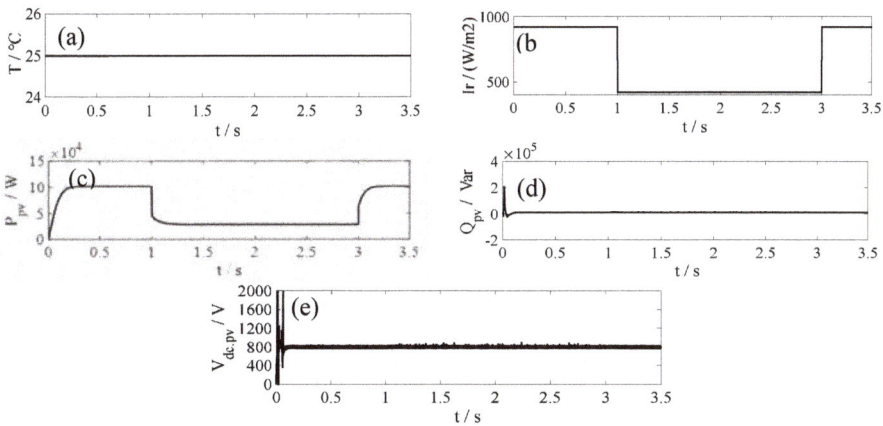

Figure 11. The simulation results of PV. (**a**) the initial temperature of the PV array. (**b**) the initial conditions of the light intensity of PV array. (**c**) the active power of PV. (**d**) the reactive power of PV. (**e**) the DC voltage of PV.

In Figure 11c, the active power output of PV array has the same change trend, falling at around 1 s and increasing back to the original state at around 3 s. Therefore, the PV array has realized the MPPT operation under the control of DC/DC. The variation of active power output is smooth and this reduces the impact on the system. The reactive power of the AC side of the PV inverter is stable at 0 Var as shown in Figure 11d. Figure 11e shows that the DC voltage of the inverter can be stabilized at around 800 V, which is the same as the instruction value of the grid side. As a power grid power source, PV has characteristics of uncertainty and randomness. The constant reactive power and constant DC voltage control strategy of DC/AC inverter used in this paper, is able to achieve the function of releasing reactive power according to the power grid instructions, stabilizing DC voltage and stabilizing the active power.

Figure 12a–f are the battery-energy storage results. The reactive power of the battery-energy storage could well track its instruction value shows Figure 12b, which is about 0 Var. As shown in Figure 12c, the DC voltage at the inverter side could track well its instruction value that is 800 V. The active power of it could stabilize the output as shown in Figure 12a. As depicted in Figure 12d–f, the bidirectional DC/DC converter of battery-energy storage adopts the constant current discharge control method, and the discharge current is basically kept at around 112 A as shown in Figure 12d. The discharge voltage is slightly reduced is shown in Figure 12e and the charge capacity of the battery decreases uniformly is shown in Figure 12f. The battery-energy storage could be controlled by constant DC voltage constant reactive power and the mode of constant current, which can realize stable output of the reactive power and active power, thus ensure the stable operation of the power grid. Figure 12g–i are the simulation results of EV. It can be seen that under the control of constant reactive power and constant DC voltage of energy router, EV could also be connected to the network stably.

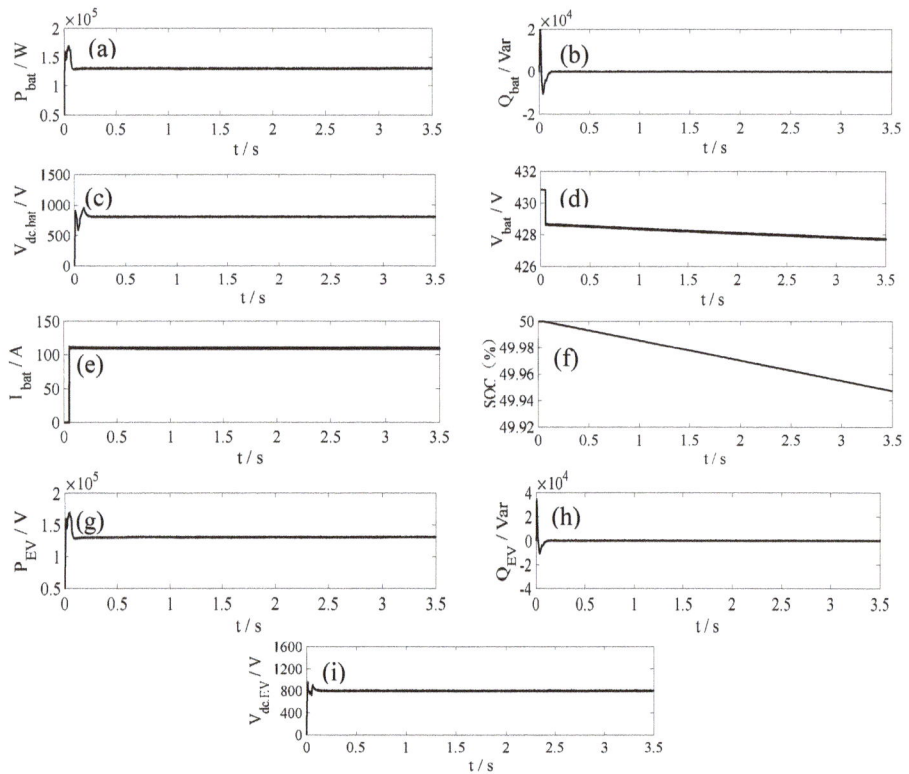

Figure 12. The simulation results of battery-energy storage and electric vehicle (EV). (**a**) the active power of battery-energy storage. (**b**) the reactive power of battery-energy storage. (**c**) the DC voltage of battery-energy storage. (**d**) the discharge current of battery-energy storage. (**e**) the discharge voltage of battery-energy storage. (**f**) the state of charge (SOC) of battery-energy storage. (**g**) the active power of EV. (**h**)the reactive power of EV. (**i**) the DC voltage of EV.

Figure 13 shows the active and reactive power generated by the grid. The active and reactive load of this system are respectively 0.06 MW and 0.06 MVar. The active power generated by the grid makes up for the demand of active power load as shown in Figure 13a Since the reactive power instruction values of the DG are all 0 Var, the reactive power of the system is provided by the grid, as shown in Figure 13b.

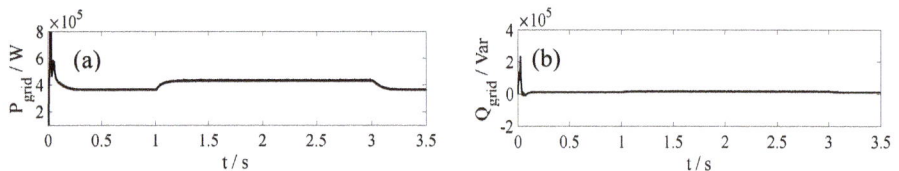

Figure 13. (**a**) active power of grid. (**b**) reactive power of grid.

5.2. Example 2: The Reactive Power Instruction Values of PV, Battery-Energy Storage and EV Are Respectively 10,000 Var, 20,000 Var and 0 Var

Figure 14 shows the PV simulation results. The reactive power instruction values of the PV, battery-energy storage and EV are respectively 10,000 Var, 20,000 Var, and 0 Var. As shown in Figure 14a, the initial temperature of the PV array is stable at about 25 °C.

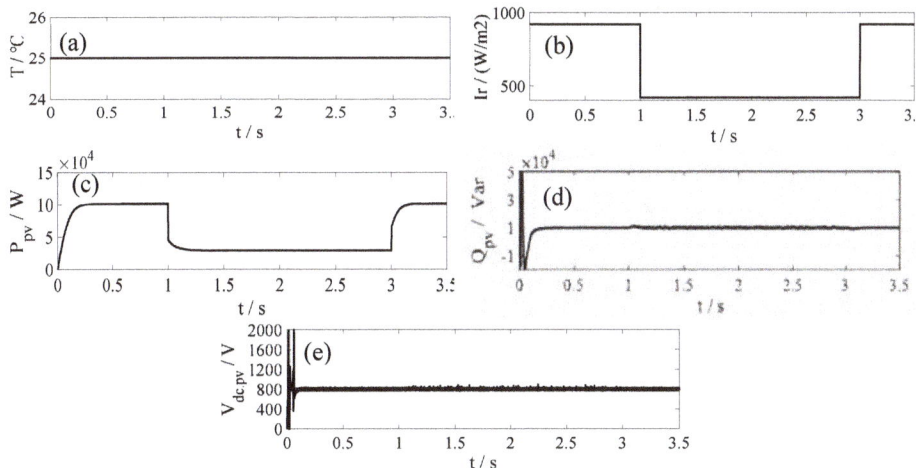

Figure 14. The simulation results of PV. (**a**) the initial temperature of the PV array. (**b**) the initial conditions of light intensity of PV array. (**c**) the active power of PV. (**d**) the reactive power of PV. (**e**) DC voltage of PV.

As depicted in Figure 14b, the light intensity of the PV array decreased at around 1 s and increased back to the original state at around 3 s. In Figure 14c, the active power output of the PV array has the same change trend, and falls at around 1 s and increases back to the original state at around 3 s, therefore the PV array has realized the MPPT operation under the control of DC/DC. The variation of active power output is smooth and reduces the impact on the system. The reactive power of the AC side of the PV inverter is stable at 10,000 Var, as shown in Figure 14d. Figure 14e shows that the DC voltage of the inverter can be stabilized at around 800 V, which is the same as the reactive power and DC voltage instruction value of the grid side. As a power source of the power grid, PV has characteristics of uncertainty and randomness. The constant reactive power constant DC voltage control strategy of DC/AC inverter used in this paper can achieve the function of releasing reactive power according to the instruction of power grid, stabilizing DC voltage, and then stabilizing the active power.

Figure 15a–f are the battery-energy storage results. As shown in Figure 15b the reactive power of the battery-energy storage could track well its instruction value, which is about 20,000 Var. As shown in Figure 15c, the DC voltage at the inverter side could track well its instruction value, which is 800 V. The active power could be stabilized as shown in Figure 15a. The bidirectional DC/DC converter of battery-energy storage adopts the constant current discharge control method, the discharge current is basically kept at around 115 A as shown in Figure 15d, the discharge voltage is slightly reduced as shown in Figure 15e and the charge capacity of the battery decreases uniformly as shown in Figure 15f. The battery-energy storage could generate stable reactive power and active power, which ensure the stable operation of the power grid under the control of constant DC voltage constant reactive power and the mode of constant current. Figure 15g,h are the EV simulation results where the reactive power of the EV is about 0 Var and the DC voltage of the EV is about 800 V which are the same as their instruction values. It can be seen that under the control of constant reactive power constant DC voltage of energy router, the EV could be connected as a distributed power source to the grid and ensure the stability of the power grid.

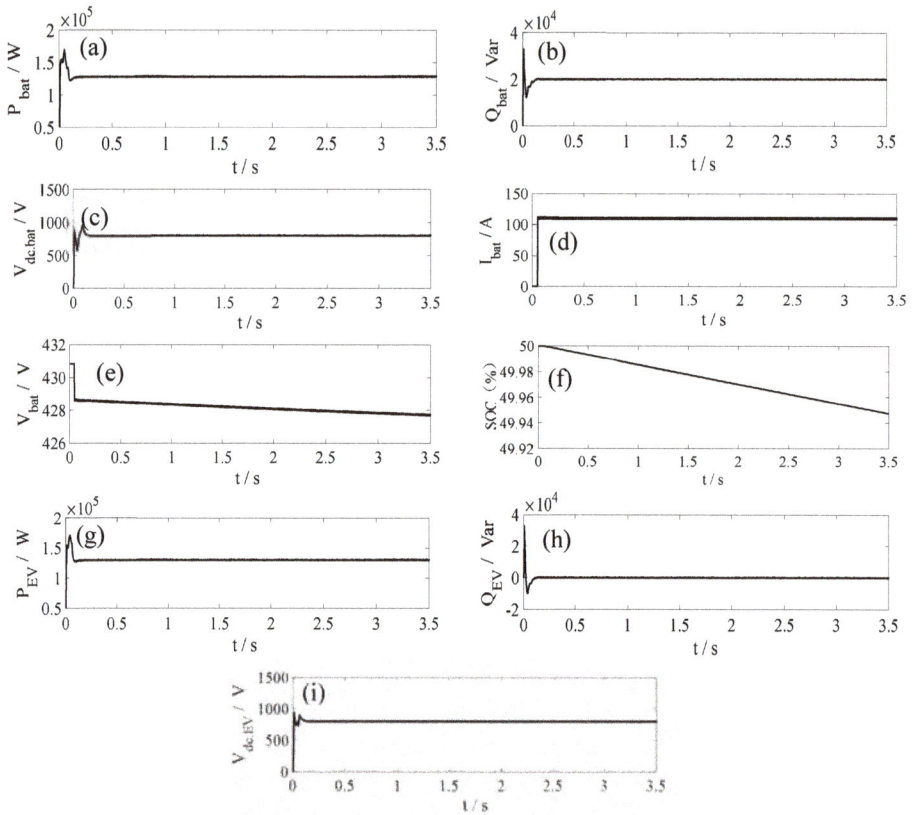

Figure 15. The simulation results of battery-energy storage and EV. (**a**) the active power of battery-energy storage. (**b**) the reactive power of battery-energy storage. (**c**) the DC voltage of battery-energy storage. (**d**) the discharge current of battery-energy storage. (**e**) the discharge voltage of battery-energy storage. (**f**) the SOC of battery-energy storage. (**g**) the active power of EV. (**h**) the reactive power of EV. (**i**) the DC voltage of EV.

Figure 16 shows the active and reactive power generated by the grid. The active and reactive load of this system are respectively 0.06 MW and 0.06 MVar. The active power generated by the grid makes up for the demand of active power load as shown in Figure 16a. Since the reactive power instruction values of the DG are respectively 10,000 Var, 20,000 Var, 0 Var., the rest of the reactive power of the system is provided by the grid. As shown in Figure 16b, the reactive power of grid is about 30,000 Var.

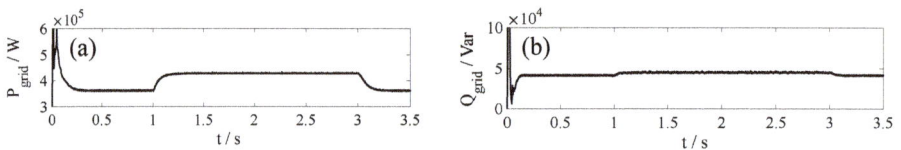

Figure 16. (**a**) active power of grid. (**b**) reactive power of grid.

5.3. Example 3: EV Charging

Figure 17 shows the simulation results of the EV when it is the load of the grid. The constant current and constant voltage are adopted when the EV is charged. The constant current charging current is 10 A, and the constant voltage charging voltage is 440 V. It can be seen from Figure 17a,b, that a constant current is adopted at the beginning stage, and this constant current is 110 A. With the deepening of charging, the battery voltage gradually increases to 440 V, and the charging machine is switched to the constant voltage charging mode. The constant current charging time is 8 s. After 8 s, the terminal voltage of the power battery remains at 440 V, the charging current gradually decreases and enters the trickling charging stage. The battery continues to be charged. It can be seen from Figure 17c,d that the charge capacity of the power battery increases faster in the constant current stage, while the charge capacity increases slowly in the constant voltage stage due to the reduction of charging current. The intermediate voltage of AC/DC inverter remains stable after charging, which verifies the effectiveness of the control strategy of constant voltage and constant current charging of the power battery.

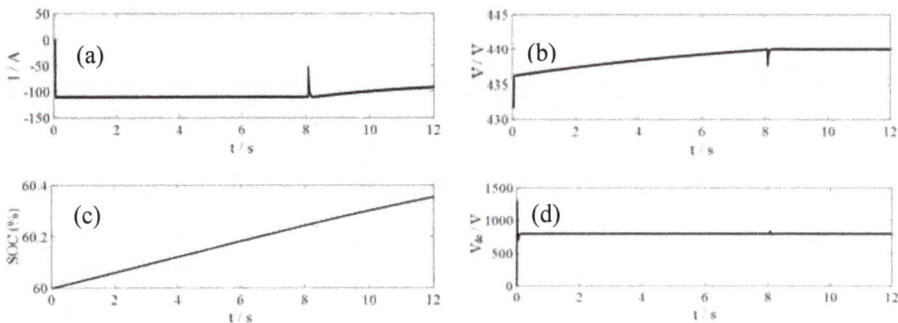

Figure 17. The simulation results of EV charging. (**a**) Charging current of EV. (**b**) Charging voltage of EV. (**c**) SOC of EV. (**d**) DC voltage of EV

6. Conclusions

This paper mainly studies the control of a LAN-level energy router, discusses the structure and components of the energy router, and focuses on the energy transmission control strategy of the energy router for different distributed generations. The models of EV, PV and battery-energy storage are established respectively by using Simulink. Three examples are simulated to verify the effectiveness of the control strategy. The examples are the reactive power instruction values of DGs are all 0 Var; the reactive power instruction values of PV, battery-energy storage and EV are respectively 10,000 Var, 20,000 Var and 0 Var; and EV charging.

The simulation results show that the control method of the PV system could realize MPPT, and the EV system could achieve constant voltage and constant current control when it is charging. The EV, PV and battery-energy storage could release reactive power according to the instructions of the grid. The DC voltages of DGs could be well controlled and stabilized. The normal operation and quality of the power grid be achieved under the control of the energy router while maintaining the reasonable flow of distributed energy and safe intelligent energy flow. The limitation of this work is that the control model of the energy router lacks any interaction with the communication and decision-making part of the power system. In fact, the control of the energy router needs to interact with the communication part in real time, which can be further studied and combined in the future.

Author Contributions: Conceptualization, Y.L. (Yingpei Liu) and Y.L. (Yan Li); Funding acquisition, Y.L. (Yingpei Liu) and H.L.; Methodology, Y.L. (Yingpei Liu), Y.L. (Yan Li) and H.L.; Software, Y.L. (Yan Li), H.L., J.H. and H.C.; Validation, Y.L. (Yingpei Liu), Y.L. (Yan Li), H.L., J.H. and H.C.; Writing—original draft, Y.L. (Yingpei Liu), Y.L. (Yan Li) and H.L.; Writing—review & editing, Y.L. (Yingpei Liu), Y.L. (Yan Li), H.L., J.H. and H.C.

Funding: This research was funded by the Project Supported by National Natural Science Foundation of China, grant number 51607069 and the Fundamental Research Funds for the Central Universities, grant number 2016MS88; 2017MS091.

Acknowledgments: An earlier version of this paper was presented at 2018 IEEE International Conference on Information, Communication and Engineering. The authors would like to express their sincere gratitude to IEEE ICICE 2018 committees for recommending it to submit to Energies.

Conflicts of Interest: The authors declare no conflict of interest.

References

1. Liang, H.; Li, Y.; Liu, Y.; He, J.; Cui, H. Grid-connected Control Strategy of PV Battery-energy Storage Hybrid Power System with Electric vehicle. In Proceedings of the 2018 IEEE International Conference on Information, Communication and Engineering, Xiamen, China, 28–30 September 2018.
2. Rifkin, J. *The Third Industrial Revolution: How Lateral Power Is Transforming Energy, the Economy, and the World*; Palgrave Macmillan: New York, NY, USA, 2011; pp. 74–114.
3. Chen, L.; Sun, Q.; Zhao, L.; Cheng, Q. Design of A Novel Energy Router and Its Application in Energy Internet. In Proceedings of the 2017 Chinese Automation Congress (CAC), Wuhan, China, 27–29 November 2017.
4. Abdella, J.; Shuaib, K. Peer to Peer Distributed Energy Trading in Smart Grids: A Survey. *Energies* **2018**, *11*, 1760. [CrossRef]
5. Panagiotis, K.; Lambros, E. (Eds.) *Electricity Distribution, Intelligent Solutions for Electricity Transmission and Distribution Networks*; Springer: Berlin/Heidelberg, Germany, 2016.
6. De Quevedo, P.M.; Contreras, J.; Mazza, A.; Chicco, G.; Porumb, R. Reliability Assessment of Microgrids With Local and Mobile Generation, Time-Dependent Profiles, and Intraday Reconfiguration. *IEEE Trans. Ind. Appl.* **2018**, *54*, 61–72. [CrossRef]
7. Hirsch, A.; Parag, Y.; Guerrero, J. Microgrids: A review of technologies, key drivers, and outstanding issues. *Renew. Sustain. Energy Rev.* **2018**, *90*, 402–411. [CrossRef]
8. Yan, B.; Wang, B.; Zhu, L.; Liu, H.; Liu, Y.; Ji, X.; Liu, D. A Novel, Stable, and Economic Power Sharing Scheme for an Autonomous Microgrid in the Energy Internet. *Energies* **2017**, *8*, 14741–14764. [CrossRef]
9. Liu, Q.; Wang, R.; Zhang, Y.; Wu, G.; Shi, J. An Optimal and Distributed Demand Response Strategy for Energy Internet Management. *Energies* **2018**, *11*, 217. [CrossRef]
10. Liu, Y.; Gao, S.; Zhao, X.; Zhang, C.; Zhang, N. Coordinated Operation and Control of Combined Electricity and Natural Gas Systems with Thermal Storage. *Energies* **2017**, *10*, 917. [CrossRef]
11. BDI Initiative. *Internet of Energy*; Federation of German Industries: Berlin, Germany, 2008.
12. Huang, A.Q.; Crow, M.L.; Heydt, G.T. The future renewable electric energy delivery and management (FREEDM)system the energy internet. *Proc. IEEE* **2011**, *99*, 133–148. [CrossRef]
13. Boyd, J. An internet-inspired electricity grid. *IEEE Spectr.* **2013**, *50*, 14. [CrossRef]
14. Zhang, J.; Gao, F.; Xu, S.; Zhang, X.; Zhang, L.; Sun, Y. Energy Internet technology architecture and case analysis. *China Electr. Power* **2008**, *51*, 24–30.
15. Liu, Y.; Fang, Y.; Li, J. Interconnecting Microgrids via the Energy Router with Smart Energy Management. *Energies* **2017**, *10*, 1497.
16. Nguyen-Van, T.; Abe, R.; Tanaka, K. Digital Adaptive Hysteresis Current Control for Multi-Functional Inverters. *Energies* **2018**, *11*, 2422. [CrossRef]
17. Cao, J.; Meng, K.; Wang, J. An energy internet and energy routers. *Sci. China Inf. Sci.* **2014**, *44*, 714–772.
18. Lin, C.; Zhao, H.; Liu, X.; Li, H.Q.; Xu, J.Q. Research on routing strategy of energy Internet. *J. Electr. Technol.* **2017**, *30*, 37–44.
19. Jiang, Y.; Ye, H.; Zhang, Q.; Wang, K.; Xu, Z.; Yang, R. Implementation of distributed power routing strategy based on Dijkstra algorithm in energy Internet. *Power Syst. Technol.* **2017**, *41*, 2071–2078.
20. Guo, H.; Wang, F.; Zhang, L.J.; Luo, J. Intelligent distributed energy network technology based on energy router. *Proc. CSEE* **2016**, *36*, 3314–3325.

Energies **2019**, *12*, 302

21. Tian, B.; Lei, J.; Guo, X.; Li, P.; Liu, Z. Main circuit structure and function simulation of multi-interface energy router. *Autom. Power Syst.* **2017**, *41*, 16–21.
22. Xu, W.; Zhou, G. Research on energy router design based on distributed power supply. *Power Syst. Prot. Control* **2017**, *45*, 64–71.
23. Tchakoua, P.; Wamkeue, R.; Ouhrouche, M.; Slaoui-Hasnaoui, F.; Tameghe, T.A.; Ekemb, G. Wind Turbine Condition Monitoring: State-of-the-Art Review, New Trends, and Future Challenges. *Energies* **2014**, *7*, 2595–2630. [CrossRef]
24. Shafiullah, G.M.; Oo, A.M.T.; Ali, A.B.M.S.; Wolfs, P.; Stojcevski, A. Experimental and simulation study of the impact of increased photovoltaic integration with the grid. *J. Renew. Sustain. Energy* **2014**, *6*, 033144. [CrossRef]
25. Li, H.; Wen, C.; Chao, K.-H.; Li, L.-L. Research on Inverter Integrated Reactive Power Control Strategy in the Grid-Connected PV Systems. *Energies* **2017**, *10*, 914. [CrossRef]
26. Wang, R.; Wang, D.; Jia, H.; Yang, Z.; Qi, Y.; Fan, M.; Sheng, W.; Hou, L. A Coordination Control Strategy of Battery and Virtual Energy Storage to Smooth the Micro-grid Tie-line Power Fluctuations. *Proc. CSEE* **2017**, *35*, 5144–5514.
27. Marcos, J.; de la Parra, I.; García, M.; Marroyo, L. Control Strategies to Smooth Short-Term Power Fluctuations in Large Photovoltaic Plants Using Battery Storage Systems. *Energies* **2014**, *7*, 6593–6619. [CrossRef]
28. Shepero, M.; Munkhammar, J.; Widén, J.; Bishop, J.D.K.; Bostrom, T. Modeling of photovoltaic power generation and electric vehicles charging on city-scale: A review. *Renew. Sustain. Energy. Rev.* **2018**, *89*, 61–71. [CrossRef]
29. Habib, S.; Khan, M.M.; Abbas, F.; Sang, L.; Shahid, M.U.; Tang, H. A Comprehensive Study of Implemented International Standards, Technical Challenges, Impacts and Prospects for Electric Vehicles. *IEEE Access* **2018**, *6*, 13866–13890. [CrossRef]
30. Nefedov, E.; Sierla, S.; Vyatkin, V. Internet of Energy Approach for Sustainable Use of Electric Vehicles as Energy Storage of Prosumer Buildings. *Energies* **2018**, *11*, 2165. [CrossRef]

![energies logo] *energies*

MDPI

Article

Virtual Oscillator Control of Equivalent Voltage-Sourced and Current-Controlled Power Converters †

Daniel F. Opila *, Keith Kintzley, Spencer Shabshab and Stephen Phillips

Department of Electrical and Computer Engineering, United States Naval Academy, Annapolis, MD 21402, USA; kintzley@usna.edu (K.K.); doubleshab@gmail.com (S.S.); stephen.phillips.m@gmail.com (S.P.)

* Correspondence: opila@usna.edu

† This paper is an extended version of our work published in the IFAC Workshop on Control of Transmission and Distribution Smart Grids, Prague, Czech Republic, 11–13 October 2016.

Received: 27 November 2018; Accepted: 9 January 2019; Published: 18 January 2019

Abstract: The dynamics of a general class of weakly nonlinear oscillators can be used to control power converters to create a self-forming AC network of distributed generators. Many control stability results for these "virtual" oscillators consider the interaction of voltage-source converters, but most practical converters use a nested current loop. This paper develops a general method to extend voltage-source stability results to current-controlled converters using a virtual admittance. A fast current control loop allows a singular perturbations analysis to demonstrate the equivalence of the two. This virtual admittance can also manipulate load sharing between converters without changing the core nonlinear dynamics. In addition, Virtual Oscillator Control is experimentally demonstrated with three-phase voltage-sourced and current-controlled inverters. This validates the equivalence of the two formulations, and extends previous single phase testing into three phases. The extension to current-controlled converters enhances safety and increases the breadth of applications for existing control methods.

Keywords: coupled oscillators; virtual impedance; synchronization; power converters; droop control; virtual admittance; distributed generation; energy; renewable energy; microgrids

1. Introduction

1.1. Motivation

The increasing penetration of power converters has raised new problems and opportunities in the control of small power systems and microgrids [1–3]. Power conversion decouples the physical dynamics of generators and loads from the rest of the system, allowing almost any set of dynamics to be substituted. Converter dynamics can be designed to facilitate load sharing, synchronization, and voltage and frequency regulation among multiple generators in both AC and DC power systems [4].

Historically, inverters would connect to an existing stiff AC power grid where a synchronous machine established the voltage and frequency. The proliferation of distributed generation and renewable energy has spurred interest in "grid forming" controls that can create a functioning AC power system using only power converters. However, the synchronous machines that comprise a standard power system have well-understood dynamics that serve to create an oscillating AC waveform and synchronize multiple power sources [5].

This innate feature of synchronous machines must be replaced by power converters. As a simple approach, the converters may have their own sinusoidal reference voltage generation, somewhat similar to the traditional Phase Lock Loop controls used with existing AC grids [6]. Alternatively, a synchronous machine's dynamics can be simulated in software to produce a "Virtual synchronous machine" that uses tunable virtual parameters [7] to create the desired system dynamics including droop [8], frequency dynamics [9,10], and damping [11].

1.2. Approach

This paper focuses on an entirely separate method, a class of weakly nonlinear Liénard oscillators that form nearly sinusoidal stable limit cycles when simulated in software. They take the general form of a resonant linear oscillator with weakly nonlinear forcing terms, as $\ddot{x} + f(x)\dot{x} + g(x) = 0$ [12,13]. Theoretical analysis has shown that coupled networks of these oscillators can produce the desired AC system behaviors including synchronization, stability, power sharing, and droop [14–16]. This general approach is termed virtual oscillator control (VOC) [12,13,17–19] because the oscillator dynamics are simulated to produce inverter commands. Two examples include the Van Der Pol Oscillator and the Nonlinear Dead Zone Oscillator (DZO) [17–20], which is the focus of the examples in this paper.

The theoretical analysis of VOC typically approximates a switching power converter as an ideal voltage source. Controller dynamics are assigned to command the output voltage as a function of output current. This is a logical approach for control of voltage-source converters. In practice, however, it is common to have a low-level current control loop to render the converter a controlled current source. A practical consideration critical to commercial utilization of power converters is the need to prevent damage to devices due to load anomalies. Semiconductors are sensitive to overloads of even short duration, and the current-controlled formulation makes it easy to enforce protective limits during faults, rapid transients, and low-voltage ride-through conditions. VOC based on the voltage-source formulation presents no obvious way to satisfy these protective limiting requirements, which presently restricts its commercial deployment.

This paper attempts to bring theoretical VOC converter control methods closer to deployment through three main contributions built on our previous work [20]. The first is a general theoretical approach that can extend a voltage-source analysis into a current-controlled version while maintaining its stability proofs, enabling more direct implementation into the current-controlled converters common in practice [20]. This method uses a simulated virtual admittance (the inverse of impedance) to create an equivalence between the two implementations under the assumption of fast current controller dynamics. A virtual admittance simulates the output current produced by a voltage difference across a component that exists only in simulation, while a virtual impedance simulates a voltage drop based on a current. The second contribution is the ability to control the power output and droop characteristics of current-controlled VOC inverters by varying their virtual output admittance, rather than the physical filter components as in the voltage-sourced derivation. Finally, we extend the experimental VOC results from the existing single-phase dead-zone oscillators [17–19] to three-phase voltage-sourced inverters, and then to three-phase current-controlled inverters using the equivalence technique. Compared to our previous work [20], this paper adds a more detailed theoretical analysis, simulations, voltage-sourced inverter testing, and considers current controllers with limited bandwidth.

1.3. Literature Review

The idea of virtual admittance is not new in converter controls [21–26]. It has been studied for DC systems to create droop phenomena and enforce power sharing in DC [23,27,28] and AC [24,29–31] microgrids, and similarly in battery management systems [22]. Another application is to regulate harmonic voltages in current-controlled inverters [29,32]. The idea of having dual voltage-sourced or current-controlled implementations has been studied for virtual synchronous machines, where the simulated dynamics already include a clear method to implement the two variants [33,34]. The novelty here and in our previous work is using virtual admittance to enable current control for VOC controllers, which have no obvious method to create a current-controlled equivalent.

When operating as a voltage-sourced virtual oscillator, tuning the effective output impedance of an inverter requires changing the core oscillator dynamics or the inverter's physical output filter. However, the effective output impedance of a current-controlled inverter can be manipulated by changing only a virtual admittance [20]. As a consequence, the relative power contributions of parallel-connected current-controlled inverters can be manipulated without changing the core oscillator control or any physical components. The idea of varying virtual impedances or admittances to enforce a desired power sharing arrangement in voltage controlled inverters was shown in [30,35], but is new for VOC applications.

To demonstrate these ideas, we implemented them experimentally using the dead zone oscillator for virtual oscillator control. A key feature of DZO control is that the parameters can be tuned such that parallel-connected inverters self-synchronize with no communication other than that inherent to their common electrical coupling [18]. Existing DZO work assumes voltage-source inverters and was tested in single-phase [17–19]. This paper extends the hardware testing to three-phase via the Clarke transformation [19]. The voltage-to-current conversion is then used to drive three-phase current-controlled inverters, as in our previous work [20]. The resulting AC grid in both cases demonstrates self-synchronization, voltage and frequency regulation, controllable load sharing, and droop characteristics, all without dedicated communication between units. The testbed parameters for both current and voltage-controlled systems were identical thus proving the stability characteristics apply under either control framework.

The proposed DZO control method is only one of many available for microgrid control. Most involve some form of low level control, combined with a hierarchical control scheme. Droop control is commonly used for the low level, in which the voltage and/or frequency reference are modified based on the measured real and reactive power outflows [6,36,37]. DZO control does not explicitly implement this phenomenon, but the nonlinear dynamics of the system inherently exhibit this same behavior [12,13]. There are many proposed algorithms for hierarchical control between multiple converters [38,39], but DZO functions in a truly distributed fashion without a higher level controller.

This paper is organized as follows: Section 2 describes the extension of voltage-source stability results to the current-controlled case. Section 3 summarizes DZO dynamics for both voltage and current-controlled variants, and Section 4.1 describes the testbed setup. Section 4.2 describes simulation results, while Section 4.3 presents the results of hardware testing which demonstrate three phase DZO synchronization, load sharing, and response to step changes in loads.

2. Voltage- and Current-Controlled Inverters

This section describes a typical model of a voltage-sourced converter and its interaction with a grid or load. It then describes how the same dynamics and control can be used with a current-controlled inverter. This concept initially appeared in [20].

2.1. Converter Models

A simple voltage-source converter with ideal semiconductor switches is illustrated in Figure 1. The current I_F and voltage V_i are measured at the output of the inverter bridge. An inductor and a parallel capacitor serve as an output filter, however, these elements may be eliminated if filtering is not required. The voltage where the output of the converter connects to the grid is labeled V_g. The converter establishes a voltage V_g given the injected current and activity in the rest of the system. The "Grid" component G represents everything external in the system: loads, impedances, and other inverters. Although Figure 1 depicts a single-phase AC converter, different switch configurations can produce three-phase or DC converters.

Assuming the switching dynamics are sufficiently fast to be neglected and by replacing the inverter bridge with an ideal voltage source, the voltage-source converter can be further simplified into Figure 2. The series-connected inductor filter has been replaced in the system by F_S which uses the voltage drop $V_i - V_g$ to calculate a current. Also, the parallel-connected capacitor filter is combined with the grid system, now denoted G'.

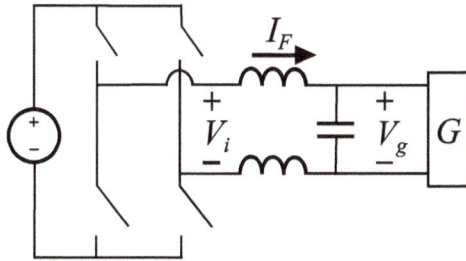

Figure 1. Simple model of a voltage-source converter [20].

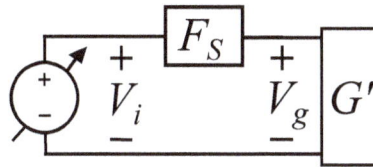

Figure 2. The converter switches can be represented as an ideal voltage source. Series filter components are represented as F_S, while parallel filter elements are lumped into the grid to become G' [20].

2.2. Voltage-Sourced Analysis

For an AC or DC converter system, stability analysis depends on assumptions about the following elements: The dynamics of the controller; the grid impedance seen at the output of the voltage source; and the switching dynamics, which are fast and often neglected.

The software-based oscillator controller C has an internal state x_C that evolves with dynamics $\dot{x}_C = C_x(x_C, I_F)$ and assigns a voltage based on the output function $V_i = C_y(x_C, I_F)$ in response to I_F, the measured output current. This current is produced by the voltage drop across the output filter F via its internal dynamics $\dot{x}_F = F_x(x_F, V_i - V_g)$ and output function $I_F = F_y(x_F, V_i - V_g)$. In a similar manner, the internal model of the grid is specified by $\dot{x}_{G'} = G'_x(x_{G'}, I_F)$ with output $V_g = G'_y(x_{G'}, I_F)$. It should be noted that the components C, F, and G' may all be nonlinear. The basic equations are thus

$$\dot{x}_C = C_x(x_C, I_F) \tag{1}$$
$$\dot{x}_F = F_x(x_F, V_i - V_g) \tag{2}$$
$$\dot{x}_{G'} = G'_x(x_{G'}, I_F) \tag{3}$$
$$V_i = C_y(x_C, I_F) \tag{4}$$
$$I_F = F_y(x_F, V_i - V_g) \tag{5}$$
$$V_g = G'_y(x_{G'}, I_F). \tag{6}$$

In the case where the F is linear, it becomes an admittance.

2.3. Current-Controlled Equivalent

To adapt the previous analysis for the case of a current-controlled converter, we assume a sufficiently fast current-control loop (PI or similar) which allows the converter to maintain a desired output current by varying the converter output voltage. Because the current controller must be able to modulate output current, it will not function with an open circuit. Hence, we require the output impedance G' to be bounded. To meet this condition even if the grid interface to the converter terminals G is unconnected, we include a parallel-connected filter capacitor as in Figure 1.

Under this ideal current-controlled model, the effects of the series filter F_S on current I_F are removed as it is within the closed-loop portion. Yet, this offers the possibility to simulate the effect of any desired output filter in software as a virtual impedance F'_S as shown in Figure 3.

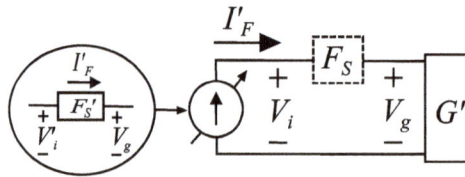

Figure 3. A current-controlled inverter with a sufficiently fast PI loop can enforce any I'_F. Thus, given a simulated output voltage V'_i, it can enforce any virtual filter current $I'_F = F'_y(x_{F'}, V'_i - V_g)$ (12). When the virtual and hardware filters do not match $F'_S \neq F_S$, the virtual (V'_i) and real (V_i) inverter bridge terminal voltages will differ $V'_i \neq V_i$ [20].

To create this system, start with the simulated controller output V'_i and subtract the measured terminal voltage V_g to create the voltage difference needed to calculate output current with F'_S. This current is then fed back as the current command I'_F. The dynamics of the current controller K are added as (10) to drive the actual current I_F to the desired I'_F and yield the modified system

$$\dot{x}_C = C_x(x_C, I'_F) \tag{7}$$

$$\dot{x}_{F'} = F'_x(x_{F'}, V'_i - V_g) \tag{8}$$

$$\dot{x}_{G'} = G'_x(x_{G'}, I_F) \tag{9}$$

$$\epsilon \dot{x}_K = K_x(x_K, I'_F - I_F) \tag{10}$$

$$V'_i = C_y(x_C, I'_F) \tag{11}$$

$$I'_F = F'_y(x_{F'}, V'_i - V_g) \tag{12}$$

$$V_g = G'_y(x_{G'}, I_F). \tag{13}$$

The "singular perturbations" theorem [40] states that for a dynamical system where a portion of the dynamics are much faster than the rest, those fast dynamics can be treated as an algebraic relationship. Assuming the current controller K is fast compared to the other dynamics and drives the current tracking error to zero, ϵ in (10) approaches zero. A singular perturbations argument [40] allows us to consider (10) as an algebraic relationship, and the converter can be treated as a current source where I_F matches exactly the desired I'_F within the bandwidth of the current controller. This reduces the current controlled system (7)–(13) to match the original voltage source system (1)–(6) with the original filter dynamics F replaced with the virtual filter F'.

For a voltage-source model, the output impedance includes the series output filter of the inverter F_S in Figure 2, which is a physical hardware component. For the current-controlled version, the physical output filter impedance is neglected due to the ideal current control loop, but a simulated filter impedance is included in the controller dynamics. This makes the full system analysis and system dynamics identical to the previous case, except for that filter dynamics F_S are now virtual rather than real.

Thus, for $F'_S = F_S$, the system dynamics are identical to the voltage-source case. This method makes no assumptions about linearity or AC vs. DC operation. The main underlying assumption is that the current loop is stable and much faster than other dynamics such that I_F converges to I'_F. A necessary condition for this assumption is the effective grid impedance G' must be finite.

Specifying the inverter's effective output filter as a virtual impedance whose characteristics are controlled via software offers several benefits. Firstly, methods exist to control power sharing, droop, etc. based on this filter, thus software control of this parameter permits unlimited flexibility as compared to changing the actual hardware values. Also, most causal filter models can be implemented independent of the actual hardware. This also permits improved analysis since the simulated filter impedance will be accurately known. This is not always the case with actual hardware.

Although presented here for the single-phase AC case, this current-controlled model similarly extends to three-phase systems for both unbalanced or nonlinear conditions.

2.4. Non-Ideal Current Amplifiers

This equivalent controller method relies on a current control loop or current source that rapidly tracks the commanded current (10). Typically, this is only true within a specific frequency range, and so long as the current controller is sufficiently fast compared to the oscillator dynamics, the singular perturbations analysis holds. For completeness, consider the behavior of the system as the base dynamics approach the bandwidth of the current controller, and the tracking performance weakens.

Consider the transfer function at the output of the controller C, which sends a commanded inverter voltage, and receives a current feedback. The voltage-source converter of Figure 4a observes an output transfer function that behaves as a simple impedance

$$\frac{I_F}{V_i} = \frac{1}{Z_F + Z_G}. \tag{14}$$

The more complex current-controlled converter of Figure 4b observes the output transfer function

$$\frac{I'_F}{V'_i} = \frac{Z_F + Z_G + K(s)}{Z'_{F'}(Z_F + Z_G) + K(s)(Z'_{F'} + Z_G)}. \tag{15}$$

When the current controller gain $K(s)$ is large, those terms dominate, and the transfer function reverts to the form of the voltage-sourced version (14), with the virtual filter $Z'_{F'}$ replacing the hardware filter Z_F, as expected. If $K(s)$ becomes very small, for example if the output is disabled, the dynamics reduce to the simple impedance of the virtual filter $Z'_{F'}$.

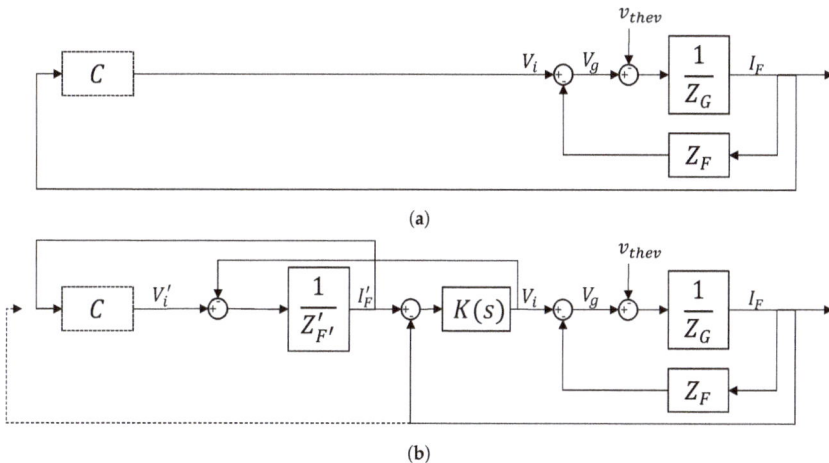

(a)

(b)

Figure 4. Diagram of the output dynamics seen by the virtual oscillator for voltage-sourced and current-controlled systems. The oscillator controller C is not modified, and is shown with a dashed border. The remainder of the system is shown as a linear transfer function block diagram. The filter impedance is Z_F for the hardware filter or $Z'_{F'}$ for the virtual filter. The grid (either G or G') is treated as its Thévenin equivalent with impedance Z_G and voltage v_{thev}. The current controller (e.g., a PI controller) is $K(s)$. The various impedances are all functions of s. (**a**) Voltage-sourced inverter block diagram. In the base derivation, the virtual oscillator control (VOC) controller C receives a current feedback I_F measured at the inverter output. (**b**) Current-controlled inverter block diagram. In the base derivation, the VOC controller C receives a virtual current feedback I'_F based on the virtual filter. Alternatively, one could use the feedback of the actual current I_F, shown as a dashed line.

As an alternative, the feedback current could be the output of the real system, rather than the commanded current, as shown by the dashed feedback line in Figure 4b. In this case, the transfer function seen by the oscillator control output is

$$\frac{I_F}{V_i'} = \frac{K(s)}{Z_{F'}'(Z_F + Z_G) + K(s)(Z_{F'}' + Z_G)}. \tag{16}$$

This configuration exhibits the same behavior when current controller gain $K(s)$ is large, again matching the form of the voltage-sourced version (14), with the virtual filter $\hat{Z}_{F'}$ replacing the hardware filter. However, as the gain $K(s)$ approaches zero, this output transfer function also approaches zero to reflect the reality that there is little real output current. This alternative method is not used in the results presented in this paper, but its behavior may be desirable in specific applications.

2.5. Synchronization and Over-Current Protection

Converters typically have output filters with impedances of 5%–20% per unit at the fundamental frequency. Directly connecting a voltage-source virtual oscillator via switch or a bridge and enabling switching without synchronizing first will usually result in a large current spike that will either damage the converter or trip the protection. This is not an issue from an analysis or stability perspective, but can cause significant damage in hardware.

Current-controlled inverters can be set to self-limit this current spike, protect the device, and continue operating. This is one reason why current control loops are used so often in practice.

Voltage-controlled inverters cannot directly limit current in the same way, and they are often protected by software trip logic that instantly ceases outputs if the measured current exceeds given threshold values. In normal operation this is not a problem, but a direct startup to an operating grid will trip the protection.

For the voltage-controlled case, one possible solution is to achieve some degree of synchronization before connecting the converter or enabling the switching action. This can be achieved with a voltage measurement on the line side of any switches or breakers. One approach is to use this voltage measurement to adjust the states of the oscillator to match the grid oscillation.

A second approach is used in the experimental section of this paper. The line-side voltage measurement is fed to the control software, and a simulated output admittance is connected between the oscillator dynamics and the grid. This process is entirely virtual and results in no real currents or voltages, but it does synchronize the oscillator dynamics in software before the physical converter is connected or enabled.

3. Application of Dead-Zone Oscillator Control to Current and Voltage-Controlled Three-Phase Inverters

DZO control was formulated in [17–19] as a voltage-source algorithm, but a current-controlled equivalent would both increase the breadth of application for DZO control and offer the advantages of controlling the effective output impedance of inverters through software. To these ends, the method developed in Section 2 of adapting voltage-source control algorithms to current-controlled converters is applied to DZO control. The basics of this method are summarized here for convenience. The various stability and convergence proofs are found in [17–19].

3.1. Dead-Zone Oscillator Control

Under DZO control, an inverter mimics the dynamics of a nonlinear DZO in a manner similar to the well-known Van der Pol oscillator. A DZO circuit equivalent model, shown in Figure 5, consists of a nonlinear voltage-dependent current source and a parallel resistor inductor capacitor (RLC) circuit.

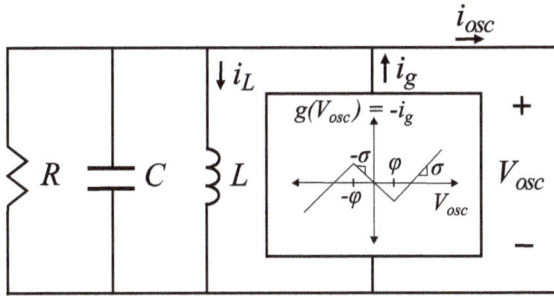

Figure 5. Circuit model of the Dead-Zone Oscillator. i_g, the output of the voltage-dependent current source, is a nonlinear function $g(V_{osc})$. The impedance of the parallel resistor inductor capacitor (RLC) circuit, together with the parameters σ and φ of $g(V_{osc})$, control the limit cycle of a single oscillator and the interaction of coupled DZOs [17,20].

The DZO [17,18,20] is a nonlinear dynamic system of two states: The DZO terminal voltage V_{osc} and the inductor current i_L with dynamics specified by

$$\frac{dV_{osc}}{dt} = \frac{1}{C}\left[(V_{osc}(\sigma - \frac{1}{R}) - f(V_{osc}) - i_L - i_{osc}\right] \tag{17}$$

$$\frac{di_L}{dt} = \frac{1}{L}V_{osc} \tag{18}$$

$$i_g = -g(V_{osc}) = \sigma V_{osc} - f(V_{osc}) \tag{19}$$

where $f(V_{osc})$ is the dead-zone function with parameters σ and φ, and $g(V_{osc})$ is the voltage-current characteristic of the voltage-dependent current source, which is related to $f(V_{osc})$ by (19).

The dynamics of the DZO, as presented in [18], are governed by the parameters of $g(V_{osc})$ and by the impedance of the parallel RLC circuit. For a standalone inverter, V_{osc} has a stable, unique limit cycle if $\sigma > 1/R$. If $\sqrt{L/C}(\sigma - 1/R) \ll 1$, it can be shown that the limit cycle is approximately sinusoidal with frequency close to the natural frequency of the RLC circuit, $1/\sqrt{LC}$.

When multiple inverters under DZO control are connected in parallel, each inverter's output current is determined by V_i, V_g, and F_s, and because each inverter's output current I_F depends on the common grid voltage, the inverters demonstrate the ability to self-synchronize. Further, due to the dependence of the inverter current I_F on its own output filter impedance F_s, load-sharing between inverters can be controlled by changing F_s.

3.1.1. Voltage-Controlled Inverter DZO Requirements

The DZO control derivation and testing was initially created for use in voltage-source inverters. The controller measures the current I_F at the terminal of the inverter and commands the bridge voltage V_i.

3.1.2. Adaptation of DZO Control to Current-Controlled Inverters

DZO control can be extended from voltage-source inverters to current-controlled inverters using the approach presented in Section 2. As illustrated in Figure 3, the DZO terminal voltage V_i' and the voltage measured at the grid terminal are used to compute I_F', which are both sent as a control signal to the current-controlled inverter and fed back into the DZO model to calculate the value of V_i' at the next time step.

3.2. Extension of DZO to Three-Phase Networks

In [19], the Clarke Transform was used to generate the balanced three-phase voltage control signal V_i from V_{osc} and i_L, the virtual inductor current of (18). For control of current-controlled inverters in the test bed described here, V_i' is generated in the same way.

3.3. Software DZO Inverter Synchronization

Voltage measurements are processed by the controller to provide a reference for multiple inverter synchronization. During initial startup of additional inverters when at least one is already running, software in the starting inverter synchronizes the output of the inverter with AC voltage across the load to decrease potential current spikes upon connecting multiple inverters across the same load. Before the inverter switches are enabled, the load side voltage measurement is fed to the simulated oscillator and connected through a virtual impedance. This triggers the natural synchronization behavior of the oscillator and causes it to synchronize with the grid before the output bridge is enabled, minimizing transients.

4. Results

4.1. Test Bed Configuration

The three-phase testbed consists of three inverter subunits connected in parallel to a Y-connected resistive load. The general configuration is identical for both current- and voltage-controlled inverters. Figure 6 shows one inverter subunit connected to the load, and displays the core functional elements of the subunit. A picture of this unit in the physical testbed is shown in Figure 7 without the attached load.

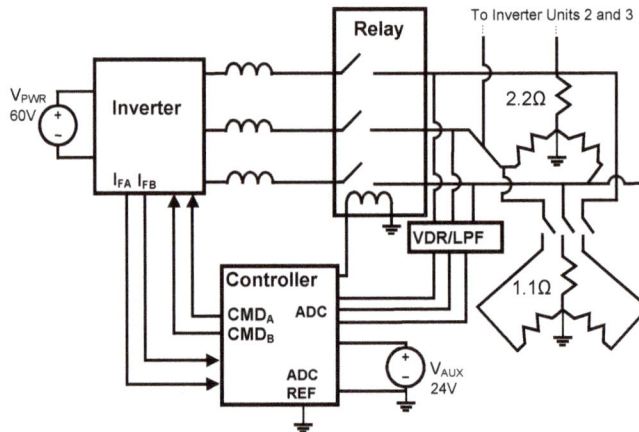

Figure 6. One inverter subunit coupled to a Y-connected load. The inverter's control signal reference and its negative DC rail are isolated. The negative DC power rail of each inverter floats in isolation. Each subunit's controller reads $V_{g,A}$ and $V_{g,B}$, the load voltages on phases A and B, and receives feedback of the amplifier output current. Additional inverter subunits are connected in parallel to the load for synchronization testing. The inverter can be either voltage-sourced and receive a voltage command, or current-controlled with a current command [20].

The controller samples $V_{i,A}$ and $V_{i,B}$, the grid voltages on the A and B phases, each at a rate of 10 kHz. Current measurements from the inverter are also fed back into the controller for processing. An analog signal conditioning board provides a voltage divider and low-pass filter with a single pole at 22 kHz.

The controller sends a reference command to the inverter, either an A- and B-phase voltage or an A- and B-phase current depending if the inverter is voltage-sourced or current-controlled.

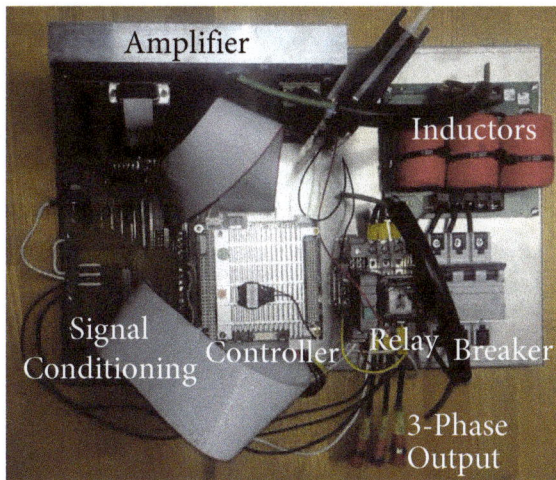

Figure 7. One power converter subunit, as shown schematically in Figure 6. The testbed comprises three subunits connected in parallel to a Y-connected three-phase load. A DC source (not shown) feeds power to the inverter in the top left of the picture.

Current and voltage measurements are fed back to the controller regardless of the controller type. The voltage-controlled variant nominally requires only current feedback, while the current-controlled variant requires only voltage feedback. However, both types use the voltage measurement for initial grid synchronization, and current measurements for over-current protection.

Each inverter subunit's controller actuates a normally-open relay to connect the subunit to the grid. For a current-controlled inverter, an open circuit can create control loop stability problems, so when the relay is open the controller holds the states of the simulated oscillator and the current commands at zero to prevent current loop saturation. The governing state equations of the dead-zone oscillator are discretized and solved in real time with a fixed step size of 100 μs.

4.2. Simulation Results

The dead zone oscillator system with both voltage-sourced and current-controlled amplifiers was simulated in the MATLAB/SIMULINK environment. The ability of the current amplifier to accurately track current commands is critical to the correct functionality of the system, as described in Section 2.4. The simulation was configured and tuned to match the test bed, with both voltage-sourced and current-controlled amplifiers connected through a 150 μH inductor to a 4 Ω Y-connected resistor bank. The current controller is a Proportional Integral (PI) type.

To understand the amplifier response, the transfer function from amplifier command to output current was measured via a frequency sweep in the amplifier reference input as shown in Figure 8.

For the voltage-sourced inverter, the voltage command was scaled by the output resistance (4 Ω) so that a 1 V command would nominally produce 1 A and measure 0 dB. This is visible in both the simulated and measured output currents until at least 1 kHz, when the current response rolls off due to the system L/R pole at 4.2 kHz. This means the voltage follows the command very well within this measurement window.

The current-controlled amplifier response is shown as the transfer function from current command to current output. In this case, both real and simulated current controllers show good tracking until a −3 dB bandwidth of about 200 Hz, when the response no longer tracks the command. So for this system, both simulation and hardware tests indicated that the current-controlled and voltage-sourced systems should be equivalent for frequencies below about 100 Hz. This should suffice for the steady-state behavior of oscillators set for 60 Hz. Rapid transients may cause the two versions to diverge somewhat.

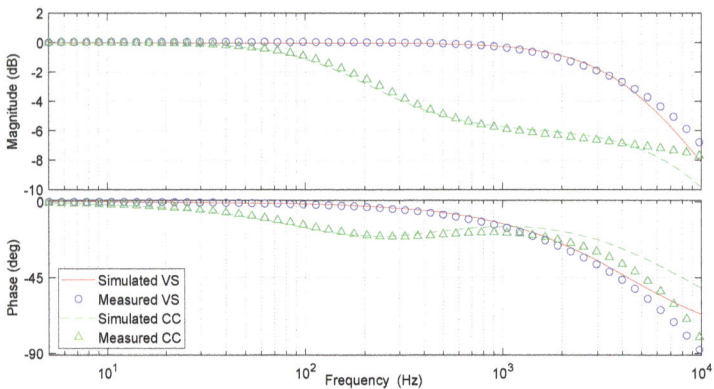

Figure 8. Amplifier Bode plots in simulation and hardware for both voltage-sourced (VS) and current-controlled (CC) inverters. In all cases, the amplifier is connected through a 150 μH inductor to a Y-connected 4 Ω resistive load. The voltage transfer function is measured as voltage command to current response, and scaled by the resistance (4 Ω) to produce 0 dB at low frequencies.

The basic operation of the virtual oscillator is tested in simulation by starting from a zero initial condition. The system starts from rest, begins oscillating with increasing amplitude, and reaches a stable limit cycle. Both voltage-sourced and current-controlled versions are shown in Figure 9, and demonstrate good agreement between the two.

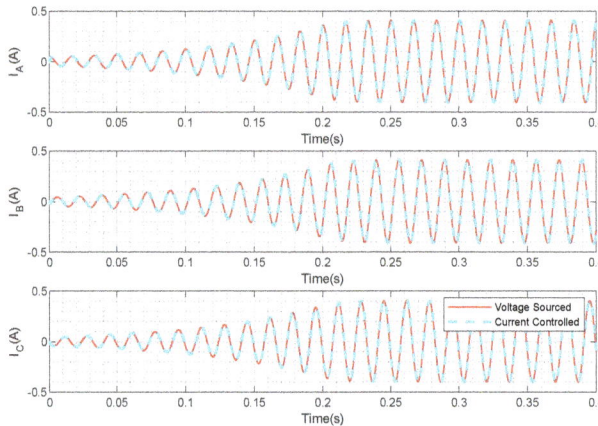

Figure 9. Simulated startup transient for both voltage-sourced and current-controlled amplifiers operated by a DZO controller. They match almost exactly, as would be expected for operations within the bandwidth of the current controller. The phase of the traces diverge slightly over longer time intervals because the two systems are not completely identical.

To study the effects of the two different implementations in a more realistic scenario, simulations were conducted to study the droop characteristics of the oscillator. As more power is drawn from the system, the output voltage decreases based on the nonlinear oscillator dynamics. This feature aids in load sharing between inverters. As shown in Figure 10, the voltage-sourced and current-controlled versions match very well during 60 Hz operation for two different nominal power ratings of the oscillator. However, when set for 400 Hz operation, more divergence appears. This validates the predictions of the current tracking response shown in Figure 8, which starts to introduce errors above about 100 Hz.

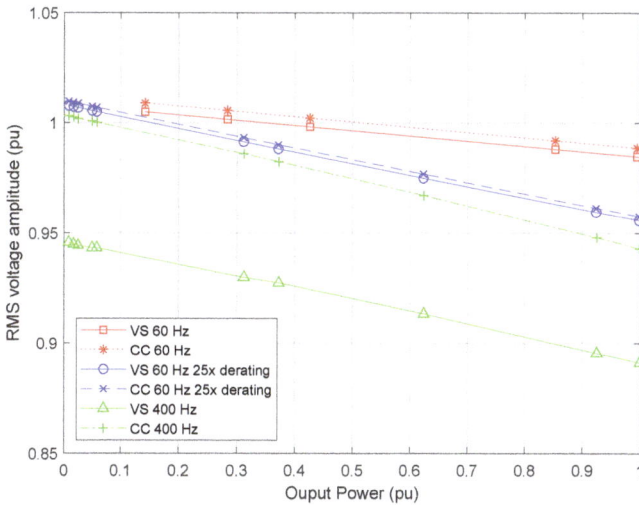

Figure 10. Simulated droop characteristics for voltage-sourced (VS) vs. current-controlled (CC) inverters for two different nominal ratings at 60 Hz and and one rating at 400 Hz. As expected, the two equivalent VS and CC formulations perform almost exactly the same for both ratings at 60 Hz, but they diverge at 400 Hz because it is beyond the current controller bandwidth shown in Figure 8. A derating by a factor of 25 is implemented by artificially scaling the current feedback to the oscillator, which creates a different droop characteristic [17–19].

4.3. Experimental Results

This section experimentally demonstrates three phenomena that are useful for practical implementation of oscillator-based converter controls. The first is the three-phase voltage-source DZO control theoretically developed and simulated (but not tested) in [19], including synchronization and the predicted power droop characteristic. The second is the equivalent performance of a current-controlled inverter to its voltage-controlled counterpart, including oscillation and synchronization of parallel-connected inverters regardless of their control framework. This validates the results of Section 2.

The final tests demonstrate the ability to dynamically change and control the power output of a current-controlled inverter by adjusting a virtual output filter admittance as described in Section 2.3. The ability to alter the load-share of an inverter by changing its *physical* output filter $Z_{F,j}$ was demonstrated for single-phase voltage-controlled inverters in [17]. Changing a *virtual* output filter as described here achieves the same result, and can be done dynamically without hardware changes. This concept is only applicable to the current-controlled implementation.

4.3.1. Voltage-Controlled Inverter

A three-phase voltage-sourced converter using DZO control was tested for self-oscillation, step changes in load, synchronization, and droop characteristics. The behavior matches expectations, confirming the theoretical results in [19].

As a first test, a single inverter self-oscillates and powers a Y-connected load of 8 Ω per phase as shown in Figure 11a. The load resistance then changes to 2.66 Ω (66% decrease) in each phase. A brief transient occurs, then settles to a steady state at higher current values as expected.

An important feature of DZO control is synchronization with other inverters. In Figure 11b, one inverter is operating and powering a load, and another inverter is then connected. They synchronize and then share the current provided to the load.

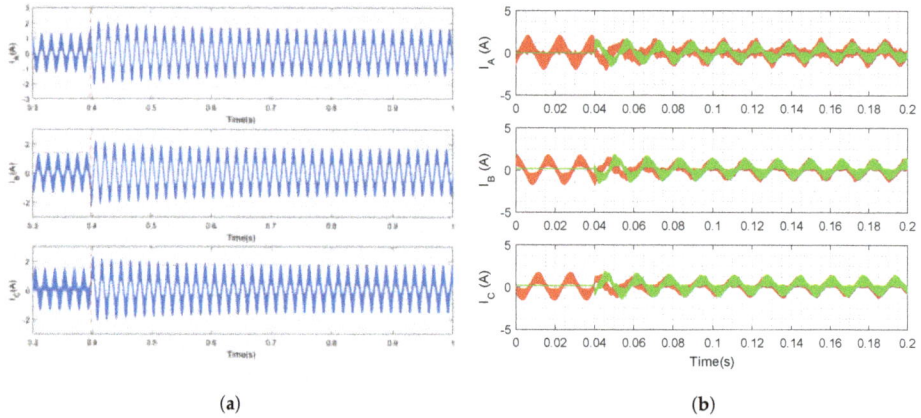

(a)   (b)

Figure 11. Experimental testing of a three phase voltage-sourced inverter which responds to step changes in load and synchronizes with another inverter. (**a**) Voltage-controlled inverter load resistance changes from 8 Ω to 2.66 Ω per phase resulting in transient response from the system with voltage and current overshoot. The DZO output then settles to nominal conditions. (**b**) Voltage-controlled inverter synchronizes when connected to another inverter.

Finally, the DZO dynamics yield a droop characteristic where the output voltage drops in the presence of increasing load. The predicted and experimental results are shown in Figure 12. The simulation results are perfectly symmetric and all three phases yield the same voltage, current, and power magnitudes. The experimental results indicate some discrepancy between phases, but the general trend is the same. The output power is calculated based on the load resistance and the nominal output voltage.

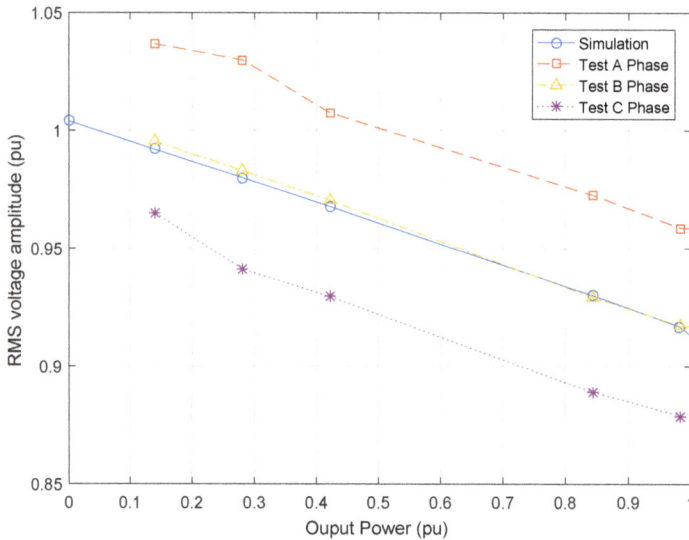

Figure 12. Three-phase voltage-controlled inverter droop characteristic demonstrated in simulation.

4.3.2. Current-Controlled Inverter

Three-phase current-controlled converters were tested for self-oscillation, synchronization, and step changes in load. These are the same physical converters as in the voltage-controlled case, but they have an internal current feedback loop such that the command from the microprocessor to the converter is a current command.

In the first test case, a single inverter self-oscillates to power a load, and a second inverter is then connected. The three phase current measurements from each inverter are shown in Figure 13a. They synchronize and begin to share load. A current spike occurs 10 ms after the synchronizing inverter is connected, reaching a maximum of 4.64 A on one phase. Out-of-phase oscillation is observed from 10–200 ms after connection, and then low-amplitude oscillation of the synchronizing inverter from 200–250 ms. Both converge to full-amplitude, in-phase oscillation after 250 ms with each providing half the load current.

As a second test case, a third inverter is added to the two that were already synchronized, as shown in Figure 13b. The same general trend occurs, with all three equally sharing the load current once they synchronize.

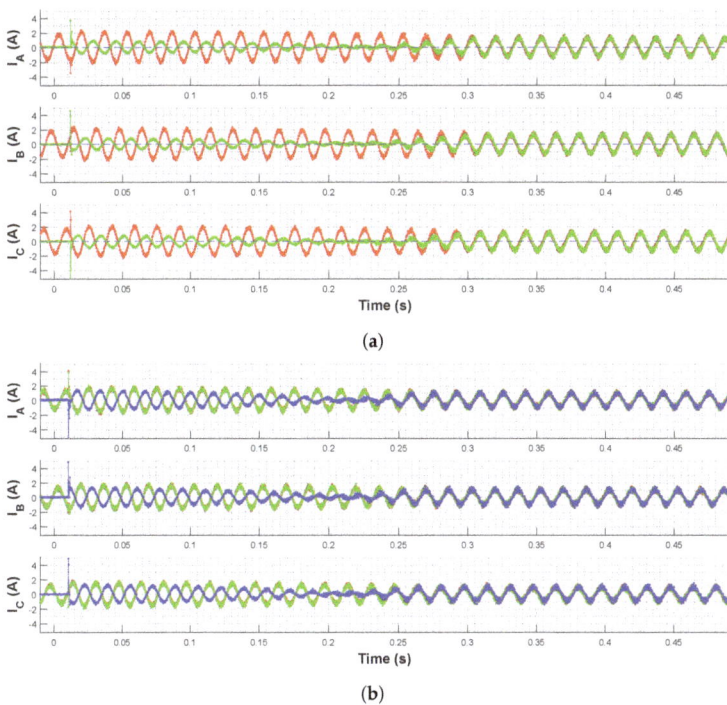

(a)

(b)

Figure 13. Experimental testing of current-controlled converter synchronization with DZO control. (**a**) One current-controlled inverter synchronizes when connected to another inverter operating at rated power. (**b**) Current-controlled inverter synchronizes when connected to two already-synchronized inverters. Three traces are shown (red, green, blue), but the red trace is directly under the green because they are already synchronized. A maximum load phase current of 4.64 A and a maximum inverter phase current of 8.48 A were observed in testing [20].

To demonstrate stability through rapid step-changes in load, two synchronized, identical inverters were equally sharing the load, and the load resistance was switched from a Y-connected 2.2 Ω to 0.733 Ω (66% decrease) on each phase, as shown in Figure 14a. These two inverters have identical virtual output impedance ($Z'_{F,1} = Z'_{F,2}$), and thus share load equally. They tolerate the change and maintain synchronization.

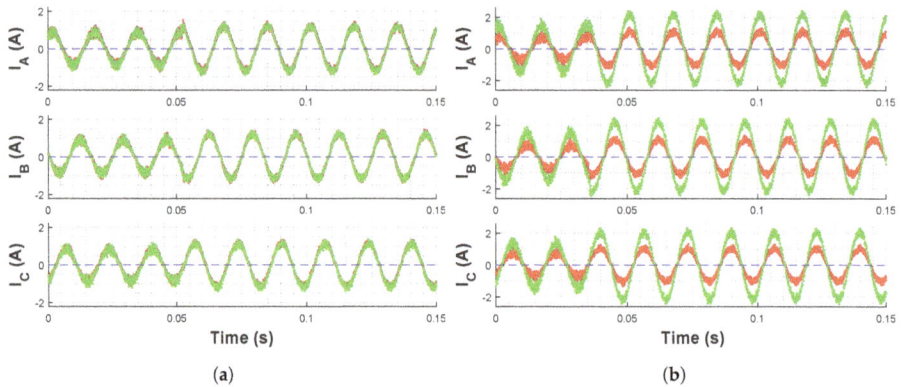

Figure 14. Hardware testing of two synchronized current-controlled inverters, with both equal and unequal load-sharing, that maintain their behavior through a step increase in load. Load resistance changes from 2.2 Ω to 0.733 Ω [20]. (**a**) Inverters with equal power sharing. (**b**) Inverters with unequal power sharing.

4.4. Inverter Load-Sharing Control through Manipulation of Virtual Output Impedance

One unique advantage of a current-controlled inverter is that the virtual output impedance can be manipulated during operation to change that inverter's share of the load (Section 2.3). It was shown in [17] that the relative power contributions P of synchronized inverters under DZO control are related to the relative output impedances κ of the inverters j and k,

$$\frac{P_k}{P_j} = \frac{\kappa_j}{\kappa_k} \qquad \forall k, j = 1...N \tag{20}$$

In Figure 15a, both inverters oscillate in synchronization with identical virtual output impedances $(Z'_{F,1} = Z'_{F,2})$ until κ_2 is doubled at time 0, increasing $|Z'_{F,2}|$ to twice $|Z'_{F,1}|$. After κ_2 is doubled, the output currents $I_{F,1}$ and $I_{F,2}$ continue to oscillate in phase but change magnitude, $|I_{F,2}|$ is approximately half of $|I_{F,1}|$. This observation is consistent with (20). The opposite process, in which κ_2 is halved rather than doubled, also yields the expected results in Figure 15b, with $|I_{F,2}|$ increasing to approximately twice $|I_{F,1}|$.

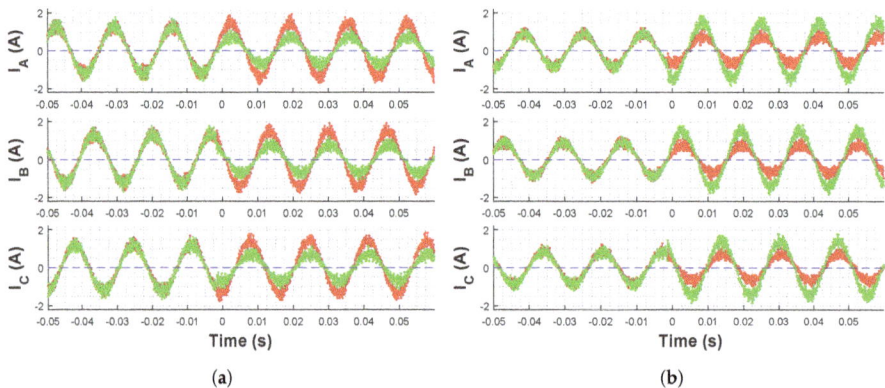

Figure 15. Experimental three phase test cases demonstrate the ability to control load sharing between two different current-controlled inverters by changing the virtual output impedance of one of them [20]. (**a**) The load carried by one inverter is decreased by increasing virtual output impedance. (**b**) The load carried by one inverter is increased by decreasing virtual output impedance.

This change in virtual output impedance does not affect the response to step changes in load. Figure 14b shows the same step change in load resistance (2.2 Ω to 0.733 Ω) applied to inverters with unequal sharing ($Z'_{F,1} = 2Z'_{F,2}$) such that the load-share of the first inverter is half that of the second inverter. It can be seen for both tests that the inverter output currents maintain both their relative amplitudes and their phase synchronization during a step change in load resistance. Recall that Figure 14a is the same test, but with identical virtual impedances ($Z'_{F,1} = Z'_{F,2}$).

5. Conclusions

This paper formulated and experimentally validated a method to bridge the gap between voltage-source and current-controlled converters for Virtual Oscillator Control applications. A virtual output admittance was used to allow a controller based on voltage output to interact with an amplifier based on a current command. If the current amplifier exhibits good command following at a bandwidth much higher than the base controller, the singular perturbations theorem allows the amplifier to be treated as an ideal current source. This paper showed that under these conditions, the dynamic behavior of the current-controlled amplifier oscillator system would match that of a voltage sourced amplifier. Simulations and hardware experiments demonstrated the effectiveness of this method for the basic oscillator performance, synchronization of multiple inverters, responses to load changes, and droop characteristics. The approach was also analyzed and simulated for cases when the amplifier has limited bandwidth. As expected, once the speed of the core oscillator dynamics approaches or exceeds the current control bandwidth, the behavior of the two implementations starts to diverge.

This VOC control was experimentally validated for the first time in a three-phase grid system with both voltage-source and current-controlled converters, and the voltage droop characteristics were verified to match predictions. The ability to tune the effective output impedances of current-controlled inverters was demonstrated by dynamically changing the load-share of parallel-connected inverters during operation. The particular variant tested, the dead-zone oscillator, demonstrated robustness to a 66% step-change in load was also with equal and unequal load-sharing. These results facilitate the deployment of VOC by extending it to three-phase and current-controlled inverter networks, and by demonstrating a new method of implementing load-sharing.

Author Contributions: Conceptualization, D.F.O.; Data curation, K.K., S.S. and S.P.; Formal analysis, D.F.O.; Investigation, K.K., S.S. and S.P.; Project administration, D.F.O.; Software, K.K., S.S. and S.P.; Supervision, D.F.O. and K.K.; Writing—original draft, D.F.O. and S.S.; Writing—review & editing, D.F.O. and K.K.

Funding: This work was partially supported by the United States Office of Naval Research under N0001415WX01739 and DURIP award N0001415WX02048.

Conflicts of Interest: The authors declare no conflict of interest. The funders had no role in the design of the study; in the collection, analyses, or interpretation of data; in the writing of the manuscript, or in the decision to publish the results.

References

1. Trudnowski, D.; Donnelly, M.; Lightner, E. Power-System Frequency and Stability Control using Decentralized Intelligent Loads In Proceedings of the 2006 IEEE/PES Transmission and Distribution Conference and Exhibition, Dallas, TX, USA, 21–24 May 2006; pp. 1453–1459.
2. Dong, B.; Li, Y.; Zheng, Z.; Xu, L. Control strategies of microgrid with Hybrid DC and AC Buses. In Proceedings of the 2011 14th European Conference on Power Electronics and Applications, Birmingham, UK, 30 August–1 September 2011; pp. 1–8.
3. Salehi, V.; Mohamed, A.; Mohammed, O.A. Implementation of real-time optimal power flow management system on hybrid AC/DC smart microgrid. In Proceedings of the 2012 IEEE Industry Applications Society Annual Meeting, Las Vegas, NV, USA, 7–11 Octorber 2012; pp. 1–8.
4. Bidram, A.; Lewis, F.; Davoudi, A. Distributed Control Systems for Small-Scale Power Networks: Using Multiagent Cooperative Control Theory. *IEEE Control Syst.* **2014**, *34*, 56–77. [CrossRef]

5. Kundur, P.; Klein, M.; Rogers, G.; Zywno, M.S. Application of power system stabilizers for enhancement of overall system stability. *IEEE Trans. Power Syst.* **1989**, *4*, 614–626. [CrossRef]

6. Guerrero, J.M.; Vasquez, J.C.; Matas, J.; de Vicuna, L.G.; Castilla, M. Hierarchical Control of Droop-Controlled AC and DC Microgrids—A General Approach Toward Standardization. *IEEE Trans. Ind. Electron.* **2011**, *58*, 158–172. [CrossRef]

7. D'Arco, S.; Suul, J.A.; Fosso, O.B. Automatic Tuning of Cascaded Controllers for Power Converters Using Eigenvalue Parametric Sensitivities. *IEEE Trans. Ind. Appl.* **2015**, *51*, 1743–1753. [CrossRef]

8. D'Arco, S.; Suul, J.A. Equivalence of Virtual Synchronous Machines and Frequency-Droops for Converter-Based MicroGrids. *IEEE Trans. Smart Grid* **2014**, *5*, 394–395. [CrossRef]

9. Torres, L.M.A.; Lopes, L.A.C.; Morán, T.L.A.; Espinoza, C.J.R. Self-Tuning Virtual Synchronous Machine: A Control Strategy for Energy Storage Systems to Support Dynamic Frequency Control. *IEEE Trans. Energy Convers.* **2014**, *29*, 833–840. [CrossRef]

10. Moeini, A.; Kamwa, I. Analytical Concepts for Reactive Power Based Primary Frequency Control in Power Systems. *IEEE Trans. Power Syst.* **2016**, *31*, 4217–4230. [CrossRef]

11. Alipoor, J.; Miura, Y.; Ise, T. Power System Stabilization Using Virtual Synchronous Generator with Alternating Moment of Inertia. *IEEE J. Emerg. Sel. Top. Power Syst.* **2015**, *3*, 451–458. [CrossRef]

12. Sinha, M.; Dörfler, F.; Johnson, B.B.; Dhople, S.V. Virtual Oscillator Control subsumes droop control. In Proceedings of the 2015 American Control Conference (ACC), Chicago, IL, USA, 1–3 July 2015.

13. Johnson, B.B.; Sinha, M.; Ainsworth, N.G.; Dörfler, F.; Dhople, S.V. Synthesizing Virtual Oscillators to Control Islanded Inverters. *IEEE Trans. Power Electron.* **2016**, *31*, 6002–6015. [CrossRef]

14. Dörfler, F.; Chertkov, M.; Bullo, F. Synchronization in complex oscillator networks and smart grids. *Proc. Natl. Acad. Sci. USA* **2013**, *110*, 2005–2010. [CrossRef]

15. Simpson-Porco, J.W.; Dörfler, F.; Bullo, F. Synchronization and power sharing for droop-controlled inverters in islanded microgrids. *Automatica* **2013**, *49*, 2603–2611. [CrossRef]

16. Dörfler, F.; Bullo, F. Synchronization in complex networks of phase oscillators: A survey. *Automatica* **2014**, *50*. [CrossRef]

17. Johnson, B.B.; Dhople, S.V.; Hamadeh, A.O.; Krein, P.T. Synchronization of Parallel Single-Phase Inverters With Virtual Oscillator Control. *IEEE Trans. Power Electron.* **2014**, *29*, 6124–6138. [CrossRef]

18. Johnson, B.B.; Dhople, S.V.; Hamadeh, A.O.; Krein, P.T. Synchronization of Nonlinear Oscillators in an LTI Electrical Power Network. *IEEE Trans. Circuits Syst. I* **2014**, *61*, 834–844. [CrossRef]

19. Johnson, B.B.; Dhople, S.V.; Cale, J.L.; Hamadeh, A.O.; Krein, P.T. Oscillator-Based Inverter Control for Islanded Three-Phase Microgrids. *IEEE J. Photovolt.* **2014**, *4*, 387–395. [CrossRef]

20. Shabshab, S.C.; Opila, D.F. Extending control stability results from voltage-source to current-controlled AC or DC power converters. *IFAC* **2016**, *49*, 60–65. [CrossRef]

21. Wang, X.; Li, Y.W.; Blaabjerg, F.; Loh, P.C. Virtual-Impedance-Based Control for Voltage-Source and Current-Source Converters. *IEEE Trans. Power Electron.* **2015**, *30*, 7019–7037. [CrossRef]

22. Chowdhury, S.M.; Badawy, M.; Sozer, Y.; Garcia, J.A.D.A. A novel battery management system using a duality of the adaptive droop control theory. In Proceedings of the 2017 IEEE Energy Conversion Congress and Exposition (ECCE), Cincinnati, OH, USA, 1–5 Octorber 2017; pp. 5164–5169.

23. Jin, Z.; Meng, L.; Han, R.; Guerrero, J.M.; Vasquez, J.C. Admittance-type RC-mode droop control to introduce virtual inertia in DC microgrids. In Proceedings of the 2017 IEEE Energy Conversion Congress and Exposition (ECCE), Cincinnati, OH, USA, 1–5 Octorber 2017; pp. 4107–4112.

24. Zhang, W.; Remon, D.; Candela, I.; Luna, A.; Rodriguez, P. Grid-connected converters with virtual electromechanical characteristics: Experimental verification. *CSEE J. Power Energy Syst.* **2017**, *3*, 286–295. [CrossRef]

25. Rodriguez, P.; Candela, I.; Citro, C.; Rocabert, J.; Luna, A. Control of grid-connected power converters based on a virtual admittance control loop. In Proceedings of the 2013 15th European Conference on Power Electronics and Applications (EPE), Lille, France, 2–6 September 2013. [CrossRef]

26. Tarrasó, A.; Candela, J.I.; Rocabert, J.; Rodriguez, P. Grid voltage harmonic damping method for SPC based power converters with multiple virtual admittance control. In Proceedings of the 2017 IEEE Energy Conversion Congress and Exposition (ECCE), Cincinnati, OH, USA, 1–5 Octorber 2017; pp. 64–68.

27. Jin, Z.; Meng, L.; Guerrero, J.M. Comparative admittance-based analysis for different droop control approaches in DC microgrids. In Proceedings of the 2017 IEEE Second International Conference on DC Microgrids (ICDCM), Nuremburg, Germany, 27–29 June 2017; pp. 515–522.

28. Chandorkar, M.C.; Divan, D.M.; Adapa, R. Control of parallel connected inverters in standalone AC supply systems. *IEEE Trans. Ind. Appl.* **1993**, *29*, 136–143. [CrossRef]

29. He, J.; Li, Y.W.; Guerrero, J.M.; Blaabjerg, F.; Vasquez, J.C. An Islanding Microgrid Power Sharing Approach Using Enhanced Virtual Impedance Control Scheme. *IEEE Trans. Power Electron.* **2013**, *28*, 5272–5282. [CrossRef]

30. Guerrero, J.M.; de Vicuna, L.G.; Matas, J.; Castilla, M.; Miret, J. Output impedance design of parallel-connected UPS inverters with wireless load-sharing control. *IEEE Trans. Ind. Electron.* **2005**, *52*, 1126–1135. [CrossRef]

31. He, J.; Li, Y.W. Analysis, Design, and Implementation of Virtual Impedance for Power Electronics Interfaced Distributed Generation. *IEEE Trans. Ind. Appl.* **2011**, *47*, 2525–2538. [CrossRef]

32. Blanco, C.; Reigosa, D.; Vasquez, J.C.; Guerrero, J.M.; Briz, F. Virtual Admittance Loop for Voltage Harmonic Compensation in Microgrids. *IEEE Trans. Ind. Appl.* **2016**, *52*, 3348–3356. [CrossRef]

33. Chen, Y.; Hesse, R.; Turschner, D.; Beck, H.P. Improving the grid power quality using virtual synchronous machines. In Proceedings of the 2011 International Conference on Power Engineering, Energy and Electrical Drives, Malaga, Spain, 11–13 May 2011; pp. 1–6.

34. Chen, Y.; Hesse, R.; Turschner, D.; Beck, H.P. Comparison of methods for implementing virtual synchronous machine on inverters. In Proceedings of the International Conference on Renewable Energies and Power Quality, Santiago de Compostela, Spain, 28–30 March 2012; pp. 1–6.

35. Guerrero, J.M.; Matas, J.; De Vicuna, L.G.; Castilla, M.; Miret, J. Wireless-Control Strategy for Parallel Operation of Distributed-Generation Inverters. *IEEE Trans. Ind. Electron.* **2006**, *53*, 1461–1470. [CrossRef]

36. Guerrero, J.M.; Chandorkar, M.; Lee, T.L.; Loh, P.C. Advanced Control Architectures for Intelligent Microgrids—Part I: Decentralized and Hierarchical Control. *IEEE Trans. Ind. Electron.* **2013**, *60*, 1254–1262. [CrossRef]

37. Guerrero, J.M.; Loh, P.C.; Lee, T.L.; Chandorkar, M. Advanced Control Architectures for Intelligent Microgrids—Part II: Power Quality, Energy Storage, and AC/DC Microgrids. *IEEE Trans. Ind. Electron.* **2013**, *60*, 1263–1270. [CrossRef]

38. Baghaee, H.R.; Mirsalim, M.; Gharehpetian, G.B. Real-time verification of new controller to improve small/large-signal stability and fault ride-through capability of multi-DER microgrids. *IET Gener. Transm. Distrib.* **2016**, *10*, 3068–3084. [CrossRef]

39. Baghaee, H.R.; Mirsalim, M.; Gharehpetan, G.B.; Talebi, H.A. Nonlinear Load Sharing and Voltage Compensation of Microgrids Based on Harmonic Power-Flow Calculations Using Radial Basis Function Neural Networks. *IEEE Syst. J.* **2018**, *12*, 2749–2759. [CrossRef]

40. Khalil, H.K. *Nonlinear Systems*, 3rd ed.; Prentice Hall: Upper Saddle River, NJ, USA, 2002.

energies

MDPI

Article

Design and Analysis of Non-Isolated Three-Port SEPIC Converter for Integrating Renewable Energy Sources

C. Anuradha [1,*], N. Chellammal [1], Md Saquib Maqsood [2] and S. Vijayalakshmi [1]

[1] Department of Electrical and Electronics Engineering, SRM Institute of Science and Technology, Kattankulathur, Kancheepuram District, Tamilnadu PIN-603203, India; chellammal.n@ktr.srmuniv.ac.in (N.C.); vijayalakshmi.s@ktr.srmuniv.ac.in (S.V.)

[2] NFTDC, Hyderabad, Telangana PIN-500059, India; mdsaquibmaqsood@gmail.com

* Correspondence: anurithvik2007@gmail.com; Tel.: +91-9840527150

Received: 26 November 2018; Accepted: 7 January 2019; Published: 11 January 2019

Abstract: An efficient way of synthesizing a three port non-isolated converter from a single-ended primary inductor converter (SEPIC) is proposed in this paper. The primary SEPIC converter is split into a source cell and a load cell. Two such source cells are integrated through direct current (DC) link capacitors with a common load cell to generate a three-port SEPIC converter. The derived converter features single-stage power conversion with reduced structural complexity and bidirectional power flow capability. For bidirectional power flow, it incorporates a battery along with an auxiliary photovoltaic source. Mathematical analyses were carried out to describe the operating principles and design considerations. Experiments were performed on an in-house-built prototype three-port unidirectional converter, and the results are presented to validate the feasibility of the designed converter.

Keywords: multiport converter (MPC); single ended primary inductor converter (SEPIC); multi-input single output (MISO); renewable power system

1. Introduction

Renewable energy resources are of potential interest nowadays due to environmental problems, high oil prices, global warming, and the depletion of fossil fuels. Though these renewable energy resources are abundant and cause zero emissions, they are intermittent in nature. When integrating, these power generators provide low voltage and require high-gain converters to meet the load demand. Thus, power electronic technology plays a significant role in interfacing hybrid renewable power systems, electric traction, and uninterrupted power supplies [1,2]. To overcome the intermittent nature of energy resources, future power systems will also require the interfacing of various energy sources using multisource technology. To enable multi-source technology, a multi-input power converter (MIPC) that can accommodate a variety of sources, as shown in Figure 1, seems to be essential. A few limitations in this structure are as follows:

1. MIPCs utilize separate direct current (DC)–DC power converters to integrate diversified energy sources to a common DC bus, which results in a higher implementation cost.
2. In the case of alternating current (AC) loads, the system needs an extra inverter, and as a result the efficiency is reduced.

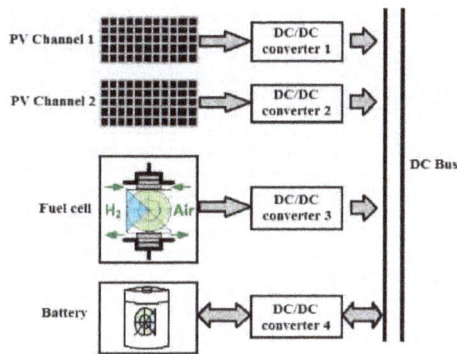

Figure 1. Conventional multi-input converter.

Due to the above disadvantages, the implementation of a MIPC is complicated.

To overcome the drawbacks of MIPCs, such as structure complications, multi-port converters (MPC) have been proposed. In MPCs, various sources are fed to the load through a single power electronic converter, as shown in Figure 2. MPCs reduce the structural complexity and the control technique.

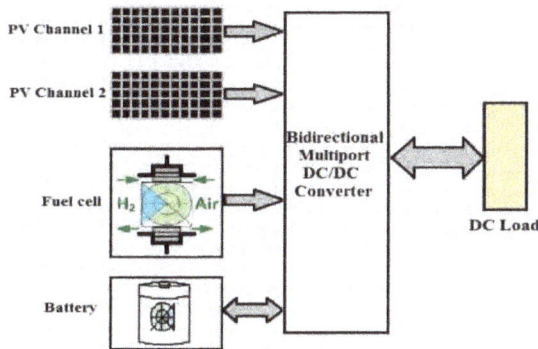

Figure 2. The multiport structure.

These power electronic converters can be classified based on:

a. Topology (in series and parallel)
b. Coupling (isolated and non-isolated)
c. Port placement (single input–single output, multi-input–single output, multi-input–multi-output, and single input–multi-output)
d. Conversion process (unidirectional and bidirectional)

Two input circuits with a current source connected in series to realize a high-efficiency, zero voltage switching dual-input converter are investigated in Reference [3]. In such series circuits, the weakest current source connected in series limits the current of the entire string. Thus, parallel-connected topologies are traditionally popular. Isolated converters use a transformer to achieve a high voltage gain, which increases the bulkiness of the system and leads to core-saturation problems. To achieve a wide output voltage, an interleaved LLC converter with voltage doublers is proposed in Reference [4]. A coupled-inductor-based bidirectional converter, as investigated in Reference [5], provides a high voltage gain and efficiency by employing soft switching and voltage clamping techniques to reduce the switching loss and achieve a high voltage gain. However, the shortcoming of this converter is that current ripples are introduced, due to an increase in the ratio of the coupled inductor. It therefore requires a filter arrangement, which in turn increases the structural complexity and the cost. To overcome the need for a filter requirement, a (non-isolated) converter without galvanic isolation is a good candidate to provide a large voltage gain with reduced size and cost. Such a single-stage Zeta single-ended primary inductor converter (SEPIC) converter is presented in Reference [6]. A high-gain, non-isolated DC–DC converter is analyzed in Reference [7]. A step-up converter combining the features of a KY converter and a buck–boost converter with a high voltage conversion ratio is presented in Reference [8]. A non-isolated, bidirectional DC–DC converter that uses four active switches is addressed in Reference [9]. A SEPIC-integrated boost converter with isolation, proposed in Reference [10], has one active switch, two inductors, and three capacitors. The above-mentioned converters are unidirectional, with a single input–single output (SISO) configuration. A systematic way of deriving MPCs using full bridge (FB) and bidirectional DC–DC converters (BDCs) is explained in Reference [11]. Various configurations of multiport converters using a DC link inductor are well-described in Reference [12]. A new multi-input DC–DC converter topology that is capable of integrating diversified energy resources of different voltage–current characteristics is proposed in Reference [13]. The setback for this converter is that only one input is allowed to transfer energy into the load at a time in boost mode. The concept of extracting pulsating source cells from basic converters like buck, boost, buck–boost, Zeta, Cuk, and SEPIC is reported in Reference [14].

A dual-input boost–buck converter with a coupled inductor is analyzed in Reference [15]. Though the presence of the coupled inductor causes current ripples, the interleaving mode operation of the converter reduces the ripples and makes the converter suitable for thermoelectric generator application. Integrated two input converters using buck and SEPIC topology are addressed in Reference [16]. A generic structure of a single-ended primary-inductor converter (SEPIC)-based multi (m)-input DC–DC converter is investigated in Reference [17]. This converter utilizes (m-2) +3 switches to interface 'm' input resources. For example, four active switches are required for realization of a three input structure of this converter. A single input–multi-output structure converter with the ability of generating buck, boost, and inverted output simultaneously is presented in Reference [18]. The drawback of this converter is that it is a single input–multi-output (SIMO) model, capable of interfacing only one input. The unified energy management scheme dealt with in Reference [19] employs different current control structures for various components of micro-grids, such as super capacitors, battery, renewable energy resources (RES), and voltage source converters (VSC). A power flow management control strategy based on sensing the battery voltage to select the operating modes of a bidirectional converter is presented in Reference [20]. A multiport power electronic interface as energy router with inductively coupled power transfer, ultra-capacitor and battery to solve the pulse charging is investigated in Reference [21]. Power budgeting using DC link voltage and current control methods have been analyzed in Reference [22].

To overcome drawbacks such as the size and control complexity due to the isolated transformer and a greater number of components, and to have a flexible integration of the diversified energy resources of different characteristics, a compact, high-profile power electronic interface is required. Most of the multi-input converter (MIC) topologies in the literature are derived from basic buck–boost topologies, and they leave scope for the further development of topological structures by using special converters. Therefore, this paper proposes a modular, non-isolated converter that is derived from basic SEPIC topology for accommodating arbitrary input sources and output loads. Due to special features, such as a better power factor from a continuous input current, non-inverting output, gracious response, and true shut down during short circuits, this SEPIC converter has wide applications in connecting flexible input voltages with stable outputs, battery-operated equipment, and lighting applications.

2. Three-Port SEPIC Converter

2.1. Synthesis of the Three-Port SEPIC Converter

A generalized structure of the proposed circuit is shown in Figure 3. It indicates that the number of ports can be further increased/decreased by connecting/disconnecting the additional pulsating voltage cells (PVC), depending on the availability of the sources.

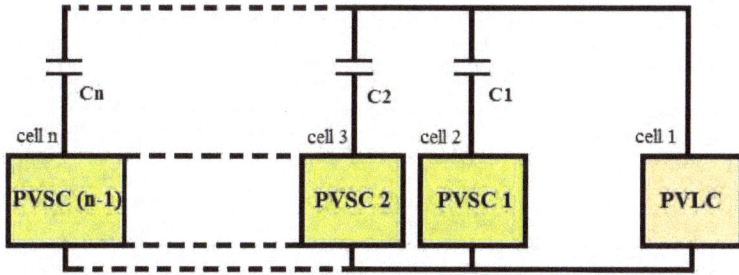

Figure 3. Generalized diagram of an n-port SEPIC converter.

In this paper, three-port (two inputs and one output) unidirectional and bidirectional SEPIC converters are proposed, as shown in Figure 4 and Figure 9. The proposed structures are a combination of PVC. PVCs can be categorized into two types; pulsating voltage source cells (PVSC) for the input side, and pulsating voltage load cell (PVLC) for the output side. Each PVSC connects with a common PVLC (as it is a MISO structure) through a coupling capacitor, and this forms a complete SEPIC structure.

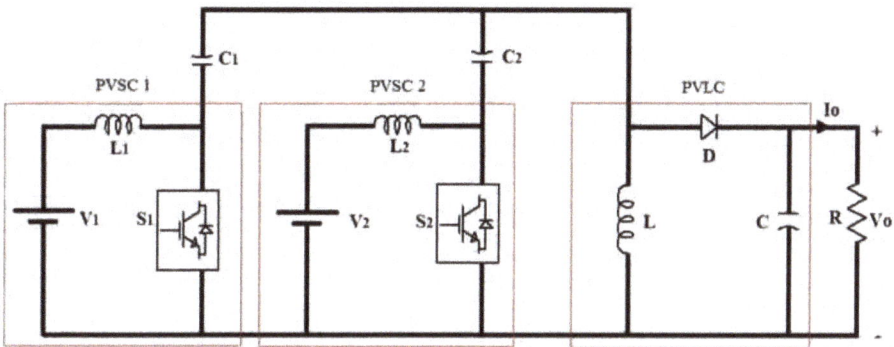

Figure 4. A three-port unidirectional SEPIC converter (Topology-1).

2.1.1. Steady State Analysis of a Three-Port Unidirectional SEPIC Converter (Topology-1)

In topology-1, if both sources are renewable DC sources; for example, if V_1 is a solar photovoltaic system and if V_2 is a fuel cell, then the converter works as a unidirectional converter, as the energy flows only from the source to the load. If both sources have an equal voltage magnitude, then the switches S_1 and S_2 operate with the same duty cycle simultaneously. If the primary source is a solar photovoltaic system and the secondary source is an energy storage device, then it works as a partially bidirectional converter. The reason for partial bidirectional is that the battery will charge from the primary source if the battery nominal voltage is less than the primary source voltage. To charge the battery, the switch in the corresponding PVSC has to be permanently turned off, and only the switches of other PVSCs will operate. Figure 4 shows the circuit diagram of a three-port unidirectional SEPIC converter.

If two DC voltage sources with different magnitudes, V_1 and V_2, are considered, then in order to use the sources effectively, the two sources must operate at different duty cycles. The source with a higher magnitude operates for the lower duty cycles, and the source with the lower magnitude operates for a higher duty cycle.

Assuming that $V_1 > V_2$, then $D_1 < D_2$. D_1 and D_2 are the respective duty cycles for PVSC1 and PVSC2. The operation of the three-port unidirectional SEPIC converter (Topology-1) is categorized into three modes, as shown in Figure 5.

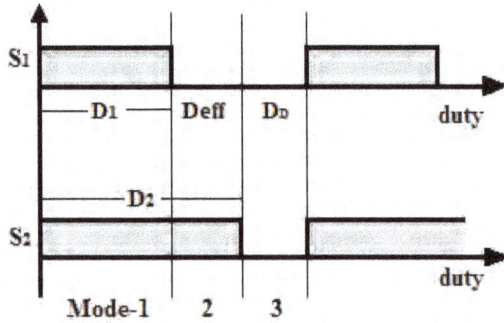

Figure 5. Modes of operation.

Where D_{eff} is the effective duty of PVSC2, $D_{eff} = D_2 - D_1$, and D_D is the duty for which the diode conducts; $D_D = 1 - D_2$.

■ Mode-1 (S_1 and S_2 ON, only S_1 conducts): The equivalent circuit of Mode-1 is shown in Figure 6. Switches S_1 and S_2 are on during this mode. In a steady-state condition, the voltages on the input capacitors C_1 and C_2 are the source voltages V_1 and V_2, respectively. Since V_1 is considered to be greater than V_2, S_2 blocks the possibility of a reverse current through the input leg of PVSC2. This mode makes S_1 conduct the current, while S_2 is reverse-biased. Since D is reverse-biased, it does not conduct, and meanwhile, the load side current is maintained by the output capacitor C.

$$L_1 \frac{di_{L1}}{dt} = V_1 \tag{1}$$

$$L_2 \frac{di_{L2}}{dt} = V_2 - V_{C2} + V_{C1} \tag{2}$$

$$L \frac{di_L}{dt} = V_{C1} \tag{3}$$

$$C_1 \frac{dV_{C1}}{dt} = -(i_L + i_{L2}) \tag{4}$$

$$C_2 \frac{dV_{C2}}{dt} = i_{L2} \tag{5}$$

$$C \frac{dV_C}{dt} = -\frac{V_C}{R} \tag{6}$$

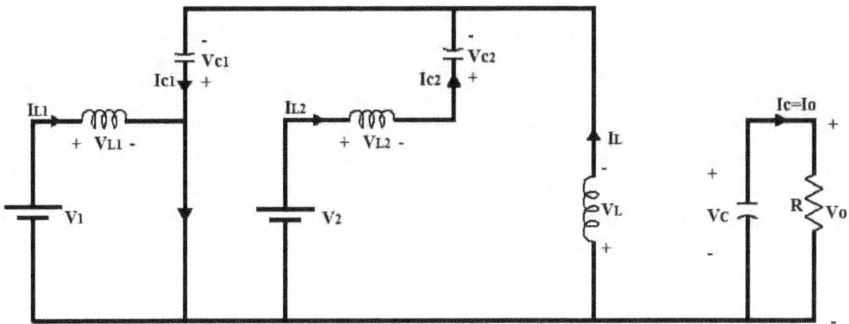

Figure 6. Mode-1 of topology-1 (S_1 and S_2 ON, only S_1 conducts).

■ Mode-2 (S_1 OFF and S_2 ON): The equivalent circuit of Mode-2 is shown in Figure 7. Once the switch S_1 is off, the switch S_2 becomes forward-biased and starts conducting, since the duty cycle of S_2 is greater than S_1. Since the diode D is still in a reverse-biased condition, the load current is again maintained by the output capacitor C.

$$L_1 \frac{di_{L1}}{dt} = V_1 - V_{C1} + V_{C2} \tag{7}$$

$$L_2 \frac{di_{L2}}{dt} = V_2 \tag{8}$$

$$L \frac{di_L}{dt} = V_{C2} \tag{9}$$

$$C_1 \frac{dV_{C1}}{dt} = i_{L1} \tag{10}$$

$$C_2 \frac{dV_{C2}}{dt} = -(i_{L1} + i_L) \tag{11}$$

$$C \frac{dV_C}{dt} = -\frac{V_C}{R} \tag{12}$$

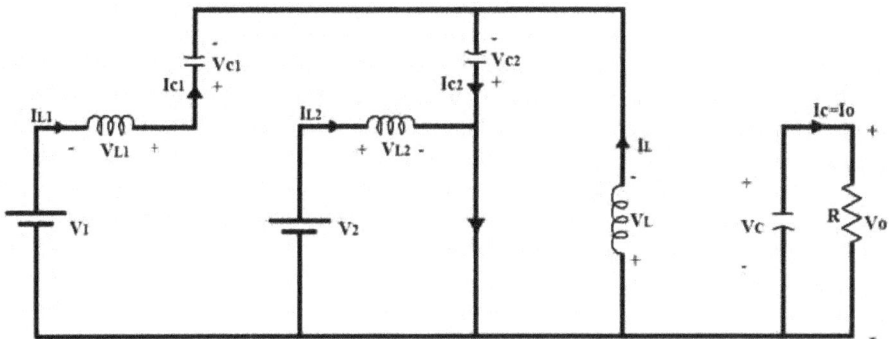

Figure 7. Mode-2 of topology-1 (S_1 OFF and S_2 ON).

■ Mode-3 (S_1 and S_2 OFF, D conducts): The equivalent circuit of Mode-3 is shown in Figure 8. Both the switches S_1 and S_2 are in the off-state. The inductor L_1 and L_2 starts discharging, and the capacitors C_1 and C_2 start charging from the sources V_1 and V_2. The diode D becomes forward-biased. The load current is now supplied by the sources V_1 and V_2, through L_1, C_1, and L_2, C_2, respectively.

$$L_1 \frac{di_{L1}}{dt} = V_1 - V_{C1} - V_C \tag{13}$$

$$L_2 \frac{di_{L2}}{dt} = V_2 - V_{C2} - V_C \tag{14}$$

$$L \frac{di_L}{dt} = -V_{C2} \tag{15}$$

$$C_1 \frac{dV_{C1}}{dt} = i_{L1} \tag{16}$$

$$C_2 \frac{dV_{C2}}{dt} = i_{L2} \tag{17}$$

$$C \frac{dV_C}{dt} = (i_{L1} + i_{L2} + i_L) - \frac{V_C}{R} \tag{18}$$

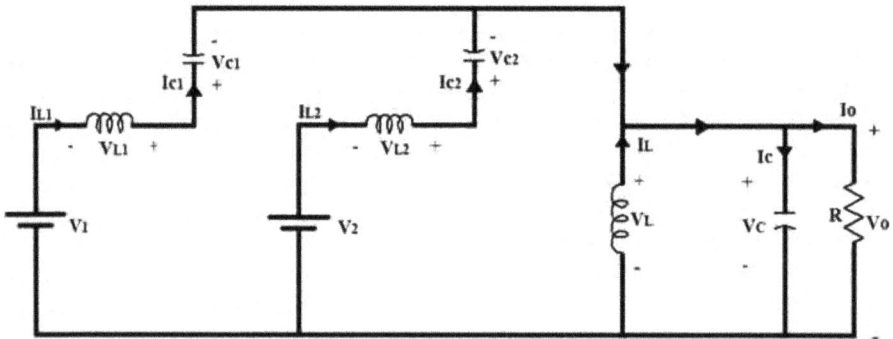

Figure 8. Mode-3 of topology-1 (S_1 and S_2 OFF, D conducts).

Combining the equations from (1) to (18) with their respective operating periods, the steady-state equations of the proposed unidirectional converter can be deduced as follows:

$$L_1 \frac{di_{L1}}{dt} = V_1 - D_{eff}(V_{C1} - V_{C2}) - D_D(V_{C1} + V_{C2}) \tag{19}$$

$$L_2 \frac{di_{L2}}{dt} = V_2 - D_1(V_{C2} - V_{C1}) - D_D(V_{C2} + V_C) \tag{20}$$

$$L \frac{di_L}{dt} = D_1 V_{C1} + D_{eff} V_{C2} - D_D V_C \tag{21}$$

$$C_1 \frac{dV_{C1}}{dt} = -D_1(i_L + i_{L2}) + (1 - D_1) i_{L1} \tag{22}$$

$$C_2 \frac{dV_{C2}}{dt} = (1 - D_{eff}) i_{L2} - D_{eff}(i_{L1} + i_L) \tag{23}$$

$$C \frac{dV_C}{dt} = D_D(i_{L1} + i_{L2} + i_L) - \frac{V_C}{R} \tag{24}$$

Assuming that the converter is operated in CCM, and neglecting the ripple voltage and ripple current, in a steady state average condition, $V_{C1} = V_1$, $V_{C2} = V_2$, and $V_C = V_O$. So, from Equation (21), we have:

$$0 = D_1 V_1 + D_{eff} V_2 - D_D V_O$$

$$V_O = \frac{D_1 V_1 + D_{\text{eff}} V_2}{1 - D_2} \text{ If } (V_1 > V_2) \tag{25}$$

$$V_O = \frac{D_2 V_2 + D_{\text{eff}} V_1}{1 - D_1} \text{ If } (V_2 > V_1) \tag{26}$$

The above Equation (25) represents the output voltage expression of topology-1. On solving the steady state Equations (19)–(24) by taking the left hand side as zero, the six state variables can be derived (i_{L1}, i_{L2}, i_L, V_{C1}, V_{C2}, and V_C):

$$i_{L1} = \frac{D_1 V_O}{(1 - D_2) R} = I_1 \tag{27}$$

$$i_{L2} = \frac{D_{\text{eff}} V_O}{(1 - D_2) R} = I_2 \tag{28}$$

$$i_L = \frac{V_O}{R} \tag{29}$$

$$V_{C1} = V_1 \tag{30}$$

$$V_{C2} = V_2 \tag{31}$$

$$V_C = V_O \tag{32}$$

2.1.2. Steady State Analysis of a Three-Port Bidirectional SEPIC Converter (Topology-2)

Topology-2 is designed to be a bidirectional converter with a solar photovoltaic system as a primary source, and a battery as an energy storage device. Figure 9 shows the circuit diagram of a three-port, bidirectional SEPIC converter. The converter possesses bidirectional power flow capability, as the battery can be charged both from the primary source and the regenerative energy, if it is available from the load side.

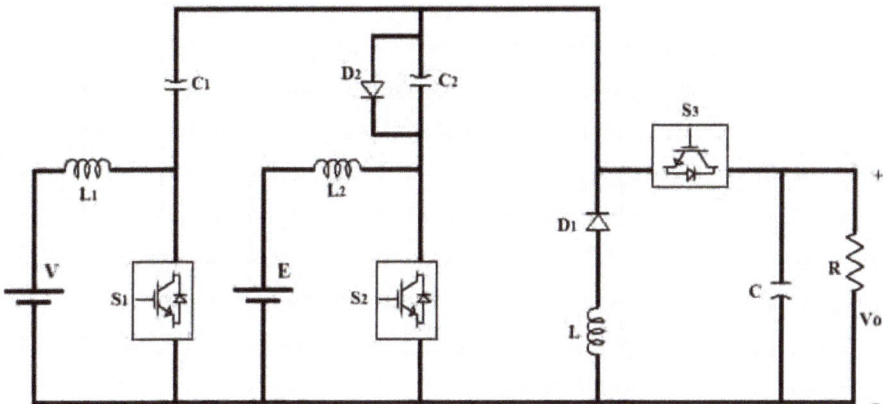

Figure 9. Three-port bidirectional SEPIC converter (Topology-2).

For bidirectional power flow with a solar photovoltaic system (V) and a battery (E), the converter operates with two possible conditions.

Case-1 (V < E, battery discharging): Switches S_1 and S_2 operate with duty cycles D_1 and D_2 (where $D_1 > D_2$) and S_3 will remain in the OFF condition. In this case, the converter operates as a unidirectional converter, i.e., in a similar fashion to that of topology-1. Figure 10 shows the modes of operation of topology-2 under this condition.

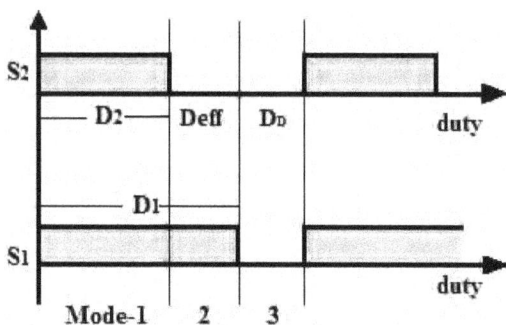

Figure 10. Modes of operation (case 1, when V < E).

Since the converter operates in a similar fashion as that of topology-1 in this case, the output voltage expression will remain the same as topology-1:

$$V_O = \frac{D_2E + D_{eff}V}{1 - D_1} \tag{33}$$

where $D_{eff} = D_1 - D_2$, V = primary source voltage, E = battery nominal voltage.

Case-2 (V > E, battery charging): Switch S_2 will remain OFF; S_1 only will operate at duty cycle D. Two modes of operation are possible in this case, as shown in Figure 11.

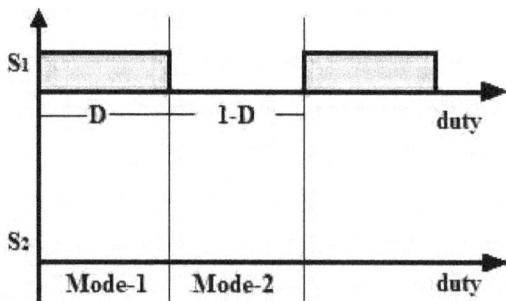

Figure 11. Modes of operation (case 2, when V > E).

■ Mode 1 (S_1 ON, S_2 and S_3 OFF): In this mode of operation, S_2 and S_3 remain in an off state and S_1 is on for the duty cycle D. C_1 discharges energy through the short path of S_1, and it flows through L and also through the anti-parallel diode of S_2, and charges C_2. The stored energy of L_2 freewheels through the battery, and the short path makes the battery charge up. L_1 charges from the source V. Meanwhile, the load is fed by the capacitor C, as the diode is reverse-biased. The equivalent circuit of this mode shown in Figure 12.

$$L_1\frac{di_{L1}}{dt} = V \tag{34}$$

$$L_2\frac{di_{L2}}{dt} = -E \tag{35}$$

$$L\frac{di_L}{dt} = V_{C1} \tag{36}$$

$$C_1\frac{dV_{C1}}{dt} = C_2\frac{dV_{C2}}{dt} + i_L \tag{37}$$

$$C\frac{dV_C}{dt} = -\frac{V_C}{R} \tag{38}$$

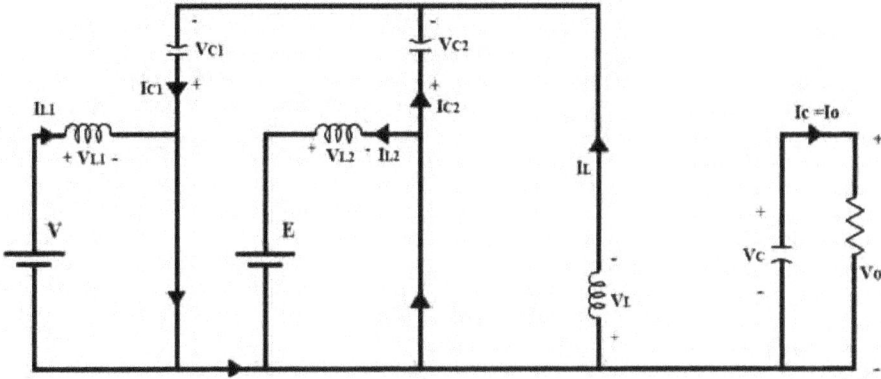

Figure 12. Mode 1 of topology-2 (case-2).

■ Mode-2 (S_1 OFF, S_2 OFF): In this mode of operation, S_2 and S_3 remain in the off state, and S_1 is also in the off state for 1-D. The capacitor C_1 charges from the voltage source V, and C_2 discharges through the inductor L_2 and the battery. This mode makes L_2 store energy, and the battery charge. L_1 releases stored energy through C_1. The anti-parallel diode of S_3 becomes forward-biased, and the load is powered up by the source and output inductor L. The equivalent circuit of this mode is shown in Figure 13.

$$L_1\frac{di_{L1}}{dt} = V - V_{C1} - V_C \tag{39}$$

$$L_2\frac{di_{L2}}{dt} = V_{C2} - E - V_C \tag{40}$$

$$L\frac{di_L}{dt} = V_{C2} - E - V_C \tag{41}$$

$$C_1\frac{dV_{C1}}{dt} = i_{L2} \tag{42}$$

$$C\frac{dV_C}{dt} = i_L + (i_{L1} - i_{L2}) - \frac{V_C}{R} \tag{43}$$

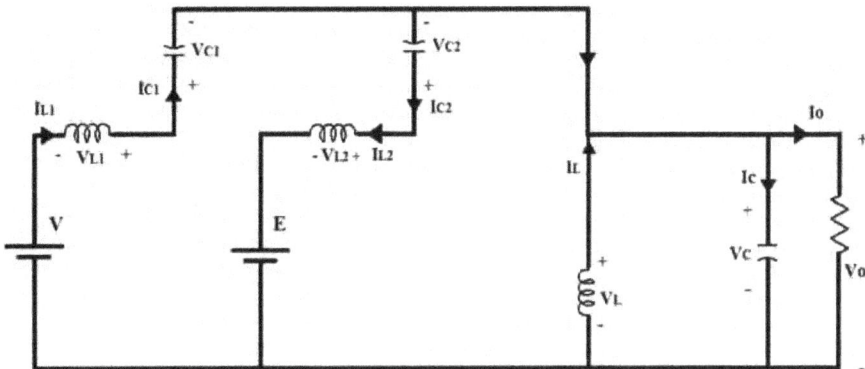

Figure 13. Mode-2 of topology-2 (case-2).

Combining the equations from (34) to (43) with their respective operating periods, the steady-state equations of the proposed bidirectional converter (case-2) can be deduced as follows:

$$L_1 \frac{di_{L1}}{dt} = DV + (1 - D)(V - V_{C1} - V_C) \tag{44}$$

$$L_2 \frac{di_{L2}}{dt} = -DE + (1 - D)(V_{C2} - E - V_C) \tag{45}$$

$$L \frac{di_L}{dt} = DV_{C1} + (1 - D)(V_{C2} - E - V_C) \tag{46}$$

Assuming that the converter is operated in CCM, and neglecting the ripple voltage and ripple current, in a steady-state average condition, $V_{C1} = V$, $V_{C2} = V$, and $V_C = V_O$. So, from Equation (46), we have:

$$0 = DV + (1 - D)(V - E - V_O)$$

$$V_O = \frac{DV + (1 - D)(V - E)}{1 - D} \tag{47}$$

The above expression (47) represents the output voltage of topology-2 while charging.

During reverse power flow, if any, the switches S_1 and S_2 will remain off, and only S_3 will operate. If S_3 is on, the current will flow from the load side to the battery (E) through D_2 and L_2, and the inductor L_2 will store energy. If S_3 is off, the inductor L_2 discharges energy to the battery (E) through the anti-parallel diode of S_2. This means the battery is charged from the regenerative energy, if it is available from the load side.

2.1.3. Small Ripple Approximation of a Three-Port SEPIC Converter

The methods of operation for topology-1 and topology-2 (case-1) are similar. The parameters used for topology-1 can also be used for topology-2. In this section, the expressions for all of the circuit parameters are described. Following the individual modes, the expressions for L_1, L_2, L, C_1, C_2, and C concerning the current and voltage ripples are described below.

Consider the three modes, (i.e., D_1, D_{eff}, and D_D) operating for t_1, t_2, and t_3 periods respectively.

In mode-1 and mode-2, the inductor current increases from a low level to high level, say, I_{L11} to I_{L12}, and in mode-3, the current falls from I_{L12} to I_{L11}. Therefore, the current ripple is considered to be $\Delta I_{L1} = I_{L12} - I_{L11}$. Thus:

$$L_1 \frac{di_{L1}}{dt} = V_1$$

$$L_1 \frac{\Delta I_{L1}}{t_1 + t_2} = V_1$$

If 'T' is the total period, then t_1, t_2, and t_3 can be represented as $D_1 T$, $D_{eff} T$, and $D_D T$, respectively. Thus:

$$\Delta I_{L1} = \frac{V_1 D_2}{f L_1} \tag{48}$$

where $f = 1/T$, assuming that the inductor L_2 charges linearly during the periods t_1 and t_2 from I_{L21} to I_{L22}. Thus, the current ripple is $\Delta I_{L2} = I_{L22} - I_{L21}$.

$$L_2 \frac{di_{L2}}{dt} = V_2$$

$$L_2 \frac{\Delta I_{L2}}{t_1 + t_2} = V_2$$

$$\Delta I_{L2} = \frac{V_2 D_2}{f L_2} \tag{49}$$

The output side inductor L discharges from a high value to low value, say I_{L2} to I_{L1}, only within period t_3. Hence the ripple is $\Delta I_L = I_{L2} - I_{L1}$.

$$L\frac{di_L}{dt} = -V_C$$

$$-L\frac{\Delta I_L}{t_3} = -V_C$$

Here $V_C = V_O$, so:

$$\Delta I_L = \frac{D_1 V_1 + D_{eff} V_2}{f\,L} \tag{50}$$

Similarly, the voltage ripples can be calculated from the steady-state equations. The voltage across C_1 rises (assuming linearly) from a low value to a high value, say V_{C11} to V_{C12}, during the time periods t_2 and t_3. This flow gives a voltage ripple $\Delta V_{C1} = V_{C12} - V_{C11}$. In this period, the capacitor is charging by the source current of V_1, i.e., $I_1 = i_{L1}$.

$$\Delta V_{C1} = \frac{1}{C_1}\int_0^{t_2+t_3} i_{L1}dt = \frac{1}{C_1}\int_0^{t_2+t_3} I_1 dt$$

$$\Delta V_{C1} = \frac{I_1}{f\,C_1}(D_{eff} + D_D)$$

$$\Delta V_{C1} = \frac{I_1}{f\,C_1}(1 - D_1) \tag{51}$$

The capacitor C_2 is charging during period t_3 and t_1 from the source current V_2, i.e., $I_2 = i_{L2}$. Therefore, a voltage ripple of $\Delta V_{C2} = V_{C22} - V_{C21}$ appears across capacitor C_2.

$$\Delta V_{C2} = \frac{1}{C_2}\int_0^{t_3+t_1} i_{L2}dt = \frac{1}{C_2}\int_0^{t_3+t_1} I_2 dt$$

$$\Delta V_{C2} = \frac{I_2}{f\,C_2}(D_D + D_1)$$

$$\Delta V_{C2} = \frac{I_2}{f\,C_2}(1 - D_{eff}) \tag{52}$$

The capacitor C discharges during periods t_1 and t_2, providing the load current. The voltage ripple is defined as $\Delta V_C = V_{C2} - V_{C1}$.

$$-\Delta V_C = -\frac{1}{C}\int_0^{t_1}\frac{V_C}{R}dt - \frac{1}{C}\int_0^{t_2}\frac{V_C}{R}dt$$

$$\Delta V_C = \frac{V_C}{R\,C}(D_1 + D_{eff})$$

$$\Delta V_C = \frac{D_1 V_1 + D_{eff} V_2}{f\,R\,C}(D_1 + D_{eff}) \tag{53}$$

Thus, the circuit parameters can be obtained from Equations (48) to (53):

$$L_1 = \frac{V_1 D_2}{f\,\Delta I_{L1}} \tag{54}$$

$$L_2 = \frac{V_2 D_2}{f\,\Delta I_{L2}} \tag{55}$$

$$L = \frac{D_1 V_1 + D_{eff} V_2}{f \, \Delta I_L} \tag{56}$$

$$C_1 = \frac{I_1}{f \, \Delta V_{C1}} (1 - D_1) \tag{57}$$

$$C_2 = \frac{I_2}{f \, \Delta V_{C2}} (1 - D_{eff}) \tag{58}$$

$$C = \frac{D_1 V_1 + D_{eff} V_2}{f \, R \, \Delta V_C} (D_1 + D_{eff}) \tag{59}$$

3. Results Analysis

The analysis of the proposed converter is further discussed and verified in this section, through simulations for both topologies, using MATLAB/Simulink software. Table 1 indicates the parameters used in the simulation of topology-1.

Table 1. Simulation parameters.

Parameters	Estimated Values	Simulation Values	Unit
L_1, L_2	14.74	15	mH
C_1, C_2	0.462	0.54	mF
L	14.74	15	mH
C	0.299	0.54	mF
R	60	60	Ohm
V_1	42	42	V
V_2	42	42	V
D_1	67	67	%
D_2	50	50	%
f (switching freq.)	10,000	10,000	Hz

The permissible value of the current and voltage ripples are assumed to be $\Delta I_{L1} = \Delta I_{L2} = \Delta I_L$ = 0.5 A, and $\Delta V_{C1} = \Delta V_{C2} = \Delta V_C = 0.5$ V. The input voltages (V_1, V_2), duty cycles (D_1, D_2), and the corresponding values of L_1, L_2, L, C_1, C_2, and C are estimated by using Equations (54)–(59), as shown in Table 1.

Figure 14 shows the waveforms of the current and voltage through and across the inductors and capacitors, respectively. The output voltage and current of the proposed converter (topology-1) shown in Figure 15 was found to be $V_O = 79$ V and $I_O = 1.3$ A. The actual current and voltage ripples estimated from the simulation were approximately equal, and within the allowed value range of the ripples. The actual values of the ripples from the simulation were $\Delta I_{L1} = 0.3$ A, $\Delta V_{C1} = 0.13$ mV, $\Delta I_{L2} = 0.35$ A, $\Delta V_{C2} = 0.4$ mV, $\Delta I_L = 0.4$ A, and $\Delta V_C = 0.4$ mV.

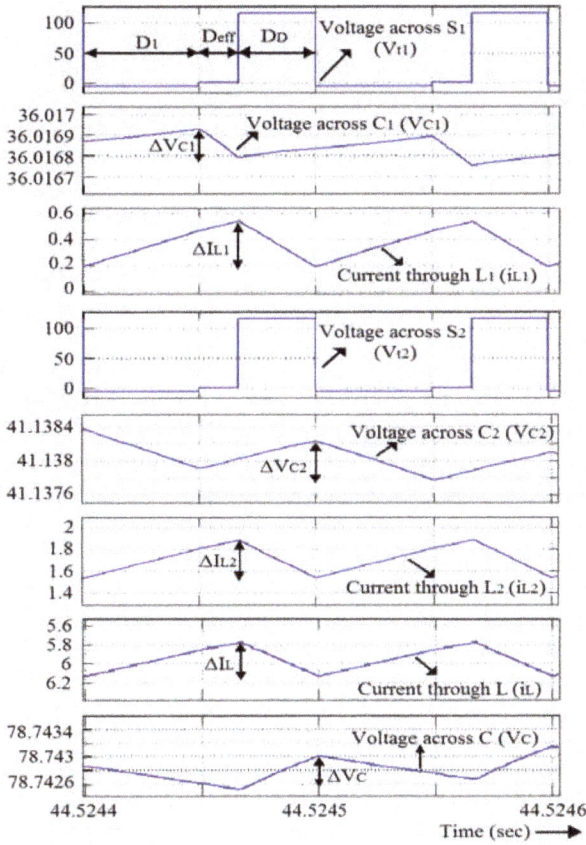

Figure 14. Current and voltage waveforms of each component (switch, inductor, and capacitor) present in PVSC1, PVSC2, and PVLC of topology-1.

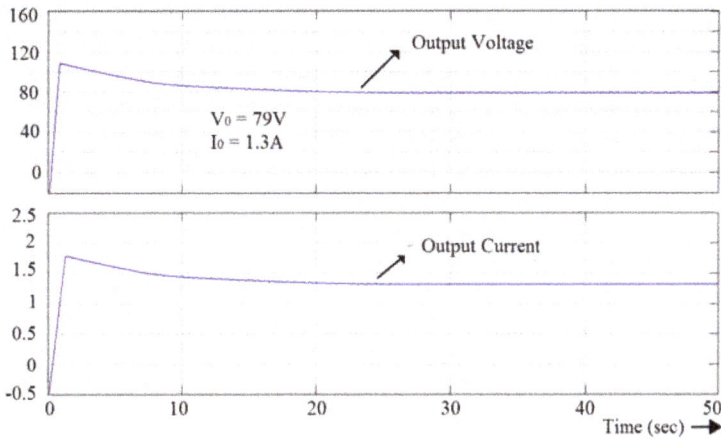

Figure 15. Output voltage and current waveforms of topology-1.

The comparisons of the simulated output voltage (from the Simulink model) and the estimated output voltage from Equations (25) and (26) for the different sets of input voltages (V_1, V_2) and duty cycles (D_1, D_2) are shown in Table 2. The simulation and the estimated results were approximately same for each individual set of inputs.

Table 2. Comparison of the simulated and estimated results of topology-1.

V_1 (Volt)	V_2 (Volt)	D_1 (%)	D_2 (%)	Sim. V_O (Volt)	Est. V_O (Volt)
24	12	30	60	28.5	27
30	15	30	60	36	33.75
25	20	55	68.75	50	52.8
30	20	50	75	77	80
36	24	40	60	52	48
36	24	50	75	92	96
35	42	67	50	79	81.66

The output voltage and current waveforms of topology-2 during discharging (case-1) are shown in Figure 16. The primary source voltage was taken as V = 16 V, and the nominal battery voltage rating was E = 24 V. An initial state of charge of the battery (SOC) was considered to be 80%. The duty cycles of the primary source and the battery were assumed to be 75% and 50%, respectively. The discharging of the battery can be seen in the SOC graph, as shown in Figure 17. The SOC was decreasing in nature.

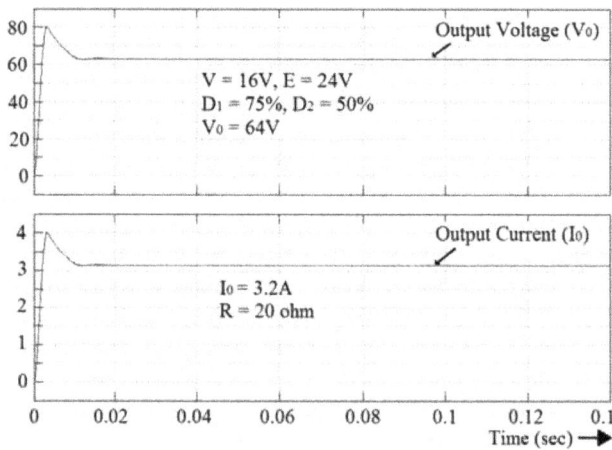

Figure 16. Output voltage and current waveforms of topology-2 during discharging (case-1).

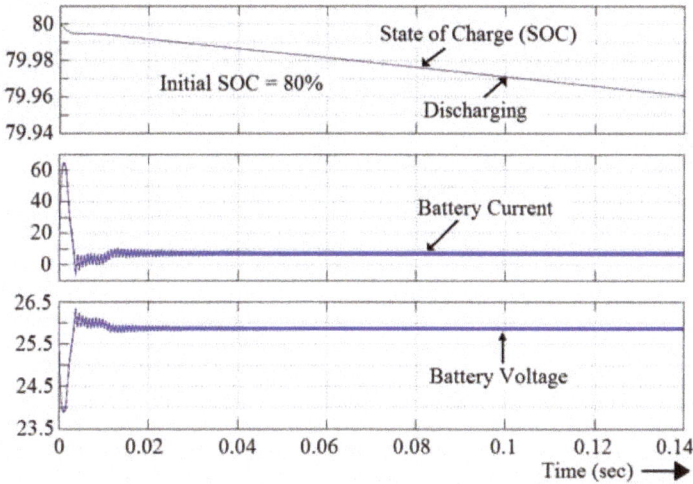

Figure 17. State of charge, current, and voltage waveforms of battery for topology-2 during discharging (case-1).

The comparison of the simulated output voltage (from the Simulink model) and the estimated output voltage from Equation (33) for the different sets of primary inputs and battery nominal voltages (V, E) and duty cycles (D_1, D_2) is shown in Table 3. The simulation and estimated results were approximately the same for each individual set of inputs.

Table 3. Comparison of the simulated and estimated results of topology-2 during discharging (case-1).

V (Volt)	E (Volt)	D_1 (%)	D_2 (%)	Sim. V_O (Volt)	Est. V_O (Volt)
8	12	82.5	55	49	50.28
10	12	62.4	52	21	19.4
20	24	72	60	61	60
16	24	75	50	64	64
30	36	75	50	104	102
20	36	72	40	75	74.28

The output voltage and current waveforms of topology-2 during charging (case-2) are shown in Figure 18. The primary source voltage was taken as V = 30 V, and the nominal battery voltage rating was E = 24 V. An initial SOC was considered at 40%. The duty cycle of the primary source was assumed to be 60%, and the switch under the battery cell remained off. The charging of the battery can be seen in the SOC graph, shown in Figure 19. The SOC was increasing in nature.

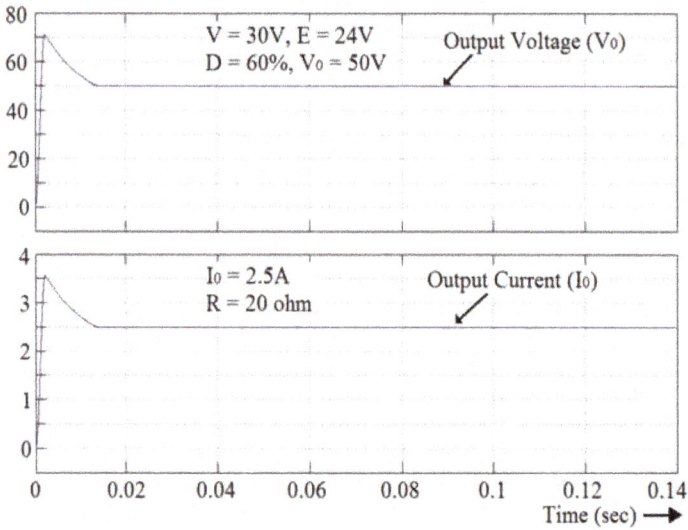

Figure 18. Output voltage and current waveforms of topology-2 during charging (case-2).

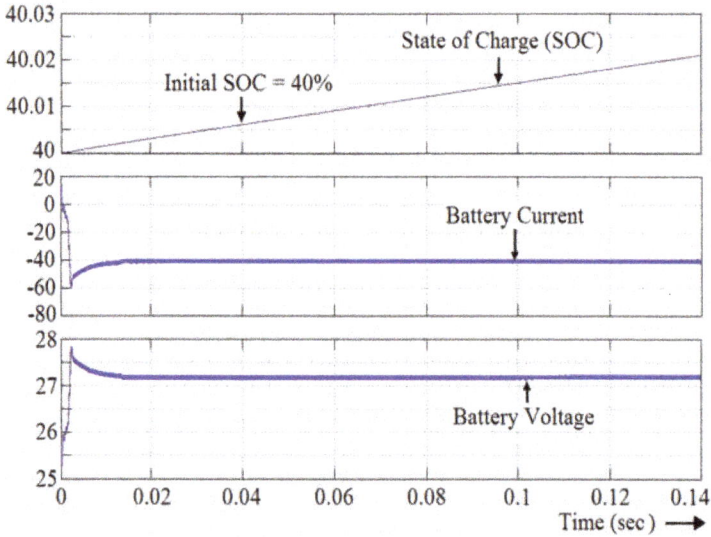

Figure 19. State of charge (SOC), current, and voltage waveforms of the battery for topology-2 during charging (case-2).

A comparison of the simulated output voltage (from the Simulink model) and the estimated output voltage from Equation (47), for different sets of primary inputs and battery nominal voltages (V, E), and duty cycles (D_1, D_2) are shown in Table 4. The simulation and estimated results were approximately same for each individual set of inputs.

Table 4. Comparison of the simulated and estimated results of topology-2 during charging (case-2).

V (Volt)	E (Volt)	D_1 (%)	D_2 (%)	Sim. V_O (Volt)
20	12	60	39.5	38
15	12	70	39.3	38
30	24	60	53.3	51
20	12	40	23	21.3
40	24	60	80	76
44	36	50	63	52
40	36	60	66.5	64

4. Design of Controllers for the Three-Port SEPIC Converter

A controller is needed for closed loop operation of the proposed converter, in order to maintain the output voltage as constant. A state-space analysis of the proposed converter was performed in order to obtain the transfer function, followed by the step response of the converter. Two controller structures were designed in this work; the first one was for unidirectional topology, and the other one was for bidirectional topology. The controller used in this work was a PI controller for both topologies, and an MPPT controller for the solar photovoltaic system was also used as an input in the bidirectional topology to track the maximum available power.

4.1. State Space Analysis of the Three-Port SEPIC Converter

State space analysis refers to the smallest set of variables whose knowledge at $t = t_0$, together with the knowledge of the input for $t > t_0$, gives complete knowledge of the behavior of the system at any time $t \geq t_0$. The state variable refers to the smallest set of variables that can help us to determine the state of the dynamic system. For the proposed converter, the current through all of the inductors (i_{L1}, i_{L2}, i_L) and the voltage across each capacitor (V_{C1}, V_{C2}, V_C) are considered to be the state variables. The state space model is represented as:

$$\dot{x} = Ax + Bu \tag{60}$$

$$y = Cx + Du \tag{61}$$

where x = state variable matrix, u = input matrix, y = output matrix.

$$x = \begin{bmatrix} i_{L1} \\ i_{L2} \\ i_L \\ V_{C1} \\ V_{C2} \\ V_C \end{bmatrix} ; u = \begin{bmatrix} V_1 \\ V_2 \end{bmatrix} ; y = \begin{bmatrix} V_o \\ I_o \end{bmatrix}$$

The transfer function (TF) of the system is defined as follows by Equation (62):

$$TF(s) = C(sI - A)^{-1}B + D \tag{62}$$

The A, B, C, and D matrices of the proposed converter are derived using the steady-state Equations (19)–(24):

$$A = \begin{bmatrix} 0 & 0 & 0 & \frac{D_1-1}{L_1} & \frac{D_2-D_1}{L_1} & \frac{D_2-1}{L_1} \\ 0 & 0 & 0 & \frac{D_1}{L_2} & -(1+D_1-D_2) & \frac{D_2-1}{L_2} \\ 0 & 0 & 0 & \frac{D_1}{L} & \frac{D_2-D_1}{L} & \frac{D_2-1}{L} \\ \frac{1-D_1}{C_1} & \frac{-D_1}{C_1} & \frac{-D_1}{C_1} & 0 & 0 & 0 \\ \frac{D_1-D_2}{C_2} & \frac{1+D_1-D_2}{C_2} & \frac{D_1-D_2}{C_2} & 0 & 0 & 0 \\ \frac{1-D_2}{C} & \frac{1-D_2}{C} & \frac{1-D_2}{C} & 0 & 0 & \frac{-1}{RC} \end{bmatrix}$$

$$B = \begin{bmatrix} \frac{1}{L_1} & 0 \\ 0 & \frac{1}{L_2} \\ 0 & 0 \\ 0 & 0 \\ 0 & 0 \\ 0 & 0 \end{bmatrix} \quad C = \begin{bmatrix} 0 & 0 & 0 & 0 & 0 & 1 \\ 0 & 0 & 0 & 0 & 0 & \frac{1}{R} \end{bmatrix} \quad D = \begin{bmatrix} 0 & 0 \\ 0 & 0 \end{bmatrix}$$

The transfer function of the proposed converter has been derived using the state-space model, and it is represented in Equations (63) and (64):

$$\frac{V_o}{V_1} = \frac{1.941e004\,s^4 + 2.023e-009\,s^3 + 4.453e009\,s^2 + 0.0003845\,s^1 + 1.451e014}{s^6 + 30.3\,s^5 + 2.057e005\,s^4 + 5.652e006\,s^3 + 8.792e009\,s^2 + 1.751e011\,s^1 + 8.866e013} \tag{63}$$

$$\frac{V_o}{V_2} = \frac{1.941e004\,s^4 + 3.724e-009\,s^3 + 1.644e009\,s^2 + 0.0001096\,s^1 + 3.493e013}{s^6 + 30.3\,s^5 + 2.057e005\,s^4 + 5.652e006\,s^3 + 8.792e009\,s^2 + 1.751e011\,s^1 + 8.866e013} \tag{64}$$

For the open-loop system in Figure 20, the step response had a steady state error of 1.5%, and the maximum overshoot was 59%, which was greater. A PI controller was designed for maintaining the output at a reference value. For the closed-loop system in Figure 21, the steady-state error and the maximum overshoot were almost 0%. The step response of the proposed converter model is shown in Figure 22.

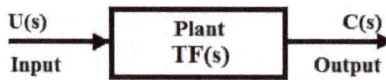

Figure 20. Open-loop system block diagram.

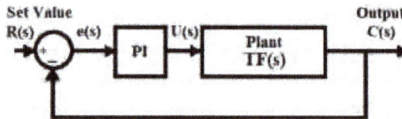

Figure 21. Closed-loop system block diagram.

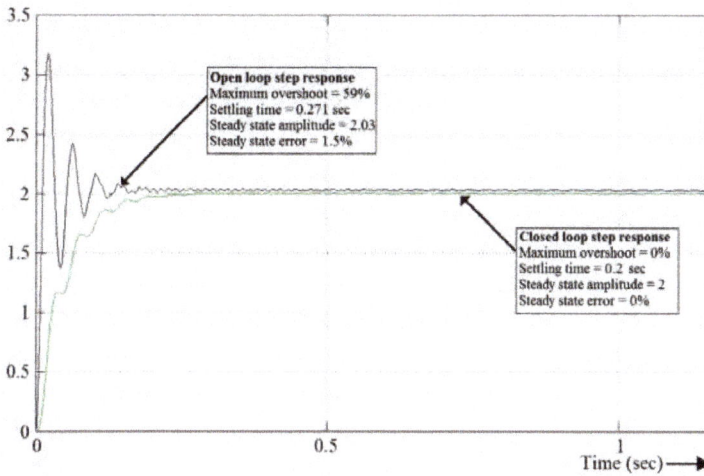

Figure 22. Step response of the proposed converter for the open-loop and closed-loop systems.

4.2. The Closed-Loop Unidirectional Controller

The output voltage V_O in this topology varied with the change in the input voltage sources V_1 and V_2. A PI controller was incorporated between the output and the input V_2, as shown in Figure 23. The actual output voltage was compared with a set value or a reference voltage, and generated an error. The error was processed through a PI controller tuned with a proportional constant (K_P) and an integral constant (K_I). The output of the PI controller was again processed through a saturation block, the maximum limit of which was set to 0.8, so that the switch did not operate beyond an 80% duty cycle for the safety of the switch. The saturated signal then passed through a PWM block, which generated the switching pulses D_2 for the voltage source V_2. With the known values of V_0, V_1, V_2, and D_2, the switching pulses for V_1, i.e., (D_1) could then be found, using the output voltage equation of topology-1 (Equation (25)).

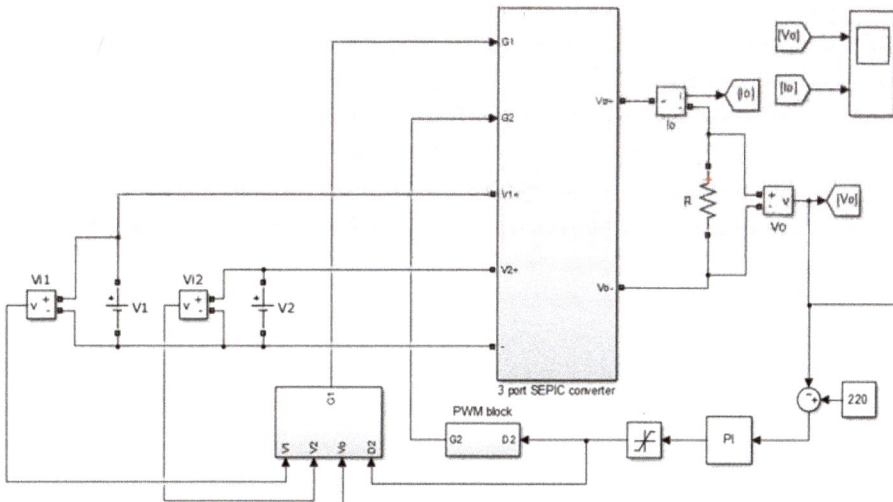

Figure 23. Unidirectional closed-loop converter.

The reference value of the output voltage was set to be 220 V. With this closed-loop topology of a unidirectional structure, the output voltage was maintained at a constant value of 220 V, even if there was a variation in the input supply V_1 and V_2. The proportional gain constant (K_P) of the PI controller was taken as 0.0001, and the integral gain constant (K_I) was taken as 0.005. Figure 24 shows the output voltage and the output current of a unidirectional closed-loop converter.

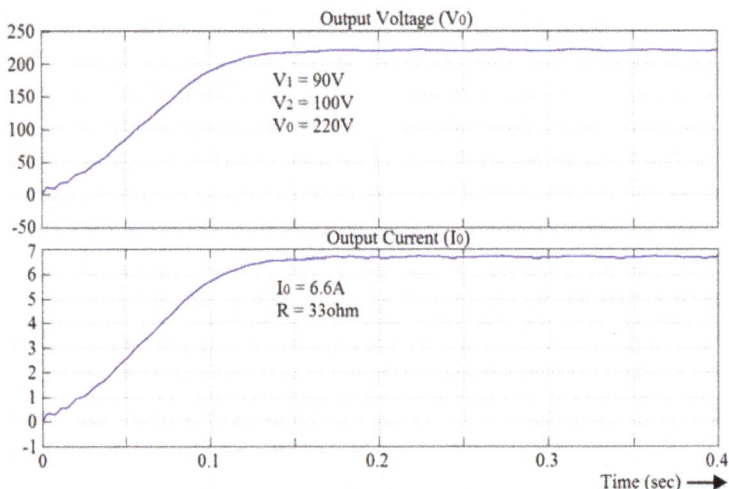

Figure 24. Output voltage and current of a unidirectional closed-loop converter.

With 220 V as the reference voltage, the generated output was V_O = 220 V with inputs V_1 = 90 V and V_2 = 100 V. From the results of the closed loop simulation, it was inferred that the output voltage of the converter remained constant with the variation of input source. The closed loop system reduced the steady state error and maximum overshoot to 0%.

4.3. The Closed-Loop Bidirectional Controller

In this design, as shown in Figure 25, two input voltage sources, a solar photovoltaic system and a battery were used. The output voltage V_O varied with the variation of the input voltage sources. A PI controller was incorporated between the output and one of the input sources, depending on the availability of the sources. A P & O algorithm-based MPPT controller was incorporated with a PV input, in order to track the maximum available power. Three operating conditions were possible for this closed-loop structure.

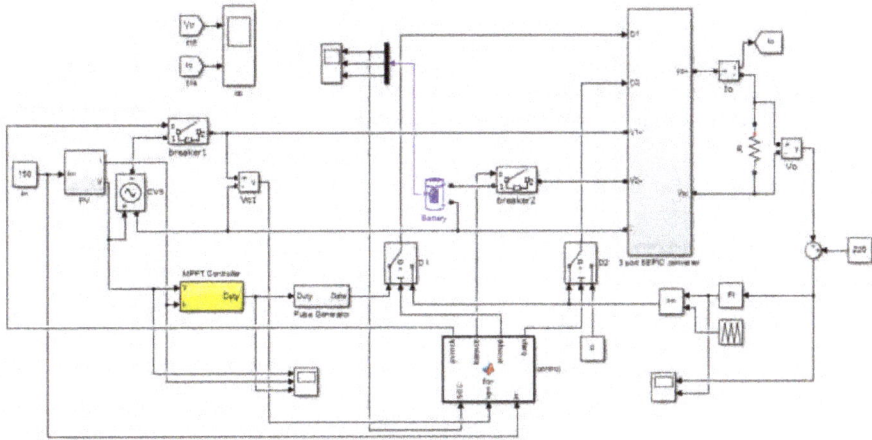

Figure 25. Bidirectional closed-loop converter.

Case-1: When both sources supply to the load: When both the sources are active, then the MPPT controller generates the switching pulse (D_1) for the cell containing the PV, and the PI controller generates the switching pulses (D_2) for the cell containing the battery. The PI controller incorporated with the battery maintains the output voltage to the set value. The actual output voltage is compared with a set value or a reference voltage, and it generates an error. The error is processed through a PI controller tuned with a proper value of a proportional constant (K_P) and an integral constant (K_I). The output of the PI controller is again processed through a saturation block, the maximum limit of which is set to 0.8, so that the switch does not operate beyond the 80% duty cycle for the safety of the switch. The signal forms the saturation block that is fed to the pulse generation circuit or the PWM block, which generates the switching pulses D_2 for the cell containing the battery source. Meanwhile, the MPPT controller transfers the maximum available power from the PV to the load.

Case-2: When sufficient solar power is available: When maximum solar power is available and it is sufficient to supply the load, then the MPPT controller has to be removed, and the PI controller will be incorporated between the load and the PV input. During this condition, if the battery is not fully charged, the switch for the battery port (S_2) can be turned off. This makes the battery charge from the PV input. If the battery is fully charged, then it can be disconnected from the system through the breaker. The PI controller with the PV input will maintain a constant output, even if there is a variation in solar irradiation to an extent. The operating conditions for the above three cases are given in Table 5. The signals 'pvcut' and 'batcut' represent the operating signals for the breakers of the PV port and battery port, respectively. When 'pvcut' or 'batcut' is 1, then the corresponding breaker will be shorted, and the respective source will be added to in the system. Similarly, if it is 0, then the corresponding breaker will be open, and the respective source will be disconnected from the system. The signals 'pvmpp' and 'batpi' are the operating signals of the selection switches, allowing the switching of pulses to two respective switches, S_1 and S_2, of the input ports from the controllers. If 'pvmpp' is 1, then the switching pulse D_1 will be generated from the MPPT controller, and if 'pvmpp' is 0, then the switching pulse D_1 will be generated from the PI controller. Similarly, if 'batpi' is 1, then the switching pulse D_2 will be generated from the PI controller, and if 'batpi' is 0, then the switching pulse D_2 will be 0 and switch S_2 turns off.

Table 5. Control logic of a closed-loop bidirectional controller.

Condition	SOC	vpv	irr	batcut	pvcut	pvmpp	batpi
1	>80%	-	-	1	1	1	1
2	<95%	>120	-	1	1	0	0
3	>99%	>120	-	0	1	0	0
4	<40%	-	-	1	1	0	0
5	-	-	0	0	1	1	1

Case-3: When solar power is not available: When PV power is not available during the night or in cloudy weather, the battery alone will supply to the load, and maintain a constant output. During this condition, the PV can be disconnected from the system through a breaker. The PI controller is incorporated between the load and battery source. The actual output voltage is compared to the reference voltage, and it generates an error. The error is processed through the PI controller, tuned with a proper value of the proportional constant (K_P) and the integral constant (K_I). The output of the PI controller is again processed through a saturation block. The signal form the saturation block is fed to the pulse generation circuit or PWM block, which generates the switching pulses D_2 for the cell containing the battery source.

In the output, the voltage shown in Figure 26 implies that it remains constant at a set value of 220 V, even if there is a variation in the PV input due to intermittent solar irradiation levels, to an extent. The proportional gain constant (KP) of the PI controller is taken as 0.0001, and the integral gain constant (KI) is taken as 0.005.

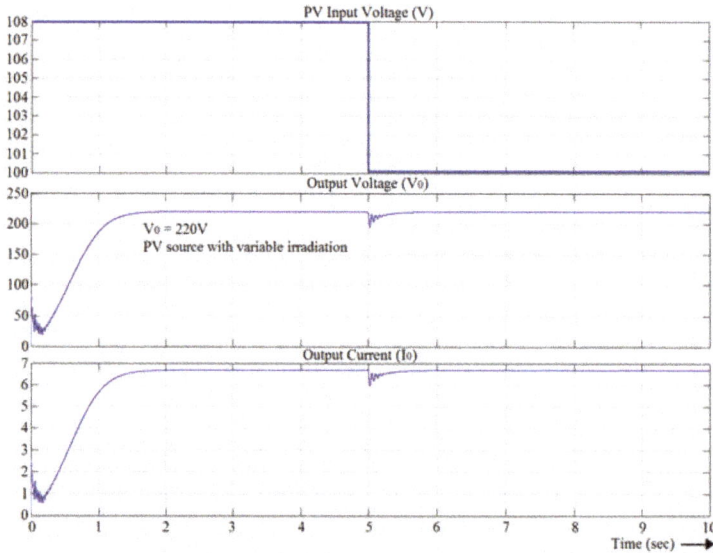

Figure 26. PV input, output voltage, and current during intermittent solar irradiation.

Condition-1 of control logic: Figure 27 shows the output voltage and output current of a bidirectional closed loop converter when both the PV and the battery are supplying. The inputs are taken to be PV = 119 V, E = 120 V, and the initial battery state of charge is assumed to be 82%, as shown in Figure 28. From the results of the closed-loop simulation for this case, it is inferred that the output voltage of the converter remains constant at a set value of 220 V. Both sources are supplying to the load, and the battery is discharging.

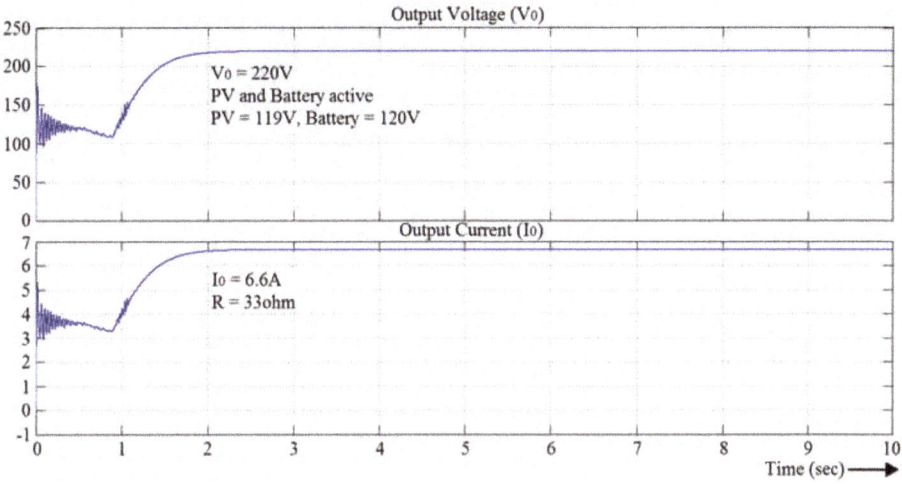

Figure 27. Output voltage and current of a bidirectional closed-loop converter when both the PV and battery are supplying (condition-1 of control logic).

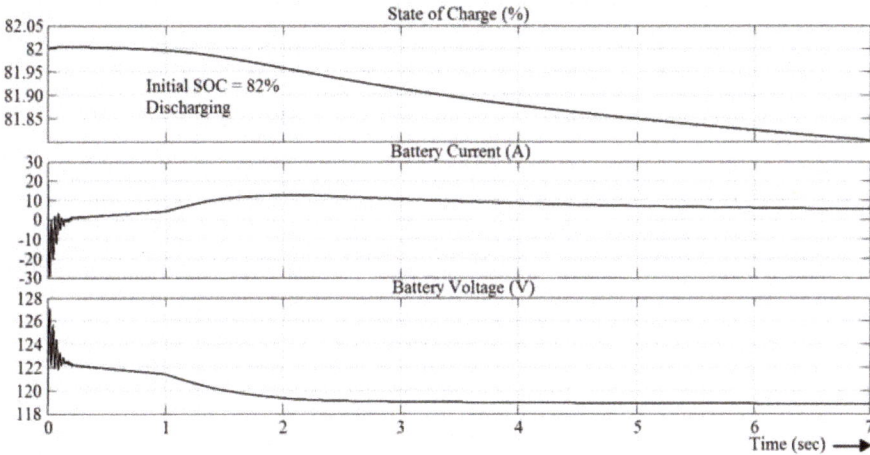

Figure 28. SOC, current, and voltage of the battery during discharging (condition-1 of control logic).

Condition-2 of control logic: Figure 29 shows the output voltage and output current of a bidirectional closed loop converter when only the PV is supplying. The PV is taken to be PV = 123 V, and the initial battery state of charge is assumed to be 60%, as shown in Figure 30. The output voltage is maintained constant at a set value of 220 V, and the battery is charging.

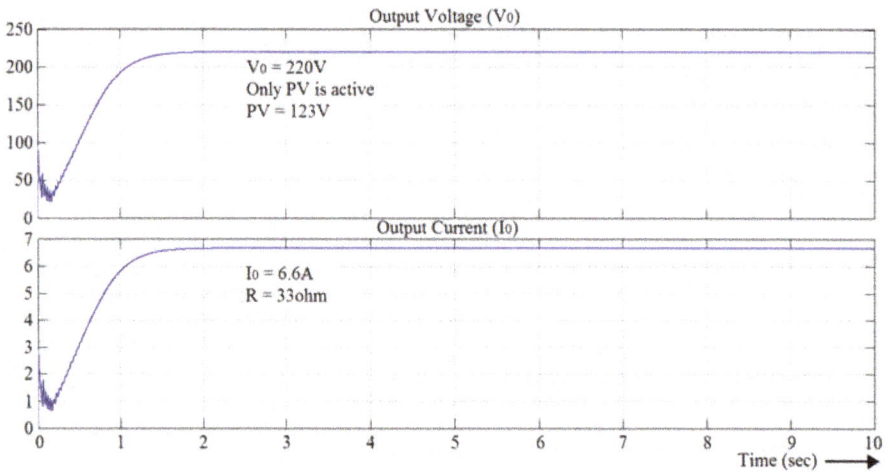

Figure 29. Output voltage and current of a bidirectional closed-loop converter when only the PV is supplying (condition-2 of control logic).

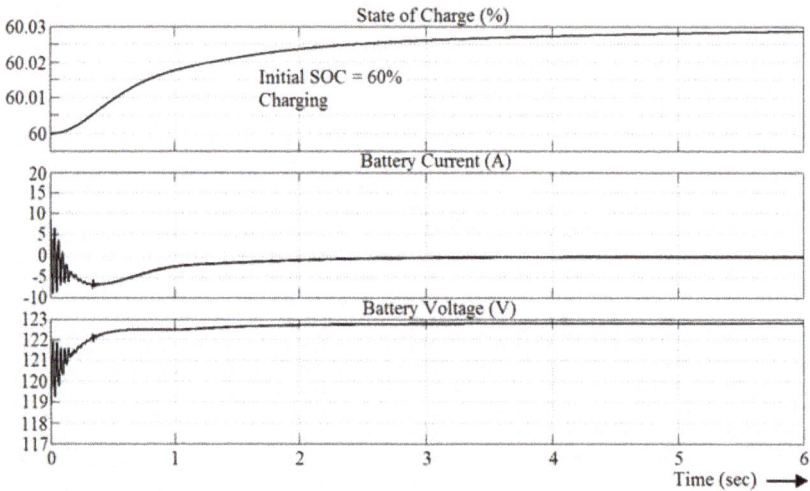

Figure 30. SOC, current, and voltage of the battery during charging (condition-2 of control logic).

Condition-3 of control logic: In this case, if the battery is fully charged and PV is sufficient to supply the load, then the battery can be disconnected. For a fully charged battery, the initial battery SOC is assumed to be 99.9%. From the results of the closed-loop simulation for this case, in Figures 31 and 32, it is inferred that the output voltage of the converter remains constant at a set value of 220 V. The PV is supplying to the load and the battery is disconnected.

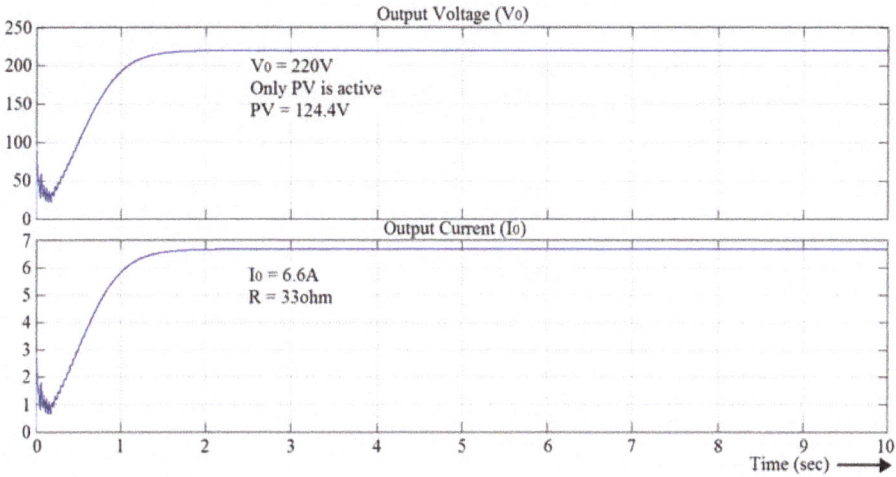

Figure 31. Output voltage and current of a bidirectional closed-loop converter when only the solar PV is supplying, and the battery is fully charged (Condition-3 of control logic).

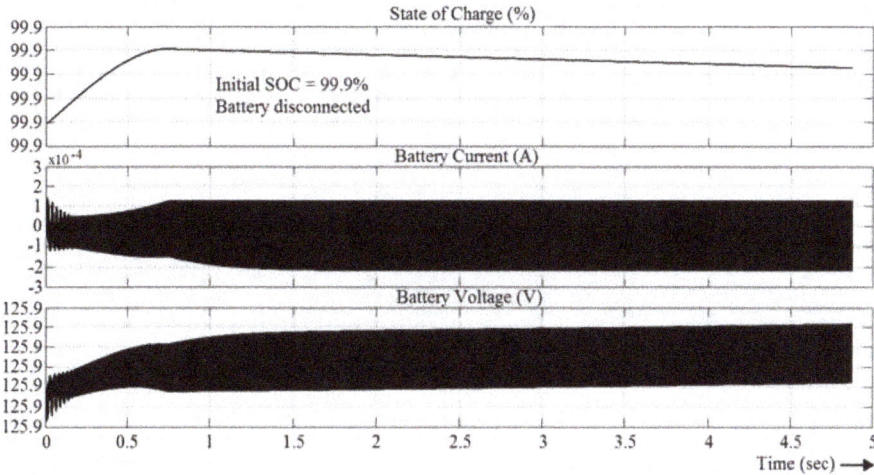

Figure 32. SOC, current, and voltage of the battery when it is fully charged (Condition-3 of control logic).

Condition-4 of control logic: In this case, the solar PV supplies the load and charges the battery. The PV is taken to be PV = 119 V, and the battery is very much less charged. The initial battery state of charge is assumed to be 35%, as shown in Figures 33 and 34. The output voltage is maintained constantly at a set value of 220 V, and the battery is charging.

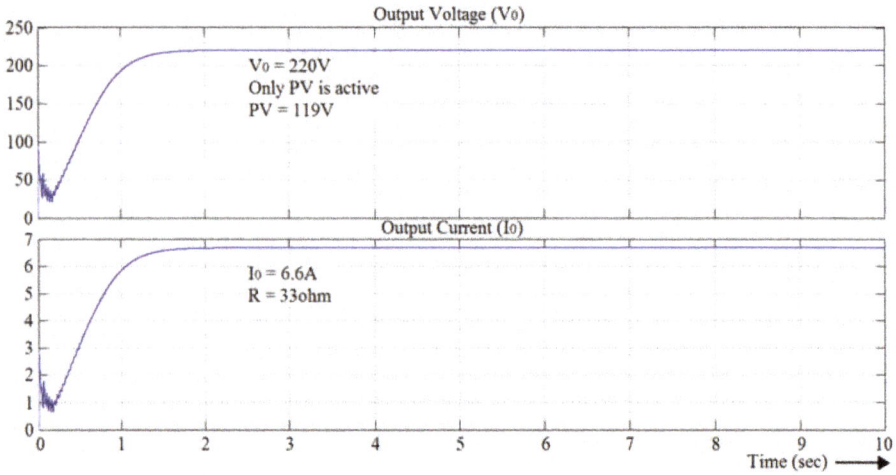

Figure 33. Output voltage and current of a bidirectional closed-loop converter when only the solar PV is supplying, and the battery is undercharged (Condition-4 of control logic).

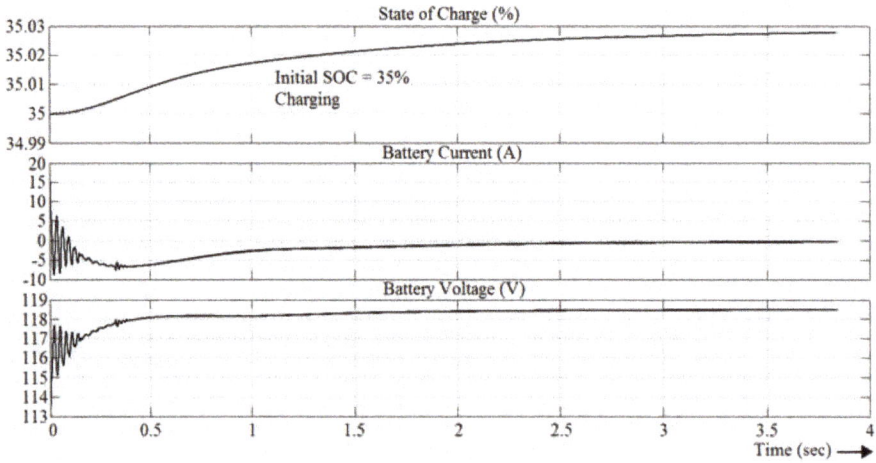

Figure 34. SOC, current, and voltage of the battery when it is undercharged (Condition-4 of control logic).

Condition-5 of control logic: In this case, if the PV is not available or if irradiation level is 0, then the battery supplies to the load. The initial battery SOC is assumed to be 65%. From the results of the closed loop simulation for this case, in Figures 35 and 36, it is inferred that the output voltage of the converter remains constant at a set value of 220 V. The PV is disconnected, and the battery supply to the load is being discharged.

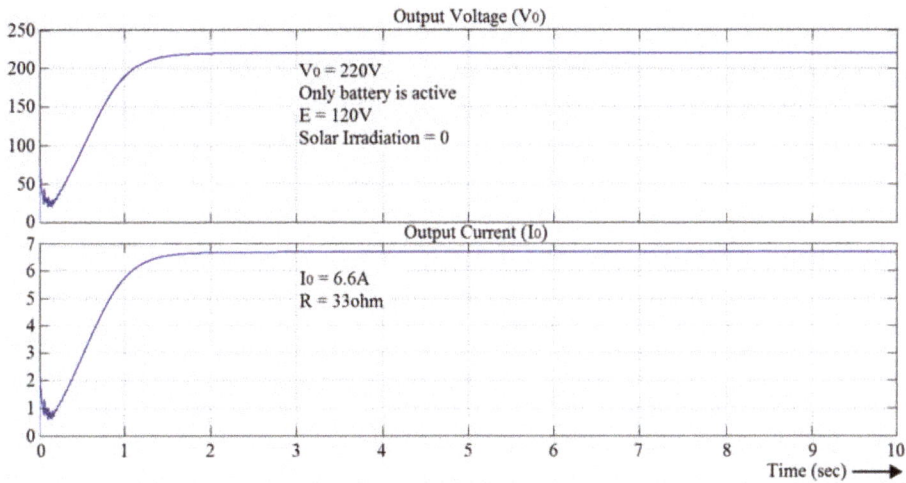

Figure 35. Output voltage and current of a bidirectional closed-loop converter when only the battery is supplying (Condition-5 of control logic).

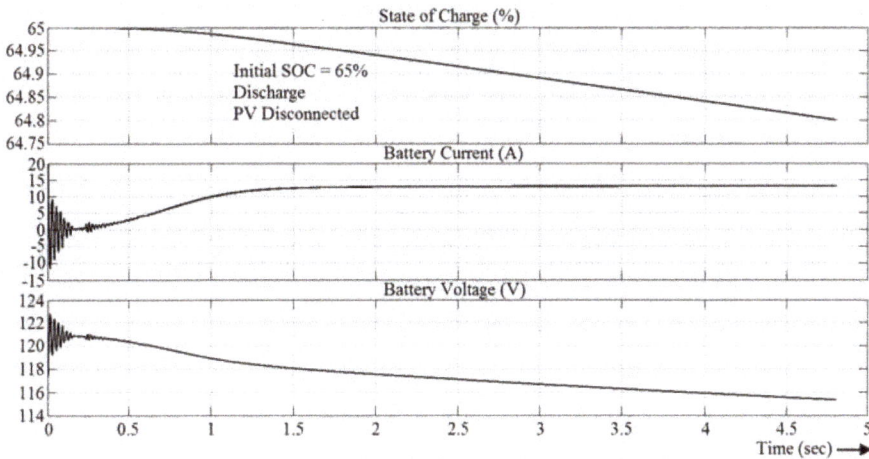

Figure 36. SOC, current, and voltage of the battery (Condition-5 of control logic).

5. Experimental Verification

The simulation analysis of the open loop topology-1 was verified with a real-time hardware setup. The hardware setup was realized with two IGBTs (FGA15N120) and a high-frequency diode (HER3006PT). A DSPIC30F2010 microcontroller was used for generating the switching pulses. In both the input ports, V_1 and V_2, supply was provided by a solar PV system of equal power rating.

The input to PVSC1 is $V_1 = 35$ V, whereas for PVSC2, $V_2 = 42$ V. The duty cycle for two corresponding switches is: $D_1 = 67\%$ and $D_2 = 50\%$. The components of the setup were designed to operate at a maximum of 1 kW. The component design parameters of the setup are given in Table 6.

Table 6. Design parameters for the hardware setup.

Description	Specification
PVSC1 inductor, L_1	15 mH
PVSC2 inductor, L_2	15 mH
PVLC inductor, L	15 mH
PVSC1 capacitor, C_1	0.54 mF
PVSC2 capacitor, C_2	0.54 mF
PVLC capacitor, C	0.54 mF
Switching frequency, f	10,000 Hz

With the given input and duty cycle, the calculated output voltage from Equation (26) was 81.66 V. The output of the MATLAB simulation of the proposed topology for the similar parameters was approximately 79 V. The output voltage of the proposed hardware setup for the given input and duty cycle was 78.8 V. This indicates that for similar values of parameters and inputs, the outputs corresponding to a mathematical analysis, MATLAB simulation, and hardware setup are approximately the same. Figure 37 shows the hardware setup of the three-port unidirectional topology.

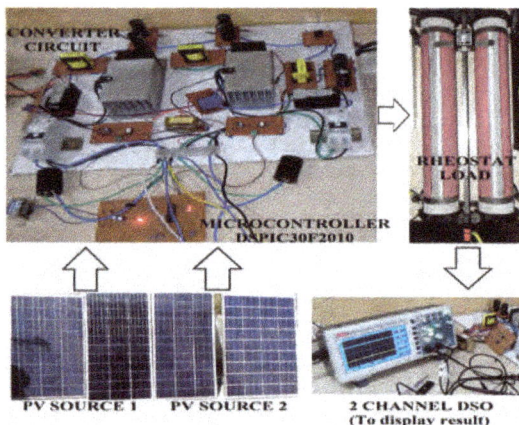

Figure 37. Hardware setup of the proposed topology-1.

Figure 38a shows the input voltage waveform, $V_1 = 37$ V and $V_2 = 42$ V. Figure 38b shows the switching pulses of the switch S_1 & S_2, i.e., $D_1 = 67\%$ and $D_2 = 50\%$ respectively. Figure 39a shows the voltage across the switches. Figure 39b shows the current through the inductors. Figure 40a,b shows the voltage across the capacitors. Figure 41 shows the output voltage of the converter $V_0 = 78.8$ V.

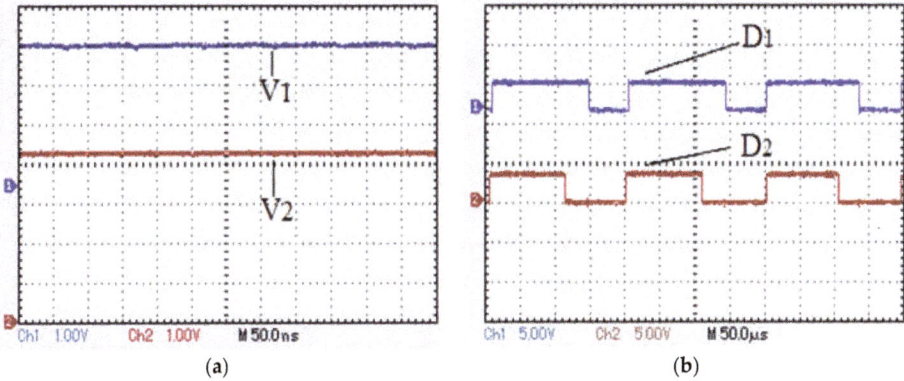

Figure 38. (a) Input voltages ($V_1 = 35$ V, $V_2 = 42$ V), (b) switching pulses for S_1 and S_2 ($D_1 = 67\%$ and $D_2 = 50\%$).

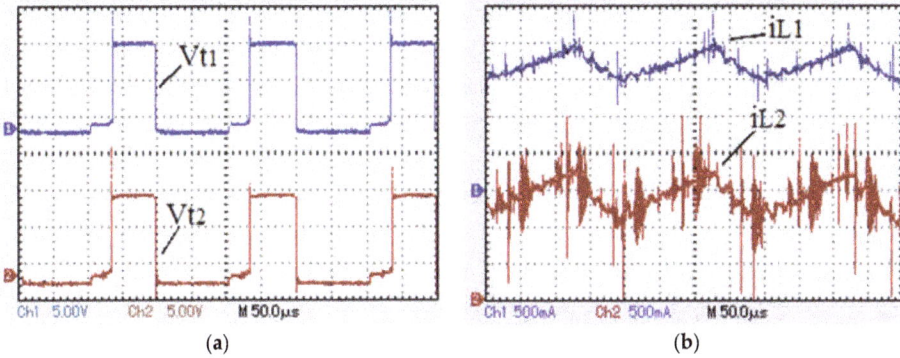

Figure 39. (a) Voltages across switches S_1 and S_2 (V_{t1} and V_{t2}), (b) currents through inductors L_1 and L_2 (i_{L1} and i_{L2}).

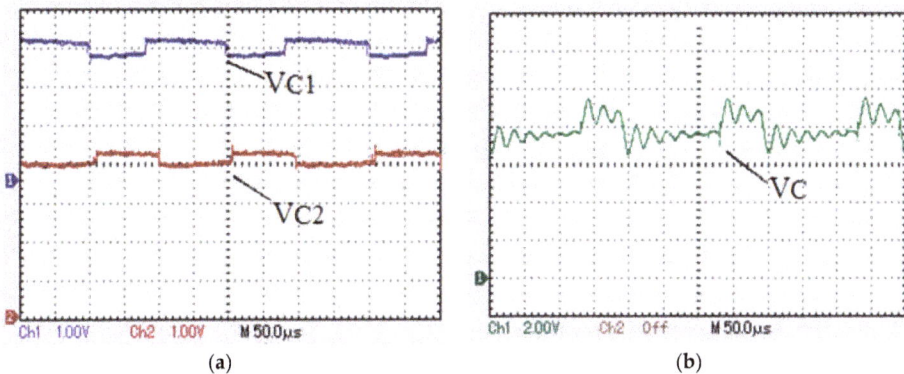

Figure 40. (a) Voltages across the capacitors C_1 and C_2 (V_{C1} and V_{C2}), (b) voltage across the output capacitor C (V_{C1}).

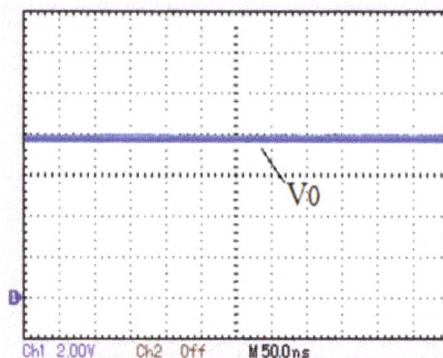

Figure 41. Output voltage (V_0 = 78.8 V).

6. Conclusions

In this paper, a unidirectional and bidirectional three-port converter based on SEPIC topology has been proposed and analyzed thoroughly. The detailed operations in various cases along with design concepts, theoretical analyses, have been cross verified through simulation and experimental results. The proposed three-port converter offers alternate solutions for integrating renewable sources with energy storage devices. This makes three port SEPIC-SEPIC converters a promising topology for electric vehicles, satellite, or DC micro-grid applications.

Author Contributions: Conceptualization done by C.A.; Supervision done by N.C., Methodology done by M.S.M.; Formal analysis done by S.V.

Funding: This research received no external funding.

Conflicts of Interest: The authors declare no conflict of interest.

References

1. Liserre, M.; Sauter, T.; Hung, J. Future Energy Systems: Integrating Renewable Energy Sources into the Smart Power Grid through Industrial Electronics. *IEEE Ind. Electron. Mag.* **2010**, *4*, 18–37. [CrossRef]
2. Carrasco, J.; Franquelo, L.G.; Bialasiewicz, J.; Galvan, E.; PortilloGuisado, R.; Prats, M.; Leon, J.I.; Moreno-Alfonso, N.; Prats, M.A.M. Power-Electronic Systems for the Grid Integration of Renewable Energy Sources: A Survey. *IEEE Trans. Ind. Electron.* **2006**, *53*, 1002–1016. [CrossRef]
3. Wai, R.J.; Lin, C.Y.; Chen, B.H. High-Efficiency DC–DC Converter with Two Input Power Sources. *IEEE Trans. Power Electron.* **2012**, *27*, 1862–1875. [CrossRef]
4. Shahzad, M.I.; Iqbal, S.; Taib, S. Interleaved LLC Converter with Cascaded Voltage Doubler Rectifiers for Deeply Depleted PEV Battery Charging. *IEEE Trans. Transp. Electrif.* **2018**, *4*, 89–98. [CrossRef]
5. Wai, R.-J.; Duan, R.-Y.; Jheng, K.-H. High-efficiency bidirectional dc–dc converter with high-voltage gain. *IET Power Electron.* **2011**, *5*, 173–184. [CrossRef]
6. Singh, A.K.; Pathak, M.K. Single-stage ZETA-SEPIC-based multifunctional integrated converter for plugin electric vehicles. *IET Electr. Syst. Transp.* **2017**, *8*, 101–111. [CrossRef]
7. Lakshmi, M.; Hemamalini, S. Nonisolated High Gain DC–DC Converter for DC Microgrids. *IEEE Trans. Ind. Electron.* **2018**, *65*, 1205–1212. [CrossRef]
8. Hwu, K.I.; Huang, K.W.; Tu, W.C. Step-up converter combining KY and buck-boost converters. *Electron. Lett.* **2011**, *47*, 722–724. [CrossRef]
9. Lin, C.-C.; Yang, L.-S.; Wu, G.W. Study of a non-isolated bidirectional DC–DC converter. *IET Power Electron.* **2012**, *6*, 30–37. [CrossRef]
10. Park, K.B.; Moon, G.W.; Youn, M.J. Nonisolated High Step-up Boost Converter Integrated with SEPIC Converter. *IEEE Trans. Power Electron.* **2010**, *25*, 2266–2275. [CrossRef]

11. Wu, H.; Xu, P.; Hu, H.; Zhou, Z.; Xing, Y. Multiport Converters Based on Integration of Full-Bridge and Bidirectional DC–DC Topologies for Renewable Generation Systems. *IEEE Trans. Ind. Electron.* **2014**, *61*, 856–869. [CrossRef]

12. Wu, H.; Zhang, J.; Xing, Y. A Family of Multiport Buck–Boost Converters Based on DC-Link-Inductors (DLIs). *IEEE Trans. Power Electron.* **2015**, *30*, 735–746. [CrossRef]

13. Khaligh, A.; Cao, J.; Lee, Y.-J. A Multiple-Input DC–DC Converter Topology. *IEEE Trans. Power Electron.* **2009**, *24*, 862–868. [CrossRef]

14. Li, Y.; Ruan, X.; Yang, D.; Liu, F.; Tse, C.K. Synthesis of Multiple-Input DC/DC Converters. *IEEE Trans. Power Electron.* **2010**, *25*, 2372–2385. [CrossRef]

15. Wang, C.; Nehrir, M.H. Power Management of a Stand-Alone Wind/Photovoltaic/Fuel Cell Energy System. *IEEE Trans. Energy Convers.* **2008**, *23*, 957–967. [CrossRef]

16. Veerachary, M. Two-Loop Controlled Buck–SEPIC Converter for Input Source Power Management. *IEEE Trans. Ind. Electron.* **2012**, *59*, 4075–4087. [CrossRef]

17. Haghighian, S.K.; Tohidi, S.; Feyzi, M.R.; Sabahi, M. Design and analysis of a novel SEPIC-based multi-input DC/DC converter. *IET Power Electron.* **2017**, *10*, 1393–1402. [CrossRef]

18. Patra, P.; Patra, A.; Misra, N. A Single-Inductor Multiple-Output Switcher with Simultaneous Buck, Boost, and Inverted Outputs. *IEEE Trans. Power Electron.* **2012**, *27*, 1936–1951. [CrossRef]

19. Tummuru, N.R.; Mishra, M.K.; Srinivas, S. Dynamic Energy Management of Renewable Grid Integrated Hybrid Energy Storage System. *IEEE Trans. Ind. Electron.* **2015**, *62*, 7728–7737. [CrossRef]

20. Rani, B.I.; Ilango, G.S.; Nagamani, C. Control Strategy for Power Flow Management in a PV System Supplying DC Loads. *IEEE Trans. Ind. Electron.* **2013**, *60*, 3185–3194. [CrossRef]

21. McDonough, M. Integration of Inductively Coupled Power Transfer and Hybrid Energy Storage System: A Multiport Power Electronics Interface for Battery-Powered Electric Vehicles. *IEEE Trans. Power Electron.* **2015**, *30*, 6423–6433. [CrossRef]

22. Tani, A.; Camara, M.B.; Dakyo, B. Energy Management Based on Frequency Approach for Hybrid Electric Vehicle Applications:Fuel-Cell/Lithium-Battery and Ultracapacitors. *IEEE Trans. Veh. Technol.* **2012**, *61*, 3375–3386. [CrossRef]

energies

MDPI

Article

Speed Control for Turbine-Generator of ORC Power Generation System and Experimental Implementation

Hyung-Seok Park [1], Hong-Jun Heo [1], Bum-Seog Choi [2], Kyung Chun Kim [3] and Jang-Mok Kim [1,*]

[1] Department of Electrical Engineering, Pusan National University, Busan 46241, Korea; hs_4451@pusan.ac.kr (H.-S.P.); hhongjun@pusan.ac.kr (H.-J.H.)
[2] Korea Institute of Machinery & Materials, Daejeon 34103, Korea; bschoi@kimm.re.kr
[3] School of Mechanical Engineering, Pusan National University, Busan 46241, Korea; kckim@pusan.ac.kr
* Correspondence: jmok@pusan.ac.kr; Tel.: +82-51-510-2366

Received: 11 December 2018; Accepted: 3 January 2019; Published: 9 January 2019

Abstract: This paper presents a rotation speed estimation and an indirect speed control method for a turbine-generator in a grid-connected 3-phase electrical power conversion system of an organic Rankine cycle (ORC) generation system. In addition to the general configuration mechanism and control techniques that are required in the grid-connected ORC power generation system, the indirect speed control method using the grid-side electric power control and the speed estimation method is proposed for the proper speed control of turbine-generators. The speed estimation method utilizes a digital phase-locked loop (PLL) method that uses a state observer to detect the positive-sequence voltages. A 10 kW system where a Motor-Generator set is used as a turbine simulator and a 23 kW actual system for the grid-connected ORC power generation were designed and manufactured, respectively. This paper includes various experimental results obtained from field tests conducted on actual installed ORC systems.

Keywords: generator speed control; electrical power generation; turbine and generator; grid-connected converter; organic Rankine cycle; renewable energy

1. Introduction

As part of the renewable energy generation system, researches on power generation systems using heat sources have been developed. An organic Rankine cycle (ORC) system with an organic compound having a low boiling point as a working fluid can obtain high-pressure steam even with a low-temperature heat source. Accordingly, there are many technical and economic advantages, and it is possible to generate high efficiency power from various heat sources [1,2].

In the ORC generation system, the output power of the turbine is converted into electric power by the generator, and it transferred to the grid network via an electric power conversion system. The generated electric power must be synchronized with the grid electric power under the constant frequency before fed into the grid [3,4]. Figure 1 shows the schematic diagram and the photographs of the ORC generation system with the grid-connected electric power conversion system.

The generator is directly coupled to the turbine expander that is designed for a high-speed drive so as to reduce the size and increase the efficiency [5–7]. Additionally, the generator rotor will operate at variable speeds according to the operating conditions of the ORC system. However, due to the ripple or fluctuation of the turbine rotation speed caused by the unpredictable nature of the ORC system, the generator is exposed to the speed ripple, which in turn causes significant vibration and noise. This means that the degradation in control performance and durability of the ORC system are inevitable. Thus, the rotation speed of the turbine-generator should be operated constantly for the stable operation of the ORC generation system [8].

The rotation speed information of the turbine-generator is needed for proper speed control and it can be measured by a speed sensor, such as encoders or resolvers [9]. However, these sensors add difficulties to the installation and maintenance, increasing system cost, and greatly reducing the reliability due to the hostile environment of high temperature and humidity of the turbine [9,10]. Thus, the estimation of the generator speed should be adopted for the ORC system [10,11].

In general, PWM switched converters or diode rectifiers are used as generator side AC/DC converters. In References [12,13], a PWM converter has high power flow management capability and can directly control the generator speed. However, the use of the PWM converter for the high-speed generator causes several problems such as high switching frequency, high device breakdown voltage, and price increase. A multilevel PWM topology to meet the voltage and power requirement is also used in MW generation systems, as in Reference [14]. However, a high-performance control system is required along with control complexity, and the price is considerably increased [15].

Many studies using a diode rectifier instead of a PWM converter on the generator side have been performed [16]. Advantages of this topology include reliability, durability, lower cost and higher rated power than PWM converters, especially in the power generation system with the high-speed driven turbine [15,17,18]. However, the usage of a diode rectifier makes it impossible to directly control the generator speed on the generator side. Furthermore, the generator terminal voltage has a distorted waveform caused by the conduction of the diode rectifier [5,15,18], which results in a significant ripple component of the estimated speed. Together, it leads to a great challenge for the improved speed control performance in the ORC system. A method of controlling the generator speed by using an additional DC-DC boost converter in a diode rectifier has been recently studied, as in references [19,20]. However, it also has the same problems as PWM converters.

Therefore, in order to solve the problem of speed estimation and control caused by the use of the generator-side diode rectifier, as shown in Figure 1, this paper presents a rotation speed estimation method under the distorted generator terminal voltage of the diode rectifier and an indirect speed control method using grid-side electric power control for the ORC power generation system. The proposed speed control system is verified by experimental results of the manufactured and installed actual grid-connected ORC system.

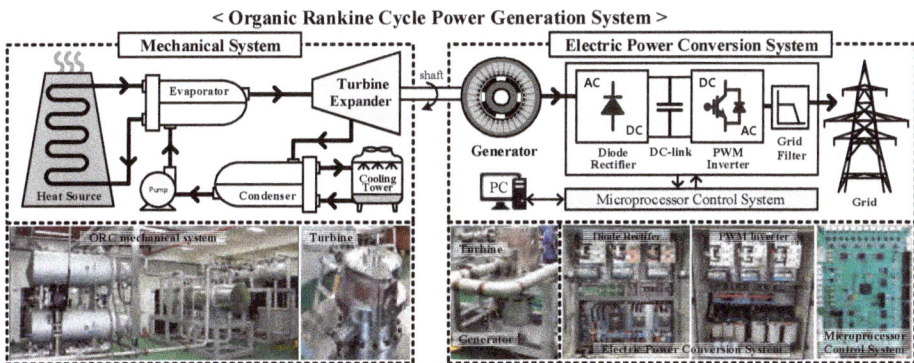

Figure 1. Schematic diagram and photographs of the whole manufactured organic Rankine cycle (ORC) generation system.

2. Grid-Connected Electric Power Conversion System

Grid-connected electric power conversion systems are commonly comprised of a generator-side rectifier, a DC-link, a grid-side inverter, a grid filter, and a control system. The grid-side three-phase PWM inverter supplies the regulated AC power from the rectified DC voltage to the grid. It is critical in grid-connected electric power conversion systems for optimized control to meet the grid interconnection and required electric power quality [21–24].

The control algorithm for the grid-side PWM inverter is illustrated in Figure 2. It represents the 'Microprocessor Control System' part of Figure 1 in detail. The control structure for the grid-side PWM inverter consists of two cascaded loops. The two current controllers in the inner loop have a fast response in a synchronous reference frame [25–27]. The DC voltage controller in one of the outer loops implements the balanced electric power of the DC- link, which enables the active power flowing to the grid.

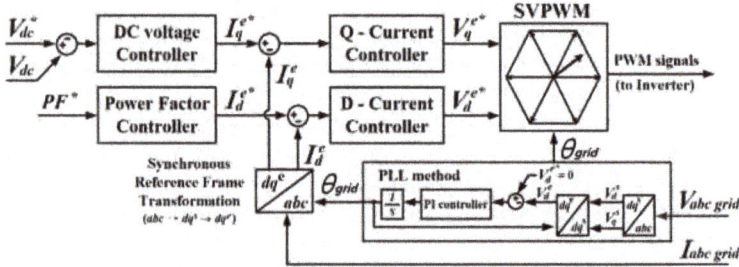

Figure 2. General control block diagram of the gird-side PWM inverter with the PLL method.

Additionally, the power factor controller in the other outer loop controls the reactive power. Therefore, the two current controllers for q-axis and d-axis obtain their references from the DC voltage and power factor controller, respectively. Additionally, the two current controllers generate the voltage references on d and q axes, for the PWM signal generation [28,29]. The current synchronized with the grid voltage should be injected into the grid as the standards required. Accurate phase angle information of the grid voltages can be detected using a phase-locked loop (PLL) method [30].

The synchronous reference frame (SRF) PLL method detects the angular position of three-phase voltage by controlling the error of the actual and the estimated [31]. The SRF-PLL is robust and has better dynamic performance than zero crossing PLL discussed in [32,33]. This method has been generally utilized various applications for the phase angle detection [34,35]. The SRF-PLL is conducted in the d and q axis synchronous reference frame as shown in Figure 2. This PLL requires reference frame transformations, namely the stationary and synchronous reference frame, and the phase locker is implemented by setting the d-axis voltage to zero. The PI controller output is the angular velocity of the grid voltage. After taking the integration of the angular velocity, the phase angle is obtained [31,34].

3. Speed Estimation and Control for Turbine and Generator

3.1. Speed Estimation Using PLL

The SRF-PLL used for the phase angle detection of the grid voltage can be also adopted in generator control to estimate the rotational speed [36,37]. In this case, the extracted three-phase terminal voltage from the generator is used instead of the grid, the generator speed can be estimated in the same way as the grid voltage [38,39]. In the case of the PWM switched converters are used as generator side AC/DC converters, since the generator, three-phase terminal voltage, has a non-distorted voltage without any harmonics, it has fast and good dynamic characteristics like in the grid voltage.

However, a diode rectifier is used instead of the PWM converter in this paper. As shown in Figure 3, the three-phase terminal voltage has a non-sinusoidal waveform due to the diode conduction of the rectifier in the generation region, and has a variable frequency of the generator. As a result, the estimated speed from the generator terminal voltage by the conventional SRF-PLL method has a significant error, which is similar to the estimated phase angle under the distorted grid angle conditions. This means that the speed control performance and overall system efficiency will be degraded [36,37].

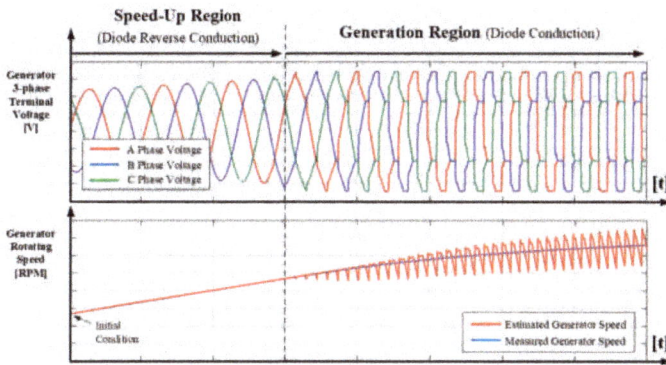

Figure 3. Waveforms of 3-phase generator voltages by diode rectifier and estimated speed using conventional SRF-PLL method.

In order to achieve satisfactory performance of the speed estimation under the non-sinusoidal distorted voltage, certain techniques such as harmonics compensation, filtering, and positive-sequence component detection can be additionally used [40–42]. An improved performance of the conventional SRF-PLL under distorted voltages is achieved by separating the positive and negative sequences and by feeding back only the positive-sequence. The SRF-PLL method with positive-sequence detection has the better tracking performance than the conventional SRF-PLL topology [43]. In this paper, a state observer is utilized as the positive-sequence voltage detection method. Based on the positive-sequence of the distorted three-phase generator terminal voltages, detected by a state observer, the generator position and speed can be obtained from the positive-sequence voltage by the SRF-PLL.

Assuming that the three-phase voltage is unbalanced and distorted with high-order harmonics, it can be decomposed into positive, negative, and homopolar (zero) sequence components. By calculating and decomposing each sequence component, only the voltage of the positive-sequence component can be extracted [44]. Using this as the input voltage of the SRF-PLL, accurate position estimation from the unbalanced and distorted three-phase voltage is possible.

The positive and negative sequence components of the three-phase voltage are given in Equation (1), where p and n represent positive and negative, respectively. The zero sequence component is not considered because of the balanced electric circuit of the generator.

$$\begin{bmatrix} V_a \\ V_b \\ V_c \end{bmatrix} = V_p \begin{bmatrix} \cos(\omega t + \varphi_p) \\ \cos(\omega t - \frac{2\pi}{3} + \varphi_p) \\ \cos(\omega t + \frac{2\pi}{3} + \varphi_p) \end{bmatrix} + V_n \begin{bmatrix} \cos(\omega t + \varphi_n) \\ \cos(\omega t + \frac{2\pi}{3} + \varphi_n) \\ \cos(\omega t - \frac{2\pi}{3} + \varphi_n) \end{bmatrix} \tag{1}$$

Equation (1) can be calculated in the d-q stationary reference frame as Equation (2).

$$\begin{bmatrix} V_d^s \\ V_q^s \end{bmatrix} = \begin{bmatrix} V_{pd}^s \\ V_{pq}^s \end{bmatrix} + \begin{bmatrix} V_{nd}^s \\ V_{nq}^s \end{bmatrix} = V_p \begin{bmatrix} \cos(\omega t + \varphi_p) \\ \sin(\omega t + \varphi_p) \end{bmatrix} + V_n \begin{bmatrix} \cos(\omega t + \varphi_n) \\ -\sin(\omega t + \varphi_n) \end{bmatrix} \tag{2}$$

The generator three-phase terminal voltage can be modeled by using a state equation with the positive and negative sequence voltage in the stationary reference frame. The equation of the general state observer is shown in Equation (3). Where x and y is the state and output vector, respectively, and L is the gain of the state observer, A and C is the system and output matrix, respectively.

$$\frac{d}{dt}\hat{x} = A\hat{x} + L(y - C\hat{x}) \tag{3}$$

By differentiating Equation (2), the state equation can be derived as Equation (4).

$$
\frac{d}{dt}
\begin{bmatrix} V_{pd}^s \\ V_{pq}^s \\ V_{nd}^s \\ V_{nq}^s \end{bmatrix}
=
\begin{bmatrix}
0 & -\omega & 0 & 0 \\
0 & 0 & 0 & 0 \\
\omega & 0 & 0 & \omega \\
0 & 0 & -\omega & 0
\end{bmatrix}
\begin{bmatrix} V_{pd}^s \\ V_{pq}^s \\ V_{nd}^s \\ V_{nq}^s \end{bmatrix}
\tag{4}
$$

The output equation for the decomposed state variables of Equation (2) can be expressed as Equation (5).

$$
y = cx =
\begin{bmatrix} V_d^s \\ V_q^s \end{bmatrix}
=
\begin{bmatrix}
1 & 0 & 1 & 0 \\
0 & 1 & 0 & 1
\end{bmatrix}
\begin{bmatrix} V_{pd}^s \\ V_{pq}^s \\ V_{nd}^s \\ V_{nq}^s \end{bmatrix}
\tag{5}
$$

The state observer equation is derived from Equations (3)–(5) and is given in Equation (6), where the angular speed variable $\hat{\omega}$ is estimated from the previous sample and the tuned observer constant should be used.

$$
\frac{d}{dt}
\begin{bmatrix} \hat{V}_{pd}^s \\ \hat{V}_{pq}^s \\ \hat{V}_{nd}^s \\ \hat{V}_{nq}^s \end{bmatrix}
=
\begin{bmatrix}
0 & -\hat{\omega} & 0 & 0 \\
0 & 0 & 0 & 0 \\
\hat{\omega} & 0 & 0 & \hat{\omega} \\
0 & 0 & -\hat{\omega} & 0
\end{bmatrix}
\begin{bmatrix} \hat{V}_{pd}^s \\ \hat{V}_{pq}^s \\ \hat{V}_{nd}^s \\ \hat{V}_{nq}^s \end{bmatrix}
+
\begin{bmatrix}
l_{11} & l_{12} \\
l_{21} & l_{22} \\
l_{31} & l_{32} \\
l_{41} & l_{42}
\end{bmatrix}
\left\{ y -
\begin{bmatrix}
1 & 0 & 1 & 0 \\
0 & 1 & 0 & 1
\end{bmatrix}
\begin{bmatrix} \hat{V}_{pd}^s \\ \hat{V}_{pq}^s \\ \hat{V}_{nd}^s \\ \hat{V}_{nq}^s \end{bmatrix}
\right\}
\tag{6}
$$

The block diagram of the speed estimation method using the SRF-PLL with a state observer is presented in Figure 4. The measured three-phase generator terminal voltage is transformed into the stationary reference frame voltage. The positive and negative sequence component is decomposed by the state observer. Then, the generator speed is estimated from the positive-sequence. The estimated electrical speed is transformed into mechanical speed depending on the pole number of the generator. The speed can be estimated with minimal error and ripple, and used for the speed control afterwards.

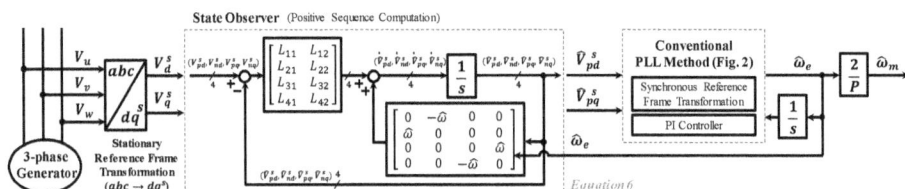

Figure 4. Speed estimation method using SRF-PLL with a state observer.

3.2. Indirect Speed Control

Since a diode rectifier is used on the generator side instead of the PWM converter, the torque and speed of turbine and generator cannot be directly controlled. Therefore, the torque of the turbine and generator should be indirectly controlled through the generated electrical power on the grid side, which allows the speed of the turbine and generator to be operated constantly.

The input power of the generator from the turbine can be expressed, as shown in Equation (7). When the output power of turbine increases under a DC-link voltage control of the grid-side PWM inverter, the turbine-generator speed increase due to the difference between turbine and generator (load) torques. Then, the operating point of the turbine and generator is changed by the torque characteristic line. Therefore, for the constant speed operation, the generator torques can be controlled according to the given output power of the turbine.

$$
P_{input} = T_M \times \omega_M
\tag{7}
$$

The mechanical generator torque is expressed in Equation (8). At the steady state, the right-side differential term of Equation (8) is zero. Neglecting the damping effect of the inertia moment and the friction term, the mechanical turbine output torque is approximately equal to the load torque of the generator.

$$T_M = J\frac{d\omega}{dt} + B\omega + T_L \tag{8}$$

The load torque can be regarded as the generator electric torque delivered to the grid, as given in Equation (9). Where K_T is the torque constant and i_{qs}^e is the generated torque component current of the q-axis synchronous reference frame. Therefore, the mechanical speed of the turbine and generator can be maintained by controlling the torque component current of the generator.

$$T_L = T_e = K_T \times i_{qs}^e \tag{9}$$

The peak value of output voltage with the constant excitation of the generator is proportional to the rotating speed of the machine. The variation of the q-axis current (the generated peak output current from the generator) depends on the difference between the peak output voltage of the generator and the DC–link voltage. This implies that it is possible to control the torque component (q-axis) current by adjusting the DC-link voltage, which further controls the mechanical speed of the turbine and generator.

The schematic diagram for speed control is shown in Figure 5. The indirect speed control system is developed based on the relationship between the DC-link voltage and the torque component current of the generator. The speed controller outputs the DC-link voltage reference from the error between the reference and the estimated speed. The DC-link voltage should be guaranteed to be the standard value at least for system stability and can be changed within the system limits.

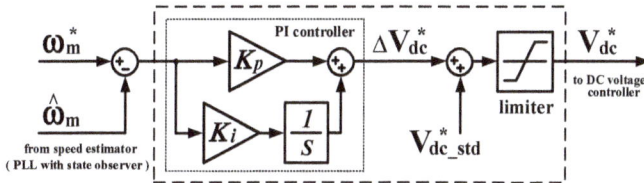

Figure 5. Block diagram for indirect speed control.

The overall proposed speed control algorithm for the turbine-generator of the ORC system is illustrated in Figure 6. The three-phase generator voltage is measured by the generator-side voltage sensor. The rotor speed information of the generator, which is used in the speed controller, is estimated using the PLL method. The rotor speed with reduced ripple components can be estimated from the positive sequence voltage. The speed controller is located in the outer loop of the DC-link voltage controller, which controls the active power generation of the generator. The torque component current control can control the turbine-generator speed indirectly.

Therefore, the proposed speed control system can control the generator speed for the stable operation by controlling the amount of transmitted electric power to the grid. All the other components of Figure 6 have been explained in detail in the Section 2.

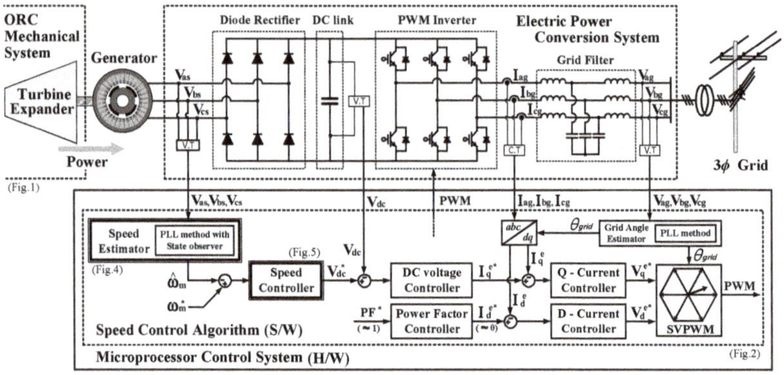

Figure 6. Proposed indirect speed control algorithm in a grid-connected ORC generation system.

4. Experimental Set-Up and Results

Two types of experiments are conducted in order to verify the proposed indirect speed control systems. The first experimental setup shown in Figure 7 is a 10 kW small-scale system, where a motor simulates the turbine and is used for the confirmation of the proposed control method during the developing stage. This system consists of two grid-connected electric power conversion systems and a motor-generator set. A 15 kW PMSM, which is directly coupled with a 12 kW PMSG, is operated as the turbine simulator. Two converter systems are used as the drive systems for turbine simulator and generator, respectively. The entire control algorithm is implemented in a micro-processor control board. The electrical parameters of the experimental setup are presented in Table 1.

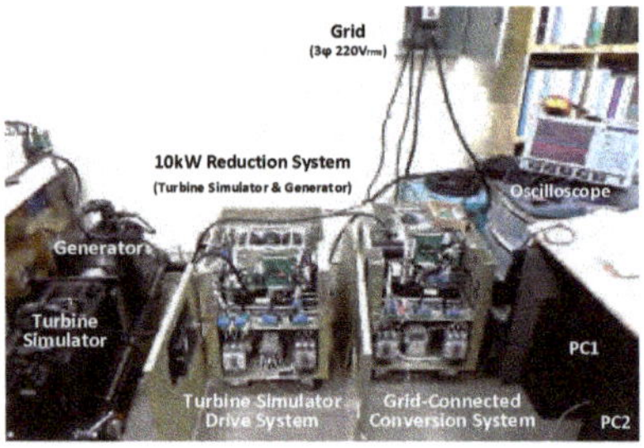

Figure 7. Experimental setup of 10 kW small-scale grid-connected ORC simulation system.

Table 1. Parameters of Experimental Setup.

System	Parameter	Value
Converter System	DC-link capacitance	3133 (uF)
	Switching frequency	10 (kHz)
	Control period	100 (us)
Grid-Network	Grid line-line voltage	55 (V_{rms}) (4:1 trans.)
	Frequency	60 (Hz)
	Filter L inductance	5 (mH)
Simulator/Generator (PMSM/G)	Rated power	15 (kW)/12 (kW)
	Rated torque	95.4 (Nm)/70 (Nm)
	Back EMF constant	109.0 ($V_{peak\ L\text{-}L}$/krpm)
	Pole number	8

Figure 8 shows the experimental results of the generator speed estimation performance under the conventional SRF-PLL and the state observer SRF-PLL. In Figure 8a, using the conventional SRF-PLL, significant ripple components are observed in waveforms of the generator voltages V_d^s, V_d^e, and the estimated speed ω_{PLL}. On the other hand, the state observer SRF-PLL eliminates the ripple components from the waveforms of the estimated speed by using the positive sequence component of generator voltage $V_{d\_positive}^s$ and $V_{d\_positive}^e$. The ripple of the estimated generator speed $\omega_{proposed\_PLL}$ is reduced by about 85% compared to that of the conventional estimation method.

Figure 8b represents the results of speed estimation under the conventional and the proposed method in a full operating range. In the region indicated as 'Generator Speed-up Region' in Figure 8b, the generated power cannot be transmitted to the grid side because the excited voltage amplitude according to the generator speed does not reach to the DC-link voltage. At the transition stage between 'Generator Speed-up Region' and 'Generation Region', the generated power starts flowing to the grid. Additionally, the generator voltage is distorted due to diode conduction. As can be seen, the speed waveform of the conventional method has much more ripple components than the proposed method. The curve $\omega_{m\_measured}$ in Figure 8b is the actual speed measured by the encoder in order to confirm the speed estimation performance.

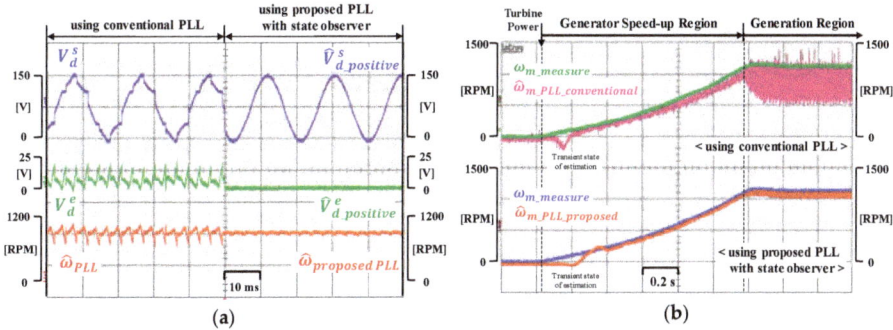

Figure 8. Experimental results of the rotation speed estimation for the turbine-generator: (**a**) Estimated speed by PLL; and (**b**) estimated speed in a full operating range.

The experimental results of the proposed indirect speed control is shown in Figure 9. From speed control in the 'Generation Region', by increasing q-axis current to the grid-side (negative direction) using the speed control action, the generator speed is controlled constantly at the reference speed of 1000RPM. The grid-side d-axis current is controlled to be zero to achieve the unity power factor. When the turbine output power increases suddenly, the q-axis current injected into the grid is increased (from −4A to −18A) by the DC-link voltage control to keep the constant speed for the safe operation of the ORC system.

The second experiments are implemented in the actual ORC power generation system, as shown in Figure 10. The configuration of the experimental setup is same as the one shown in Figure 1. And the ORC loop configurations, the working fluid, the operation process, performance curves, sensors and system parameters used in this experiment are the same as that previously designed and manufactured in Reference [45]. Generator (PMSM) is directly coupled to the expander shaft.

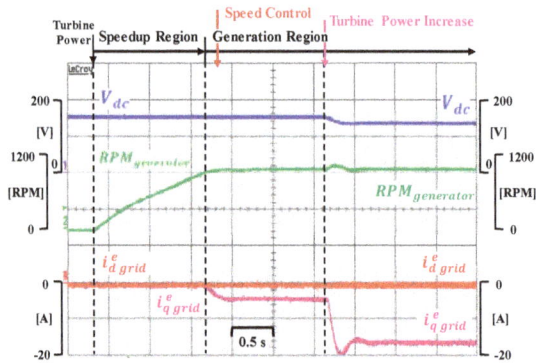

Figure 9. Experimental results of the proposed indirect generator speed control.

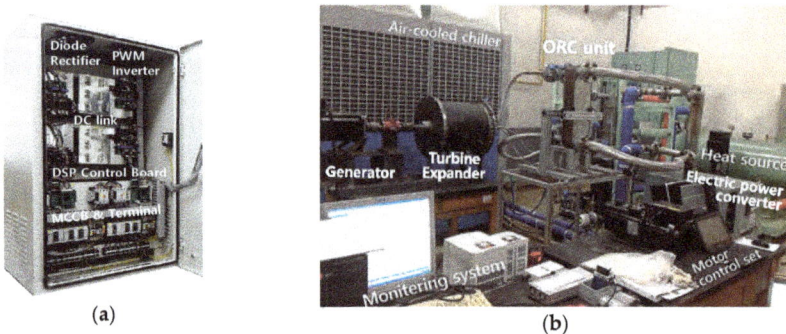

Figure 10. Experimental setup for the actual ORC power generation system: (**a**) Gird-connected electric power conversion system; and (**b**) 23 kW ORC power generation system.

The experimental results of the proposed indirect speed control are presented in Figure 11a–d. It shows the waveforms for the mass flow rate of the working fluid and the turbine in/outlet pressure (a), the pressure ratio and the turbine output power (b), the rotation speed and torque of the turbine-generator (c), and the DC-link voltage and the grid-side torque component current (d) in detail. During the experiment, the turbine output power is increased from 0 to 1.4 kW to confirm the control performance.

Initially (before 80s), when the turbine output power is supplied, the turbine and generator speeds increase until reaching the start point of the 'Generation Region'. At this point (from 80s), the rotation speed control is started, and the generator starts supplying the power to the grid system and the constant speed control is implemented (3000RPM). As presented in Figure 11d, by controlling the DC-link voltage, the injection of the generated q-axis torque component current gets regulated, accordingly. Then, the generator torque is indirectly controlled through the injection of grid-side current, which further maintains the rotor speed, as shown in Figure 11c.

At time 1050s, the turbine output power increases due to the increase in the mass flow rate of the working fluid as shown in Figure 11a,b. For the generator speed to be controlled constantly in this region, the DC-link voltage is controlled to have a lower value, and the grid-side q-axis current correspondingly increases (Figure 11d). Accordingly, the generator (load) torque increases, and the generator speed is controlled to maintain constantly (Figure 11c).

In conclusion, as observed in the experimental waveforms, the proposed speed control system can achieve the desired control performance under various supplied turbine output due to the power fluctuation of the ORC system. Since the noise and vibration of turbine-generator are caused by the speed ripple from the ORC power output ripple, constant speed control reduces noise and vibration. This means that, without the use of indirect speed control, the ORC output ripple cause to the noise and vibration of the generator. Therefore, from the experimental results shown in Figure 11, the ORC turbine output ripple waveform and the indirectly controlled speed waveform can be considered as the performance comparison.

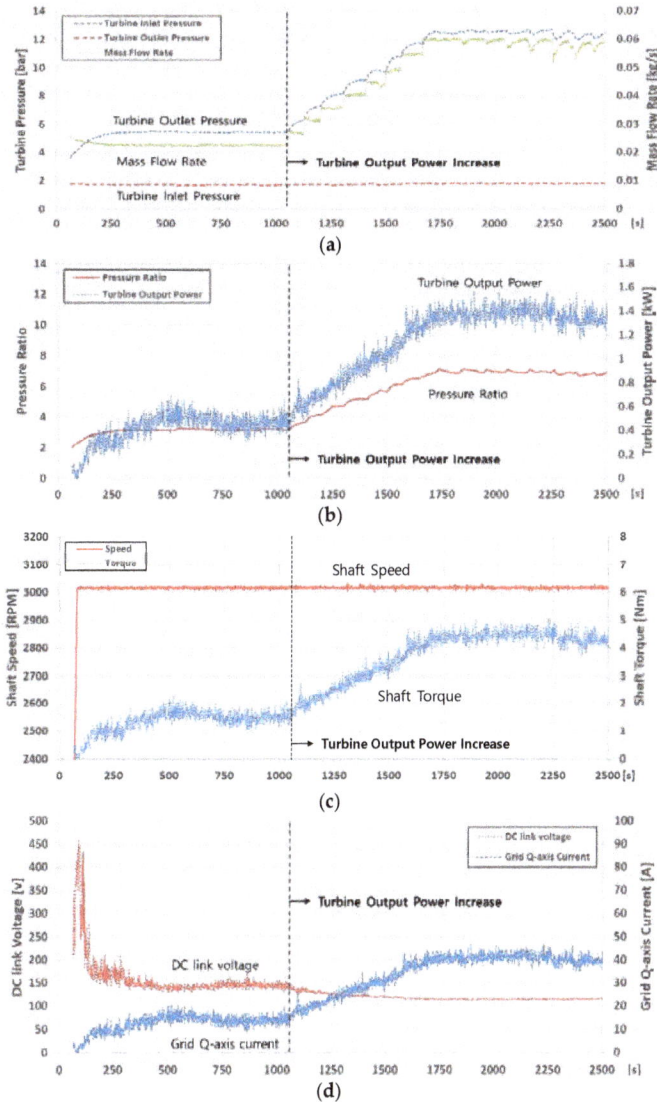

Figure 11. Experimental results of proposed speed control in the actual ORC generation system: (**a**) Mass flow rate of working fluid and turbine in/outlet pressure; (**b**) pressure ratio and turbine output power; (**c**) Rotation speed and torque of the turbine-generator; and (**d**) DC-link voltage and grid-side torque component (q-axis) current.

5. Conclusions

This paper presented an indirect speed control method for a turbine-generator in a grid-connected electric power conversion system of an ORC power generation system. The rotor speed information for proper speed control was estimated from the positive-sequence voltage of the generator by using the state observer PLL method. Additionally, the constant speed control could be guaranteed by controlling the grid-side generated electrical power. Accordingly, the proposed control method improves stability and durability of the ORC generation system by making it possible to estimate the rotating speed under the distorted generator terminal voltage due to the diode rectifier. It also realizes the constant rotation speed control despite the ripple or fluctuation of the turbine power. This paper also presented the information about the configuration mechanism and control techniques that are required in the grid-connected generation system. For the experimental verification, 10 kW and 23 kW grid-connected ORC power generation experimental systems with a turbine simulator and actual turbine, respectively, were designed and manufactured. The field test was conducted in the installed actual grid-connected ORC system. The effectiveness of the proposed control system has been verified through extensive experimental results. As a future plan, the proposed control system will be applied to the 220 kW large-scale ORC power generation system (photographs in Figure 1), which was manufactured and installed previously in Reference [46].

Author Contributions: Methodology, H.-S.P., H.-J.H. and J.-M.K.; Software, H.-S.P.; Validation, H.-J.H.; Formal analysis, H.-S.P. and H.-J.H.; Data Curation, B.-S.C. and K.C.K.; Writing-Original Draft Preparation, H.-S.P.; Writing-Review and Editing, H.-S.P., K.C.K. and J.-M.K.; Visualization, B.-S.C.; Project Administration, J.-M.K.

Funding: This research was funded by the National Research Foundation of Korea (NRF) grant funded by the Korean government (MSIT) through GCRC-SOP (No. 2011-0030013).

Conflicts of Interest: The authors declare no conflict of interest.

References

1. Larjola, J. Electricity from industrial waste heat using high-speed organic Rankine cycle (ORC). *Int. J. Prod. Econ.* **1995**, *41*, 227–235. [CrossRef]
2. Hung, T.C.; Shai, T.Y.; Wang, S.K. A review of organic Rankine cycles (ORCs) for the recovery of low-grade waste heat. *Energy* **1997**, *22*, 661–667. [CrossRef]
3. Curtiss, P.S.; Kreider, J.F. Recent developments in the control of distributed electrical generation systems. *ASME J. Sol. Energy Eng.* **2003**, *125*, 352–358. [CrossRef]
4. Lawrence, R.; Middlekauff, S. The new guy on the block. *IEEE Ind. Appl. Mag.* **2005**, *11*, 54–59. [CrossRef]
5. Papini, F.; Bolognesi, P. Preliminary design and analysis of a high speed permanent magnets synchronous generator. In Proceedings of the IEEE MELECON 2010, Valletta, Malta, 26–28 April 2010.
6. Quoilin, S.; Broek, M.V.D.; Declaye, S.; Dewallef, P.; Lemort, V. Techno-economic survey of Organic Rankine Cycle (ORC) systems. *Renew. Sustain. Energy Rev.* **2013**, *22*, 168–186. [CrossRef]
7. Colonna, P.; Casati, E.; Trapp, C.; Mathijssen, T.; Larjola, J.; Turunen-Saarest, T.; Uusitalo, A. Organic Rankine cycle power systems: From the concept to current technology, applications, and an outlook to the future. *ASME J. Eng. Gas Turbines Power* **2015**, *137*, 100801. [CrossRef]
8. Petrovic, V.; Ortega, R.; Stankovic, A.M.; Tadmor, G. Design and implementation of an adaptive controller for torque ripple minimization in PM synchronous motors. *IEEE Trans Power Electron.* **2000**, *15*, 871–880. [CrossRef]
9. Bojoi, R.; Pastorelli, M.; Bottomley, J.; Giangrande, P.; Gerada, C. Sensorless Control of PM Motor Drives—A Technology Status Review. In Proceedings of the IEEE WEMDCD 2013, Paris, France, 11–12 March 2013.
10. Holtz, J. Developments in sensorless AC drive technology. In Proceedings of the IEEE PEDS 2005, Kuala Lumpur, Malaysia, 28 November–1 December 2005.
11. Islam, M.N.; Seethaler, R.J. Sensorless position control for piezoelectric actuators using a hybrid position observer. *IEEE/ASME Trans. Mechatron.* **2014**, *19*, 667–675. [CrossRef]
12. Xu, G.; Liu, F.; Hu, J.; Bi, T. Coordination of wind turbines and synchronous generators for system frequency control. *Renew. Energy* **2018**, *129*, 225–236. [CrossRef]

13. Alcalá, J.; Cárdenas, V.; Espinozac, J.; Duránaa, M. Investigation on the limitation of the BTB-VSC converter to control the active and reactive power flow. *Electr. Power Syst. Res.* **2017**, *143*, 149–162. [CrossRef]

14. Bunjongjit, K.; Kumsuwan, Y.; Sriuthaisiriwong, Y. An implementation of three-level BTB NPC voltage source converter based-PMSG wind energy conversion system. In Proceedings of the IEEE TENCON 2014, Bangkok, Thailand, 22–25 October 2014.

15. Vilathgamuwa, D.M.; Jayasinghe, S.D.G. Rectifier systems for variable speed wind generation—A review. In Proceedings of the IEEE ISIE 2012, Hangzhou, China, 28–31 May 2012.

16. Tan, K.; Islam, S. Optimum control strategies in energy conversion of PMSG wind turbine system without mechanical sensors. *IEEE Trans. Energy Convers.* **2004**, *19*, 392–399. [CrossRef]

17. Binder, A.; Schneider, T. Permanent magnet synchronous generators for regenerative energy conversion—A survey. In Proceedings of the IEEE EPE 2005, Dresden, Germany, 11–14 September 2005.

18. Rodriguez, J.; Franquelo, L.G.; Kouro, S.; Leon, J.I.; Portillo, R.C.; Prats, M.A.M.; Perez, M.A. Multilevel converters: An enabling technology for high-power applications. *Proc. IEEE* **2009**, *97*, 1786–1817. [CrossRef]

19. Rahimi, M. Modeling, control and stability analysis of grid connected PMSG based wind turbine assisted with diode rectifier and boost converter. *Electr. Power Energy Syst.* **2017**, *93*, 84–96. [CrossRef]

20. Aziz, A.; Mto, A.; Stojcevski, A. Full converter based wind turbine generator system generic modelling: Variations and applicability. *Sustain. Energy Technol. Assess.* **2016**, *14*, 46–62. [CrossRef]

21. Kazmierkowski, M.P.; Malesani, L. Current control techniques for three-phase voltage-source PWM converters: A survey. *IEEE Trans. Ind. Electron.* **1998**, *45*, 691–703. [CrossRef]

22. Pozzebon, G.G.; Goncalves, A.F.Q.; Pena, G.G.; Mocambique, N.E.M.; Machado, R.Q. Operation of a three-phase power converter connected to a distribution system. *IEEE Trans. Ind. Electron.* **2013**, *60*, 1810–1818. [CrossRef]

23. Song, S.-H.; Kang, S.-I.; Hahm, N.-K. Implementation and control of grid connected AC-DC-AC power converter for variable speed wind energy conversion system. In Proceedings of the IEEE APEC 2003, Miami Beach, FL, USA, 9–13 February 2003.

24. Prodanovic, M.; Green, T.C. Control and filter design of three-phase inverters for high power quality grid connection. *IEEE Trans. Power Electron.* **2013**, *18*, 373–380. [CrossRef]

25. Twining, E.; Holmes, D.G. Grid current regulation of a three-phase voltage source inverter with an LCL input filter. *IEEE Trans. Power Electron.* **2003**, *18*, 888–895. [CrossRef]

26. Blaabjerg, F.; Teodorescu, R.; Liserre, M.; Timbus, A.V. Overview of control and grid synchronization for distributed power generation systems. *IEEE Trans. Ind. Electron.* **2006**, *53*, 1398–1409. [CrossRef]

27. Dai, J.C.; Liu, D.; Hu, Y.; Shen, X. Research on joint power and loads control for large scale directly driven wind turbines. *ASME J. Sol. Energy Eng.* **2013**, *136*, 021015. [CrossRef]

28. Svensson, J.; Lindgren, M. Vector current controlled grid connected voltage source converter-influence of non-linearities on the performance. In Proceedings of the IEEE PESC Record, Fukuoka, Japan, 22–22 May 1998.

29. Van der Broeck, H.W.; Skudelny, H.C.; Stanke, G.V. Analysis and realization of a pulse width modulator based on voltage space vectors. *IEEE Trans. Ind. Appl.* **1988**, *24*, 142–150. [CrossRef]

30. Chung, S.-K. Phase-locked loop for grid-connected three-phase power conversion systems. *IEE Proc. Electr. Power Appl.* **2000**, *147*, 213–219. [CrossRef]

31. Kaura, V.; Blasko, V. Operation of a phase locked loop system under distorted utility conditions. *IEEE Trans. Ind. Appl.* **1997**, *33*, 58–63. [CrossRef]

32. Nash, G. Phase-Locked Loop Design Fundamentals. Motorola Application Note AN-535 1994.

33. Hsieh, G.; Hung, J.C. Phase-Locked Loop Techniques—A Survey. *IEEE Trans. Lnd. Electron.* **1996**, *43*, 609–615. [CrossRef]

34. Chung, S.-K. A phase tracking system for three phase utility interface inverters. *IEEE Trans. Power Electron.* **2000**, *15*, 431–438. [CrossRef]

35. Amuda, L.N.; Cardoso Filho, B.J.; Silva, S.M.; Silva, S.R.; Diniz, A.S.A.C. Wide bandwidth single and three-phase PLL structures for grid-tied PV systems. In Proceedings of the IEEE PVSC, Anchorage, AK, USA, 15–22 September 2000.

36. Nozari, F.; Mezs, P.A.; Julian, A.L.; Sun, C.; Lipo, T.A. Sensorless synchronous motor drive for use on commercial transport airplanes. *IEEE Trans. Ind. Appl.* **1995**, *31*, 850–859. [CrossRef]

37. Caliskan, V.; Perreault, D.J.; Jahns, T.M.; Kassakian, J.G. Analysis of three-phase rectifiers with constant-voltage loads. *IEEE Trans. Circuits Syst.* **2003**, *50*, 1220–1225. [CrossRef]
38. Ketzer, M.B.; Jacobina, C.B. Sensorless control technique for PWM rectifiers with voltage disturbance rejection and adaptive power factor. *IEEE Trans. Ind. Electron.* **2015**, *62*, 1140–1151. [CrossRef]
39. Blasko, V.; Moreira, J.C.; Lipo, T.A. A new field oriented controller utilizing spatial position measurement of rotor end ring current. In Proceedings of the IEEE PESC Record, Milwaukee, WI, USA, 26–29 June 1989.
40. Song, H.; Nam, K. Dual current control scheme for PWM converter under unbalanced input voltage conditions. *IEEE Trans. Ind. Electron.* **1999**, *46*, 953–959. [CrossRef]
41. Limongi, L.R.; Bojoi, R.; Pica, C.; Profumo, F.; Tenconi, A. Analysis and comparison of phase locked loop techniques for grid utility applications. In Proceedings of the IEEE PCCON 2007, Nagoya, Japan, 2–5 April 2007.
42. Ko, Y.; Park, K.; Lee, K.-B.; Blaabjerg, F. A new PLL system using full order observer and PLL system modeling in a single phase grid-connected inverter. In Proceedings of the IEEE ICPE 2011, Jeju, South Korea, 30 May–3 June 2011.
43. Freijedo, F.D.; Doval-Gandoy, J.; López, Ó.; Acha, E. A generic open-loop algorithm for three-phase grid voltage/current synchronization with particular reference to phase, frequency, and amplitude estimation. *IEEE Trans. Power Electron.* **2009**, *24*, 94–107. [CrossRef]
44. Fortescue, C. Method of symmetrical coordinates applied to the solution of polyphase networks. *Trans. AIEE* **1918**, *37*, 1027–1140.
45. Yun, E.; Kim, D.; Yoon, S.Y.; Kim, K.C. Experimental investigation of an organic Rankine cycle with multiple expanders used in parallel. *Appl. Energy* **2015**, *145*, 246–254. [CrossRef]
46. Sung, T.; Yun, E.; Kim, H.D.; Yoon, S.Y.; Choi, B.S.; Kim, K.; Kim, J.; Jung, Y.B.; Kim, K.C. Performance characteristics of a 200-kW organic Rankine cycle system in a steel processing plant. *Appl. Energy* **2016**, *183*, 623–635. [CrossRef]

energies

MDPI

Article

Study of Inertia and Damping Characteristics of Doubly Fed Induction Generators and Improved Additional Frequency Control Strategy

Xiangwu Yan *, Zijun Song, Yun Xu, Ying Sun, Ziheng Wang and Xuewei Sun

Key Laboratory of Distributed Energy Storage and Micro-Grid of Hebei Province, North China Electric Power University, Baoding 071003, China; 2172213052@ncepu.edu.cn (Z.S.); xuyun2009054@163.com (Y.X.); syingmeer@163.com (Y.S.); wangzh0306@126.com (Z.W.); sun_xw@126.com (X.S.)
* Correspondence: xiangwuy@ncepu.edu.cn; Tel.: +86-139-0336-5326

Received: 20 November 2018; Accepted: 17 December 2018; Published: 23 December 2018

Abstract: Large-scale wind farms connect to the grid and deliver electrical energy to the load center. When a short-circuit fault occurs on the transmission line, there will be an excess of electric power, but the power demand will increase instantaneously once the fault is removed. The conventional additional frequency control strategies of wind farms can effectively reduce the frequency fluctuation caused by load mutation, but still there are some limitations for the frequency fluctuation caused by the whole process of occurrence, development and removal of a short-circuit fault on the transmission line. Therefore, this paper presents an improved additional frequency control strategy for wind farms. According to the variation law of system frequency during the whole process of a short-circuit fault, the proposed strategy revises the parameters in conventional additional frequency control of the doubly-fed induction generator (DFIG) to have effective damping characteristics throughout the entire process from failure to removal, thereby the output power of DFIGs could respond to frequency fluctuation rapidly. MATLAB/ Simulink is used to build a four-machine two-area model for simulation analysis. The results show that the control strategy can effectively reduce the frequency fluctuation of DFIGs, and enhance the stability of the system.

Keywords: doubly-fed induction generator; short-circuit fault; frequency regulation; variable power tracking control; improved additional frequency control; variable coefficient regulation; inertia and damping characteristics

---

1. Introduction

There is abundant wind energy in Northwest China and large-scale wind farms are connected to the grid. Wind energy is transported to the load center through long distance high voltage transmission lines. Since 2009, IEA Wind TCP member countries have increased their wind share in the energy mix at an average rate of 0.44% per year. Wind-generated electricity met almost 5.6% of the world's demand in 2017 [1]. The steady increase of wind turbines brings new challenges to the safe and stable operation of the system [2–4]. Because of the high power generation efficiency, small capacity of the converter and decoupling control of active power and reactive power, doubly-fed induction generator (DFIG) has become the main model of large-scale wind farms. DFIGs have become a dominant wind turbine (WT) technology, therefore, this paper takes DFIGs as the research object [5,6]. However, the rotor speed is decoupled from the system frequency when DFIGs adopt the converter control mode [7–9], which reduces the equivalent inertia of the system. When the permeability of wind turbines increases to a certain extent, the dynamic response ability to the system frequency will be greatly declined [10]. In fact, the rotational speed range of DFIGs is 0.7 pu~1.2 pu, and the rotational kinetic energy stored in the rotor (the blades, hubs, gearboxes and generator rotors of WTs are equivalent to one mass) is much larger than that of a synchronous machine [11–13]. If the rotational speed of DFIGs and the system frequency can be coupled, the frequency regulation ability will be greatly improved.

A certain control strategy can make the WTs have a frequency adjustment characteristic similar to that of the synchronous generator, which can control the WTs to participate in the system frequency regulation. The currently applied methods include virtual inertial control and droop control [14]. Virtual inertial control [15–17] and droop control [18–20] are additional frequency control modules in the rotor side control system of WTs. WTs usually operate in the Maximum Power Point Tracking (MPPT) mode. Therefore, due to the lack of reserve capacity of WTs, the WTs can only participate in the system frequency regulation in a short time by increasing the virtual inertia. In order to expand the time scale of frequency regulation, the common method is to increase the over speed method [21–24] or the pitch angle control method [25,26] on the basis of additional frequency control, so WTs obtain a certain reserve capacity to participate in the primary frequency regulation. Although the above methods solve the basic problem of DFIGs participating in system frequency modulation, there are still many detailed problems to be considered for additional frequency control. Due to the droop control coefficient is not easy to determine, the excess coefficient makes it difficult for the system to reach a stable state [27]. In order to avoid this problem, the relationship among the virtual inertia of DFIGs, the speed regulation and the frequency variation of the grid is discussed in [28], and the wind power tracing curve is adjusted according to the frequency fluctuation of the system. However, the scheme is slow to respond at the initial time of frequency fluctuation. Therefore, [29,30] adopt active power control instead of phase-locked loop (PLL) technology to realize synchronous operation between DFIGs and the grid. Although the response speed and the stability are improved, the control strategy proposed in the above studies only regulates the frequency fluctuation caused by load mutation and the frequency fluctuation in an event of a short-circuit fault is not taken into consideration. There is a clear difference in the frequency fluctuation trend between the load mutation and the short-circuit fault. Therefore, the above methods are not applicable to the frequency regulation when the short-circuit fault occurs.

In view of this situation, the inertia damping characteristics of DFIGs during the whole process of occurrence, development and removal in an event of a short-circuit fault are studied in this paper. On this basis, this paper proposes an improved frequency control strategy for DFIGs. The parameters of additional frequency control are modified according to the variation of the system frequency during the process from fault occurrence to complete removal. The output power of DFIGs is adjusted rapidly with the frequency fluctuation of the system, and the transient stability of the system is improved. Finally, a four-machine two-area simulation model is built in MATLAB/Simulink to verify the effectiveness of the proposed strategy. The results show that after adopting the improved additional frequency control, DFIGs can adjust the rotational speed in time to release the stored kinetic energy to increase the output power once the fault is removed. The proposed method could also effectively reduce the frequency offset, and the improved control method is suitable for different operating conditions.

2. Principle of Conventional Additional Frequency Control for DFIGs

2.1. Conventional Additional Frequency Control Strategy

The power generation and power consumption of the power system are real-time balanced. When the system frequency fluctuates greatly, synchronous generators can respond in time and release or absorb the rotational kinetic energy to damp the system frequency fluctuation, because the rotating speed is closely coupled with the system frequency. Especially in the early stage of the event, the inertia of the generator directly affects the rate of frequency fluctuation and even the stability of the system [31]. The rotor motion equation of synchronous generators is shown as Equation (1):

$$2H\omega\frac{d\Delta\omega}{dt} = P_M - P_E - D\Delta\omega \tag{1}$$

where, H is the inertia constant of the generator; $\triangle\omega = \omega - \omega_0$, ω is the actual angular velocity, ω_0 is the rated angular velocity; P_M is the mechanical power; P_E is the electromagnetic power; D is the damping coefficient.

DFIGs usually operate in MPPT mode without the capability of frequency response. When the permeability of WTs is high, in order to improve the frequency dynamic characteristics of the system, additional frequency control such as virtual inertial control is usually added to DFIGs to increase the system inertia.

The conventional additional frequency control method is to add frequency regulation auxiliary power on the basis of MPPT control of WTs. The additional power comes from the kinetic energy released or absorbed by the rotor, as shown in Figure 1.

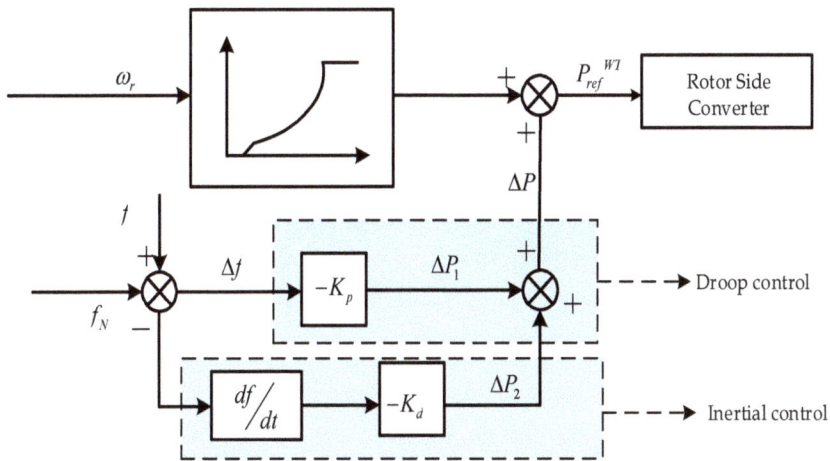

Figure 1. Conventional additional frequency control.

where ω_r is the rotor speed; P_{ref} is the active power reference value of the rotor side converter under the maximum power tracking mode; $\triangle f$ is the deviation between the system frequency f and the rated frequency f_N. Additional active power can be expressed as Equation (2):

$$\Delta P = \Delta P_1 + \Delta P_2 = -(K_p \Delta f + K_d \frac{d\Delta f}{dt}) \tag{2}$$

The form of the equation can be expressed as the follow, which is similar to the rotor motion equation of a conventional synchronous generator:

$$K_d \frac{d\Delta f}{dt} = \Delta P - K_p \Delta f \tag{3}$$

By comparing Equations (1) and (3), it can be seen that when the differential control coefficient $K_d > 0$, the rotational inertia is similar to that of a synchronous machine, so differential control is also called inertia control; When the proportional control coefficient $K_p > 0$, it can increase the damping coefficient and improve the frequency dynamic response ability, so the proportional control is also called droop control.

Because of the different feedback signals, the adjustment processes of inertial control and droop control are different. Inertial control is a transient process, which uses the rate of frequency fluctuation as the feedback signal. It is mainly used to damp the abrupt fluctuation of frequency. Therefore, it can provide a larger active power support at the initial time of the event. But near the frequency extreme point, the rate of frequency fluctuation is close to zero, and the active power support is weak.

In contrast, the additional signal of droop control is related to the frequency deviation. In most cases, it is a steady-state process and is mainly used to reduce the frequency deviation of the system. In the process of transient frequency fluctuation caused by various events in the grid, droop control mainly plays a damping role and provides strong active power support near the peak frequency, but the supporting effect is weak at the initial time, so that the control speed is slower than that of inertial control. Effective combination of the rapidity of inertial control and the persistence of droop control can make the system have good dynamic frequency characteristics.

2.2. Limitation Analysis of Conventional Additional Frequency Control under A Short-Circuit Fault

DFIGs are connected to the grid and deliver electrical energy to the load center through high voltage transmission lines. When an instantaneous three-phase short-circuit fault occurs on high voltage transmission lines, the variation of relevant parameters of DFIGs under the condition of conventional additional frequency control is shown in Figure 2.

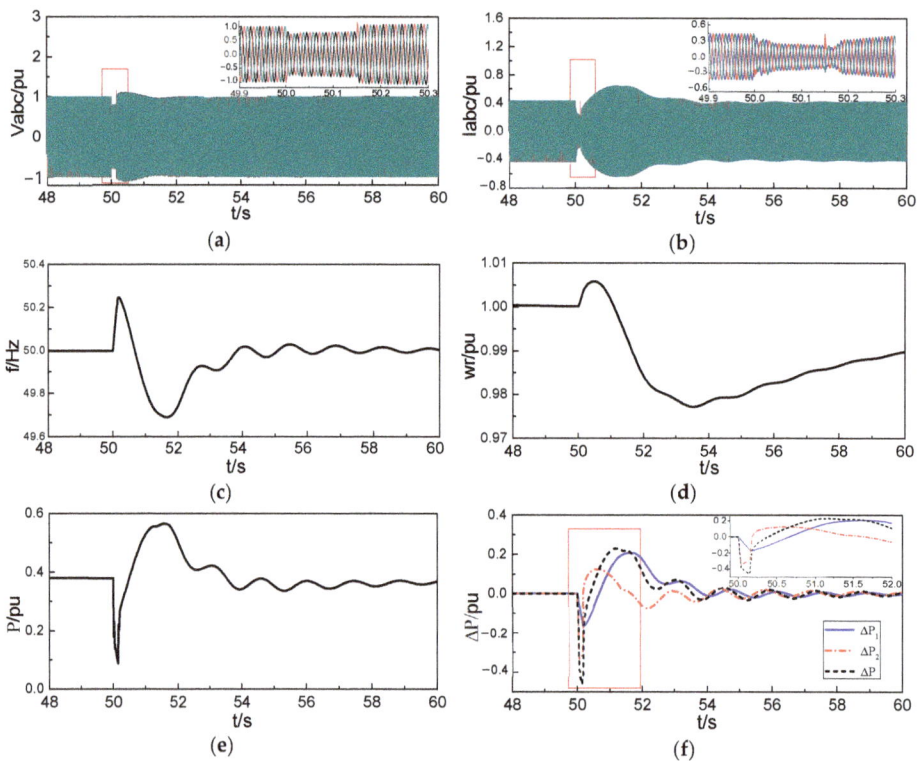

Figure 2. Variation of parameters under a short-circuit fault using conventional additional frequency control (**a**) Voltage of the DFIGs; (**b**) Current at the common connection point between the DFIGs and the grid; (**c**) System frequency; (**d**) Rotor speed of DFIGs; (**e**) Active power provided by DFIGs; (**f**) Additional power.

In Figure 2, the short-circuit fault occurs in 50 s. At 50.15 s, the fault is removed. During the period, the lowest terminal voltage of DFIGs falls to 0.75 pu, and the frequency of the system rises sharply during the fault period and falls rapidly after the fault is removed. In this paper, the frequency variation curve is obtained through the simulation in the whole process of fault occurrence, development and removal. The frequency variation trend is similar to that of the permanent-magnet direct-drive wind turbines in the short-circuit fault at the common connection point [32]. The increase or decrease of load results in the single decrease or increase of the system frequency. But the short-circuit fault will go through two processes of frequency rise and decrease continuously from occurrence to removal.

During the period of frequency rise, the additional power $\triangle$P1 and$\triangle$P2 make the reference value of active power on the rotor side smaller. This could control the acceleration of the rotor and absorb the excess active power, which can reduce the output power of the WTs to damp the system frequency rise. Inertial control plays a leading role in the initial time of the short-circuit fault and provides strong power support, while droop control plays a weak role. With the increase of frequency deviation, the additional power of droop control gradually increases. After 0.15 seconds, the fault is removed and the frequency rises to the maximum and then goes down. The active power demand of the system increases instantaneously, and the generators are required to replenish the shortage of active power in time. During this period, the power $\triangle$P2 produced by inertial control becomes positive, which helps to increase the reference value of active power on the rotor side, but the inertial support function is weak. At this time, the frequency deviation is large, and the power $\triangle$P1 produced by droop control plays a strong role. But the value is negative and the total additional power is negative, the reference value of active power on the rotor side is still low. The output electromagnetic power of WTs is less than the captured mechanical power, so the rotor keeps the acceleration trend and cannot release kinetic energy to meet the instantaneous surge of active power demand. As the frequency gradually decreases, the additional power eventually becomes positive and the rotor starts to decelerate.

3. Improved Additional Frequency Control Strategy

According to the analysis in Section 2.2, the instantaneous increase of active power demand causes the system frequency to drop dramatically after the short-circuit fault is removed. At this time, if the advantages of fast response of speed regulation of DFIGs can be brought into play, the rotating speed of DFIGs can be regulated in time to release kinetic energy. This can supplement the active power shortage of the system, and alleviate the frequency regulation pressure of synchronous generators. In order to reduce the rotating speed of DFIGs, the reference value of active power on the rotor side should be increased firstly, so that the output electromagnetic power of WTs is greater than the captured mechanical power.

As can be seen from Figure 2, after the short-circuit fault is removed, the frequency begins to decrease, and the frequency deviation starts to decrease from the positive maximum value. At this time, the droop control which plays a dominant role is still negative, so that the output electromagnetic power is still lower than the captured mechanical power. The rotor of WTs is in the acceleration trend, which not only hinders the rapid release of kinetic energy, but also enhance the rapid drop of the system frequency.

Based on the analysis of inertia damping characteristics of conventional additional frequency control strategy for DFIGs under the condition of short-circuit fault exists in Section 2.2, this section proposes an improved additional frequency control strategy to overcome the limitations. The strategy modifies the droop control coefficient according to the frequency variation during the occurrence, development and removal of a short-circuit fault. The DFIGs have effective inertia damping characteristics throughout the whole process, so that the output active power will be adjusted in time with the frequency fluctuation. The implementation method of the improved additional frequency control strategy is shown in Figure 3, and the specific implementation process is shown in Figure 4.

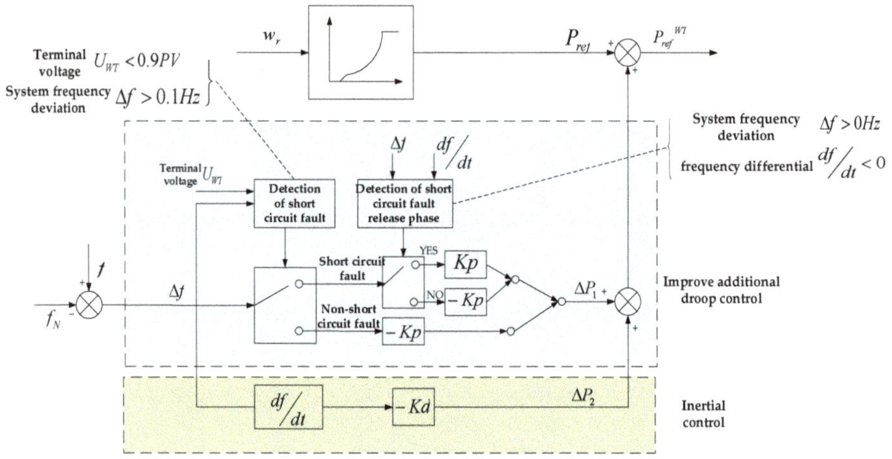

Figure 3. Improved additional frequency control.

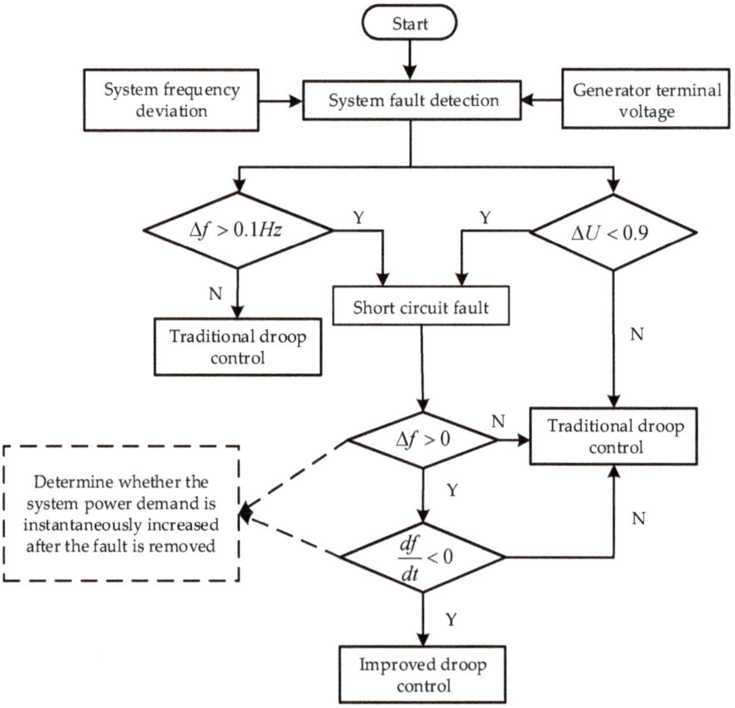

Figure 4. Implementation process of improved additional frequency control algorithm.

When the frequency fluctuates due to a short-circuit fault, the WTs adopt the improved additional frequency control strategy shown in Figure 3. The event that can cause the increase of system frequency are mainly load reduction or a short-circuit fault. Load reduction will cause a short-term increase in terminal voltage, while a short-circuit fault will cause a large drop in terminal voltage. In order to distinguish the two kinds of events, the variation of terminal voltage and the system frequency are introduced to judge whether a short-circuit fault occurs. When the terminal voltage amplitude is lower than 0.9 pu and the system frequency is higher than 50.1 Hz, the short-circuit fault of the system occurs. After the short-circuit fault is removed, the frequency starts to decrease and the frequency is still higher than the rated value, the value of additional active power produced by droop control remains unchanged when the symbol changes. Not until the frequency is below the rated value that the WTs are restored to the conventional additional frequency control. The detailed adjustment process can be seen in Figure 4.

4. Results

In this paper, a simulation model of four-machine two-area system including DFIGs is built in MATLAB/Simulink. The four-machine two-area test system of Kundur has been used to investigate the impact of the DFIGs on the power oscillation damping of the power system [33]. This system is specifically used to study low frequency electromechanical oscillation modes in the interconnected power system. The single line diagram of the two-area test system is shown in Figure 5. The two-area system is linked together by 230 kV lines of 220 km length. Each area has two round-rotor synchronous generators each with 20 KV/900 MVA rated power. All conventional synchronous generators are steam turbine driven, round-rotor synchronous generators provided with governor and excitation control. The synchronous generators are identical. The loads are modeled as constant impedances.

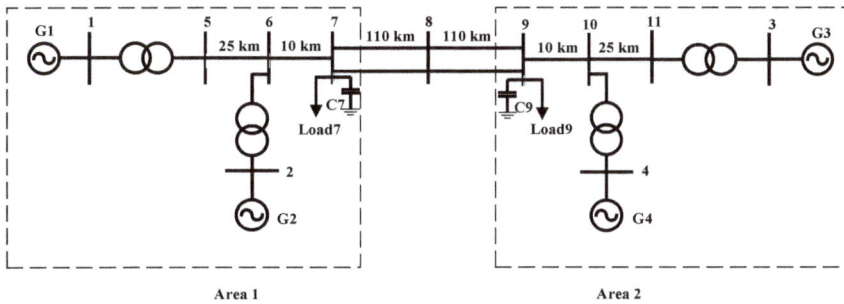

Figure 5. Four-machine two-area test system.

In order to examine the impact of DFIG-WPP on inertial damping characteristics of the power system, the synchronous generator G1 is replaced by the DFIG-based WPP as shown in Figure 6.

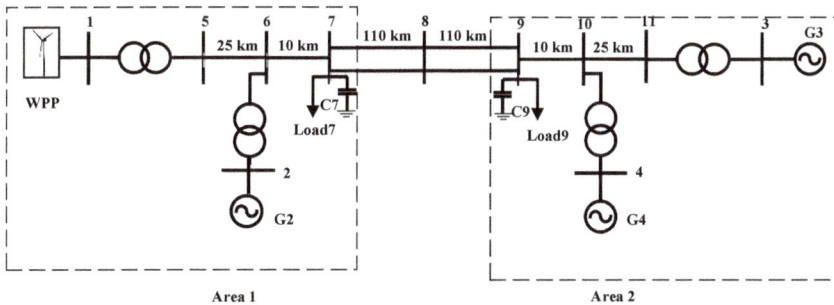

Figure 6. Two-area test system with G1 replaced by WPP.

To simplify the analysis, the above model is simplified as Figure 7a, G1 is a doubly-fed wind farm with 300 doubly-fed wind generators. G2~G4 are power plants with governor and excitation regulator. The load L1 and L2 are constant active load respectively, C1 and C2 are reactive power compensation devices and the three-phase short-circuit fault occurs on the single circuit.

The typical DFIG configuration is considered, as illustrated in Figure 7b. Its model includes the main electrical components, the mechanical and aerodynamic subsystems, and the controllers. In the converter control system, shown in Figure 7c, the rotor-side converter (RSC) controls the active and reactive power delivered to the grid, while the grid-side converter (GSC) regulates the dc-bus voltage, operating at unity power factor.

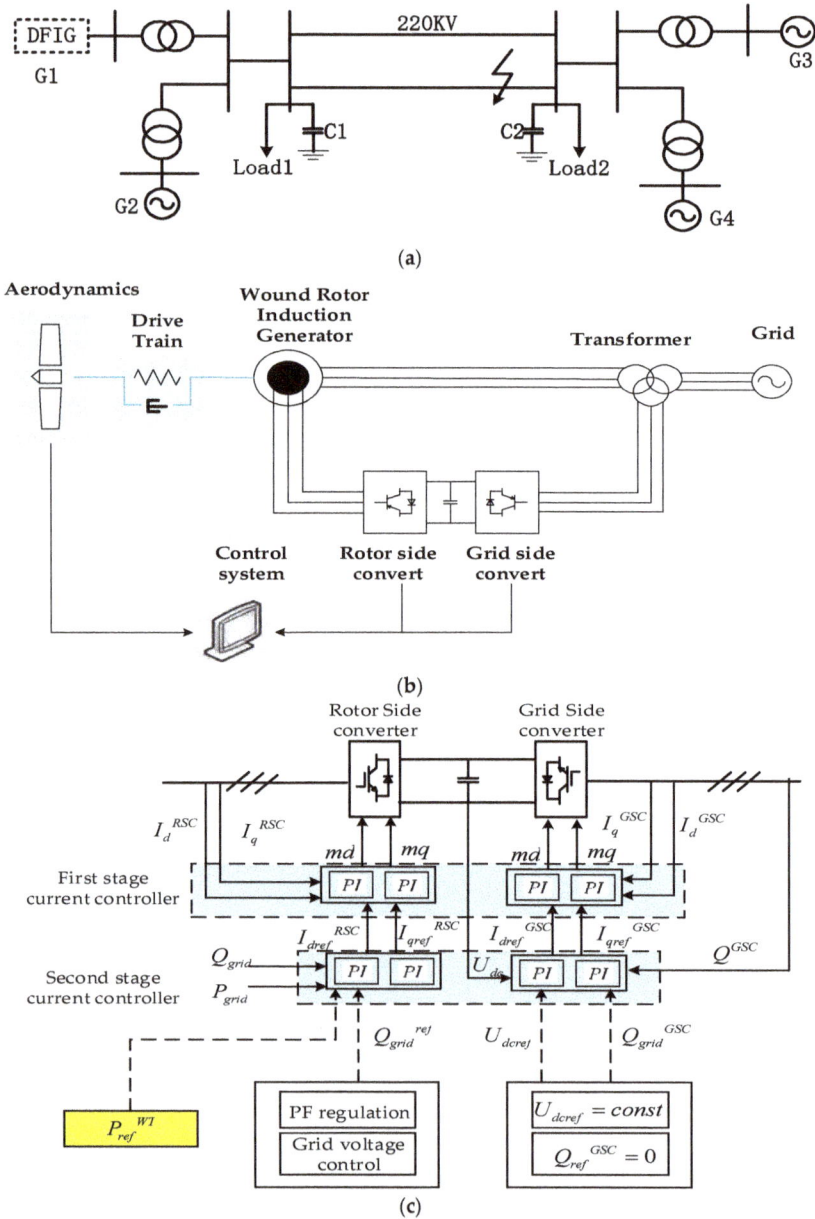

(a)

(b)

(c)

Figure 7. Simplified model of four-machine two-area system and control scheme (**a**) Simplified model of four-machine two-area system; (**b**) DFIGs configuration; (**c**) Converter control scheme for the DFIGs.

The controller structure is based on cascaded loops: a fast inner current controller regulates current to the reference values, specified by external power control loops. The control scheme are presented based on the phasor model of the DFIGs system. In this section, an outline of the models used is only presented for the sake of completeness. Details are provided in [34]. The load flow data is shown in Table 1 and the main component parameters are reported in Table 2.

Table 1. Load flow data.

Percentage Penetration of Wind (%)	Area-1		Area-2		Total Load + Losses (MW)
	Wind Generation G1 (MW)	Synchronous Machine G2 (MW)	Synchronous Machine G3 (MW)	Synchronous Machine G4 (MW)	
10	160	484	483	483	1610

Table 2. Simulation parameters.

Header Components	Parameters	Values
Wind turbine parameters	Pitch angle, β	0
	Tip speed ratio constant, λ	8.1
	Power coefficient, C_p	0.48
	Wind speed, v	10 m/s
DFIG parameters	Stator resistance, R_s	0.00706
	Stator leakage inductance, L_{ls}	0.171
	Rotor resistance, $R_{r'}$	0.005
	Rotor leakage inductance, $L_{lr'}$	0.156
	Magnetizing inductance, L_m	2.9
	Inertia constant, H	5.04
	Viscous friction factor, F	0
	Number of pole pairs, p	3
	Nominal DC bus voltage, U_{dc}	1200 V
GSC controllers parameters	DC voltage proportional gains, k_p,	0.002
	DC voltage integral gains, k_i	0.05
	GSC current proportional gains, k_p,	1
	GSC current integral gains, k_i	100
RSC controllers parameters	Power proportional gains, k_p	1,
	Power integral gains, k_i	5
	RSC current proportional gains, k_p,	0.3,
	RSC current integral gains, k_i	8
Synchronous generator parameters	The d-axis synchronous reactance, X_d	1.8
	The d-axis transient reactance, $X_{d'}$	0.3
	The d-axis subtransient reactance, $X_{d''}$	0.25
	The q-axis synchronous reactance, X_q	1.7
	The q-axis transient reactance, $X_{q'}$	0.55
	The q-axis subtransient reactance, $X_{q''}$	0.25
	The stator resistance, R_s	0.0025

At 50 seconds, a three-phase short-circuit fault occurs in one circuit of 220 kV high voltage transmission line. After the fault continues 0.15 s, the system is restored to the normal operation. In the additional frequency control, the virtual inertial control coefficient is 10.08 and the droop control coefficient is 1/0.03.

Under the three modes of no additional frequency control, conventional additional frequency control and improved additional frequency control proposed in this paper, the terminal voltage of DFIGs, the grid-connected point current, the voltage of the converter on DC side, the transmission frequency, the active power provided by WTs and the rotor speed of DFIGs are shown in Figure 8.

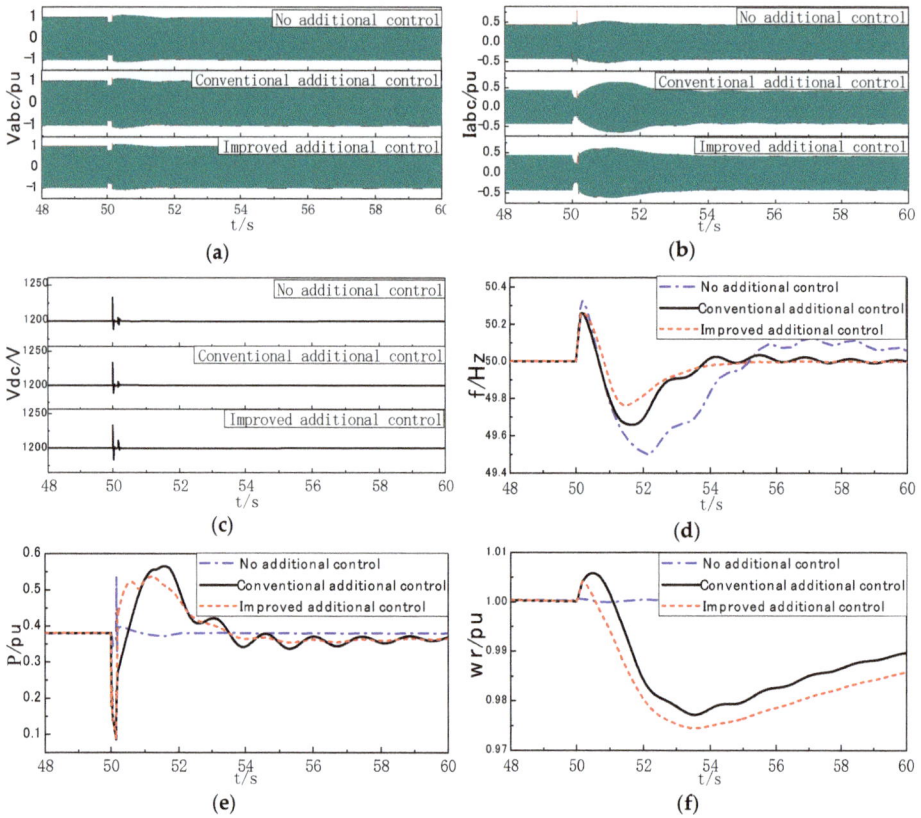

Figure 8. Variation of parameters under a short-circuit fault by three different control methods (**a**) Voltage of the DFIGs; (**b**) Current at the common connection point between the DFIGs and the grid; (**c**) DC voltage of converter; (**d**) System frequency; (**e**) Active power provided by DFIGs; (**f**) Rotor speed of DFIGs

4.1. Analysis of System Frequency Change

According to the comparative analysis of frequency variation under three different control strategies in Figure 8d, it can be concluded that when the system adopts MPPT control without additional frequency control for DFIGs, the frequency fluctuation amplitude is the largest. The maximum frequency increased to 50.32 Hz. After the fault is removed, the frequency decreases to 49.50 Hz. By contrast, the maximum frequency offset is reduced to a certain extent by using conventional frequency control in addition. The maximum frequency amplitude is reduced from 50.32 Hz to 50.26 Hz during the period of a fault, and the maximum frequency deviation is reduced by 18.7% during the period of frequency rise. After the fault removal, the minimum amplitude of frequency is improved from 49.50 Hz to 49.68 Hz. The maximum frequency deviation decreased by 36% in the period of frequency drop. The DFIGs play an obvious role of inertia support in the process of frequency fluctuation. After adopting the improved additional frequency control, the DFIGs can adjust the rotational speed in time. This can release the stored kinetic energy in the rotor after the fault is removed and resumed normal operation, and the output active power increases, which effectively reduces the frequency offset. The minimum amplitude is 49.76 Hz during frequency decline, which is 25% higher than that of conventional additional frequency control. Compared with the former two methods, we can see that this method significantly weakens the trend of frequency oscillation and improves the stability of the system in the later period of frequency regulation.

4.2. Analysis of additional power variation

Figure 9 shows the variation of the additional power of DFIGs under the three control modes.

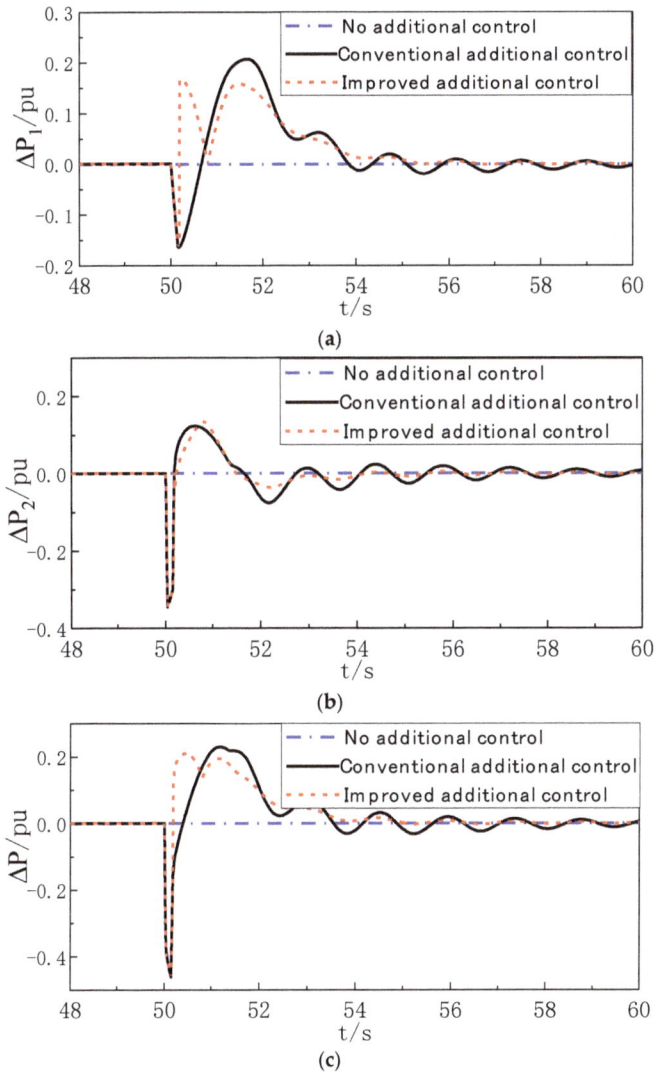

Figure 9. Change of additional power under a short-circuit fault by three different control methods (**a**) Additional active power produced by droop control of DFIGs, $\triangle P_1$; (**b**) Additional active power produced by inertial control of DFIGs, $\triangle P_2$; (**c**) Total additional power, $\triangle P$.

As shown in Figure 9, the additional power is 0 without additional frequency control. In the case of conventional additional frequency control, the analysis of additional power changes is detailed in Section 2.1. In the case of improved additional frequency control, after the short-circuit fault is removed, the frequency deviation begins to fall from the maximum positive value to zero. During this period, the additional power $\triangle P_1$ produced by droop control is changed from positive to negative. The total additional power is increased from -0.1 pu of the conventional additional to 0.2 pu at the time of the fault removal. The increase of active power reference value enables the DFIGs to adjust the kinetic energy and the rotational speed timely and increase the output active power.

4.3. Analysis of Output Active Power of DFIGs

By comparing and analyzing the change of output active power of DFIGs in the process of frequency mutation in Figure 8e, it can be seen that when the DFIGs operate in MPPT mode without additional frequency control, the output active power fluctuates in a small range near 0.38 pu, and almost has no response to system frequency. With conventional additional frequency control, the output power of DFIGs can rapidly follow the change of active power reference value through additional power signal after a short-circuit fault occurs, and instantaneously reduces to 0.13 pu. The reduction of output power leads to the increase of speed, which makes the WTs deviate from the MPPT mode and overspeed. The reserve of rotor kinetic energy is increased. After the fault is removed, the additional power signal makes the output power instantly rise to 0.28 pu, which is still lower than the mechanical power. It is unable to meet the instantaneous active power demand of the system. Then the output active power gradually increases until the electromagnetic power is greater than the captured mechanical power. The speed begins to drop and the power is released. When the additional frequency control is improved, after the fault is removed, the additional power signal makes the output power rise to 0.43 pu instantaneously, which is larger than the captured mechanical power. The rotational speed of the WTs decreases and releases kinetic energy to replenish the active power in time.

4.4. Analysis of Speed Variation of DFIGs

Through the change of rotational speed of DFIGs in Figure 8f, we can see that when DFIGs have no additional frequency control, in order to maintain the optimal tip speed ratio, the rotational speed only adjusts with the change of wind speed, and cannot respond to the frequency change of the system. The rotational speed is always maintained at 1pu. Under the condition of conventional additional frequency control, the WTs can adjust the rotational speed of the system to response to the frequency change of the system. The rotational speed of DFIGs increases with the increase of the system frequency and absorbs excess active power in the period of a fault. However, after the fault is removed and returns to normal operation, the frequency shows a downward trend, but still higher than the rated value. Droop control plays a leading role, and the additional power is related to frequency deviation. In a short period after the fault is removed, the value is opposite to the frequency change rate. The rotational speed of the WTs still has an upward trend, which is not conducive to the rapid stability of the system frequency. Aiming at the problems of conventional additional frequency control, improving additional frequency control can make the rotational speed of WTs adjust rapidly according to the change trend of system frequency. From the figure, it can be seen that the improved additional frequency control can reduce the rotational speed in time after the frequency starts to decrease, and release kinetic energy to provide effective support for the system.

4.5. Analysis at Different Wind Speeds

In order to verify the effectiveness of the improved strategy, this section performs simulation analysis under different wind speeds, which is to prove that the improved strategy can effectively improve the frequency regulation of the system under different operating points. Figure 10 shows the variation of the system frequency, the output active power of DFIGs, the additional power of DFIGs and the rotational speed of DFIGs under three different control modes in the wind speed of 8 m/s. and Figure 11 shows the variation in the wind speed of 11 m/s.

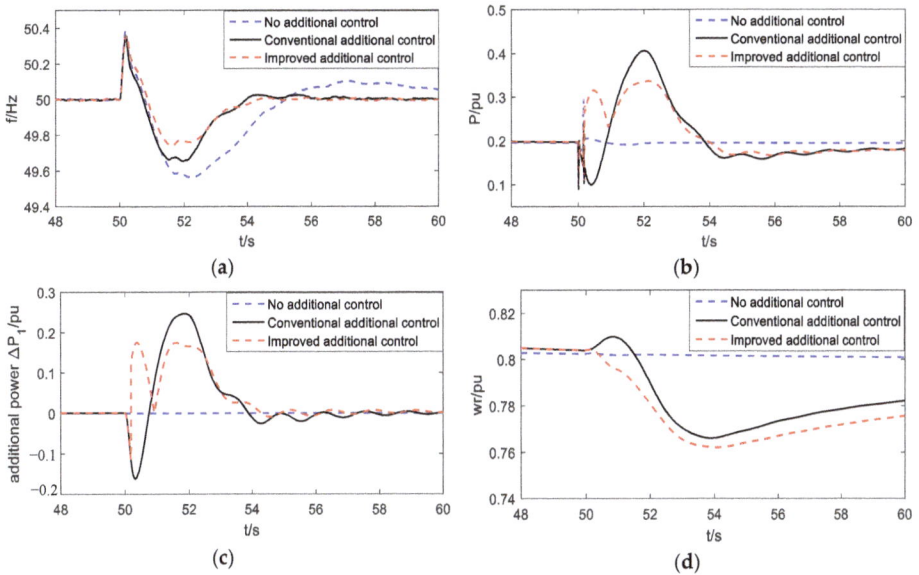

Figure 10. Change of parameters under a short-circuit fault by three different control methods in the wind speed of 8 m/s (**a**) System frequency; (**b**) Output active power of DFIGs; (**c**) Additional power of DFIGs; (**d**) Rotational speed of DFIGs.

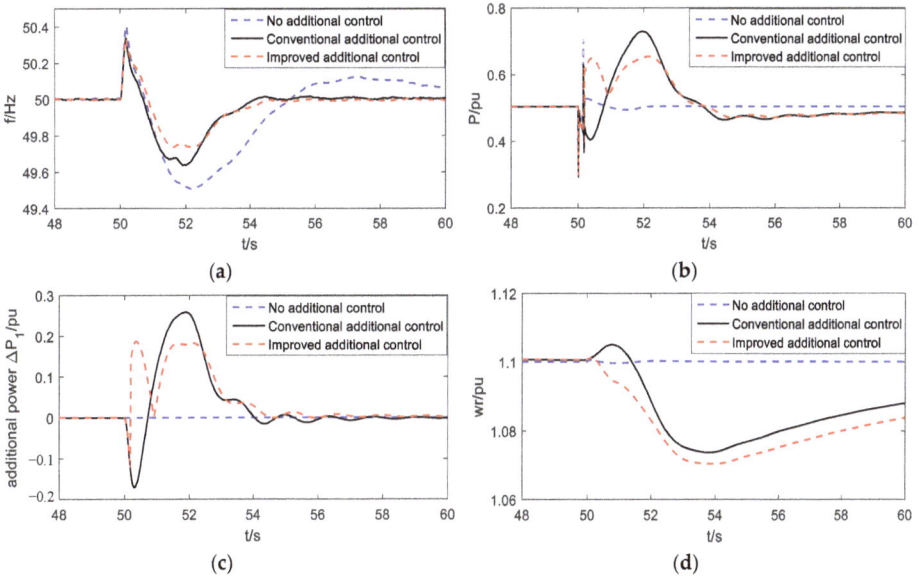

Figure 11. Change of parameters under a short-circuit fault by three different control methods in the wind speed of 11 m/s (**a**) System frequency; (**b**) Output active power of DFIGs; (**c**) Additional power of DFIGs; (**d**) Rotational speed of DFIGs.

It can be seen from Figures 10 and 11 that the variation trends of each variable are similar under different wind speed conditions. After adopting the improved additional frequency control, the DFIGs can adjust the rotational speed in time to release the stored kinetic energy in the rotor after the fault is removed and resumed normal operation, and the output active power increases, which effectively reduces the frequency offset. Comparing with the former two methods, we can see that this method significantly weakens the trend of frequency oscillation and improves the stability of the system in the later period of frequency regulation. Therefore the improved control method is suitable for different operating conditions.

5. Conclusions

The large-scale grid connection of DFIGs has a significant influence on the stability of the power system. The additional frequency control strategy of DFIGs has a vital effect on the frequency fluctuation caused by the sudden change of load, but the control has some limitations on the frequency fluctuation caused by the short-circuit fault. In this paper, the variation law of the system frequency during the whole process from the occurrence to the removal of the short-circuit fault has been deeply studied, and the parameters in the additional frequency control of the DFIGs have been modified, so that the output power of the DFIGs can be quickly adjusted with the change of the system frequency. And the transient stability of the system is improved. The experimental results have shown that after adopting the improved additional frequency control, the DFIGs release the stored kinetic energy after the fault removal and resumes normal operation, which effectively reduces the frequency offset, and the improved control method is suitable for different operating conditions, which has a highly applicability to the frequency regulation of wind power grid-connected systems in the future.

Author Contributions: X.Y. proposed the research direction, based on the limitation of conventional additional frequency control proposed improved additional frequency control strategy. Z.S. completed the improved additional frequency control strategy algorithm. X.Y., Z.S., Y.X., Y.S., Z.W. and M.W. performed the verification and analyzed the results. Z.S. wrote the paper.

Acknowledgments: This paper was supported by the Natural Science Foundation of Hebei Province (E2018502134); Liaoning Power Grid Corporation's 2018 Science and Technology Project "Research on the Reactive Power and Voltage Optimization Strategy and Evaluation Index Considering Source and Load Fluctuation Characteristics".

Conflicts of Interest: The authors declare no conflict of interest.

References

1. International Energy Agency Wind TCP 2017 Annual Report. September 2018. Available online: https://community.ieawind.org/publications/ar (accessed on 7 December 2018).
2. Ayodele, T.R.; Jimoh, A.; Munda, J.L.; Tehile, A.J. Challenges of Grid Integration of Wind Power on Power System Grid Integrity: A Review. *Int. J. Renew. Energy Res.* **2012**, *2*, 618–626.
3. Kowli, A.; Gross, G. Evaluation of the Impacts of Deep Penetration of Wind Resources on Transmission Utilization and system stability. *IEEE Power Energy Mag.* **2010**, 10–12. [CrossRef]
4. Zhang, G.; Yang, J.; Sun, F. Primary Frequency Regulation Strategy of DFIG Based on Virtual Inertia and Frequency Droop Control. *Trans. China Electrotech. Soc.* **2017**, *32*, 225–232.
5. He, Y.; Hu, J. Several Hot-spot Issues Associated With the Grid-connected Operations of Wind-turbine Driven Doubly Fed Induction Generators. *Proc. CSEE* **2012**, *32*, 1–15.
6. Rini Ann Jerin, A.; Kaliannan, P.; Subramaniam, U.; Shawky El Moursi, M. Review on FRT solutions for improving transient stability in DFIG-WTs. *IET Renew. Power Gener.* **2018**, *12*, 1786–1799. [CrossRef]
7. Li, D.D.; Chen, C. Decoupled control of speed and reactive power of doubly-fed induction generator. In Proceedings of the 2004 International Conference on Power System Technology, Singapore, 21–24 November 2004; Volume 1, pp. 21–24.
8. Abad, G.; Rodriguez, M.A.; Iwanski, G.; Poza, J. Direct Power Control of Doubly Fed Induction Generator based Wind Turbines under Unbalanced Grid Voltage. *IEEE Trans. Power Electron.* **2010**, *25*, 442–452. [CrossRef]
9. Jiang, J.; Chao, Q.; Chen, J.; Chang, X. Simulation study on frequency response characteristic of different wind turbines. *Renew. Energy Resour.* **2010**, *28*, 24–28.
10. Tang, X.; Miao, F.; Qi, Z. Survey on Frequency Control of Wind Power. *Proc. CSEE* **2014**, *34*, 4304–4314.

11. De Haan, S.W.H.; Morren, J.; Ferreira, J.A.; Kling, W.L. Wind Turbines Emulating Inertia and Supporting Primary Frequency Control. *IEEE Trans. Power Syst.* **2006**, *21*, 433–434.

12. Xue, Y. Wind power grid-connected operation control and frequency regulation research. *Shanghai Jiaotong Univ.* **2013**.

13. Xue, Y.; Yan, N.; Liu, L.; Yang, X.; Jin, N.; Xiong, N. A New Method for Frequency Response of Doubly-Fed Wind Turbines Participating in System. *High Volt. Eng.* **2009**, *35*, 2839–2845.

14. Fan, L.; Miao, Z.; Osborn, D. Wind Farm with HVDC Delivery in Inertial and Primary Frequency Response. *Green Energy Technol.* **2012**, *78*, 465–483.

15. Tian, X.; Wang, W.; Chi, Y. Variable Parameter Virtual Inertia Control Based on Effective Energy Storage of DFIG-based Wind Turbines. *Autom. Electr. Power Syst.* **2015**, *39*, 20–26, 33.

16. Kayikci, M.; Milanovic, J.V. Dynamic Contribution of DFIG-Based Wind Plants to System Frequency Disturbances. *IEEE Trans. Power Syst.* **2009**, *24*, 859–867. [CrossRef]

17. Mauricio, J.M.; Marano, A.; Gomez-Exposito, A. Frequency Regulation Contribution Through Variable-Speed Wind Energy Conversion Systems. *IEEE Trans. Power Syst.* **2009**, *24*, 173–180. [CrossRef]

18. Vidyanandan, K.V.; Senroy, N. Primary Frequency Regulation by Deloaded Wind Turbines Using Variable Droop. *IEEE Trans. Power Syst.* **2013**, *28*, 837–846. [CrossRef]

19. You, R.; Chai, J.; Sui, X. Variable Speed Wind Turbine Micro-grid Frequency Regulation Control Based on Variable Droop. *Proc. CSEE* **2016**, *36*, 6751–6758.

20. Pan, W.; Quan, R.; Wang, F. A Variable Droop Control Strategy for Doubly-fed Induction Generators. *Autom. Electr. Power Syst.* **2015**, *39*, 126–131, 186.

21. Chang-Chien, L.R.; Lin, W.T.; Yin, Y.C. Enhancing Frequency Response Control by DFIGs in the High Wind Penetrated Power Systems. *IEEE Trans. Power Syst.* **2011**, *26*, 710–718. [CrossRef]

22. Wu, Z.; Yu, J.; Peng, X. DFIG's Frequency Regulation Method only for High Wind Speed with Suboptimal Power Tracking. *Trans. China Electrotech. Society.* **2013**, *28*, 112–119.

23. Ding, L.; Yin, S.; Wang, T. Integrated Frequency Control Strategy of DFIGs Based on Virtual Inertia and Over-Speed Control. *Power Syst. Technol.* **2015**, *39*, 2385–2391.

24. Zhao, J.; Lv, X.; Fu, Y. Frequency Regulation of the Wind/Photovoltaic/Diesel Microgrid Based on DFIG Cooperative Strategy with Variable Coefficients Between Virtual Inertia and Over-speed Control. *Trans. China Electrotech. Soc.* **2015**, *30*, 59–68.

25. Zhao, J.; Lv, X.; Fu, Y. Dynamic Frequency Control Strategy of Wind Photovoltaic Diesel Microgrid Based on DFIG Virtual Inertia Control and Pitch Angle Control. *Proc. CSEE* **2015**, *35*, 3815–3822.

26. Fu, Y.; Wang, Y.; Zhang, X. Analysis and Integrated Control of Inertia and Primary Frequency Regulation for Variable Speed Wind Turbines. *Proc. CSEE* **2014**, *34*, 4706–4716.

27. Zhou, T.; Sun, W. Study on Virtual Inertia Control for DFIG-based Wind Farms with High Penetration. *Proc. CSEE* **2017**, *37*, 486–496.

28. Li, H.; Zhang, X.; Wang, Y. Virtual Inertia Control of DFIG-based Wind Turbines Based on the Optimal Power Tracking. *Proc. CSEE* **2012**, *32*, 32–39.

29. Wang, S.; Hu, J.; Yuan, X. Virtual Synchronous Control for Grid-Connected DFIG-Based Wind Turbines. *IEEE J. Emerg. Sel. Top. Power Electron.* **2015**, *3*, 932–944. [CrossRef]

30. Wang, S.; Hu, J.; Yuan, X. On Inertial Dynamics of Virtual-Synchronous-Controlled DFIG-Based Wind Turbines. *IEEE Trans. Energy Convers.* **2015**, *30*, 1691–1702. [CrossRef]

31. Ding, L.; Yin, S.; Wang, T. Active Rotor Speed Protection Strategy for DFIG-based Wind Turbines with Inertia Control. *Autom. Electr. Power Syst.* **2015**, *39*, 29–34, 95.

32. Han, G.; Zhang, C.; Cai, X. Mechanism of Frequency Instability of Full-Scale Wind Turbines Caused by Grid Short-circuit Fault and Its Control Method. *Trans. China Electrotech. Soc.* **2017**, 1–9. [CrossRef]

33. Kundur, P. *Power System Stability and Control*; Balu, N.J., Lauby, M.G., Eds.; McGraw-Hill: New York, NY, USA, 1994; Volume 7.

34. Han, Y. Control Strategies and Digital Simulation of DFIG-based Wind Generators for Transient Stability Studies. *Przegląd Elektrotechniczny* **2013**, *89*, 43–46.

energies

MDPI

Article

Modeling and Analysis of a PCM-Controlled Boost Converter Designed to Operate in DCM

Teuvo Suntio

Laboratory of Electrical Energy Engineering, Tampere University of Technology, 33720 Tampere, Finland;
teuvo.suntio@tut.fi; Tel.: +358-400-828-431

Received: 15 November 2018; Accepted: 17 December 2018; Published: 20 December 2018

Abstract: Peak current-mode (PCM) control has been a very popular control method in power electronic converters. The small-signal modeling of the dynamics associated with PCM control has turned out to be extremely challenging. Most of the modeling attempts have been dedicated to the converters operating in continuous conduction mode (CCM) and just a few to the converters operating in discontinuous operation mode (DCM). The DCM modeling method published in 2001 was proven recently to be very accurate when applied to a buck converter. This paper provides the small-signal models for a boost converter and analyses for the first time its real dynamic behavior in DCM. The objectives of this paper are as follows: (i) to provide the full-order dynamic models for the DCM-operated PCM-controlled boost converter; (ii) to analyze the accuracy of the full and reduced-order dynamic models; and iii) to verify the validity of the high-frequency extension applied in the DCM-operated PCM-controlled buck converter in the case of the boost converter. It is also shown that the DCM-operated boost converter can operate only in even harmonic modes, similar to all the CCM-operated PCM-controlled converters. In the case of the DCM-operated PCM-controlled buck converter, its operation in the odd harmonic modes is the consequence of an unstable pole in its open-loop power-stage dynamics.

Keywords: boost converter; peak-current-mode control; dynamic modeling; discontinuous operation mode

1. Introduction

The concept of peak current-mode (PCM) control was launched publicly in 1978 [1,2], and it has become a very popular control method in DC–DC converters. The popularity is a consequence of the properties it provides, such as virtually first-order control dynamics, inherent overcurrent limiting of the switching elements, and high input-to-output voltage–noise attenuation in buck-derived converters [3]. The dynamic modeling of PCM control has turned out to be quite challenging. A large number of different modeling attempts has been published for the converters operating in continuous conduction mode (CCM), as discussed in [4], but just a few for the converters operating in discontinuous conduction mode (DCM), such as [5–9]. The DCM models in [5] (pp. 478–480) were given implicitly as equivalent circuits for buck, boost, and buck-boost converters, but the low-frequency control-to-output-voltage transfer function was given explicitly in a generalized form applicable to the named converters, which seemed to predict quite well the location of the load-resistor-affected low-frequency pole in the case of a buck converter [9]. The models are load-resistor affected, and the effect cannot be removed to obtain the required unterminated models. The dynamic models in [6] were derived assuming that there was no internal feedback from the inductor current even if it was used for generating the duty ratio. The author of [6] promoted earlier, in the case of CCM operation, the influence of the sampling effect on the dynamic behavior of PCM-controlled converters but forgetting it in the case of DCM-operated converters. The basic assumption in the modeling method in [6] was wrong, and therefore, the dynamic models were inaccurate as well. A discrete-time modeling method was applied to PCM-controlled DCM-operated buck converter in [7], but no explicit dynamic models were given. The modeling method proposed in [8] in the early 2000s and applied to buck converters was proven to be very accurate in [9]. The dynamic models in [8] were load-resistor affected, which hid the true, unterminated dynamic behavior, as discussed in [9,10]. The reduced-order (i.e., no parasitic elements are considered), unterminated small-signal state spaces applicable to DCM-operated PCM control in buck, boost, and buck-boost converters were also given in [11] (pp. 222–224), but the transfer functions were not solved for comparison.

It was recently proven that the small-signal models of PCM control in CCM contain infinite duty-ratio gain at the mode limit (i.e., the maximum duty ratio after which the converter enters into the harmonic mode of operation), which forces the converter to enter into the second-harmonic mode of operation, as discussed and demonstrated explicitly in [4]. The modeling method introduced in [8] also includes the infinite duty-ratio gain, which takes place at the boundary between the DCM and CCM operations. This implies that the existence of the infinite duty-ratio gain at the mode limit between the operations at the switching frequency and its harmonics are characteristic features of PCM control.

It is widely assumed that PCM-controlled converters are very sensitive to bifurcation phenomena and chaotic operation regardless of the operation mode (i.e., CCM or DCM) [12,13]. The harmonic operation modes are assumed to be the consequence of the bifurcation phenomena, which can take place in open and closed conditions as well. The infinite duty-ratio gain at the mode limit forces the converter to enter into the harmonic operation mode, as discussed explicitly in [4,9]. The harmonic operation mode can also take place as a consequence of a high-switching frequency ripple applied to the duty-ratio process, as discussed in [12,13]. Actually, a proper controller design will eliminate the appearance of the harmonic modes; when the control bandwidth is limited to 1/5th of the switching frequency, the controller is provided with a high-frequency noise filtering, and the operation is limited to the duty ratios less than the mode-limit duty ratio, as discussed and demonstrated in [4]. Both Reference [12] (plain proportional controller) and Reference [13] (proportional-integral (PI) controller without high-frequency pole) demonstrate the appearing of the bifurcation phenomena by using highly impractical controllers.

The main objective of this paper is to provide a comprehensive analysis of the dynamic behavior of a PCM-controlled boost converter operating in DCM including all the relevant parasitic circuit elements. The corresponding analyses are not published earlier in the literature. The investigations of this paper show clearly that the modeling technique introduced in [8] will also accurately predict the dynamic behavior of the DCM-operated PCM-controlled boost converter, when the relevant parasitic elements are considered [14] and the load-resistor effect is removed [10]. In addition, the boost converter is shown to operate only in even harmonic modes, where the averaged duty ratio equals approximately the duty ratio in boundary conduction mode (BCM). The validation of the proposed models was performed in a simulated environment, where all the components with parasitic elements were exactly known, and there were no extra source or load interactions.

The rest of the paper is organized as follows: Section 2 introduces the modeling method briefly and provides the relevant full and reduced-order state spaces, as well as the corresponding explicit transfer functions. Section 3 provides the validation of the developed small-signal models by utilizing Matlab™ Simulink-based switching models [11] (pp. 279–291) and the pseudorandom-binary-sequence-based frequency-response measurement technique introduced in [15,16]. The conclusions are presented in Section 4.

2. Modeling of PCM Control in DCM

The PCM small-signal state space can be obtained from the corresponding state space of a direct-duty-ratio (DDR) or voltage-mode (VM) controlled converter [4,11]. The transformation can be performed by replacing the perturbed duty ratio ($\hat{d}$) with the duty-ratio constraints given in Equation

(1), where F_m denotes the duty-ratio gain, x_c is the control variable, x_i is the state and input variable, and q_i is the different feedback and feedforward gains related to x_i.

$$\hat{d} = F_m \left(\hat{x}_c - \sum_{i=1}^{n} q_i \hat{x}_i \right). \tag{1}$$

The DCM dynamic modeling under DDR control was established in the late 1990s [17], and it was later elaborated into a more convenient form in [14], providing the possibility of adding the effect of parasitic circuit elements and performing mixed-conduction-mode modeling as well. The models in [15] were load-resistor affected, and therefore, the modeling technique introduced in [14] is applied in this paper to obtain the unterminated state space, which accurately represents the DDR control dynamics in DCM. The power stage of the boost converter is given in Figure 1 with the open-loop PCM control system that is analyzed in this paper. The selection of the inductor was performed in such a manner that the converter would operate in DCM, when the duty ratio varied from approximately 0.039 to 0.786. The given duty-ratio range can be computed based on $K_{crit} = DD'^2$ and $K = 2L/T_s/R_L$ (i.e., $K_{crit} = K$), as instructed in detail in [5] (pp. 107–125) (Note: The method based on the application of K_{crit} can yield quite inaccurate values for the corresponding minimum and maximum duty ratios, because it omits the effect of the circuit parasitic elements).

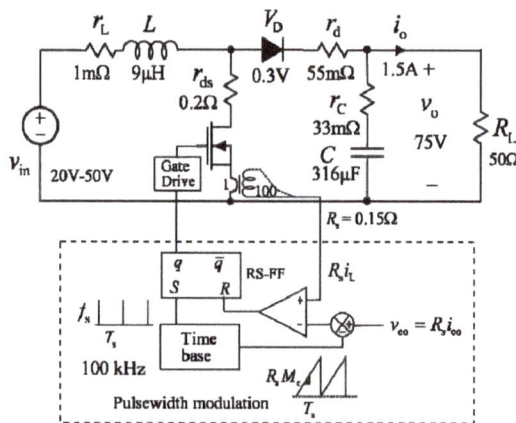

Figure 1. The power stage of the peak current-mode (PCM)-controlled boost converter in open loop with definitions of the component values and the operational conditions.

In principle, the accuracy of the DDR small-signal models in DCM is important for obtaining accurate PCM small-signal models due to the method used to develop the latter models. The frequency response of the control-to-output-voltage transfer function of the DDR-controlled boost converter in DCM (cf. Figure 2) is extracted from the Simulink-based switching model by applying the pseudorandom-binary-sequence method described in [15,16]. The construction of the Simulink-based switching models is described in detail in [11] (pp. 279–291).

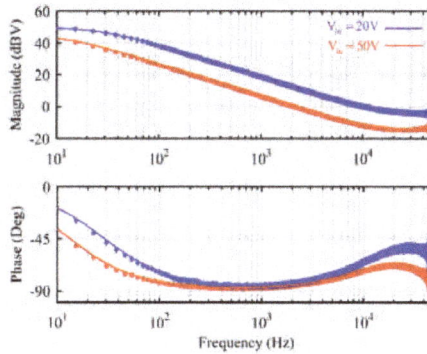

Figure 2. The frequency responses of the control-to-output-voltage transfer function of the direct-duty-ratio (DDR)-controlled boost converter at the input voltages of 20 V (red) and 50 V (blue). The solid lines denote the predicted responses, and the squares (20 V) and diamonds (50 V) denote the simulated responses.

Figure 2 shows explicitly that the predicted (solid lines: red at $V_{in} = 20$ V and blue at $V_{in} = 50$ V) and simulated (squares marked at 20 V and diamonds marked at 50 V) frequency responses matched each other perfectly. The predictions were based on the full-order models, which were computed according to the method introduced in [12]. The used modulator gain (i.e., $1/V_M$) equaled $1/1$ V (cf. [5]).

2.1. Small-Signal State Space of DDR-Controlled Boost Converter in DCM

The average state space of the DDR-controlled boost converter can be derived from Figure 1 by applying the methods described in detail in [14], which yields the following:

$$\frac{d\langle i_L \rangle}{dt} = \frac{d((R_1 - R_2)\langle i_L \rangle + \langle v_C \rangle - r_C \langle i_o \rangle + V_D)}{L} - \frac{2\langle i_L \rangle}{dT_s} \cdot \frac{R_1 \langle i_L \rangle + \langle v_C \rangle - r_C \langle i_o \rangle + V_D - \langle v_{in} \rangle}{\langle v_{in} \rangle - R_2 \langle i_L \rangle}$$

$$\frac{d\langle v_C \rangle}{dt} = \frac{\langle i_L \rangle}{C} - \frac{d^2 T_s}{2LC}(\langle v_{in} \rangle - R_2 \langle i_L \rangle) - \frac{\langle i_o \rangle}{C}$$

$$\langle i_{in} \rangle = \langle i_L \rangle \tag{2}$$

$$\langle v_o \rangle = \langle v_C \rangle + r_C C \frac{d\langle v_C \rangle}{dt}$$

$$R_1 = r_L + r_d + r_C \quad R_2 = r_L + r_{ds}$$

The corresponding small-signal state space can be derived from Equation (2) by linearizing the averaged state space at a certain operating point by applying a partial-derivative-based method, which was introduced in detail in [11] (pp. 60–61). This procedure yields the following:

$$\frac{d\hat{i}_L}{dt} = -\frac{A_1}{L}\hat{i}_L - \frac{A_2}{L}\hat{v}_C + \frac{A_3}{L}\hat{v}_{in} + \frac{A_4}{L}\hat{i}_o + \frac{V_e}{L}\hat{d}$$

$$\frac{d\hat{v}_C}{dt} = \frac{(1 + B_1 R_3)}{C}\hat{i}_L - \frac{B_1}{C}\hat{v}_{in} - \frac{\hat{i}_o}{C} - \frac{I_e}{C}\hat{d} \tag{3}$$

$$\hat{i}_{in} = \hat{i}_L$$

$$\hat{v}_o = \hat{v}_C + r_C C \frac{d\hat{v}_C}{dt}$$

where A_{1-4}, B_1, V_e, and I_e are given in Equation (4), as well as V_{1-3} and $R_{1,2}$ in Equation (5), respectively.

$$A_1 = D(R_2 - R_1) + \frac{2L}{DT_s}\left(\frac{V_1 + R_1 I_L}{V_3} + \frac{R_2 I_L V_1}{V_3^2}\right)$$

$$A_2 = -D + \frac{2LI_L}{DT_s}\frac{1}{V_3} \quad A_3 = \frac{2LI_L}{DT_s} \cdot \frac{V_2}{V_3^2}$$

$$A_4 = \left(\frac{2I_L L}{DT_s} \cdot \frac{1}{V_3} - D\right)r_C \quad B_1 = \frac{D^2 T_s}{2L} \tag{4}$$

$$V_e = V_2 + \frac{2LI_L V_1}{D^2 T_s V}\quad I_e = \frac{DT_s V_3}{L}$$

$$V_1 = V_o - V_{in} + V_D + R_1 I_L - r_C I_o$$
$$V_2 = V_o + V_D + (R_1 - R_2)I_L - r_C I_o \quad V_3 = V_{in} - R_2 I_L \ . \tag{5}$$
$$R_1 = r_L + r_d + r_C \quad R_2 = r_L + r_{ds}$$

The operating points (i.e., I_L and D) with the parasitic circuit elements can be computed by the following:

$$I_L = \frac{I_o + \frac{D^2 T_s}{2L}V_{in}}{1 + \frac{D^2 T_s}{2L}R_2} \tag{6}$$

$$f(D) = a_4 D^4 + a_2 D^2 + a_0 = 0$$

where

$$a_4 = -\frac{T_s^2}{2L}\left(V_{in}I_o R_2(R_1 - R_2) + V_4 I_o R_2^2\right)$$

$$a_2 = -T_s\left(R_2(R_1 - R_2)I_o^2 + V_{in}I_o R_1 - V_{in}^2 + 2V_4 I_o R_2\right) \tag{7}$$

$$a_0 = 2LI_o(V_{in} - V_4 - R_1 I_o)$$

$$V_4 = V_o + V_D - r_C I_o$$

and without the parasitic circuit elements by

$$I_L = MI_o$$
$$D = \sqrt{KM(M-1)} \tag{8}$$

where $M = V_o/V_{in}$. The duty ratio (D) in Equation (6) (i.e., $f(D) = 0$) can be solved by Matlab^{TM} as follows:

$$D = \min(\text{abs}(\text{roots}(f(D)))). \tag{9}$$

2.2. Averaged Comparator Equation for PCM-Controlled DCM Boost Converter

The development of the generalized form of the duty-ratio constraints, which is applicable to the second-order converters, was given explicitly in [9]. The comparator equation applicable to the conventional boost converter can be given by the following:

$$\langle i_{co}\rangle - m_c d T_s = \langle i_L\rangle + m_1 d T_s\left(1 - \frac{d}{2}\cdot\frac{m_1 + m_2}{m_2}\right) \tag{10}$$

where i_{co} denotes the control current (cf. Figure 1: $R_s i_{co}$), d denotes the duty ratio, m_1 and m_2 denote the inductor-current up and down slopes as absolute values, and m_c denotes the inductor-current compensation slope (cf. Figure 1: $R_s M_c$), respectively.

The averaged comparator equation in Equation (10) can be given as a function of D in a steady state with $M_c = 0$ [11] (p. 199) according to Equation (10) by the following:

$$D^2 - \frac{2M_2}{M_1 + M_2}\cdot D + \frac{2M_2\Delta I_L}{M_1(M_1 + M_2)T_s} = 0 \tag{11}$$

where $\Delta I_L = I_{co} - I_L$. Equation (11) can be developed further in terms of M and K (i.e., $R_{eq} = R_L$), when the inductor-current slopes M_1 and M_2 are substituted with their physical values (Note: the parasitic circuit elements are omitted) corresponding to the actual converter as follows:

$$2D - \frac{KR_L I_{co}}{V_{in}} = 0 \tag{12}$$

which is applicable for boost and buck-boost converters. The final form of Equation (12) can be obtained for a boost converter by substituting D with $\sqrt{KM(M-1)}$ (cf. (8)) yielding

$$M^2 - M - K\left(\frac{I_{co}R_L}{2V_{in}}\right)^2 = 0. \tag{13}$$

Equation (13) has two real roots, which means that there are no right-half-plane (RHP) poles in the open-loop dynamics of a boost converter (i.e., the converter is stable in open loop). The corresponding buck converter incorporates one RHP pole, and therefore, it is unstable in open loop, as discussed and demonstrated in [9].

As discussed in [9], the duty-ratio gain (F_m) will become infinite when the converter enters into the boundary between DCM and CCM operation, i.e., into the boundary-conduction-mode (BCM) operation, which will take place in the case of the boost converter in Figure 1, at approximately $V_{in} \approx 17.5$ V with $D \approx 0.76$. Figure 3 shows the behavior of the inductor current, when the converter enters into the mode boundary at $V_{in} \approx 17.5$ V and goes deeper into the harmonic operation mode.

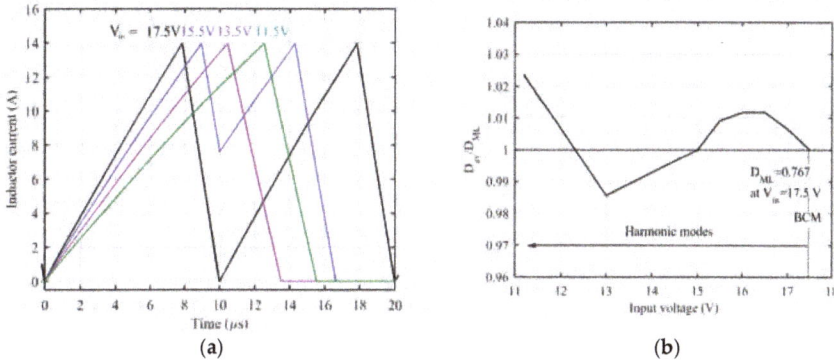

Figure 3. The behavior of (**a**) the inductor current and (**b**) the average duty ratio in the mode limit (boundary conduction mode (BCM)) and in harmonic operation modes.

Figure 3a shows that the converter entered into the second harmonic mode when the operating point passed through the BCM mode of operation (i.e., the black line), and it kept operating in DCM and in the second harmonic mode as well (i.e., the blue, magenta, and green lines). Figure 3b shows that the averaged duty ratio, which was computed based on $D_{av} = 1 - V_{in}/V_o$, stays approximately at the value of $D \approx 0.76$, equaling the value in BCM. This kind of behavior equals the behavior of D_{av} in CCM, as discussed in [4]. The origin of such a behavior is the infinite duty-ratio gain at the mode limit, maintaining the average derivative of inductor current at zero.

The DCM-operated PCM-controlled buck converter can adopt both odd and even harmonic operation modes, when the operating point passes through the mode limit, as demonstrated in [5]. The existence of the odd harmonic operation modes is the consequence of the open-loop instability at $M \approx 2/3$ at the resistive load. In Reference [8], the operation at odd and even harmonics was assumed to be a characteristic feature of the DCM-operated PCM-controlled converters in general. The analyses performed with the boost converter, in this paper, show clearly that the even harmonic operation is a general characteristic of PCM control.

2.3. Small-Signal Duty-Ratio Constraints for PCM-Controlled DCM Boost Converter

The coefficients of the duty-ratio constraints in Equation (1) can be solved by replacing m_1 and m_2 with their physical values, which are given for a boost converter in Equation (14), and by linearizing the averaged comparator equation in Equation (10), including all the parasitic circuit elements.

$$m_1 = \frac{\langle v_{in} \rangle - R_2 \langle i_L \rangle}{L}$$
$$m_2 = \frac{R_1 \langle i_L \rangle + \langle v_C \rangle - r_C \langle i_o \rangle + V_D - \langle v_{in} \rangle}{L} \qquad (14)$$

The duty-ratio constraints for a boost converter can be given in general by

$$\hat{d} = F_m(\hat{i}_{co} - q_L \hat{i}_L - q_C \hat{v}_C - q_{in} \hat{v}_{in} - q_o \hat{i}_o) \qquad (15)$$

and the corresponding full-order coefficients by

$$F_m = \frac{1}{T_s \left(M_c + \frac{V_3 (D'(V_o + V_D) - V_{in} + (D'R_1 + DR_2)i_L - D'r_C i_o)}{L V_1} \right)}$$
$$q_L = 1 - \frac{DT_s}{L} R_2 - \frac{D^2 T_s}{2L V_1} \left((R_1 - R_2) V_{in} - R_2 (V_2 + (R_1 - R_2) I_L)) - \frac{R_1 V_2 V_3}{V_1} \right)$$
$$q_C = \frac{D^2 T_s}{2L} \frac{V_3^2}{V_1^2} \qquad (16)$$
$$q_{in} = \frac{DT_s}{L} \left(1 - \frac{D}{2} \cdot \frac{V_2^2}{V_1^2} \right)$$
$$q_o = \frac{D^2 T_s r_C}{2L} \frac{V_3^2}{V_1^2}$$

where V_{1-3} and $R_{1,2}$ are defined in Equation (5).

The duty-ratio constraint coefficients in Equation (16) can also be given in a reduced-order form as a function of M and K by omitting all the parasitic circuit elements, as in [8,11], as given in Equation (17).

$$F_m = \frac{1}{T_s \left(M_c + \frac{V_{in}(D'M-1)}{L(M-1)} \right)}$$
$$q_L = 1$$
$$q_C = \frac{M}{R_{eq}(M-1)} \qquad (17)$$
$$q_{in} = \frac{1}{R_{eq}} \left(2\sqrt{\frac{M(M-1)}{K}} - \frac{M}{M-1} \right)$$
$$q_o = 0$$

2.4. Full-Order PCM State-Space for PCM-Controlled DCM Boost Converter

The full-order PCM state space can be obtained by replacing the perturbed duty ratio in Equation (3) with Equation (15) yielding

$$\frac{d\hat{i}_L}{dt} = -\frac{A_1 + F_m q_L V_e}{L} \hat{i}_L - \frac{A_2 + F_m q_C V_e}{L} \hat{v}_C + \frac{A_3 - F_m q_{in} V_e}{L} \hat{v}_{in} + \frac{A_4 - F_m q_o V_e}{L} \hat{i}_o + \frac{F_m V_e}{L} \hat{d}$$
$$\frac{d\hat{v}_C}{dt} = \frac{(1 + B_1 R_2 + F_m q_L I_e)}{C} \hat{i}_L + \frac{F_m q_C I_e}{C} \hat{v}_C$$
$$- \frac{B_1 - F_m q_{in} I_e}{C} \hat{v}_{in} - \frac{1 - F_m q_o I_e}{C} \hat{i}_o - \frac{F_m I_e}{C} \hat{d} \qquad (18)$$
$$\hat{i}_{in} = \hat{i}_L$$
$$\hat{v}_o = \hat{v}_C + r_C C \frac{d\hat{v}_C}{dt}$$

where A_{1-4}, B_1, V_e, and I_e are given in Equations (4) and (5).

The reduced-order state space can be obtained from Equation (18) by applying the reduced-order duty-ratio constraints in Equation (17) and replacing A_{1-4}, B_1, V_e, and I_e with

$$A_1 = R_{eq}\sqrt{\tfrac{K(M-1)}{M}} \quad A_2 = \sqrt{\tfrac{KM}{M-1}}$$

$$A_3 = M^2\sqrt{\tfrac{KM}{M-1}} \quad A_4 = 0 \quad B_1 = \tfrac{M(M-1)}{R_{eq}} \quad . \tag{19}$$

$$V_e = 2V_o \quad I_e = 2I_o\sqrt{\tfrac{M-1}{KM}}$$

2.5. Full-Order PCM Transfer Functions for PCM-Controlled DCM Boost Converter

The full-order transfer functions can be solved from the linearized state space in Equation (18) with the complete duty-ratio constraints in Equation (16) and the complete DDR elements (i.e., A_{1-4}, B_1, V_e, and I_e) in Equation (4) by formulating the state space into a matrix form and applying matrix manipulation techniques (cf. [11] (pp. 57–64).

The determinant (Δ) of the transfer functions can be given by

$$s^2 + s\left(\frac{A_1 + F_m q_L V_e}{L} - \frac{F_m q_C I_e}{C}\right) + \frac{(A_2 + F_m q_C V_e)(1 + B_1 R_2) + F_m I_e(A_2 q_L - A_1 q_C)}{LC} \tag{20}$$

and the transfer functions representing the input dynamics by

$$\Delta G_{ci\text{-}o} = F_m\left(s\tfrac{V_e}{L} + \tfrac{A_2 I_e}{LC}\right)$$

$$\Delta T_{oi\text{-}o} = s\tfrac{A_4 - F_m q_o V_e}{L} + \tfrac{A_2 + F_m q_C V_e}{LC} - \tfrac{F_m I_e(A_4 q_C + A_2 q_o)}{LC} \tag{21}$$

$$\Delta Y_{in\text{-}o} = s\tfrac{A_3 - F_m q_{in} V_e}{L} + \tfrac{A_2 B_1 - F_m I_e(A_2 q_{in} + A_3 q_C)}{LC} + \tfrac{B_1 F_m q_C V_e}{LC}$$

where $G_{ci\text{-}o} = \hat{i}_{in}/\hat{i}_{co}$, $T_{oi\text{-}o} = \hat{i}_{in}/\hat{i}_o$, and $Y_{in\text{-}o} = \hat{i}_{in}/\hat{v}_{in}$ and the transfer functions represent the output dynamics as follows:

$$\frac{\Delta G_{co\text{-}o}}{(1+sr_C C)} = \frac{F_m(-sLI_e + (1+B_1 R_2)V_e - A_1 I_e)}{LC}$$

$$\frac{\Delta G_{io\text{-}o}}{1+sr_C C} = \frac{sL(F_m q_{in} I_e - B_1) + A_3(1+B_1 R_2 + F_m q_L I_e) - A_1(B_1 - F_m q_{in} I_e) + F_m V_e(B_1(q_L + q_{in} R_2) + q_{in})}{LC}$$

$$\frac{\Delta Z_{o\text{-}o}}{1+sr_C C} = \frac{sL(1 - F_m q_o I_e) + A_1(1 - F_m q_o I_e) - A_4(1+B_1 R_2 + F_m q_o I_e) + F_m V_e(q_L + (1+B_1 R_2)q_o)}{LC} \tag{22}$$

where $G_{co\text{-}o} = \hat{v}_o/\hat{i}_{co}$, $G_{io\text{-}o} = \hat{v}_o/\hat{v}_{in}$, and $Z_{o\text{-}o} = \hat{v}_o/\hat{i}_o$.

2.6. Reduced-Order PCM Transfer Functions for PCM-Controlled DCM Boost Converter

The denominator (Δ) of the reduced-order transfer functions can be given by

$$s^2 + s\left(\frac{R_{eq}D}{D'M-1}\left(\frac{M-1}{LM} - \frac{KM}{R_{eq}^2 C}\right)\right) + \frac{1}{LC}\left(\sqrt{\frac{KM}{M-1}} + \frac{KM^2}{D'M-1}\right). \tag{23}$$

The transfer functions representing the input dynamics can be given by

$$\Delta Y_{in\text{-}o} = s\left(\frac{M^2\sqrt{\tfrac{KM}{M-1}} - 2F_m q_{in} V_o}{L}\right) + \frac{\tfrac{M(M-1)}{R_{eq}}\left(\sqrt{\tfrac{KM}{M-1}} - 2F_m q_C V_o\right) + 2F_m I_o(q_{in} + M^2 q_C)}{LC}$$

$$\Delta T_{oi\text{-}o} = \frac{\sqrt{\tfrac{KM}{M-1}} + 2F_m q_C V_o}{LC} \tag{24}$$

$$\Delta G_{ci\text{-}o} = \tfrac{2F_m I_o}{LC}(sR_{eq}C + 1)$$

and the transfer functions representing the output dynamics can be given by

$$\frac{\Delta Z_{o\text{-}o}}{1+sr_C C} = \frac{(sL+R_{eq}(\sqrt{\frac{K(M-1)}{M}}+\frac{K(M-1)M}{D'M-1}))}{LC}$$

$$\frac{\Delta G_{io\text{-}o}}{1+sr_C C} = s\frac{2F_m q_{in} I_o \sqrt{\frac{M-1}{KM}}-\frac{M(M-1)}{R_{eq}}}{C}+$$

$$\frac{M^2\sqrt{\frac{KM}{M-1}}-(M-1)\sqrt{KM(M-1)}+2F_m I_o(q_{in} R_{eq}\frac{M-1}{M}+M^2)}{LC}$$

$$\frac{\Delta G_{co\text{-}o}}{1+sr_C C} = \frac{2F_m V_{in}(1-s\cdot\frac{DT_s}{2})}{LC}$$

(25)

2.7. Load-Resistor-Affected Dynamics

The load resistor effect on the control-to-output-voltage transfer function ($G_{co\text{-}o}$) can be computed by [11] (pp. 38–39)

$$G_{co\text{-}o}^{R_L} = \frac{G_{co\text{-}c}}{1+\frac{Z_{o\text{-}o}}{R_L}}$$

(26)

where $G_{co\text{-}o}$ and $Z_{o\text{-}o}$ (i.e., output impedance) are the unterminated transfer functions of the converter given in Equation (22) or in Equation (25). Here, we treat only the load effect on the denominator of the transfer functions, which represents the dynamics of the system. The load-resistor-affected full-order denominator is given in Equation (27), and the reduced-order denominator with $M_c = 0$ is given in Equation (28), respectively.

$$s^2 + s\left(\frac{1}{R_L C}+\frac{A_1+F_m q_L V_e}{L}-\frac{F_m I_e(q_C+\frac{q_o}{R_L})}{C}\right)+\frac{(A_2+F_m q_C V_e)(1+B_1 R_2))+\frac{A_1-A_4(1+B_1 R_2)}{R_L}}{LC}$$

$$+\frac{+F_m V_e\frac{q_L+q_o(1+B_1 R_2)}{R_L}+F_m I_e(A_2 q_L-A_1 q_C-\frac{A_1 q_o+A_4 q_L}{R_L})}{LC}$$

(27)

$$s^2 + s\left(\frac{1}{R_L C}+\frac{R_{eq}D}{D'M-1}(\frac{M-1}{LM}-\frac{KM}{R_{eq}^2 C})\right)+\frac{1}{LC}\left(\frac{2M-1}{M}\sqrt{\frac{KM}{M-1}}+\frac{KM(2M-1)}{D'M-1}\right).$$

(28)

The damping in the DCM-operated PCM-controlled converters is very high [9], and therefore, the system poles are well separated, where the low-frequency pole ($\omega_{p\text{-LF}}$) is located close to origin and the high-frequency pole ($\omega_{p\text{-HF}}$) is located close to the switching frequency or beyond it. Therefore, the system poles can be approximated by utilizing the properties of a second-order polynomial $s^2 + sa + b$ as follows:

$$\omega_{p\text{-LF}} \approx \frac{b}{a} \quad \omega_{p\text{-HF}} \approx a.$$

(29)

According to Equation (29), the poles of the unterminated system in Equation (23) can be given by

$$\omega_{p\text{-LF}} \approx \frac{\sqrt{\frac{KM}{M-1}}+\frac{KM^2}{D'M-1}}{\frac{R_{eq}D}{D'M-1}(\frac{M-1}{M}C-\frac{KM}{R_{eq}^2}L)}$$

$$\approx \frac{(D'M-1)M\sqrt{\frac{KM}{M-1}}+KM^3}{CR_{eq}D(M-1)}$$

(30)

and

$$\omega_{p\text{-HF}} \approx \frac{R_{eq}D}{D'M-1}(\frac{M-1}{LM}-\frac{KM}{R_{eq}^2 C}) \approx \frac{R_{eq}D(M-1)}{L(D'M-1)M}.$$

(31)

In the case of the load-resistor-affected system in Equation (28), the poles can be given by

$$\omega_{p\text{-LF}}^{R_L} \approx \frac{(2M-1)(D'M-1)\sqrt{\frac{KM}{M-1}}+KM^2(2M-1)}{R_L CD(M-1)}$$

$$\approx \frac{2M-1}{R_L C(M-1)} \cdot \frac{(D'M-1)\sqrt{\frac{KM}{M-1}}+KM^2}{D}$$

(32)

and

$$\begin{aligned}
\omega_{\text{p-HF}}^{R_L} &\approx \frac{1}{R_L C} + \frac{R_{eq} D}{D' M - 1}\left(\frac{M-1}{LM} - \frac{KM}{R_{eq}^2 C}\right) \\
&\approx \frac{R_{eq} D(M-1)}{LM(D'M-1)}
\end{aligned} \tag{33}$$

According to Equations (31) and (33), we can conclude that the load resistor does not affect significantly the location of $\omega_{\text{p-HF}}$. Equation (32) indicates that the low-frequency pole given in [5] (p. 480) (i.e., Equation (34)) resembles the pole given in Equation (32) but it does not equal it.

$$\omega_{\text{p-LF}}^{R_L} \approx \frac{2M-1}{R_L C(M-1)}. \tag{34}$$

The denominator of the transfer function in [5] can be computed to be

$$s^2 + s\left(\frac{2}{R_L C} + \frac{R_L}{L} \cdot \frac{M-1}{M^2}\right) + \frac{1}{LC} \cdot \frac{2M-1}{M^2} \tag{35}$$

from which the low and high-frequency poles can be approximated to be

$$\begin{aligned}
\omega_{\text{p-LF}} &\approx \frac{2M-1}{2M^2\frac{L}{R_L} + R_L C(M-1)} \approx \frac{2M-1}{R_L C(M-1)} \\
\omega_{\text{p-HF}} &\approx \frac{2}{R_L C} + \frac{R_L}{L} \cdot \frac{M-1}{M^2} \approx \frac{R_L}{L} \cdot \frac{M-1}{M^2}
\end{aligned}. \tag{36}$$

The low-frequency pole in Equation (34) can be obtained from Equation (36) by setting $L = 0$, as instructed in [5].

Table 1 shows the computed values for $\omega_{\text{p-LF}}$ and $\omega_{\text{p-HF}}$ at the input voltages of 20 V and 50 V. The values in parenthesis equal the load-resistor-affected values. Table 1 shows clearly that the system poles were well separated. It shows also that the effect of the load resistor was quite small. In addition, it shows that the low-frequency pole $\omega_{\text{p-LF}}^{R_L}$ in Equation (34) and Equation (36) predicted quite well the location of the load-resistor-affected low-frequency pole as well. The high-frequency pole $\omega_{\text{p-HF}}^{R_L}$ in Equation (36) predicted quite inaccurately the load-resistor-affected pole.

Table 1. The location of the system poles $\omega_{\text{p-LF}}$ and $\omega_{\text{p-HF}}$, as well as the location of the right-half-plane RHP zero ($\omega_{\text{z-RHP}}$) of $G_{\text{co-o}}$ at $V_o = 75$ V and $I_o = 1.5$ A.

Input Voltage	$\omega_{\text{p-LH}}$/Hz	$\omega_{\text{p-HF}}$/Hz	$\omega_{\text{z-RHP}}$/Hz
20 V	14 (24)	1 M (864 k)	53 k
50 V	29 (40)	193 k (192 k)	195 k
20 V [5]	(24)	(648 k)	-
50 V [5]	(40)	(295 k)	-

Equation (25) indicates that $G_{\text{co-o}}$ contains a right-half-plane (RHP) zero approximately at $2/DT_s$. Table 1 (i.e., $\omega_{\text{z-RHP}}$) shows that the RHP zero is located at much higher frequencies than in the corresponding CCM boost converter (i.e., $\omega_{\text{z-RHP}}^{\text{DCM}} \geq f_s/2$ vs. $\omega_{\text{z-RHP}}^{\text{CCM}} \geq f_s/100$; cf. [9] (p. 153). Thus, the main contributions of the RHP zero on the control design are reflected via the high-frequency phase behavior in DCM, which allows for the use of higher bandwidth controllers than in CCM.

2.8. Generalized Transfer Functions

The control engineering block diagrams, from which the generalized transfer functions in Equation (37) (output dynamics) and in Equation (38) (input dynamics) are defined, are given explicitly in [9]:

$$G_{io\text{-}o}^{PCM} = \frac{\hat{v}_o}{\hat{v}_{in}} = \frac{(1+\frac{Fmq_L^B H_{sr}B}{A})G_{io\text{-}o}^{DDR} - Fm(q_{in}^B + \frac{q_L^B H_{sr}C}{A})G_{co\text{-}o}^{DDR}}{1+L_c+L_v}$$

$$Z_{o\text{-}o}^{PCM} = \frac{\hat{v}_o}{\hat{i}_o} = \frac{(1+\frac{Fmq_L^B H_{sr}B}{A})Z_{o\text{-}o}^{DDR} + \frac{Fmq_L^B H_{sr}}{A}G_{co\text{-}o}^{DDR}}{1+L_c+L_v} \qquad (37)$$

$$G_{co\text{-}c}^{PCM} = \frac{\hat{v}_o}{\hat{i}_{co}} = \frac{F_m^B G_{co\text{-}o}^{DDR}}{1+L_c+L_v}$$

$$Y_{in\text{-}o}^{PCM} = \frac{\hat{i}_{in}}{\hat{v}_{in}} = Y_{in\text{-}o}^{DDR} - \frac{F_m^B G_{ci\text{-}o}^{DDR}}{1+L_c+L_v}(q_{in}^B + \frac{q_L^B H_{sr}C}{A} + (q_o^B + \frac{q_L^B H_{sr}}{AZ_C})G_{io\text{-}o}^{DDR})$$

$$T_{oi\text{-}o}^{PCM} = \frac{\hat{i}_{in}}{\hat{i}_o} = T_{oi\text{-}o}^{DDR} + \frac{F_m G_{ci\text{-}o}^{DDR}}{1+L_c+L_v}((q_o^B + \frac{q_L^B H_{sr}}{AZ_C})Z_{o\text{-}o}^{DDR} - \frac{q_L^B H_{sr}}{A}) \qquad (38)$$

$$G_{ci\text{-}o}^{PCM} = \frac{\hat{i}_{in}}{\hat{i}_{co}} = \frac{F_m G_{ci\text{-}o}^{DDR}}{1+L_c+L_v}$$

where L_c and L_v denote the inductor-current and output-voltage feedback-loop gains as given in Equation (39), $G_{cL\text{-}o}^{DDR}$ and $G_{co\text{-}o}^{DDR}$ denote the control-to-inductor-current and control-to-output-voltage transfer functions of the corresponding DDR-controlled converter

$$L_c = F_m q_L^B H_{sr} G_{cL\text{-}o}^{DDR}$$
$$L_v = F_m q_o^B G_{co\text{-}c}^{DDR} \qquad (39)$$

and Z_C denotes the impedance of the output capacitor, H_{sr} denotes the high-frequency extension for correcting the phase behavior given in Equation (40), and A, B, and C are defined in Equation (41) for a boost converter, respectively.

$$H_{sr} = 1 + s\frac{2\zeta}{\omega_{sr}} + \frac{s^2}{\omega_{sr}^2}. \qquad (40)$$

$$A = 1 - \frac{D^2 T_s}{2L}(r_L + r_{ds}) \quad B = \frac{DT_s}{L}V_{in} \quad C = \frac{D^2 T_s}{2L}. \qquad (41)$$

The duty-ratio constraints applicable for Equations (37) and (38) in the case of a boost converter can be derived from Equation (16) by setting $r_C = 0$ as

$$F_m^B \approx F_m$$
$$q_L^B \approx q_L \quad q_{in}^B \approx q_{in} \quad q_o^B \approx q_C \qquad (42)$$

and $G_{cL\text{-}o}^{DDR}$ and $G_{co\text{-}o}^{DDR}$ for computing $G_{co\text{-}o}^{PCM}$ of the boost converter in Equation (43) can be given by

$$G_{co\text{-}c} = \frac{-sLI_e + (1+B_1 R_2)V_e - A_1 I_e}{LC(s^2 + s\frac{A_1}{L} + \frac{A_2(1+B_1 R_2)}{LC})}(1 + sr_C C)$$

$$G_{cL\text{-}o} = \frac{sV_e C + A_2 I_e}{LC(s^2 + s\frac{A_1}{L} + \frac{A_2(1+B_1 R_2)}{LC})} \qquad (43)$$

where $A_{1,2}$, B_1, V_e, and I_e are given in Equation (4), respectively.

3. Simulink-Based Model Validation

The validation was performed in such a manner that the simulated frequency responses were extracted from the Simulink-base switching model corresponding exactly to the boost converter given in Figure 1. The pseudo-random-binary-sequence-based method to extract the frequency responses is described in [13,14]. The predicted frequency responses were computed based on the complete transfer functions given in Equation (22), where the effect of the high-frequency extension H_{sr} in Equation (40) is added with $\zeta = 0.5$ and $\omega_{sr} = 2\pi f_s$ (i.e., $f_s = 100$ kHz) (cf. [9]).

Figure 4 shows the predicted (solid lines) and simulated (diamond and square marks) output impedance at the input voltages of 20 V and 50 V. The figure shows that there were no high-frequency effects visible at the magnitude or phase, which were actually removed by the output capacitor, as also discussed in [5]. This means that the average-model-based transfer function given in Equation (22) predicted exactly the dynamic behavior of the output impedance as such.

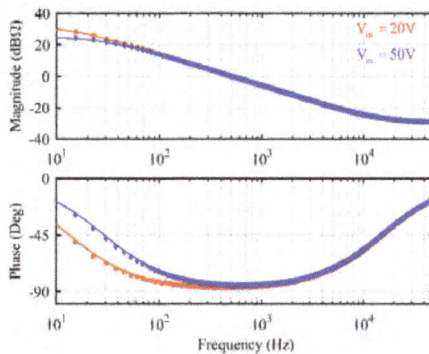

Figure 4. The predicted (solid lines) and simulated (diamond and square marks) frequency responses of the output impedance at the input voltages of 20 V (red) and 50 V (blue).

Figure 5 shows the predicted (solid black lines) and simulated (red square marks at 20 V and blue diamond marks at 50 V) frequency responses of the control-to-output-voltage transfer functions. The figure shows that the predicted and simulated responses with the application of H_{sr} in Equation (38) had very good matches with each other.

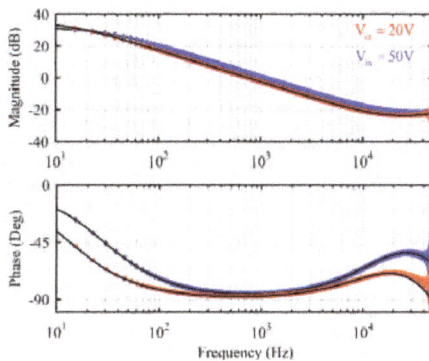

Figure 5. The predicted (solid black lines) and simulated (diamond and square marks) frequency responses of the control-to-output-voltage transfer functions at the input voltages of 20 V (red) and 50 V (blue).

Figure 6 shows the comparison of the predicted responses of the control-to-output-voltage transfer functions, where the solid black lines denote the responses computed by using the full-order transfer functions, and the dashed lines denote the responses by using the reduced-order transfer functions, respectively. The figure shows that the responses coincided. This implies that the reduced-order models can be used for different design purposes, but it is, however, recommended to verify the situation with the actual design in a similar manner as shown in Figure 6.

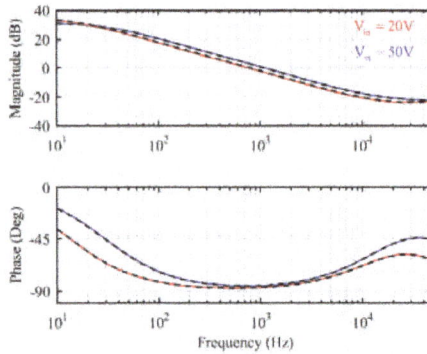

Figure 6. The black solid lines represent the frequency responses of the control-to-output transfer function predicted by the full-order model, and the dashed lines represent the corresponding frequency responses predicted by the reduced-order model at 20 V (red) and 50 V (blue).

Figure 7 shows the unterminated (dashed lines) and load-resistor-affected (solid lines) frequency responses of the control-to-output-voltage transfer function at the input voltages of 20 V (red) and 50 V (blue), respectively. The figure confirms that the load-resistor effects were concentrated at the low frequencies, as discussed in Section 2.6. It is obvious that the load-resistor-affected control-to-output-voltage responses can be utilized in the control design, when the feedback-loop crossover frequency is designed to be at high enough frequencies (i.e., >1 kHz in this case).

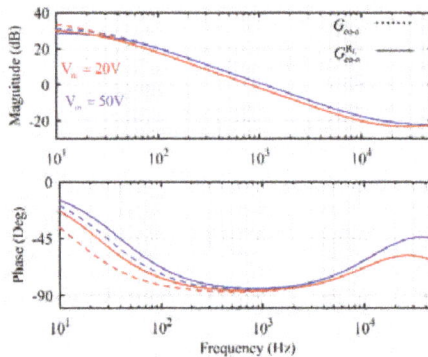

Figure 7. The dashed lines denote the predicted unterminated control-to-output-voltage frequency responses, and the solid lines denote the predicted load-resistor-affected frequency responses at the input voltages of 20 V (red) and 50 V (blue).

The output-voltage feedback loop of the boost converter was designed at the input voltage of 20 V, where the RHP zero is at its minimum frequency (cf. Table 1). The phase behavior of the control-to-output-voltage transfer function is such that a proportional-integral (PI) controller can be used (cf. Figure 7) as given in

$$G_{cc} = K_{cc}\frac{1+s/\omega_z}{s(1+s/\omega_p)}. \tag{44}$$

The controller zero (ω_z) was placed at 1 kHz, the controller pole (ω_p) was placed at 5 kHz, and the controller gain (K_{cc}) was set to 88,614 to obtain the feedback-loop crossover frequency of 10 kHz and the phase margin of 60 degrees, as shown in Figure 8. If looking carefully the behavior of the high-frequency magnitude of the feedback-loop gain (solid red line) at the input voltage of 20 V, then it would be clear that the sufficient gain margin (i.e., 6 dB at least) would determine the obtainable crossover frequency. The dashed lines denote the PI-controller design, where the high-frequency pole was omitted, as in [17]. According to the corresponding high-frequency-magnitude behavior (dashed red line), it would be obvious that the design would be easily sensitive to the high-frequency ripple effects in the duty-ratio generation process.

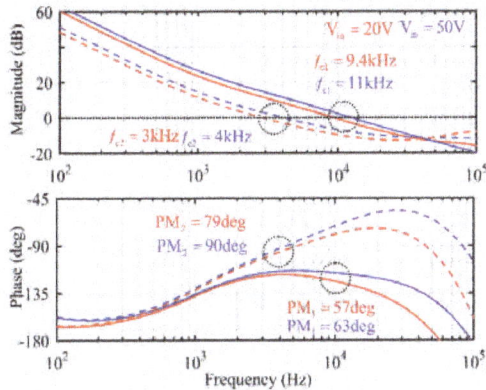

Figure 8. The output-voltage feedback-loop gains, when the PI controller with a high-frequency pole is used (solid lines) and when the PI controller without the high-frequency pole is used (dashed lines).

Figure 9 shows the output-voltage response to a load current change of 1.5 A at the input voltages of 20 V (red) and 50 V (blue). The responses were quite close to each other, because the feedback-loop gains do not change much when the input voltage varies. The same responses in the case of the CCM-operated PCM-controlled boost converter can be seen in [11] (p. 324, Figure 6.53), where the voltage dipped at 20 V and 50 V to 1.37 V and 0.56 V, respectively, for a steep change in the output current of 1.3 A. This is a very good indication of why the DCM operation is preferred in converters having a RHP zero in their control dynamics.

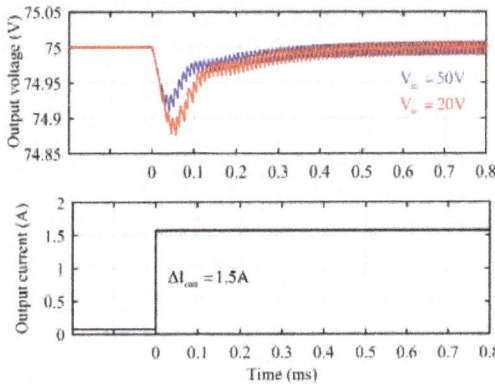

Figure 9. The output-voltage response to a load current change of 1.5 A at the input voltages of 20 V (red) and 50 V (blue).

4. Conclusions

The full-order dynamic modeling and comprehensive analysis of a DCM-operated PCM-controlled boost converter are presented for first time in the literature in this paper. The investigations show that the PCM-modeling technique introduced in [8] and validated in the case of a buck converter in [9] also produces accurate models for a boost converter, when the load-resistor effect is removed [10] and the high-frequency extension introduced in [9] is added. In addition, this paper shows that the reduced-order models also quite accurately predicted the dynamic behavior in this particular case. This phenomenon should not be generalized until the validity of the reduced-order models in each specific case is verified by means of the full-order models.

Suntio [8] stated explicitly that DCM-operated PCM-controlled converters will operate in harmonic operation modes at both even and odd harmonic frequencies. The investigations in this paper show explicitly that the even harmonic-mode operation is a characteristic feature of PCM-controlled converters in general because of the infinite duty-ratio gain at the mode limit. The even and odd harmonic-mode operations of the buck converter are the consequence of the open-loop RHP pole.

It was also shown that the output-voltage transient response of the DCM-operated PCM-controlled boost converter is outstanding compared with the corresponding CCM-operated PCM-controlled boost converter.

Funding: The research received no external funding.

Conflicts of Interest: The author declares no conflict of interest.

References

1. Deisch, C.W. Simple switching control method changes power converter into a current source. In Proceedings of the IEEE power Electron. Specialists Conf. (IEEE PESC), Syracuse, NY, USA, 13–15 June 1978; pp. 300–306.
2. Capel, A.; Ferrante, G.; O'Sullivan, D.; Weinberg, A. Application of injected current model for the dynamic analysis of switching regulators with the new concept of LC^3 modulator. In Proceedings of the IEEE power Electron. Specialists Conf. (IEEE PESC), Syracuse, NY, USA, 13–15 June 1978; pp. 135–147.
3. Redl, R.; Sokal, N.O. Current-mode control, five different types, used with the three basic classes of power converters: Small-signal ac and large-signal DC characterization, stability requirements, and implementation of practical circuits. In Proceedings of the IEEE Power Electron. Specialists Conf. (IEEE PESC), Toulouse, France, 24–28 June 1985; pp. 771–785.
4. Suntio, T. On dynamic modeling of PCM-controlled converters—Buck converter as an example. *IEEE Trans. Power Electron.* **2018**, *33*, 5502–5518. [CrossRef]
5. Erickson, R.W.; Maksimović, D. *Fundamentals of Power Electronics*, 2nd ed.; Kluwer Academic Publishers: Norwell, MA, USA, 2001.
6. Ridley, R.B. A new continuous-time model for current-mode control with constant frequency, constant on-time and constant off-time in CCM and DCM. In Proceedings of the IEEE Power Electron. Specialists Conf. (IEEE PESC), San Antonio, TX, USA, 11–14 June 1990; pp. 382–389.
7. Fang, C.-C. Unified discrete-time modeling of buck converter in discontinuous mode. *IEEE Trans. Power Electron.* **2011**, *26*, 2335–2342. [CrossRef]
8. Suntio, T. Analysis and modeling of peak-current-mode controlled buck converter in DICM. *IEEE Trans. Ind. Electron.* **2001**, *48*, 127–135. [CrossRef]
9. Suntio, T. Dynamic modeling and analysis of PCM-controlled DCM-operated buck converter—A reexamination. *Energies* **2018**, *11*, 1267. [CrossRef]
10. Suntio, T. Load-resistor-affected dynamic models in control design of switched-mode converters. *EPE J.* **2018**, *28*, 159–168. [CrossRef]
11. Suntio, T.; Messo, T.; Puukko, J. *Power Electronic Converters—Dynamics and Control in Conventional and Renewable Energy Applications*; Wiley VCH: Weinheim, Germany, 2017.
12. Gurbina, M.; Ciresan, A.; Lascu, D.; Lica, S.; Pop-Calimanu, I.M. A new exact mathematical approach for studying bifurcation in DCM operated dc-dc switching converters. *Energies* **2018**, *11*, 663. [CrossRef]
13. Fang, C.-C.; Redl, R. Subharmonic instability limits for the peak-current-controlled buck converter with closed voltage feedback loop. *IEEE Trans. Power Electron.* **2015**, *30*, 1085–1092. [CrossRef]
14. Suntio, T. Unified average and small-signal modeling of direct-on-time control. *IEEE Trans. Ind. Electron.* **2006**, *53*, 287–295. [CrossRef]

15. Roinila, T.; Helin, T.; Vilkko, M.; Suntio, T.; Koivisto, H. Circular correlation based identification of switching power converter with uncertainty analysis using fuzzy density approach. *J. Simul. Model. Pract. Theory* **2009**, *17*, 1043–1058. [CrossRef]

16. Roinila, T.; Vilkko, M.; Suntio, T. Fast loop gain measurement of switched-mode converter using binary signal with specified Fourier amplitude spectrum. *IEEE Trans. Power Electron.* **2009**, *24*, 2746–2755. [CrossRef]

17. Sun, J.; Mitchell, D.M.; Greuel, M.F.; Krein, P.T.; Bass, R.M. Averaged modeling of PWM converters operating in discontinuous mode. *IEEE Trans. Power Electron.* **2001**, *16*, 482–492.

energies

MDPI

Article

Oscillation Suppression Method by Two Notch Filters for Parallel Inverters under Weak Grid Conditions

Ling Yang [1], Yandong Chen [1,*], Hongliang Wang [1], An Luo [1] and Kunshan Huai [2]

[1] College of Electrical and Information Engineering, Hunan University, Changsha 410082, China; yangling_1992@163.com (L.Y.); liangliang-930@163.com (H.W.); an_luo@126.com (A.L.)
[2] Guangzhou Power Supply Co., Ltd., Guangzhou 510620, China; huaikunshan@126.com
[*] Correspondence: yandong_chen@hnu.edu.cn; Tel.: +86-151-1626-8089

Received: 4 November 2018; Accepted: 5 December 2018; Published: 8 December 2018

Abstract: With plenty of parallel inverters connected to a weak grid at the point of common coupling (PCC), the impedance coupling interactions between the inverters and the grid are enhanced, which may cause high-frequency harmonic oscillation and further aggravate the system instability. In this paper, a basic technique for inverter output impedance is proposed to suppress the oscillation, showing that the inverter output impedance should be designed relatively high at the harmonic oscillation frequency, while relatively low at other frequencies. On the basis of the proposed technique, two virtual impedances are added to be in parallel and in series with the original inverter output impedance, respectively. Thus, an oscillation suppression method by two notch filters is proposed to realize the virtual impedances and increase the whole system damping. The implementation forms of the virtual impedances are presented by the proposed PCC voltage feedforward and grid-side inductor current feedback with two notch filters. Finally, simulation and experimental results are provided to verify the validity of the proposed control method.

Keywords: weak grid; parallel inverters; oscillation suppression; notch filter; impedance reshaping

1. Introduction

With the increasing energy crisis and environmental problems, renewable energy generation, mostly in power plants or microgrids, have grown rapidly [1–3]. Due to the distributed locations of renewable energy generators, multiple transformers and long transmission lines are utilized to connect the systems to the public grid [4]. Thus, the public grid exhibits the feature of a weak grid in which grid impedance cannot be ignored [5].

Especially, in a large-scale power plant or microgrid, renewable energies are mostly connected to the grid via parallel inverters [6,7]. By this way, the power plant or microgrid can easily expand the output power capacity, and it is also convenient to connect plenty of renewable energies into the grid [8]. Under weak grid conditions, the impedance coupling interactions between the inverters and the grid are enhanced further due to grid impedance [9]. If control parameters and device selection of all inverters are the same, the grid impedance seen by each inverter will become n times of the real grid impedance [10]. Therefore, these interactions may cause oscillation if inverters are improperly designed or controlled.

There are two methods to analyze the system's stability. The first is the eigenvalue-based analysis [11], which is usually utilized to evaluate the system's stability. The eigenvalue-based stability analysis studies the eigenvalues of a system's state space model matrix, which requires the physical features and control parameters in the system [12]. The second one is the impedance-based stability criteria [13,14], which is well built to adjudicate the system stability. The system will be stable if two requirements are satisfied [13]. The first requirement is that the grid-connected inverter is stable in the public grid, and the other is that the product of grid impedance and inverter output admittance satisfies the Nyquist criterion. The industrial and academia community have widely accepted the theory of impedance-based stability analysis [15,16].

Some strategies were proposed to suppress the oscillation, which can be divided into two cases. One is to introduce additional hardware equipment to the PCC [17], the other is to reshape the inverter output impedance [18–23]. Adding additional equipment was adopted to stabilize the paralleled multi-inverter system, in which structures and control parameters of the inverters are unknown. Reference [17] installed an active damper at the PCC to suppress the oscillation. However, extra cost may exist owing to the demand of additional circuitry [24].

Different impedance reshaping methods need to be adopted to suppress the oscillation. The existing impedance reshaping methods mainly include: capacitance-current-feedback methods [18,19], capacitance-voltage-feedback methods [20,21], and virtual resistor methods [22,23]. In Reference [18], the real-time computational method was proposed to decrease the computational delay, which can simplify the design and enhance the performance. However, if the grid voltage has much harmonics, it will make the duty cycle of the inverters change sharply. Similar to the capacitance-current-feedback methods, useful active damping was induced by the derivative feedback of the capacitance voltage [20,21]. However, the capacitance-voltage-feedback methods should handle the challenge of grid voltage variation. In References [22,23], the virtual resistor methods made the inverter output impedance show high impedance at all frequencies, which can suppress the oscillation. However, due to the high impedance characteristics of the inverters at the fundamental frequency, the methods will cause the change of the fundamental current, and then influence the tracking precision of the grid-connected power. Therefore, the fundamental impedance and high-frequency harmonic impedance of inverters are separately considered. The fundamental impedance should be designed as the low impedance, which does not affect the grid-connected power tracking. However, the high-frequency harmonic impedance is designed as the high impedance to suppress the oscillation.

In this paper, the oscillation suppression method by two notch filters is proposed to increase the whole system damping. This paper is organized as follows. Section 2 analyzes the oscillation mechanism of a paralleled multi-inverter system. Section 3 presents the demand for the inverter output impedance for the purpose of oscillation suppression, and the method of adding the virtual impedances to meet the demand is proposed, which introduces two notch filters to the PCC voltage feedforward and grid-side inductor current feedback. Section 4 compares and analyzes the system's stability in two cases. Sections 5 and 6 provide the simulation and experimental results to verify the validity of the proposed control method. Finally, Section 7 gives the conclusion.

2. Oscillation Mechanism of Paralleled Multi-Inverter System

2.1. System Description

The structure of a paralleled multi-inverter system is shown in Figure 1. The left and right side are the inverter subsystem and the grid subsystem. $j = 1, 2, \ldots, n$. U_{dc} is the DC voltage. u_{invj}, u_{C1j}, and u_{PCC} are the inverter output voltage, filter capacitor voltage, and PCC voltage. u_g is the grid voltage. Z_g is the grid impedance. Inductor-capacitor-inductor-type (LCL-type) filter is constituted by the inverter-side inductor L_{1j}, grid-side inductor L_{2j}, and filter capacitor C_{1j}. R_{L1j} and R_{L2j} are parasitic resistances of L_{1j} and L_{2j}. i_{L1j} is the inverter-side inductor current. i_{C1j} is the filter capacitor current. i_{oj} is the grid-side inductor current. i_g is the grid-connected current.

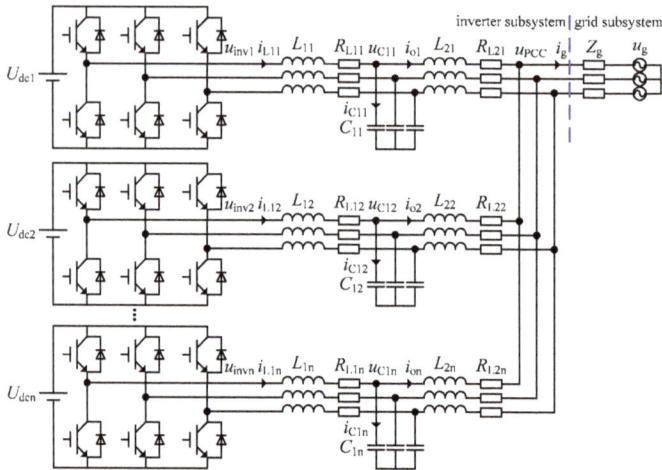

Figure 1. Structure of paralleled multi-inverter system.

2.2. Oscillation Mechanism

In the grid-connected mode, single inverter is equivalent to the current source i_j in parallel with equivalent admittance Y_j, which is the Norton equivalent circuit [23]. The grid is equivalent to the grid voltage u_g in series with grid impedance Z_g. From the PCC, the Norton equivalent circuit of paralleled multi-inverter system is shown in Figure 2a. From Figure 2a, the circuit relationship about u_{PCC} is shown in Equation (1) by the nodal analysis method.

$$(Y_1 + Y_2 + \cdots + Y_n)u_{PCC} + u_{PCC}/Z_g = (i_1 + i_2 + \cdots + i_n) + u_g/Z_g \tag{1}$$

where i_j is the current source of a single inverter, and Y_j is the equivalent admittance of a single inverter.

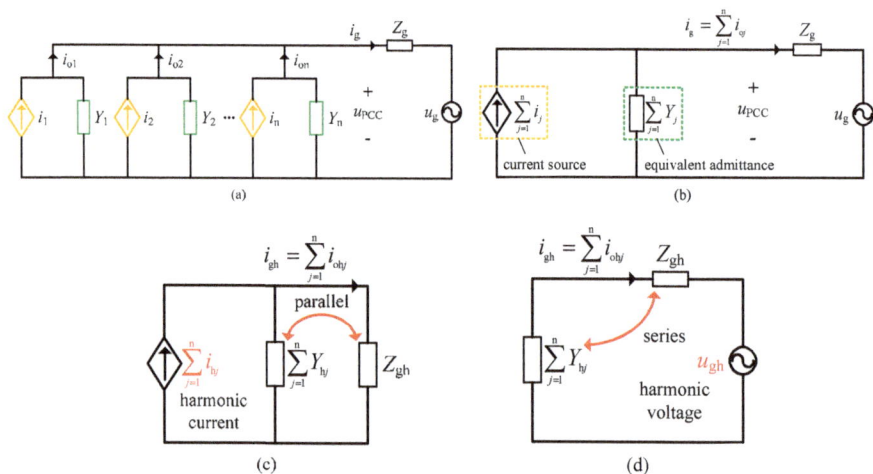

Figure 2. Output admittance model and oscillation mechanism of paralleled multi-inverter system. (**a**) Norton equivalent circuit; (**b**) Output admittance model; (**c**) Parallel oscillation; (**d**) Series oscillation.

From Equation (1), the output admittance model of paralleled multi-inverter system satisfies the Equation (2), as depicted in Figure 2b.

$$u_{PCC} \sum_{j=1}^{n} Y_j + u_{PCC}/Z_g = \sum_{j=1}^{n} i_j + u_g/Z_g \tag{2}$$

The frequency of harmonic current i_{hj} is caused by the system nonlinear factor. If it is equal to or close to the parallel resonance frequency of impedance network, it will cause the parallel oscillation, as shown in Figure 2c. The frequency of harmonic voltage u_{gh} is caused by the grid distortion. If it is equal to or close to the series resonance frequency of impedance network, it will result in the series oscillation, as shown in Figure 2d.

From Figure 2b, the grid-connected current i_g can be derived as

$$i_g = (\sum_{j=1}^{n} i_j - u_g \sum_{j=1}^{n} Y_j) \cdot (\frac{1}{1 + Z_g \cdot \sum_{j=1}^{n} Y_j}) \tag{3}$$

where the product of grid impedance and inverter output admittance is defined as the impedance ratio K, which can be expressed as

$$K = Z_g \cdot \sum_{j=1}^{n} Y_j \tag{4}$$

From Equation (3), it can be assumed that the grid voltage is stable in the absence of the inverter, and the inverter will be stable if the grid impedance is zero. Nevertheless, when the grid impedance is not negligible, the system will be stable only if the impedance ratio K satisfies the Nyquist criterion in Reference [12]. In other words, the system will be stable, only if the Nyquist curve of the impedance ratio does not surround $(-1, j0)$.

Thus, there are impedance coupling interactions between the inverters and the grid, which will aggravate the harmonic distortion of grid-connected current, lead to the oscillation in paralleled multi-inverter system, and even cause the system to be unstable.

3. Oscillation Suppression Method by Two Notch Filters for Parallel Inverters

3.1. Demand for the Inverter Output Impedance

To suppress the oscillation, the inverter output impedance should be designed relatively high at the harmonic oscillation frequency while relatively low at other frequencies. For this purpose, the virtual impedances are added to be connected with the original output impedance, as shown in Figure 3. Figure 3a–c presents parallel, series, and parallel-series virtual impedances, respectively. The above three forms can reach the same performance for oscillation suppression. For the first and second forms, it is relatively complicated to introduce one virtual impedance to realize that the inverter output impedance shows high at the harmonic oscillation frequency while relatively low at other frequencies. However, the third form with two virtual impedances is relatively easy to achieve this purpose. Thus, the third form is mainly discussed in this paper.

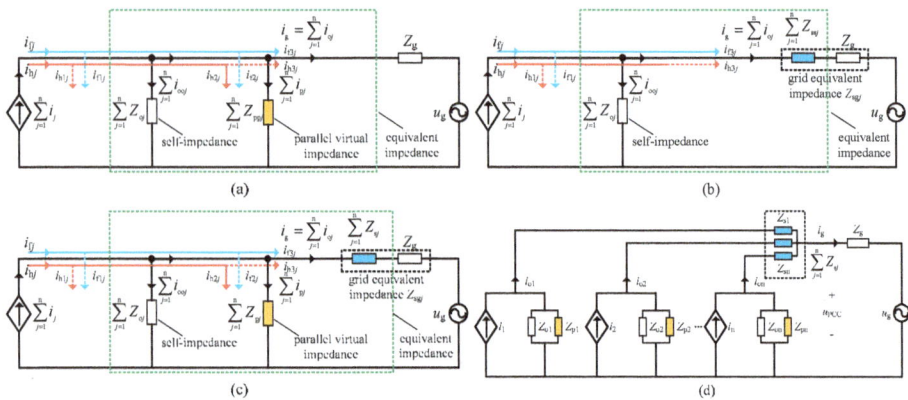

Figure 3. Equivalent schematic diagram. (**a**) Parallel virtual impedance; (**b**) series virtual impedance; (**c**) parallel-series virtual impedances; (**d**) equivalent output impedance model.

In Figure 3c, Z_{oj} is the self-impedance when the parallel-series virtual impedances are not added. The equivalent impedance Z_j ($Z_j = 1/Y_j$) is formed by Z_{oj} in parallel with Z_{pj}, and then in series with Z_{sj}. The grid equivalent impedance Z_{sgj} consists of the virtual impedance Z_{sj} and the grid impedance Z_g connected in series. $i_{f/hj}$, $i_{f/h1j}$, $i_{f/h2j}$, and $i_{f/h3j}$ are total fundamental/high-frequency harmonic current, fundamental/high-frequency harmonic current of Z_{oj} branch, fundamental/high-frequency harmonic current of Z_{pj} branch, and fundamental/high-frequency harmonic current of Z_{sgj} branch, respectively. $i_{f/hj} = i_{f/h1j} + i_{f/h2j} + i_{f/h3j}$. The total fundamental frequency current i_{fj} and the total high-frequency harmonic current i_{hj} are determined by the shunt circuit, consisting of the self-impedance Z_{oj}, parallel virtual impedance Z_{pj}, and grid equivalent impedance Z_{sgj}. From the PCC, the Norton equivalent circuit of Figure 3c is refined into the forms of Figure 3d.

On the basis of the proposed basic technique, the parallel virtual impedance Z_{pj} should be designed to show low impedance at the harmonic oscillation frequency. By doing so, most high-frequency harmonic current will flow into the parallel virtual impedance Z_{pj} branch; it effectively suppresses the oscillation of paralleled multi-inverter system. In the meantime, to improve the power quality of grid-connected current, the series virtual impedance Z_{sj} should be designed to display low impedance at the fundamental frequency. This way, most fundamental frequency current flows into the grid branch with relatively low impedance.

3.2. Oscillation Suppression Method

The parallel-series virtual impedances in Figure 4 can be realized by introducing two notch filters. In Figure 4, two virtual impedances are added in parallel and series with inverter output impedance, respectively. G_i is the grid-connected current loop proportional resonant (PR) controller, G_{PWM} is the equivalent gain of the inverter, $Z_{L1j} = sL_{1j} + R_{L1j}$, $Z_{C1j} = 1/sC_{1j}$, $Z_{L2j} = sL_{2j} + R_{L2j}$.

The parallel virtual impedance Z_{pj} and series virtual impedance Z_{sj} can be expressed as

$$\begin{cases} Z_{pj} = r_1/G_N \\ Z_{sj} = r_2 G_N \end{cases} \tag{5}$$

where r_1 and r_2 are the proportional coefficient and G_N is the notch filter.

The grid-connected current loop PR controller G_i can be expressed as

$$G_i = k_p + \frac{2k_{i1}\omega_c s}{s^2 + 2\omega_c s + \omega_0^2} \tag{6}$$

where k_p is the proportional coefficient of quasi-proportional resonant controller, k_{i1} is the resonance gain of quasi proportional resonance controller, ω_c is the cut-off angular frequency, and ω_o is the fundamental angular frequency.

The effects of dead-time of switching devices in paralleled multi-inverter system are regarded as a disturbance, which have a constant amplitude and an alternative direction depending on the inverter-side inductor current i_{L1j} [25]. It is notable that the disturbance can be seen as the controlled current source i_{dj} in Norton equivalent circuit, which can be presented as

$$i_{dj} = \frac{U_{ej}}{Z_{pj}A + Z_{sj}B} \cdot \text{sign}(i_{L1j}) \tag{7}$$

where $A = Z_{L1j}Z_{L2j} + Z_{C1j}(G_iG_{PWM} + Z_{L1j} + Z_{L2j})$ and $B = Z_{L1j}Z_{C1j} + (Z_{L1j} + Z_{C1j})(Z_{L2j} + Z_{pj})$.

According to Reference [25], U_{ej} in Equation (7) can be expressed as

$$U_{ej} = 2(U_{dcj} + U_{Dj} - U_{Tj})\frac{t_{dj} + t_{onj} - t_{offj}}{T_{sj}} - U_{Dj} - U_{Tj} \tag{8}$$

where U_{Tj} and U_{Dj} are the on-state voltage drop of switching devices and diodes, T_{sj}, t_{dj}, t_{onj} and t_{offj} are switching period, dead-time, turn-on time, and turn-off time of switching devices.

Meanwhile, $\text{sign}(i_{L1j})$ in Equation (7) can be expressed as

$$\text{sign}(i_{L1j}) = \begin{cases} 1 & i_{L1j} > 0 \\ -1 & i_{L1j} < 0 \end{cases} \tag{9}$$

From Figure 4, the closed-loop transfer function of the system can be expressed as

$$i_{oj} = G_j i_{refj} + i_{dj} - Y_j u_{PCC} = \frac{G_i G_{PWM} Z_{C1j} Z_{pj}}{Z_{pj}A + Z_{sj}B} i_{refj} + i_{dj} - \frac{Z_{L1j}Z_{C1j} + (Z_{L1j} + Z_{C1j})(Z_{pj} + Z_{L2j})}{Z_{pj}A + Z_{sj}B} u_{PCC} \tag{10}$$

where i_{refj} is the reference current of single inverter, G_j is the current source equivalent coefficient of single inverter, and Y_j is the equivalent admittance of single inverter.

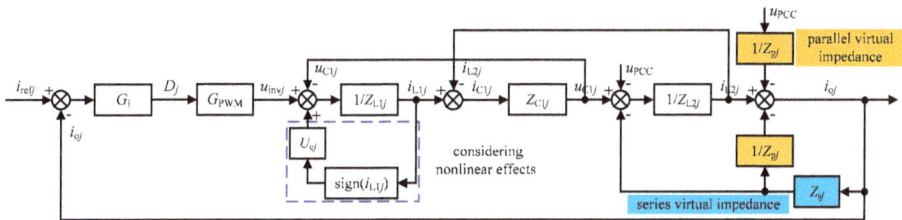

Figure 4. Equivalent control block diagram of oscillation suppression method by two notch filters for parallel inverters.

According to Equation (10), the refined equivalent output impedance model is shown in Figure 5. The current source i_j is equivalent to the current source i_{1j} in parallel with the current source i_{dj}. Two virtual impedances are added to be in parallel and in series with the original inverter output impedance, respectively. Thus, Figure 5 is equivalent to the refinement of Figure 3d, which can achieve the proposed approach.

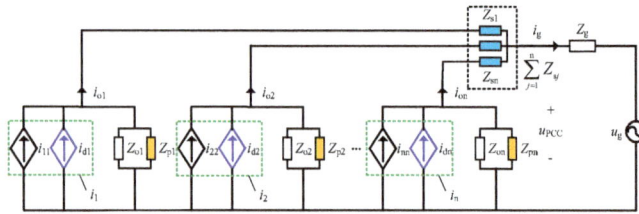

Figure 5. Refined equivalent output impedance model.

From Figure 5, the equivalent model of the m-th inverter can be expressed as

$$
\begin{aligned}
i_{om} &= G_m i_{refm} + i_{dm} - Y_m u_{PCC} \\
&= \frac{G_i G_{PWM} Z_{C1m} Z_{pm}}{Z_{pm} A + Z_{sm} B} i_{refm} + i_{dm} - \frac{Z_{L1m} Z_{C1m} + (Z_{L1m} + Z_{C1m})(Z_{pm} + Z_{L2m})}{Z_{pm} A + Z_{sm} B} u_{PCC}
\end{aligned}
\tag{11}
$$

where i_{refm}, G_m, Y_m, i_{dm}, Z_{L1m}, Z_{C1m}, Z_{L2m}, Z_{pm}, and Z_{sm} are the variables of the m-th inverter. Substitute Equation (1) into Equation (11), and Equation (11) can be rewritten as

$$
\begin{aligned}
i_{om} &= G_{selfm} i_{refm} + \sum_{j=1, j \neq m}^{n} G_{paralm,j} i_{refj} - G_{serim} u_g + i_{dm} \\
&= \left(G_m - \frac{Z_g Y_m G_m}{1 + Z_g \sum_{j=1}^{n} Y_j} \right) i_{refm} + \sum_{j=1, j \neq m}^{n} \left(-\frac{Z_g Y_m}{1 + Z_g \sum_{j=1}^{n} Y_j} G_j \right) i_{refj} - \frac{Y_m}{1 + Z_g \sum_{j=1}^{n} Y_j} u_g + i_{dm}
\end{aligned}
\tag{12}
$$

where G_{selfm} is the transfer relationship between the grid-side inductance current i_{om} of the m-th inverter and reference current i_{refm} of the m-th inverter, $G_{paralm,j}$ is the transfer relationship between the grid-side inductance current i_{om} of the m-th inverter and reference current i_{refj} of the jth inverter, and G_{serim} is the transfer relationship between grid-side inductance current i_{om} of the m-th inverter and grid voltage u_g.

When a similar analysis method is adopted to other inverters, the system can be expressed using a closed-loop transfer function matrix with reference currents, grid voltage and controlled current sources as inputs and grid-side inductor currents as outputs

$$
\begin{bmatrix}
i_{o1} \\
i_{o2} \\
\vdots \\
i_{on}
\end{bmatrix}
=
\begin{bmatrix}
G_{self1} & G_{paral1,2} & \cdots & G_{paral1,n} \\
G_{paral2,1} & G_{self2} & \cdots & G_{paral2,n} \\
\vdots & \vdots & \vdots & \vdots \\
G_{paraln,1} & G_{paraln,2} & \cdots & G_{selfn}
\end{bmatrix}
\begin{bmatrix}
i_{ref1} \\
i_{ref2} \\
\vdots \\
i_{refn}
\end{bmatrix}
-
\begin{bmatrix}
G_{seri1} \\
G_{seri2} \\
\vdots \\
G_{serin}
\end{bmatrix}
u_g +
\begin{bmatrix}
i_{d1} \\
i_{d2} \\
\vdots \\
i_{dn}
\end{bmatrix}
\tag{13}
$$

It is obvious that the strong coupling between the inverters and the grid may exist and introduce harmonic oscillation currents under weak grid condition.

According to Figure 4, Figure 6 gives the control block diagram of the oscillation suppression method by two notch filters for parallel inverters. The first notch filter is introduced into the feedforward path of PCC voltage and the second notch filter is introduced into the feedback path of the grid-side inductor current. The feedback path of the grid-side inductor current with the notch filter is equivalent to a virtual impedance in series with inverter output impedance, which can effectively improve the power quality of the grid-connected current. The feedforward path of PCC voltage with the notch filter equals to a virtual impedance in parallel with inverter output impedance. This can effectively restrain the parallel inverters' harmonic current from flowing into the grid and avoid the oscillation phenomenon. H_{1j} is the feedback coefficient of the grid-side inductor current, and H_{2j} is the feedforward coefficient of PCC voltage.

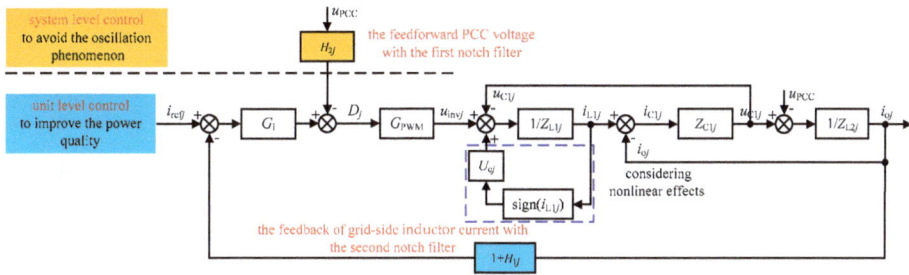

Figure 6. Control block diagram of the oscillation suppression method by two notch filters for parallel inverters.

From Figure 6, the equivalent closed-loop transfer function of the system can be expressed as

$$i_{oj} = G_{jeq}i_{refj} + i_{dj} - Y_{jeq}u_{PCC} = \frac{G_i G_{PWM} Z_{C1j}}{C+D+E}i_{refj} + i_{dj} - \frac{Z_{L1j} + Z_{C1j} + H_{2j}G_{PWM}Z_{C1j}}{C+D+E}u_{PCC} \quad (14)$$

where G_{jeq} is the equivalent coefficient of the current source after the single inverter transformation, Y_{jeq} is the inverter equivalent admittance after the single inverter transformation, $C = Z_{L1j}Z_{L2j}$, $D = (1 + H_{1j})G_i G_{PWM} Z_{C1j}$, and $E = Z_{C1j}(Z_{L1j} + Z_{L2j})$.

In order to achieve the same purpose of Figures 4 and 6, the equivalent coefficient of the current source and the inverter equivalent admittance in Equation (10) are equal to those in Equation (14). Thus, it can be expressed as

$$\begin{cases} \frac{G_i G_{PWM} Z_{C1j} Z_{pj}}{Z_{pj}A + Z_{sj}B} = \frac{G_i G_{PWM} Z_{C1j}}{C+D+E} \\ \frac{Z_{L1j}Z_{C1j} + (Z_{L1j} + Z_{C1j})(Z_{pj} + Z_{L2j})}{Z_{pj}A + Z_{sj}B} = \frac{Z_{L1j} + Z_{C1j} + H_2 G_{PWM} Z_{C1j}}{C+D+E} \end{cases} \quad (15)$$

From Equation (15), the feedback coefficient of the grid-side inductor current H_{1j} and the feedforward coefficient of PCC voltage H_{2j} can be expressed as

$$\begin{cases} H_{1j} = \frac{Z_{sj}(Z_{L1j}Z_{C1j} + (Z_{L1j} + Z_{C1j})(Z_{L2j} + Z_{pj}))}{G_i G_{PWM} Z_{C1j} Z_{pj}} \\ H_{2j} = \frac{Z_{L1j}Z_{C1j} + (Z_{L1j} + Z_{C1j})Z_{L2j}}{G_{PWM} Z_{C1j} Z_{pj}} \end{cases} \quad (16)$$

At the specific frequency, the amplitude of the notch filter is greatly attenuated while the amplitude at other frequencies is almost non-destructive. The notch filter G_N can be expressed as

$$G_N = \frac{(\frac{s}{2\pi f_o})^2 + 1}{(\frac{s}{2\pi f_o})^2 + \frac{s}{Q2\pi f_o} + 1} \quad (17)$$

where f_o is the fundamental frequency and Q is the quality factor of the notch filter.

The analysis diagrams of the notch filter with $Q = 0.25$, 0.5, and 1 are shown in Figure 7. From Figure 7a, the larger Q is, the better the notch characteristic of the notch filter, but the worse the frequency adaptability. From Figure 7b, when $Q = 0.25$, the characteristic equation has two unequal real poles on the negative real axis of the s plane, which is over-damping. When $Q = 0.5$, the characteristic equation has two equal real poles on the negative real axis of the s plane, which is critical-damping. When $Q = 1$, the characteristic equation has the conjugate complex poles on the left half plane, which is under-damping. From Figure 7c, the tuning time of notch filter with $Q = 0.5$ is better than $Q = 0.25$, 1. Therefore, considering the notch characteristics and dynamics comprehensively, the Q value was selected to be 0.5.

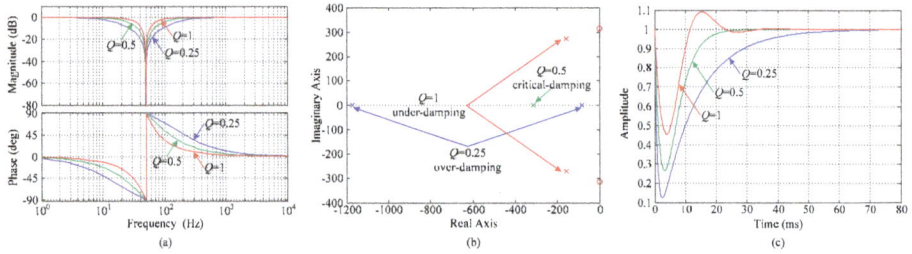

(a)   (b)   (c)

Figure 7. Analysis diagrams of notch filter with $Q = 0.25, 0.5, 1$. (**a**) The Bode diagram of notch filter; (**b**) Pole-zero plot of notch filter; (**c**) Unit step dynamic responses of notch filter.

The Bode diagrams of parallel virtual impedance Z_{pj} and series virtual impedance Z_{sj} are shown in Figure 8. At the fundamental frequency, the parallel virtual impedance Z_{pj} shows a high impedance, and the series virtual impedance Z_{sj} displays a low impedance, so that the fundamental current flows into the grid. At the high frequency, the parallel virtual impedance Z_{pj} shows a low impedance, and the series virtual impedance Z_{sj} displays a high impedance, so that the high-frequency harmonic current flows into the parallel virtual impedance Z_{pj} branch.

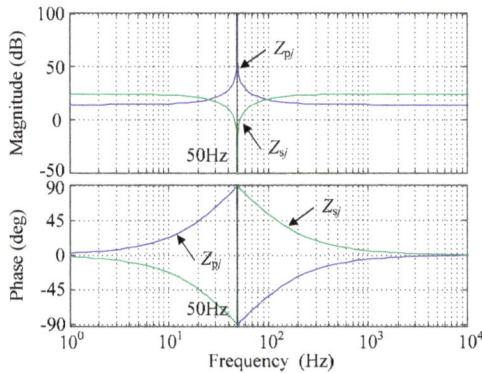

Figure 8. The Bode diagrams of parallel virtual impedance Z_{pj} and series virtual impedance Z_{sj}.

The Bode diagrams of the inverter self-impedance Z_{oj}, parallel virtual impedance Z_{pj}, grid equivalent impedance Z_{sgj}, and grid impedance Z_g are shown in Figure 9. Combined with Figure 5, the grid equivalent impedance Z_{sgj} is much lower than the self-impedance Z_{oj} and the parallel virtual impedance Z_{pj} at the fundamental frequency, thus most fundamental frequency current flows into the grid branch with relatively low impedance, which improves the power quality of grid-connected current. At the high frequency, the parallel virtual impedance Z_{pj} is much lower than the self-impedance Z_{oj} and grid equivalent impedance Z_{sgj}, thus most high-frequency harmonic current flows into the parallel virtual impedance Z_{pj} branch with relatively low impedance, and it effectively suppresses the oscillation of the paralleled multi-inverter system.

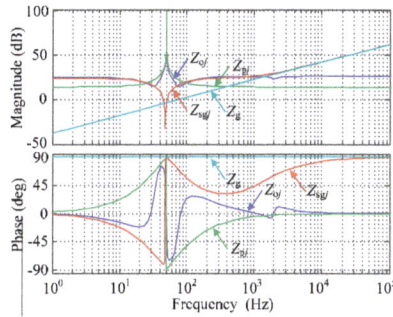

Figure 9. The Bode diagrams of the inverter self-impedance Z_{oj}, parallel virtual impedance Z_{pj}, grid equivalent impedance Z_{sgj}, and grid impedance Z_g.

4. Contrast Analysis of System Stability under Weak Grid Condition

In order to compare and analyze the system stability, different conditions of adding parallel-series virtual impedances are shown in Table 1. Parallel virtual impedance $Z_{pj} = r_1$ and series virtual impedance $Z_{sj} = r_2$, abbreviated as case I. Parallel virtual impedance $Z_{pj} = r_1/G_N$ and series virtual impedance $Z_{sj} = r_2 G_N$, abbreviated as case II (proposed control method).

Table 1. Different conditions of adding parallel-series virtual impedances.

Number	Parallel Virtual Impedance Z_{pj}	Series Virtual Impedance Z_{sj}	Control Method
I	r_1	r_2	case I
II	r_1/G_N	$r_2 G_N$	case II

To verify that the system satisfies the precondition of using the Nyquist criterion, the open-loop Bode diagrams and closed-loop pole-zero diagrams of the equivalent coefficient of current source G_{jeq} for single inverter in different cases are shown in Figure 10. From Figure 10a,b, it can be seen that the gain margin (GM) and the phase margin (PM) are greater than 0, no right half-plane pole exists, and the single-inverter system is in a stable state. Therefore, the system satisfies the precondition of using the Nyquist criterion. The stability condition of the paralleled multi-inverter system is that the Nyquist curve of the impedance ratio does not surround $(-1, j0)$.

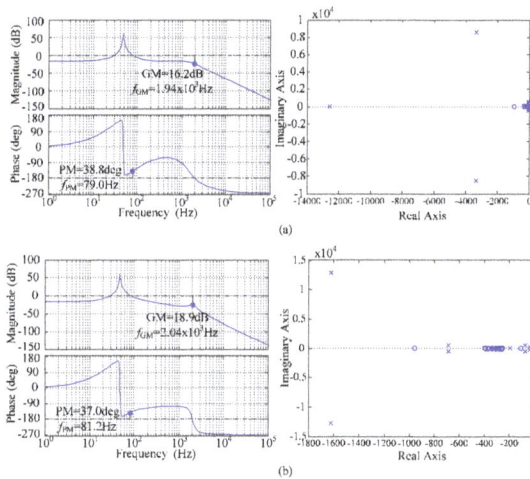

Figure 10. The open-loop Bode diagrams and closed-loop pole-zero diagrams of the equivalent coefficient of current source G_{jeq} for single inverter in different cases. (**a**) Case I; (**b**) Case II.

Figure 11 shows the Nyquist diagrams of impedance ratio *K* for the paralleled multi-inverter system. From Figure 11a, when the number of inverters is 2, 3, 4, 5, and 6, respectively, the Nyquist curve does not surround $(-1, j0)$ in case I, and the system is in a stable state. However, when the number of inverters is 7, 8, 9, 10, and 11, respectively, the Nyquist curves all surround $(-1, j0)$, and the system is in an unstable state. However, from Figure 11b, regardless of the number of inverters, the Nyquist curves would never wrap around $(-1, j0)$ in case II, and the system is in a stable state. Therefore, compared with case I, the proposed oscillation suppression method by two notch filters can effectively restrain parallel inverters' harmonic current from flowing into the grid, and avoid the oscillation phenomenon.

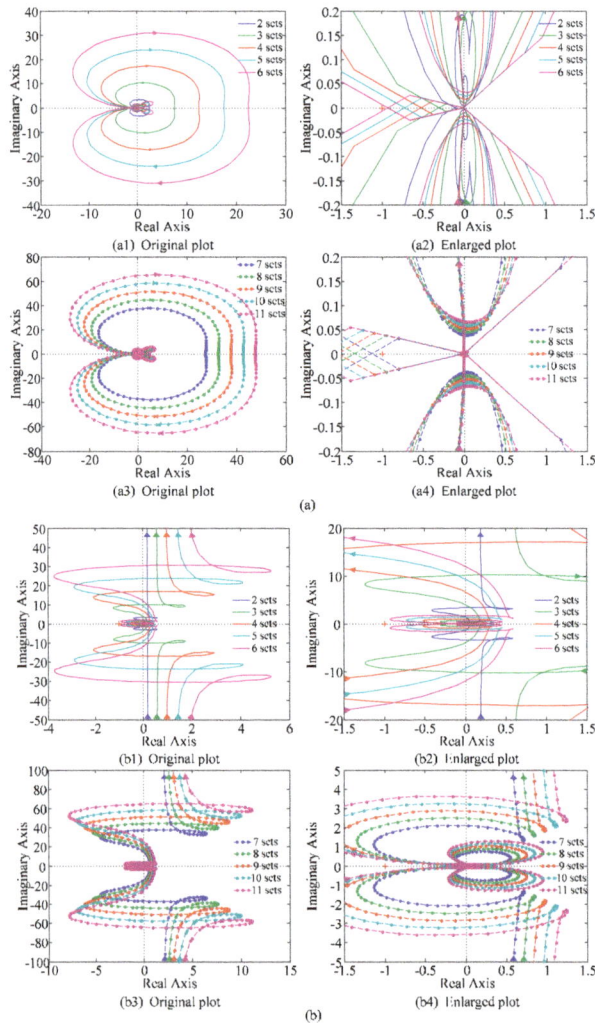

Figure 11. Nyquist diagrams of impedance ratio *K* for the paralleled multi-inverter system. (**a**) Case I; (**b**) Case II.

5. Simulation Verification

To verify the correctness of theoretical analysis, the simulation model of a paralleled multi-inverter system was built based on Figure 1. Several cases including two and seven inverters were simulated by PSIM (Powersim Inc., Rockville, MD, USA) simulation. The system parameters are shown in Table 2.

When two inverters were connected to a weak grid in case I and case II, the root mean square (RMS) value of the reference grid-side inductor current i_{refm} for each grid-connected inverter increased from 18.75 A to 37.5 A at 0.405 s. Therefore, the RMS of the reference grid-connected current i_{gref} increased from 37.5 A to 75 A for the two parallel inverters. Simulation results of the grid-connected current with two parallel inverters are shown in Table 3.

Table 2. System parameters.

Parameter/Unit	Value	Parameter/Unit	Value
DC voltage U_{dc}/V	720	Switching frequency f_s/kHz	10
Grid phase voltage U_g/V	220	Proportional coefficient k_p	2.1
Amplitude of triangular carrier U_{tri}/V	1	Resonance gain k_{i1}	175
Inverter-side inductor L_1/mH	2.2	Cut-off angular frequency ω_c/rad/s	6.28
Parasitic resistance of L_1 R_{L1}/Ω	0.002	Fundamental angular frequency ω_o/rad/s	314
Filter capacitor C_1/μF	10	Fundamental frequency f_o/Hz	50
Grid-side inductor L_2/mH	0.8	Quality factor Q	0.5
Parasitic resistance of L_2 R_{L2}/Ω	0.001	Proportional coefficient r_1	5
Grid inductor L_g/mH	2	Proportional coefficient r_2	15

Table 3. Simulation results of grid-connected current with two parallel inverters. THD: total harmonic distortion.

Number	Case	Before Reference Current Transient		After Reference Current Transient	
		THD	Resonance Point and Resonance Peak	THD	Resonance Point and Resonance Peak
2	I	6.07%	25th harmonic 1.77 A	3.16%	25th harmonic 1.81 A
	II	3.27%	39th harmonic 0.88 A	1.51%	39th harmonic 0.79 A

In case I, the transient simulation waveforms of PCC voltage u_{PCC} and grid-connected current i_g are depicted in Figure 12a,b. It can be seen that the distortion rate of PCC voltage u_{PCC} is 1.57%. The grid-connected current i_g and corresponding spectrum from 0.30 s to 0.36 s are shown in Figure 12c,d, which describes the situation before the reference current surges. The distortion rate of grid-connected current i_g is 6.07%, the resonance point is near the 25th harmonic (1250 Hz), and the resonance peak is 1.77 A. The grid-connected current i_g and corresponding spectrum after the reference current suddenly increases are shown in Figure 12e,f. The distortion rate of grid-connected current i_g decreases to 3.16%, the resonance point is still near the 25th harmonic (1250 Hz), and the resonance peak value is 1.81 A. At this time, the major high-frequency harmonics of the grid-connected current are 25th harmonics.

In case II, the transient simulation waveforms of PCC voltage u_{PCC} and the grid-connected current i_g are depicted in Figure 13a,b. From Figure 13a, the distortion rate of PCC voltage u_{PCC} is 0.82%. The grid-connected current i_g and corresponding spectrum before the reference current surges are shown in Figure 13c,d. The distortion rate of the grid-connected current i_g is 3.27%, the resonance point is near the 39th harmonic (1950 Hz), and the resonance peak is 0.88 A. The grid-connected current i_g and corresponding spectrum after the reference current suddenly increases are shown in Figure 13e,f. The distortion rate of the grid-connected current i_g decreases to 1.51%, the resonance point is still near the 39th harmonic (1950 Hz), and the resonance peak value is 0.79 A. Meanwhile, the power quality of the grid-connected current can be significantly improved, and the resonance phenomenon has obviously decreased.

When seven inverters are connected to a weak grid in cases I and II, the RMS of reference grid-side inductor current i_{refm} for each grid-connected inverter is 10.71 A. Therefore, the RMS of the reference grid-connected current i_{gref} is 75 A for seven parallel inverters. The simulation waveforms and spectrograms of PCC voltage u_{PCC} and the grid-connected current i_g are shown in Figures 14 and 15. The simulation results of the grid-connected current with seven parallel inverters are shown in Table 4. In case I, the system is in an unstable state. The resonance phenomenon is obvious. The reason is that the high-frequency harmonic current frequency is equal to or close to the parallel resonance frequency of self-impedance, resulting in a parallel resonance or quasi-resonance of the impedance network. The grid-connected current is still severely polluted by the impedance coupling interactions between the inverters and the grid. In case II, the distortion rate of the grid-connected current i_g is 3.38%, the resonance point is near the 39th harmonic (1950 Hz), and the resonance peak is 1.51 A. Due to the sufficient resistive damping introduced to the impedance network, the system can operate stably. Therefore, case II can effectively improve the power quality of the grid-connected current and suppress the oscillation of the paralleled multi-inverter system.

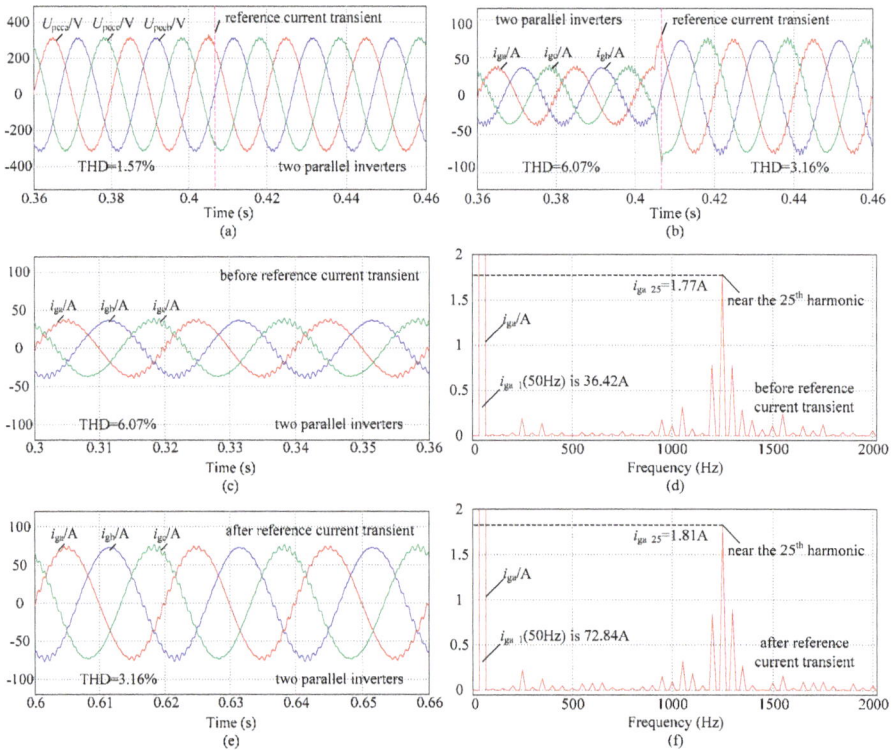

Figure 12. Simulation waveforms of point of common coupling (PCC) voltage u_{PCC} and grid-connected current i_g in case I during reference current transient (two parallel inverters). (**a**) PCC voltage u_{PCC}; (**b**) grid-connected current i_g; (**c**) grid-connected current i_g (before); (**d**) spectrogram of grid-connected current i_g (before); (**e**) grid-connected current i_g (after); (**f**) spectrogram of grid-connected current i_g (after).

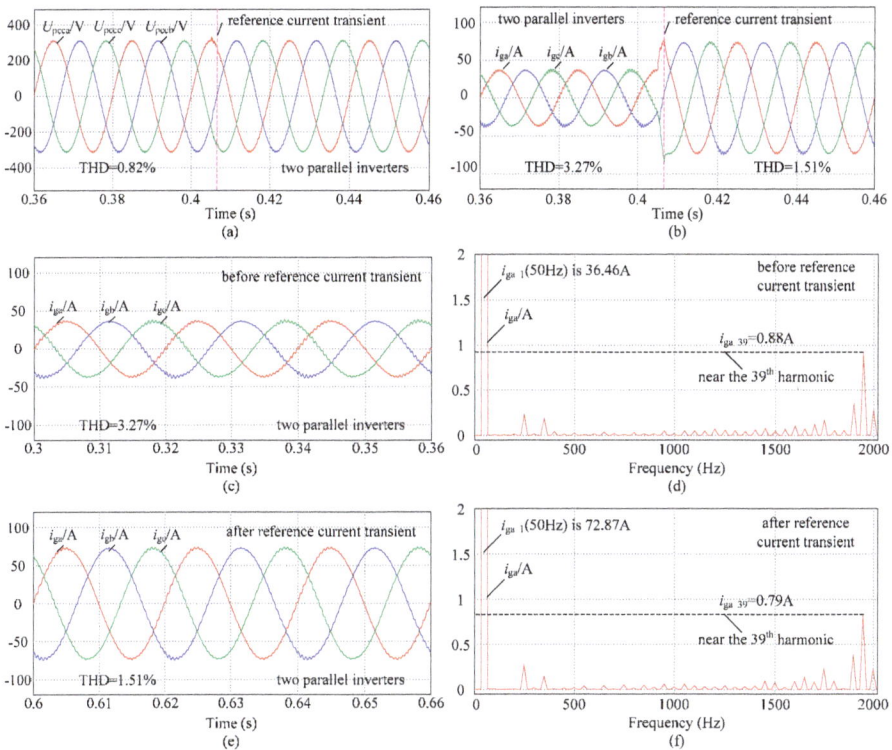

Figure 13. Simulation waveforms of PCC voltage u_{PCC} and the grid-connected current i_g in case II during reference current transient (two parallel inverters). (**a**) PCC voltage u_{PCC}; (**b**) grid-connected current i_g; (**c**) grid-connected current i_g (before); (**d**) spectrogram of grid-connected current i_g (before); (**e**) grid-connected current i_g (after); (**f**) spectrogram of grid-connected current i_g (after).

Table 4. Simulation results of grid-connected current with seven parallel inverters.

Number	Case	THD	Resonance Point and Resonance Peak
7	I	(unstable)	(unstable)
	II	3.38%	39th harmonic 1.51 A

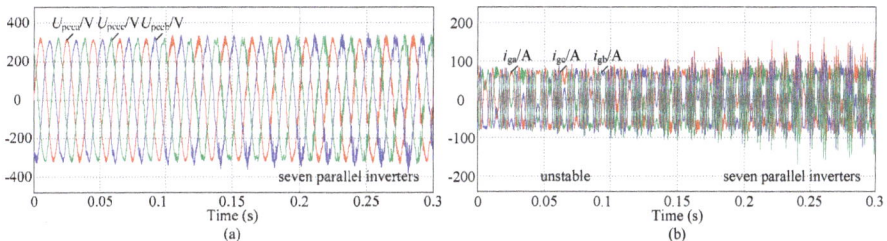

Figure 14. Simulation waveforms of PCC voltage u_{PCC} and the grid-connected current i_g in case I (seven parallel inverters). (**a**) PCC voltage u_{PCC}; (**b**) grid-connected current i_g.

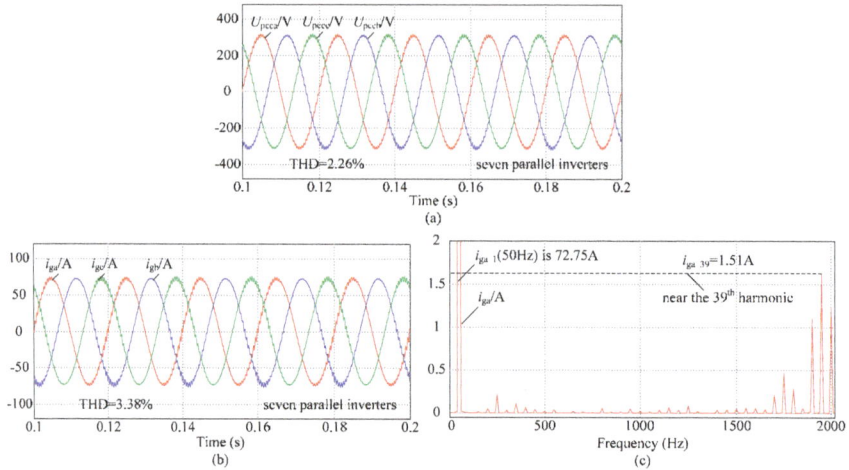

Figure 15. Simulation waveforms of PCC voltage u_{PCC} and the grid-connected current i_g in case II (seven parallel inverters). (**a**) PCC voltage u_{PCC}; (**b**) grid-connected current i_g; (**c**) spectrogram of grid-connected current i_g.

6. Experimental Verification

To verify the validity of simulation analysis, a hardware-in-the-loop experimental platform based on a real controller TMS320F2812 digital signal processor (DSP) (Texas Instruments, Inc, Dallas, TX, USA) and a real-time laboratory (RT-LAB) (Opal-RT Technologies, Montreal, QC, Canada) was built [26], as shown in Figure 16a. The system parameters are shown in Table 2. The hardware-in-the-loop experimental platform mainly includes an RT-LAB simulator OP5700 (Opal-RT Technologies, Montreal, QC, Canada), real controllers of grid-connected inverters, a host computer as a real-time control interface, and an oscilloscope, as shown in Figure 16b. The main circuit model of the system was established in the host computer, which was loaded to the OP5700. When the model was running, the OP5700 sent the analog signals (voltages and currents) to the real controllers through the input/output (I/O) ports in real time. After data processing, the real controllers transmitted the digital control signals (pulses) to OP5700 through the I/O ports in real time. The data interacted in real time during the simulation process to ensure the normal operation of the system. The measurement information during the simulation were converted into analog signals through the I/O ports, which could be observed on the oscilloscope.

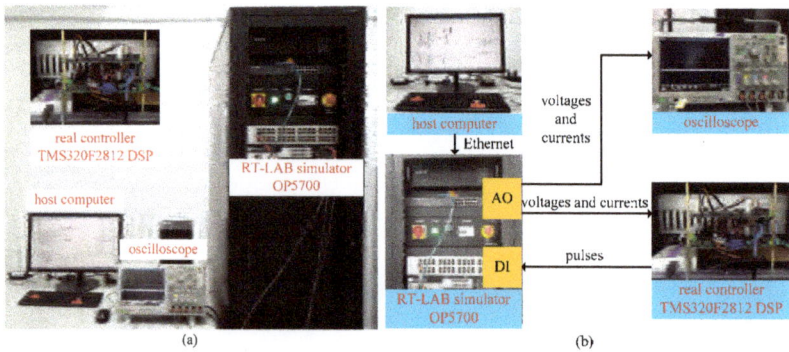

Figure 16. Experimental platform for paralleled multi-inverter system based on hardware-in-loop simulation. (**a**) Whole; (**b**) Structure.

When two inverters operated in parallel, as in case I and case II, the RMS values of reference for the grid-side inductor current i_{refm} for each grid-connected inverter increased from 18.75 A to 37.5 A at 0.405 s. Therefore, the RMS of reference grid-connected current i_{gref} increased from 37.5 A to 75 A for two parallel inverters. The transient experimental waveforms and spectrograms in case I and case II are depicted in Figures 17 and 18. Experimental results of the grid-connected current with two parallel inverters are shown in Table 5. Before the reference current surged in Figure 17b, the distortion rate of the grid-connected current i_g was 7.01%, the resonance point was near the 25th harmonic (1250 Hz), and the resonance peak was 2.66 A. After the reference current suddenly increased in Figure 17c, the distortion rate of the grid-connected current i_g decreased to 4.22%, the resonance point was still near the 25th harmonic (1250 Hz), and the resonance peak value was 2.89 A. Thus, the grid-connected current apparently contained high-frequency ripples, and it is obvious that the major harmonics are 25th harmonics.

Before the reference current surged in Figure 18b, the distortion rate of the grid-connected current i_g was 4.82%, the resonance point was near the 39th harmonic (1950 Hz), and the resonance peak was 2.06 A. After the reference current suddenly increased in Figure 18c, the distortion rate of grid-connected current i_g decreased to 2.79%, the resonance point was still near the 39th harmonic (1950 Hz), and the resonance peak value was 2.04 A. Meanwhile, the power quality of grid-connected current can be significantly improved, and the resonance phenomenon has obviously decreased. Therefore, the distortion rate was smaller after the reference current surge in the same case. Moreover, before and after the reference current surged, the distortion rate in case II was less than case I.

Table 5. Experimental results of the grid-connected current with two parallel inverters.

Number	Case	Before Reference Current Transient		After Reference Current Transient	
		THD	Resonance Point and Resonance Peak	THD	Resonance Point and Resonance Peak
2	I	7.01%	25th harmonic 2.66 A	4.22%	25th harmonic 2.89 A
	II	4.82%	39th harmonic 2.06 A	2.79%	39th harmonic 2.04 A

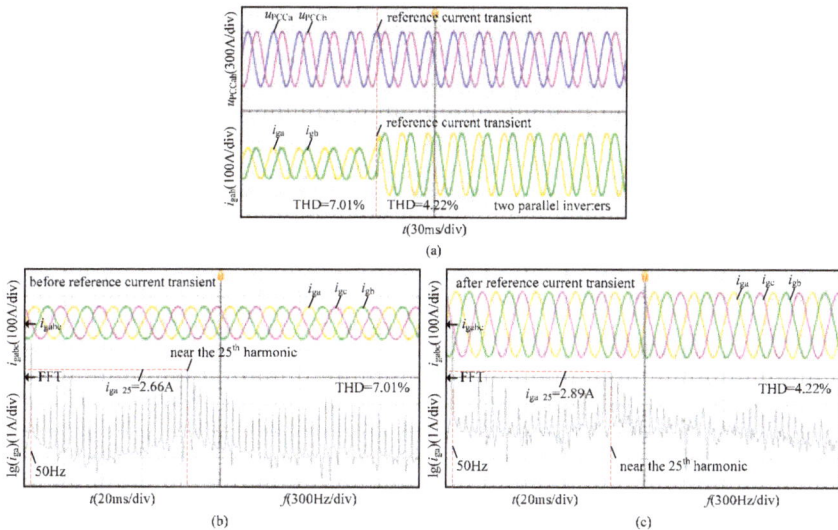

Figure 17. Experimental waveforms of PCC voltage u_{PCC} and grid-connected current i_g in case I during reference current transient (two parallel inverters). (**a**) PCC voltage u_{PCC} and grid-connected current i_g; (**b**) grid-connected current i_g and spectrogram (before); (**c**) grid-connected current i_g and spectrogram (after).

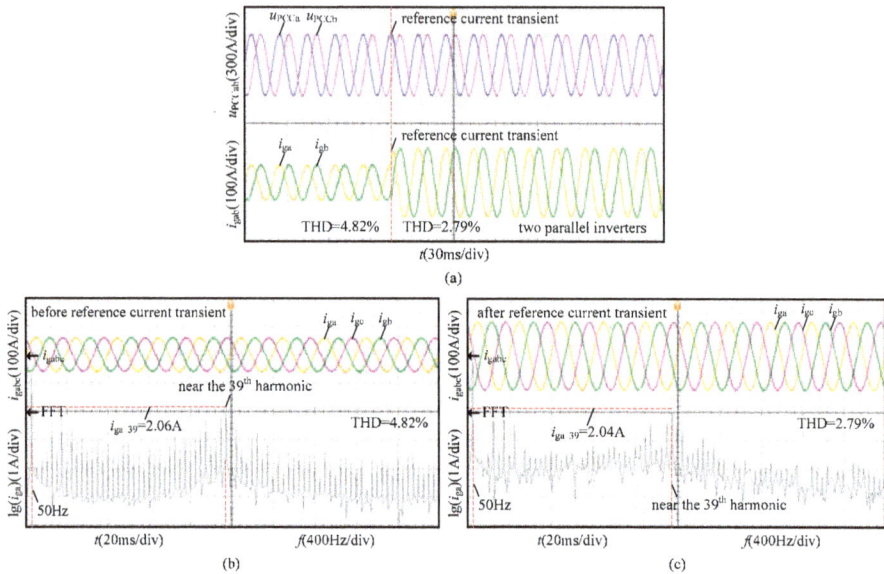

Figure 18. Experimental waveforms of PCC voltage u_{PCC} and grid-connected current i_g in case II during reference current transient (two parallel inverters). (**a**) PCC voltage u_{PCC} and grid-connected current i_g; (**b**) grid-connected current i_g and spectrogram (before); (**c**) grid-connected current i_g and spectrogram (after).

When seven inverters operate in parallel in cases I and case II, the RMS of the reference grid-side inductor current i_{refm} for each grid-connected inverter was 10.71 A. Therefore, the RMS of the reference grid-connected current i_{gref} was 75 A for the seven parallel inverters. The experimental waveforms and spectrograms in cases I and case II are shown in Figures 19 and 20. The experimental results of the grid-connected current with the seven parallel inverters are shown in Table 6. As can be seen in Figure 19, the system is in an unstable state. The resonance phenomenon is obvious. The reason is that the high-frequency harmonic current frequency is equal to or close to the parallel resonance frequency of self-impedance, resulting in a parallel resonance or quasi-resonance of the impedance network. The grid-connected current is still severely polluted by the impedance coupling interactions between the inverters and the grid. From Figure 20, it can be seen that the distortion rate of the grid-connected current i_g is 4.36%, the resonance point is near the 39th harmonic (1950 Hz), and the resonance peak is 2.99 A. Due to the sufficient resistive damping introduced to the impedance network, the system can operate stably. Therefore, case II can effectively improve the power quality of the grid-connected current and suppress the oscillation of the paralleled multi-inverter system.

Table 6. Experimental results of the grid-connected current with seven parallel inverters.

Number	Case	THD	Resonance Point and Resonance Peak
7	I	(unstable)	(unstable)
	II	4.36%	39th harmonic 2.99 A

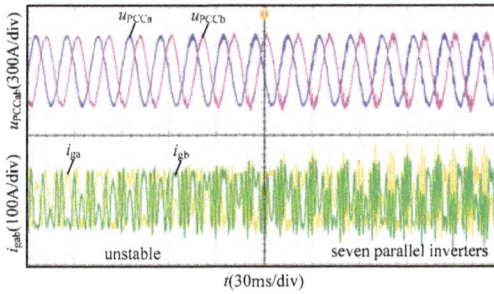

Figure 19. Experimental waveforms of PCC voltage u_{PCC} and the grid-connected current i_g in case I (seven parallel inverters).

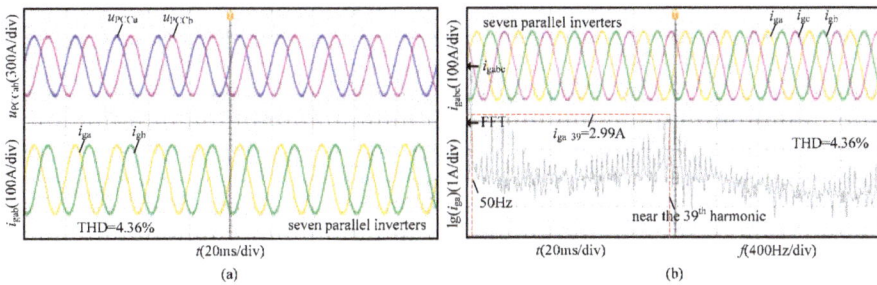

Figure 20. Experimental waveforms of PCC voltage u_{PCC} and the grid-connected current i_g in case II (seven parallel inverters). (**a**) PCC voltage u_{PCC} and grid-connected current i_g; (**b**) grid-connected current i_g and spectrogram.

7. Conclusions

In this paper, the oscillation suppression method by two notch filters is proposed to realize the virtual impedances and increase the whole system damping. The implementation form of the virtual impedances is presented by the proposed PCC voltage feedforward and grid-side inductor current feedback with two notch filters. The feedforward path of PCC voltage with the notch filter equals to a virtual impedance in parallel with inverter output impedance, which is designed to show low impedance at the harmonic oscillation frequency. By doing so, most high-frequency harmonic current will flow into the parallel virtual impedance branch, and it effectively suppresses the oscillation. Meanwhile, the feedback path of the grid-side inductor current with the notch filter is equivalent to a virtual impedance in series with inverter output impedance, which is designed to display low impedance at the fundamental frequency. This way, most of the fundamental frequency current flows into the grid branch with relatively low impedance. In addition, it improves the power quality of the grid-connected current. Finally, simulation and experimental results are provided to verify the validity of proposed control method.

Author Contributions: L.Y. and Y.C. provided the original idea for this paper. L.Y., H.W., A.L. and K.H. organized the manuscript and attended the discussions when analysis and verification were carried out. All the authors gave comments and suggestions on the writing and descriptions of the manuscript.

Funding: This research was supported by the National Key R&D Program of China under Grant No. 2017YFB0902000, and the Science and Technology Project of State Grid under Grant No. SGXJ0000KXJS1700841.

Conflicts of Interest: The authors declare no conflict of interest.

References

1. Yu, Y.; Konstantinou, G.; Hredzak, B.; Agelidis, V.G. Power balance optimization of cascaded H-bridge multilevel converters for large-scale photovoltaic integration. *IEEE Trans. Power Electron.* **2016**, *31*, 1108–1120. [CrossRef]
2. Guo, X.; Yang, Y.; Zhu, T. ESI: A novel three-phase inverter with leakage current attenuation for transformerless PV systems. *IEEE Trans. Ind. Electron.* **2018**, *65*, 2967–2974. [CrossRef]
3. Bouloumpasis, I.; Vovos, P.; Georgakas, K.; Vovos, N.A. Current harmonics compensation in microgrids exploiting the power electronics interfaces of renewable energy sources. *Energies* **2015**, *8*, 2295–2311. [CrossRef]
4. Hany, M.H. Whale optimisation algorithm for automatic generation control of interconnected modern power systems including renewable energy sources. *IET Gener. Transm. Distrib.* **2018**, *12*, 607–614.
5. Li, X.; Fang, J.; Tang, Y.; Wu, X. Robust design of LCL filters for single-current-loop-controlled grid-connected power converters with unit PCC voltage feedforward. *IEEE J. Emerg. Sel. Top. Power Electron.* **2018**, *6*, 54–72. [CrossRef]
6. Guo, X.; Yang, Y.; Wang, X. Advanced control of grid-connected current source converter under unbalanced grid voltage conditions. *IEEE Trans. Ind. Electron.* **2018**, *65*, 9225–9233. [CrossRef]
7. Yang, Y.; Ye, Q.; Tung, L.J.; Greenleaf, M.; Li, H. Integrated size and energy management design of battery storage to enhance grid integration of large-scale PV power plants. *IEEE Trans. Ind. Electron.* **2018**, *65*, 394–402. [CrossRef]
8. Rohner, S.; Bernet, S.; Hiller, M.; Sommer, R. Modulation, losses, and semiconductor requirements of modular multilevel converters. *IEEE Trans. Ind. Electron.* **2010**, *57*, 2633–2642. [CrossRef]
9. Remon, D.; Cantarellas, A.M.; Mauricio, J.M.; Rodriguez, P. Power system stability analysis under increasing penetration of photovoltaic power plants with synchronous power controllers. *IET Renew. Power Gener.* **2017**, *11*, 733–741. [CrossRef]
10. Malinowski, M.; Gopakumar, K.; Rodriguez, J.; Perez, M.A. A survey on cascaded multilevel inverters. *IEEE Trans. Ind. Electron.* **2010**, *57*, 2197–2206. [CrossRef]
11. Amin, M.; Molinas, M. Small-signal stability assessment of power electronics based power systems: A discussion of impedance- and eigenvalue-based methods. *IEEE Trans. Ind. Appl.* **2017**, *53*, 5014–5030. [CrossRef]
12. Bakhshizadeh, M.K.; Yoon, C.; Hjerrild, J.; Bak, C.L.; Kocewiak, Ł.H.; Blaabjerg, F.; Hesselbæk, B. The application of vector fitting to eigenvalue-based harmonic stability analysis. *IEEE J. Emerg. Sel. Top. Power Electron.* **2017**, *5*, 1487–1498. [CrossRef]
13. Sun, J. Impedance-based stability criterion for grid-connected inverters. *IEEE Trans. Power Electron.* **2011**, *26*, 3075–3078. [CrossRef]
14. Yang, L.; Chen, Y.; Luo, A.; Chen, Z.; Zhou, L.; Zhou, X.; Wu, W.; Tan, W.; Guerrero, J.M. Effect of phase-locked loop on small-signal perturbation modelling and stability analysis for three-phase LCL-type inverter connected to weak grid. *IET Renew. Power Gener.* **2008**. [CrossRef]
15. Xu, J.; Xie, S.; Tang, T. Improved control strategy with grid-voltage feedforward for LCL-filter-based inverter connected to weak grid. *IET Power Electron.* **2014**, *7*, 2660–2671. [CrossRef]
16. Harnefors, L. Modeling of three-phase dynamic systems using complex transfer functions and transfer matrices. *IEEE Trans. Ind. Electron.* **2007**, *54*, 2239–2248. [CrossRef]
17. Wang, X.; Blaabjerg, F.; Liserre, M.; Chen, Z.; He, J.; Li, Y. An Active Damper for Stabilizing Power-Electronics-Based AC Systems. *IEEE Trans. Power Electron.* **2014**, *29*, 3318–3329. [CrossRef]
18. Yang, D.; Ruan, X.; Wu, H. A real-time computation method with dual sampling modes to improve the current control performances of the LCL-type grid-connected inverter. *IEEE Trans. Ind. Electron.* **2015**, *62*, 4563–4572. [CrossRef]
19. Li, X.; Wu, X.; Geng, Y.; Yuan, X.; Xia, C.; Zhang, X. Wide damping region for LCL-type grid-connected inverter with an improved capacitor-current-feedback method. *IEEE Trans. Power Electron.* **2015**, *30*, 5247–5259. [CrossRef]
20. Dannehl, J.; Fuchs, F.W.; Hansen, S.; Thogersen, P.B. Investigation of active damping approaches for PI-based current control of grid-connected pulse width modulation inverters with LCL filters. *IEEE Trans. Ind. Appl.* **2010**, *46*, 1509–1517. [CrossRef]

21. Komurcugil, H.; Altin, N.; Ozdemir, S.; Sefa, I. Lyapunov-function and proportional-resonant-based control strategy for single-phase grid-connected VSI with LCL filter. *IEEE Trans. Ind. Electron.* **2016**, *63*, 2838–2849. [CrossRef]

22. Chen, Z.; Chen, Y.; Guerrero, J.M.; Kuang, H.; Huang, Y.; Zhou, L.; Luo, A. Generalized coupling resonance modeling, analysis, and active damping of multi-parallel inverters in microgrid operating in grid-connected mode. *J. Mod. Power Syst. Clean Energy* **2016**, *4*, 63–75. [CrossRef]

23. He, J.; Li, Y.; Bosnjak, D.; Harris, B. Investigation and active damping of multiple resonances in a parallel-inverter-based microgrid. *IEEE Trans. Power Electron.* **2013**, *28*, 234–246. [CrossRef]

24. Zheng, C.; Zhou, L.; Xie, B.; Zhang, Q.; Li, H. A stabilizer for suppressing harmonic resonance in multi-parallel inverter system. In Proceedings of the IEEE Transportation Electrification Conference and Expo, (ITEC Asia-Pacific), Harbin, China, 1–6 August 2017.

25. Jeong, S.-G.; Park, M.-H. The analysis and compensation of dead-time effects in PWM inverters. *IEEE Trans. Ind. Electron.* **1991**, *38*, 108–114. [CrossRef]

26. Shuai, Z.; Huang, W.; Shen, C.; Ge, J.; Shen, Z.J. Characteristics and restraining method of fast transient inrush fault currents in synchronverters. *IEEE Trans. Ind. Electron.* **2017**, *64*, 7487–7497. [CrossRef]

![energies logo] *energies*

MDPI

Article

A Grid-Supporting Photovoltaic System Implemented by a VSG with Energy Storage

Huadian Xu, Jianhui Su, Ning Liu * and Yong Shi

School of Electrical Engineering and Automation, Hefei University of Technology, Hefei 23009, China;
xuhuadian@mail.hfut.edu.cn (H.X.); su_chen@hfut.edu.cn (J.S.); shiyong@hfut.edu.cn (Y.S.)
* Correspondence: Ning.Liu@unb.ca; Tel.: +86-551-6290-4042

Received: 24 October 2018; Accepted: 8 November 2018; Published: 14 November 2018

Abstract: Conventional photovoltaic (PV) systems interfaced by grid-connected inverters fail to support the grid and participate in frequency regulation. Furthermore, reduced system inertia as a result of the integration of conventional PV systems may lead to an increased frequency deviation of the grid for contingencies. In this paper, a grid-supporting PV system, which can provide inertia and participate in frequency regulation through virtual synchronous generator (VSG) technology and an energy storage unit, is proposed. The function of supporting the grid is implemented in a practical PV system through using the presented control scheme and topology. Compared with the conventional PV system, the grid-supporting PV system, behaving as an inertial voltage source like synchronous generators, has the capability of participating in frequency regulation and providing inertia. Moreover, the proposed PV system can mitigate autonomously the power imbalance between generation and consumption, filter the PV power, and operate without the phase-locked loop after initial synchronization. Performance analysis is conducted and the stability constraint is theoretically formulated. The novel PV system is validated on a modified CIGRE benchmark under different cases, being compared with the conventional PV system. The verifications demonstrate the grid support functions of the proposed PV system.

Keywords: coordination control; energy storage; grid support function; inertia; photovoltaic; virtual synchronous generator

---

1. Introduction

This paper proposes a grid-supporting photovoltaic system, including implementation and performance analysis. In this section, the background, literature review, formulation of the problem of interest for this investigation, scope and contribution of this study, and organization of the paper are presented.

1.1. Background and Significance

Synchronous generators (SGs), which take responsibility for frequency regulation in electric power systems (EPS), operate as inertial voltage sources, providing the inertia to slow down frequency dynamics and moderate the power imbalance between generation and consumption in an autonomous fashion. Driven by issues such as potential exhaustion of conventional fossil fuel based energies (e.g., coal, oil, and natural gas) and increasing environmental concerns, the quantity of renewable energy sources (RES) integrated into EPS is escalating [1,2]. In consequence, SGs are gradually being replaced by inverters with high penetration of RES.

Among RES, solar energy via photovoltaic (PV) systems is one of the most promising, and has largely penetrated the global energy market [3,4]. A decrease of investment costs, technological development, and governmental support has led to the significant increase in PV systems that has been seen in recent years, and there is still significant need for growth [1,5].

However, grid-connected inverters are controlled as current sources with phase-locked loops (PLL) in conventional PV systems [6], which turns conventional PV systems into power injectors and grid-following units [7,8]. As power injectors, conventional PV systems inject the power extracted from the PV array to EPS without the capability to mitigate the power imbalance between generation and consumption. Meanwhile, conventional PV systems, as grid-following units, provide little of the inertia that plays an essential role in short-term system stability [9], thus the increasing penetration of PV generators reduces the inertia of EPS, which exacerbates the system's stability [9–11]. Therefore, implementing the function of a supporting grid in PV systems, which offers inertia and participates in frequency regulation, is significantly beneficial to the stability enhancement of EPS.

1.2. Formulation of the Problem of Interest for This Study

Existing research mainly concentrates on the realization of a virtual synchronous generator (VSG) and the improvement of the performance of inverters that are assumed to be supplied by stiff DC voltage sources. However, rare attention has been paid to the realization of a VSG in practical PV systems. It is a challenge to introduce inertia, the indispensable property for VSGs, into an inverter only powered by a PV system. To enable the inverter to emulate the inertia of SGs, an energy buffer, whose function is identical to a rotor for kinetic energy, needs to be installed at the dc link of the inverter.

The main objective of this paper is therefore to equip PV systems with the function of supporting the grid through emulating SG characteristics, then analyze the performance, formulate the stability constraint, and corroborate the implemented function with numerical experiments.

1.3. Literature Review

Virtual synchronous generator (VSG) technology, which controls inverters to mimic the characteristics of SGs to provide inertia and participate in the frequency regulation of EPS, emerged in response to issues addressed in Section 1.2 [12–15]. The core of VSG technology is to present the various energy sources interfaced to the grid through power electronic converters as SGs [16]. Researches on VSGs in References [12–15] are devoted to realizing the basic function of emulating inertia using converters. Recently, VSG research focus towards developing novel control strategies, and improving the performance of these strategies from the point-of-view of enhanced dynamics, stability, and so on. Compared with the method investigated in Reference [14], the filter inductance of the synchronverter studied in Reference [17] is virtually increased to improve the stability. In the early Ise lab's topology presented in Reference [15], active power oscillation becomes one of the major concerns during the emulation of inertia [18]. In Reference [18], an alternating moment of inertia is proposed to suppress such oscillation. Furthermore, the self-adaptive inertia and damping approach is presented to improve the dynamics in Reference [19]. To smooth transitions and reduce frequency excursions, a particle swarm optimization technique was developed in Reference [20], a self-tuning VSG was investigated in Reference [21] and an auxiliary loop is proposed to adjust the dynamic response speed through correct the damping in Reference [22].

In References [14,23], inverters that mimic SGs were studied with an assumption that inverters are supplied by stiff DC voltage sources, but inverters only energized by PV systems cannot satisfy this assumption. In Reference [24], a VSG was realized by a battery/ultracapacitor hybrid ES system, but the inverters based on RES were not competent in emulating the characteristics of SGs. In References [15,25], it is pointed out that energy storage (ES) should be installed to emulate the kinetic energy stored in the rotating rotors of SGs, but the detailed system topology is not considered, nor is the control strategy coordinating ES and the renewable energy generator. In References [26,27], ES is applied to a PV system to smooth power fluctuation, and this system possesses no characteristics of SGs. In Reference [28], a battery is used as the backup for a PV inverter that employs only PQ control or droop control, causing the inverter to operate unlike SGs.

1.4. Scope and Contribution of This Study

Motivated by the above observations, this paper presents a novel grid-supporting PV system, achieving emulation of SG characteristics. Consequently, the grid-supporting PV system, behaving as SGs, contributes to supporting the grid by autonomously mitigating the power imbalance between generation and consumption, and slowing down frequency dynamics with virtual inertia. Accordingly, the proposed grid-supporting PV system is superior to the conventional PV system, while the conventional PV system cannot moderate the deficits and surplus of power in the grid, and is unable to provide inertia.

1.5. Organization of the Paper

The content of this paper is organized as follows: Section 2 introduces the topology and control scheme. In Section 3, performance analysis is conducted, and the stability constraint is obtained through the established small-signal model. Results of case studies conducted on a modified CIGRE LV network benchmark are presented and discussed in Section 4. Section 5 draws the conclusions of this paper and discusses future research directions.

2. Topology and Control Scheme

Figure 1 shows the topology and control scheme of the grid-supporting PV system. In order to mimic the kinetic energy stored in the rotating rotors of SGs, an energy storage (ES) unit equipped with a bidirectional DC-DC converter was installed in the conventional PV system. Accordingly, the hardware consists of an ES unit, a bidirectional DC-DC converter, an inverter, and a PV array. The PV array was tied directly to the dc link, sharing the same dc bus with the DC-DC converter and the inverter. A buck/boost converter is adopted in this paper.

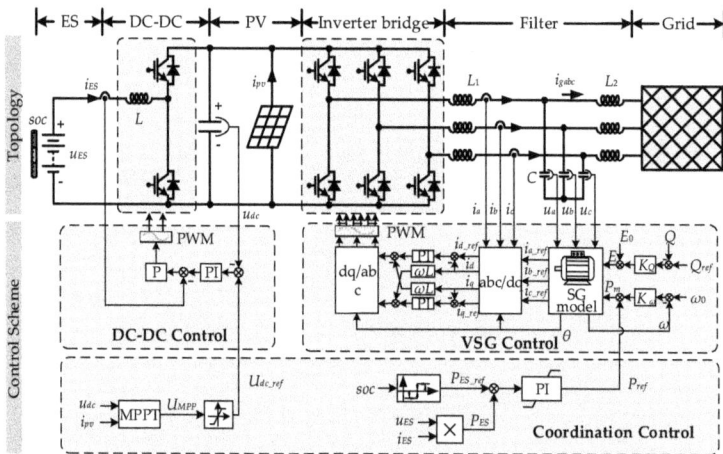

Figure 1. Topology and control scheme of the grid-supporting PV system.

As Figure 1 depicts, the overall control scheme of the grid-supporting PV system comprises three strategies: DC-DC control, VSG control, and coordination control. The coordination control is designed to attune the system with two tasks: (1) One is to ascertain the value of the dc link voltage reference U_{dc_ref}, utilizing a maximum power point tracking (MPPT) algorithm to draw maximum power from the PV array. U_{dc_ref} is provided for DC-DC control, which performs the regulation of dc link voltage u_{dc}. (2) Another is to constrain the state of charge (SOC) of the ES unit through regulating the inverter active power reference P_{ref}, which capacitates the buck/boost to control u_{dc} for the inverter emulating SGs. P_{ref} is delivered to the VSG control that drives the inverter to emulate the characteristics of SGs.

The conventional PV system is only energized by PV input, and its interface inverter is controlled by a voltage-oriented control method, with an outer dc link voltage control loop and an inner current control loop [29]. As shown in Figure 1, the proposed grid-supporting PV system however, is energized by a PV array and ES unit, and the interface inverter is driven by the VSG control. Benefiting from the topology and control scheme, which are different from those of the conventional PV system, the grid-supporting PV system is able to provide inertia and participate in frequency regulation as SGs.

2.1. Coordination Control

To behave as an energy buffer like a rotating rotor, the ES unit must have not only energy to release, but also capacity to store absorbed energy. Therefore, the SOC of the ES unit must be kept within a proper range. Meanwhile, it is necessary to constrain the SOC to control the dc link voltage u_{dc}, so that the ES unit can release energy when u_{dc} falls, and store the absorbed energy when u_{dc} rises.

To constrain the SOC, the exchanged power P_{ES} (positive for discharge and negative for charge) between the ES unit and the dc link must be regulated. However, P_{ES} cannot be directly controlled by the buck/boost converter, which is resulted from that the DC-DC control performs the regulation of u_{dc}.

According to the law of conservation of energy, the following equation is obtained when the energy change of the capacitor at the dc link is ignored:

$$P_{ES} + P_{pv} = P_e \tag{1}$$

where P_{pv} is the power generated by the PV array, and P_e is the output active power of the inverter in the system.

Since the PV array operates at the maximum power point, P_{pv} in Equation (1) fails to adjust. Thus, regulating P_e is the only way to control P_{ES}. Due to the emulated SG characteristics of the inverter, P_e can be controlled with coordination control through regulating the inverter active power reference P_{ref}. To track the exchanged power reference P_{ES_ref}, a proportional-integral (PI) regulator, whose input is the error between P_{ES_ref} and P_{ES}, is used for generating P_{ref}. Then, P_{ref} can be expressed as

$$P_{ref} = G_{PI}(s)\left(P_{ES\_ref} - P_{ES}\right) \tag{2}$$

where $G_{PI}(s) = K_p + K_i/s$ is the transfer function of the PI regulator, K_p and K_i are the proportional coefficient and integral coefficient of the PI regulator, respectively, and s is the Laplace operator.

The relationship of exchanged power reference P_{ES_ref} with respect to the SOC is designed as shown in Figure 2, where SOC_M is the mean of lower limit SOC_L and higher limit SOC_H, and P_0 is the absolute value of the charge power and the discharge power. The ES unit starts charging once the SOC is less than SOC_L, and discharging when the SOC is more than SOC_H. Both charging and discharging are terminated when the SOC reaches SOC_M. Applying the curve shown in Figure 2 to specify P_{ES_ref}, frequent operations of charge/discharge near SOC_L/SOC_H can be avoided by the coordination control. As the charge power and the discharge power of the ES unit depend on P_0, the rated power of the inverter and the charge-discharge rate (C-rate) of the ES unit need to be taken into account when determining P_0. First, the charge power and the discharge power of the ES unit should not be more than the rated power of the inverter to protect the inverter from over-current. Second, P_0 should ensure the charge current and the discharge current do not exceed the maximum C-rate so that the cycling life and the capacity of the ES unit are not significantly affected.

The dc link voltage reference U_{dc_ref} is generally equal to U_{MPP}, which is calculated by a maximum power point tracking (MPPT) algorithm, to ensure that the PV array operates at the point where it can output maximum power. Incremental Conductance [30,31], a classical MPPT algorithm, is employed in this paper.

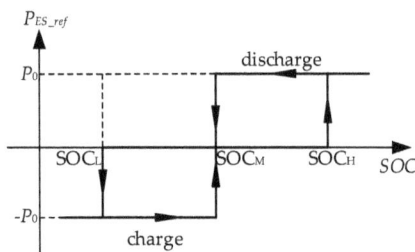

Figure 2. Relationship between exchanged power reference and SOC.

2.2. VSG Control

The VSG control aims to equip the inverter with the characteristics of a SG so that the inverter is capable of behaving like the SG. To realize the inverter emulating the characteristics of the SG, there are three sub-processes to implement during one carrier cycle, as illustrated in Figure 1. The voltages of the filter capacitors, u_a, u_b and u_c, are measured in real time, and virtualize the phase terminal voltages of the stator windings and serve as the input variables of a SG model. Through solving the model, the stator currents of the SG, i_{a_ref}, i_{b_ref}, and i_{c_ref}, are obtained as the reference currents for the inverter. To complete the emulation of the SG characteristics, the inverter output currents should be driven to track the reference currents. Thus, proportional-integral (PI) regulators in the rotating frame are employed to control the inverter.

The SG model adopted in this work comprises third-order electrical equations and second-order mechanical equations. The electrical equation set, which reproduces the stator circuit of the SG, is given by

$$L_s \frac{di_{abc\_ref}}{dt} = e_{abc} - u_{abc} - R_s i_{abc\_ref} \tag{3}$$

where $u_{abc} = [u_a, u_b, u_c]^T$ denotes the phase terminal voltages of the stator windings; R_s and L_s are respectively the stator resistance and the stator inductance; $i_{abc\_ref} = [i_{a\_ref}, i_{b\_ref}, i_{c\_ref}]^T$ represents the currents of the stator windings, which serve as reference currents for the inverter; and $e_{abc} = [e_a, e_b, e_c]^T = E[\sin\theta, \sin(\theta - 2\pi/3), \sin(\theta + 2\pi/3)]^T$ denotes the induced phase electromotive forces in the stator windings.

The electromechanical characteristics of the SG, neglecting the mechanical losses and considering the effect of damper windings, can be described as

$$2H\frac{d\omega}{dt} = \frac{P_m}{\omega} - \frac{P_e}{\omega} - D(\omega - \omega_0) \tag{4}$$

$$\frac{d\theta}{dt} = \omega \tag{5}$$

where H is the inertia constant, P_m is the mechanical power, D is the damping coefficient, ω and ω_0 are the actual and the nominal angular frequency, respectively, and θ is the electrical rotation angle.

To emulate the droop characteristics of the primary frequency control (PFC) and primary voltage control (PVC), P_m and E can be expressed as

$$P_m = P_{ref} + K_\omega(\omega_0 - \omega) \tag{6}$$

$$E = E_0 + K_Q(Q_{ref} - Q) \tag{7}$$

where K_ω is the unit power regulation, E_0 is the no-load electromotive force (EMF), Q_{ref} and Q are the reference value and the actual value of the inverter output reactive power, respectively, and K_Q is the voltage droop coefficient.

2.3. DC-DC Control

The objective of the DC-DC control is to keep the actual value of the dc link voltage u_{dc} equal to the voltage reference U_{dc_ref} provided by the MPPT algorithm of the coordination control. Regulating u_{dc} to track u_{dc_ref} enables the PV array to operate at the maximum power point. The DC-DC control strategy incorporates an outer voltage loop with an inner current loop, as depicted in Figure 1. Through DC-DC control, a stiff dc link voltage, which is required to emulate the inertia, is provided for the inverter.

The buck/boost converter works in BUCK mode and charges the ES unit to prevent u_{dc} from rising when $P_{pv} > P_e$. Under the condition of $P_{pv} < P_e$, the buck/boost converter operates in BOOST mode and discharges the ES unit to stop u_{dc} from dropping. This indicates that the ES unit behaves as an energy buffer, emulating a rotating rotor through complementing the deficit, or absorbing the surplus, of PV production.

3. Performance Analysis and Stability Constraint Formulation

In this section, a small signal per unit (pu) model that considers the Q-E droop control is established. Utilizing the model, performance analysis is conducted, the impact of the Q-E droop control on stability is investigated, and the stability constraint is obtained.

3.1. Small-Signal Modelling

Figure 3 depicts the equivalent circuit of the inverter when connected to the grid. In this figure, δ is the power angle; U_g is the amplitude of the grid voltage; $R_s + jX_s$ is the virtual stator impedance implemented by the VSG control; $R_g + jX_g$ is the grid impedance, which includes the line impedance; and the grid-side filter impedance $j\omega L_2$ and $P_e + jQ$ is the apparent power measured for the control scheme.

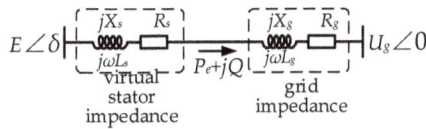

Figure 3. Equivalent circuit of the inverter when connected to the grid.

When R_s and R_g are neglected due to $X_s \gg R_s$ and $X_g \gg R_g$, P_e and Q can be expressed according to Figure 3 as follows:

$$P_e = \frac{EU_g}{X_s + X_g} \sin \delta \tag{8}$$

$$Q = \frac{1}{(X_s + X_g)^2} \left[X_g E^2 - X_s U_g^2 + (X_s - X_g) EU_g \cos \delta \right] \tag{9}$$

Linearizing P_e and Q with respect to E and δ, the deviations of P_e and Q are given by

$$\Delta P_e = k_{P\delta} \Delta \delta + k_{PE} \Delta E \tag{10}$$

$$\Delta Q = k_{Q\delta} \Delta \delta + k_{QE} \Delta E \tag{11}$$

where Δx ($x = P_e, Q, E,$ and δ) represents the deviation of x, and

$$k_{P\delta} = \frac{\partial P_e}{\partial \delta} = \frac{EU_g}{X_s + X_g} \cos \delta \tag{12}$$

$$k_{PE} = \frac{\partial P_e}{\partial E} = \frac{U_g}{X_s + X_g} \sin \delta \tag{13}$$

$$k_{Q\delta} = \frac{\partial Q}{\partial \delta} = \frac{(X_g - X_s)EU_g}{(X_s + X_g)^2} \sin \delta \tag{14}$$

$$k_{QE} = \frac{\partial Q}{\partial E} = \frac{1}{(X_s + X_g)^2} \left[2X_g E + (X_s - X_g) U_g \cos \delta \right] \tag{15}$$

The small signal model of the Q-E droop control described by Equation (7) is given by

$$\Delta E = -K_Q \Delta Q \tag{16}$$

Solving Equations (10), (11), and (16), ΔP_e and ΔQ can be further derived as

$$\Delta P_e = (k_{p\delta} + \Delta K_S)\Delta \delta = K_S \Delta \delta \tag{17}$$

$$\Delta E = -\frac{K_Q k_{Q\delta}}{1 + K_Q k_{QE}} \Delta \delta = K_{\delta E} \Delta \delta \tag{18}$$

$$\Delta K_S = -\frac{K_Q k_{PE} k_{Q\delta}}{1 + K_Q k_{QE}} \tag{19}$$

Normalizing and linearizing Equations (4) and (5) yields [32]

$$\frac{d\Delta \omega}{dt} = \frac{1}{2H} \left[\Delta P_m - \Delta P_e - D\Delta \omega \right] \tag{20}$$

$$\frac{d\Delta \delta}{dt} = \frac{d\Delta \theta}{dt} - \omega_0 = \omega_0 \Delta \omega \tag{21}$$

where $\Delta \omega$ is the angular frequency deviation, ΔP_m is the deviation of P_m, D is the damping coefficient, H is the inertia constant in seconds, and ω_0 is the nominal angular frequency in rad/s.

The incremental Equations of (1), (2) and (6) are

$$\Delta P_{pv} + \Delta P_{ES} = \Delta P_e \tag{22}$$

$$\Delta P_{ref} = G_{PI}(s)\left(\Delta P_{ES\_ref} - \Delta P_{ES}\right) \tag{23}$$

$$\Delta P_m = \Delta P_{ref} - K_\omega \Delta \omega \tag{24}$$

By combining Equations (17)–(24), the small-signal model considering the Q-E droop control is established in Figure 4, and $\Delta \omega$ is correspondingly derived as Equation (25).

$$\Delta \omega = G(s)\left(\Delta P_{ES\_ref} + \Delta P_{pv}\right) \tag{25}$$

where the transfer function $G(s)$ is given by

$$G(s) = \frac{(K_p s + K_i)s}{2Hs^3 + (D + K_\omega)s^2 + \omega_0 K_S (K_p + 1)s + \omega_0 K_S K_i} \tag{26}$$

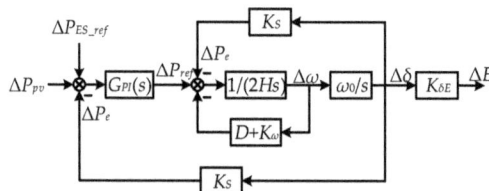

Figure 4. Small signal model of the grid-supporting PV system.

The expected performance of the grid-supporting PV system can be achieved if H, $D + K_\omega$, K_p, and K_i are properly selected in such a way that the poles of Equation (26) are located at desired locations.

3.2. Performance Analysis

The root loci family of the proposed PV system is shown in Figure 5a, where $2H = 1$ s, 3 s, and 15 s, and $D + K_\omega$ changes from 10 pu to 200 pu. It is clear that H plays an important role in determining the settling time of the proposed PV system. As $D + K_\omega$ increases, the damping of the system rises and the stability is improved.

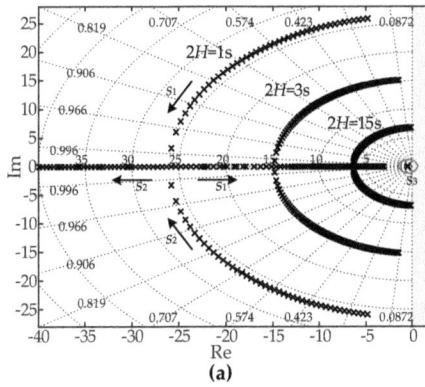

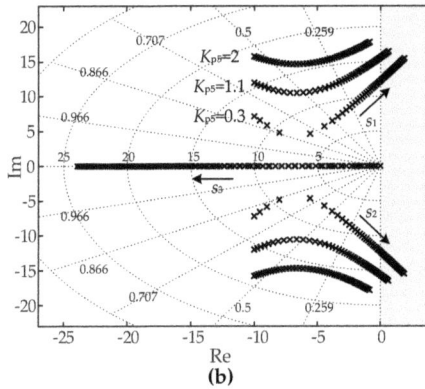

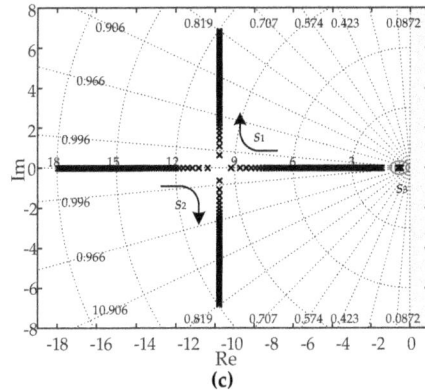

Figure 5. Family of root loci for variations of: (**a**) H and $D + K_\omega$; (**b**) K_p and K_i; (**c**) K_S.

Figure 5b depicts the root loci family considering variations of K_i from 0.1 pu to 50 pu, and K_p = 0.3 pu, 1.1 pu, and 2 pu. As Figure 5b shows, the damping drops, the overshoot rises, and the undamped natural frequency increases when K_p becomes larger. The damped frequency and the stability margin mainly depend on K_i. Instability may happen with an excessively large value of K_i. To ensure the stability of the system, it is necessary to formulate the constraint explicitly on H, $D + K_\omega$, K_p, and K_i to guide the tuning of parameters.

Figure 5c depicts the root locus where K_S varies from 0.5 pu to 2 pu. Among the three poles depicted in the plane, s_3 is the real root and its position depends on K_p and K_i. As K_S increases, the conjugate complex roots s_1 and s_2 evolve in the direction of the arrows. A larger K_S increases the real parts of s_1 and s_2, which improves the stability.

3.3. Impact of Q-E Droop Control on Stability

As Figure 5c illustrates, K_S is a factor that affects the stability of the grid-supporting PV system. According to Equation (17), K_s consists of two parts, $k_{P\delta}$ and ΔK_S. It is indicated in Equation (19) that ΔK_S is related to the coefficient K_Q of the Q-E droop control and embodies the impact of the Q-E droop control on the stability. When the Q-E droop control is invalidated (i.e., K_Q = 0) and only P-ω is considered, ΔK_S vanishes identically, and K_S in Equation (17) is equal to $k_{P\delta}$.

By substituting Equations (13)–(15) into Equation (19), ΔK_S can be obtained as Equation (27), where $r = X_g/X_S$. On the condition that the Q-E droop control works, K_Q is set to be a positive number. The power angle δ lies in the range from 0° to 90°, and E is generally larger than U_g. Thus, $k_{P\delta}$, k_{PE}, and k_{QE} are all positive. Accordingly, the curve of ΔK_S with respect to the ratio X_g/X_S is shown in Figure 6.

$$\Delta K_S \overset{r=X_g/X_S}{=} K_Q \cdot \frac{1-r}{1+r} \cdot \frac{EU_g^2 \sin^2 \delta}{X_S^2(1+r)^2 + K_Q X_S \left[r(2E - U_g \cos \delta) + U_g \cos \delta\right]} \tag{27}$$

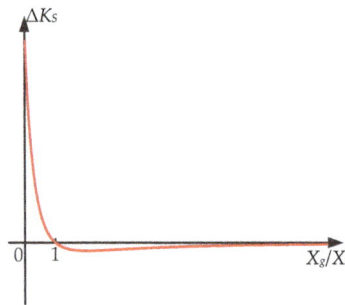

Figure 6. Curve of ΔK_S with respect to the ratio X_g/X_S.

As Figure 6 illustrates, ΔK_S is positive if the ratio X_g/X_S is less than 1, while ΔK_S is negative if the ratio X_g/X_S is greater than 1. Accordingly, K_S is greater than $k_{P\delta}$ under the condition of $X_g < X_S$, which means the Q-E droop control improves the stability, since a larger K_S improves the stability. Conversely, K_S is less than $k_{P\delta}$ when $X_g > X_S$, which indicates that the Q-E droop control worsens the stability in the weak grid. Besides, K_S is zero in the case of $X_g = X_S$, implying that the Q-E droop control has no effect on the stability.

3.4. Stability Constraint Formulation

The closed-loop system of Equation (26) is a third-order linear time-invariant. To analyze the stability, the Routh–Hurwitz stability criterion is used. The system characteristic equation is obtained from Equation (26) as

$$D(s) = a_0 s^3 + a_1 s^2 + a_2 s + a_3 = 0 \tag{28}$$

where $a_0 = 2H$, $a_1 = D + K_\omega$, $a_2 = \omega_0 K_S(K_p + 1)$, and $a_3 = \omega_0 K_S K_i$.

Through applying the Routh–Hurwitz stability criterion to Equation (28), the system stability discriminant is yielded as

$$\begin{cases} a_i > 0, i = 0, 1, \ldots, 3 \\ a_1 a_2 - a_0 a_3 > 0 \end{cases} \tag{29}$$

Since H, $D + K_\omega$, K_p, and K_i are positive real numbers, the discriminant can be simplified as

$$\begin{cases} K_S > 0 \\ a_1 a_2 - a_0 a_3 > 0 \end{cases} \tag{30}$$

Substituting a_i into Equation (30) gives

$$\begin{cases} K_S > 0 \\ \frac{D + K_\omega}{2H} > \frac{K_i}{K_p + 1} \end{cases} \tag{31}$$

Equation (31) presents the stability constraint for the grid-supporting PV system, which H, $D + K_\omega$, K_p, K_i, and K_S must satisfy to guarantee the stability of the system.

4. Results and Discussion

The proposed grid-supporting PV system was verified on the CIGRE benchmark of the European LV distribution network elaborated in Reference [33]. The topology of the benchmark is shown in Figure 7, and the line parameters of the benchmark are given in Table 1. All loads were configured to be balanced for simplicity. The apparent power and power factor (PF) of the loads are described in Figure 7. The 20 kV medium voltage grid in this benchmark was equated with a SG system with inertia constant of $H = 9$ s [32]. The PFC of the SG system reacts in 5 s when there is a frequency deviation, and the speed regulation of the PFC is 3.33%. Six cases were considered when disturbances occur, as listed in Table 2.

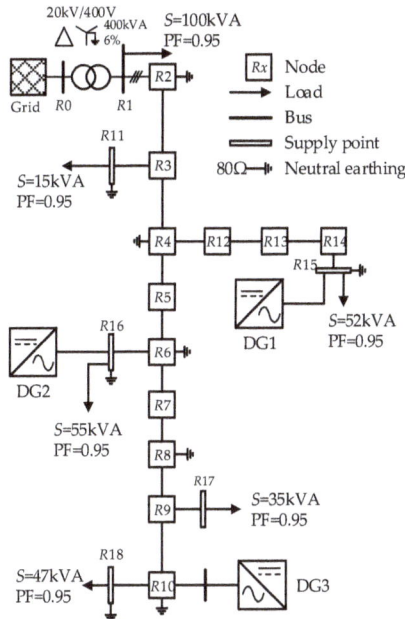

Figure 7. Modified benchmark of the European LV distribution network.

Table 1. Line parameters of the benchmark.

Node (From–To)	Length (m)	Resistance (mΩ)	Inductance (µH)
R1–R2	35	10.045	18.6052
R2–R3	35	10.045	18.6052
R3–R4	35	10.045	18.6052
R4–R6	70	20.09	37.2104
R6–R9	105	31.135	55.8156
R9–R10	35	10.045	18.6052
R4–R15	135	155.52	196.811
R6–R16	30	34.56	43.7358
R9–R17	30	34.56	43.7358
R10–R18	30	34.56	43.7358
Transformer	-	3.2	40.7437

Table 2. Cases Description.

Case	Disturbance	DG1–DG3
A	Sudden Load Variation	Conventional PV System
B		Proposed PV System
C	Short Circuit Fault	Conventional PV System
D		Proposed PV System
E	Step of Solar Irradiance	Conventional PV System
F		Proposed PV System

Case A and Case B considered sudden load variation by switching a load of 25 kW in R11 at 2 s. Case C and Case D considered a three-phase short circuit fault occurring at 2 s, the fault in each case was located at R17 and was cleared at 3 s. A step of solar irradiance from 1000 W/m^2 to 1050 W/m^2 was exerted on the PV array of DG3 at 2 s in Case E and Case F. After the disturbance occurred at 2 s in each case, the PFC was activated in 5 s; that is, at 7 s.

In Case A, Case C, and Case E, three conventional single-stage PV systems were applied to the benchmark as DG1−DG3. In comparison with the conventional PV system, three proposed grid-supporting PV systems, with parameters listed in Table 3, were connected to the feeder as DG1−DG3 in Case B, Case D, and Case F, and each ES unit was comprised of 25 Powersonic PS-121100 batteries in series. The SOC of each ES unit was set to 50%, which leads to $P_{ES\_ref} = 0$.

Table 3. Inverter Parameters of the Proposed PV System.

Meaning, Symbol, and Unit		No.1	No.2	No.3
Rated Power S_n (kVA)		20	20	50
Nominal frequency f_n (Hz)			50	
Nominal voltage V_n (V)			380	
Carrier frequency f_c (kHz)			10	
Filter values	L_1 (mH)	1.2	1.2	0.72
	C (µF)	20	20	50
	L_2 (mH)	0.8	0.8	0.18
Current loop controller gains	K_{cp} (pu)		2	
	K_{ci} (pu)		1500	
K_p (pu)			0.05	
K_i (pu)			0.3	
Parameter values of SG model	$2H$ (s)	10	14	10
	D (pu)	30	40	40
	R_s (pu)		0.08	
	L_s (pu)		0.8	
No-load EMF	E_0 (pu)		1.22	
Droop gains	K_ω (pu)	20	10	30
	K_Q (pu)		0	

To verify the functions of smoothing the power and tracking the maximum power point of the conventional PV system and the proposed grid-supporting PV system, solar irradiance steps were used in Case E and Case F to provide the most severe condition, although there is little possibility that the solar irradiance step would occur in a real case.

4.1. Case A: Sudden Load Variation—Conventional PV System

Figure 8a gives the resultant frequency of the grid, DG1, DG2, and DG3, respectively. All frequencies decrease consistently between 2 s and 5 s. With the phase-lock loop, the conventional PV system tracks the grid frequency (i.e., the LV distribution network frequency), but fails to provide the inertia due to operating as a grid-following unit. As depicted in Figure 8b, the output power of the conventional PV system injects power into the LV distribution network without change after the load variation, and thus is incapable of mitigating the power imbalance between generation and consumption. Figure 8c illustrates the dc link voltages, which are always regulated by the inverter of the conventional PV systems to perform MPPT.

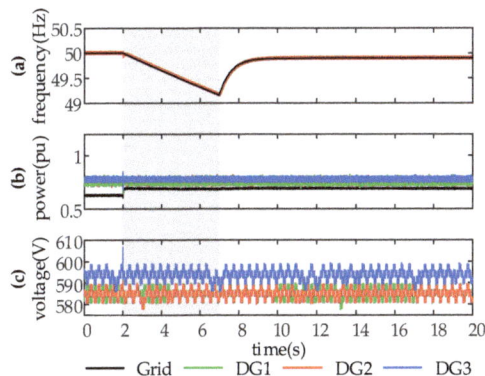

Figure 8. Responses to sudden load variation when conventional PV systems are applied as DG1–DG3: (**a**) Frequency response; (**b**) output active power; (**c**) dc link voltage of the inverter.

4.2. Case B: Sudden Load Variation—Proposed PV System

Figure 9 presents the responses when three proposed PV systems are integrated into the LV distribution network as DG1–DG3. As depicted in Figure 9a, all frequencies decrease between 2 s and 5 s, with a smaller rate of change of frequency (ROCOF) and higher frequency nadir when compared with Figure 8a of Case A. Through emulating the inertia of SGs, the proposed PV system is able to slow down frequency response and allow decent time for frequency control. Figure 9b shows that the proposed PV system mitigates the power imbalance between generation and consumption by increasing the output active power P_e autonomously, and thus supports the grid, mimicking the SG. Figure 9c plots the dc link voltage of the interface inverter, which is maintained at U_{MPP} by the buck/boost converter with the ES unit, even if there is an imbalance between P_{pv} and P_e. It is demonstrated that a stiff dc link voltage can be provided in the proposed PV system for the inverter to emulate the inertia. Figure 9d illustrates the exchanged power P_{ES} between the ES unit and the dc link. After the sudden load variation, the incremental of P_{ES} is consistent with that of P_e shown in Figure 9b. It is indicated that the ES unit balances the power between the PV array and the inverter, which emulates the behavior of a rotating rotor releasing kinetic energy when it's frequency drops. After PFC activation, the exchanged power P_{ES}, regulated by coordination control, gradually converges to P_{ES_ref} to constrain the SOC.

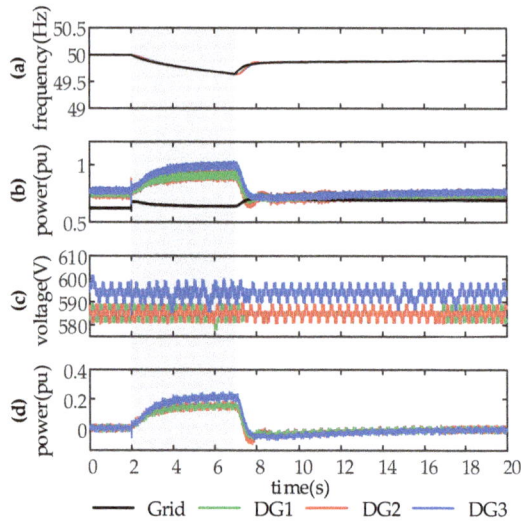

Figure 9. Responses to sudden load variation when the proposed PV systems are applied as DG1–DG3: (**a**) Frequency response; (**b**) output active power; (**c**) dc link voltage of the inverter; (**d**) exchanged power between the ES unit and dc link.

4.3. Case C: Short Circuit Fault—Conventional PV System

As Figure 10a depicts, the grid frequency, which is tracked by the conventional PV system, decreases until the fault is cleared and reaches the nadir of 49.82 Hz at 3 s. However, after the clearance of the fault, frequencies of the grid, DG1, DG2, and DG3 hardly change until the PFC is activated at 5 s. This shows that the conventional PV system fails to participate in frequency regulation. As Figure 10b depicts, the power imbalance resulting from the fault is counteracted only by the grid, while the conventional PV system is incapable of responding to the power imbalance, and outputs power without change after the fault occurs. The dc link voltage, illustrated in Figure 10c, is regulated during the fault to draw maximum power from the PV array.

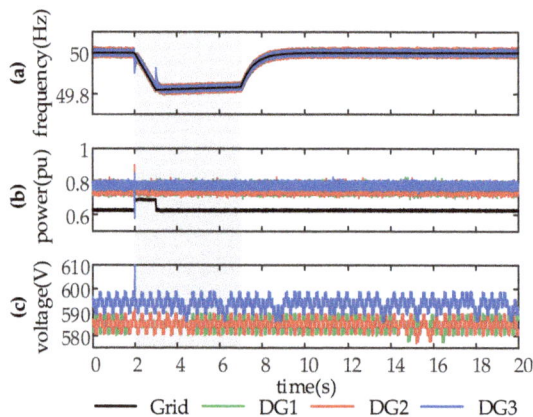

Figure 10. Responses to a short circuit fault when conventional PV systems are applied as DG1–DG3: (**a**) Frequency response; (**b**) output active power; (**c**) dc link voltage of the inverter.

4.4. Case D: Short Circuit Fault—Proposed PV System

System responses to a three-phase short circuit fault, where three proposed PV systems are integrated as DG1–DG3, are studied in this case, and given in Figure 11.

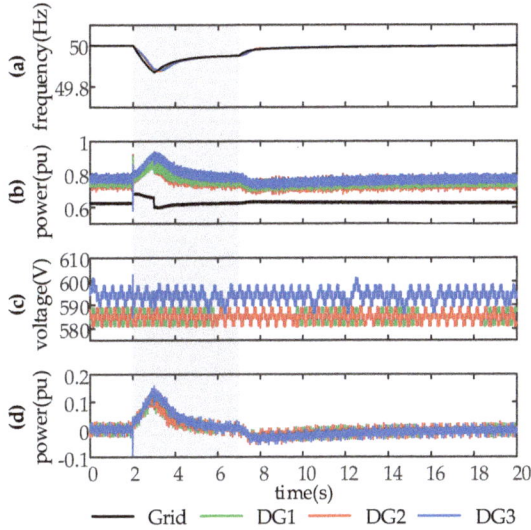

Figure 11. Responses to a short circuit fault when the proposed PV systems are applied as DG1–DG3: (**a**) Frequency response; (**b**) output active power; (**c**) dc link voltage of the inverter; (**d**) exchanged power between the ES unit and dc link.

The grid frequency, as shown in Figure 11a, decreases from the normal value of 50 Hz to the nadir of 49.87 Hz lasting from 2 s to 3 s. In comparison with Case C, a smaller ROCOF and slighter frequency deviation are caused by the fault in this case. It is perceived in Figure 11a that the grid frequency is regulated by the proposed PV system in the absence of PFC action during 3–7 s. As depicted in Figure 11b, the proposed PV system increases its output active power and moderates the power imbalance, while the conventional PV system fails to implement this function, as shown in Figure 10b.

Figure 11c illustrates the dc link voltage of the proposed PV system, which indicates that the proposed PV system is able to provide a stiff dc link voltage to emulate the inertia and perform MPPT. As depicted in Figure 11d, the incremental of P_{ES} is consistent with that of P_e shown in Figure 11b, verifying that the ES unit balances the power between the PV array and the inverter, which emulates the behavior of a rotor releasing kinetic energy when it's frequency declines.

4.5. Case E: Step of Solar Irradiance—Conventional PV System

As Figure 12a shows, the step of solar irradiance exerted on the PV array of DG 3 causes a sudden change of the power generated from the PV array in DG3. The dc link voltage of the inverter in DG 3 increases, as depicted in Figure 12b, to track U_{MPP} specified by the MPPT algorithm in the coordination control. Due to the lack of an energy buffer in the conventional PV system, DG3 injects the fluctuant PV power resulting from the solar irradiance step into the LV distribution network, and the output active power rises suddenly, as shown in Figure 12c. Figure 12d shows that the grid frequency deviates after the solar irradiance step until the PFC is activated.

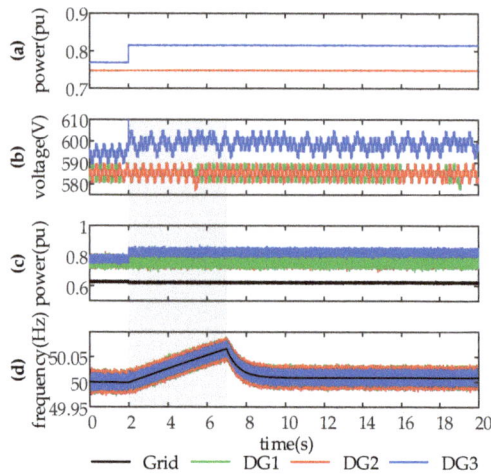

Figure 12. Responses to a step of solar irradiance when conventional PV systems are applied as DG1~DG3: (**a**) Power generated from the PV array; (**b**) dc link voltage of the inverter; (**c**) output active power; (**d**) frequency response.

4.6. Case F: Step of Solar Irradiance—Proposed PV System

Responses to the step of solar irradiance exerted on DG3 are studied in Figure 13 when three proposed PV systems are applied as DG1−DG3. After the solar irradiance steps, the power generated from the PV array in DG3 rises suddenly, as shown in Figure 13a, and the dc link voltage of DG3 increases to perform MPPT, as shown in Figure 13b. In comparison with *Case E*, the active power that DG3 feeds to the LV distribution network is filtered and rises smoothly, as depicted in Figure 13c, and the grid frequency deviates with a smaller ROCOF and lower zenith, as depicted in Figure 13d. Figure 13e shows that the exchanged power P_{ES} bètween the ES unit and the dc link decreases suddenly, and the ES unit of DG3 absorbs the surplus of PV production after the solar irradiance steps. P_{ES} eventually returns to zero, tracking the reference P_{ES_ref} to constrain the SOC, and the output active power P_e is finally equal to the power generated by the PV array P_{pv}.

It is demonstrated that the proposed PV system is able to smooth the power fed to the LV distribution network, even if the power generated by the PV array fluctuates suddenly.

Combining with the results in Cases A–F, Table 4 shows the following advantageous features of the proposed grid-supporting PV system as compared with the conventional PV system:

(1) The grid-supporting PV system, presenting as SGs from the point-of-view of the grid by mimicking SG characteristics with VSG control, can support the grid through mitigating the power imbalance between generation and consumption, and slowing down frequency dynamics with virtual inertia.

(2) Through emulating the droop characteristics of PFC and PVC, the grid-supporting PV system can participate in frequency regulation and voltage regulation.

(3) The proposed PV system has the functions of filtering the PV power and smoothing the power fed to the grid, which leads to a reduced impact of PV fluctuation on the grid.

(4) The grid-supporting PV system synchronizes with the grid through mimicking the synchronization mechanism of SGs, and thus the PLL, in which the delay may cause instability [34], is discarded in the proposed PV system.

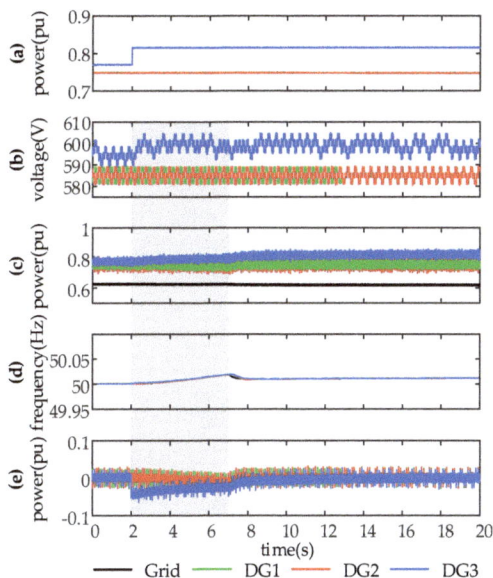

Figure 13. Responses to a step of solar irradiance when the proposed PV systems are applied as DG1–DG3: (**a**) Power generated from the PV array; (**b**) dc link voltage of the inverter; (**c**) output active power; (**d**) frequency response; (**e**) exchanged power between the ES unit and dc link.

Table 4. Advantages of the proposed grid-supporting PV system as compared with the conventional PV system.

Features	Conventional PV System	Proposed PV System
Emulating the characteristics of SGs to support the grid	✕	✓
Primary frequency control and primary voltage control	✕	✓
Smoothing the fluctuation of the power fed to the grid	✕	✓
PLL-less operation after initial synchronization	✕	✓

✕: operating without the feature; ✓: operating with the feature.

5. Conclusions

A novel grid-supporting PV system, which operates as an inertia voltage source by emulating the characteristics of SGs, is proposed in this paper.

To present the PV system as a SG from the point-of-view of the grid, both the topology and the control scheme were investigated. An ES unit equipped with a bidirectional DC-DC converter was installed, which can mimic the function of a rotating rotor for kinetic energy and buffer the imbalance of the PV power. On the other hand, the coordination control, DC-DC control, and VSG control were employed in the proposed system. The coordination control is able to constrain the SOC of the ES unit and calculate the voltage at the maximum power point of the PV array. The DC-DC control can perform the regulation of the dc link voltage to realize MPPT, and the VSG control is capable of equipping the interface inverter with SG characteristics.

To guide the tuning of the parameters, the system performance was analyzed with the variation parameter values. It was found that the inertia constant H plays an important role in determining the settling time, and that the system damping mainly depends on $D + K_\omega$. Furthermore, the stability constraint was formulated as Equation (31), which should be satisfied to guarantee the stability of the system.

Results of case studies conducted on the modified CIGRE LV network benchmark verify the grid-supporting PV system. Compared with the conventional PV system, the grid-supporting PV system is advantageous in the following areas:

(1) The VSG control and the ES unit capacitate the interface inverter of the proposed inverter to mimic SG characteristics, slowing down frequency response with inertia, and moderating the power imbalance between generation and consumption.

(2) The emulation of the droop characteristics of PFC and PVC capacitates the grid-supporting PV system to contribute to frequency regulation and voltage regulation of EPS.

(3) In the conventional PV system, the power fed to the grid must be equal to the power generated from the PV array. However, the installation of the ES unit in the grid-supporting PV system spares this embarrassment. Through the installed ES unit and the control scheme, the function of filtering the PV power and smoothing the PV system output power, which reduces the impact of PV fluctuations on the grid, is implemented in the proposed grid-supporting PV system.

(4) With the VSG control, the interface inverter of the grid-supporting PV system mimics not only the inertia and frequency damping of SGs, but also the synchronization mechanism of SGs. Beneficially, the impact of the PLL on the stability is eliminated, which leads to the proposed PV system being more compatible with EPS than the conventional PV system that employs the PLL.

The topology, the control scheme, and the stability constraint for parameters are presented in this paper. In future, the method for optimizing the capacity of the ES unit under different conditions, where the values of H, D, and K_ω, and the capacity of the PV array vary, need to be obtained in order to reduce costs. As the analysis indicates, the grid impedance has impacts on the stability of the system with the Q-E droop control. How to reduce or avoid these impacts is also set as one of our future research directions.

Author Contributions: H.X. and J.S. provided the original idea for this paper. H.X., N.L., and Y.S. organized the manuscript and attended the discussions when analysis and verification were carried out. All the authors gave comments and suggestions on the writing and descriptions of the manuscript.

Funding: This research was supported partly by the National Key Research and Development Program of China (2017YFB0903503), the National Science Foundation of China (51677050) and Double First Class Project for Independent Innovation and Social Service Capabilities (45000-411104/012).

Conflicts of Interest: The authors declare no conflict of interest.

Abbreviations

The following abbreviations are used in this manuscript:

EMF	Electromotive force
EPS	Electric power systems
ES	Energy storage
MPPT	Maximum power point tracking
PFC	Primary frequency control
PLL	Phase-locked loops
PV	Photovoltaic
PVC	Primary voltage control
RES	Renewable energy sources
ROCOF	Rate of change of frequency
SG	Synchronous generator
SOC	State of charge
VSG	Virtual synchronous generator

References

1. Pintér, G.; Baranyai, N.H.; Wiliams, A.; Zsiborács, H. Study of Photovoltaics and LED Energy Efficiency: Case Study in Hungary. *Energies* **2018**, *11*, 790. [CrossRef]
2. Egwebe, A.M.; Fazeli, M.; Igic, P.; Holland, P.M. Implementation and stability study of dynamic droop in islanded microgrids. *IEEE Trans. Energy Convers.* **2016**, *31*, 821–832. [CrossRef]
3. Zsiborács, H.; Baranyai, N.H.; András, V.; Háber, I.; Pintér, G. Economic and Technical Aspects of Flexible Storage Photovoltaic Systems in Europe. *Energies* **2018**, *11*, 1445. [CrossRef]
4. Radwan, A.A.A.; Mohamed, Y.A.R.I. Power synchronization control for grid-connected current-source inverter-based photovoltaic systems. *IEEE Trans. Energy Convers.* **2016**, *31*, 1023–1036. [CrossRef]
5. Nižetić, S.; Papadopoulos, A.M.; Tina, G.M.; Rosa-Clot, M. Hybrid energy scenarios for residential applications based on the heat pump split air-conditioning units for operation in the Mediterranean climate conditions. *Energy Build.* **2017**, *140*, 110–120. [CrossRef]
6. Rocabert, J.; Luna, A.; Blaabjerg, F.; Rodríguez, P. Control of power converters in AC microgrids. *IEEE Trans. Power Electron.* **2012**, *27*, 4734–4749. [CrossRef]
7. Delille, G.; Francois, B.; Malarange, G. Dynamic frequency control support by energy storage to reduce the impact of wind and solar generation on isolated power system's inertia. *IEEE Trans. Sustain. Energy* **2012**, *3*, 931–939. [CrossRef]
8. Kroposki, B.; Johnson, B.; Zhang, Y.; Gevorgian, V.; Denholm, P.; Hodge, B.; Hannegan, B. Achieving a 100% renewable grid: Operating electric power systems with extremely high levels of variable renewable energy. *IEEE Power Energy Mag.* **2017**, *15*, 61–73. [CrossRef]
9. Wesenbeeck, M.P.N.; de Haan, S.W.H.; Varela, P.; Visscher, K. Grid tied converter with virtual kinetic storage. In Proceedings of the IEEE Bucharest PowerTech, Bucharest, Romania, 28 June–2 July 2009.
10. Eftekharnejad, S.; Vittal, V.; Heydt, G.T.; Keel, B.; Loehr, J. Impact of increased penetration of photovoltaic generation on power systems. *IEEE Trans. Power Syst.* **2013**, *28*, 893–901. [CrossRef]
11. Ulbig, A.; Borsche, T.S.; Andersson, G. Impact of low rotational inertia on power system stability and operation. In Proceedings of the IFAC World Congress, Capetown, South Africa, 24–29 August 2014.
12. Beck, H.P.; Hesse, R. Virtual synchronous machine. In Proceedings of the 9th International Conference on Electrical Power Quality Utilizations, Barcelona, Spain, 9–11 October 2007.
13. VSG Control Algorithms: Present Ideas. Available online: http://www.vsync.eu.ProjectVSYNC (accessed on 22 March 2016).
14. Zhong, Q.C.; Weiss, G. Synchronverters: Inverters that mimic synchronous generators. *IEEE Trans. Ind. Electron.* **2011**, *58*, 1259–1267. [CrossRef]
15. Sakimoto, K.; Miura, Y.; Ise, T. Stabilization of a power system with a distributed generator by a virtual synchronous generator function. In Proceedings of the IEEE 8th International Conference Power Electronics ECCE Asia (ICPE & ECCE), Jeju, Korea, 30 May–3 June 2011.
16. Bevrani, H.; Ise, T.; Miura, Y. Virtual synchronous generators: A survey and new perspectives. *Int. J. Electr. Power Energy Syst.* **2014**, *54*, 244–254. [CrossRef]
17. Natarajan, V.; Weiss, G. Synchronverters with better stability due to virtual inductors, virtual capacitors and anti-windup. *IEEE Trans. Ind. Electron.* **2017**, *64*, 5994–6004. [CrossRef]
18. Alipoor, J.; Miura, Y.; Ise, T. Power system stabilization using virtual synchronous generator with alternating moment of inertia. *IEEE J. Emerg. Sel. Top. Power Electron.* **2015**, *3*, 451–458. [CrossRef]
19. Li, D.; Zhu, Q.; Lin, S.; Bian, X.Y. A self-adaptive inertia and damping combination control of VSG to support frequency stability. *IEEE Trans. Energy Convers.* **2017**, *32*, 397–398. [CrossRef]
20. Alipoor, J.; Miura, Y.; Ise, T. Stability Assessment and Optimization Methods for Microgrid with Multiple VSG Units. *IEEE Trans. Smart Grid* **2016**, *9*, 1462–1471. [CrossRef]
21. Torres, L.M.A.; Lopes, L.A.C.; Morán, T.L.A.; Espinoza, C.J.R. Self-tuning virtual synchronous machine: A control strategy for energy storage systems to support dynamic frequency control. *IEEE Trans. Energy Convers.* **2014**, *29*, 833–840. [CrossRef]
22. Dong, S.; Chen, Y.C. Adjusting synchronverter dynamic response speed via damping correction loop. *IEEE Trans. Energy Convers.* **2017**, *32*, 608–619. [CrossRef]
23. Suul, J.A.; D'Arco, S.; Guidi, G. Virtual synchronous machine-based control of a single-phase bi-directional battery charger for providing vehicle-to-grid services. *IEEE Trans. Ind. Appl.* **2016**, *52*, 3234–3244. [CrossRef]

24. Fang, J.; Tang, Y.; Li, H.; Li, X. A battery/ultracapacitor hybrid energy storage system for implementing the power management of virtual synchronous generators. *IEEE Trans. Power Electron.* **2018**, *33*, 2820–2824. [CrossRef]

25. Liu, J.; Miura, Y.; Bevrani, H.; Ise, T. Enhanced virtual synchronous generator control for parallel inverters in microgrids. *IEEE Trans. Smart Grid* **2017**, *8*, 2268–2277. [CrossRef]

26. Alam, M.J.E.; Muttaqi, K.M.; Sutanto, D. A novel approach for ramp-rate control of solar PV using energy storage to mitigate output fluctuations caused by cloud passing. *IEEE Trans. Energy Convers.* **2014**, *29*, 507–518. [CrossRef]

27. Thang, T.V.; Ahmed, A.; Kim, C.I.; Park, J.H. Flexible system architecture of stand-alone PV power generation with energy storage device. *IEEE Trans. Energy Convers.* **2015**, *30*, 1386–1396. [CrossRef]

28. Fuente, D.V.; Trujillo Rodríguez, C.L.; Garcerá Figueres, G.E.; Gonzalez, R.O. Photovoltaic power system with battery backup with grid-connection and islanded operation capabilities. *IEEE Trans. Ind. Electron.* **2013**, *60*, 1571–1581. [CrossRef]

29. Kadri, R.; Gaubert, J.; Champenois, G. An improved maximum power point tracking for photovoltaic grid-connected inverter based on voltage-oriented control. *IEEE Trans. Ind. Electron.* **2011**, *58*, 66–75. [CrossRef]

30. Esram, T.; Chapman, P.L. Comparison of photovoltaic array maximum power point tracking techniques. *IEEE Trans. Energy Convers.* **2007**, *22*, 439–449. [CrossRef]

31. Yazdani, A.; Fazio, A.R.D.; Ghoddami, H.; Russo, M.; Kazerani, M.; Jatskevich, J.; Strunz, K.; Leva, S.; Martinez, J.A. Modeling guidelines and a benchmark for power system simulation studies of three phase single-stage photovoltaic systems. *IEEE Trans. Power Deliv.* **2011**, *26*, 1247–1264. [CrossRef]

32. Kundur, P.; Balu, N.J.; Lauby, M.G. *Power System Stability and Control*; McGraw-Hill Inc.: New York, NY, USA, 1994; ISBN 9780070359581.

33. Strunz, K.; Abbasi, E.; Fletcher, R.; Hatziargyriou, N.D. *Benchmark Systems for Network Integration of Renewable and Distributed Energy Resources*; CIGRÉ TF C6.04.02, Technical Report 575; CIGRE: Paris, France, 2014.

34. Norouzi, A.H.; Sharaf, A.M. Two control scheme to enhance the dynamic performance of the STATCOM and SSSC. *IEEE Trans. Power Deliv.* **2005**, *20*, 435–442. [CrossRef]

energies

MDPI

Article

Coordinated Control of Multiple Virtual Synchronous Generators in Mitigating Power Oscillation

Pan Hu [1,*], Hongkun Chen [2], Kan Cao [1], Yuchuan Hu [1], Ding Kai [1], Lei Chen [2] and Yi Wang [1]

[1] State Grid Hubei Electric Power Research Institute, Wuhan 430072, China; cao_kan@foxmail.com (K.C.); huyuchuan91@foxmail.com (Y.H.); dingkay@sina.com (D.K.); yi.wang@bath.edu (Y.W.)
[2] School of Electrical Engineering, Wuhan University, Wuhan 430072, China; chkinsz@163.com (H.C.); stclchen1982@163.com (L.C.)
* Correspondence: hpan2@ualberta.ca; Tel.: +86-027-8856-4097

Received: 10 September 2018; Accepted: 11 October 2018; Published: 17 October 2018

Abstract: Virtual synchronous generators (VSGs) present attractive technical advantages and contribute to enhanced system operation and reduced oscillation damping in dynamic systems. Traditional VSGs often lack an interworking during power oscillation. In this paper, a coordinated control strategy for multiple VSGs is proposed for mitigating power oscillation. Based on a theoretical analysis of the parameter impact of VSGs, a coordinated approach considering uncertainty is presented by utilizing polytopic linear differential inclusion (PLDI) and a D-stable model to enhance the small-signal stability of system. Subsequently, the inertia and damping of multiple VSGs are jointly exploited to reduce oscillation periods and overshoots during transient response. Simulation, utilizing a two-area four-machine system and a typical microgrid test system, demonstrates the benefits of the proposed strategy in enhancing operation stability and the anti-disturbing ability of multiple VSGs. The results conclusively confirm the validity and applicability of the method.

Keywords: multiple VSGs; oscillation mitigation; coordinated control; small-signal and transient stability

1. Introduction

The high penetration of renewable energy sources (RESs) reduces rotational inertia significantly and hence lowers the frequency support and damping to a power system [1–3]. To address the challenges, many scholars introduce a virtual synchronous generator (VSG) control strategy to resemble the operation of the SG with its inertia behavior [4–6]. By introducing rotor motion equations, a VSG-based converter integrates the inertia and damping functions in one single term [7]. Different from traditional voltage control, this "synchronverter concept" controller enjoys a better frequency response during a disturbance and provides voltage and frequency support in the weak grid. However, the unsuitable parameters of controllers may deteriorate oscillation suppressing ability and reduce the stability margin of a system when multiple VSGs operate in parallel [8].

Since the time-scale of converter controllers is inconsistent with the mechanical adjustment of synchronous machines (SMs), the authors of [9–11] proved that the integration of converter-based generators exerts little impact on the original electromagnetic oscillation mode (EOM) of a power system. Though the collateral impact of RESs replacing SMs reduces the overall inertia, the small-signal stability of the system improves, especially displacing SMs, thereby affecting the modes. However, the VSG-based converters will be involved in the original electromagnetic oscillation mode due to the implementation of rotor motion behavior [12]. Consequently, the small-signal stability of a power system may be deteriorated if the parameters of the VSG are not coordinated with other SMs and VSGs.

On the other hand, as the VSG unit is not a real synchronous machine, the parameters can be adaptively updated to operate faster and more stably during disturbances [13]. This characteristic provides outstanding flexible and convenient performances of VSGs in oscillation mitigation. Some studies [14,15] have addressed this idea and designed a bang-bang control strategy according to four intervals of the oscillation cycle. The inertial is set to be a big or small value when the product of $|d\omega/dt|$ and $\Delta\omega$ is positive or negative, respectively [14]. However, this work lacks the detailed design of the inertial parameters during each cycle and does not consider the damping factor. To overcome this drawback, the authors of [15] introduce a parameter design method of rotor inertia combined with damping factors. The thresholds of J and D are set, and $|d\omega/dt|$ and $\Delta\omega$ are multiplied by two experiential coefficients to add as a correction term. However, this parameter design can still be enriched to better suppress power oscillation.

Dealing with the power oscillation issue, this paper aims to contribute a coordinated control for multiple VSGs. The novelty of this paper mainly focuses on the following:

(a) The mechanism of low-frequency oscillations caused by the interaction between VSGs and SMs is quantitatively investigated. A coordinated method is then put forward to keep robustness and damping under disturbances and uncertainty.

(b) The possibility of reducing intervals in one oscillation cycle is expounded. Subsequently, an optimized issue is built to fulfill the coordinate-adaptive update of inertial and damping parameters during transient disturbances.

The article is organized in the manner as follows. Section 2 theoretically presents the impact of VSG parameters on the small-signal stability. Subsequently, the coordinated design method is demonstrated for parameter optimization. Section 3 presents an advanced control strategy for multiple VSGs in reducing oscillation periods. Section 4 is devoted to simulation analysis. Conclusions and future works are summarized in Section 5.

2. Coordinated Parameters Optimization of Multiple VSGs for Small-Signal Stability Improvement

2.1. The Mechanism of Low-Frequency Oscillations Caused by VSGs and SMs

The impact of the VSG-based converters participating in the EOM of a system is investigated quantitatively below using traditional small-signal analysis method. The derivation is given by utilizing a simple two-machine system. Figure 1a is the control block of VSGs and Figure 1b is the system network. In Figure 1b, VSG control is introduced to DG. Assume the phase angle of the load is zero, the classical small-signal model of SG is expressed as:

$$J_G\omega_0\frac{d\Delta\omega}{dt} = -\Delta P_e - D_G\Delta\omega - \Delta P_{DG} \tag{1}$$

$$P_L = \frac{U_1U_2}{Z_L}\cos\delta. \tag{2}$$

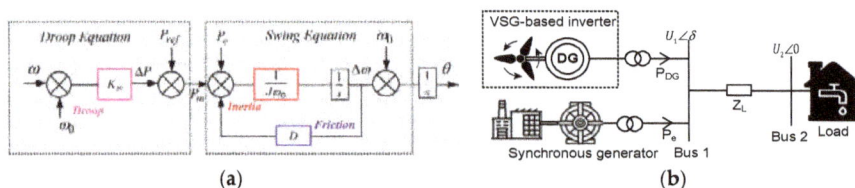

Figure 1. Virtual synchronous generator (VSG) control block and two-machine infinite-bus system: (a) the control block of VSG; (b) the two-machine infinite-bus system.

The linearized model of power balance equation with an initial state δ_0 at Bus 1 meets:

$$\Delta P_L = \frac{U_1U_2}{Z_L}\cos\delta_0\Delta\delta = \Delta P_e + \Delta P_{DG}. \tag{3}$$

As traditional inverters lack inertia, the VSG control algorithm shown in Figure 1a is introduced to provide frequency and voltage support during power fluctuations. Combined with a $\omega - P$ droop strategy, the model for a VSG under yields:

$$\begin{cases} \frac{d\delta}{dt} = \omega - \omega_0 \\ J_{dg}\omega_0 \frac{d\omega}{dt} = P_{ref} - P_{DG} - (K_\omega + D_{dg})(\omega - \omega_0) \end{cases}.$$ (4)

Since traditional control of the DG is often working on a maximum power point tracking (MPPT) state, the active power output does not vary with a small disturbance on the grid side, indicating $\Delta P_{DG} = 0$ in Equation (1). When VSG control strategy is introduced, the output of the active power responses as grid frequency varies. Combined with Equations (1), (3) and (4), the linearized model with/without VSG control is expressed:

$$\begin{cases} J_G\omega_0 \frac{d\Delta\omega}{dt} + D_G\Delta\omega + I\Delta\delta = 0 \\ (J_G + J_{dg})\omega_0 \frac{d\Delta\omega}{dt} + (D_G + K_\omega + D_{dg})\Delta\omega + I\Delta\delta = 0 \end{cases}$$ (5)

where $I = [(U_1U_2)/Z_L]\cos\delta_0$. For the original system, the eigenvalues and damping ratio are calculated as:

$$p_{1,2} = \frac{-D'_G \pm \sqrt{D'^2_G - 4J'_G\omega_0^2 I}}{2J'_G\omega_0}$$ (6)

$$\xi = \frac{-D'_G}{2}\sqrt{\frac{1}{IJ'_G\omega_0}}.$$ (7)

For VSG, $D'_G = D_G + K_\omega + D_{dg}$, $J'_G = J_G + J_{dg}$. The inertia and droop coefficient of the system increase when VSG control is introduced. According to Equations (6) and (7), the damping ratio may decrease, and the eigenvalues may move from left to right when the parameters of multiple VSGs and SGs do not complement each other well. The unsuitable parameters of the system deteriorate stability margin and increase angle instability risk, which is unfavorable to the grid power oscillation.

Figure 2 gives a simple example of two VSG controllers operating in parallel. The eigenvalues of this simple system are calculated as the inertia and damping varies from their rated values to their limitations. The results show that the coordinated design needs to be introduced for multiple VSGs to obtain a better operational performance.

(a)

(b)

(c)

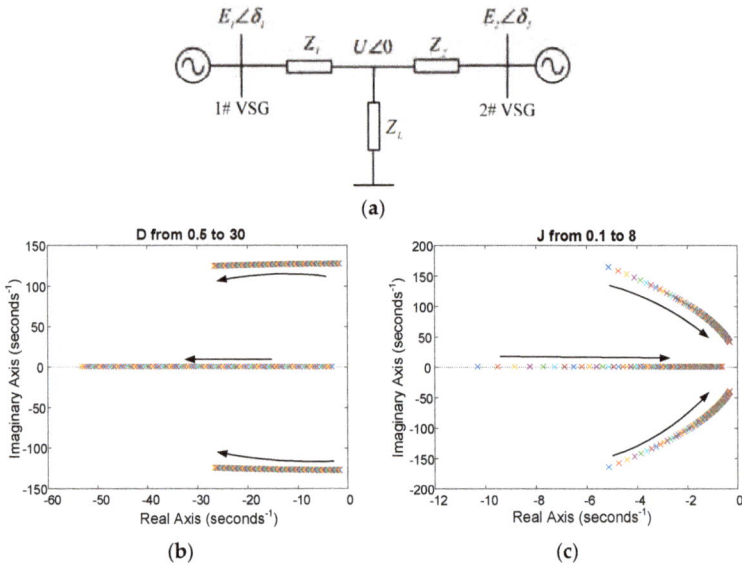

Figure 2. Small-signal stability analysis results of two VSGs operating in parallel. (**a**) The network of two VSG system, (**b**) The eigenvalues of system as the inertia varies, (**c**) The eigenvalues of system as the damping varies.

2.2. Parameter Design Method for Multiple VSG in Improving Small-Signal Stability of System

Quite often, we want the system to operate with well-damped oscillations and keep robustness under disturbances [16]. For this motivation, the D-stable region [17] is presented here to let the eigenvalues of typical operating points lie in an area of secure operation. To enhance the damping ratio and stability margin of the system, this region is used to define a criterion for the controller design. Modes with higher damping behaviors stand on a complex plane shown in Figure 3. The performance robustness constraints for the design methodology is presented as follows:

$$AQ + QA^T + 2\alpha Q < 0 \tag{8}$$

$$\begin{bmatrix} \sin\theta(A^TQ + QA) & \cos\theta(A^TQ - QA) \\ \cos\theta(QA - QA^T) & \sin\theta(A^TQ - QA) \end{bmatrix} < 0. \tag{9}$$

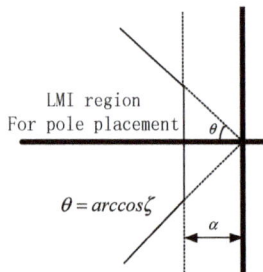

Figure 3. D-stable area.

The proof of this theorem can be found in [17]. As the output of renewable generation is stochastic, the system under different operation states indicates the weak adaptability of the traditional certainty model. To guarantee the robustness of the converter and enhance dynamic response, a polytopic linear differential inclusion model is presented here for better expressing uncertainty in RES. Instead of precisely predicting system state in a period, the LDI only requires a motion-changed region of the uncontrolled resource in time [18]. This advantage provides a possibility to apply a stochastic forecast model to deal with the stability analysis with systematical prediction errors. According to its mathematical definition, the system based on PLDI is expressed as:

$$\dot{x}(t) \in co(A_i x + B_i u). \tag{10}$$

For a typical PLDI, the system has globally uniform stability when it satisfies an exponentially stable theorem [19]. This characteristic provides a possibility to apply PDLI theory to qualitatively judge the stability of the stochastic time-variant system and simplifies the analysis process of uncertainty disturbances. According to [19,20], the stability criterion for PLDI is given as follows:

Theorem 1. *Consider a composite positive function $V_c(x)$ in Equation (11). A PDLI system is exponentially stable if and only if positive–definite matrices $Q_k(x) \in \mathbb{R}^{n \times n}$, $F_k(x) \in \mathbb{R}^{m \times n}$, and $Q(\gamma^*) = \sum_{k=1}^{N} \gamma_k^* Q_k$, $F(\gamma^*) = \sum_{k=1}^{N} \gamma_k^* F_k$ exist, such that*

$$V_c(x) = \min_{\gamma \in \Gamma} x^{\mathsf{T}} P(\gamma) x = \min_{\gamma \in \Gamma} x^{\mathsf{T}} \left(\sum_{i=1}^{N} \gamma_i Q_i \right)^{-1} x \tag{11}$$

$$A_i Q_k + Q_k A_i^{\mathsf{T}} + B_i F_k + B_i^{\mathsf{T}} F_k^{\mathsf{T}} < -\beta Q_k. \tag{12}$$

According to the Schur complement rule, Equation (12) is equal to:

$$V_c x = \min \vartheta$$
$$\text{subject to } \begin{bmatrix} \vartheta & x^{\mathsf{T}} \\ x & Q(\gamma) \end{bmatrix} \geq 0, \ \gamma \in \Gamma, \ \sum_{i=1}^{N} \gamma_i = 1 \tag{13}$$

The detail proving process of Equations (12) and (13) is found in [19]. In a practical circuit, K_ω is determined by the system operation requirement. The parameters D_{dg} and J_{dg}, according to [21], should satisfy:

$$0 \leq J \leq \frac{(D_{dg} + K_\omega)}{2\pi f_{cpmin}} \cot PM_{req}, \ (D_{dg} + K_\omega / \omega_0) \leq D_{max}. \tag{14}$$

Combined with the small-signal model of the VSG-based converter, the optimized issue considered stochastic excitation yields

$$\min \vartheta$$
$$\text{subject to } (6), (8), (9), (10), (12), (13), (14) \tag{15}$$

In the practical project, we always want a ξ_0 larger than 0.05, which is called a strong damping mode. Then, let $\xi_0 = 0.05$, and $\alpha = \xi_0 \omega_n$. The polytopic linear differential matrices A_i is:

$$A_i \in \begin{bmatrix} A_{sysi} & 0 \\ 0 & A_{vsg} \end{bmatrix} A_{sysi} \text{ and } A_{vsg} \text{ satisfies (1) and (5).} \tag{16}$$

When a polytopic model is considered, Equations (8) and (9) should be rewritten for each vertex system, and the resulting set of inequalities should be solved simultaneously to ensure all eigenvalues that lie in a designed region. The optimized issue expressed by Equation (15) consists of a family of bilinear matrix inequities (BMIs) that contain bilinear terms as the product of a full matrix and a scalar. To effectively solve the BMI problem and improve the applicability for large-scale systems, the path-following method [22] is adopted here to update all parameters.

3. Improved Coordinate-Adaptive *J/D* Control Strategy of Multiple VSGs in Mitigating OSC

3.1. The Mechanism of Improved Bang-Bang Control Strategy in Improving the Transient Stability of a System

Generally, the system needs to go through three intervals during transient disturbances before converging into a steady state [16]. This transition (i.e., a–b–c–b) shown in Figure 4a often inevitably causes power oscillation and deteriorates frequency damping. In contrast, as the rotor does not exist, the controller parameters of the VSG can be more flexible during the OSC cycle, hence accelerating the response of the VSG in tracking the steady state. For example, during the acceleration modes (i.e., a–b and c–b), high inertial parameters are adapted to resist disturbances, while small inertial parameters need to be chosen during the deceleration period (i.e., b–c and b–a) to accelerate convergence. This strategy was first introduced by [14,15] and represents one step further on how to detail the coordinate design of adaptive parameters (i.e., *J* and *D*) during each stage.

In fact, the intervals during the OSC cycle can be reduced into one interval (i.e., a–b) if the appropriate control strategy is utilized. For example, on the right side of Figure 4b, we usually want the frequency converges into a rated value at the same time t_1 that the output of VSG increases to its final required power. This process requires the angular velocity to grow first and then decrease. If this scene happens, the other stages in the OSC cycle no longer exist, and the other transient adjustments of the system are eliminated. Specifically, when a large disturbance occurs, the angular frequency quickly increases and obtains its upper limit at time t_u. According to Equation (4), during this period, the inertial term at the left side of the equation is larger than the damping term. The ω then remains with the fastest velocity and waits for the decreasing order. When the decline time t_d comes, the ω begins to decrease and the inertial term at the left side of the equation is smaller than the damping term. Finally, the ω will converge into its rated value at time t_1, and the system operates steadily without any further adjustment. During this whole cycle, the output of the VSG remains increasing and reaches P_2 at the same time t_1. It needs to be emphasized that the output of the VSG rises linearly during periods between t_u and t_d, and the ω is continuous but nondifferentiable during the whole time. Compared to SM, only the VSG is flexible enough to operate in this idea.

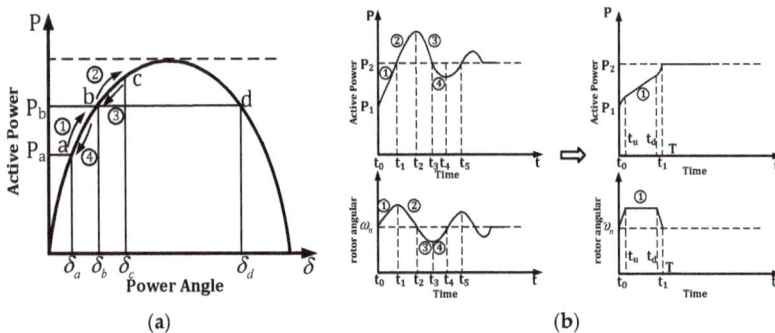

Figure 4. The power-angle curve of a synchronous machine (SM) and the proposed VSG: (**a**) the SM response during transient disturbances; (**b**) the proposed VSG response during transient disturbances.

3.2. Parameter Design Method for Multiple VSG Corresponding to Oscillation Cycle

To begin with, we should set a trigger threshold that enables transient self-adaptive control. The threshold is determined by the endurance capacity and of the VSG-based converter. When the disturbance $\Delta P > L$ happens, the controller starts. Then, to fulfil the expectation in Figure 4b, the parameter-optimization issue, for one VSG, yields

$$\min_{\omega} T_{osc}$$

$$\text{subject to } \Delta P \approx \frac{U_{vsg}U_{grid}}{X_s} \int_0^{T_{osc}} [\omega(t) - \omega_0] dt, \; \omega(t)\Big|_{t=0} = \omega(t)\Big|_{t=T_{osc}} = \omega_n \tag{17}$$

$$|\omega(t) - \omega_0| \leq 2\pi\Delta f_{max}, \; |d\omega/dt| \leq 2\pi k_{max}$$

The first equality constraint ΔP represents the required power deviation that is calculated by the terminal voltage U_{vsg} and U_{grid} at the converter and grid side and the impedance X_s between these two sides. The second equality term represents the steady-state constraint. The third and fourth constraints consider the limitation of the frequency change-rate threshold k_{max}, e.g., the charge–discharge rate of the battery in the VSG-based converter and the maximum frequency deviation Δf_{max}. According to the model in Equation (17) at a particular time $t \in (0, T_{osc})$, the constrains $\omega(t)$ at the time-span T_{osc} in Figure 4b satisfies:

$$\omega(t) - \omega_0 = \begin{cases} 2\pi k_{max}t & 0 \leq t \leq T_1 \\ 2\pi\Delta f_{max} & T_1 \leq t \leq T_2 \\ 2\pi k_{max}(T_{osc} - t) & T_2 \leq t \leq T_{osc} \end{cases} . \tag{18}$$

It is obvious that $T_1 = \Delta f_{max}/k_{max}$, and $T_2 = T_{osc} - \Delta f_{max}/k_{max}$. Assume the same decline rate of frequency. Based on Equations (17) and (18), we have

$$\Delta P \approx \frac{U_{vsg}U_{grid}}{X_s} \int_0^{T_{osc}} [\omega(t) - \omega_0] \, dt \leq \frac{2U_{vsg}U_{grid}\pi}{X_s} \Delta f_{max}\left(T_{osc} - \frac{\Delta f_{max}}{k_{max}}\right). \tag{19}$$

If we know the final state of the system (i.e., ΔP is obtained), then T_{osc} is identified by Equation (19). Then, according to Equation (18), the term $d\omega/dt$ should meet the following condition:

$$d\omega/dt = \begin{cases} 2\pi k_{max} & 0 \leq t \leq T_1 \\ 0 & T_1 \leq t \leq T_2 \\ -2\pi k_{max} & T_2 \leq t \leq T_{osc} \end{cases} . \tag{20}$$

Combined with Equations (4) and (20), parameters J and D during three periods should meet the following conditions:

(a) In a period $0 < t < T_1$, the frequency linearly increases as a rate of k_{max}. Then at this stage, the parameters J should meet the following condition:

$$J_1 = (P_{ref} + k_\omega(\omega - \omega_0) - P_e)/2\pi k_{max}\omega. \tag{21}$$

To accelerate this transient process, $D_{dg} = 0$ at this period.

(b) When the system meets $d\omega/dt = 0$, the system transits to Stage 2, and parameters D_{dg} should have

$$D_{dg} = \frac{[P_{ref} + k_\omega(\omega - \omega_0) - P_e]}{2\pi\Delta f_{max}(2\pi\Delta f_{max} + \omega_0)}. \tag{22}$$

During this period, J is set as the original value.

(c) Then, in $T_1 < t < T_2$, the frequency linearly decreases as a rate of k_{max}, and $D_{dg} = 0$ while J meets the following condition:

$$J_3 = -(P_{ref} + k_\omega(\omega - \omega_0) - P_e)/2\pi k_{max}\omega. \tag{23}$$

This control strategy can be applied to a single VSG-based system owing to the power fluctuation, and the adjustment is mainly provided by this converter. From Equations (17)–(19), to obtain the adjust time T_2, the adjusting power $\Delta P = P_2 - P_1$ should be identified. This situation indicates that the coordinated controller needs to obtain the next system steady state. However, for a multiple VSG-based system, the steady-state output of each VSG is not easy to achieve. For this concern, a system state estimation algorithm is needed.

Considering a typical AC network, the oscillation happens due to the power matching process between each SMs when a disturbance happens. When detecting the imbalance frequency between each bus, the SMs adjust their output for system synchronization. Subsequently, the frequency recovers to its rated value. This characteristic indicates that each VSG at the same bus should synchronize at any moment to eliminate OSC. Therefore, if each VSG introduces the proposed parameter-optimization issue, according to Equation (18), the deviation of the power angle is the same. Hence, the state estimation algorithm can utilize this characteristic to simplify the calculation procedure. Since the controller needs to obtain the final system state variable (i.e., voltage and active power), this estimation is somehow similar to static security analysis [23] after a disturbance.

For a typical network contained n bus, the system at each bus meets the following power balance constraints [23]:

$$P_i = U_i \sum_{j \in i} U_j (G_{ij} \cos \delta_{ij} + B_{ij} \sin \delta_{ij}), \quad Q_i = U_i \sum_{j \in i} U_j (G_{ij} \sin \delta_{ij} - B_{ij} \cos \delta_{ij}). \tag{24}$$

In traditional power flow analysis, SM is set as a PV node, and a $V\delta$ node is needed for power balancing. However, a VSG-based converter is not appropriate to be considered as a PV node, especially V/f control is introduced. When system operates in an island model, the power balance between load and energy resource is maintained by adjusting converter output. Therefore, according to Equation (24), the VSG-based converter meets the following conditions:

$$\begin{aligned}
\frac{U_{vsg} U_i}{X'_L} \sin \delta_{iv} &= U_i \sum_{j \in i} U_j (G_{ij} \cos \delta_{ij} + B_{ij} \sin \delta_{ij}) \\
\frac{U_{vsg} U_i}{X'_L} \frac{U_{vsg} \cos \delta_{iv} - U_i}{U_{vsg}} &= U_i \sum_{j \in i} U_j (G_{ij} \sin \delta_{ij} - B_{ij} \cos \delta_{ij})
\end{aligned} \tag{25}$$

The left side of Equation (25) represents the output of the VSG when line impedance meets $X \gg R$. This requirement can be fulfilled by adding virtual impedance [24]. Hence, in Equation (25), $X'_L = X_L + X_{vir}$, U_{vsg} is set as the nominal integrated voltage. The equation contains two unknown variables (i.e., U_i and δ_{ij}), which are consistent with traditional PQ node constraint. Then, based on Newton-Raphson, the Jacobian matrix yields

$$\begin{bmatrix} \Delta P \\ \Delta Q \end{bmatrix} = \begin{bmatrix} H & N \\ J & L \end{bmatrix} \begin{bmatrix} \Delta \delta \\ \Delta V / V \end{bmatrix}. \tag{26}$$

When $\Delta P > L$, the next operation state is obtained by Equations (25) and (26) and the equality constraint of power angle on the same bus.

Equation (26) can be solved by the Newton-Raphan method in this case. The calculation speed can be faster if the system satisfies the constraints of the PQ decoupled method. Moreover, if the grid structure is fixed, the iterative matrix in Equation (26) is determined and only needs to generate once according to the PQ algorithm.

Noted that this calculation will cause a time-delay issue. From Equation (18), the controller satisfies $T_1 = \Delta f_{max} / k_{max}$, which is constant and identified by controller response ability. Therefore, the time-span at Stage 1 (T_1) is determined. This characteristic indicates that the time-delay issue can be eliminated if the measurement and calculation process can be done before time T_2. From a practical perspective, the speed requirement can be fulfilled, especially for a large-scale power system containing Wide Area Measurement System (WAMS) equipment, e.g., Power Management Unit (PMU) [25].

The maximal power margin of each VSG is different, so the proposed method is limited by the minimum of the maximal adjust-margin of VSG-based converters. Let $\Delta\delta = \int_0^{T_{osc}} [\omega(t) - \omega_0]\, dt$. Then, for a given system,

$$\Delta\delta_{max} = \min\{\Delta\delta_{1max}, \Delta\delta_{2max}, \ldots\ldots \Delta\delta_{nmax}\}. \tag{27}$$

Equation (27) represents the maximal self-adjustment ability of a system, which is limited by each $\Delta\delta_{imax}$. If the strength of the disturbances exceeds the limitation of the system, the strategy in [15] is utilized in this paper.

In all, the steps for coordinate-adaptive J/D control of multiple VSGs is summarized as follows:

(a) Measure the bus disturbances. If $\Delta P > L$, go to Step (b);
(b) Set the J/D value to satisfy the first-period requirement in Equation (21), while estimating the final state of the system by utilizing Equations (25) and (26) and equality constraint;
(c) Transit J/D to the values in Equation (22) at time T_1; meanwhile, calculate T_{osc} and T_2;
(d) At time T_2, let the J/D value satisfy Equation (23) and return to their rate values.

4. Case Study

To validate the proposed approach, two test cases are utilized. The first case is a four-machine two-area test system shown in Figure 5a. In this case, each generator model has six generator states, and the additional control is not considered. The SMs in Areas 1 and 2 are replaced by two VSG-based converters with the equivalent MVA rating of the original SM. The specification and setting of VSGs are shown in Table 1, and the typical system operation data is supplied by [16]. To be mentioned, the VSG parameters J and D before optimization are consistent with the original SM in the same bus. The purpose of this setting is to prove that the VSG-based converter participates in the original electromechanical oscillation modes.

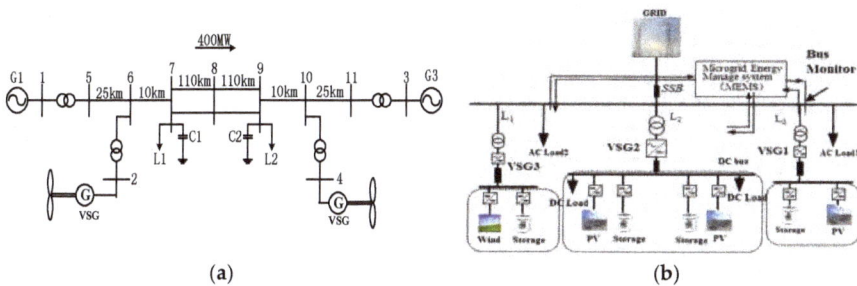

(a)   (b)

Figure 5. Test cases system: (**a**) four-machine two-area system containing VSG; (**b**) a typical microgrid containing three VSGs.

Table 1. Case 1: Main simulation parameters of the VSG model.

Controller Constraints of VSG	
Minimum PM_{req}, f_{cpmin}/f_{cpmin}, D_{max}	45°, 8.5/12.3 Hz, 30 N·m/rad
J_{max}, DC Voltage, DC Capacitor	20 kg/m², 1150 V, 10,000 μF
Control parameters of VSG at bus 2 and 4	
VSG at Bus 2: virtual, inertia/damping	11.571 kg/m², 11.773 N·m/rad
VSG at Bus 4: virtual, inertia/damping	12.418 kg/m², 10.89 N·m/rad

The second case is a typical Microgrid system shown in Figure 5b. The operation parameters, which were optimized by the PLDI and the D-stable model, are given in Table 2. The validation was carried out on a real-time digital simulator (RTDS) to validate the controller presented in Section 2. The model of the VSG and the proposed controller were built on RSCAD, which is accompanying software of RTDS. The signal connection of the RTDS simulation is shown in Figure 6. The digital output card (GTDO) and the digital input card (GTDI) was provided to generate a D/A signal for communication between the RTDS and controller platform. The Field Programmable Gate Array (FPGA, XC3S200-AN) and Digital Signal Processing (DSP, TMS-320F28335) constitute the main hardware architectures of a platform. The PWM pulse signal was provided by this platform.

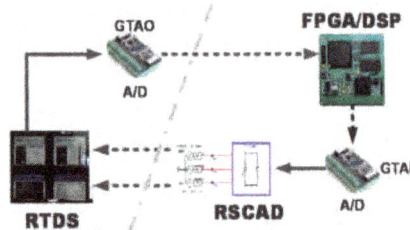

Figure 6. The semi-physical platform of the microgrid based on the RTDS system.

Table 2. Case 2: Main simulation parameters of the microgrid system.

Main Network System	
Inertia/damping/DC voltage/Droop coefficient/Virtual impedance of VSG	0.2 kg/m^2, 2·N m/rad, 700 V, 100 kW/Hz, $j2\Omega$
Equivalent power source	10.5 kV, 50 Hz
AC Load 1/AC Load 3, DC Load1/DC Load2	(8 + j2) kVA, (11 + j2) kVA, 6 kW/5 kW
Transmission Line	j0.347 Ω/km, L_1 = 15 km, L_2 = 10 km, L_3 = 20 km
Virtual Synchronous Generator Parameters	
VSG1: PV generation/Storage	12 kW, 10 kW
VSG2: PV generation/Storage, PV generation/Storage	8 kW/4 kW, 6 kW/4 kW
VSG 3: Wind power/Storage	14 kW/8 kW

4.1. Case 1 Study

Small-signal stability is of great importance for a system to perform well. To mitigate the power oscillation caused by VSGs, the Case 1 test system was mainly introduced to validate the coordinated parameter design method in Section 3. Compared with the traditional VSG controller [4], the stable margin and the dynamic response of the multiple-VSG system before and after optimization are presented here to show the advantages.

The dominant eigenvalues of the original two-area system, before and after optimization, are shown in Figure 7. The oscillation decays in all cases when a disturbance happens. However, the level of damping ratio of the original interarea model, i.e., ξ = 0.0341), is too small to be accepted. After replacing the SMs with a VSG-based converter station, the dominant eigenvalues shown in Figure 7 are changed and improved but still quite close to the original oscillation model. Since the VSG parameters J and D are consistent with the SM, the similarity of oscillation indicates that both VSGs and SMs are coupled, and parameter optimization of multiple VSGs needs to be considered. To satisfy operation requirement, the minimum ξ of the system is 0.05, and the physical constraints of the VSG converter are shown in Table 1. Based on the proposed designed method in Section 2.1, the dominant eigenvalues after optimization is also shown in Figure 7. The dynamic response to step changes of 2% with/without optimization are shown in Figures 8 and 9.

In Figure 7, the eigenvalues are located in a specified complex plane that satisfies the design requirement, and the damping ratios of the system are all larger than 0.05. According to Figures 8 and 9, the generator speed responses to changes in the mechanical torque decays fast when J/D parameters are designed. In contrast, the oscillation of the original system with insufficient ξ decays for more than 30 s. In addition, the simulation result shows the high tracking speed and small overshoot of the proposed algorithm.

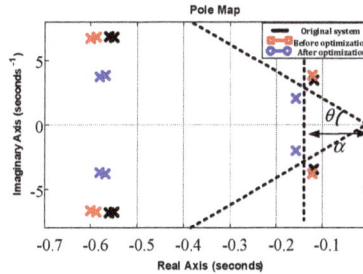

Figure 7. The eigenvalues of the system before/after optimization.

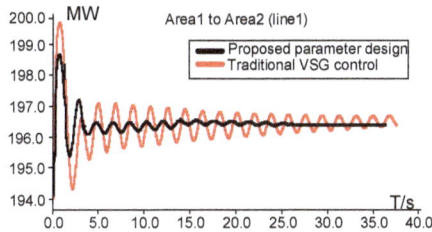

Figure 8. Dynamic response of inter-area power before/after optimization.

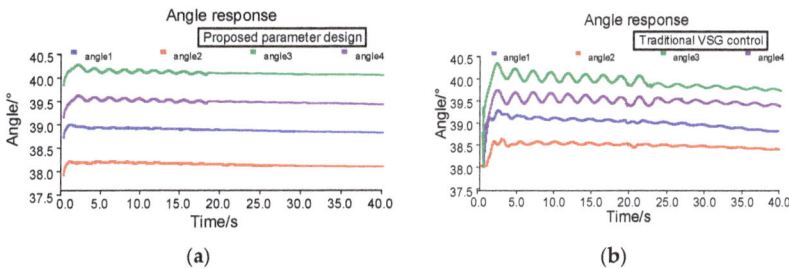

Figure 9. Dynamic response of power angle before/after optimization. (**a**) After optimization, (**b**) before optimization.

4.2. Case 2 Study

To fully utilize the outstanding flexible and convenient performances of VSGs during oscillation, Case 2 was performed to validate the coordinate-adaptive method in Section 2. Compared with the constant J/D controller [4] and the traditional adaptive control [6], the effectiveness and advantage of the improved coordinate-adaptive J/D control strategy are verified.

Figure 10 gives the root locus of this test system as the inertia varies. The disturbance fluctuation margin is set as 10 kW, and related analyses are performed to assess the dynamic transient performance behaviors of the VSGs with the proposed coordinated strategy. The description of this case is as follows: At 0.5 s, there is a step change from 8 to 18 kW in the active power of AC Load1 in Figure 5b. Then, 2 s later, the load decreases to its rated value. The transient response of VSG-based converters is shown in Figures 11–14 as follows.

Figures 11–13 shows the comparison of current, frequency, and active power among different control strategies at each VSG. Figure 14 is the changing process of adaptive virtual inertia and damping coefficients during transient response. According to Figures 11–13, the coordinated controller is well able to control the system response during the sharp rise of power. During simulation, the overshoot under coordinated control is restricted to a small area and the settling time of the system is less than 0.01 s. The total transient period is less than 0.05 s. In contrast, the constant J and D control and traditional adaptive control [6] took longer to converge into the rated frequency. Meanwhile, the overshoot of the system under these controllers is higher than the proposed method. Moreover, the oscillating amplitude of frequency under constant J/D is larger than 1 Hz. The dynamic response of current and active power shown in Figures 11 and 13 also demonstrates the advantages of the proposed method over the other controllers. The overshoot and total transient period under the proposed method are constricted to a small region, and the active power experiences a smooth transition to the final state.

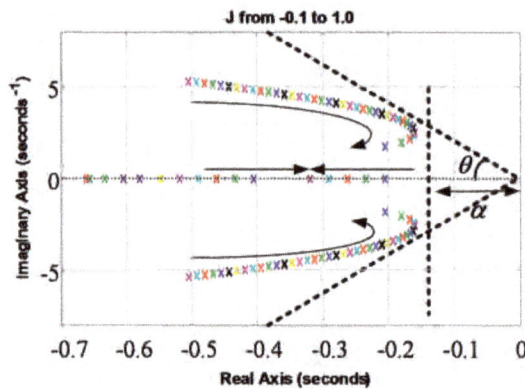

Figure 10. The root locus of this test system with the J from -0.1 to 1.0.

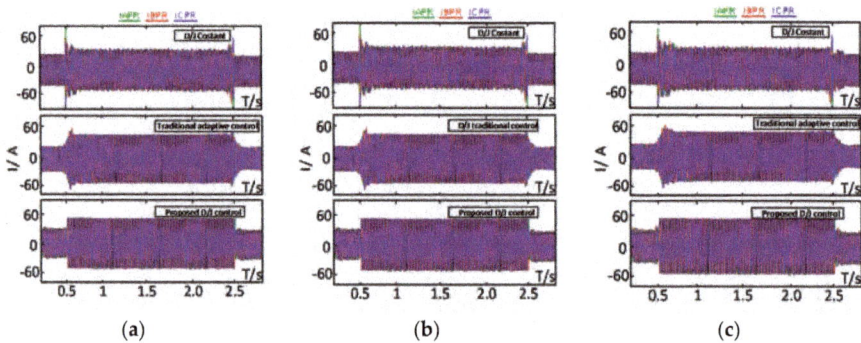

Figure 11. Current responses among different controllers: (a) VSG1; (b) VSG2; (c) VSG3.

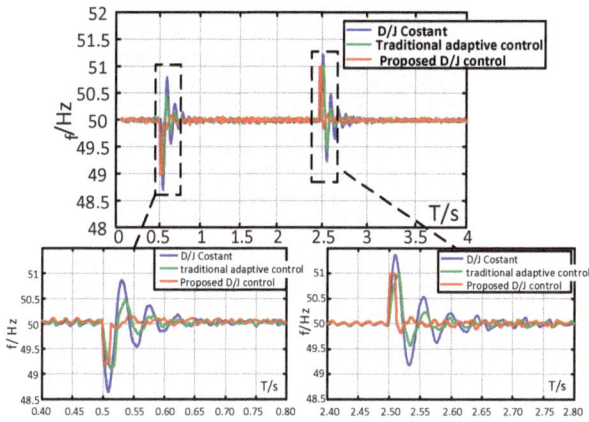

(a)

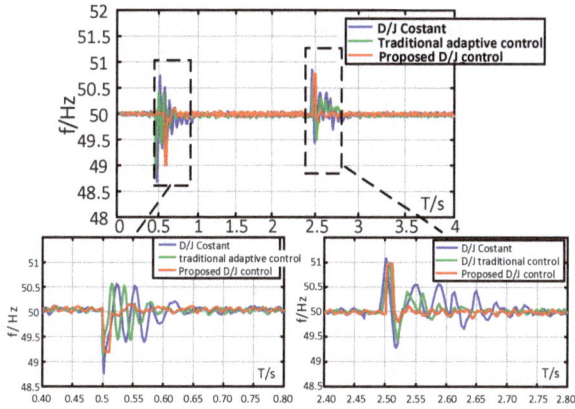

(b)

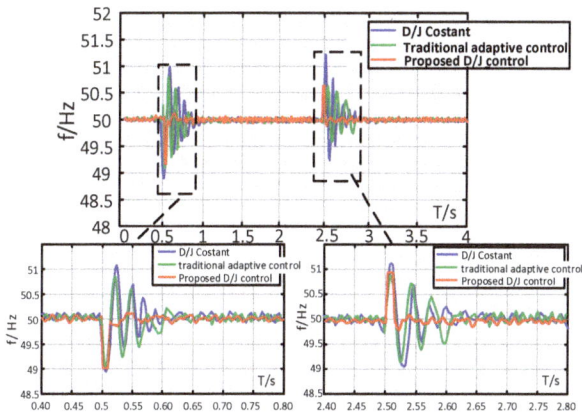

(c)

Figure 12. Frequency response among different controllers: (**a**) VSG1; (**b**) VSG2; (**c**) VSG3.

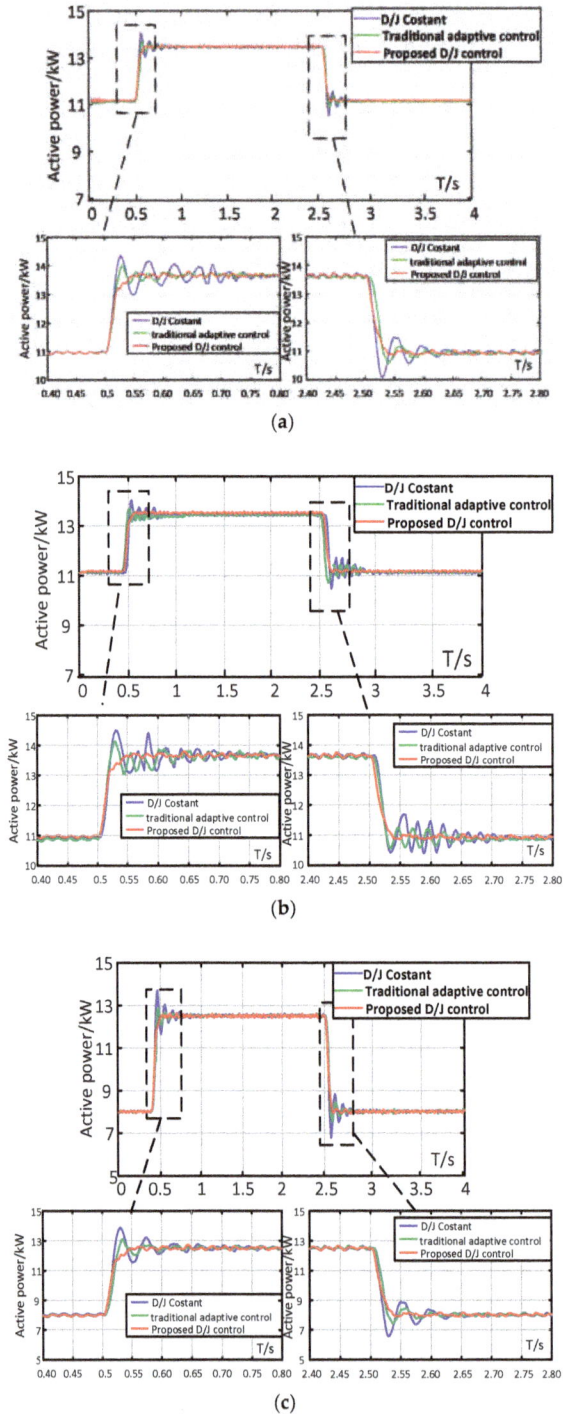

Figure 13. Active power response among different controllers: (**a**) VSG1; (**b**) VSG2; (**c**) VSG3.

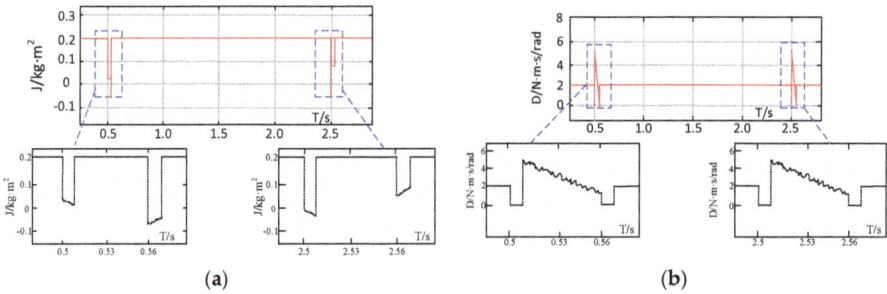

Figure 14. The variation of virtual inertia and damping coefficient: (**a**) virtual inertia; (**b**) damping coefficient.

In Figure 14, the adaptive dynamic process of parameters *J* and *D* are fitted to Equations (21)–(23). During the first period, according to Equation (21), parameter *J* satisfies a linearly decreasing function, while the ω linearly increases. The parameter *D* at Stage 2 is obviously linearly increasing. At Stage 3, parameter *J* shows a contrary tendency with a change in ω. It should be noted that the inertial parameter *J* is negative during this period to fast convergence.

Note that, since the structure of the microgrid is fixed and the network impedance satisfies $x \gg R$, the *PQ* decoupled method can be utilized and the $\Delta\delta$ and ΔV can be directly and quickly identified using Equation (26). The proposed estimation algorithm is fast enough to calculate the threshold time T_2 (in this case, it is 6 ms, according to Figure 14) before the controller requires. From the viewpoints of increasing the adaptability of the proposed controller and reducing the requirement of the measurement of the device's ability, some follow-up work can be done, such as the improvement of the estimation algorithm with deep learning as well as reduced-order models of real-sized power systems.

5. Conclusions

This paper introduces a coordinated control for multiple virtual synchronous generators to improve small-signal stability. Simulation, utilizing a two-area four-machine system and a typical microgrid test system, validates the proposed strategy. Based on theoretical analysis and simulation study, the major conclusions include the following:

(1) The theoretical analysis and related simulation prove that the VSG-based converters are involved in the original electromagnetic oscillation mode. Hence, the damping ratio may decrease and the eigenvalues will move from left to right. The simulation result in Case 1 shows the high tracking speed and small overshoot and thus shows the effectiveness of the proposed algorithm.

(2) To further reduce dynamic periods during transient disturbances, an optimization algorithm for the coordinated controlling of multiple VSG-based converters is presented. Compared to the constant *J* and *D* control and traditional adaptive control, the advanced controller can well reduce overshoot and oscillating amplitude in Case 2.

The results effectively confirm the applicability of the method and indicates the benefits of the proposed strategy in enhancing operation stability and the anti-disturbing ability of multiple VSGs. Concerning the proposed coordinated control, a few aspects should be enriched, and these include reducing and optimizing the investment of a detection system. Moreover, a coping strategy responding to asymmetric faults is necessary. These tasks will be carried out in the future.

Author Contributions: P.H. helped design the study, analyze the data, and write the manuscript. H.C. helped conduct the study. K.C. helped analyze the data. Y.H. helped draw the figure. D.K. helped draft the manuscript. L.C. helped conduct the simulation analysis. Y.W. helped edit the manuscript.

Funding: This research was funded by the National Science Foundation of China, grant number 510507117

Nomenclature

Subscript i, j, k	integer index = $\{1, 2, 3, \ldots \ldots \}$.
Subscript G	presents the synchronous machine (SM).
Subscript dg, DG	presents the distribution generation (DG).
U_1, U_2	the voltage at bus 1 and bus 2 respectively.
Z_L	impedance between bus 1 and bus 2.
P_e, P_L	the electromagnetic power and supply power of SM.
ω, ω_0	operational and rated rotor angular velocity.
J_{dg}, D_{dg}, K_ω	virtual inertia, damping coefficient, and droop coefficient of VSG.
$Q \in \mathbb{R}^{n \times n}$	symmetric matrix of linear matrix inequality (LMI) variables.
$\gamma^*(x) = \arg\min\limits_{\gamma \in \Gamma^N} x^T \left(\sum\limits_{j=1}^{N} \gamma_j Q_j \right)^{-1} x$	a typical convex hull function which meets $0 \leq \gamma_i \leq 1$, $\sum\limits_{i=1}^{N} \gamma_i = 1$.
α, θ	designer specified scalar values shown in Figure 3.
$x \in R^n, u \in R^n, A_i \in R^{n \times n}, B_i \in R^{n \times m}$	the state, input variable and constant matrices of the system.
$PM_{req}, f_{cpmin}, D_{max}$	the required phase margin, cut-off frequency and max droop of VSG.
ξ_0, T_{osc}, θ	the damping ratio threshold and response time-span of the system, $\theta = \arccos\xi_0$.
G_{ij}, B_{ij}	the real and imaginary parts of an element in nodal admittance matrix.
P_i, Q_i	the net active and reactive power at node i.

References

1. Kozlova, M. Real option valuation in renewable energy literature: Research focus, trends and design. *Renew. Sustain. Energy Rev.* **2017**, *80*, 180–196. [CrossRef]
2. Miller, N.; Loutan, C.; Shao, M.; Clark, K. Emergency response: US system frequency with high wind penetration. *IEEE Power Energy Mag.* **2013**, *6*, 63–71. [CrossRef]
3. Collins, S.; Deane, J.P.; Poncelet, K.; Panos, E.; Pietzcker, R.C. Integrating short term variations of the power system into integrated energy system models: A methodological review. *Renew. Sustain. Energy Rev.* **2017**, *76*, 839–856. [CrossRef]
4. Zhong, Q.C.; Hornik, T. Synchronverters: Grid-friendly inverters that mimic synchronous generators. *IEEE Trans. Ind. Electron.* **2011**, *58*, 1259–1267. [CrossRef]
5. Karimi-Ghartemani, M.; Khajehoddin, S.A.; Piya, P.; Ebrahimi, M. Universal controller for three-phase inverters in a microgrid. *IEEE J. Emerg. Sel. Top. Power Electron.* **2016**, *4*, 1342–1353. [CrossRef]
6. Shintai, T.; Miura, Y.; Ise, T. Oscillation damping of a distributed generator using a virtual synchronous generator. *IEEE Trans. Power Deliv.* **2014**, *29*, 668–676. [CrossRef]
7. Guan, M.; Pan, W.; Zhang, J.; Hao, Q.; Cheng, J. Synchronous generator emulation control strategy for voltage source converter (vsc) stations. *IEEE Trans. Power Syst.* **2015**, *30*, 3093–3101. [CrossRef]
8. Xi, X.; Geng, H.; Yang, G. Small signal stability of weak power system integrated with inertia tuned large scale wind farm. In Proceedings of the IEEE PES Innovative Smart Grid Technologies (ISGT), Washington, DC, USA, 19–22 February 2014; pp. 514–518.
9. Eftekharnejad, S.; Vittal, V.; Heydt, G.T.; Keel, B.; Loehr, J. Impact of increased penetration of photovoltaic generation on power systems. *IEEE Trans. Power Syst.* **2013**, *28*, 893–901. [CrossRef]
10. Gautam, D.; Vittal, V.; Harbour, T. Impact of increased penetration of DFIG-based wind turbine generators on transient and small signal stability of power systems. *IEEE Trans. Power Syst.* **2009**, *24*, 1426–1434. [CrossRef]
11. Quintero, J.; Vittal, V.; Heydt, G.T.; Zhang, H. The impact of increased penetration of converter control-based generators on power system modes of oscillation. *IEEE Trans. Power Syst.* **2015**, *29*, 2248–2256. [CrossRef]
12. Ma, J.; Qiu, Y.; Li, Y.; Zhang, W.; Song, Z. Research on the impact of dfig virtual inertia control on power system small-signal stability considering the phase-locked loop. *IEEE Trans. Power Syst.* **2017**, *32*, 2094–2105. [CrossRef]
13. Torres L., M.A.; Lopes, L.A.C.; Morán T., L.A.; Espinoza C., J.R. Self-tuning virtual synchronous machine: A control strategy for energy storage systems to support dynamic frequency control. *IEEE Trans. Energy Convers.* **2014**, *29*, 833–840. [CrossRef]

14. Alipoor, J.; Miura, Y.; Ise, T. Power system stabilization using virtual synchronous generator with alternating moment of inertia. *IEEE J. Emerg. Sel. Top. Power Electron.* **2015**, *3*, 451–458. [CrossRef]

15. Li, D.; Zhu, Q.; Lin, S.; Bian, X.Y. A self-adaptive inertia and damping combination control of vsg to support frequency stability. *IEEE Trans. Energy Convers.* **2017**, *32*, 397–398. [CrossRef]

16. Rogers, G. *Power System Oscillations*; Kluwer Academic Publishers: Norwell, MA, USA, 2000.

17. Chilali, M.; Gahinet, P. H∞ design with pole placement constraints: A LMI approach. *IEEE Trans. Autom. Control* **1996**, *4*, 358–367. [CrossRef]

18. Smirnov, G.V. Introduction to the theory of differential inclusions. *Am. Math. Soc.* **2002**, *255*, 114–139.

19. Hu, T.; Lin, Z. Properties of the composite quadratic Lyapunov functions. *IEEE Trans. Autom. Control* **2004**, *49*, 1162–1167. [CrossRef]

20. Pham, H.; Jung, H.; Hu, T. State-space approach to modeling and ripple reduction in ac–dc converters. *IEEE Trans. Control Syst. Technol.* **2013**, *21*, 1949–1955. [CrossRef]

21. Wu, H.; Ruan, X.; Yang, D.; Chen, X.; Zhao, W. Small-signal modeling and parameters design for virtual synchronous generators. *IEEE Trans. Ind. Electron.* **2016**, *63*, 4292–4303. [CrossRef]

22. Hassibi, A.; How, J.; Boyd, S. A path-following method for solving BMI problems in control. In Proceedings of the American Control Conference, Anchorage, AK, USA, 8–10 May 2002; Volume 2, pp. 1385–1389.

23. Sauer, P.W.; Pai, M.A. *Power System Dynamics and Stability*; Prentice-Hall: Englewood Cliffs, NJ, USA, 1998.

24. Pogaku, N.; Prodanovic, M.; Green, T.C. Modeling, analysis and testing of autonomous operation of an inverter-based microgrid. *IEEE Trans. Power Electron.* **2007**, *22*, 613–625. [CrossRef]

25. Bevrani, H.; Watanabe, M.; Mitani, Y. *Power System Monitoring and Control*; John Wiley & Sons: Hoboken, NJ, USA, 2014.

![energies logo]

Article

Ultra-Short-Term Wind Power Prediction Based on Multivariate Phase Space Reconstruction and Multivariate Linear Regression

Rongsheng Liu [1], Minfang Peng [1,*] and Xianghui Xiao [2]

[1]   College of Electrical and Information Engineering, Hunan University, Changsha 410082, China;
    lrs0623@hnu.edu.cn
[2]   College of Automation, Foshan University, Foshan 52800, China; xiaoxianghui@fosu.edu.cn
*   Correspondence: pengminfang@hnu.edu.cn; Tel.: +86-189-7332-6940

Received: 15 September 2018; Accepted: 9 October 2018; Published: 15 October 2018

Abstract: In order to improve the accuracy of wind power prediction (WPP), we propose a WPP based on multivariate phase space reconstruction (MPSR) and multivariate linear regression (MLR). Firstly, the multivariate time series (TS) are constructed through reasonable selection of wind power and weather factors, which are closely associated with wind power. Secondly, the phase space of the multivariate time series is reconstructed based on the chaos theory and C-C method. Thirdly, an auto regression model for multivariate phase space is created by regarding phase variables as state variables, and the very-short-term wind power is predicted by using a multi-linear regression algorithm. Finally, a parallel algorithm based on map/reduce is presented to improve computing speed. A cloud computing platform, Hadoop consisting of five nodes, is established as a matter of convenience, followed by the prediction of wind power of a wind farm in the Hunan province of China. The experimental results show that the model based on MPSR and MLR is more accurate than both the continuous method and the simple approximation method, and the parallel algorithm based on map/reduce effectively accelerates the computing speed.

Keywords: wind power prediction; phase space reconstruction; multivariate linear regression; cloud computing; time series

---

1. Introduction

In the past decades, with the increasing population, industrial need, and energy need [1], a large amount of fuels such as fossil oil, coal, and natural gas have been consumed. However, the fossil fuels can discharge a large amount of greenhouse gas and pollute the environment. What is more, the fossil fuels are non-renewable and diminishing day-by-day. Therefore, researchers have focused on the renewable energy sources, among which wind power generation is one of the most mature renewable energies with lower pollution and greenhouse gas emissions [2]. Wind power generation is affected by wind speed, wind direction, temperature, turbine type, terrain roughness, air density, and so on [3]. The wind power is random and intermittent. To reduce the risk that is caused by the wind power's fluctuation, both a one-day-ahead (0–24 h) and a real-time (15 min–4 h) wind power predicting report should be submitted to the Grid Dispatch Center in China. Actually, the ultra-short-term wind power prediction (UST-WPP) has been extensively employed in many fields, such as balancing load, optimal operation of reserves [4], and wind farm control [5]. The wind farm owners, power users, and facilities benefit from an improved wind power prediction (WPP).

According to the current law in China, the wind power prediction with horizons of 1–4 h and 24 h are necessary. This paper focuses on forecasting wind power with short horizons. Specifically, we propose a distributed model for ultra-short-term WPP based on multivariate phase space reconstruction

(MPSR) and multivariate linear regression (MLR). Compared with other approaches for short-term wind power prediction, such as artificial neural networks (ANNs), support vector machine (SVM) and single-variable phase space reconstruction, the proposed wind power prediction model is more precise and quicker.

The remainder of this paper is organized as follows. The background of our work is introduced in Section 1. Section 2 reviews the related work. The prepared knowledge is given and the parallel model is proposed in Section 3. Section 4 expresses the key points of the parallel algorithm based on map/reduce. Section 5 describes the prediction experiments. Section 6 outlines the conclusions.

2. Related Works

According to the time horizons, WPP is divided into four categories: long-term prediction, medium-term prediction, short-term prediction, and ultra-short-term prediction [6]. The time horizon and application of the special wind power prediction is shown as Table 1.

Table 1. Classification of wind power prediction based on time horizon.

No.	Method	Time Horizon	Application
1	Long-term	Year	- planning wind farms - Planning of annual generation
2	Medium-term	Week or Month	- Scheduling maintenance
3	Short-term	3 days	- Reducing the discarded wind power - Optimizing the maintenance scheduling - Optimizing the generation scheduling
4	Ultra-short-term	4 h	- Optimizing the frequency - Optimizing the spinning reserve capacity - Optimizing the unit commitment online

The long-term and medium-term wind power predictions do not require very high forecasting accuracy. The short-term WPP forecasts the wind power in next three days and needs more precise results [7], while the ultra-short-term WPP, whose temporal resolution is 15 min, predicts the wind power in next 4 h and requires the highest precision [1]. It is very difficult to accurately forecast wind power because of its chaotic and stochastic characteristics. Additionally, compared to other WPP, the ultra-short-term WPP is more difficult due to its shorter time frames [7]. In the past decades, extensive efforts have focused on WPP, and a large number of wind power prediction methods, models, and tools have been developed. Generally, the WPP methods includes five categories: (a) physical methods, (b) statistic methods, (c) artificial intelligent methods, (d) hybrid methods, and (e) spatio-temporal methods [8].

The physical methods forecast the wind power in terms of the meteorological parameters such as topography, temperature, pressure, wind speed, and wind direction. The well-known physical wind power prediction systems are Prediktor system [9], Previento, and eWind. However, the physical methods are suitable to predict the wind power in 6–72 h or long-term wind power due to the low update frequency of numerical weather prediction (NWP). Because of the high computational cost of NWP, the physical models' application to ultra-short-term WPP are limited. The hybrid methods, which improve the WPP by making use of the advantages of physical methods, statistical methods, and intelligent methods, are gaining attention. Mehmet Baris Ozkan and Pinar Karagoz presented a novel wind power forecast model: Statistical Hybrid Wind Power Forecast Technique (SHWPFT). Compared with other statistical and physical models such as ANNs and SVM, SHWPFT requires less historical data to establish a model [10]. A hybrid forecasting model based on grey relational analysis and wind speed distributional features is presented to improve the effects of the ultra-short-term WPP [11]. Reference [12] provides an approach of short-term WPP with multiple observation points. The speed and direction of wind are used to forecast the wind power; unfortunately, the error of numeric weather prediction (NWP) must be considered. Since the hybrid models also require the NWP, their application to ultra-short-term WPP is limited too. The statistical methods are widely used for the shorter-term wind power prediction because of the lower computational complexity and cost.

The statistical models can be modified based on the allocation and features of wind farms, and statistical models are widely used to forecast short-term wind power. Generally, the statistical models include the direct prediction and the indirect prediction. The wind speed is forecasted first and then is mapped to the wind power according to the wind power curve in indirect wind power prediction. However, the wind power curve is influenced by the environmental and meteorological variables, which are not considered in indirect methods. Consequently, the accuracy of indirect methods is limited. In direct methods, the wind power is predicted directly according to the historical data, and the wind power predictions have greater accuracy [13]. The statistical models are divided into time series (TS) methods and artificial intelligence (AI) methods. Persistence method is a classical TS method, and it assumes that the wind power at time '$t + \Delta t$' equals the wind power at time 't'. Persistence method is more accurate than most physical and other statistical methods in short-term or ultra-short-term WPP. Hence, any new predicting method should be tested against the classical benchmark (persistence method). The other well-known TS models includes autoregressive (AR), autoregressive moving average (ARMA), autoregressive integrated moving average (ARIMA), and so on. However, these TS models are linear and it is difficult to capture the non-linear patterns. It is too difficult to directly create precise mathematical models for wind power prediction because of wind's complexity, so the artificial intelligence and machine learning methods are employed to improve WPP. Artificial neural network (ANN), Bayesian network (BN) and least square support vector regression (LSSVR) are widely used to improve the ultra-short-term WPP. Duehee Lee and Ross Baldick presented an ensemble model for short-term WPP based on Gaussian processes and neural networks [14] from which the predicted wind power and distribution can be obtained. The ensemble model is relatively accurate in the whole forecasting time horizon, and the predicted distribution objectively reflects the real distribution of wind power. The comparative study shows that the neural networks (NNs) outperform the ARIMA in [15]. A short-term WPP model based on wavelet support vector machine is proposed in reference [16], in which a wavelet support vector machine-based (WSVM) approach effectively reduces the negative effects on predicting results induced by errors of NWP data. Additionally, wind power can be forecasted when fewer training samples are accessed. Nevertheless, the universality should be further verified. The least square support vector machine (LSSVM) outperformed the SVM and NNs in terms of the computational complexity and the global convergence [17] in WPP. It is difficult to directly model the wind power time series that is composed of several components with different characteristics. To improve WPP, the wind power time series are decomposed by the wavelet transform (WT), empirical mode decomposition (EMD), and ensemble empirical mode decomposition (EEMD) in the preprocessing stage [18]. Some prior knowledge and assumptions are necessary to determine the mother wavelet in WT. EMD is a data-driven decomposition method, and outperforms WT. Compared with WT, EMD and its variant EEMD have the greatest accuracy. Although the non-linearity of wind power time series can be dealt with using decomposition methods, the chaotic nature of wind power time series can lead to prediction errors [19]. A method of wind power prediction based on chaotic theory and phase space reconstruction is proposed in reference [20]. Reference [13] proposed a short-term WPP framework based on chaotic time series analysis and singular spectrum analysis, and the accuracy is improved by distinguishing the chaotic and non-chaotic components in both decomposition and prediction stages. However, only the historical wind power data are used to forecast wind power in the phase-space-reconstruction-based WPP model.

According to the output, the statistical WPP is divided into probabilistic prediction and point prediction. To deal with wind ramp dynamics, a support-vector-machine-enhanced Markova model (SVMEM) for short-term WPP was presented by Lei Yang, Miao He, and Junshan Zhang et al. in which both distributional predictions and point predictions are derived [21]. Reference [14] has proved that the uncertainty of wind speed satisfies Gaussian distribution with zero mean and heteroscedasticity.

An optimal loss function for heteroscedastic regression and a new framework of v-support vector regression (v-SVR) for learning tasks of Gaussian noise with heteroscedasticity were thus developed to improve the situation. Both the wind speed and wind power, however, satisfy other distribution but not the proposed model in reference [22]. The uncertainty of the wind power is modeled by the probabilistic prediction and prediction interval, which are the extension of point prediction. The point prediction is regarded as the most likely wind power. The probabilistic prediction has been applied to power system security [23], various unit commitment strategies [24], and energy storage sizing [25], etc.

The wind power data at each site are self-correlated individually, but also the wind power data at different sites are spatially cross-correlated. A sparsity-controlled vector autoregressive model is introduced to obtain the spatio-temporal wind power prediction in reference [26]. Although, the predicting methods are discussed in batch learning mode, a general overview of wind power prediction is presented in Table 2.

Table 2. The wind power prediction methods. WPP: wind power prediction.

No.	WPP Methods	Remarks
1	Persistence Method	- benchmark method - Very accurate for ultra-short and short term prediction - Low accuracy for drastically fluctuating wind power series
2	Physical Method	- Uses the meteorological and environmental data - Applied to WPP whose time horizon is more than 6 h
3	Time-Series Models	- Accurate for short-term predictions - Cannot model the non-linearity
4	Artificial Intelligence	- Accurate for short-term predictions - Models the non-linearity - Cannot process the chaotic nature
5	Hybrid Method	- Accurate for medium and long term predictions
6	Spatial-Temporal Method	- Accurate for ultra-short-term predictions - Batch learning mode - Cannot update using the latest and real-time information

The high penetration of wind power will definitely affect the reliability and security of the grid due to the fluctuation and randomness of wind power. Compared with the short-term WPP, the ultra-short-term WPP is more effective to solve these problems. Although the above mentioned works contribute a lot to the development of WPP, the ultra-short-term WPP is still a young area and needs more contributions from different areas.

3. Prepare Knowledge and System Modeling

3.1. Time-Series Similarity

Time-series similarity [27,28] is measured with many methods, such as Minkowski distance, Euclidean distance, Pearson coefficient, and dynamic time warping (DTW) distance [29,30]. Compared to other methods, DTW is not sensitive to synchronization and can measure the similarity of time-series with different lengths. Thus, DTW is employed to measure the similarity of numerical weather prediction series. DTW is defined as follows:

Definition 1. *Dynamic time warping (DTW) distance.* $X = \{x_1, x_2, \ldots, x_l\}$ *and* $Y = \{y_1, y_2, \ldots, y_n\}$ *are 2 time-series, where l and n are the lengths of time series X and Y, respectively. The DTW between X and Y is defined recursively as Equation (1).*

$$D_{dtw}(X,Y) = \begin{cases} 0, & if\ X = Y = \{\} \\ +\infty, & if\ X = \{\}\ and\ Y \neq \{\} \\ +\infty, & if\ X \neq \{\}\ and\ Y = \{\} \\ d(x_1,y_1) + min \begin{cases} D_{dtw}(rest(X),Y) \\ D_{dtw}(rest(X),rest(Y)) \\ D_{dtw}(X,rest(Y)) \end{cases} \end{cases} \tag{1}$$

where {} indicates empty time-series, $d(x,y) = ||x,y||_2$ is the 2-norm [31] *of x and y, and rest(X) = $\{x_2, \ldots, x_l\}$.*

Definition 2. *Time-series similarity. X and Y are 2 time-series, and the time-series similarity between X and Y is defined as Equation (2).*

$$S(X,Y) = \begin{cases} 0, & if\ \theta < D_{dtw}(X,Y) \\ \theta - D_{dtw}(X,Y), & else \end{cases} \tag{2}$$

where θ is a threshold, $D_{dtw}(X,Y)$ is the DTW between X and Y.

3.2. Multi-Variable Phase Space Reconstruction

According to Takens-theorem [32,33], the phase space of chaotic time series $x_1, x_2, \ldots, x_n$ is reconstructed with appropriate embedding dimension m and time delay τ; the original system and the trajectory of reconstructed phase space are differential homeomorphism. The phase space reconstruction theory [34] of single-variable chaotic time series could be extended to multi-variable chaotic time series. The multi-variable phase space reconstruction is defined as follows.

Definition 3. *Multi-variable phase space reconstruction. X^1, X^2, ... , X^K are K chaotic time-series, where $X^i = \{x_1^i, x_2^i, \ldots, x_n^i\}$, and i = 1, 2, ... , K. The q-th phase point of reconstructed phase space for multi-variable chaotic time series X^1, X^2, ... , X is shown as Equation (3).*

$$V_q = \Big\{ x_q^1, x_{q-\tau_1}^1, \ldots, x_{q-(m_1-1)\tau_1}^1, \\ x_q^2, x_{q-\tau_2}^2, \ldots, x_{q-(m_2-1)\tau_2}^2, \ldots, \\ x_q^K, x_{q-\tau_K}^K, \ldots, x_{q-(m_K-1)\tau_K}^K \Big\} \tag{3}$$

where $q = j, j+1, \ldots n, j = \max\limits_{1 \leq i \leq K} \{(m_i - 1)\tau_i\} + 1$ and m_i and τ_i are embedding dimension and time delay of the chaotic time-series X^i, respectively. The embedding dimension m_i and time delay τ_i are obtained using C-C algorithm [33].

3.3. Ultra-Short-Term WPP Modeling

In this paper the ultra-short-term WPP model is based on multi-variable phase space reconstruction, similarity of time-series, and linear regression. Our forecasting model includes steps steps, which are specifically presented in Figure 1.

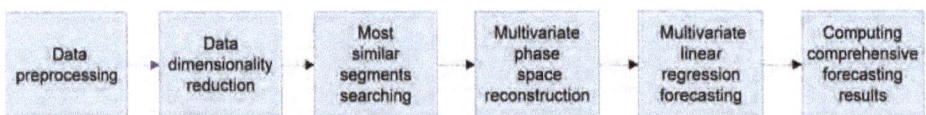

Figure 1. Flow diagram of ultra-short-term wind power prediction (WPP).

More details about the proposed ultra-short-term WPP model are listed below.

Step 1: Data preprocessing. The historical wind power data and weather data may be missing for some reason, for instance, equipment failure, network interruption, etc. The continuous missing data series Md, whose length is more than 5, is removed from historical data series. Otherwise, data is specified by simple interpolation, as expressed in Equation (4).

$$x_i = x_0 + \frac{i \times (x_e - x_s)}{N+1} \tag{4}$$

where x_s is the data point before Md, x_e is the data point behind Md, $i = 1, 2, \ldots, N$, and N is the length of missing data series Md.

Step 2: Data dimensionality reduction. NWP includes some important parameters, for instance, wind speed v, wind direction θ, atmospheric pressure p, and temperature t. The speed and direction of wind are shown in Figure 2.

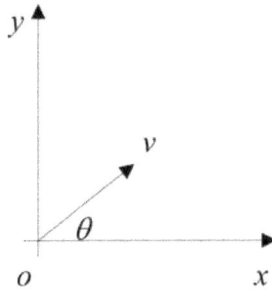

Figure 2. The decomposition of wind speed.

In Figure 2, the x-axis is the latitude line of the given location, and the east is the positive direction. The y-axis is the longitude of the given location, and the north is the positon direction. In order to conveniently forecast wind power, wind speed v is decomposed to v_x and v_y.

$$\begin{aligned} v_x &= v \times \cos\theta \\ v_y &= v \times \sin\theta \end{aligned} \tag{5}$$

where θ is the angle between wind direction v and the x-axis.

Only the most important parameters (v_x, v_y and p) of NWP are selected in the presented model. The historical NWP data can be expressed as a $N \times 3$ matrix M, which is clearly shown as Equation (6).

$$M = \begin{pmatrix} u_1, v_1, p_1 \\ u_2, v_2, p_2 \\ \ldots \\ u_N, v_N, p_N \end{pmatrix} \tag{6}$$

where N is the number of NWP data points. The matrix M is converted to a $N \times 1$ matrix using the approach in reference [2], as displayed in Equation (8).

$$C = M^T \times M \tag{7}$$

$$X = M \times e = (w_1, w_2, \ldots, w_N)^T \tag{8}$$

where e is an eigenvector, whose corresponding Eigen value is the maximal Eigen value of matrix C.

Step 3: The most similar NWP series segments searching: NWP_0 is the NWP data series during $2L$ hours, which consist of two equal parts, namely, the former L hours and the latter L hours. NWP0 includes 8L data points (4 points per hour). NWP_0 is converted to X^0, and the whole historical NWP data series is converted to X according to Equations (6)–(8). X^{is} is the top K sub-series of X, which are most similar to X^0, and S_i is the similarity between X^0 and X^{is}, $i = 1, 2, \ldots, K$.

Step 4: The phase space reconstruction for multivariate time series. τ_X and τ_P are the delay time of X and P, respectively, and m_X and m_P are the embedding dimensions of X and P, respectively. τ_X, τ_P, m_X and m_P are achieved by C-C method. P^i and X^i $(i = 0, 1, \ldots, K)$ are the wind power sub-series and reduced NWP sub-series at the same time period, and Ph^i is the reconstructed phase space of P^i and X^i.

Step 5: Multivariate linear regression. The multivariate linear regressions are shown as Equation (9)

$$
\begin{pmatrix} y_1 \\ y_2 \\ \cdots \\ y_m \end{pmatrix} = \begin{pmatrix} c_{1,1}, c_{1,2}, \ldots, c_{1,m} \\ c_{2,1}, c_{2,2}, \ldots, c_{2,m} \\ \cdots \\ c_{m,1}, c_{m,2}, \ldots, c_{m,m} \end{pmatrix} * \begin{pmatrix} x_1 \\ x_2 \\ \cdots \\ x_m \end{pmatrix} + \begin{pmatrix} c_{1,0} \\ c_{2,0} \\ \cdots \\ c_{m,0} \end{pmatrix} + \begin{pmatrix} \xi_1 \\ \xi_2 \\ \cdots \\ \xi_m \end{pmatrix} \tag{9}
$$

where $\xi = (\xi_1, \xi_2, \ldots, \xi_m)^T$ is residual error, $(x_1, x_2, \ldots, x_m)^T$ is the input, and $(y_1, y_2, \ldots, y_m)^T$ is the output. The coefficient matrix C and constant term C_0 are shown as Equation (10).

$$
C = \begin{pmatrix} c_{1,1}, c_{1,2}, \ldots, c_{1,m} \\ c_{2,1}, c_{2,2}, \ldots, c_{2,m} \\ \cdots \\ c_{m,1}, c_{m,2}, \ldots, c_{m,m} \end{pmatrix} \text{ and } C_0 = \begin{pmatrix} c_{1,0} \\ c_{2,0} \\ \cdots \\ c_{m,0} \end{pmatrix} \tag{10}
$$

The training process of linear regression models is to search for appropriate coefficient matrix C and constant term C_0, so that residual error ξ minimizes.

Given the residual error equals 0, then Equation (9) is converted into Equation (11).

$$
\begin{pmatrix} y_1 \\ y_2 \\ \cdots \\ y_m \end{pmatrix} = \begin{pmatrix} c_{1,0}, c_{1,1}, c_{1,2}, \ldots, c_{1,m} \\ c_{2,0}, c_{2,1}, c_{2,2}, \ldots, c_{2,m} \\ \cdots \\ c_{m,0}, c_{m,1}, c_{m,1}, \ldots, c_{m,m} \end{pmatrix} * \begin{pmatrix} 1 \\ x_1 \\ x_2 \\ \cdots \\ x_m \end{pmatrix} \tag{11}
$$

The Equation (11) is further expressed as follows:

$$
Y = (C_0, C) * \begin{pmatrix} 1 \\ X \end{pmatrix} \tag{12}
$$

Given a group of training samples $< X_i, Y_i >$, $(i = 1, 2, \ldots, K)$, the goal of the training process is to find appropriate C and C_0, so that the mean error, namely, Equation (13) minimizes.

$$
\sum_{i=1}^{n} \frac{(\| Y_i^* - Y_i \|_2)}{n} \tag{13}
$$

where $Y_i = (C_0, C) * \begin{pmatrix} 1 \\ X_i \end{pmatrix}$.

After the linear regression model is trained, Y is obtained from Equation (12) by assuming X is given.

$Ph(i)$ is the i-th phase point of reconstructed phase space Ph. Considering that the reconstructed phase space is locally linear, $Ph(i)$ and $Ph(i+1)$ are linearly related as follows.

$$Ph(i+1) = (C_0, C) \begin{pmatrix} 1 \\ Ph(i) \end{pmatrix} \tag{14}$$

The Equation (15) is obtained by substituting all the phase points of Ph into Equation (14).

$$(Ph(i+1), Ph(i), \ldots, Ph(2)) = C' \begin{pmatrix} 1, & 1, & \ldots, & 1 \\ Ph(i), & Ph(i-1), & \ldots, & Ph(1) \end{pmatrix} \tag{15}$$

where $C' = (C_0, C)$. C' is achieved from the following Equation.

$$C' = (Ph(i+1), Ph(i), \ldots, Ph(2)) \times \begin{pmatrix} 1, & 1, & \ldots, & 1 \\ Ph(i), & Ph(i-1), & \ldots, & Ph(1) \end{pmatrix}^{-1} \tag{16}$$

where $\begin{pmatrix} 1, & 1, & \ldots, & 1 \\ Ph(i), & Ph(i-1), & \ldots, & Ph(1) \end{pmatrix}^{-1}$ is the generalized inverse matrix of $\begin{pmatrix} 1, & 1, & \ldots, & 1 \\ Ph(i), & Ph(i-1), & \ldots, & Ph(1) \end{pmatrix}$. When C' is given, the next phase point is obtained from the current phase point and Equation (14).

Step 6: The comprehensive forecasting results. If $Ph^1, Ph^2, \ldots, Ph^k$ are the reconstructed phase space of the top K most similar time-series segments respectively, according to step 4, then C'_k is achieved by substituting Ph^k into Equation (16). The multivariate linear regression modes are shown as Equation (17).

$$Ph^k(i+1) = C'_k \begin{pmatrix} 1 \\ Ph^k(i) \end{pmatrix} \tag{17}$$

where $Ph^k(i)$ and $Ph^k(i+1)$ are the input and output of the k-th linear regression model, $k = 1, 2, \ldots, k$. The predicted phase point is achieved when the phase points of the current NWP and wind power data are input into Equation (17). The predicted value of the k-th WPP model is shown as Equation (18).

$$P_k = Ph^k(m_P, i+1) \tag{18}$$

where P_k is the predicted value of wind power by the k-th regression model and $Ph^k(m_P, i+1)$ is the m_P-th element of the predicted phase point $Ph^k(i+1)$, $k = 1, 2, \ldots, K$. The comprehensive forecasting sums the weighed P_k as follows.

$$P = \frac{S_k}{\sum\limits_{i=1}^{K} S_k} P_k, k = 1, 2, \ldots, K \tag{19}$$

where S_k is the similarity between X^0 and X^{is}. The forecasting wind power series will be obtained when the above comprehensive forecasting process is executed iteratively.

4. Parallel Algorithm Based on Map/Reduce

To improve the ultra-short-term WPP, the multivariate phase space reconstruction, the similarities of time-series and the multi-variate linear regression are employed to model wind power prediction. However, the increasing wind power and NWP data increase the time and space complexity and affect the predicting accuracy of the ultra-short-term WPP. In order to accelerate the computing speed, we present a parallel algorithm of ultra-short-term WPP, which is based on the map/reduce programming model.

4.1. Map/Reduce Programming Model

Map/reduce is an extendable parallel programming model, which is widely used in the parallel computation of big data. The diagrammatic layout of map/reduce is shown in Figure 3.

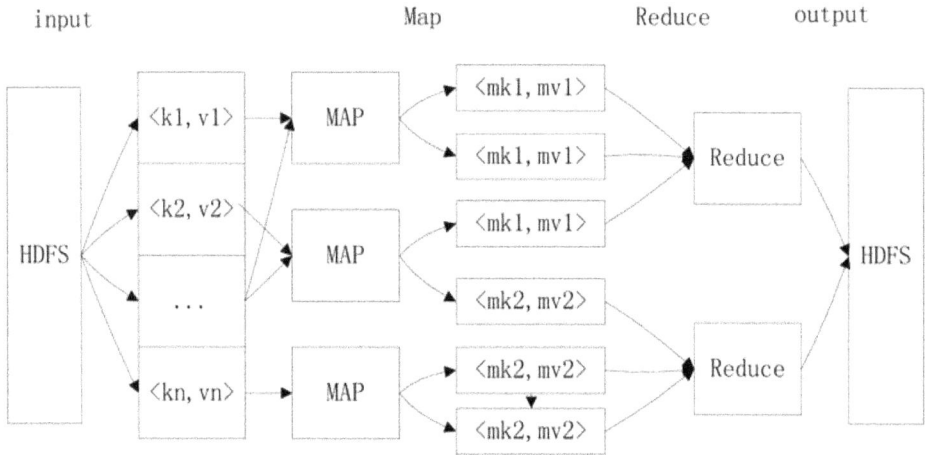

Figure 3. Diagrammatic layout of map/reduce. MAP: the mapping process of hadoop; HDFS: hadoop distributed file system.

Map/reduce includes two phases: mapping and reducing. Every sub-process is highly parallel in these two phases. The specific procedures are as follows:

1. The raw data are input into the key/value pairs (key, value), and the data are processed as much as possible without communication. The intermediate data created in map phase are also saved as key/value pairs (intermediate-key, intermediate-value).
2. The intermediate pairs with the same intermediate-key are transferred to the same reducing process with the completion of the mapping process. The reducing process starts when all intermediate data are transferred. When both mapping and reducing processes are completed, the final results are achieved.

4.2. The Algorithm of Ultra-Short-Term WPP

The map/reduce model is widely used to process big data, however, a single map/reduce job only can complete simple jobs. To solve complex problems, the complex process is usually divided into some sub-processes, and the work is completed together with many map/reduce jobs. Due to the complexity of the ultra-short-term WPP, which is proposed in Section 3.2, the work is completed by five map/reduce jobs (see Algorithm 1).

Algorithm 1. Parallel Ultra-short-term wind power prediction based on Map/reduce.

Job 1: Reducing dimension of NWP matrix.

Map: The NWP matrix M is separated into N sub-matrixes M_i, where $i = 1, 2, \ldots, N$. Each sub-matrix M_i is a reduced dimension based on the method in Section 3.2.

(1) Each map process reads sub-matrix M_i of M.

(2) The intermediate data pairs $(0,' i, X_i')$ is achieved by reducing the data dimension of each sub-matrix M_i, according to Equation (8).

Reduce: The reducing dimension series X is created by connecting in order all X_i, which are from map process.

(1) $X = \{\}$;

(2) For $i = 1{:}N$

$\quad\quad X_i$ is added to the end of X;

$\quad\quad$ end for

(3) Output X;

$\quad\quad$ For the sake of convenience, X_0 is the reduced NWP data series during $2L$ hours, which consists of two equal parts, namely, the former L hours and the latter L hours. X_0 is removed from X, and the remainder of X is divided into N sub-series, $X_1, X_2, \ldots, X_N$, where $|X_i|| \geq 8L$, the head of X_1 is the first node in X, and the head of X_{i+1} is the $(8L - 1)$-th node of X_i from the bottom, $i = 1, 2, \ldots, N - 1$.

Job 2: Searching top k sub-series, which are the most similar to X_0.

Map:

(1) Each map process reads the pairs $(i, < X_i, X_0 >)$;

(2) For $j = 1{:}||Xi|| - L\%$ where $||Xi||$ is the length of Xi [31].

(1) We compute the DTW between X_0 and X_{ij}, which is X_i's sub-series including L elements and starting with the j-th element of Xi;

(2) The intermediate data are key/value pairs $(i, 'j, DTW')$

End for

Reduce:

(1) $k = (\sum\limits_{k=1}^{i-1} ||X_i||) + j$;

(2) $S_k = \left| \theta - DTW_{ij} \right|$;

(3) Output (k, S_k).

$\quad\quad$ The sub-series are sorted by S_k according to the method in reference [35], and the top k sub-series of NWP and wind power are selected. The top k most similar sub-series of NWP and wind power are denoted as X_k and P_k, where $k = 1, 2, \ldots, k$.

$\quad\quad$ The sub-series are sorted by S_k according to the method in reference [35], and the top k sub-series of NWP and wind power are selected. The top k most similar sub-series of NWP and wind power are denoted as X_k and P_k where $k = 1, 2, \ldots, k$.

Job 3: The embedding dimension and delaying time of P and X are computed in this job.

$\quad\quad$ For the sake of convenience, the time series P and X in our algorithm are denoted as P_1 and P_2, respectively.

Map:

(1) $(i, P_i) \overset{C-C\ algorithm}{\rightarrow} (m_i, \tau_i)$; $\quad$ % m_i and τ_i are obtained by C-C algorithm

(2) $J_i = (m_i - 1)\tau_i$; $\quad\quad\quad\quad$ % calculating the starting index J_i of P_i

(3) The intermediate data of this map are data pairs $(0, ''m_i, \tau_i, J_i'')$, where m_i, τ_i and J_i are the embedding dimension, delaying time, and starting point of P_i, respectively.

Reduce:

(1) $J = \max\limits_{1 \leq i \leq 2} \{J_i\} + 1$

(2) $m_P = m_1, \tau_P = \tau_1, m_X = m_2, \tau_X = \tau_2$;

(3) Output $m_P, \tau_P, m_X, \tau_X, J$.

Job 4: Multivariate phase space reconstruction

Map:

(1) $(k,' J\#m_X\#\tau_X\#m_P\#\tau_P\#X_k\#P_k') \xrightarrow{reconstructed} (0, \aleph_k)$, where X_k and P_k are the most similar sub-series of top K achieved from Job 2, and $k = 1, 2, \ldots, k$.

Reduce:

(1) Output reconstructed phase space $\aleph_k$.

Job 5: Multi-variables linear regression

Map:

(1) Input $< i,' \aleph_k\#S_k\#\aleph_0' >$, where $\aleph_0$ is the reconstructed phase space during 2L hours.

(2) The linear model is trained by $< \aleph_{k,j}, \aleph_{k,j+1} >$, where $\aleph_{k,j}$ is the j-th phase point of reconstructed phase space $\aleph_k, j = 1, 2, \ldots, \|\aleph_k\| - 1$;

(3) After the linear model has been trained, the next phase point is obtained according to Equation (12) and the last point of $\aleph_0$.

(4) The step 3 is executed iteratively, and then the forecasting reconstructed phase space $\aleph_k^P$ is achieved.

(5) The forecasting power sequence P_{kj} is obtained based on an inverse process of the phase space reconstruction and reconstructed phase space $\aleph_k^P$.

(6) The intermediate data are $< 0,' P_{kj}, S_k' >$.

Reduce:

The comprehensive wind power is given as follows.

$$P_j = \frac{S_k}{\sum\limits_{k=1}^{k} S_k} P_{kj}, k = 1, 2, \ldots, k; j = 1, 2, \ldots, B \qquad (20)$$

where $j = 1, 2, \ldots, B$, and B is the forecasting time-scale

5. Application and Case Study

5.1. The Experimental Data and Environments

The experimental data includes 2 sections: NWP data and wind power data. The NWP data ranged from 1 September 2012 to 31 August 2013 and were from the key laboratory of regional numerical weather predictions in a province of China. NWP data include temperature, humidity, wind speed, wind direction, atmosphere pressure, and so on. The NWP is from the mesoscale numerical prediction model. The temporal resolution of NWP is 1 h. The horizontal resolution of NWP is 3×3 km, and it is improved to 1×1 km by the dynamic downscaling model [36]. In our experiment, only the data of wind speed, wind direction, and atmosphere pressure were selected. The wind speed and wind direction were first converted into x components and y components, where the east and north are regarded as the positive direction of x-axis and y-axis, respectively. The converted NWP data are shown in Table 3.

Table 3. Historical numerical weather prediction (NWP) data of a wind farm.

No.	Time 1 February 2013	Wind Speed X-Direction (m/s)	Wind Speed Along Y Direction (m/s)	Atmosphere Pressure (atm)
1	1:00	7.43	6.45	0.98
2	2:00	8.6	5.36	0.98
3	3:00	6.07	9.05	0.986
...	...	...	...	...

In Table 3, each row is an NWP data record, which is composed of the time, the wind speed in the x-component, the wind speed in the y-component, and the atmosphere pressure. simple interpolation method was used to get the 15-min NWP data. The wind power (15-min) from 1 September 2012 to 31 August 2013 was used to train the model. The Figure 4 shows the daily peak wind power in each month.

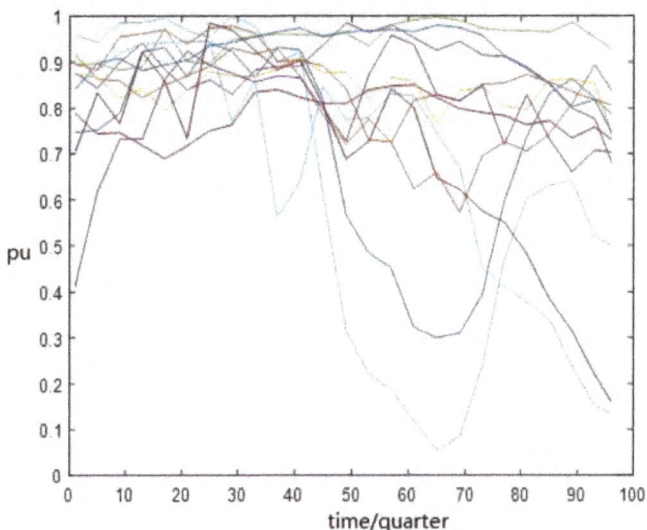

Figure 4. Historical data of wind power.

In Figure 4, the *x* axis is time, *y* axis is the per-unit value of wind power. Because the wind power fluctuates severely at different times, the data are normalized based on Equation (18).

$$P_N = P/P_{max} \tag{21}$$

where P_N is the normalized wind power, P is wind power, and P_{max} is the peak daily wind power from 1 September 2012 to 31 August 2013.

To implement the experiment, we tried to establish an experimental cloudy computing platform Hadoop, which is composed of five nodes including a 4-core central processing unit (CPU) (Intel Core i5) and an Ubuntu operation system. The ultra-short-term wind power was individually predicted by continuous prediction method, simple approximation method, and other methods proposed in this paper.

In terms of continuous prediction method, the nearest observed values were regarded as the next predicting values, namely $x'_{i+1} = x_i$. The simple approximation method is another kind of continuous method based on phase space reconstruction, and the nearest observed phase points were regarded as the values of next phase point *i*, namely, $X'_{i+1} = X_{j+1}$, where X_j is the nearest phase point of X_i. The error index was the normal mean absolute error (NMAE) in this paper.

$$NMAE = \sum_{i=1}^{N} \frac{|x_i - y_i|}{N \times C} \tag{22}$$

where x_i is the measured wind power, y_i is the forecasting wind power, C is the installed capacity of wind farm, and N is the number of periods being forecasted. More details about NMAE are described in reference [26].

5.2. Results and Analysis of the Experiment

The embedding dimension and delaying time of wind power series P and NWP series X were computed by C-C method with 4000 points and 8000 points, respectively. The embedding dimension and delaying time of wind power series P were 5 and 18, respectively. The embedding dimension and delaying time of NWP series P were 6 and 17, respectively.

The number of the most similar sub-series could affect the ultra-short-term wind power prediction. To optimize the number of most similar sub-series, the number of most similar sub-series ranged from 1 to 10 in the experiment. The experimental results are shown as Figure 5, when the number of most similar sub-series ranges from 1 to 10.

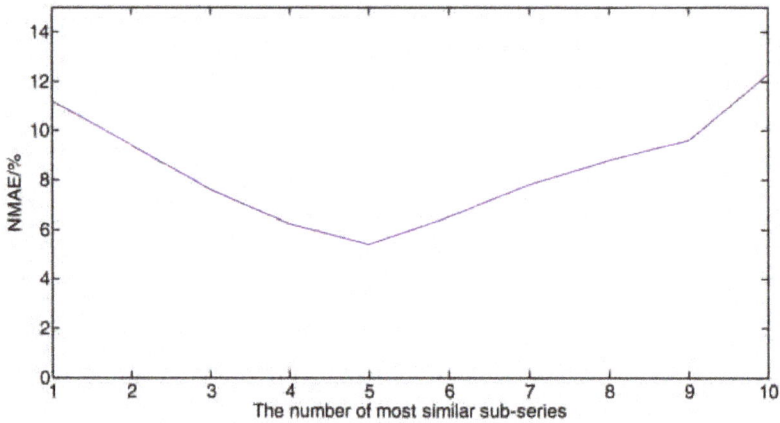

Figure 5. Relation between the number of most similar sub-series and normal mean absolute error.

Figure 5 shows the relation between NMAE and the number of most similar sub-series. When fewer of the most similar segments (especially 1) are used to train a multi-variate linear regression model, the multi-time weighted linear regression becomes a single variable linear regression, and the results have bigger errors. With increasing K, the errors are slowly reduced. When K is near to the threshold, the NMAE keeps getting smaller. However, when the K is larger than a certain number, the errors increase slowly. The reason for this phenomenon is that the dissimilar series could increase the errors of ultra-short-term WPP. A good result can be achieved when K is between 4 and 8.

Figure 6 demonstrates the relation between forecasting time-scale and errors. When the predicting time-scale is less than 6 h, the predicting errors are stable and smaller. With an increasing forecasting time-scale, the average predicting error increases gradually. Figure 6 illustrates that the proposed model is suitable for ultra-short-term WPP, due to its smaller predicting time-scale.

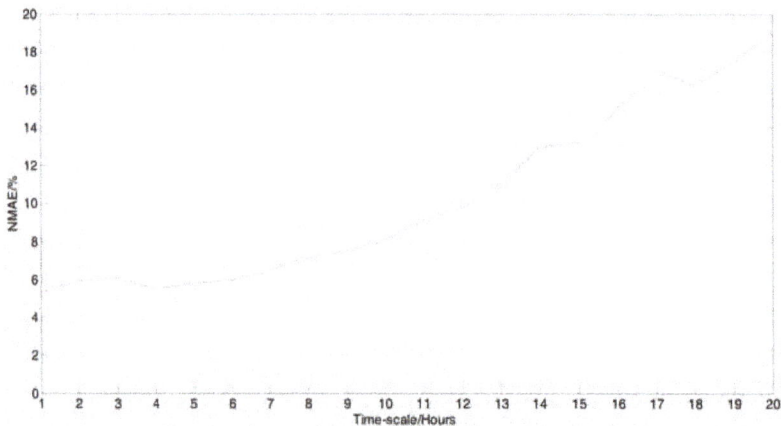

Figure 6. Relation between time-scale and NMAE.

Our model is trained with the historical wind power and NWP data from 00:00 of 1 September 2012 to 24:00 of 31 August 2013. When the five most similar segments are used to train the linear regression model, the experimental results are shown in Figure 7.

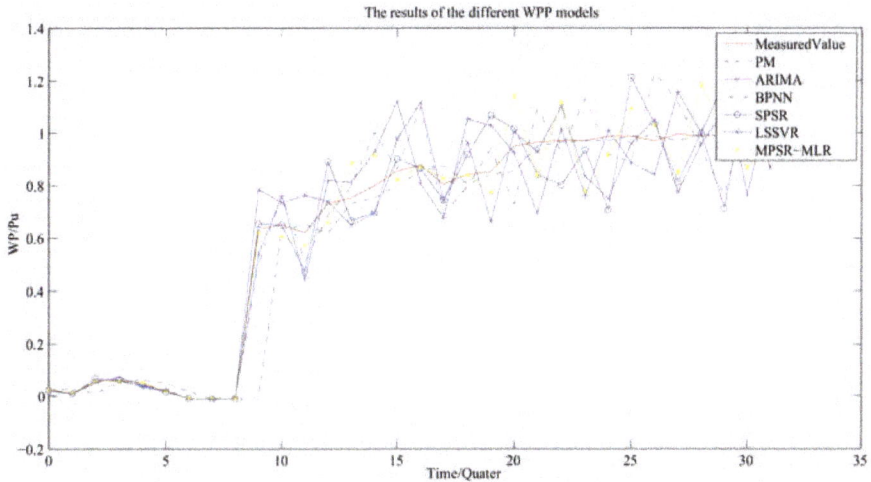

Figure 7. Results of wind power predictions. 'Measured Value' is the measurement of wind power. 'PM' is the forecasting results of persistence method. 'ARIMA' is the forecasting results of the autoregressive integrated moving average. 'BPNN' is the forecasting result of the back propagation neural networks [37]. 'SPSR' is the forecasting of the WPP based on single-variable phase space reconstruction. 'LSSVR' is the forecasting results of the WPP based on LSSVR. 'MPSR-MLR' is the forecasting results of the proposed model.

Figure 7 shows that most of the forecasting results of the proposed model are near to the actual measurements. It proves that MPSR and MLR are suitable for ultra-short-term WPP. The error of the different models are shown as Figure 8.

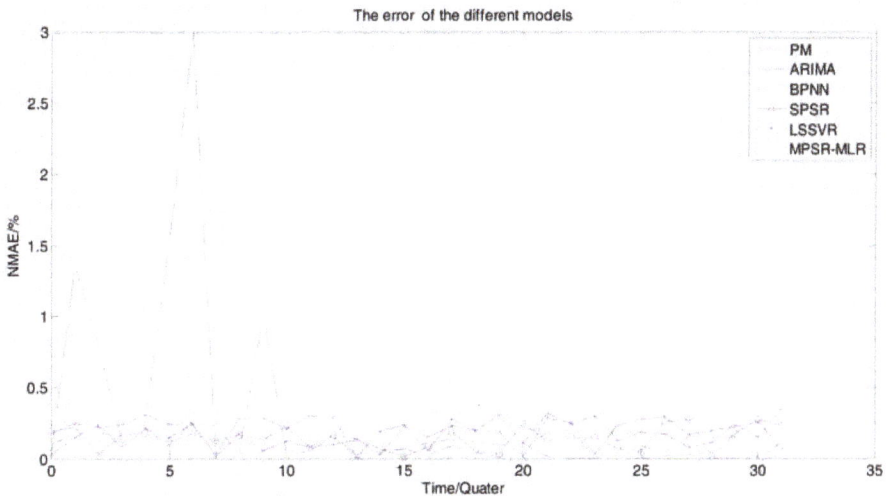

Figure 8. The errors of different wind power predictions.

In Figure 8, 'PM' is the errors of the persistence method. 'ARIMA' is the errors of the WPP based on autoregressive integrated moving average. 'BPNN' is the errors of the WPP based on back propagation neural networks [37]. 'SPSR' is the errors of the WPP based on the single-variable phase space reconstruction. 'LSSVR' is the errors of the WPP based on LSSVR. 'MPSR-MLR' is the errors of the proposed model. The statistical results of errors for different methods are listed in Table 4.

Table 4. Historical data of a wind farm.

No.	Methods	Max Error (%)	Min Error (%)	Avg Error (%)
1	PM	389.64	0.75	24.25
2	ARIMA	26.00	0.0114	12.76
3	BPNN	25.16	0.0089	10.40
4	LSSVR	21.24	0.0071	9.34
5	SPSR	22.36	0.0016	11.86
6	MPSR-MLR	14.38	0.0042	6.46

Figure 8 and Table 4 show that PM is not applicable to the seriously fluctuating wind power series. MPSR-MLR can reduce the predicting errors in the seriously fluctuating wind power series. The error of our model is lower than that of other WPP methods. MPSR-MLR outperforms SPSR by mining the correlation of different time series.

6. Conclusions

Various models and methods have been extensively employed in the field of ultra-short-term WPP as no model is suitable for all conditions and all wind farms. In order to improve the ultra-short-term WPP, we have proposed a novel model of the ultra-short-term WPP based on the multi-variates phase space reconstruction and the linear regression in this paper. The proposed model improves the ultra-short-term WPP by mining the correlation of different time series. The performance is particularly good, when the wind power series fluctuate seriously. A map/reduce-based parallel algorithm is implemented on a cloudy computing platform, Hadoop. The research results allow for three conclusions:

1. Wind power is a chaotic time series, and multi-variate phase space reconstruction can improve the ultra-short-term WPP by mining the correlation of these wind power series.
2. It is very difficult to forecast wind power, especially when the wind power series fluctuates seriously. The proposed model improves the performance during the wind ramp drastically.
3. The forecasting speed is accelerated by adopting both the map/reduce-based parallel algorithm and the cloudy computing platform.

Author Contributions: Conceptualization, R.L. and M.P.; methodology, R.L.; software, R.L.; validation, X.X. and M.P.; formal analysis, R.L.; investigation, R.L.; resources, R.L.; data curation, R.L.; writing—original draft preparation, R.L.; writing—review and editing, R.L. and M.P.; visualization, R.L.; supervision, R.L.; project administration, M.P.; funding acquisition, M.P.

Acknowledgments: The authors would like to acknowledge the support provided by the Natural Science Foundation of China No. 61472128 and No. 61173108.

Conflicts of Interest: The authors declare no conflict of interest.

References

1. Xue, Y.; Yu, C.; Zhao, J.; Li, K.; Liu, X.; Wu, Q.; Yang, G. A Review on Short-term and Ultra-short-term Wind Power Prediction. *Autom. Electric Power Syst.* **2015**, *39*, 141–151.
2. Rafique, M.M.; Rehman, S.; Alam, M.M.; Alhems, L.M. Feasibility of a 100MW Installed CapacityWind Farm for Different Climatic Conditions. *Energies* **2018**, *11*, 2147. [CrossRef]

3. Xie, W.; Zhang, P.; Chen, R.; Zhou, Z. A Nonparametric Bayesian Framework for Short-Term Wind Power Probabilistic Forecast. *IEEE Trans. Power Syst.* **2018**. [CrossRef]
4. Khorramdel, B.; Chung, C.Y.; Safari, N.; Price, G.C.D. A Fuzzy Adaptive Probabilistic Wind Power Prediction Framework Using Diffusion Kernel Density Estimators. *IEEE Trans. Power Syst.* **2018**, 958–965. [CrossRef]
5. Wang, Y.; Liao, W.; Chang, Y. Gated Recurrent Unit Network-Based Short-Term Photovoltaic Forecasting. *Energies* **2018**, *11*, 2163. [CrossRef]
6. Foley, A.; Leahy, P.G.; Marvuglia, A.; McKeogh, E.J. Current methods and advances in forecasting of wind power generation. *Renew. Energy* **2012**, *31*, 1–8. [CrossRef]
7. Khodayar, M.; Wang, J.; Manthouri, m. Interval Deep Generative Neural Network for Wind Speed Forecasting. *IEEE Trans. Smart Grid* **2018**, 1–16. [CrossRef]
8. Pimkumwong, N.; Wang, M.-S. Online Speed Estimation Using Artificial Neural Network for Speed Sensorless Direct Torque Control of Induction Motor based on Constant V/F Control Technique. *Energies* **2018**, *11*, 2176. [CrossRef]
9. Lu, P.; Ye, L.; Sun, B.; Zhang, C.; Zhao, Y.; Teng, J. A New Hybrid Prediction Method of Ultra-Short-Term Wind Power Forecasting Based on EEMD-PE and LSSVM Optimized by the GSA. *Energies* **2018**, *11*, 697. [CrossRef]
10. Ozkan, M.B.; Karagoz, P. A Novel Wind Power Forecast Model: Statistical Hybrid Wind Power Forecast Technique (SHWIP). *IEEE Trans. Ind. Inform.* **2015**, *11*, 375–387. [CrossRef]
11. Shi, J.; Ding, Z.; Lee, W.-J.; Yang, Y.; Liu, Y.; Zhang, M. Hybrid Forecasting Model for Very-Short-Term Wind Power Forecasting Based on Grey Relational Analysis and Wind Speed Distribution Features. *IEEE Trans. Smart Grid* **2014**, *5*, 521–526. [CrossRef]
12. Xu, Q.; He, D.; Zhang, N.; Kang, C.; Xia, Q.; Bai, J.; Huang, J. A Short-Term Wind Power Forecasting Approach with Adjustment of Numerical Weather Prediction Input by Data Mining. *IEEE Trans. Sustain. Energy* **2015**, *6*, 1283–1291. [CrossRef]
13. Safari, N.; Chung, C.Y.; Price, G.C.D. Novel Multi-Step Short-Term Wind Power Prediction Framework Based on Chaotic Time Series Analysis and Singular Spectrum Analysis. *IEEE Trans. Power Syst.* **2018**, *33*, 590–601. [CrossRef]
14. Lee, D.; Baldick, R. Short-Term Wind Power Ensemble Prediction Based on Gaussian Processes and Neural networks. *IEEE Trans. Smart Grid* **2014**, *5*, 501–510. [CrossRef]
15. Chen, L.; Lai, X. Comparision between ARIMA and ANN models used in short-term wind speed forecasting. In Proceedings of the 2011 Asia-Pacific Power and Energy Engineering Conference, Wuhan, China, 25–28 March 2011; pp. 1–4.
16. Zeng, J.; Qiao, W. Short-Term Wind Power Prediction Using a Wavelet Support Vector Machine. *IEEE Trans. Sustain. Energy* **2012**, *3*, 255–264. [CrossRef]
17. Giorgi, M.G.D.; Campilongo, S.; Ficarella, A.; Congedo, P.M. Comparison between wind power prediction models based on wavelet decomposition with least-squares support vector machine (LSSVM) and artificial neural network (ANN). *Energies* **2014**, *7*, 5251–5272. [CrossRef]
18. Wu, Q.; Peng, C. Wind power generation forecasting using least square support machine combined with ensemble empirical model decomposition, principal component analysis and a bat algorithm. *Energies* **2016**, *9*, 261. [CrossRef]
19. An, X.L.; Jiang, D.X.; Zhao, M.H.; Liu, C. Short-term preciton of wind power using EMD and chaotic theory. *Commun. Nonlinear Sci. Numer. Simul.* **2012**, *17*, 1036–1042. [CrossRef]
20. Zhang, Y.; Lu, J.; Meng, Y.; Yan, H.; Li, H. Wind power short-term forecasting Based on Empirical Mode Decomposition and Chaotic Phase Space Recontruction. *Autom. Electric Power Syst.* **2012**, *36*, 24–28.
21. Yang, L.; He, M.; Zhang, J.; Vittal, V. Support-Vector-Machine-Enhanced Markov Model for Short-Term Wind Power Forecast. *IEEE Trans. Sustain. Energy* **2015**, *6*, 791–799. [CrossRef]
22. Morshedizadeh, M.; Kordestani, M.; Carriveau, R.; Ting, D.S.-K.; Saif, M. Power production prediction of wind turbines using a fusion of MLP and ANFIS networks. *IET Renew. Power Gener.* **2018**, *12*, 1025–1033. [CrossRef]
23. Rarki, R.; Thapa, S.; Billinto, R. A Simplified risk-based method for short-term wind power commitment. *IEEE Trans. Sustain. Energy* **2012**, *3*, 498–505.
24. Wen, Y.; Li, W.; Hunag, G.; Liu, X. Frequency dynamaics constrained unit commitment with battery energy storage. *IEEE Trans. Power Syst.* **2016**, *31*, 5115–5125. [CrossRef]

25. Bitaraf, H.; Rahman, S.; Pipattanasomporn, M. Sizing energy storage to mitigate wind power forecast error impacts by signal processing techniques. *IEEE Trans. Sustain. Energy* **2015**, *6*, 1457–1465. [CrossRef]

26. Zhao, Y.; Ye, L.; Pinson, P.; Tang, Y.; Lu, P. Correlation-Constrained and Sparsity-Controlled Vector Autoregressive Model for Spatio-Temporal Wind Power Forecasting. *IEEE Trans. Power Syst.* **2018**, *33*, 5029–5040. [CrossRef]

27. Cui, B.; Zhao, Z.; Tok, W.H. A Framework for similarity search of Time Series Cliques with Natural Relations. *IEEE Trans. Knowl. Data Eng.* **2012**, *24*, 385–398. [CrossRef]

28. Schramm, R.; Jung, C.R.; Miranda, E.R. Dynamic Time Warping for Music Conducting Gestures Evaluation. *IEEE Trans. Multimed.* **2015**, *17*, 243–255. [CrossRef]

29. Yin, H.; Yang, S.; Ma, S.; Liu, F.; Chen, Z. A novel parallel scheme for fast similarity search in large time series. *China Commun.* **2015**, *12*, 129–140. [CrossRef]

30. Han, M.; Zhang, R.; Qiu, T. Multivariate Chaotic Time Series Prediction Based on Improved Grey Relational Analysis. *IEEE Trans. Syst. Man Cybern. Syst.* **2018**, 1–11. [CrossRef]

31. Mercorelli, P. Denoising and Harmonic Detection Using Nonorthogonal Wavelet Packets in Industrial Applications. *J. Syst. Sci. Complex.* **2007**, *20*, 325–343. [CrossRef]

32. Johnson, M.T.; Povinelli, R.J.; Lindgren, A.C.; Ye, J.; Liu, X.; Indrebo, K.M. Time-domain isolated phoneme classification using reconstructed phase spaces. *IEEE Trans. Speech Audio Proc.* **2005**, *13*, 458–466. [CrossRef]

33. Zhao, F.; Sun, B.; Zhang, C. Cooling, Heating and Electrical Load Forecasting Method for CCHP System Based on Multivariate Phase Space Reconstruction and Kalman Filter. *Proc. CSEE* **2016**, *36*, 399–406.

34. Jin, J. Research on Optimization of Sorting Algorithm under MapReduce. *Comput. Sci.* **2014**, *41*, 155–159.

35. Chen, N.; Qian, Z.; Nabney, I.T.; Meng, X. Wind Power Forecasts Using Gaussian Processes and Numerical Weather Prediction. *IEEE Trans Power Syst.* **2014**, *29*, 656–665. [CrossRef]

36. Han, Y. *The Design and Implementation of a Wind Power Forecasting System for Wind Farm in Gansu Province*; University of Electronic Science and Technology of China: Chengdu, China, 2017.

37. Castellani, F.; Astolfi, D.; Mana, M.; Burlando, M.; Meißner, C.; Piccioni, E. Wind power forecasting techniques in complex terrain: ANN vs. ANN-CFD hybrid approach. *J. Phys. Conf. Ser.* **2016**, *735*, 082002. [CrossRef]

energies

MDPI

Article

Leveraging Hybrid Filter for Improving Quasi-Type-1 Phase Locked Loop Targeting Fast Transient Response

Yunlu Li [1], Junyou Yang [1,*], Haixin Wang [1], Weichun Ge [2] and Yiming Ma [1]

[1]   School of Electrical Engineering, Shenyang University of Technology, Shenyang 110870, China;
      liyunlu@sut.edu.cn (Y.L.); sutxny_whx@126.com (H.W.); lnmayiming@126.com (Y.M.)
[2]   Liaoning Province Electric Power Company, Shenyang 110006, China; gwc@ln.sgcc.com.cn
*   Correspondence: junyouyang@sut.edu.cn

Received: 4 September 2018; Accepted: 16 September 2018; Published: 17 September 2018

Abstract: In renewable energy generation applications, phase locked loop (PLL) is one of the most popular grid synchronization technique. The main objective of PLL is to rapidly and precisely extract phase and frequency especially when the grid voltage is under non-ideal conditions. This motivates the recent development of moving average filters (MAFs) based PLL in a quasi-type-1 system (i.e., QT1-PLL). Despite its success in certain applications, the transient response is still unsatisfactory, mainly due to the fact that the time delay caused by MAFs is still large. This has significantly limited the utilization of QT1-PLL, according to common grid codes such as German and Spanish grid codes. This challenge has been tackled in this paper. The basic idea is to develop a new hybrid filtering stage, consisting of adaptive notch filters (ANFs) and MAFs, arranged at the inner loop of QT1-PLL. Such an idea can greatly improve the transient response of QT1-PLL, owing to the fact that ANFs are utilized to remove the fundamental frequency negative voltage sequence (FFNS) component while other dominant harmonics can be removed by MAFs with a small time delay. By applying the proposed technique, the settling time is reduced to less than one cycle of grid frequency without any degradation in filtering capability. Moreover, the proposed PLL can be easily expanded to handle dc offset rejection. The effectiveness is validated by comprehensive experiments.

Keywords: synchronization; adaptive notch filter (ANF); phase-locked loop (PLL)

---

1. Introduction

With the development of renewable energy system, PLL is widely used in most of grid-connected power converter applications owing to its simple structure [1]. Synchronous-reference-frame based PLL (SRF-PLL) is a standard PLL in three-phase grid connected applications, as shown in Figure 1. Since the open-loop transfer function of its model has two poles at the origin, SRF-PLL can be treated as a type-2 PLL (a type-N system has N poles at origin). When grid voltages are unpolluted, SRF-PLL can provide zero phase-error under phase jump and frequency deviation [2,3]. However, its phase-tracking performance degrades under non-ideal grid conditions owing to the existence of disturbances voltage components such as FFNS component and harmonics [4]. This motivates the work [5] to integrate low-pass filters into SRF-PLL, together with moving average filter (MAF) or delay signal cancellation (DSC) operator, to attenuate disturbances. Despite its success in completely removing harmonics, it incurs significant degradation of the dynamic performance. This has significantly limited its applications due to the restriction of common grid codes such as German and Spanish grid codes [6,7].

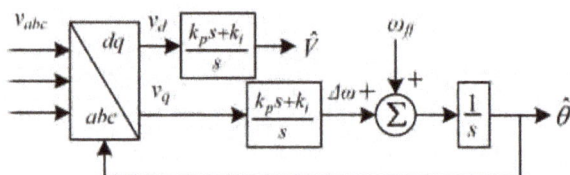

Figure 1. Block schematic of SRF-PLL.

In recent years, many PLLs were developed to make improvements in phase tracking performance. Some type-2 PLLs, such as multiple complex-coefficient-filter-based PLL (MCCF-PLL) [8], dual second-order generalized integrator based PLL (DSOGI-PLL) [9] and multiple reference frame based PLL (MRF-PLL) [10], provide a higher bandwidth by using hybrid filtering but high order harmonics cannot be totally removed. Although some MAF and DSC based hybrid filtering stage can solve this problem, the time delay induced by MAF or DSC is large. In Reference [11], a differential MAF-PLL (DAMF-PLL) was proposed. Although the window length (T_ω) of its MAF is narrowed, the settling time is still over one cycle, which can hardly satisfy the requirements in some grid code [12,13]. Recently, a novel PLL structure with rapid transient response was proposed in Reference [14], which is named quasi-type-1 PLL (QT1-PLL). Some advanced PLLs also improve their dynamic performance by using QT1 structure. But the filtering capability of these existing quasi-type-1 structure based PLLs is still unsatisfactory. A brief literature review related to some advanced PLLs mentioned above is given in Section 2.

To tackle the above technical challenge, this paper develops a new hybrid filtering stage, consisting of adaptive notch filters (ANFs) and MAFs, arranged at the inner loop of QT1-PLL. Such an idea can reduce the settling time of QT1-PLL, because ANFs are employed to eliminate the FFNS while other dominant harmonics can be removed by MAFs with a narrowed window length. By using the proposed method, the convergence time is reduced to a short time within one grid period without any degradation in filtering capability. The propose method is motivated from [11]. Moreover, compared with author's other two papers [15,16], besides the main difference in filtering technique, this paper also studies the digital implementation of the proposed method in more detail. The computational burden is evaluated and the adaptive implementation of MAFs is also discussed. On the contrary, lacking this part of the discussion [15,16], it is difficult to implement in a practical embedded system.

Our contribution is as follows.

- A novel hybrid filtering stage with narrowed T_ω of MAFs is proposed, which can remove dominant disturbances completely without degrading the dynamic performance.
- In conventional QT1-PLL, only phase margin (PM) is considered in design procedure [14]. In this paper, the settling times under two adverse grid conditions are directly taken into account in design guidelines. In addition, PM does not decrease yet.
- To validate the effectiveness of the proposed method, a comprehensive experimental study is performed which considers many cases such as phase and frequency change, voltage sag and harmonic polluted grid voltages. The results show that the transient behavior lasts for nearly one grid period. Compared with QT1-PLL, the settling time is reduced by nearly 40%, which makes it fulfill the stringent transient response requirement in most grid standard [6,7].
- As a byproduct, the proposed PLL can also handle dc offset by using two more ANFs. It is examined under dc offset injection condition. This advantage is confirmed by comparative experiments.

Literature review is presented in Section 2. The suggested hybrid filtering technique and PLL is introduced and analyzed in detailed in Section 3. The mathematical modeling of the proposed PLL is derived in Section 4. Section 4 also provides parameter design guidelines and evaluates the stability of open-loop system. In Section 5, comparative experiments are implemented to validate performance of the proposed method.

2. Literature Review

To achieve a satisfactory performance under adverse conditions, many advanced PLLs were suggested recently. Almost all these PLLs evolved from SRF-PLL. According to control structure, a general classification of some typical advanced PLLs is depicted in Figure 2. All of these PLLs, except MAF-PLL and QT1-PLL, use hybrid filter based on low pass filter (LPF) or MAF.

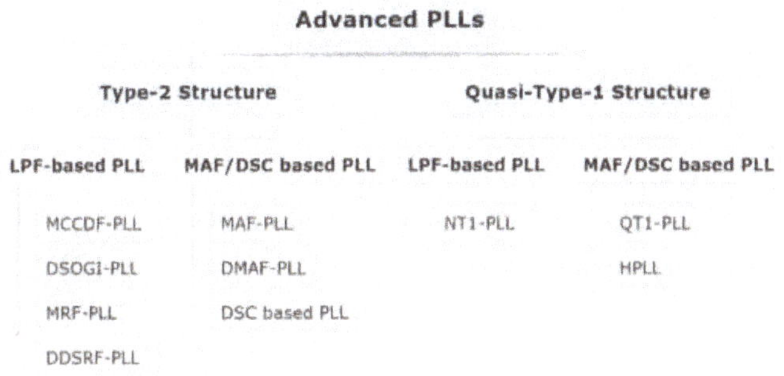

Advanced PLLs

Type-2 Structure		Quasi-Type-1 Structure	
LPF-based PLL	MAF/DSC based PLL	LPF-based PLL	MAF/DSC based PLL
MCCDF-PLL	MAF-PLL	NT1-PLL	QT1-PLL
DSOGI-PLL	DMAF-PLL		HPLL
MRF-PLL	DSC based PLL		
DDSRF-PLL			

Figure 2. A general classification of typical advanced PLLs.

To achieve a better performance, LPF-based hybrid filtering stage is employed in many advanced PLLs, such as MCCF-PLL [8], DSOGI-PLL [9] and MRF-PLL [10]. The hybrid filtering stage, which are usually arranged at the front of Park transformation, can be divided into two parts. One part is responsible for eliminating FFNS, which can be considered as a notch filter. Another, act as a LPF, is used to reject other dominant harmonics. Although the bandwidth of these PLLs open-loop system becomes higher, high order harmonics still remain owing to the usage of LPFs. MAFs and DSCs can be adopted to overcome this disadvantage. They can act as ideal filters to remove harmonic disturbance completely. Nevertheless, the transient behavior is slowed down by time delays of MAFs and DSCs [17]. MAF-based hybrid filtering technique is also developed in recent years. It makes MAF can fulfill disturbances rejection requirement with a narrowed T_ω. In Reference [11], DMAF-PLL is successful in reducing T_ω. However, a big deviation of frequency estimation occurs under a phase step change condition owing to the differential proportional component, which may bring about an unexpected tripping operation in some power generation applications [18,19].

Another approach to improve PLL's transient performance is to reduce the type of a PLL system. In Reference [20], a hybrid type-1/type-2 PLL with a so-called reconstructor unit was proposed. This PLL performs as a type-1 PLL under normal grid condition. When grid frequency is off-nominal, the reconstructor is activated to make the system behave as a type-2 system. Owing to this variable-structure, this hybrid type1/2 PLL provides a fast response. Motivated by [20], QT1-PLL was proposed in Reference [14]. The block scheme of QT1-PLL is depicted in Figure 3. Unlike type-2 PLL structure, QT1-PLL has a feed-forward control path. It makes QT1-PLL similar to type-1 PLL structure. However, from the control viewpoint, it is actually a type-2 system. Compared with MAF-PLL, it not only provides almost same filtering capability but also achieves less settling time under a frequency step change condition. Although the disturbance rejection capability is not better than that of MAF-PLL when grid frequency drifts, it can be simply solved by making the MAFs' window length adaptive adjust with grid frequency. To expand its application conditions, a hybrid filter based PLL named HPLL was proposed in Reference [21]. Its filtering stage consists of DSCs and MAFs. With the employment of DSCs, HPLL offers a dc-offset filtering capability for QT1-PLL. The disadvantage of QT1-PLL and HPLL is the large delay in MAFs and DSCs. It is a common defect of many MAF/DSC based PLLs. In Reference [22], a novel-type-1 PLL (NT1-PLL) implemented in QT1-PLL structure is proposed. Its filtering stage act as same as that in MCCF-PLL. Although NT1-PLL provides a much better dynamic performance, the utilization of LPF makes its filtering stage unable to completely remove high order harmonics disturbance. Consequently, phase-error of NT1-PLL still exists under distorted grid conditions.

A performance comparison between some of typical PLLs mentioned above is listed in Table 1. As discussed above, the transient behaviors of most of type-2 PLLs and MAF/DSC based PLLs are slow. The quasi-type-1 structure shows its advantage in dynamic performance.

Table 1. Performance comparison between some typical PLLs.

Control Structure	Sub-Classification		Ideal Filtering Capability	Dynamic Response
Type-2 Structure	LPF-based PLLs	MCCF-PLL	No	Slow
		DSOGI-PLL	No	Slow
		MRF-PLL	No	Slow
		DDSRF-PLL	No	Slow
	MAF/DSC based PLLs	MAF-PLL	Yes	Slow
		DSC-based PLLs	Yes	Slow
		DMAF-PLL	Yes	Average
Quasi-Type-1 Structure	MAF based PLLs	QT1-PLL	Yes	Average
	LPF-based PLLs	NT1-PLL	No	Fast

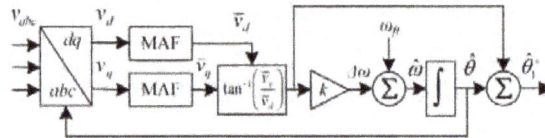

Figure 3. Block scheme of QT1-PLL.

3. The Hybrid Filtering Stage and Proposed PLL

To reduce the settling time of PLLs as small as possible, a hybrid cascaded filtering stage is incorporated into QT1-PLL structure. The window length of MAFs is narrowed. The proposed PLL enhances the advantage of QT1-PLL in dynamic performance.

3.1. The Proposed PLL

The scheme of QT1-PLL structure is depicted in Figure 3. v_{abc} is three-phase grid voltage, $v_{d,q}$ is d,q-axis voltage of v_{abc}. $\bar{v}_d$ and $\bar{v}_q$ are dc components in v_d and v_q. ω_{ff} is the nominal frequency value of fundamental frequency positive voltage sequence (FFPS). $\Delta\omega$ denotes the deviation of input frequency from ω_{ff}. The estimated values of FFPS's frequency and phase is represented by $\hat{\omega}$ and $\hat{\theta}_1^+$.

In three-phase grid applications, unbalanced grid voltages can be decomposed into FFPS, FFNS and non-triplen odd harmonic sequences [22]. Since these dominant disturbances in $\alpha\beta$-frame turn to be even harmonics in dq-frame after using Park transformation [23,24], frequency of the lowest order harmonic which is FFNS component turns to be −100 Hz. FFPS turns to be DC components. Thus, T_w of MAF is selected to be half a cycle (0.01 s for 50 Hz grid) in QT1-PLL. Table 2 lists the dominant component in the most practical conditions [25]. Since these components represent grid voltage vectors, some of their signs are negative to represent that the negative sequence vectors rotate in counterclockwise direction.

Table 2. Dominant voltage disturbances of grid voltages.

Harmonic order	...	−11	−5	−1	+1	+7	+13	...
$\alpha\beta$-frame (Hz)	...	−550	−250	−50	50	350	650	...
Harmonic order	...	−12	−6	−2	0	+6	+12	...
dq-frame (Hz)	...	−600	−300	−100	0	300	600	...

Figure 4 illustrates the block scheme of the proposed PLL. ANFs are embedded into QT1-PLL. All disturbances are filtered by the proposed hybrid filtering stage in dq-frame. ANFs are utilized to eliminate −100Hz FFNS. The rest of harmonics (±300 Hz, ±600 Hz, etc.) are removed by MAFs. Since the lowest harmonic order turns to be ±6 (±300 Hz components) rather than −2 (−100 Hz component) in QT1-PLL, T_ω of MAF is reduced to be 0.0033 s (1/6 grid cycle).

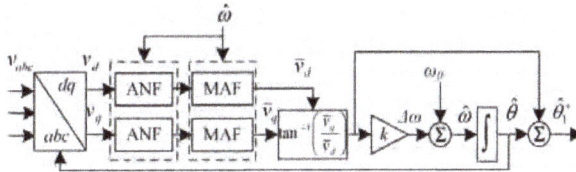

Figure 4. Block scheme of the proposed PLL.

3.2. Hybrid Cascaded Filtering Stage

As introduced above, the proposed filtering stage is composed of ANFs and MAFs. It can provide an ideal filtering capability and improve the dynamic performance. To achieve this goal, the parameters in the hybrid filtering stage needs to be properly designed in this part.

Since T_ω of MAFs is already set to be 0.0033 s as mentioned above, ANF is the only component to be designed, which is written as

$$\text{ANF}(s) = \frac{s^2 + (2\hat{\omega})^2}{s^2 + 2\hat{\omega}\zeta s + (2\hat{\omega})^2} \tag{1}$$

where ζ is the damping factor and $\hat{\omega}$ is the estimation of grid frequency. For a 50 Hz power system under normal condition, $\hat{\omega}$ equals to $2\pi 50$ rad/s. Figure 5 shows the bode plot of ANF part of filtering stage with different values of ζ. It is observed that FFNS (−100 Hz) component is eliminated and FFPS (0 Hz) remains without any change in magnitude or phase. ζ is determined by step response simulations of ANF(s). The results are depicted in Figure 6. A trade-off is made between the transient response and peak deviation. Therefore, ζ is selected to be 0.7. Since ANF is an adaptive filter and its notch frequency depends on the $\hat{\omega}$, the frequency adaptive structure of ANF is necessary and depicted in Figure 7.

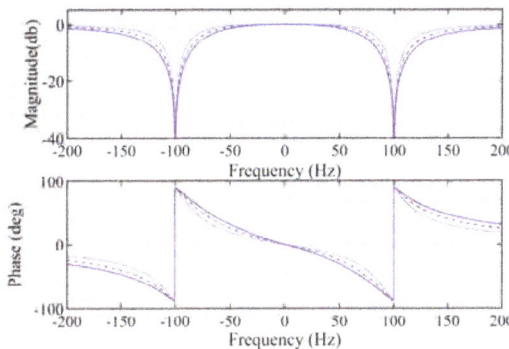

Figure 5. Bode plot of ANF part in filtering stage. ζ = 0.5 (dotted lines), 0.7 (dashed lines), 0.9 (solid lines).

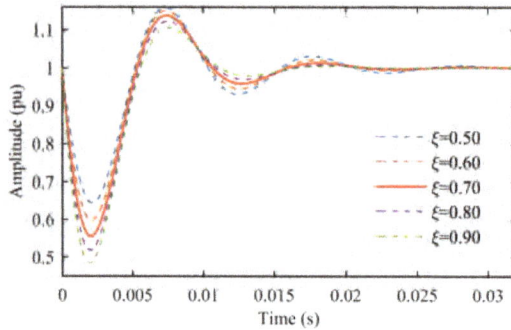

Figure 6. Step response of ANF(*s*).

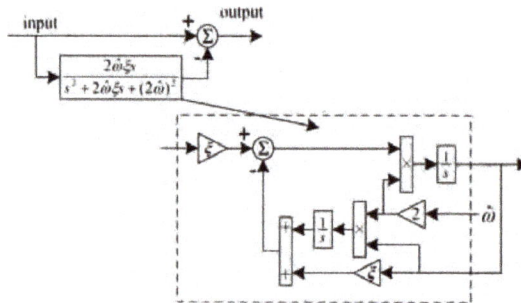

Figure 7. The adaptive structure of ANF.

As studied in many literature [26,27], the MAF can be expressed as

$$\text{MAF}(s) = \frac{1 - e^{-T_\omega s}}{T_\omega s} \tag{2}$$

Then, the proposed hybrid filtering stage can be expressed as

$$\text{H}(s) = \text{ANF}(s)\text{MAF}(s) = \frac{s^2 + (2\hat{\omega})^2}{s^2 + 2\hat{\omega}\xi s + (2\hat{\omega})^2} \frac{1 - e^{-T_\omega s}}{T_\omega s} \tag{3}$$

where $\xi = 0.7$ and $T_\omega = 0.0033$ s.

Figure 8 depicts the frequency characteristic of H(*s*). Observing Figure 8, H(*s*) has a unity gain and zero phase shift at 0 Hz. It means H(*s*) has no impact on FFPS in dq-frame. H(*s*) also provides zero gain at frequencies of the dominant disturbances (−100 Hz, ±300 Hz, ±600 Hz, etc.). The dominant disturbances listed in Table 2 can be totally removed by the proposed hybrid filtering stage.

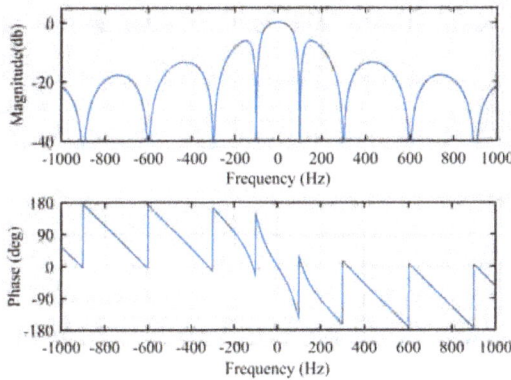

Figure 8. Frequency response of the entire hybrid cascaded filtering stage.

3.3. Proposed Hybrid Filter with DC Offset Rejection Capability

In some cases, with DC injection from power converters or A/D conversion, DC offset may present at the input of PLL. If DC offset rejection is required and necessary in some applications, an ANF designed to remove DC offset can be included in the inner loop of the proposed method. The DC offset occurs as −50 Hz voltage sequence vector in dq-frame. Hence, the ANF used for DC offset removal is expressed as

$$\text{ANF}_{\text{dc}}(s) = \frac{s^2 + (\hat{\omega})^2}{s^2 + 2\hat{\omega}\xi s + (\hat{\omega})^2} \tag{4}$$

The whole hybrid filtering stage with ANF_{dc} can be written as

$$H_{\text{dc}}(s) = \text{ANF}(s)\text{ANF}_{\text{dc}}(s)\text{MAF}(s) = \frac{s^2 + (2\hat{\omega})^2}{s^2 + 2\hat{\omega}\xi s + (2\hat{\omega})^2} \frac{s^2 + (\hat{\omega})^2}{s^2 + 2\hat{\omega}\xi s + (\hat{\omega})^2} \frac{1 - e^{-T_\omega s}}{T_\omega s} \tag{5}$$

where ξ is also 0.7. The frequency response of $H_{\text{dc}}(s)$ is illustrated in Figure 9. It shows that $H_{\text{dc}}(s)$ can completely remove −50 Hz voltage sequence caused by dc offset and other disturbance components listed in Table 2.

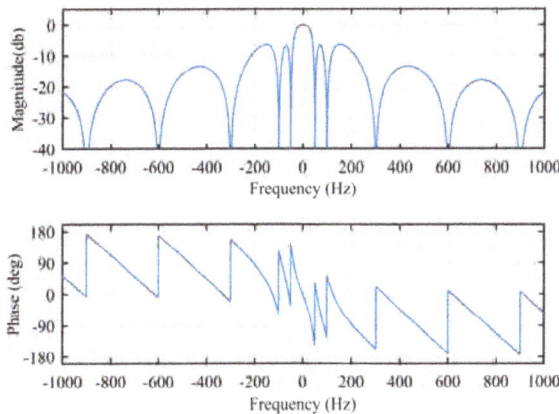

Figure 9. Frequency response of $H_{\text{dc}}(s)$.

4. Small-Signal Modeling and Design Procedure

Based on a small-signal model, parameters design procedures are suggested in Section 4. The transient response under two abnormal conditions are the key factors concerned in the procedure. The stability analysis is also presented in this section.

4.1. Small-Signal Model

Since the filtering stage of QT1-PLL and the proposed PLL is the major difference, the model of the proposed method can be simply obtained from that of QT1-PLL (as shown in Figure 10) [14], by substituting H(s) for MAF. The small-signal model of the proposed PLL is shown in Figure 11. D(s) represents all disturbances. To be brief, the detailed derivation of the model is not presented. A simulation is implemented in Section 3 to assess the modeling accuracy.

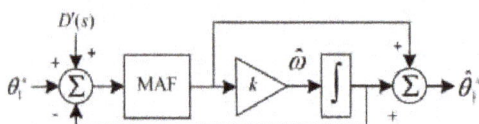

Figure 10. Small-signal model of QT1-PLL.

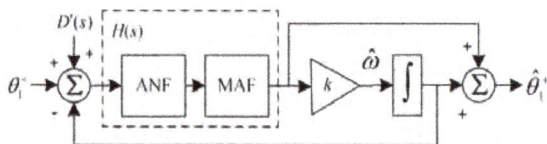

Figure 11. Small-signal model of the proposed PLL.

4.2. Parameter Design Guidelines

As already mentioned above, the parameters in the proposed hybrid filter, T_w and ξ are already selected to be 0.0033 s and 0.7, respectively. Therefore, only k is left to be designed.

Applying block diagram algebra to Figure 11, a simplified small-signal model is achieved in Figure 12, which is a typical close-loop system. Its open-loop transfer function is

$$G_{ol}(s) = \frac{\hat{\theta}_1^+(s)}{\theta_1^+(s) - \hat{\theta}_1^+(s)} = \left(\frac{\text{ANF}(s)\text{MAF}(s)}{1 - \text{ANF}(s)\text{MAF}(s)} \right) \left(\frac{s+k}{s} \right) \tag{6}$$

The phase tracking error transfer function is

$$G_e(s) = \frac{\theta_e(s)}{\theta_1^+(s)} = \frac{\theta_1^+(s) - \hat{\theta}_1^+(s)}{\theta_1^+(s)} = \frac{1}{1 + G_{ol}(s)} \tag{7}$$

where $\theta_e(s)$ represents phase error. In response to a phase jump $(\Delta\theta)$, phase-error can be expressed in s-domain as

$$\Theta_e^{\Delta\theta}(s) = \frac{\Delta\theta}{s} G_e(s) \tag{8}$$

In response to a frequency jump $(\Delta\omega)$, phase-error is

$$\Theta_e^{\Delta\theta}(s) = \frac{\Delta\omega}{s^2} G_e(s) \tag{9}$$

To provide a rapid transient response under both phase and frequency jump conditions, the settling time is examined by applying inverse Laplace transform to phase-error. As a function of k, the variations of 2% settling time for phase jump (solid line) and frequency jump (dashed line) are depicted in Figure 13. It is obvious that choosing k to be 150 is an optimal value considering both conditions.

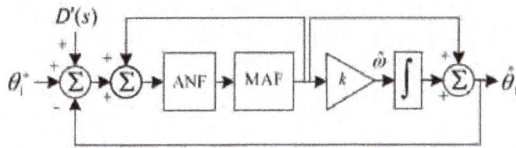

Figure 12. The simplified model of the proposed structure.

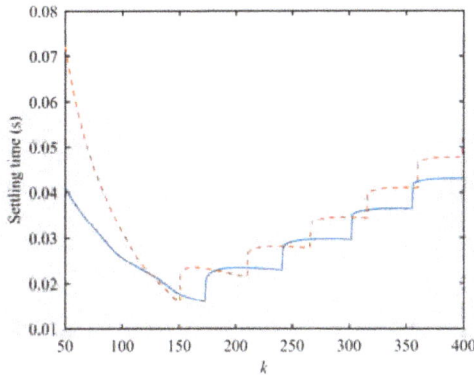

Figure 13. 2% settling time as a function of k.

Since all the parameters are given, the open-loop bode diagram is depicted in Figure 14. PM of the proposed PLL and QT1-PLL is 45.3 degree and 44.8 degree, respectively. It is enough to ensure their stability. Compared with QT1-PLL, the crossover frequency of the suggested PLL is bigger. It also illustrates that all dominant disturbance components listed in Table 2 can be eliminated, completely.

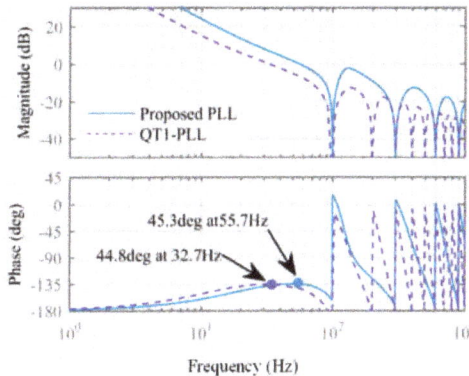

Figure 14. Bode plot of open-loop transfer function in proposed PLL and QT1-PLL.

When dc offset exists in grid voltages and is required to be eliminated in some applications, $H_{dc}(s)$ is recommended for this task, which is proposed in the previous section. The design procedure is similar to the design guidelines mentioned above. For the sake of brevity, it is not presented here. According to Figure 15, k can be chosen to be 76.5. The corresponding open-loop bode diagram is depicted in Figure 16. DC component (50 Hz component in dq-frame) is removed by the proposed method. The phase margin is 31 degrees at 32.5 Hz, which illustrates that the system is stable. Compared with QT1-PLL, the bandwidth is bigger.

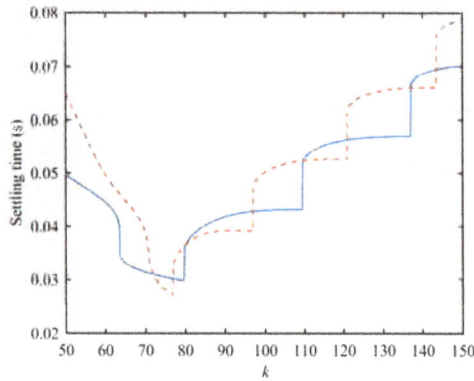

Figure 15. 2% settling time of the proposed PLL (with dc rejection capability) under phase and frequency jump.

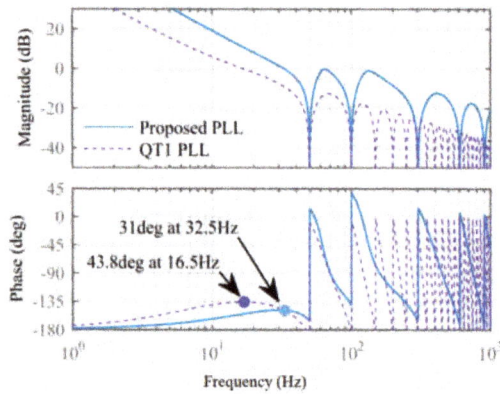

Figure 16. Open-loop bode plot of the proposed PLL and QT1-PLL (with dc rejection capability).

4.3. Assessment of Small-Signal Model

To assess the accuracy of the small-signal model, a simulation is carried out under phase jump ($+40°$) and frequency step change ($+5$ Hz) simulations. As depicted in Figure 17, the model can precisely predict the transient response of the proposed PLL. For brevity, the assessment of the proposed method with DC rejection capability is not presented.

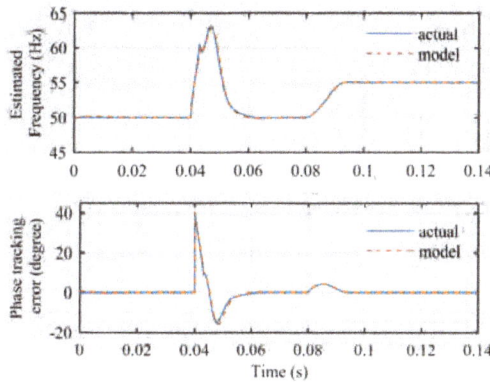

Figure 17. Dynamic behavior of the actual proposed PLL and its model.

4.4. Digital Implementation

The proposed PLL is designed in the continuous-time domain. However, a discrete-time realization is required in practice. The main difference between the proposed method and QT1-PLL is ANFs. To discretize ANFs, Back Euler method is used to approximate the integrals in ANF as follow,

$$\frac{1}{s} \Leftrightarrow \frac{T_S}{1 - z^{-1}} \tag{10}$$

Figure 18 shows the discrete-time realization of ANF. It can be observed that ANF requires five multipliers, three adders, two subtractors and two stored samples. To assess the computational burden, Table 3 lists the math operator required in the hybrid filtering stage. Although the proposed PLL require a little more math operation, it takes much less storage space than QT1-PLL.

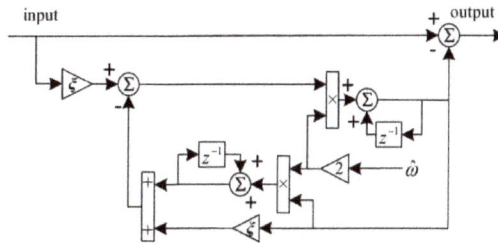

Figure 18. Discrete-time realization of ANF.

Table 3. Calculation operators in filtering stage.

Operator	+/−	×/÷	Sorted Samples
Proposed PLL	10	10	66
QT1-PLL	2	4	202

The impact of frequency variation of MAF is another thing need to be noticed. If the grid frequency drifts away from its nominal value, MAF with a fixed window length cannot completely remove harmonics. Hence, MAFs are frequency-adaptive in the proposed PLL. They can change their window length based on the online estimation of grid frequency. Several methods can be implemented to realize a frequency-adaptive MAF. A simplest form to adjust the window length of MAF is to round-down or round-up T_ω / T_s to the nearest integer. However, this method introduces discretization error. To reduce this error, linear interpolation method is used in this paper, which increases computational burden.

MAFs used in QT1-PLL, DMAF-PLL are also adaptive in this paper. The detailed linear interpolation method and frequency MAF implementation can be found in References [20,28].

Figure 19 illustrates the discretization effect of the proposed PLL in simulation. The sample time used in discretization is 10 kHz. As depicted in Figure 19, the dynamic behaviors in continuous domain and discrete domain are almost same.

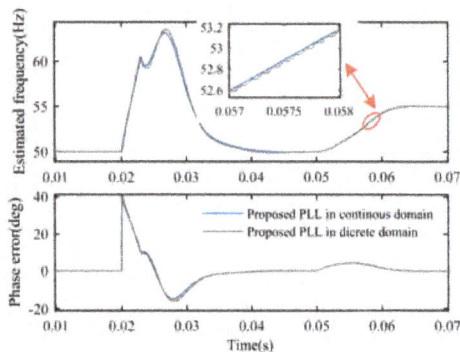

Figure 19. The discretization effect of the proposed PLL.

5. Experimental Results

To validate performance, experimental results are provided and analyzed here. The proposed PLL is realized in a digital signal processor. The sampling frequency is 10 kHz. A programmable arbitrary waveform generator, which is built by PC and acquisition board, is utilized to obtain 50 Hz three-phase voltages signals. DSP board exports the estimated frequency and phase-error through DA conversion circuit. All waveforms are captured by oscilloscope. Figure 20 shows the experimental setup.

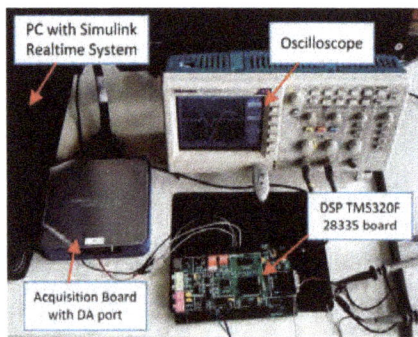

Figure 20. Experimental setup.

For comparison, several advanced PLLs are also implemented in the experiments. QT1-PLL [14] and Novel Type-1 PLL (NT1-PLL) [22] are carried out since they both have quasi-type-1 structure. DMAF-PLL [11], MCCF-PLL$_{PID}$ [29], MCCF-PLL$_{PI}$ [8] are implemented since their filtering stages are also hybrid. These PLLs were proposed in recent three years. Their parameters used in experiment can be found in the literature mentioned above.

5.1. Phase Jump

All PLLs are examined under a +40° phase jump voltages condition. Observing the waveforms in Figure 21, the settling time of the proposed PLL is shortest. Its 2% settling time is about 0.9 grid period. NT1-PLL and DMAF-PLL also provide satisfactory dynamic performance. However, an over 30 Hz overshoot occurs in the estimated frequency of DMAF-PLL. It may violate some restriction in some grid standard [30]. Unexpected tripping operation may be triggered [31]. The settling time of other three PLLs is almost 35 ms. According to the requirement in transient response mentioned in many grid standard [6,7,12,13], the estimation of voltage parameters need to be finished within 25 ms. Hence, QT1-PLL, MCCF-PLL$_{PID}$, MCCF-PLL$_{PI}$ are not eligible under such condition.

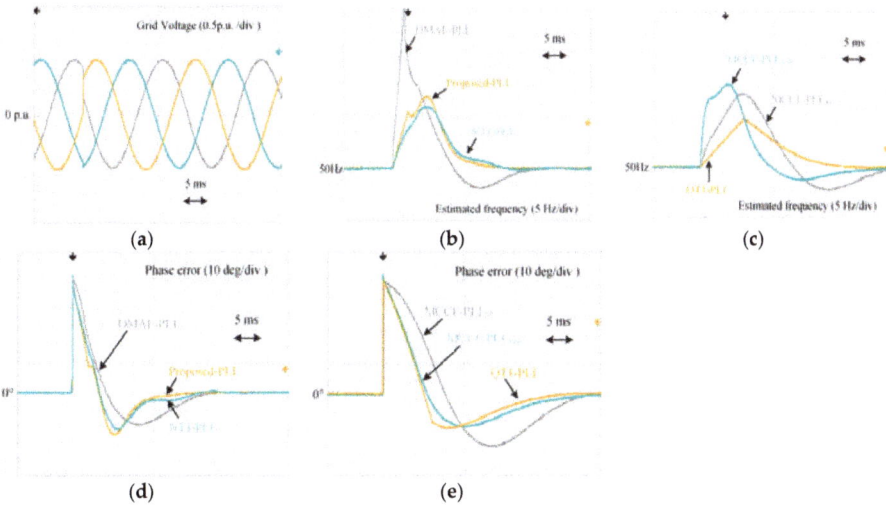

(a)   (b)   (c)

(d)   (e)

Figure 21. Experimental waveforms under a +40° phase jump: (**a**) Three-phase voltages; (**b**,**c**) Estimated frequency; (**d**,**e**) Phase error.

5.2. Frequency Step Change

Figure 22 illustrates the waveforms under a frequency step change grid condition. As shown, the proposed PLL and NT1-PLL track the grid frequency in 14 ms and 17 ms, respectively. Furthermore, within 15 ms, the phase-error of the suggested PLL converges to zero. The frequency tracking transient response of QT1-PLL and DMAF-PLL are also acceptable. But, the settling time of MCCF-PLL$_{PID}$, MCCF-PLL$_{PI}$ is over 30 ms, which cannot meet the requirement of the grid code. On the other hand, DMAF-PLL, MCCF-PLL$_{PID}$, MCCF-PLL$_{PI}$ have over +1.5 Hz peak frequency deviation. On the contrary, other three PLLs have no frequency overshoot.

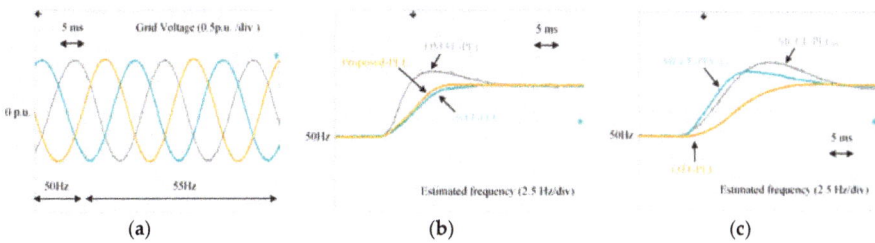

(a)   (b)   (c)

Figure 22. *Cont.*

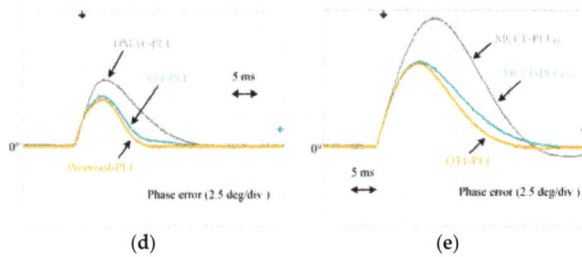

(d)　　　　　　　　　　　　　　(e)

Figure 22. Experimental waveforms under a +5 Hz frequency step change: (**a**) Three-phase voltages; (**b,c**) Estimated frequency; (**d,e**) Phase error.

5.3. Frequency Ramp Change

To evaluate the effectiveness during frequency ramp change, voltage frequency is increased from 50 Hz to 55 Hz in 50 ms. The ramp rising rate is +100 Hz/s. As depicted in Figure 23, during the transient behavior, the suggested PLL has minimum phase-tracking error of 0.7°. The phase-error of NT1-PLL is 0.8°, which is also a small error. Compared with these two PLLs, the phase-errors of other PLLs are relatively large.

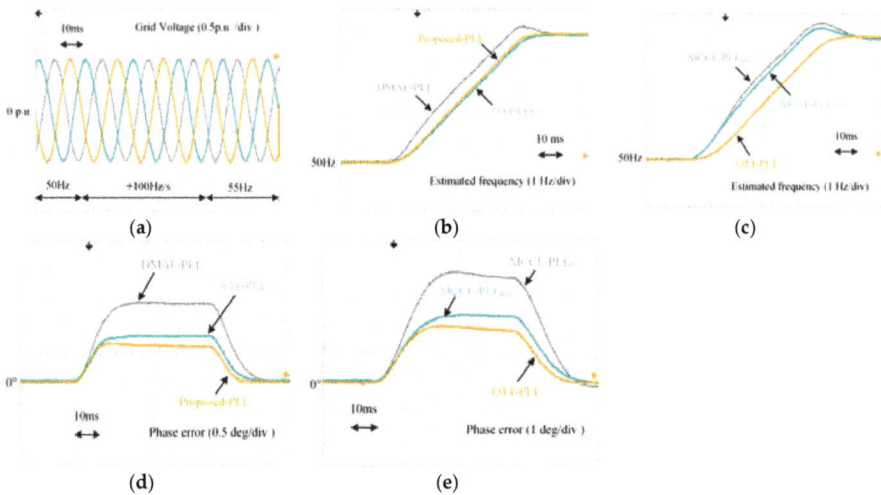

(a)　　　　　　　　　　(b)　　　　　　　　　　(c)

(d)　　　　　　　　　　　　　　(e)

Figure 23. Experimental waveforms under a +100 Hz/s frequency ramp change: (**a**) Three-phase voltages; (**b,c**) Estimated frequency; (**d,e**) Phase error.

5.4. Voltage Sag

A test case when three-phase voltages undergo a voltage sag is also carried out. The waveforms are shown in Figure 24. Owing to the utilization of arc tangent operation, the output of PLL cannot be influenced by voltage amplitude. Hence, the performance of the proposed PLL, NT1-PLL and QT1-PLL are not deteriorated. Similar to the result under phase jump condition, DMAF-PLL also has undesired +13 Hz overshoot in estimated frequency under voltage sag condition.

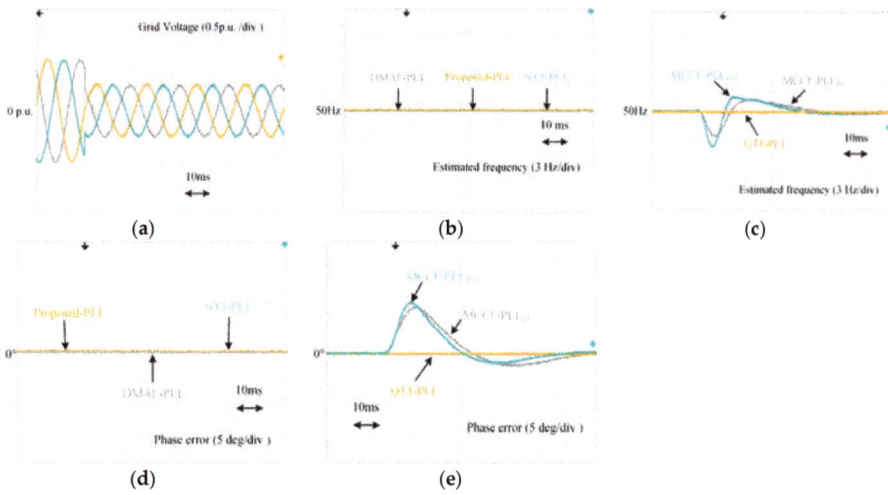

Figure 24. Experimental waveforms under a 0.5 p.u. voltage sag: (**a**) Three-phase voltages; (**b**,**c**) Estimated frequency; (**d**,**e**) Phase error.

5.5. Distorted Grid Voltages

To examine the filtering capability, an experiment under distorted grid voltage condition is implemented. To validate the filtering capability under different grid frequency, the grid voltage undergoes a +5 Hz frequency jump. Table 4 lists the parameters of grid voltages. The experimental waveforms are depicted in Figure 25.

Table 4. Distorted grid voltage components.

Voltage Sequences (in $\alpha\beta$-Frame)	Amplitude (p.u.)
FFPS (+50 Hz)	1
FFNS (−50 Hz)	0.1
−5th voltage sequence (−250 Hz)	0.1
+7th voltage sequence (+350 Hz)	0.05
−11th voltage sequence (−550 Hz)	0.05
+13th voltage sequence (+650 Hz)	0.05

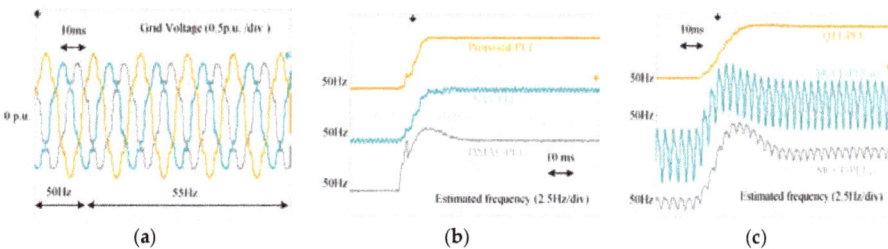

Figure 25. *Cont.*

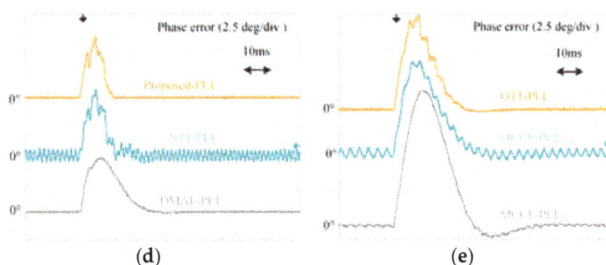

Figure 25. Experimental waveforms under distorted voltage condition with a +5 Hz frequency jump: (a) Three-phase voltages; (b,c) Estimated frequency; (d,e) Phase error.

With the help of adaptive MAF and ANF, the oscillations under 50 Hz and 55 Hz are totally eliminated in the proposed PLL. The steady-state phase-errors of DMAF-PLL and QT1-PLL are also zero under both grid frequency conditions. On the contrary, for the case of NT1-PLL, 0.4 Hz and 1 oscillations error occur MCCF-PLL$_{PID}$, MCCF-PLL$_{PI}$ also have the same trouble. The oscillations in phase-error of MCCF-PLL$_{PID}$, MCCF-PLL$_{PI}$ are 0.7 and 0.2.

5.6. Voltages with DC Offset

To validate the dc rejection capability when grid voltages are polluted by dc offset, 0.2 p.u., 0.1 p.u. and −0.2 p.u. DC components are suddenly injected to the phases A, B and C, respectively. Since NT1-PLL, MCCF-PLL$_{PID}$ and MCCF-PLL$_{PI}$ did not consider DC offset in their design procedure, only DMAF-PLL and QT1-PLL are implemented for comparison. The DC elimination structure and corresponding parameters of DMAF-PLL can be found in Reference [11]. To eliminate the DC component, T_ω of MAF and k used in QT1-PLL are selected to be 0.02 s and 50 in Reference [21], respectively.

Figure 26 shows the performance of DMAF-PLL, QT1-PLL and the proposed PLL. When DC offsets are suddenly injected into grid voltages, tracking errors occur in both estimated frequency and phase error. According to existing grid code [6,7,12,13], 0.2 Hz frequency deviation is selected as the criterion to define settling time. Compared with other two PLLs, the proposed PLL has a shorter settling time when DC offsets are suddenly injected into grid voltage.

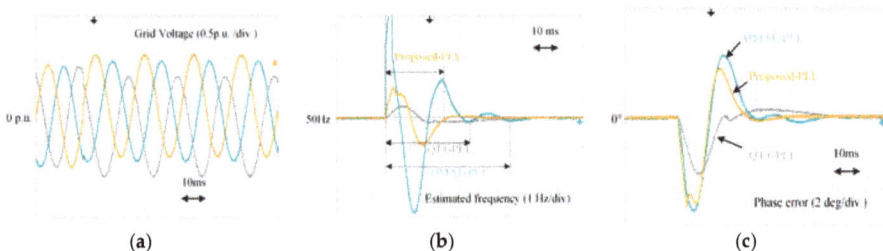

Figure 26. Experimental waveforms under a dc offset injection condition: (a) Three-phase voltages; (b) Estimated frequency; (c) Phase error.

5.7. Summary

Table 5 lists all the experimental results. A comprehensive assessment from different perspective is given in this section.

Table 5. Summary of results.

Advanced PLL	MCCF-PLL$_{PID}$	MCCF-PLLPI	QT1-PLL	DMAF-PLL	NT1-PLL	Proposed PLL
Phase jump (+40°)	-	-	-	-	-	-
Settling time (2%)	≈1.81 cycles	≈2.5 cycles	≈1.5 cycles	≈1.25 cycles	≈1.1 cycles	**≈0.92 cycles**
Peak phase-error	**11.4°** **(28.5%)**	18.74° (46.8%)	12.2° (30.6%)	11.5° (28.8%)	13.2° (33%)	14.8° (37%)
Peak frequency deviation	16.1 Hz	14.2 Hz	**8.9 Hz**	33.8 Hz	11.5 Hz	13.1 Hz
Frequency jump (+5 Hz)	-	-	-	-	-	-
Settling time of estimated frequency (2%)	≈1.74 cycles	≈2.6 cycles	≈1.6 cycles	≈1.3 cycles	≈0.85 cycles	**≈0.7 cycles**
Peak phase-error	7.9°	12.4°	7.6°	6.1°	4.5°	**4.1°**
Peak frequency deviation	1.6 Hz (32%)	2.5 Hz (50%)	**0 Hz (0%)**	1.6 Hz (32%)	0 Hz (0%)	0 Hz (0%)
Frequency ramp change (+100 Hz/s)	-	-	-	-	-	-
Phase-error	2.3°	3.8°	1.9°	1.4°	0.8°	**0.7°**
Voltage sag (0.5 p.u.)	-	-	-	-	-	-
Settling time of phase-error (1°)	≈2.7 cycles	≈3.3 cycles	**0 cycle**	≈2 cycles	0 cycle	0 cycle
Peak phase-error	9.2°	8.2°	**0°**	−12.7°	0°	0°
Peak frequency deviation	−4.4 Hz	−3.1 Hz	**0 Hz**	13.5 Hz	0 Hz	0 Hz
Distorted grid voltage	-	-	-	-	-	-
Peak-to-peak phase-error	0.7°	0.2°	**0°**	0°	1°	0°
Peak-to-peak frequency error	4.5 Hz	1.1 Hz	**0 Hz**	0 Hz	0.4 Hz	0 Hz
DC offset injection	-	-	-	-	-	-
0.2 Hz settling time of estimated frequency	-	-	≈1.7 cycles	≈2.5 cycles	-	**≈1.2 cycles**
Phase margin	55.4°	39.3°	45°	43°	**69.9°**	45.3°

With a smaller T_ω in the hybrid filter, the transient response of the proposed PLL is satisfactory. It provides fastest transient response in every test cases. Its filtering stage can eliminate disturbances completely. The dynamic behavior is not affected by voltage sag. In addition, its peak frequency deviation under +40° phase jump is also small.

NT1-PLL also provides a fast transient response. Voltage sag also cannot degrade its phase tracking performance. However, a major drawback is that it cannot completely eliminate disturbance. The peak-to-peak phase error of NT1-PLL is biggest. Another imperfection is that NT1-PLL in Reference [22] did not consider DC offset injection condition.

Compared with the proposed PLL and NT1-PLL, the transient behavior of DMAF-PLL is longer. Moreover, its frequency deviation during phase jump is large, which is a real risk to be considered in practical application. This undesired behavior may arise from its differential operation.

The peak deviation of QT1-PLL is the smallest. As same as the proposed PLL and NT1-PLL, its performance has no impact of voltage sag. But, the dynamic behavior is comparatively slow. MCCF-PLL$_{PID}$ and MCCF-PLL$_{PI}$ cannot provide satisfactory performance in dynamic behavior and filtering capability, which is not suitable for grid-connected applications.

6. Conclusions

A new PLL leveraging the hybrid cascaded filtering is proposed in this paper. Through incorporating well-designed notch filters into the filtering stage of QT1-PLL in cascaded way, the window length of MAF is reduced and the dynamic performance is significantly improved. Theoretical analysis and bode diagram demonstrates that new PLL provides a much better dynamic performance and filtering capability, while the stability margin is still sufficient. Moreover, it is insensitive to the change of voltage amplitude. The experimental comparisons with the state-of-the-art advanced PLL designs clearly confirm its effectiveness.

Author Contributions: The original ideal was provided by Y.L. Some theoretical analysis was given by J.Y. and W.G. Experiments were implemented by H.W. and Y.M. The manuscript was completed by Y.L.

Funding: This research was funded by Application Technology Research and Engineering Demonstration Program of National Energy of China grant number [NY20150303].

Acknowledgments: The fund of Application Technology Research and Engineering Demonstration Program of National Energy of China (NY20150303) supports this work.

Conflicts of Interest: The authors declare no conflicts of interest.

References

1.   Golestan, S.; Guerrero, J.M.; Vidal, A.; Yepes, A.G.; Doval-Gandoy, J.; Freijedo, F.D. Small-signal modeling, stability analysis and design optimization of single-phase delay-based plls. *IEEE Trans. Power Electr.* **2016**, *31*, 3517–3527. [CrossRef]

2.   Kulkarni, A.; John, V. Analysis of bandwidth-unit vector distortion trade off in PLL during abnormal grid conditions. *IEEE Trans. Ind. Electron.* **2013**, *60*, 5820–5829. [CrossRef]

3.   Batista, Y.N.; de Souza, H.E.P.; Neves, F.A.S.; Dias Filho, R.F.; Bradaschia, F. Variable-structure generalized delayed signal cancellation pll to improve convergence time. *IEEE Trans. Ind. Electr.* **2015**, *62*, 7146–7150. [CrossRef]

4.   Eodorescu, R.; Liserre, M.; Rodríguez, P. Grid Synchronization in Three-Phase Power Converters. *Grid Convert. Photovolt. Wind Power Syst.* **2010**, 169–204. [CrossRef]

5.   Han, Y.; Xu, L.; Khan, M.M.; Yao, G.; Zhou, L.-D.; Chen, C. A novel synchronization scheme for grid-connected converters by using adaptive linear optimal filter based pll (alof–pll). *Simul. Model. Pract. Theory* **2009**, *17*, 1299–1345. [CrossRef]

6.   E. On Netz Gmbh. *Grid Code-High and Extra High Voltage*; E. On Netz Gmbh: Bayreuth, Germany, 2006. Available online: http://www.pvupscale.org/IMG/pdf/D4_2_DE_annex_A-3_EON_HV_grid_ _connection_requirements_ENENARHS2006de.pdf (accessed on 16 April 2018).

7.   De Espana, R.E. *PO-12.3. Requisitos de Respuesta Frente a Huecos de Tension de las Instalaciones Eolicas*; Comisión Nacional de Energía: Madrid, Spain, 2006.

8.   Guo, X.; Wu, W.; Chen, Z. Multiple-complex coefficient-filter-based phase-locked loop and synchronization technique for three-phase grid-interfaced converters in distributed utility networks. *IEEE Trans. Ind. Electron.* **2011**, *58*, 1194–1204. [CrossRef]

9.   Rodriguez, P.; Timbus, A.V.; Teodorescu, R.; Liserre, M.; Blaabjerg, F. Flexible Active Power Control of Distributed Power Generation Systems During Grid Faults. *IEEE Trans. Ind. Electr.* **2007**, *54*, 2583–2592. [CrossRef]

10.   Xiao, P.; Corzine, K.; Venayagamoorthy, G. Multiple reference frame-based control of three-phase PWM boost rectifiers under unbalanced and distorted input conditions. *IEEE Trans. Power Electron.* **2008**, *23*, 2006–2017. [CrossRef]

11.   Wang, J.; Liang, J.; Gao, F.; Zhang, L.; Wang, Z. A method to improve the dynamic performance of moving average filter-based pll. *IEEE Trans. Power Electr.* **2015**, *30*, 5978–5990. [CrossRef]

12.   IEEE Std. *IEEE Standard for Interconnecting Distributed Resources with Electric Power Systems*; IEEE: New York, NY, USA, 2003. [CrossRef]

13.   *The Grid Code: Revision 31*; National Grid Electricity Transmission: Warwick, UK, 2008.

14.   Golestan, S.; Freijedo, F.D.; Vidal, A.; Guerrero, J.M.; Doval-Gandoy, J. A quasi-type-1 phase-locked loop structure. *IEEE Trans. Power Electr.* **2014**, *29*, 6264–6270. [CrossRef]

15.   Li, Y.; Wang, D.; Han, W.; Tan, S.; Guo, X. Performance improvement of quasi-type-1 pll by using a complex notch filter. *IEEE Access* **2016**, *4*, 6272–6282. [CrossRef]

16.   Li, Y.; Wang, D.; Han, W.; Sun, Z.; Yuan, T. A hybrid filtering stage based quasi-type-1 pll under distorted grid conditions. *J. Power Electron.* **2017**, *17*, 704–715. [CrossRef]

17.   Golestan, S.; Freijedo, F.D.; Vidal, A.; Yepes, A.G.; Guerrero, J.M.; Doval-Gandoy, J. An efficient implementation of generalized delayed signal cancellation pll. *IEEE Trans. Power Electr.* **2016**, *31*, 1085–1094. [CrossRef]

18. Hadjidemetriou, L.; Kyriakides, E.; Blaabjerg, F. An adaptive tuning mechanism for phase-locked loop algorithms for faster time performance of interconnected renewable energy sources. *IEEE Trans. Ind. Appl.* **2015**, *51*, 1792–1804. [CrossRef]

19. Hadjidemetriou, L.; Kyriakides, E.; Blaabjerg, F. A robust synchronization to enhance the power quality of renewable energy systems. *IEEE Trans. Ind. Electr.* **2015**, *62*, 4858–4868. [CrossRef]

20. Wang, L.; Jiang, Q.; Hong, L. A novel three-phase software phase-locked loop based on frequency-locked loop and initial phase angle detection phase-locked loop. In Proceedings of the IECON 2012—38th Annual Conference on IEEE Industrial Electronics Society, Montreal, QC, Canada, 25–28 October 2012. [CrossRef]

21. Golestan, S.; Guerrero, J.M.; Abusorrah, A.M.; Al-Turki, Y. Hybrid synchronous/stationary reference-frame-filtering-based pll. *IEEE Trans. Ind. Electr.* **2015**, *62*, 5018–5022. [CrossRef]

22. Kanjiya, P.; Khadkikar, V.; Moursi, M.S.E. A novel type-1 frequency-locked loop for fast detection of frequency and phase with improved stability margins. *IEEE Trans. Power Electr.* **2016**, *31*, 2550–2561. [CrossRef]

23. Gonzalez-Espin, F.; Figueres, E.; Garcera, G. An adaptive synchronous-reference-frame phase-locked loop for power quality improvement in a polluted utility grid. *IEEE Trans. Ind. Electron.* **2012**, *59*, 2718–2731. [CrossRef]

24. Golestan, S.; Monfared, M.; Freijedo, F.; Guerrero, J. Advanced and challenges of a type-3 PLL. *IEEE Trans. Power Electron.* **2013**, *28*, 4985–4997. [CrossRef]

25. Golestan, S.; Monfared, M.; Freijedo, F. Design-oriented study of advanced synchronous reference frame phase-locked loops. *IEEE Trans. Power Electron.* **2013**, *28*, 765–778. [CrossRef]

26. Golestan, S.; Guerrero, J.; Abusorrah, A. MAF-PLL with phase-lead compensator. *IEEE Trans. Ind. Electron.* **2015**, *62*, 3691–3695. [CrossRef]

27. Golestan, S.; Guerrero, J.M.; Vidal, A.; Yepes, A.G.; Doval-Gandoy, J. Pll with maf-based prefiltering stage: Small-signal modeling and performance enhancement. *IEEE Trans. Power Electr.* **2016**, *31*, 4013–4019. [CrossRef]

28. Golestan, S.; Ramezani, M.; Guerrero, J.M.; Freijedo, F.D.; Monfared, M. Moving average filter based phase-locked loops: Performance analysis and design guidelines. *IEEE Trans. Power Electr.* **2014**, *29*, 2750–2763. [CrossRef]

29. Golestan, S.; Monfared, M.; Freijedo, F.; Guerrero, J. Performance improvement of a prefiltered synchronous-reference-frame pll by using a PID-type loop filter. *IEEE Trans. Power Electron.* **2014**, *61*, 3469–3479. [CrossRef]

30. Tsili, M.; Papathanassiou, S. A review of grid code technical requirements for wind farms. *IET Renew. Power Gener.* **2009**, *3*, 308. [CrossRef]

31. Luna, A.; Rocabert, J.; Candela, J.; Hermoso, J.; Teodorescu, R.; Blaabjerg, F.; Rodriguez, P. Grid voltage synchronization for distributed generation systems under grid faults conditions. *IEEE Trans. Ind. Appl.* **2015**, *51*, 3414–3425. [CrossRef]

![energies logo] *energies*

MDPI

Article

THD Reduction in Wind Energy System Using Type-4 Wind Turbine/PMSG Applying the Active Front-End Converter Parallel Operation

Nadia Maria Salgado-Herrera [1,*], David Campos-Gaona [1], Olimpo Anaya-Lara [1], Aurelio Medina-Rios [2], Roberto Tapia-Sánchez [2] and Juan Ramon Rodríguez-Rodríguez [3]

[1] Institute for Energy and Environment, University of Strathclyde, Glasgow G1 1XW, UK;
    d.campos-gaona@strath.ac.uk (D.C.-G.); olimpo.anaya-lara@eee.strath.ac.uk (O.A.-L.)
[2] Facultad de Ingenieria Electrica, División de estudios de posgrado, Universidad Michoacana de San Nicolas
    de Hidalgo, Morelia 58030, Mexico; amedinr@gmail.com (A.M.-R.); rtsanchez@dep.fie.umich.mx (R.T.-S.)
[3] Facultad de Ingeniería; Departamento de Energía, Eléctrica Universidad Nacional Autónoma de México,
    Coyoacán, Ciudad de México 04510, Mexico; Jr_rodriguez@fi-b.unam.mx
* Correspondence: nadia.salgado-herrera@strath.ac.uk; Tel.: +52-144-3101-9237

Received: 29 August 2018; Accepted: 15 September 2018; Published: 16 September 2018

Abstract: In this paper, the active front-end (AFE) converter topology for the total harmonic distortion (THD) reduction in a wind energy system (WES) is used. A higher THD results in serious pulsations in the wind turbine (WT) output power and several power losses at the WES. The AFE converter topology improves the capability, efficiency, and reliability in the energy conversion devices; by modifying a conventional back-to-back converter, from using a single voltage source converter (VSC) to use pVSC connected in parallel, the AFE converter is generated. The THD reduction is achieved by applying a different phase shift angle at the carrier of digital sinusoidal pulse width modulation (DSPWM) switching signals of each VSC. To verify the functionality of the proposed methodology, the WES simulation in Matlab-Simulink® (Matlab r2015b, Mathworks, Natick, MA, USA) is analyzed, and the experimental laboratory tests using the concept of rapid control prototyping (RCP) and the real-time simulator Opal-RT Technologies® (Montreal, QC, Canada) is achieved. The obtained results show a type-4 WT with a total output power of 6 MVA, generating a THD reduction up to 5.5 times of the total WES current output by Fourier series expansion.

Keywords: active front-end converter; back-to-back converter; permanent magnet synchronous generator (PMSG); THD; type-4 wind turbine; wind energy system; Opal-RT Technologies®

1. Introduction

Nowadays, the number of wind energy systems (WES) has increased dramatically, as evidence of this; in 2013, WES were installed in more than 80 countries, generating a power of 240 GW [1], in 2014, the generation reached a capacity of 369.9 GW [2], in 2015, a production of 432.883 GW was generated [3]. By the end of 2016 a global generation of 487 GW was installed [4], and in 2021 the installed capacity is expected to exceed 800 GW [5]. Within the types of variable speed wind turbines (WT) there are three types: Type-2 (squirrel-cage induction generator (SCIG)), type-3 (double-fed induction generator (DFIG)) and type-4 (squirrel-cage induction generator (SCIG)/permanent magnet synchronous generator (PMSG) with full-scale back-to-back converter); in which, type-2 has a 10% variability in the rotor, type-3 has a 30% variability, and type-4 has 60% in the variability of the rotor speed [6]. The type-3 (DFIG) wind turbine schemes constitute the majority of variable speed commerce applications; however, the type-4 WT with a PMSG (WT-PMSG) is an attractive and the best option since this is not directly connected to the grid, presenting advantages such as: High efficiency, increased reliability, major variable speed operation, and low cost in maintenance and installation, due the absence of gearboxes [7]. In the type-4 WT-PMSG installation the important aspects to prevent are associated problems with the wind-nature fluctuations. For example: The flicker generation is mainly caused by load flow changes, due to its continuous operation [8]; a power factor not unity, this characteristic happens as the modulation index of the back-to-back converter is not high [9]. Voltage sags occur by the sudden changes in the rotor speed of the type-4 WT-PMSG and cause a decrement in the transferred power from the dc-link to the grid [10]. A higher total harmonic distortion (THD) is mainly produced by the power converters switching, this results in serious pulsations in the type-4 WT-PMSG output power and in several power losses at the WES [11,12]. All these problems can be mitigated through the full-scale back-to-back converter in the type-4 WT-PMSG scheme, and this generates the following advantages [13–16]: (i) Bidirectional power flow; (ii) adjustable dc-link voltage; (iii) a sinusoidal grid-side current with an exchange of active and reactive power. These advantages are possible because the generated whole power by the type-4 WT-PMSG on the AC grid is supplied through the back-to-back converter.

However, its implementation is very difficult, since this must handle very high powers of up to 6 MVA. Notwithstanding, the Active Front-End (AFE) converter topology provides a viable and efficient solution to improve the power transfer capacity and reliability of the WES quality; the AFE converter is generated by modifying a conventional back-to-back converter, from using a single voltage source converter (VSC) to use *p*VCS connected in parallel, as shown in Figure 1. As evidence, in [3] the authors describe the principal WT manufacturers, those in low voltage and medium voltage technologies are classified, generating power ratings of >3 MVA and <3 MVA, respectively. In the open literature there exists some research works that address the AFE converter topology applied to WES; for example, in [17] the authors analytically and experimentally present the control method for the current balance in an AFE power converter of 600 kVA, this is a very important topic in the parallel connection of power converters, however, the authors make the AFE converter analysis connecting only two VSCs in parallel, generating: A THD of 4.32% (three times higher than in our research work with THD of 1.23%); in addition, they use the space vector modulation for the switching of VSCs, which generates a more complex control if *p*VSC in parallel are connected.

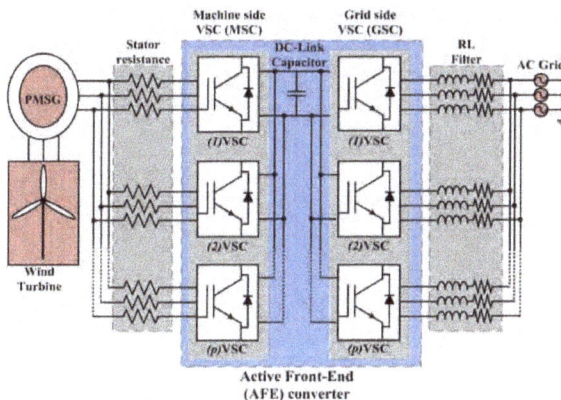

Figure 1. Type-4 wind turbine (WT) connected at wind energy system (WES) through the active front-end (AFE) converter parallel topology.

The main goal of this work is the AFE converter topology application for the THD reduction in a WES and the increase of power transfer between the WT and the AC grid; generating greater capability, efficiency, and reliability in the energy conversion at the WES.

Contributions from this Work

In this paper, the AFE converter topology applied in the THD reduction at WES is made. Through the AFE converter parallel topology the following advantages were possible:

(i)   Increased the converter power capacity.
(ii) Minimized size of each VSC unit, which manages a portion of the total nominal power.
(iii) A reduced ripple on the injected current, which improves the voltage quality at the Point of Common Coupling (PCC).
(iv) An increased equivalent switching frequency, generating a smaller passive filter on the AC-side.
(v)   The possibility of THD Reduction at the WES, modifying the Digital sinusoidal pulse width modulation (DSPWM) switching signals in each VSC.

To verify the functionality and robustness of the proposed methodology, an AFE converter formed with three VSCs connected in parallel is incorporated, as shown in Figure 1. The WES simulation in Matlab-Simulink® is analyzed, and the experimental laboratory tests using the concept of rapid control prototyping (RCP) and the real-time simulator Opal-RT® is achieved. The obtained results show a WES prototyping that incorporates a type-4 wind turbine with a total output power of 6 MVA and a THD reduction of up to 5.5 times.

This paper is organized as follows: Section 2 details the modeling of the Type-4 WT-PMSG; first, the modeling power transfer control between the WT-PMSG and AFE converter is generated; subsequently, the modeling of the machine-side VSC control, DC-link control and the grid-side VSC control of the AFE converter is analyzed, and finally, the design of the AFE converter system parameters is presented. Section 3 presents the modeling of the DSPWM Technique Applied in the THD Reduction. Section 4 shows the simulated results of a study case for WES. Section 5 presents the real-time simulation results of a study case for WES using Opal-RT Technologies®. Finally, in Section 6, the conclusions are presented.

2. Modeling of the Type-4 WT-PMSG

The AFE converter structure consists in two power electronics converters: A machine-side VSC (MSC) to provide power conversion between medium AC voltage and low DC voltage levels, and a grid-side VSC (GSC) to generate the voltages required by the consumers [18], for which, the next sections describe the control modeling of MSC and GSC and these are shown in Figure 2.

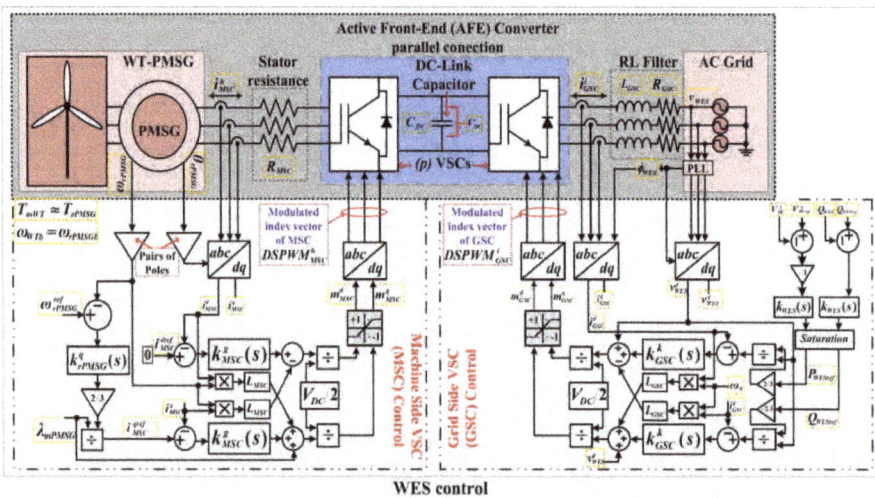

Figure 2. Modeling control of WES.

2.1. Modeling of the Machine-Side VSC Control at AFE Converter

The MSC provides the rotor flux frequency control, thus enabling the rotor shaft frequency to optimally track wind speed [19]. The time-domain relationship of the VSC AC-side is given by:

$$\left[d\left(i_{MSC}^h(t)\right)/dt\right] = -\left(R_{MSC}^h/L_{MSC}^h\right)\left[i_{MSC}^h(t)\right] + \left(1/L_{MSC}^h\right)\left[v_{MSC}^h(t)\right] - \left(1/L_{MSC}^h\right)\left[v_{WT-PMSG}^h(t)\right] \quad (1)$$

where *h* is the MSC three-phase vector (*a,b,c*), L_{MSC} is the PMSG armature inductance, R_{MSC} is the PMSG stator phase resistance, v_{MSC} and i_{MSC} are the MSC voltage and current, respectively, $v_{WT-PMSG}$ is the generated WT-PMSG voltage.

Then, the *dq* reference frame model derived from the AC-side of the MSC, including the inductances cross-coupling, is described as:

$$\left[\frac{d(i_{MSC}^d(t))}{dt}\right] = -\left(\frac{R_{MSC}^d}{L_{MSC}^d}\right)\left[i_{MSC}^d(t)\right] + \left(\frac{\omega_{rPMSG}L_{MSC}^q}{L_{MSC}^d}\right)\left[i_{MSC}^q(t)\right] + \left[\frac{v_{MSC}^d(t)}{L_{MSC}^d}\right] - \left[\frac{v_{WT-PMSG}^d(t)}{L_{MSC}^d}\right] \quad (2a)$$

$$\left[\frac{d(i_{MSC}^q(t))}{dt}\right] = -\left(\frac{R_{MSC}^q}{L_{MSC}^q}\right)\left[i_{MSC}^q(t)\right] - \left(\frac{\omega_{rPMSG}L_{MSC}^d}{L_{MSC}^q}\right)\left[i_{MSC}^d(t)\right] - \left[\frac{(\lambda_{mPMSG})(\omega_{rPMSG})}{L_{MSC}^q}\right] + \left[\frac{v_{MSC}^q(t)}{L_{MSC}^q}\right] - \left[\frac{v_{WT-PMSG}^q(t)}{L_{MSC}^q}\right] \quad (2b)$$

where ω_{rPMSG} is the PMSG rotor angular velocity; λ_{mPMSG} is the maximum flux linkage generated by the PMSG rotor magnets and transferred to the stator windings.

The generated MSC voltage is given by:

$$\left[v_{MSC}^g(t)\right] = (1/2)\left[m_{MSC}^g(t) * V_{DC}(t)\right] \quad (3)$$

where *g* is the *dq* components reference frame vector of the MSC, V_{DC} is the DC-link voltage, m_{MSC}^g is the modulated index vector.

Making $L_{MSC} = L_{MSC}^d = L_{MSC}^q$, the presence of $\omega_{rPMSG}L_{MSC}$ in (2) indicates the coupled dynamics between i_{MSC}^d and i_{MSC}^q. To decouple these dynamics, the i_{MSC}^q vector signals are changed, based in the *dq* reference frame, i.e.,

$$\left[m_{MSC}^d(t)\right] = (2/V_{DC}(t))\left[_{MSC}^d(t) - \left((\omega_{rPMSG}\cdot L_{MSC})i_{MSC}^q(t)\right) + v_{WT-PMSG}^d(t)\right] \quad (4a)$$

$$\left[m_{MSC}^q(t)\right] = (2/V_{DC}(t))\left[_{MSC}^q(t) + \left((\omega_{rPMSG}\cdot L_{MSC})i_{MSC}^d(t)\right) + \lambda_{mPMSG}\omega_{rPMSG} + v_{WT-PMSG}^q(t)\right] \quad (4b)$$

where $E_{MSC}^d(t)$ and $E_{MSC}^q(t)$ are two additional control inputs.

The MSC plant is obtained by substituting (4) into (3), subsequently, (3) is substituting into (2) generating a first order lineal system that, in Equation (5) is described.

$$\left[_{MSC}^g(t)\right] = L_{MSC}\left[di_{MSC}^g(t)/dt\right] + R_{MSC}\left[i_{MSC}^g(t)\right] \quad (5)$$

Equation (5) in the time domain is represented; its representation in the frequency domain is shown in (6); which describes a decoupled and first-order linear system, controlled through $E_{MSC}^g(s)$.

$$\left[_{MSC}^g(s)\right] = (sL_{MSC} + R_{MSC})\left[i_{MSC}^g(s)\right] \quad (6)$$

Rewriting equation (6), the transfer function representing the MSC plant is given, i.e.,

$$\left[i_{MSC}^g(s)\right] = \left[_{MSC}^g(s)\right](sL_{MSC} + R_{MSC})^{-1} \quad (7)$$

With the purpose of tracking the $i^g_{MSC}(s)$ reference commands in the loop, the proportional-integral (PI) compensators are used, obtaining:

$$\left[k^g_{MSC}(s)\right] \approx \left[k^g_{MSC}(s)\right] = \left[\frac{(\alpha_{MSC}skp^g_{MSC} + \alpha_{MSC}ki^g_{MSC})}{\alpha_{MSC}s}\right] = \left[\left(\frac{\alpha_{MSC}}{s}\right)\left(\frac{(skp^g_{MSC} + ki^g_{MSC})}{\alpha_{MSC}}\right)\right] \quad (8)$$

where kp^g_{MSC} and ki^g_{MSC} are the proportional and integral gains, respectively, $\alpha_{MSC}=2.2/\tau_{MSC}$ is the MSC bandwidth of the closed loop control and τ_{MSC} is compensator response time.

Substituting Equation (8) into (7), the closed-loop transfer function $\left[i^g_{MSC}(s)\right]$ is formed:

$$\left[i^g_{MSC}(s)\right] \approx \left[i^{gref}_{MSC}(s) - i^g_{MSC}(s)\right] = \left[\left(\frac{\alpha_{MSC}}{s}\right)\left(\frac{skp^g_{MSC}(s) + ki^g_{MSC}(s)}{\alpha_{MSC}}\right)\left(\frac{1}{sL_{MSC}(s) + R_{MSC}(s)}\right)\right] \quad (9)$$

If in open loop the expression (9) tends to be ∞ when $s = j\omega \to 0$, this guarantees that, in closed loop the system will not have a phase shift delay.

Based on (9), the relation between the plant pole and the PI compensator zero is obtained through (10), generating the kp^g_{MSC} and ki^g_{MSC} control gains.

$$\left[kp^g_{MSC}\right] = \left[\alpha_{MSC}L_{MSC}\right] = \left[(2.2/\tau_{MSC})L_{MSC}\right] \quad (10a)$$

$$\left[ki^g_{MSC}\right] = \left[\alpha_{MSC}R_{MSC}\right] = \left[(2.2/\tau_{MSC})R_{MSC}\right] \quad (10b)$$

Compensator response time, τ_{MSC}, in the range from 5 to 0.5 ms is selected, in this case a $\tau_{MSC} = 2.2$ ms is designated.

2.2. Modeling Power Transfer Control between the WT-PMSG and AFE Converter

In the WT-PMSG power transfer modeling the following power-speed characteristics are considered [20]: (i) The base angular velocity of the WT is determined by the base rotor angular velocity of the PMSG, $\omega_{WTb} = \omega_{rPMSGb}$; (ii) the WES base power is determined by the WT-PMSG nominal power, $P_{WESb} = P_{WT-PMSGb}$; iii) the output base power of the AFE converter is determined by the base WES power, $P_{AFEb} = P_{WESb}$; this power is transferred from WT to PMSG through the electric torque, this is represented by:

$$[T_{ePMSG}] = (3/2)\left[\left((L^d_{MSC} - L^q_{MSC})i^d_{MSC}i^q_{MSC}\right) + (\lambda_{mPMSG}i^q_{MSC})\right] \quad (11)$$

where T_{ePMSG} is the PMSG electrical torque, L^d_{MSC} and L^q_{MSC} are the dq reference frame components of the PMSG armature inductance.

However, considering that the rotor has a cylindrical geometry, then it is established that, $L^d_{MSC} = L^q_{MSC}$ [21], generating (12):

$$[T_{ePMSG}] = ((3/2)\lambda_{mPMSG})\left[i^q_{MSC}\right] \quad (12)$$

Then, to realize the WT-PMSG variable speed control, it is necessary to generate the plant model that represents it. Therefore, in (13) the dynamic characteristics are shown as a time function so that it represents:

$$\left[\frac{d(\omega_{rPMSG})}{dt}\right] = \frac{1}{2H}[T_{mWT} - T_{ePMSG} - D\omega_{rPMSG}(t)] \quad (13)$$

where D is the PMSG viscous damping, H is the inertia constant (s), T_{mWT} is the WT mechanical torque.

Equation (13) analyzes the WT-PMSG in the time domain; however, the WT-PMSG plant representation requires a transfer function to design the ω_{rPMSG} control. By using Laplace transformation, the WT-PMSG plant in the frequency domain is represented, i.e.,

$$[\omega_{rPMSG}(s)] = \left[(T_{mWT} - T_{ePMSG})(2Hs + D)^{-1}\right] \tag{14}$$

Equation (14) shows a multiple inputs single output system (MISO); however, because in steady state it is valid that $T_{mWT} \approx T_{ePMSG}$, then, in the control design it is considered that $T_{mWT} = 0$; generating a single input single output system (SISO), as shown in (15).

$$\left[\begin{matrix} \omega_{rPMSG}(s) \\ -T_{ePMSG} \end{matrix}\right] = \left[\frac{1}{2Hs + D}\right] \tag{15}$$

With the purpose of tracking the ω_{rPMSG} reference commands in the closed-loop transfer function, the proportional-integral (PI) compensators are used. The feedback loop $\left[i_{rPMSG}^{q}(s)\right]$ is:

$$\left[i_{rPMSG}^{q}(s)\right] = \left[\omega_{rPMSG}^{ref}(s) - \omega_{rPMSG}(s)\right] = \left[\left(\frac{\alpha_{PMSG}}{s}\right)\left(\frac{(skp_{rPMSG}^{q} + ki_{rPMSG}^{q})}{\alpha_{PMSG}}\right)\left(\frac{1}{(2Hs+D)}\right)\right] \tag{16}$$

where kp_{rPMSG}^{q} and ki_{rPMSG}^{q} are the proportional and integral gains, respectively.

From (16), the relation between the plant pole and PI compensator zero is obtained and the control gains using the next expression are generated:

$$\left[kp_{rPMSG}^{q}\right] = [2H\alpha_{PMSG}] = [(2.2/\tau_{PMSG})2H] \tag{17a}$$

$$\left[ki_{rPMSG}^{q}\right] = [\alpha_{PMSG}D] = [(2.2/\tau_{PMSG})D] \tag{17b}$$

where the subscript τ_{PMSG} is the response time by the closed loop of the WT-PMSG first-order transfer function. This is selected according to the WT-PMSG transferred power and this must be at least ten times higher than τ_{MSC}.

2.3. Modeling of the Grid-Side VSC Control of the AFE Converter

The GSC is used to keep the DC-link constant, transferring the generated power between the WT-PMSG and AC grid. The time-domain relationship of the VSC AC-side is given by:

$$\left[d\left(i_{GSC}^{l}(t)\right)/dt\right] = -\left(R_{GSC}^{l}/L_{GSC}^{l}\right)\left[i_{GSC}^{l}(t)\right] + \left(1/L_{GSC}^{l}\right)\left[v_{GSC}^{l}(t)\right] - \left(1/L_{GSC}^{l}\right)\left[v_{WES}^{l}(t)\right] \tag{18}$$

where l is the VSC three-phase vector (a,b,c), L_{GSC} and R_{GSC} are the RL filter parameters through which the AFE converter is connected to the grid, v_{GSC} and i_{GSC} are the GSC voltage and current, respectively; v_{WES} is the generated WES voltage.

Then, from (18) the derived dq model is described as:

$$L_{GSC}\left(di_{GSC}^{d}(t)/dt\right) = (\omega_0 \cdot L_{GSC})\left[i_{GSC}^{q}(t)\right] - (R_{GSC})\left[i_{GSC}^{d}(t)\right] + \left[v_{GSC}^{d}(t)\right] - \left[v_{WES}^{d}(t)\right] \tag{19a}$$

$$L_{GSC}\left(di_{GSC}^{q}/dt\right) = -(\omega_0 \cdot L_{GSC})\left[i_{GSC}^{d}(t)\right] - (R_{GSC})\left[i_{GSC}^{q}(t)\right] + \left[v_{GSC}^{q}(t)\right] - \left[v_{WES}^{q}(t)\right] \tag{19b}$$

where ω_0 is the WES angular frequency; the generated GSC voltages are given by:

$$v_{GSC}^{k}(t) = (V_{DC}/2)\left[m_{GSC}^{k}(t)\right] \tag{20}$$

where k is the dq components reference frame vector of the grid-side VSC, m_{GSC}^{k} is the modulated index vector.

Making $L_{GSC} = L_{GSC}^d = L_{GSC}^q$, the presence of $\omega_0 L_{GSC}$ in (19) indicates the coupled dynamics between i_{GSC}^d and i_{GSC}^q. Decoupling these dynamics changes m_{GSC}^d and m_{GSC}^q, based in the dq reference frame, i.e.,

$$\left[m_{GSC}^d(t)\right] = (2/V_{DC}(t))\left[_{GSC}^d(t) - \left((\omega_0 \cdot L_{GSC})i_{GSC}^q(t)\right) + v_{WES}^d(t)\right] \tag{21a}$$

$$\left[m_{GSC}^q(t)\right] = (2/V_{DC}(t))\left[_{GSC}^q(t) + \left((\omega_0 \cdot L_{GSC})i_{GSC}^d(t)\right) + v_{WES}^q(t)\right] \tag{21b}$$

where $E_{GSC}^d(t)$ and $E_{GSC}^q(t)$ are two additional control inputs.

The GSC plant is obtained by substituting (21) into (20), subsequently, (20) is substituting into (19) generating a first order lineal system, this in Equation (22) is described as:

$$L_{GSC}\left[di_{GSC}^k(t)/dt\right] = \left[_{GSC}^k(t)\right] - R_{GSC}\left[i_{GSC}^k(t)\right] \tag{22}$$

The frequency domain of the Equation (22) is shown in (23); which describes a decoupled, first-order, linear system, controlled through $E_{GSC}^k(s)$. Also, Equation (23) represents the grid-side VSC plant.

$$\left[i_{GSC}^k(s)\right] = \left[_{GSC}^k(s)\right](sL_{GSC} + R_{GSC})^{-1} \tag{23}$$

With the purpose of tracking the $i_{GSC}^k(s)$ reference commands in the closed loop, the proportional-integral (PI) compensators are used, obtaining:

$$\left[_{GSC}^k(s)\right] \approx \left[k_{GSC}^k(s)\right] = \left[\frac{(\alpha_{GSC}skp_{GSC}^k + \alpha_{GSC}ki_{GSC}^k)}{\alpha_{GSC}s}\right] = \left[\left(\frac{\alpha_{GSC}}{s}\right)\left(\frac{(skp_{GSC}^k + ki_{GSC}^k)}{\alpha_{GSC}}\right)\right] \tag{24}$$

where kp_{GSC}^k and ki_{GSC}^k are the proportional and integral gains, respectively.

The feedback loop $i_{GSC}^k(s)$ is:

$$\left[i_{GSC}^k(s)\right] = \left[i_{GSC}^{kref}(s) - i_{GSC}^k(s)\right] = \left[\left(\frac{\alpha_{GSC}}{s}\right)\left(\frac{(skp_{GSC}^k + ki_{GSC}^k)}{\alpha_{GSC}}\right)\left(\frac{1}{sL_{GSC} + R_{GSC}}\right)\right] \tag{25}$$

The relation between the plant pole and the PI compensator zero is obtained in (26), generating the kp_{GSC}^k and ki_{GSC}^k control gains and $\alpha_{GSC}=2.2/\tau_{GSC}$ is the GSC bandwidth of the closed-loop control.

$$\left[kp_{GSC}^k\right] = [\alpha_{GSC}L_{GSC}] = [(2.2/\tau_{GSC})L_{GSC}] \tag{26a}$$

$$\left[ki_{GSC}^k\right] = [\alpha_{GSC}R_{GSC}] = [(2.2/\tau_{GSC})R_{GSC}] \tag{26b}$$

where τ_{GSC} is selected from 5 to 0.5 ms based on the transferred power.

2.4. The DC-Side Control of the AFE Converter

GSC improves the DC-link control. The time-domain relationship of the DC-link of the AFE converter is given by:

$$[dV_{DC}(t)/dt] = [I_{DC}(t)/C_{DC}] - [V_{DC}(t)/(C_{DC} \cdot R_{DC})] \tag{27}$$

The sum of currents entering the capacitor is:

$$[I_{DC}(t)] = \frac{1}{2}\sum_{l=a}^{c} m_{GSC}^l(t)\left[i_{GSC}^l(t)\right] \tag{28}$$

The functionality of the AFE converter requires that:

$$V_{DC} \geq |2(v_{WESL-L})| \tag{29}$$

The DC-link control is calculated through the stored energy in the capacitor, that is,

$$[U_{DC}(s)] = (C_{DC}/2)\left[V_{DC}^2(s)\right] \tag{30}$$

where U_{DC} is the stored energy in the capacitor and C_{DC} is the DC-link capacitance.

Considering that $U_{DC}(s) \approx P_{GSCref}(s)$, and using the d reference frame component of grid-side VSC plant described in (22) the DC-link control is made, generating the active power control, that is:

$$\left[P_{GSCref}(s)\right] = (C_{DC}/2)\left[V_{DCref}^2(s) - V_{DC}^2(s)\right]\left[E_{GSC}^d(s)\right] \tag{31}$$

The reactive power control is made with the q reference frame component of the GSC plant described in (22), that is,

$$\left[Q_{GSCref}(s)\right] = \left[Q_{WESref}(s) - Q_{WES}(s)\right]\left[_{GSC}^d(s)\right] \tag{32}$$

where Q_{WES} is the presented reactive power at the WES.

It is important to consider that, the subscript τ_{WES} presented in (32) must be at least ten times higher than τ_{GSC}.

2.5. System Parameters Design of the AFE Converter

The correct operation of the type-4 WT control depends on the precise design of the AFE converter parameters; thus, the element's values of the MSC are obtained from the WT-PMSG nominal power, $P_{WT-PMSG}$, that is: the current is $i_{MSC} = (2/3)(P_{WT-PMSG}/v_{MSC})$; the machine-side impedance is $Z_{MSCt} = v_{MSC}/i_{MSC}$, thus, the MSC works with 15% of the total WT-PMSG impedance, i.e., $Z_{MSC} = (0.15)Z_{MSCt}$; from the WT-PMSG characteristics the following parameters are taken: L_{MSC}, R_{MSG}, D, H. The element's values of the GSC are obtained from the WES nominal power, but to achieve $P_{WES} = P_{WT-PMSG}$ i_{GSC} is generated using $i_{MSC} = (2/3)(P_{WES}/v_{GSC})$; the grid-side impedance is $Z_{GSCt}=v_{GSC}/i_{GSC}$ the GSC works with 15% of the total WES impedance, i.e., $Z_{GSC} = (0.15)Z_{GSCt}$; therefore, L_{GSC} is calculated with $L_{GSC} = Z_{GSC}/\omega_0$, the R_{GSC} value varies according to the transferred power, in a range from 0.1 Ω to 0.5 Ω; the base WES capacitance C_{WES} is calculated with $C_{WES} = 1/(Z_{GSC}\omega_0)$. Then, a better time response in the WES feedback is achieved, since the L_{MSC} and R_{MSC} values are used in (10), H and D values are used in (17), L_{GSC} and R_{GSC} values are used in (26), to obtain the system feedback gains. It is important to establish that from the generated active power by the GSC, v_{WES} is kept constant in the presence of any perturbation; for which, it is essential to calculate the correct capacitance value that maintains the DC-link compensation. This is determined from the base DC-link capacitance, i.e., $C_{DC} = (3/8)C_{WES}$, determining the store energy in Equation (30).

3. Modeling of the DSPWM Technique Applied in the THD Reduction

Digital modulation techniques are the most generalized framework in the control of modern power electronics converters applications. Digital sinusoidal pulse width modulation (DSPWM) is a modulation technique created by the internal generation of the modulated and carrier signals using a digital controller [22].

THD reduction is achieved by modifying the DSPWM switching signals in each VSC. This is carried out by applying a different phase shift angle in each carrier signal of each VSC; the modulated signal angle is not changed. Then, the output signals (voltage or current) of each VSC are added. In this paper, the AFE converter is built with three VSC connected in parallel. Figure 3 shows the comparison between the modulated (without phase shift angle) and carrier (with phase shift angle) signals, generating the DSPWM signal (phase a) corresponding to each VSC connected in parallel. The correct phase shift angle between each carrier signal is established putting up different values of total phase shift angle at the WES, see Figure 2. The analysis is shown in Table 1.

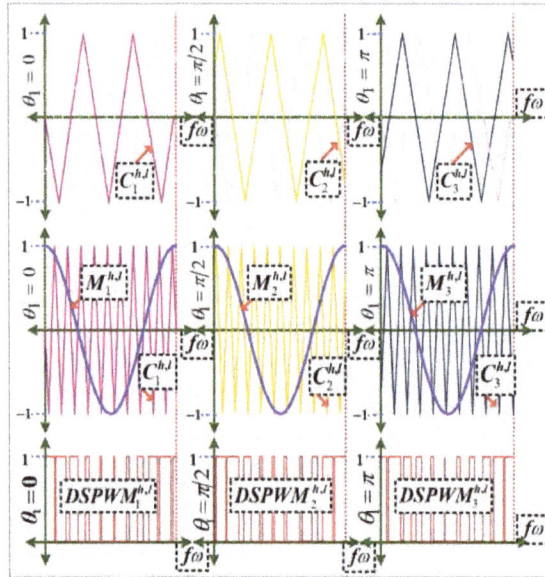

Figure 3. Digital sinusoidal pulse width modulation (DSPWM) signal applied to each voltage source converter (VSC) connected in parallel (phase a).

In Table 1 it is observed that the angle that generates a lower THD is $3\pi/2$; hence, this angle divides the number of VSCs placed in parallel, i.e.,

$$\theta_p = (3\pi/2)/p \tag{33}$$

where p is the number of VSC connected in parallel and θ_p is the carrier signal phase shift angle of each VSC.

Table 1. Analysis of different phase shift at the carrier signal.

Total Phase Shift (θ_p)	Carrier Phase Shift in Each VSC			% Total Harmonic Distortion (THD)
	θ_1	θ_2	θ_3	
0	0	0	0	6.8%
$\pi/6$	0	$\pi/18$	$\pi/9$	4.33%
$\pi/3$	0	$\pi/9$	$2\pi/9$	1.99%
$\pi/2$	0	$\pi/6$	$\pi/3$	2.054%
$2\pi/3$	0	$2\pi/9$	$4\pi/9$	1.271%
$5\pi/6$	0	$5\pi/18$	$5\pi/9$	1.608%
π	0	$\pi/3$	$2\pi/3$	4.616%
$7\pi/6$	0	$7\pi/18$	$7\pi/9$	5.635%
$4\pi/3$	0	$4\pi/9$	$8\pi/9$	2.864%
$3\pi/2$	**0**	**$\pi/2$**	**π**	**1.239%**
$5\pi/3$	0	$5\pi/9$	$10\pi/9$	1.36%
$11\pi/6$	0	$11\pi/18$	$11\pi/9$	1.867%
2π	0	$2\pi/3$	$4\pi/3$	2.756%

The *n*-harmonics content is calculated through the Fourier series expansion, i.e.,

$$F(t) = C_0 + \sum_{n=1}^{\infty} \left(C_{MSC,GSC}^n \cos(nw_0 t + \sigma) \right) \tag{34}$$

where n is the harmonic number, $C_{MSC,GSC}^n = \sqrt{(a_{MSC,GSC}^n)^2 + (b_{MSC,GSC}^n)^2}$, $\sigma = \tan^{-1}(b_{MSC,GSC}^n / a_{MSC,GSC}^n)$ and $C_0 = a_0/2$.

The magnitude of each harmonic is calculated by,

$$a_{MSC,GSC}^n = \frac{2}{T} \left(\int_{-T/2}^{T/2} F(t) \cos(nw_0 t) dw_0 t \right) \tag{35}$$

$$b_{MSC,GSC}^n = \frac{2}{T} \left(\int_{-T/2}^{T/2} F(t) \sin(nw_0 t) dw_0 t \right) \tag{36}$$

To calculate the THD in the AFE converter, the individually equivalent circuit of each three-phase VSC is analyzed.

A three-phase VSC equivalent circuit is shown in Figure 4.

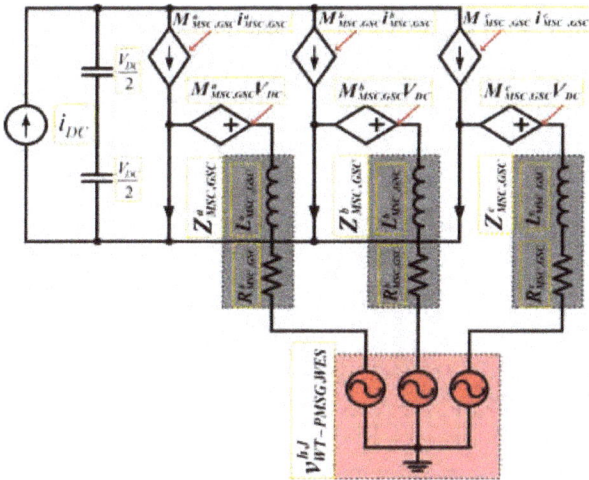

Figure 4. Three-phase VSC equivalent circuit.

The three-phase VSC is represented by the next equation,

$$\begin{bmatrix} 2(Z * i^a) & -(Z * i^b) & -(Z * i^c) \\ -(Z * i^a) & 2(Z * i^b) & -(Z * i^c) \\ -(Z * i^a) & -(Z * i^b) & 2(Z * i^c) \end{bmatrix}_{MSC,GSC} = \left[\begin{pmatrix} v^a - v^b \\ v^b - v^c \\ v^c - v^a \end{pmatrix}_{MSC,GSC} - \begin{pmatrix} v^a - v^b \\ v^b - v^c \\ v^c - v^a \end{pmatrix}_{WT-PMSG,WES} \right] \tag{37}$$

Using Kirchhoff's current law (KCL), the currents flowing towards the MSC or/and GSC node must be equal to the currents leaving the MSC or/and GSC node, i.e.,

$$i_{MSC,GSC}^c = -\left(i_{MSC,GSC}^a + i_{MSC,GSC}^b \right) \tag{38}$$

Replacing equation (38) in (37) gives line-to-line current of the MSC or/and GSC, i.e.,

$$
\begin{bmatrix} i_{MSC,GSC}^{ab} \\ i_{MSC,GSC}^{bc} \\ i_{MSC,GSC}^{ca} \end{bmatrix} = \left(\frac{1}{3Z_{MSC,GSC}^{h,l}} \right) \left(\begin{bmatrix} v^a - v^b \\ v^b - v^c \\ v^c - v^a \end{bmatrix}_{WT-PMSG,WES} - \begin{bmatrix} v^a - v^b \\ v^b - v^c \\ v^c - v^a \end{bmatrix}_{MSC,GSC} \right) \tag{39}
$$

where $v_{WT-PMSG}$ represents the WT-PMSG voltage, v_{WES} exemplifies the WES voltage, $v_{MSC,GSC}$ is the VSC AC-side output voltage of MSC or/and GSC, and $Z_{MSC,GSC}^{h,l}$ is the AC-side filter of MSC or/and GSC.

The $v_{MSC,GSC}$ value depends on $M_{MSC,GSC}^{h,l}$ signal modulation. The modulated and carrier signals implement the DSPWM technique of Figure 3; these have modulation frequencies of 60Hz (ω_0) and 7kHz ($f\omega$), respectively.

The carrier signal is composed by an up-slope and a down-slope, calculated as,

$$
C_{t_1 p} = 1 - ((4/f\omega)(\omega_0 t_1 - \theta_p)) \tag{40}
$$

$$
C_{t_2 p} = ((4/f\omega)(\omega_0 t_2 - (f\omega/2) - \theta_p)) - 1 \tag{41}
$$

where $C_{t1,t2p}$ is the composed carrier signal, θ_p is phase shift angle of each VSC, $f\omega$ is switching frequency of the carrier signal, t_1 is the time for the up-slope, t_2 is the time for the down-slope.

Time t_1 for up-slope is:

$$
\theta_p \leq t_1 \leq ((f\omega/2) + \theta_p) \tag{42}
$$

Time t_2 for down-slope is:

$$
((f\omega/2) + \theta_p) \leq t_2 \leq (f\omega + \theta_p) \tag{43}
$$

Modulated signals in each VSC are described by the carrier signal time, that is:

$$
\begin{aligned} M_{t_1 p}^{h,l} &= \cos(t_1 + \varphi) \\ M_{t_2 p}^{h,l} &= \cos(t_2 + \varphi) \end{aligned} \tag{44}
$$

where $h,l = a,b,c$ the VSC phases in MSC and GSC, respectively, and φ is the corresponding angle of each phase in the modulated signal.

The comparison between modulated and carrier signals defines the *DSPWM* signal, its representation is:

$$
\begin{aligned} DSPWM_{t_1 p}^{h,l} &= \left| M_{t_1 p}^{h,l} \leq C_{t_1 p} \right| \\ DSPWM_{t_2 p}^{h,l} &= \left| M_{t_2 p}^{h,l} \leq C_{t_2 p} \right| \end{aligned} \tag{45}
$$

Multiplying the *DSPWM* signal and DC voltage amplitude generates the VSCs output voltage for each phase value in MSC and GSC, i.e.,

$$
v_{MSC,GSC}^{h,l} = V_{DC} * DSPWM_{MSC,GSC}^{h,l} \tag{46}
$$

The WT-PMSG voltage $v_{WT-PMSG}^{h}$ is generated by,

$$
\begin{aligned} v_{t_1 WT-PMSG}^{h} &= PMSG(\cos(\omega_{rPMSG} t_1 + \theta_{rPMSG})) \\ v_{t_2 WT-PMSG}^{h} &= PMSG(\cos(\omega_{rPMSG} t_2 + \theta_{rPMSG})) \end{aligned} \tag{47}
$$

where *PMSG* is the WT-PMSG amplitude voltage and ϕ is the corresponding angle of each phase in the three-phase WT-PMSG.

And the WES voltage v_{WES}^l is produced by,

$$v_{t_1 WES}^l = WES(\cos(\omega_0 t_1 + \phi_{WES}))$$
$$v_{t_2 WES}^l = WES(\cos(\omega_0 t_2 + \phi_{WES}))$$

(48)

where *WES* is the AC grid amplitude voltage and ϕ_{WES} is the corresponding angle of each phase in the three-phase WES grid.

The output current in each VSC is calculated as,

$$\begin{bmatrix} i_{t_1 MSC,GSC}^{h,l} \\ i_{t_2 MSC,GSC}^{h,l} \end{bmatrix} = (1/(3Z_{MSC,GSC})) \left(\begin{bmatrix} v_{t_1 WT-PMSG,WES}^{h,l} \\ v_{t_2 WT-PMSG,WES}^{h,l} \end{bmatrix} - \begin{bmatrix} v_{t_1 MSC,GSC}^{h,l} \\ v_{t_2 MSC,GSC}^{h,l} \end{bmatrix} \right)$$

(49)

The harmonic content spectrum to obtain the THD is required. By using (35), (36), and (49) the spectrum is calculated as,

$$a_{MSC,GSC}^n = \left(\tfrac{2}{T}\right) \left[\left(\int_{\theta_p}^{(f\omega/2)+\theta_p} \left(i_{t_1 MSC,GSC}^{h,l} \cos(n\omega_0 t_1) \right) d\omega_0 t_1 \right) + \left(\int_{(f\omega/2)+\theta_p}^{f\omega+\theta_p} \left(i_{t_1 MSC,GSC}^{h,l} \cos(n\omega_0 t_2) \right) d\omega_0 t_2 \right) \right]$$

(50)

$$b_{MSC,GSC}^n = \left(\tfrac{2}{T}\right) \left[\left(\int_{\theta_p}^{(f\omega/2)+\theta_p} \left(i_{t_1 MSC,GSC}^{h,l} \sin(n\omega_0 t_1) \right) d\omega_0 t_1 \right) + \left(\int_{(f\omega/2)+\theta_p}^{f\omega+\theta_p} \left(i_{t_2 MSC,GSC}^{h,l} \sin(n\omega_0 t_2) \right) d\omega_0 t_2 \right) \right]$$

(51)

For the harmonic content of the output current signal, the magnitude of the individual harmonics is calculated for each VSC connected in parallel to the MSC and GSC and these are added, i.e.,

$$a_{MSC,GSC}^{n1} + a_{MSC,GSC}^{n2} + \ldots + a_{MSC,GSC}^{np}$$

(52)

$$b_{MSC,GSC}^{n1} + b_{MSC,GSC}^{n2} + \ldots + b_{MSC,GSC}^{np}$$

(53)

where p is the number of VSCs placed in parallel and n is the number of harmonics.

The THD in the AFE converter output current is,

$$THDi_{MSC,GSC}^{h,l} = \left| \left(\frac{1}{C_{MSC,GSC}^{1p}} \right) \sqrt{\sum_{n=2}^{\infty} \left(C_{MSC,GSC}^{np} \right)^2} \right| * 100$$

(54)

where $C_{MSC,GSC}^{1p}$ is the fundamental harmonic magnitude and $C_{MSC,GSC}^{np}$ is the nth harmonic magnitude.

Finally, the lower THD content in the output current of the AFE converter is generated when the output current signals of each VSC are added, i.e.,

$$i_{MSC,GSC}^{h,l} = i_{MSC1,GSC1}^{h,l} + i_{MSC2,GSC2}^{h,l} + \ldots + i_{MSCp,GSCp}^{h,l}$$

(55)

Figure 5 shows the flow diagram that describes the generated method for a lower harmonic content, represented from Equations (33) to (55).

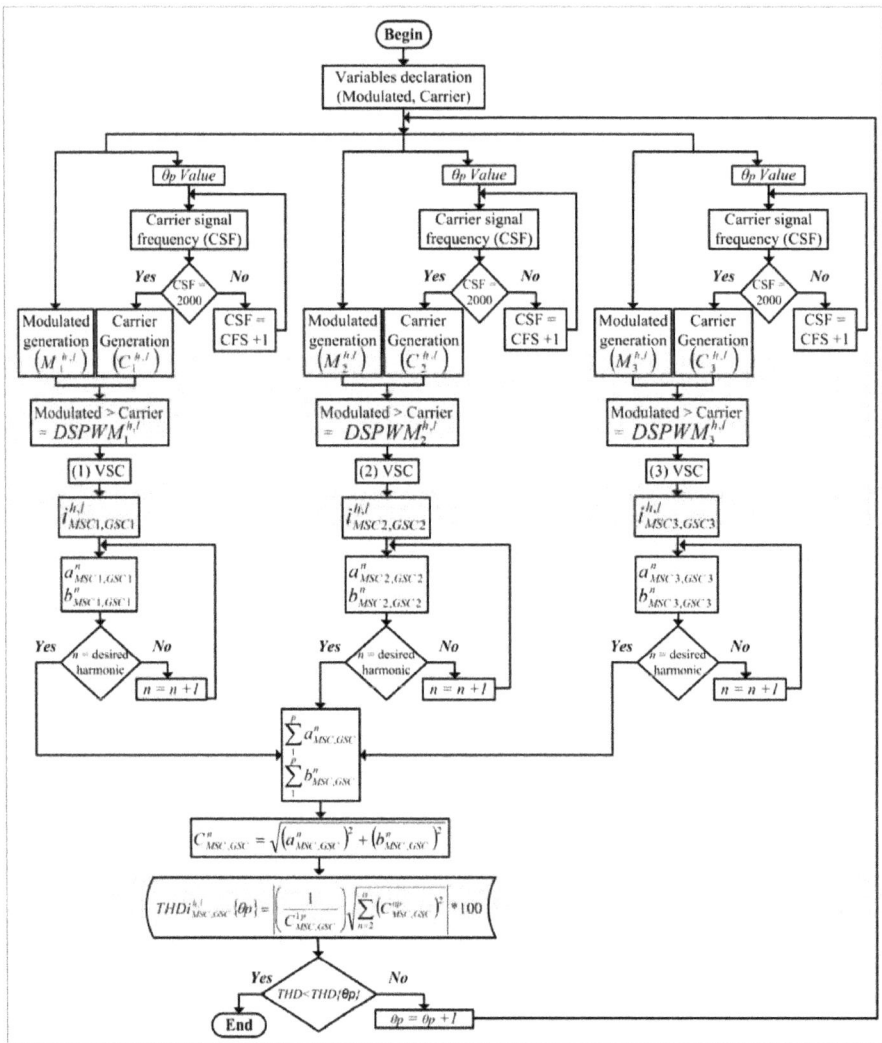

Figure 5. Flow diagram for the lower total harmonic distortion (THD) generation.

4. Simulation Results: Study Case for WES

In this paper, Matlab-Simulink® (Matlab r2015b, Mathworks, Natick, MA, USA) and Opal-RT Technologies® module (OP-5600) (Montreal, QC, Canada) are the main elements in the WES real-time simulation, since the OP-5600 module uses the rapid control prototyping (RCP) concept, which allows testing of the control law without the need for any programming code.

In Figure 2, the simulated WES is shown. It contains a WT-PMSG to supply the MSC, the AFE parallel converter and the infinite bus (considered as an ideal voltage source) to supply the GSC. The MSC and GSC are connected to WT-PMSG and the AC grid through RL filters, both converters are formed by three VSCs connected in parallel and each one is designed to possess a power and voltage of 2 MVA and 2.5 kV, respectively. The characteristics of the WT-PMSG are described in Table 2.

Table 2. Wind turbine-permanent magnet synchronous generator (WT-PMSG) characteristics.

Wind Turbine (WT)			
Nominal output power	2 MW	Base wind speed	12 m/s
Pitch angle	45 deg	base generator speed	1.2 pu
Permanent Magnet Synchronous Generator (PMSG)			
Mechanical input	-8.49×10^5 N.m.	Stator resistance	8.2×10^{-4} Ω
Armature inductance	1.6×10^{-3} H	Flux linkage	5.82
Viscous damping	4.04×10^3 N.m.s	Inertia	2.7×10^6 kg.m^2
Pole pairs	4	Rotor type	Round

To verify the correct WES operation in Figure 2, in Figure 6 the behavior of the WT mechanical torque and the PMSG electric torque are analyzed.

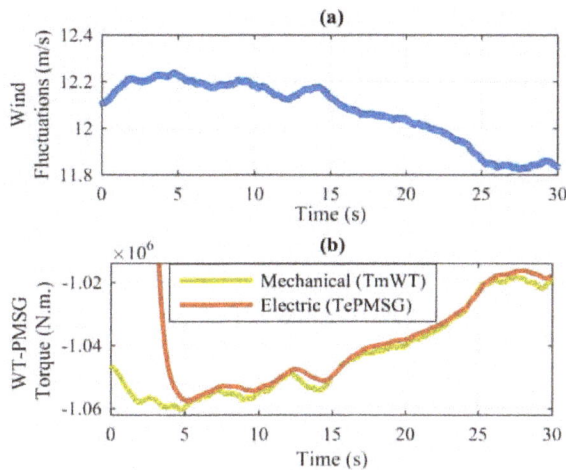

Figure 6. The behavior of the WT mechanical torque and the permanent magnet synchronous generator (PMSG) electric torque in the presence of wind fluctuations. (**a**) Wind fluctuations; (**b**) mechanical and electric torque.

Figure 6a details the wind fluctuations applied to the WT, which are generated in Matlab-Simulink® by a rotor wind model developed by RISOE National Laboratory based on Kaimal spectra. Figure 6b shows the behavior of the WT mechanical torque and the PMSG electric torque in the presence of wind fluctuations. It is possible to observe that the electric torque follows the mechanical torque behavior, due to the effective structure of the MSC closed-loop control.

Figure 7 shows the generated current by the WT-PMSG, which is controlled through the MSC of the AFE parallel converter. Because the MSC is formed using the parallel connection of three VSCs, each VSC can handle one third of the total current generated by the WT; Figure 7a–c illustrates the current in the (1), (2), and (3) VSCs, respectively, and in Figure 7d the MSC total current is shown.

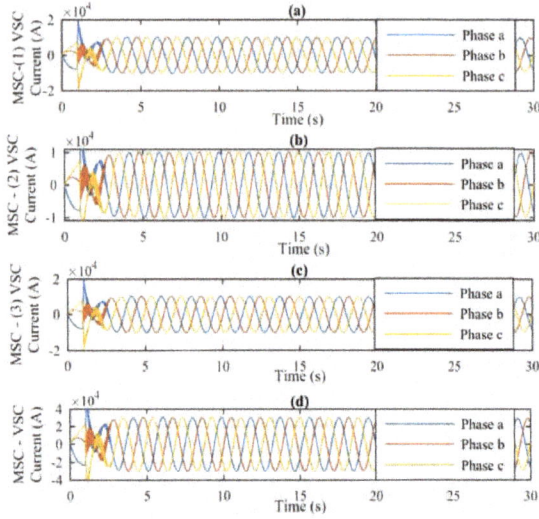

Figure 7. Current present in machine-side VSC (MSC) of Active Front-End (AFE) parallel converter. (**a**) (1) VSC; (**b**) (2) VSC; (**c**) (3) VSC; (**d**) total current.

While, the main MSC function is the rotor flux frequency control, generating the power conversion between medium AC voltage and low DC voltage levels, the most important GSC function is to keep the DC-link constant, transferring the generated power between the WT-PMSG and AC grid in the voltages required by the consumers.

Figure 8a shows that the DC-link remains constant at 5kV, because, when the MSC requires a reactive power exchange, due to the wind fluctuations of Figure 6a, the GSC restores the DC-Link, and at the same time injects the needed reactive power, as shown in Figure 8b.

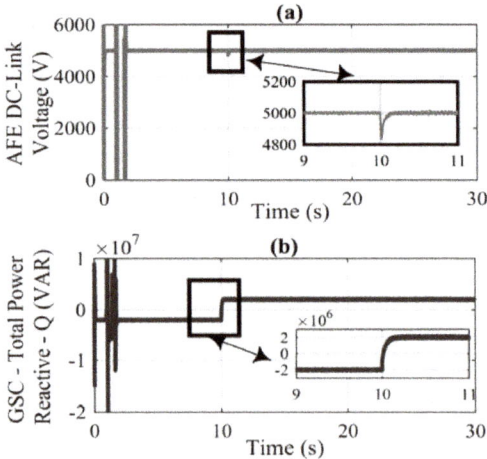

Figure 8. DC-Link and Reactive Power controlled by the grid-side VSC (GSC). (**a**) DC-link voltage; (**b**) exchange of reactive power in WES.

Figures 9–11 detail the applied DSPWM to each of the VSCs connected in parallel for the correct operation of the GSC, at the stability time from 4.5 to 4.509 ms. In Figure 9, it can be seen that both the carrier signal of Figure 9a and the modulated signal of Figure 9b start at the same time, i.e., the carrier signal does not present any phase shift, generating the DSPWM signal in Figure 9c, this is applied to the first VSC connected in parallel in the GSC. In Figure 10, the DSPWM generation applied to the second VSC connected in parallel to the GSC is shown; in Figure 10a, a phase shift of $\pi/2$ (rad/s) in the carrier signal is observed. This is compared with the modulated signal of Figure 10b, originating the DSPWM with the phase shift of Figure 10c. Finally, in Figure 11, the DSPWM signal applied to the third VSC connected in parallel of the GSC is presented; in Figure 11a the carrier is observed with a phase shift of π (rad/s) with respect to the modulated signal of Figure 11b, generating the DSPWM of Figure 11c.

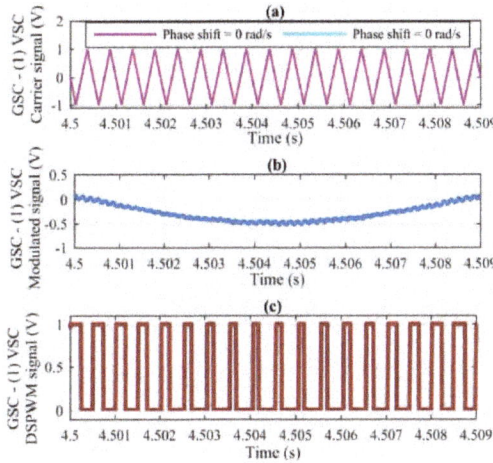

Figure 9. DSPWM signal applied to the control of the first VSC connected in parallel in GSC. (**a**) Carrier signal; (**b**) modulated signal; (**c**) DSPWM.

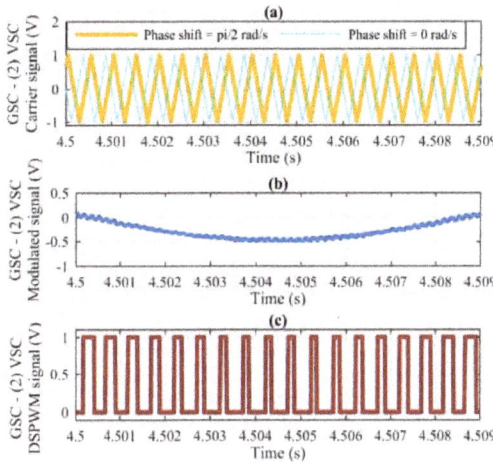

Figure 10. DSPWM signal applied to the control of the second VSC connected in parallel in GSC. (**a**) Carrier signal; (**b**) modulated signal; (**c**) DSPWM.

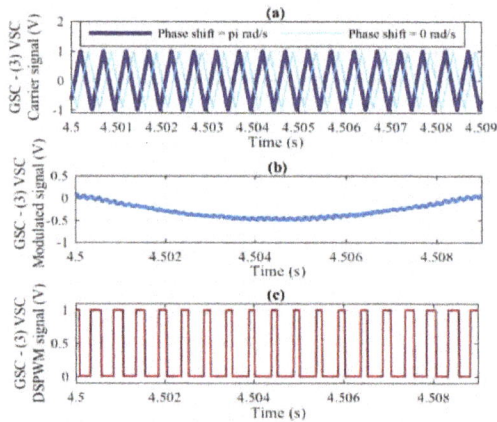

Figure 11. DSPWM signal applied to the control of the third VSC connected in parallel in GSC. (**a**) Carrier signal; (**b**) modulated signal; (**c**) DSPWM.

Figure 12 shows the electrical variables present at the GSC when the corresponding phase shift in the carriers of each VSC connected in parallel is performed, according to Equation (33). Figure 12a shows the (1) VSC current generated due to the phase shift at the carrier of Figure 9a; in which, a zoom in time is made from 9.9 to 10.1 s, observing the current magnitude and behavior in the presence of the reactive power exchange at Figure 8b. Figure 12b shows the (2) VSC current generated due the phase shift at the carrier of Figure 10a; Figure 12c shows the (3) VSC current generated due the phase shift at the carrier of Figure 11a; in Figure 12a–c, each current magnitude is 330 A, generating a total GSC current of 990 A, as seen in Figure 12d; Figure 12e details a zoom in time from 9.9 to 10.1 s, observing the generated voltage at the GSC, the magnitude of which corresponds to 2500 V.

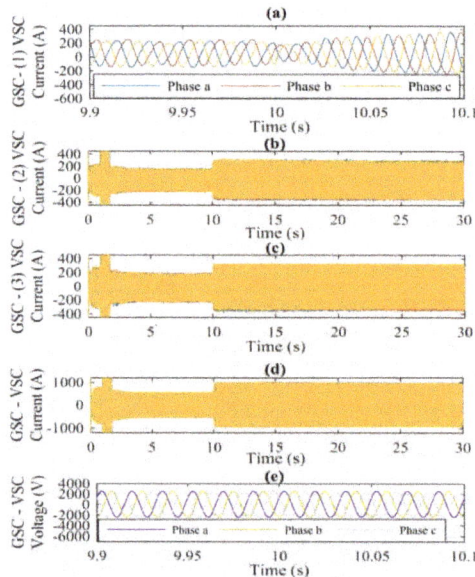

Figure 12. Electrical variables generated by the GSC. (**a**) Zoom of the handled current at the (1) VSC; (**b**) the handled current at the (2) VSC; (**c**) the handled current at the (3) VSC; (**d**) total current; (**e**) zoom at the magnitude voltage.

Finally, the current THD is shown in Figure 13; Figure 13a contents the THD without any phase shift between carriers of each VSC of the AFE converter, which corresponds to 6.8%. Please observe that, in Figure 13b, when the corresponding phase shift is performed in the carriers, the current THD is reduced to 1.239%, as specified in Table 1. The Figure 13 shows the harmonics magnitude reduction or even their elimination, once the phase shift between carriers has been made. The THD was reduced by approximately 5.5 times.

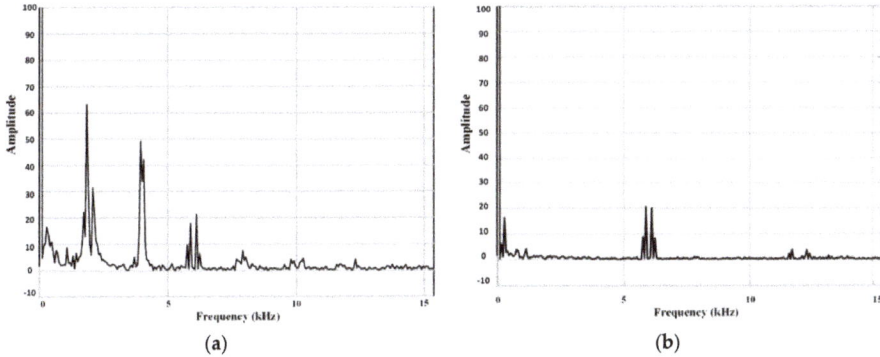

(a) (b)

Figure 13. THD present at the WES. (a) Without phase shift between carriers of each VSC; (b) with phase shift between carriers of each VSC.

5. Real Time Simulation Results: Study Case for WES using Opal-RT Technologies®

To verify the robustness of the applied control in the AFE converter and the THD reduction at the WES, the grid of Figure 2 in real time using the Opal-RT Technologies® is simulated; generating an RCP concept that tests the WES dynamics without the need for any programming code. Specifically, the VSC of the AFE converter is composed by the insulated gate bipolar transistor (IGBTs), these use a switching frequency of 7 kHz. Figure 14a shows the wind fluctuations generated by a rotor wind model developed by RISOE National Laboratory based on Kaimal spectra. Figure 14b contains the mechanical torque behavior generated by the wind turbine, and in response to the applied control at the MSC, the PMSG electric torque is able to follow the same behavior.

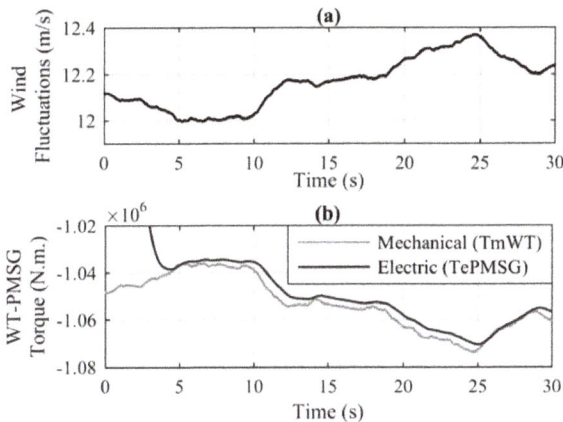

Figure 14. Behavior of the WT mechanical torque and the PMSG electric torque in the presence of wind fluctuations simulated in the Opal-RT Technologies®. (a) Wind fluctuations; (b) Mechanical and Electric torque.

Figure 15 presents the main electrical variables of the WES simulated in real time by OPAL-RT®. Figure 15a contains the current portion that handles the first VSC connected in parallel; as can be seen, as only three VSCs are connected in parallel, each one handles only a third of the total current generated by the MSC. The total current is presented in Figure 15b, and this is transferred by the WT-PMSG to the AC grid through the AFE converter. In Figure 15c, the generated voltage by the MSC is observed. It is important to mention that the main objective of the GSC is to support the constant DC-link in the presence of any disturbance (such as voltage/current variations due to wind fluctuations or reactive power exchanges by the behavior of the WT). This is evidenced in Figure 15d and is possible due to the applied control robustness. Figure 15e shows the GSC ability to exchange reactive power, that is, the ability of the injection/absorption of 6 MVA into the AC grid. Figure 15f contains the handled current portion by the first VSC connected in parallel at the GSC; similarly, as only three VSCs are connected in parallel, each one handles only a third of the total current generated by the GSC; the total current is presented in Figure 15g. Finally, in Figure 15h, the handled voltage by the GSC is observed, this is taken from the PCC attached to the AC grid. The THD of the handled total current by the GSC is generated through the OPAL-RT®. The generated THD without phase shift between the carriers of each VSC connected in parallel corresponds to 8.85%. The produced THD once the phase shift between the carriers of each VSC is made corresponds to 2.18%, and the phase shift from equation (33) is calculated; therefore, it is demonstrated that making the WES real-time simulation and applying the phase shift between the carriers of each VSC, the THD can be reduced up to four times.

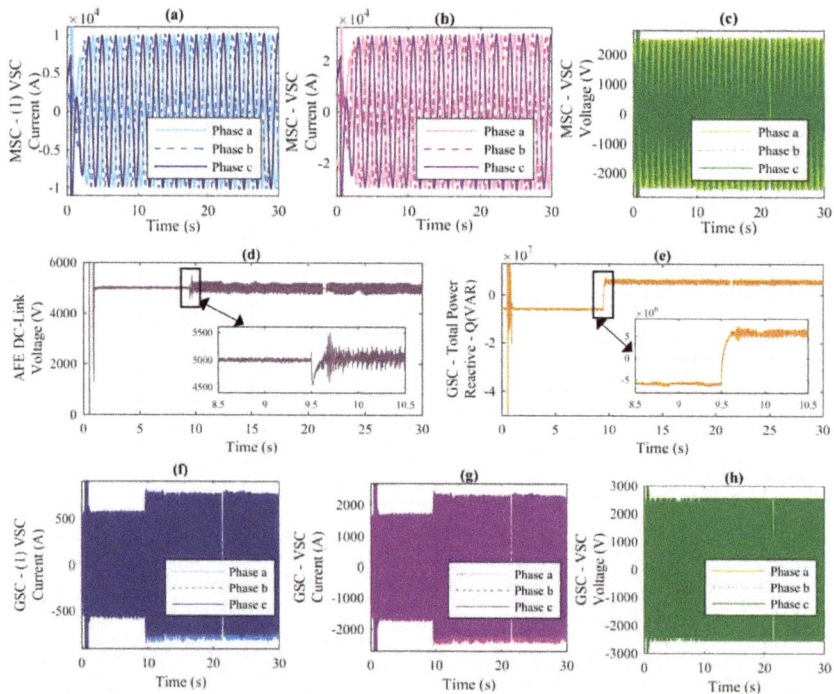

Figure 15. Electrical variables generated at the WES simulated in the Opal-RT Technologies®. (**a**) The handled current by the (1) VSC of MSC; (**b**) total current handled by the MSC; (**c**) voltage present at the MSC; (**d**) DC-Link voltage controlled by the GSC; (**e**) reactive Power controlled by the GSC; (**f**) the handled current by the (1) VSC of GSC; (**g**) total current handled by the GSC; (**h**) voltage present at the GSC.

Finally, it is important to mention that, the block *–written to file–* produces the results of Figures 14 and 15 in the MATLAB–Simulink® interface; this allows plotting the variables in MATLAB windows in order to have a better presentation.

6. Conclusions

In this paper, the AFE converter topology has been analyzed for the THD reduction in a WES. The WES has been formed by a WT-PMSG connected to the AC grid through an AFE converter. The AFE converter topology has been made from the use of a single VSC to use *p*VCS connected in parallel.

The effective THD reduction has been made through the variation in the DSPWM technique applied to each VSC, that is, applying a different phase shift angle at the carrier signals of each VSC connected in parallel, while the modulated signal angle has been kept constant.

To verify the robustness to the applied control, the WES control law has been simulated in real time using the Opal-RT Technologies®, generating an RCP concept, which tests the WES dynamics without the need for any programming code.

The obtained results have shown a type-4 WT with total output power of 6MVA generates a THD reduction up to 5.5 times of the total WES current output by Fourier series expansion.

Author Contributions: Conceptualization and Investigation, N.M.S.-H.; Methodology, D.C.-G.; Resources, R.T.-S.; Supervision, O.A.-L. and A.M.-R.; Project Administration, A.M.-R.; Funding Acquisition, J.J.R.-R.

Funding: This research was funded by the program DGAPA-PAPIIT-Project TA100718, granted by the National Autonomous University of Mexico (UNAM), to carry out the developments of this work.

Acknowledgments: This work was supported in part by The National Council of Science and Technology (CONACYT) in Mexico, through the Scholarship awarded under the Called *"Estancias Posdoctorales Vinculadas a la Consolidación de Grupos de Investigación"*.

Conflicts of Interest: The authors declare no conflict of interest.

Nomenclature

α_{GSC}	GSC bandwidth of the closed-loop control
α_{MSC}	MSC bandwidth of the closed-loop control
φ	Modulated signal angle
ϕ	WT-PMSG three-phase angle
ϕ_{WES}	WES three-phase angle
λ_{mPMSG}	PMSG maximum flux linkage
θ_p	*Phase shift angle of each VSC*
τ_{GSC}	GSC compensator response time
τ_{MSC}	MSC compensator response time
τ_{PMSG}	PMSG compensator response time
ω_{rPMSG}	PMSG rotor angular velocity
ω_{rPMSGb}	PMSG base rotor angular velocity
ω_{WTb}	WT base angular velocity
ω_o	WES angular frequency
AFE	Active Front-End
$C_{t1,t2p}$	Composed carrier signal
C_{DC}	DC-link capacitance
C_{WES}	WES capacitance
D	PMSG viscous damping
DFIG	Double-fed induction generator
DSPWM	Digital sinusoidal pulse width modulation
$DSPWM_{GSC}$	Modulated index vector at GSC
$DSPWM_{MSC}$	Modulated index vector at MSC
E_{GSC}	GSC control input

E_{MSC}	PMSC control input
$f\omega$	Switching frequency
GSC	Grid-side VSC
H	Inertia constant
i_{GSC}	GSC current
i_{MSC}	MSC current
i_{rPMSG}	PMSG rotor current
I_{DC}	DC-link current
ki_{GSC}	GSC integral compensator gain
ki_{MSC}	MSC integral compensator gain
ki_{rPMSG}	PMSG integral compensator gain
kp_{GSC}	GSC proportional compensator gain
kp_{MSC}	MSC proportional compensator gain
kp_{rPMSG}	PMSG proportional compensator gain
L_{GSC}	GSC inductance
L_{MSC}	WT-PMSG armature inductance
MSC	Machine-side VSC
p	Number of VSC in parallel
PCC	Point of Common Coupling
P_{GSCref}	GSC active power reference
PMSG	Permanent magnet synchronous generator
P_{WESb}	WES base power
$P_{WT-PMSGb}$	WT-PMSG base power
P_{AFEb}	AFE converter base power
Q_{GSCref}	GSC reactive power reference
Q_{WESref}	WES reactive power reference
Q_{WES}	WES reactive power
RCP	Rapid control prototyping
R_{DC}	DC-link resistance
R_{GSC}	GSC resistance
R_{MSC}	MSC resistance
s	Laplace operator
SCIG	squirrel-cage induction generator
Superscript d	d axis of dq reference frame
Superscript g	MSC dq components vector
Superscript h	MSC three-phase vector
Superscript k	VSC dq components vector
Superscript l	VSC three-phase vector
Superscript n	Harmonic number
Superscript q	q axis of dq reference frame
Superscript ref	Corresponding Reference value
t_1	up-slope time
t_2	down-slope time
T_{ePMSG}	PMSG electrical torque
THD	Total Harmonic Distortion
T_{mWT}	WT mechanical torque
U_{DC}	Energy capacitor
v_{GSC}	GSC voltage
v_{WES}	WES voltage
v_{WESL-L}	WES line to line voltage
v_{MSC}	WT-PMSG voltage
v_{WT}	Wind turbine voltage

$v_{WT\text{-}PMSG}$	Generated WT-PMSG voltage
V_{DC}	DC-link voltage
V_{DCref}	DC-link voltage reference
VSC	voltage source converter
WES	Wind Energy System
WT	Wind Turbine
Z_{GSCt}	GSC impedance
Z_{GSC}	Total WES impedance
Z_{MSC}	Total WT-PMSG impedance
Z_{MSCt}	MSC impedance

References

1. Wang, S.; Wang, S. Impacts of wind energy on environment: A review. *Renew. Sustain. Energy Rev.* **2015**, *49*, 437–443. [CrossRef]
2. Salgado-Herrera, N.M.; Medina-Rios, A.; Tapia-Sánchez, R. Reactive Power Compensation in Wind Energy Systems through Resonant Corrector in Distributed Static Compensator. *J. Electr. Power Compon. Syst.* **2017**, *45*, 1859–1869. [CrossRef]
3. Yaramasu, V.; Dekka, A.; Durán, M.J.; Kouro, S.; Wu, B. PMSG-based wind energy conversion systems: survey on power converters and controls. *IET Electr. Power Appl.* **2017**, *11*, 956–968. [CrossRef]
4. Renewable Energy (REN21): Global Status Report, Renewables 2017. Available online: http://www.ren21.net/status-of-renewables/global-status-report/ (accessed on 17 November 2017).
5. Global Wind Energy Council (GWEC): Global Wind Report: Annual Market Update. April 2017. Available online: http://www.gwec.net (accessed on 14 January 2018).
6. Salgado-Herrera, N.M.; Medina-Ríos, A.; Tapia-Sánchez, R.; Anaya-Lara, O. Reactive power compensation through active back to back converter in type-4 wind turbine. In Proceedings of the 2016 IEEE International Autumn Meeting on Power, Electronics and Computing (ROPEC), Ixtapa, Mexico, 9–11 November 2016; pp. 1–6.
7. Jlassi, I.; Estima, J.O.; Khil, S.K.E.; Bellaaj, N.M.; Cardoso, A.J.M. Multiple Open-Circuit Faults Diagnosis in Back-to-Back Converters of PMSG Drives for Wind Turbine Systems. *IEEE Trans. Power Electron.* **2015**, *30*, 2689–2702. [CrossRef]
8. Hu, W.; Chen, Z.; Wang, Y.; Wang, Z. Flicker Mitigation by Active Power Control of Variable-Speed Wind Turbines with Full-Scale Back-to-Back Power Converters. *IEEE Trans. Energy Convers.* **2009**, *24*, 640–649.
9. Lee, J.S.; Lee, K.B.; Blaabjerg, F. Open-Switch Fault Detection Method of a Back-to-Back Converter Using NPC Topology for Wind Turbine Systems. *IEEE Trans. Ind. Appl.* **2015**, *51*, 325–335. [CrossRef]
10. Nasiri, M.; Mohammadi, R. Peak Current Limitation for Grid Side Inverter by Limited Active Power in PMSG-Based Wind Turbines During Different Grid Faults. *IEEE Trans. Sustain. Energy* **2017**, *8*, 3–12. [CrossRef]
11. Juan, Y.L. Single switch three-phase ac to dc converter with reduced voltage stress and current total harmonic distortion. *IET Power Electron.* **2014**, *7*, 1121–1126. [CrossRef]
12. Ackermann, T. *Wind Power in Power Systems*, 2nd ed.; John Wiley & Sons Ltd.: Hoboken, NJ, USA, 2012; pp. 203–204.
13. Salgado-Herrera, N.M.; Mancilla-David, F.; Medina-Ríos, A.; Tapia-Sánchez, R. THD mitigation in type-4 Wind Turbine through AFE Back to back converter. In Proceedings of the 2015 North American Power Symposium (NAPS), Charlotte, NC, USA, 4–6 October 2015; pp. 1–6.
14. Hou, C.C.; Cheng, P.T. Experimental Verification of the Active Front-End Converters Dynamic Model and Control Designs. *IEEE Trans. Power Electron.* **2011**, *26*, 1112–1118. [CrossRef]
15. Fioretto, M.; Raimondo, G.; Rubino, L.; Serbia, N.; Marino, P. Evaluation of current harmonic distortion in wind farm application based on Synchronous Active Front End converters. In Proceedings of the IEEE Africon'11, Livingstone, Zambia, 13–15 September 2011; pp. 1–6.
16. Shen, L.; Bozhko, S.; Asher, G.; Patel, C.; Wheeler, P. Active DC-Link Capacitor Harmonic Current Reduction in Two-Level Back-to-Back Converter. *IEEE Trans. Power Electron.* **2016**, *31*, 6947–6954. [CrossRef]

17. Cai, X.; Zhang, Z.; Cai, L.; Kennel, R. Current balancing control of high power parallel-connected AFE with small current ripples. In Proceedings of the 2015 9th International Conference on Power Electronics and ECCE Asia (ICPE-ECCE Asia), Seoul, Korea, 1–5 June 2015; pp. 624–630.

18. Liu, C.; Sun, P.; Lai, J.S.; Ji, Y.; Wang, M.; Chen, C.L.; Cai, G. Cascade dual-boost/buck active-front-end converter for intelligent universal transformer. *IEEE Trans. Ind. Electron.* **2012**, *59*, 4671–4680. [CrossRef]

19. Hiskens, I.A. Dynamics of Type-3 Wind Turbine Generator Models. *IEEE Trans. Power Syst.* **2012**, *27*, 465–474. [CrossRef]

20. Yazdani, A.; Iravani, R. *Voltage-Sourced Converters in Power Systems: Modeling, Control and Applications*; John Wiley & Sons, Inc.: Hoboken, NJ, USA, 2010; pp. 385–412.

21. Orlando, N.A.; Liserre, M.; Mastromauro, R.A.; Dell' Aquila, A. A Survey of Control Issues in PMSG-Based Small Wind-Turbine Systems. *IEEE Trans. Ind. Inform.* **2013**, *9*, 1211–1221. [CrossRef]

22. Salgado-Herrera, N.M.; Medina-Ríos, J.A.; Tapia-Sánchez, R.; Anaya-Lara, O.; Rodríguez-Rodríguez, J.R. DSPWM multilevel technique of 27-levels based on FPGA for the cascaded DC/AC power converter operation. *Int. Trans. Electr. Energy Syst.* **2018**, *28*. [CrossRef]

Article

Improved Droop Control with Washout Filter

Yalong Hu and Wei Wei *

College of Electrical and engineering, Zhejiang University, Hangzhou 310027, China; 11310059@zju.edu.cn
* Correspondence: wwei@zju.edu.cn; Tel.: +86-138-6718-2377

Received: 30 August 2018; Accepted: 10 September 2018; Published: 12 September 2018

Abstract: In this paper, a droop washout filter controller (DWC), composed of a conventional droop controller and a washout filter controller, is proposed. The droop controller is used to ensure the "plug-and-play" capability, and the droop gain is set small. The washout filter is introduced to compensate the active power dynamic performance (APDP). Compared to the droop controller, the DWC can achieve accurate active power sharing and smaller frequency difference without losing the APDP. Additionally, a novel modeling technology is proposed, using which a small-signal model for an island microgrid (MG) is constructed as a singular system. The system's stability is analyzed and the DWC is verified using real-time (RT-LAB) simulation with hardware in the loop (HIL).

Keywords: microgrid (MG); droop control; washout filter; hardware in the loop (HIL)

1. Introduction

Due to the environmental pollution of fossil energy, distributed generators (DGs), such as photovoltaic panels, have attracted great attention and their use is increasing rapidly. To effectively integrate DGs, microgrid (MG) is introduced [1]. In an MG, the DG units, such as photovoltaic panels, are always installed through power electronic units in parallel, which make them adjustable. An MG should remain stable in island mode. The load should be shared by each DG proportionally when an MG operates in island, where all DGs are connected in parallel.

By imitating the operations of synchronous in power system, the droop control strategy is applied to achieve power sharing in an AC MG for its advantages such as no need for communication; however, it also has many disadvantages such as frequency difference and poor reactive power sharing, which many papers have analyzed [2,3]. To solve its defect, a secondary control strategy is widely adopted, which can find the global information of an MG [4]. However, the secondary control always needs additional communication links. When there are no communication links or the communication fails, an MG must operate stably and meet the system needs.

To improve the active power dynamic performance (APDP) of a DG embedded with the droop strategy, various control strategies have been put forward [5–9]. Another DOF (degree of freedom) is added in [5], in which the derivative term is introduced to achieve a better APDP [7]. In [6], the coefficients of the derivative control loop among DGs are set proportionally. The APDP is improved by introducing derivative control with an adaptive coefficient which is small [8]. In [9], the angle and frequency droop control strategies are combined to improve the performance of active power output. A washout, i.e., the lack of low-frequency component of output power, filter control strategy is proposed in [10]. It is actually a band-pass filter (BPF) to restore the voltage and frequency without communication. However, the over dependence on the initial state makes it weak on "plug-and-play". In [11], a secondary controller based on washout filter is proposed which analyzes the parameter setting conditions of the secondary controller. In this paper, a droop washout filter controller (DWC) is proposed which combines the conventional droop controller and the washout filter. Compared with the washout filter controller presented above, the DWC maintains important advantage of "plug-and-play" in the droop control loop. Compared with the controller that only contains droop controller, the droop coefficient could be set smaller, which results in a smaller frequency difference in the steady state, and the APDP can be compensated by the washout filter control loop.

The stability of the MG embedded with the DWC is analyzed using small signal method in this paper. In [12], the entire model of an inverter based MG is established in state-space form, which has been adopted by many articles [11,13–20]. The active load is modeled and its characteristics are analyzed in [13]. In [14], the accurate model of an islanded MG with the phase-locked loop (PLL) is built and discussed. Using the singular perturbation technique, the states of the inductor-capacitor-inductor (LCL) filter and PLL block are divided as fast states, which reduced the system order and calculation burden [15]. In [16], a system with the internal model controller is modeled and discussed. To find the optimal set of proportional parameters in inner controllers and droop gains, an objective function is designed on the basis of the small signal model of an MG using genetic algorithm [17]. An MG which contains current source DGs and voltage source DGs are modeled in [18]. In these articles, a key technology named virtual resistor is used, using which each component in an MG could be modeled together, and it is first presented in [12]. By analyzing the system, we found that the virtual resistor technology works by introducing several poles which are away from the imaginary axis when the virtual resistor value is very large. In this paper, the virtual resistor technology is abandoned and an islanded MG is remodeled as a singular system [21].

The structure of this paper is as follows. The conventional droop controller and the washout filter controller are analyzed in Section 2. In Section 3, the DWC, which combines the droop controller and washout filter controller, is presented. In Section 4, the model of an MG is constructed embedded with the DWC. The stability of an islanded MG, which is composed of two inverters and an impedance load, is analyzed in Section 5. The hardware in the loop (HIL) simulation results are presented in Section 6 to show the validity of the DWC. Finally, the conclusions are summarized in Section 7.

2. Frequency and Voltage Amplitude Deviations Analysis

2.1. Conventional Droop Controller

The line impedances are assumed to be mainly inductive in this paper. The power flow between two nodes can be expressed as:

$$P = \frac{E_1 E_2}{X} \delta \tag{1}$$

$$Q = \frac{E_1 (E_1 - E_2)}{X}, \tag{2}$$

where E_1 and E_2 are the voltage amplitudes, δ is the phase angle difference, and X is the line impedance. From the two equations, it can be informed that the real power is proportional to δ; and the reactive power is determined by the difference between E_1 and E_2 with fixed line impedance.

Using the dq theory, the instantaneous output power p and q, are given by

$$p = v_{od} i_{od} + v_{oq} i_{oq} \tag{3}$$

$$q = v_{od} i_{oq} - v_{oq} i_{od}. \tag{4}$$

The power controller received the measured output power through a low pass filter (LPF), which can be expressed as:

$$P = \frac{\omega_f}{s + \omega_f} p \tag{5}$$

$$Q = \frac{\omega_f}{s + \omega_f} q, \tag{6}$$

where ω_f is the cut-off frequency.

The conventional droop control scheme can be expressed as:

$$\omega = \omega^* - m_p(P - P^*) \tag{7}$$

$$V = V^* - n_q(Q - Q^*), \tag{8}$$

where ω and V are the angular velocity and amplitude of the output voltage, ω^* and V^* are the reference values, P^* and Q^* are the reference values of active and reactive power, and m_p and n_q are the active and reactive power droop coefficients, respectively.

The active power output is adjusted by the frequency. However, it is directly determined by the phase angular and not the frequency from Equation (1). The relation between them is

$$\Delta\delta = \frac{1}{s}\Delta\omega. \tag{9}$$

From Equation (8), it can be seen that the amplitude difference is necessary for reactive power sharing. However, $\Delta\omega$ is not necessary theoretically. The relationship between them in the steady state can be expressed as:

$$\Delta\omega = -m_p\Delta P \tag{10}$$

$$\Delta V = -n_q\Delta Q, \tag{11}$$

where $\Delta V = V - V^*$, $\Delta P = P - P^*$ and $\Delta Q = Q - Q^*$.

2.2. Washout Filter

The washout filter controller is a BPF without the low frequency component and can eliminate the frequency and amplitude deviations of output voltage in theory. As explained in [10], the control mechanism of the washout filter controller could be expressed as:

$$\omega = \omega^* - m_p \cdot \frac{s}{s + \omega_h} \cdot (P - P^*) \tag{12}$$

$$V = V^* - n_q \cdot \frac{s}{s + \omega_h} \cdot (Q - Q^*), \tag{13}$$

where ω_h is the cut-frequency of the high pass filter (HPF). By examining Equations (5), (6), (12) and (13), it can be derived that the following equation should be satisfied.

$$\omega_h < \omega_f. \tag{14}$$

From Equations (5), (6), (12) and (13), the washout filter is a BPF indeed, as explained in [11]. When a load change happens, the selected frequency signals can be used by each DG to adjust its output power which is proportional to the droop coefficients m_p and n_q. To remove the low-frequency signals, the last parts of Equations (12) and (13) should be zero theoretically in the steady state. Compared to the droop controller, the washout filter will not lead to frequency and amplitude deviation. It can achieve active power sharing at the same time. However, the lack of low-frequency signals leads to its inability to "plug and play", which is shown in Section 6.

Equation (2) shows that the reactive power sharing relies on voltage amplitudes deviation between different nodes. If the amplitude restores to the rated value for each DG, there will be no amplitude deviation between each DG, and the reactive power output will be inversely proportional to the line impedance value if the network topology is star [22,23], which indicates that the washout filter controller does nothing on reactive power sharing in the steady state.

3. Proposed Control Strategy

As explained in Section 2, the conventional droop can achieve power sharing and "plug and play" but with frequency and amplitude differences of the output voltage. The washout filter can easily eliminate the frequency and amplitude differences, but with poor reactive power sharing and cannot realize "plug and play". Inspired by this, a novel control strategy which combines the two control strategies together is proposed in this part, as shown in Figure 1. The droop control loop consists of an LPF, which makes it maintain the "plug and play" advantage, and the washout filter control loop aims to compensate the APDP caused by the small droop coefficient. Thus, the control scheme of the DWC can be derived as:

$$\omega = \omega^* - m_l \cdot (P_1 - P^*) - m_h \cdot \frac{s}{s + \omega_h} \cdot (P_2 - P^*), \tag{15}$$

where $P_1 = \frac{\omega_{l1}}{s + \omega_{l1}} p$, $P_2 = \frac{\omega_{l2}}{s + \omega_{l2}} p$, m_l is the droop coefficient, and m_h is the washout filter coefficient. The DWC consists of two independent frequency bands whose frequency characteristics are shown in Figure 2. It should be noted that there is no size relationship between ω_{l1} and ω_h. When $\omega_h = \omega_{l1}$ and $\omega_{l2} \gg \omega_{l1}$, the DWC degenerates to a PD (proportional–differential) control method [6,9], where the differential coefficient is $\frac{m_h}{\omega_h}$.

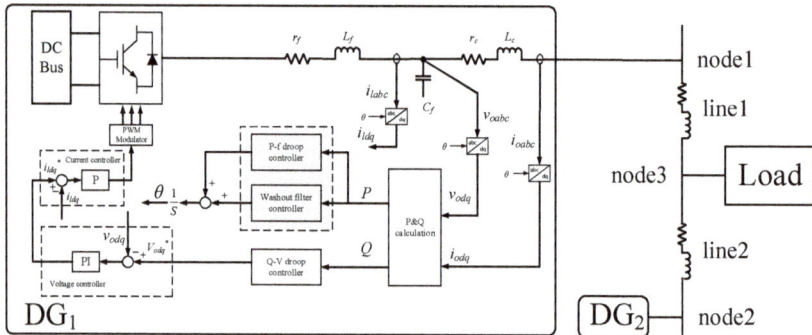

Figure 1. The study microgrid (MG) and the droop washout filter controller (DWC). DC: direct-current; PWM: pulse width modulation; PI: proportional integral.

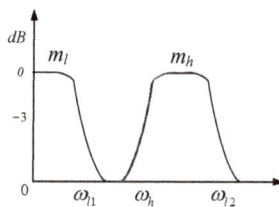

Figure 2. Frequency characteristics of the DWC.

Regardless of the difference between ω_{l1} and ω_{l2} and combined with Equation (1), Equation (15) can be rewritten as:

$$\omega - \omega^* = -m_l \cdot (P_1 - P^*) - \frac{1}{\omega_h} \cdot \frac{d\omega}{dt} - \frac{m_l + m_h}{\omega_h} \cdot \frac{E_1 E_2}{X} \cdot (\omega - \omega_g), \tag{16}$$

where ω_g is the frequency of the MG. Comparing Equation (16) and the virtual synchronous generator control equation in [24], which can be expressed as:

$$\omega - \omega^* = -m_l \cdot (P_{out} - P^*) - J m_l \omega \frac{d\omega}{dt} - D m_l \cdot (\omega - \omega_g), \tag{17}$$

where J is the virtual inertia and D is the damping factor, it can be derived that

$$\begin{cases} J = \frac{1}{\omega m_l \omega_h} \\ D = \frac{1+\frac{m_h}{m_l}}{\omega_h} \cdot \frac{E_1 E_2}{X} \end{cases} . \tag{18}$$

Thus, the APDP of the DWC can be adjusted by tuning ω_h and m_h [25].

With the analysis above, the correspondence between different control strategies can be summarized as:

1. The "P-f" droop control strategy is equivalent to $J = 0$ and $D = 0$ [24].
2. The "PD" control strategy [5–9] is equivalent to $J = 0$ with $D = \frac{m_h}{m_l} \cdot \frac{E_1 E_2}{X}$.
3. The washout filter control [10] is equivalent to removing the parameter m_l.
4. The DWC is equivalent to the "PD" controller with $\omega_h = \omega_{l1}$ and $\omega_{l2} \gg \omega_{l1}$.
5. The DWC is equal to the virtual synchronous generator control strategy with $\omega_{l2} = \omega_{l1}$.

Since the washout filter cannot reduce amplitude difference, the reactive power controller adopts the conventional $Q - V$ droop controller as follows:

$$V = V^* - n_l \cdot (Q - Q^*), \tag{19}$$

where $Q = \frac{\omega_{l1}}{s+\omega_{l1}} q$, and n_l is the reactive power control loop coefficient.

The coefficient m_l of each inverter should be set as [26]:

$$m_{l,i}\Delta P_i = m_{l,j}\Delta P_j. \tag{20}$$

Considering the APDP of the inverters, the same rule applies to coefficient m_h

$$m_{h,i}\Delta P_i = m_{h,j}\Delta P_j. \tag{21}$$

Since the frequency difference in the steady state has nothing to do with washout filter, it only relies on the droop controller. To reduce the frequency difference in the steady state, the value of m_l should be set smaller compared to the conventional droop controller. However, small droop gain always leads to slow dynamic adjustment process, which is not desired. As analyzed above, the DWC can be seen as a virtual synchronous generator controller, so the APDP can be compensated by regulating ω_h and m_h.

The inner control loop includes a voltage controller which is a PI (proportional–integral) regulator and a current controller which is a P regulator.

4. Small Signal Model

The small signal model of an islanded MG embedded with the DWC is constructed as a singular system in this section. Based on the model, the stability is analyzed. Before modeling, some symbols need to be defined. Suppose the system has "m" nodes, "s" inverters, "n" lines, and "p" load points [12]. The complete model consists of a differential algebraic part and an algebraic part. Since the differential equations has been discussed by many articles [4,12–15], the details of some matrices are given in Appendix A and not repeat in each section.

4.1. Differential Algebraic Equations

4.1.1. Load and Network Models

The state equations of loads which are general resistive and inductive loads are expressed as follows:

$$E_{LOAD}[\Delta \dot{i}_{loadDQ}] = A_{LOAD}[\Delta i_{loadDQ}] + B_{1LOAD}[\Delta v_{bDQ}] + B_{2LOAD}[\Delta \omega]$$
$$[\Delta i_{loadDQ}] = C_{LOAD}[\Delta i_{loadDQ}]$$
$$[\Delta v_{bDQ}] = \begin{bmatrix} \Delta v_{bDQ,1} & \Delta v_{bDQ,2} & \cdots & \Delta v_{bDQ,m} \end{bmatrix}, \tag{22}$$

where E_{LOAD} is a unit matrix of 2p-dimensions and C_{load} is a unit matrix of 2p-dimensions, too. The model of the network can be represented by the following equations:

$$E_{NET}[\Delta \dot{i}_{lineDQ}] = A_{NET}[\Delta i_{lineDQ}] + B_{1NET}[\Delta v_{bDQ}] + B_{2NET}[\Delta \omega]$$
$$[\Delta i_{lineDQ}] = C_{line}[\Delta i_{lineDQ}], \tag{23}$$

where E_{NET} and C_{line} are two unit matrixes of 2n-dimensions.

4.1.2. Singular Inverter Model

Power Controller: To model the power controller, Equation (15) needs to be rewritten as:

$$\Delta \omega = \Delta \omega_1 + \Delta \omega_2, \tag{24}$$

where $\Delta \omega_1 = -m_l \cdot \Delta P_1$, and $\Delta \omega_2 = -m_h \cdot \frac{s}{s+\omega_h} \cdot \Delta P_2$. Combining with Equation (9), the relations between the power angles and active power variables can be expressed as [27]:

$$\begin{cases} \Delta \delta_1 = -\frac{m_l}{s} \cdot \Delta P_1 \\ \Delta \delta_2 = -m_h \cdot \frac{1}{s+\omega_h} \cdot \Delta P_2 \end{cases}. \tag{25}$$

By linearizing Equations (19) and (25), the model of the power controller can be expressed as:

$$E_P[\Delta \dot{x}_p] = A_P[\Delta x_p] + B_P \begin{bmatrix} \Delta i_{ldq} \\ \Delta v_{odq} \\ \Delta i_{oDQ} \end{bmatrix} + B_{P\omega com}[\Delta \omega_{com}]$$
$$[\Delta v_{odq}^*] = C_{PV}[\Delta x_p] \tag{26}$$
$$[\Delta \omega] = C_{P\omega}[\Delta x_p]$$

In Equation (26),

$$[\Delta x_p] = \begin{bmatrix} \Delta \delta_1 & \Delta_2 & \Delta P_1 & \Delta P_2 & \Delta Q \end{bmatrix}^T. \tag{27}$$

Voltage and Current Controllers: As the voltage controller is a PI regulator and the current is a P regulator, they are formulated together for convenience:

$$E_{VC}[\Delta \dot{\varnothing}_{dq}] = [0][\Delta \varnothing_{dq}] + B_{VC1}[\Delta v_{odq}^*] + B_{VC2} \begin{bmatrix} \Delta i_{ldq} \\ \Delta v_{odq} \\ \Delta i_{oDQ} \end{bmatrix}$$
$$[\Delta v_{idq}^*] = C_{VC}[\Delta \varnothing_{dq}] + D_{VC1}[\Delta v_{odq}^*] + D_{VC2} \begin{bmatrix} \Delta i_{ldq} \\ \Delta v_{odq} \\ \Delta i_{oDQ} \end{bmatrix}, \tag{28}$$

LCL Filter: The LCL filter can be modeled as follows:

$$E_{LCL}[\Delta \dot{x}_{lcl}] = A_{LCL}[\Delta x_{lcl}] + B_{LCL1}[\Delta v_{idq}] + B_{LCL2}[\Delta v_{bDQ}] + B_{LCL3}[\Delta \omega] \tag{29}$$

In Equation (29),

$$[\Delta x_{lcl}] = \begin{bmatrix} \Delta i_{ldq} & \Delta v_{odq} & \Delta i_{odq} \end{bmatrix}^T. \tag{30}$$

Common Frame Transformation: For the convenience of system modeling, the small signal model of each inverter can be built separately. Every DG's dq transformation is on its local reference frame $(d - q)$. However, the output variables of each component should be converted to the common reference frame $(D - Q)$ to construct a whole system, and the transformation equations for these variables could be written as [12]:

$$[\Delta i_{oDQ}] = T_S \begin{bmatrix} \Delta i_{odq} \end{bmatrix} + T_C[\Delta \delta] \tag{31}$$

$$\begin{bmatrix} \Delta v_{bdq} \end{bmatrix} = T_S^{-1}[\Delta v_{oDQ}] + T_V^{-1}[\Delta \delta]. \tag{32}$$

Complete Model of an Individual Inverter: The complete model of an individual inverter consists of the circuit part and the controller part which has 13 state variables. The complete model of an inverter can be expressed as:

$$E_{INVi}[\Delta \dot{x}_{invi}] = A_{INVi}[\Delta x_{invi}] + B_{INVi}[\Delta v_{bDQi}] + B_{i\omega com}[\Delta \omega_{com}]$$
$$\begin{bmatrix} \Delta \omega \\ \Delta i_{oDQi} \end{bmatrix} = \begin{bmatrix} C_{INVwi} \\ C_{INVci} \end{bmatrix} [\Delta x_{invi}], \tag{33}$$

where

$$[\Delta x_{invi}] = \begin{bmatrix} \Delta x_{pi} & \Delta \varnothing_{dqi} & \Delta x_{lcli} \end{bmatrix}^T. \tag{34}$$

4.1.3. Combined Model of All Inverters

As an individual model of an inverter has been built, the combined model of all inverters in an islanded MG can be expressed as follows:

$$E_{INV}[\Delta \dot{x}_{INV}] = A_{INV}[\Delta x_{INV}] + B_{INV}[\Delta v_{bDQ}]$$
$$[\Delta i_{oDQ}] = C_{INVc}[\Delta x_{INV}], \tag{35}$$

where

$$[\Delta x_{INV}] = \begin{bmatrix} \Delta x_{inv,1} & \Delta x_{inv,2} & \cdots & \Delta x_{inv,s} \end{bmatrix}^T. \tag{36}$$

4.2. Algebraic Equations

Using Kirchhoff's current law (KCL) for each node, it is easy to derive that

$$[0][\Delta v_{bDQ,i}] = [\Delta i_{oDQ,i}] - [\Delta i_{loadDQ,i}] + \sum_{j \in G} [\Delta i_{lineDQ,ji}], \tag{37}$$

where G is a set which contains the nodes connected to node i, and $\Delta v_{bDQ,i}$ are algebraic variable.

Applying this relation to all nodes in the system, it can be obtained that

$$M_{INV}[\Delta i_{oDQ}] + M_{NET}[\Delta i_{lineDQ}] + M_{LOAD}[\Delta i_{loadDQ}] = 0, \tag{38}$$

where M_{INV}, M_{NET} and M_{LOAD} are the mapping matrix of the network structure of the system and are detailed defined in [12].

4.3. Complete Microgrid (MG) Model

By combing all of the inverters, loads, distribution lines and the relations of coupling states, the complete state matrix of an MG can be obtained as:

$$E_{sys}\left[\Delta \dot{x}_{sys}\right] = A_{sys}\left[\Delta x_{sys}\right], \tag{39}$$

where

$$\left[\Delta x_{sys}\right] = \left[\begin{array}{cccc} \Delta x_{INV} & \Delta i_{lineDQ} & \Delta i_{loadDQ} & \Delta v_{bDQ} \end{array}\right] \tag{40}$$

$$E_{sys} = \begin{bmatrix} E_{INV} & 0 & 0 & 0 \\ 0 & E_{NET} & 0 & 0 \\ 0 & 0 & E_{LOAD} & 0 \\ 0 & 0 & 0 & 0_{2m \times 2m} \end{bmatrix}$$

$$A_{sys} = \begin{bmatrix} A_{INV} & 0 & 0 & B_{INV} \\ B_{2NET}C_{INV\omega} & A_{NET} & 0 & B_{1NET} \\ B_{2LOAD}C_{INV\omega} & 0 & A_{LOAD} & B_{1LOAD} \\ M_{INV}C_{INVc} & M_{NET}C_{line} & M_{LOAD}C_{LOAD} & 0_{2m \times 2m} \end{bmatrix}. \tag{41}$$

The matrix E_{sys} is singular obviously, which indicates that the system is a singular system.

5. Stability Analysis

The small signal model built in Section 4 is singular. To analyzed the system, a determinant is defined as:

$$\Delta(s) := \left|sE_{sys} - A_{sys}\right|. \tag{42}$$

The stability of the system is determined by the roots of $\Delta(s)$. A simple method to observe the stability of the system is checking whether the real parts of all its finite eigenvalues are negative. If all are negative, the system is stable [28]. The pencil $(sE_{sys} - A_{sys})$ is regular when $\Delta(s)$ is not identically zero. The model of an MG is always regular since it is a physical dynamical system [29].

In this paper, DG$_1$ is chosen as the common reference frame, so $\Delta\delta_1$ and $\Delta\delta_2$ in DG$_1$ are ignored in calculation [15]. The eigenvalues can be easily calculated with function "eig(A,B)" in MATLAB (MathWorks, Natick, MA, USA), where "A" corresponds to A_{sys}, and "B" corresponds to E_{sys}.

In this section, the MG shown in Figure 1 is analyzed. Its complete model is constructed using the procedure shown in Section 4. The eigenvalues of the system can be found using the method described above. The parameters of the MG are shown in Table 1, and the steady points, which are measured from a MATLAB/SIMULINK simulation, of the MG are summarized in Table 2.

Table 1. Test System Parameters.

Parameter	Value	Parameter	Value
V_{DC}	800 V	f_s	5 kHz
m_{l1}	6.3×10^{-6}	V_g	380 V (line-line)
m_{h1}	6×10^{-5}	r_{L_f}	0.1 Ω
n_{l1}	0.001	L_f	1 mH
ω_{l1}	$20\,\pi$ rad/s	C_f	800 μF
ω_{l2}	$60\,\pi$ rad/s	r_{L_c}	0.03 Ω
ω_h	$40\,\pi$ rad/s	L_c	0.3 mH
K_{pv}	0.1	ω_0	50 Hz
K_{iv}	150	$r_{line,1}$	0.12 Ω
K_{pc}	0.002	$L_{line,1}$	1.2 mH
R_{load}	10 Ω	$r_{line,2}$	0.08 Ω
L_{load}	5 mH	$L_{line,2}$	0.8 mH

Table 2. Initial Conditions.

Parameter	Value	Parameter	Value
V_{od}	(10.0 309.5)	V_{oq}	(0 0)
I_{od}	(20.0 10.1)	I_{oq}	(−1.2 −4.2)
I_{ld}	(20.0 10.1)	I_{lq}	(76.6 73.2)
V_{bd}	(309.6 309.2 307.5)	V_{bq}	(29.6 27.6 23.9)
I_{lined}	(20.0 10.1)	I_{lineq}	(−0.81 −3.4)
δ_0	(−0.2 0)	I_{loadq}	(30.3 −2.6)

The root loci of the system are shown in Figure 3. Figure 3a is the eigenvalues for the parameters given in Table 2. The eigenvalues labeled Load are mainly affected by the load parameters. The modes labeled Inner controllers are sensitive to the voltage and current controllers. In the group labeled LCL filter, the modes are associated with the LCL filter parameters. The modes shown in Power controller group are sensitive to the power controller parameters. The modes labeled Washout filter are associated with the frequency parameters of the washout filter. Since the parameter ω_{l2} has little effect on the dominant poles, the root locus is not shown here.

Figure 3b shows the trajectory of the three dominant poles as the droop coefficients m_{l1} changes from 6×10^{-6} to 6×10^{-4}, where m_{l2} is always two times of m_{l1}. The influence of λ_1 and λ_2 is easily analyzed. λ_3 could be seen as an inertial link. When droop coefficients increase, the eigenvalues λ_1 and λ_2 move towards vertical axis and the eigenvalue λ_3 moves away from it, which improve the dynamic performance of the system, but oversized droop coefficients cause system oscillation and instability.

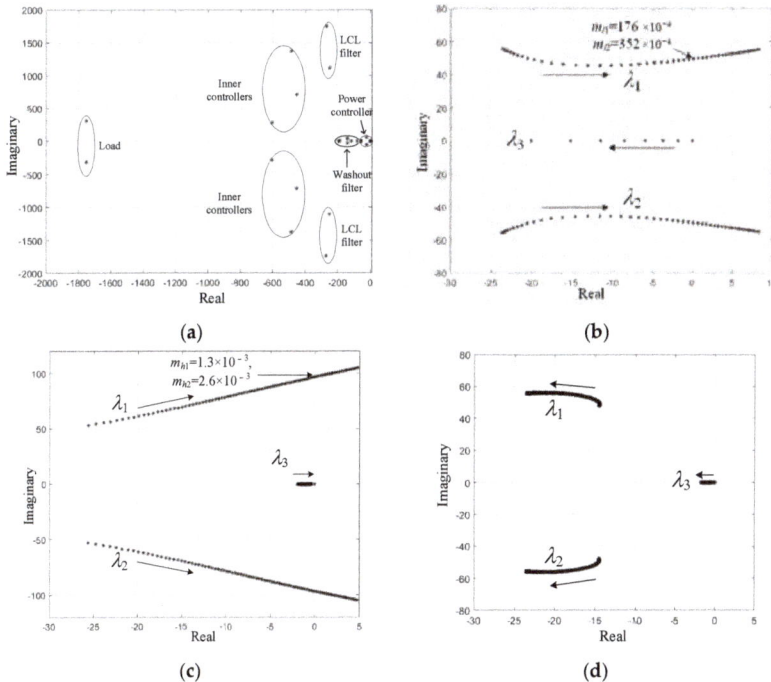

Figure 3. Dominant root locus of the system: (**a**) eigenvalue spectrum of MG state matrix; (**b**) dominant root locus as $m_{l1} \in [6 \times 10^{-6}, 6 \times 10^{-4}]$; (**c**) dominant root locus as $m_{h1} \in [0, 1.4 \times 10^{-3}]$; and (**d**) dominant root locus as $\omega_h \in [0, 125.6]$. LCL: inductor-capacitor-inductor.

Figure 3c shows the root locus of m_{h1} as it moves from 0 to 1.4×10^{-3}, where m_{h2} is always two times m_{h1}. When m_{h1} increases, the three dominate poles move close to the imaginary axis. On the one hand, λ_1 and λ_2 increase the dynamic performance of the system but make it more oscillatory. On the other hand, λ_3 suppresses the oscillation but increases the adjustment time of the system. To improve the dynamic performance, λ_1 and λ_2 should be set close to the imaginary axis.

Figure 3d shows the root locus of ω_h as it moves from 0 to 125.6, which results in the dominant poles moving away from the imaginary axis. Considering Figure 3c,d, the λ_1 and λ_2 could have more flexible assignment by adjusting parameters ω_h and m_h. Thus, the larger ω_h is, the faster the dynamic performance is. However, considering the impact on the inner control loop, it is set as 40π in this paper. There is another design shown in the next section. In that design, ω_h should be set very small, and m_h is several times m_l.

To sum up, the droop coefficient m_l could be set smaller compared to the conventional principle to reduce the frequency difference in the steady state, and the washout filter could be added to compensate the APDP.

6. Real Time Simulation Results

The DWC is verified in real-time simulations with HIL [30]. The MG shown in Figure 1 is simulated in RT-LAB (Opal-RT, Montreal, QC, Canada) and the controllers run in STM32F407 MCUs (STMicroelectronics, Geneva, Switzerland). The real-time simulation apparatus is shown in Figure 4.

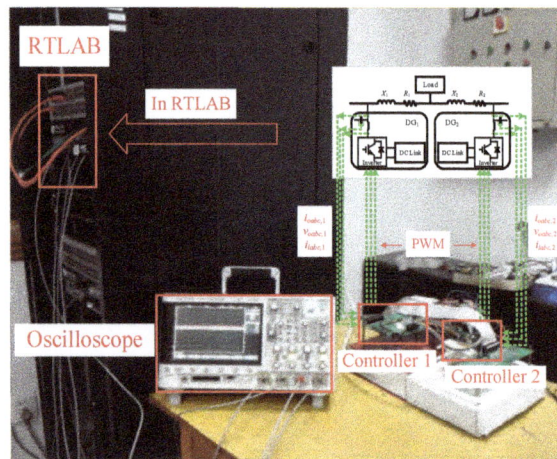

Figure 4. Hardware in the loop (HIL) simulation setup for the study system.

6.1. Performance Comparison with the Conventional Droop Controller

To verify the DWC on APDP and frequency deviation compared to the conventional droop control strategy, the performances of the two control strategies are compared in this section. Different frequency ranges are selected of the washout filter to shown the performance of the DWC. The active power performance is presented in Figure 5 and the corresponding frequency performance is presented in Figure 6, which is defined as:

$$\Delta\omega = \omega_0 - \omega. \tag{43}$$

The parameters are listed in Table 1. The droop coefficients of the active power control loop for the two DGs are set as 6.3×10^{-6} and 12.6×10^{-6} in the three cases. Initially, the MG is in steady state. At $t = 1.5$ s, a resistive load of 15 kW is connected to the MG. Figure 5a shows the performance of droop control method. The droop coefficients in Figure 5b is two times of Figure 5a. Figure 5c shows the performance in which the washout filters coefficients ω_h and ω_{l2} are set as 40π and 60π, m_{h1} and

m_{h2} are set as 5×10^{-4} and 1×10^{-3}, respectively. In Figure 5d, ω_h and ω_{l2} are set as 0.4π and 20π and m_{h1} and m_{h2} are set as 1.9×10^{-5} and 3.8×10^{-5}. The results are listed in Table 3. In Figure 5a,b, it can be seen that the APDP can be improved by increasing the droop coefficient, but it also increases the frequency deviation in Figure 6a,b.

Table 3. Dynamic performance.

Figures 5 and 6	a	b	c	d
Adjustment Time (ms)	600	350	350	300
Overshoot (%, DG$_2$)	23	15	10	15
Frequency deviation (rad)	0.08	0.16	0.08	0.08

In Figure 5c, the washout filter is designed in high-frequency band. As analyzed in Section 5, the eigenvalues are close to the imaginary axis in this design, which means the dynamic performance of the system is fast. In Figure 5d, the washout filter is designed in low-frequency band. Although the eigenvalue distribution is similar, the virtual inertia and damping factor are bigger compared with Figure 5c. Figure 6c,d shows the corresponding frequency dynamic performances. In Figure 6c, the frequency changes rapidly, as parameter ω_h is large (small virtual inertia). Conversely, the frequency changes slowly in Figure 6d, as ω_h is small (large virtual inertia). In Figures 5 and 6, it can be seen that the system using he DWC can achieve a better APDP with the same frequency deviation in the steady state compared to the droop controller.

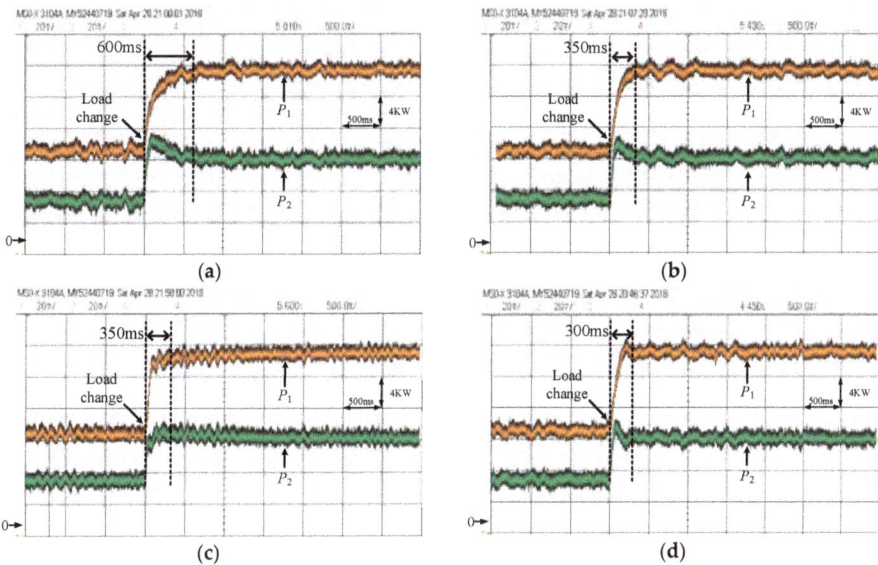

Figure 5. Active power sharing performance: (**a**) droop controller with $m_{l1} = 6.3 \times 10^{-6}$, $m_{l2} = 1.26 \times 10^{-5}$; (**b**) droop controller with $m_{l1} = 1.26 \times 10^{-5}$, $m_{l2} = 2.52 \times 10^{-5}$; (**c**) DWC with large ω_h; and (**d**) DWC with small ω_h.

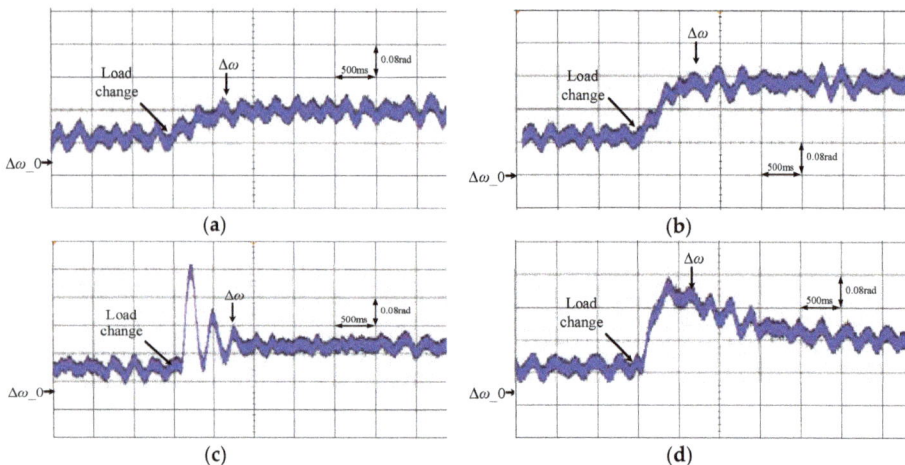

Figure 6. Frequency deviations performance: (**a**) droop controller with $m_{l1} = 6.3 \times 10^{-6}$, $m_{l2} = 1.26 \times 10^{-5}$; (**b**) droop controller with $m_{l1} = 1.26 \times 10^{-5}$, $m_{l2} = 2.52 \times 10^{-5}$; (**c**) DWC with large ω_h; and (**d**) DWC with small ω_h.

6.2. Performance Comparison with the Washout Filter Controller

The "plug-and-play" capabilities of the washout filter controller and the DWC are shown in Figures 7 and 8. The parameters of the washout filter are $\omega_h = 0.4\pi$ and $\omega_{l2} = 20\pi$. The coefficients are $m_{l1} = 1 \times 10^{-6}$, $m_{l2} = 2 \times 10^{-6}$, $m_{h1} = 2 \times 10^{-6}$, and $m_{h2} = 4 \times 10^{-6}$.

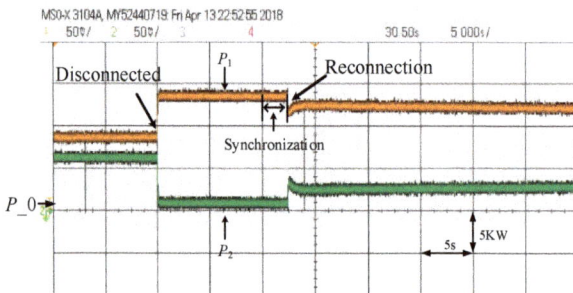

Figure 7. Performance of the washout controller under feeder disturbance.

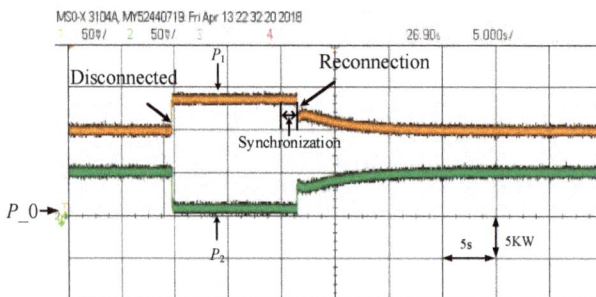

Figure 8. Performance of the DWC under feeder disturbance.

In Figure 7, there is only a washout filter control loop in the power controller. From $t = 0$ s to $t = 10$ s, the system is under steady state. The output power of the two DGs are 9 kW and 6 kW, respectively, the ratio of which does not meet the designed 2:1. At $t = 10$ s, DG$_2$ is disconnected from the MG and, at $t = 20$ s, the synchronization process starts. When the angular and voltage amplitude meets the requirements, DG$_2$ is reconnected to the MG. At this moment, the active power outputs are 13 kW and 2 kW, respectively. The operating point does not meet the requirement and is different with the operating point from 0 s to 10 s. Figure 7 shows the washout filter is weak on "plug-and-play" for the lack of static component. Figure 8 shows that the output power of the two DGs are 10 kW and 5 kW from 0 s to 10 s and from 30 s to 50 s, respectively, which indicates the sharing is accurate both before the line disconnection and after the line reconnection. Figure 8 verifies the "plug-and-play" capabilities of the DWC.

7. Conclusions

In this paper, an improved droop control strategy that combines the droop controller and washout filter controller is proposed, obtaining the advantages of both. The DWC can achieve "plug-and-play" with the droop control loop, and the washout filter controller is used to improve the APDP. The droop gains could be small to reduce the frequency difference in the steady state, and the washout filter is to compensate the APDP loss in this paper. In addition, a complete singular small signal model of an MG using the DWC is rebuilt. Using the singular model, the stability of the system is discussed. Finally, an MG embedded with the DWC has been tested with HIL, which demonstrates the effectiveness of the DWC on frequency deviation and APDP.

Author Contributions: Conceptualization, Y.H.; Methodology, Y.H.; Software, Y.H.; Validation, Y.H.; Formal Analysis, Y.H.; Investigation, Y.H.; Resources, W.W.; Data Curation, Y.H.; Writing-Original Draft Preparation, Y.H.; Writing-Review & Editing, W.W.; Visualization, W.W.; Supervision, W.W.; Project Administration, W.W.; Funding Acquisition, W.W.

Funding: This research was funded by National Key Research and Development Program of China (2016YFB0900500); State Grid Corporation Science and Technology Project (SGZJ0000BGJS1600312); Key Research Program of Zhejiang province (2017C01039).

Conflicts of Interest: The authors declare no conflict of interest.

Appendix A

The state space matrices of the load model are described here.

$$A_{LOAD} = \begin{bmatrix} A_{load,1} & 0 & \cdots & 0 \\ 0 & A_{load,2} & \cdots & 0 \\ \vdots & \vdots & \ddots & \vdots \\ 0 & 0 & \cdots & A_{load,p} \end{bmatrix}_{2p \times 2p}$$

$$B_{1LOAD} = \begin{bmatrix} B_{1load,1} \\ B_{1load,2} \\ \vdots \\ B_{1load,p} \end{bmatrix}_{2p \times 2m}$$

$$B_{2LOAD} = \begin{bmatrix} B_{2load,1} \\ B_{2load,2} \\ \vdots \\ B_{2load,p} \end{bmatrix}_{2p \times 2m}$$

$$A_{load,i} = \begin{bmatrix} \frac{-R_{load,i}}{L_{load,i}} & \omega_0 \\ -\omega_0 & \frac{-R_{load,i}}{L_{load,i}} \end{bmatrix}$$

$$B_{1load,i} = \begin{bmatrix} \cdots & \frac{1}{L_{load,i}} & 0 & \cdots \\ \cdots & 0 & \frac{1}{L_{load,i}} & \cdots \end{bmatrix}$$

$$B_{2load,i} = \begin{bmatrix} I_{loadQi} \\ -I_{loadDi} \end{bmatrix}$$

State space matrices of the network are listed below.

$$A_{NET} = \begin{bmatrix} A_{NET,1} & 0 & \cdots & 0 \\ 0 & A_{NET,2} & \cdots & 0 \\ \vdots & \vdots & \ddots & \vdots \\ 0 & 0 & \cdots & A_{NET,n} \end{bmatrix}_{2n \times 2m}$$

$$B_{1NET} = \begin{bmatrix} B_{1NET,1} \\ B_{1NET,2} \\ \vdots \\ B_{1NET,p} \end{bmatrix}_{2p \times 2m}$$

$$B_{2NET} = \begin{bmatrix} B_{2NET,1} \\ B_{2NET,2} \\ \vdots \\ B_{2NET,n} \end{bmatrix}_{2n \times 1}$$

$$A_{NET,i} = \begin{bmatrix} \frac{-r_{line,i}}{L_{line,i}} & \omega_0 \\ -\omega_0 & \frac{-r_{line,i}}{L_{line,i}} \end{bmatrix}$$

$$B_{1NET,i} = \begin{bmatrix} \cdots & \frac{1}{L_{line,i}} & 0 & \cdots & \frac{-1}{L_{line,i}} & 0 & \cdots \\ \cdots & 0 & \frac{1}{L_{line,i}} & \cdots & 0 & \frac{-1}{L_{line,i}} & \cdots \end{bmatrix}_{2 \times 2m}$$

$$B_{2NET,i} = \begin{bmatrix} I_{lineQi} \\ -I_{lineDi} \end{bmatrix}$$

The state space matrices of the power controllers are listed as follows. E_p is a unit matrix of five dimensions.

$$A_P = \begin{bmatrix} 0 & 1 & -m_l & 0 & 0 \\ 0 & -\omega_h & 0 & -m_h & 0 \\ 0 & 0 & -\omega_{l1} & 0 & 0 \\ 0 & 0 & 0 & -\omega_{l2} & 0 \\ 0 & 0 & 0 & 0 & -\omega_{l1} \end{bmatrix}$$

$$B_P = \begin{bmatrix} 0_{5 \times 2} & B_{p1} & B_{p2} \end{bmatrix}$$

$$B_{P1} = \begin{bmatrix} 0 & 0 \\ 0 & 0 \\ \omega_{l1}I_{od} & \omega_{l1}I_{oq} \\ \omega_{l2}I_{od} & \omega_{l2}I_{oq} \\ -\omega_{l1}I_{oq} & \omega_{l1}I_{od} \end{bmatrix}$$

$$B_{P2} = \begin{bmatrix} 0 & 0 \\ 0 & 0 \\ w_{l1}V_{od} & w_{l1}V_{oq} \\ w_{l2}V_{od} & w_{l2}V_{oq} \\ w_{l1}V_{oq} & -w_{l1}V_{od} \end{bmatrix}$$

$$B_{P\omega com} = \begin{bmatrix} -1 & 0 & 0 & 0 & 0 \\ 0 & -1 & 0 & 0 & 0 \end{bmatrix}^{T}$$

$$C_{PV} = \begin{bmatrix} 0 & 0 & 0 & 0 & -n_q \\ 0 & 0 & 0 & 0 & 0 \end{bmatrix}$$

$$C_{P\omega} = \begin{bmatrix} 0 & 0 & -m_l & 0 & 0 \\ 0 & -w_h & 0 & -m_h & 0 \end{bmatrix}$$

The state space matrices of the voltage and current controllers are listed as follows. E_{VC} is a unit matrix of two dimensions.

$$B_{VC1} = \begin{bmatrix} 1 & 0 \\ 0 & 1 \end{bmatrix}$$

$$B_{VC2} = \begin{bmatrix} 0 & 0 & -1 & 0 & 0 & 0 \\ 0 & 0 & 0 & -1 & 0 & 0 \end{bmatrix}$$

$$C_{VC} = \begin{bmatrix} K_{pc}K_{iv} & 0 \\ 0 & K_{pc}K_{iv} \end{bmatrix}$$

$$D_{VC1} = \begin{bmatrix} K_{pc}K_{pv} & 0 \\ 0 & K_{pc}K_{pv} \end{bmatrix}$$

$$D_{VC2} = \begin{bmatrix} -1 & 0 & -K_{pc} & 0 & 0 & 0 \\ 0 & -1 & 0 & -K_{pc} & 0 & 0 \end{bmatrix}$$

The state space matrices of the output LCL filter model are presented here. E_{lcl} is a unit matrix of six dimensions.

$$A_{LCL} = \begin{bmatrix} -\frac{r_{Lf}}{L_f} & w_0 & -\frac{1}{L_f} & 0 & 0 & 0 \\ -w_0 & -\frac{r_{Lf}}{L_f} & 0 & -\frac{1}{L_f} & 0 & 0 \\ \frac{1}{c_f} & 0 & 0 & w_0 & -\frac{1}{c_f} & 0 \\ 0 & \frac{1}{c_f} & -w_0 & 0 & 0 & -\frac{1}{c_f} \\ 0 & 0 & -\frac{1}{L_c} & 0 & -\frac{r_{Lc}}{L_c} & w_0 \\ 0 & 0 & 0 & -\frac{1}{L_c} & -w_0 & -\frac{r_{Lf}}{L_f} \end{bmatrix}$$

$$B_{LCL1} = \begin{bmatrix} \frac{1}{L_f} & 0 & 0 & 0 & 0 & 0 \\ 0 & \frac{1}{L_f} & 0 & 0 & 0 & 0 \end{bmatrix}^{T}$$

$$B_{LCL2} = \begin{bmatrix} 0 & 0 & 0 & 0 & -\frac{1}{L_c} & 0 \\ 0 & 0 & 0 & 0 & 0 & -\frac{1}{L_c} \end{bmatrix}^{T}$$

$$B_{LCL3} = \begin{bmatrix} I_{lq} & -I_{ld} & V_{oq} & -V_{od} & I_{oq} & -I_{od} \\ I_{lq} & -I_{ld} & V_{oq} & -V_{od} & I_{oq} & -I_{od} \end{bmatrix}^{T}$$

The state space matrices of the reference frame transformation are presented here.

$$T_S = \begin{bmatrix} \cos(\delta_0) & -\sin(\delta_0) \\ \sin(\delta_0) & \cos(\delta_0) \end{bmatrix}$$

$$T_C = \begin{bmatrix} -I_{od}\sin(\delta_{10}) - I_{oq}\cos(\delta_{10}) & -I_{od}\sin(\delta_{20}) - I_{oq}\cos(\delta_{20}) \\ I_{od}\cos(\delta_{10}) - I_{oq}\sin(\delta_{10}) & I_{od}\cos(\delta_{20}) - I_{oq}\sin(\delta_{20}) \end{bmatrix}$$

$$T_V^{-1} = \begin{bmatrix} -V_{bD}\sin(\delta_{10}) + V_{bQ}\cos(\delta_{10}) & -V_{bD}\sin(\delta_{20}) + V_{bQ}\cos(\delta_{20}) \\ -V_{bD}\cos(\delta_{10}) - V_{bQ}\sin(\delta_{10}) & -V_{bD}\cos(\delta_{20}) - V_{bQ}\sin(\delta_{20}) \end{bmatrix}$$

$$T_V^{-1} = \begin{bmatrix} -V_{bD}\sin(\delta_0) + V_{bQ}\cos(\delta_0) \\ -V_{bD}\cos(\delta_0) - V_{bQ}\sin(\delta_0) \end{bmatrix}$$

The state space matrices of the complete model of and individual inverter are presented here.

$$E_{INVi} = \begin{bmatrix} E_{pi} & 0 & 0 \\ 0 & E_{vci} & 0 \\ 0 & 0 & E_{lcli} \end{bmatrix}_{13 \times 13}$$

$$A_{INVi} =$$
$$\begin{bmatrix} A_{Pi} & 0 & B_{Pi} \\ B_{VC1i}C_{PVi} & 0 & B_{VC2i} \\ \begin{array}{l} B_{LCL1i}D_{VC1i}C_{PVi}+ \\ B_{LCL2i}\begin{bmatrix} T_V^{-1} & 0_{2\times3} \end{bmatrix} \\ B_{LCL3i}C_{Pwi} \end{array} & B_{LCL1i}C_{VCi} & \begin{array}{l} A_{LCLi}+ \\ B_{LCL1i}D_{VC2i} \end{array} \end{bmatrix}_{13 \times 13}$$

$$B_{INVi} = \begin{bmatrix} 0_{7\times2} \\ B_{LCL2}T_S^{-1} \end{bmatrix}$$

$$B_{iwcom} = \begin{bmatrix} B_{Pwcom} \\ 0_{8\times2} \end{bmatrix}$$

$$C_{INVwi} = \begin{cases} \begin{bmatrix} C_{Pw} & 0_{2\times8} \end{bmatrix} & i = 1 \\ \begin{bmatrix} 0_{2\times13} \end{bmatrix} & i \neq 1 \end{cases}$$

$$C_{INVci} = \begin{bmatrix} T_C & 0_{2\times9} & T_S \end{bmatrix}$$

The state space matrices of the combined model of all inverters are presented here.

$$E_{INV} = \begin{bmatrix} E_{INV,1} & \cdots & 0 \\ \vdots & \ddots & \vdots \\ 0 & \cdots & E_{INV,s} \end{bmatrix}_{13s \times 13s}$$

$$A_{INV} = \begin{bmatrix} A_{INV1} + B_{1wcom}C_{INVw1} & 0 & \cdots & 0 \\ B_{2wcom}C_{INVw1} & A_{INV2} & \cdots & 0 \\ \vdots & \vdots & \ddots & \vdots \\ B_{swcom}C_{INVw1} & 0 & \cdots & A_{INVs} \end{bmatrix}_{13s \times 13s}$$

$$B_{INV} = \begin{bmatrix} B_{INV1} & 0 & \cdots & 0 & \cdots & 0 \\ 0 & B_{INV2} & \cdots & 0 & \cdots & 0 \\ \vdots & \vdots & \ddots & \vdots & \cdots & 0 \\ 0 & 0 & \cdots & B_{INVs} & \cdots & 0 \end{bmatrix}_{13s \times 2m}$$

$$C_{INVc} = \begin{bmatrix} C_{INVc1} & 0 & \cdots & 0 \\ 0 & C_{INVc2} & \cdots & 0 \\ \vdots & \vdots & \ddots & \vdots \\ 0 & 0 & \cdots & C_{INVcs} \end{bmatrix}_{13s \times 13s}$$

$$C_{INV\omega} = \begin{bmatrix} C_{INVc1} & 0 & \cdots & 0 \\ 0 & C_{INVc2} & \cdots & 0 \\ \vdots & \vdots & \ddots & \vdots \\ 0 & 0 & \cdots & C_{INVcs} \end{bmatrix}_{13s \times 13s}$$

References

1. Hossain, A.M.; Pota, R.H.; Issa, W.; Hossain, J.M. Overview of AC Microgrid Controls with Inverter-Interfaced Generations. *Energies* **2017**, *10*, 1300. [CrossRef]
2. Li, D.; Zhao, B.; Wu, Z.; Zhang, X.; Zhang, L. An Improved Droop Control Strategy for Low-Voltage Microgrids Based on Distributed Secondary Power Optimization Control. *Energies* **2017**, *10*, 1347. [CrossRef]
3. Dou, C.; Zhang, Z.; Yue, D.; Gao, H. An Improved Droop Control Strategy Based on Changeable Reference in Low-Voltage Microgrids. *Energies* **2017**, *10*, 1080. [CrossRef]
4. Zhang, H.; Kim, S.; Sun, Q.; Zhou, J. Distributed Adaptive Virtual Impedance Control for Accurate Reactive Power Sharing Based on Consensus Control in Microgrids. *IEEE Trans. Smart Grid* **2016**. [CrossRef]
5. Mohamed, Y.A.R.I.; El-Saadany, E.F. Adaptive Decentralized Droop Controller to Preserve Power Sharing Stability of Paralleled Inverters in Distributed Generation Microgrids. *IEEE Trans. Power Electron.* **2008**, *23*, 2806–2816. [CrossRef]
6. Kim, J.; Guerrero, J.M.; Rodriguez, P.; Teodorescu, R.; Nam, K. Mode Adaptive Droop Control With Virtual Output Impedances for an Inverter-Based Flexible AC Microgrid. *IEEE Trans. Power Electron.* **2011**, *26*, 689–701. [CrossRef]
7. Mohamed, Y.A.-R.I.; Radwan, A.A. Hierarchical Control System for Robust Microgrid Operation and Seamless Mode Transfer in Active Distribution Systems. *IEEE Trans. Smart Grid* **2011**, *2*, 352–362. [CrossRef]
8. Chen, J.; Wang, L.; Diao, L.; Du, H.; Liu, Z. Distributed Auxiliary Inverter of Urban Rail Train—Load Sharing Control Strategy under Complicated Operation Condition. *IEEE Trans. Power Electron.* **2016**, *31*, 2518–2529. [CrossRef]
9. Xia, Y.; Peng, Y.; Wei, W. Triple droop control method for ac microgrids. *IET Power Electron.* **2017**, *10*, 1705–1713. [CrossRef]
10. Yazdanian, M.; Mehrizi-Sani, A. Washout Filter-Based Power Sharing. *IEEE Trans. Smart Grid* **2015**. [CrossRef]
11. Han, Y.; Li, H.; Xu, L.; Zhao, X.; Guerrero, J. Analysis of Washout Filter-Based Power Sharing Strategy—An Equivalent Secondary Controller for Islanded Microgrid without LBC Lines. *IEEE Trans. Smart Grid* **2017**. [CrossRef]
12. Pogaku, N.; Prodanovic, M.; Green, T.C. Modeling, Analysis and Testing of Autonomous Operation of an Inverter-Based Microgrid. *IEEE Trans. Power Electron.* **2007**, *22*, 613–625. [CrossRef]
13. Bottrell, N.; Prodanovic, M.; Green, T.C. Dynamic Stability of a Microgrid with an Active Load. *IEEE Trans. Power Electron.* **2013**, *28*, 5107–5119. [CrossRef]
14. Rasheduzzaman, M.; Mueller, J.A.; Kimball, J.W. An Accurate Small-Signal Model of Inverter- Dominated Islanded Microgrids Using dq Reference Frame. *IEEE J. Emerg. Sel. Top. Power Electron.* **2014**, *2*, 1070–1080. [CrossRef]
15. Rasheduzzaman, M.; Mueller, J.A.; Kimball, J.W. Reduced-Order Small-Signal Model of Microgrid Systems. *IEEE Trans. Sustain. Energy* **2015**, *6*, 1292–1305. [CrossRef]

16. Leitner, S.; Yazdanian, M.; Mehrizi-Sani, A.; Muetze, A. Small-Signal Stability Analysis of an Inverter-Based Microgrid with Internal Model–Based Controllers. *IEEE Trans. Smart Grid* **2017**. [CrossRef]
17. Yu, K.; Ai, Q.; Wang, S.; Ni, J.; Lv, T. Analysis and Optimization of Droop Controller for Microgrid System Based on Small-Signal Dynamic Model. *IEEE Trans. Smart Grid* **2015**. [CrossRef]
18. Mohammadi, F.D.; Keshtkar, H.; Feliachi, A. State Space Modeling, Analysis and Distributed Secondary Frequency Control of Isolated Microgrids. *IEEE Trans. Energy Convers.* **2017**. [CrossRef]
19. Peng, Y.; Shuai, Z.; Shen, J.; Wang, J.; Tu, C.; Cheng, Y. Reduced order modeling method of inverter-based microgrid for stability analysis. In Proceedings of the 2017 IEEE Applied Power Electronics Conference and Exposition (APEC), Tampa, FL, USA, 26–30 March 2017; pp. 3470–3474.
20. Egwebe, A.M.; Fazeli, M.; Igic, P.; Holland, P.M. Implementation and Stability Study of Dynamic Droop in Islanded Microgrids. *IEEE Trans. Energy Convers.* **2016**, *31*, 821–832. [CrossRef]
21. Rosenbrock, H.H. Structural properties of linear dynamical systems. *Int. J. Control* **2007**, *20*, 191–202. [CrossRef]
22. Mahmood, H.; Michaelson, D.; Jiang, J. Reactive Power Sharing in Islanded Microgrids Using Adaptive Voltage Droop Control. *IEEE Trans. Smart Grid* **2015**, *6*, 3052–3060. [CrossRef]
23. Zhou, J.; Kim, S.; Zhang, H.; Sun, Q.; Han, R. Consensus-based Distributed Control for Accurate Reactive, Harmonic and Imbalance Power Sharing in Microgrids. *IEEE Trans. Smart Grid* **2016**. [CrossRef]
24. Liu, J.; Miura, Y.; Ise, T. Comparison of Dynamic Characteristics between Virtual Synchronous Generator and Droop Control in Inverter-Based Distributed Generators. *IEEE Trans. Power Electron.* **2016**, *31*, 3600–3611. [CrossRef]
25. Nguyen, C.-K.; Nguyen, T.-T.; Yoo, H.-J.; Kim, H.-M. Improving Transient Response of Power Converter in a Stand-Alone Microgrid Using Virtual Synchronous Generator. *Energies* **2018**, *11*, 27. [CrossRef]
26. Guerrero, J.M.; Hang, L.; Uceda, J. Control of Distributed Uninterruptible Power Supply Systems. *IEEE Trans. Ind. Electron.* **2008**, *55*, 2845–2859. [CrossRef]
27. Sun, Y.; Hou, X.; Yang, J.; Han, H.; Su, M.; Guerrero, J.M. New Perspectives on Droop Control in AC Microgrid. *IEEE Trans. Ind. Electron.* **2017**, *64*, 5741–5745. [CrossRef]
28. Lewis, F.L. A survey of linear singular systems. *Circuits Syst. Signal Process.* **1986**, *5*, 3–36. [CrossRef]
29. Milano, F.; Dassios, I. Primal and Dual Generalized Eigenvalue Problems for Power Systems Small-Signal Stability Analysis. *IEEE Trans. Power Syst.* **2017**, *32*, 4626–4635. [CrossRef]
30. Cai, H.; Xiang, J.; Wei, W.; Chen, M.Z.Q. V-dp/dv Droop Control for PV Sources in DC Microgrids. *IEEE Trans. Power Electron.* **2017**. [CrossRef]

MDPI

Article

Adaptive Sliding Mode Speed Control for Wind Energy Experimental System

Adel Merabet

Division of Engineering, Saint Mary's University, Halifax, NS B3H 3C3, Canada; adel.merabet@smu.ca;
Tel.: +1-902-420-5712

Received: 19 July 2018; Accepted: 23 August 2018; Published: 26 August 2018

Abstract: In this paper, an adaptive sliding mode speed control algorithm with an integral-operation sliding surface is proposed for a variable speed wind energy experimental system. In the control design, an estimator is designed to compensate for the uncertainties and the unknown turbine torque. In addition, the bound of the sliding mode is investigated to deal with uncertainties. The stability of the system can be guaranteed in the sense of the Lyapunov stability theorem. The laboratory size DC generator wind energy system is controlled using a buck-boost DC-DC converter interface. The control system is validated by experimentation and results demonstrate the achievement of favorable speed tracking performance and robustness against parametric variations and external disturbances.

Keywords: adaptive control; sliding mode control; speed control; wind energy system

1. Introduction

Small-scale wind energy systems can be considered a solution to the low-to-medium energy demands for renewable sources. The efficient operations of such systems, in all wind speed regions, depends on the methods that regulate speed and power [1]. There are different configurations of wind turbine systems—fixed pitch, variable pitch and interfaced power electronics. Based on the wind system configuration, the speed and power can be regulated to optimize the operation [1–3].

In a below-rated wind speed region, the objective is to regulate turbine rotor speed and track desired optimum speed to extract maximum power from the wind, while minimizing the effects of uncertainties. In an above-rated wind speed region, the rotor speed can be controlled to follow a constant nominal speed, while the power is regulated to track the nominal value through the pitch angle controller. This control strategy is efficient compared to the use of only the pitch angle, as the generator torque cannot be assumed constant and requires to be controlled and maintained at its rated value [4,5].

Power electronics converters can be implemented to run the wind turbine in different wind regions (optimum below-rated wind speed and nominal above-rated wind speed) in power electronics interfaced wind energy systems [6–8]. The power electronics interface is regulated by a control system that maintains the desired levels of output voltages and currents, therefore controlling the torque and speed of wind turbines [9–12].

Sliding mode control (SMC) is used in the control field due to its good performance and simple structure. Furthermore, it handles uncertainties caused by un-modeled quantities, parametric variations, and modeling approximations [13,14]. The SMC control input includes a model control law based on model dynamics and a commutation control that attracts the state trajectory to the sliding surface. Various strategies for different generators in wind energy applications have been proposed. For doubly-feed induction machines, a second-order SMC is applied to optimize the power conversion [15,16]. A sliding mode control, combined with the field oriented control (FOC), is used for dual stator induction generator wind energy systems [17]. Furthermore, artificial intelligence (fuzzy, neural networks) based sliding mode control [18–20] and sliding mode control are used in DC machines [21]. In all these procedures, it is assumed that the turbine torque is known, by either measurement or calculation, which is not practically accurate as it randomly varies with change in wind speed. Furthermore, the compensation of unknown torque, unmodeled quantities and parametric variations is not guaranteed, as there is no adaptation mechanism to reduce their effects. In [22,23], SMC has been developed for power tracking with knowledge about wind turbine characteristics and aerodynamic torque observation. In [24], a second-order SMC, with fixed control gains, is applied to a DC generator based wind turbine experimental system.

The purpose of this paper is to design an SMC scheme that offers an adaptive mechanism to overcome the uncertainties found in traditional control systems. The proposed controller includes an estimator that deals with the unknown turbine torque and inaccuracies in the mathematical model of the system and attempts to achieve zero steady-state error. Furthermore, to overcome the problem of finding a suitable gain for the sliding mode control, a bound estimation algorithm is investigated, which reduces the chattering control effort and enhances the response of the overall control system through online adjustment. Investigation of the control strategy with the estimation is carried out using the Lyapunov theorem in order to guarantee closed loop system stability.

2. DC Generator Wind Turbine

The DC generator is modelled by the following equation:

$$\frac{di(t)}{dt} = \frac{1}{L}[-Ri(t) + K_b\omega(t) - V(t)] \tag{1}$$

where, i is the armature current, ω is the rotor speed, R is the resistance, L is the inductance, and K_b is back-emf constant.

The wind turbine is modelled by the mechanical dynamics of the rotor, such that:

$$\frac{d\omega(t)}{dt} = \frac{1}{J}(T_t(t) - T_{em}(t)) - \frac{B}{J}\omega(t) + \frac{1}{J}\zeta \tag{2}$$

where, ω is the rotor speed, T_t is the turbine torque to drive the rotor, T_{em} is the torque developed by the generator to oppose the driven torque, J is the total inertia (the rotor and the turbine), B is the friction coefficient, and ζ includes uncertainties related to unmodeled quantities and external disturbances.

In this work, only rotor inertia is used in control implementation. The inaccuracy in total inertia can be seen as a disturbance to be compensated for by the control system.

The generator torque is given by:

$$T_{em}(t) = K_i i(t) \tag{3}$$

where, K_i is the torque constant.

When operating the DC machine as a generator, the developed electric torque T_{em} opposes the mechanical torque developed by the wind turbine T_t. Therefore, by controlling the generator current i, the torque can be regulated, as shown in (3), to control rotor speed ω following the dynamics in (2).

The torque at the generator-turbine shaft, produced by the wind, is expressed as:

$$T_t(t) = \frac{1}{2}\pi\rho C_t r^3 v_w^2 \tag{4}$$

where, ρ is the air density, C_t is the torque coefficient, r is the radius of the turbine blade, v_w is the wind speed.

In practical applications, the torque coefficient is unknown and the wind speed measurement is affected by noise. Therefore, the turbine torque can be considered as an unknown disturbance to be compensated for by the control system.

3. Sliding Mode Control for Speed Control

The SMC for speed control is developed using the dynamics of the rotor and the generator. Using Equations (2) and (3), rotational speed dynamics is expressed as:

$$\frac{d\omega(t)}{dt} = -A_p\omega(t) + f(t) - B_pi(t) \tag{5}$$

where ω is the output, i is the control input, $A_p = \frac{B}{J}$, $B_p = \frac{K_i}{J}$, and $f(t) = \frac{1}{J}T_t(t)$ is the disturbance input related to the turbine torque.

Considering uncertainties, such as unmodeled quantities, parametric variations and external disturbances, the dynamics (5) can be upgraded to:

$$\frac{d\omega(t)}{dt} = -A_p\omega(t) + f(t) - B_pi(t) + d(t) \tag{6}$$

where d is the disturbance input that includes all system uncertainties.

The control objective is to minimize the speed tracking error dynamics:

$$\frac{de(t)}{dt} = \frac{d\omega(t)}{dt} - \frac{d\omega_{ref}(t)}{dt} = -A_pe(t) + u(t) + d(t) \tag{7}$$

where $e(t) = \omega(t) - \omega_{ref}(t)$ is the speed tracking error, ω_{ref} is the speed reference to be carried out using mechanisms for maximum or limited power extraction, and $u(t)$ is a new control signal and expressed by:

$$u(t) = f(t) - B_pi(t) - A_p\omega_{ref}(t) - \frac{d\omega_{ref}(t)}{dt} \tag{8}$$

Compensation of the uncertainties that are present in the system can be achieved by using SMC. In this work, a sliding surface, based on an integral operation, is used and expressed as:

$$S(t) = e(t) + \int ke(\tau)d\tau \tag{9}$$

where k is a positive gain.

Controlling the shaft speed can be achieved using the following speed control law:

$$u(t) = -(k - A_p)e(t) - \beta\text{sgn}(S(t)) \tag{10}$$

where β is the switching gain and sgn($\cdot$) is the sigmoid function.

The term $(k-A_p)$ is positive. Therefore, the choice of the gain k must satisfy the condition: $k > A_p$.

The choice of the gain β is based on the condition: $\beta \geq |d(t)|$ always. It is assumed that system uncertainties are bounded magnitudes with known upper bounds.

The stability of the closed loop control system, defined using (6)–(10), is based on the Lyapunov theorem. A Lyapunov function candidate can be defined such as:

$$V(t) = \frac{1}{2}S^2(t) \tag{11}$$

Using (7)–(10), the time derivative of the function (11) is expressed as:

$$\begin{aligned}
\frac{dV(t)}{dt} &= S(t)\cdot\frac{dS(t)}{dt} \\
&= S(t)\cdot\left[\frac{de(t)}{dt} + ke(t)\right] \\
&= S(t)\cdot\left[-A_pe(t) + u(t) + d(t) + ke(t)\right] \\
&= S(t)\cdot[d(t) - \beta\text{sgn}(S(t))] \\
&\leq -(\beta - |d(t)|)|S(t)| \leq 0
\end{aligned} \tag{12}$$

The stability condition is satisfied based on (12). Therefore, the closed loop system is stable based on the Lyapunov theorem.

Finally, from (8) and (10), the control law, which is the current command i^*, is given by:

$$i^*(t) = \frac{1}{B_p} \cdot \left[f(t) - A_p \omega_{\text{ref}}(t) - \frac{d\omega_{\text{ref}}(t)}{dt} + (k - A_p)e(t) + \beta \text{sgn}(S(t)) \right] \qquad (13)$$

The sliding mode control provides satisfactory tracking performance and robustness to system uncertainties. A drawback of conventional SMC is the appearance of chattering in the control input due to its discontinuity across the sliding surface. Chattering may excite high frequency dynamics, neglected during the system modeling. In this work, the chattering can be eliminated by smoothening the control discontinuity using the sigmoid function.

4. Adaptive Sliding Mode Control for Speed Control

4.1. Adaptation Based on Torque Estimation

The implementation of the control law (13) requires knowledge about the term $f(t)$, which is related to the turbine torque T_t, as observed in (4). Practically, the turbine torque T_t is related to the exact knowledge of the torque coefficient C_t, which is not always available, and wind speed v_w, which changes randomly and is affected by noise [1,2]. Therefore, the estimation of turbine torque in f can be a solution, as it takes into consideration the behavior of the closed loop system to improve the performance of the controller.

Using the torque estimation, the control law (13) becomes:

$$i^*(t) = \frac{1}{B_p} \cdot \left[\hat{f}(t) - A_p \omega_{\text{ref}}(t) - \frac{d\omega_{\text{ref}}(t)}{dt} + (k - A_p)e(t) + \beta \text{sgn}(S(t)) \right] \qquad (14)$$

where $\hat{f}$ is the estimation of f.

Torque estimation can be defined based on the following error dynamics:

$$\frac{d\tilde{f}(t)}{dt} + \eta_1 \tilde{f}(t) = 0 \qquad (15)$$

where $\tilde{f} = f - \hat{f}$ is the estimated error and η_1 is a positive constant.

From (5) and (14), and considering slow dynamics of the turbine torque compared to the electric system, the estimation (15) is reorganized as:

$$\begin{aligned}
\frac{d\hat{f}(t)}{dt} &= \eta_1 \left(f(t) - \hat{f}(t) \right) \\
&= \eta_1 \left(\frac{de(t)}{dt} + A_p \omega(t) + B_p i^*(t) - \hat{f}(t) \right) \\
&= \eta_1 \left(\frac{de(t)}{dt} + (k - A_p)e(t) + \beta \text{sgn}(S(t)) \right)
\end{aligned} \qquad (16)$$

The torque estimator can be expressed, by integrating (16), as:

$$\hat{f}(t) = \eta_1 \left(e(t) + (k - A_p) \int e(\tau)d\tau + \beta \int \text{sgn}(S(\tau))d\tau \right) \qquad (17)$$

The integral action of the speed tracking error allows the elimination of the disturbance in a steady state. Furthermore, the structure of the estimator (17), integrated with the control law (14), enhances the robustness of the controlled system with respect to uncertainties and disturbance rejection.

4.2. Adaptation Based on Bound Estimation

The dynamics of the control law (10) depends on the selection of the gain β, which can be conducted using trial and error to achieve good tracking performance. In this work, estimation is used for an automatic gain update to achieve high performance.

Replacing β by its estimated bound $\hat{\beta}$, the speed control law (10) becomes:

$$u(t) = -(k - A_p)e(t) - \hat{\beta}(t)\text{sgn}(S(t)) \tag{18}$$

Estimated gain error of the bound value can be defined as:

$$\tilde{\beta}(t) = \beta - \hat{\beta}(t) \tag{19}$$

The bound estimation law is based on the following Lyapunov candidate:

$$V_1(t) = \frac{1}{2}S^2(t) + \frac{1}{2\eta_1}\tilde{f}^2(t) + \frac{1}{2\eta_2}\tilde{\beta}^2(t) \tag{20}$$

where, η_2 is a positive constant.

On differentiating (20), with respect to time, and using (15) and (19), it is given that:

$$
\begin{aligned}
\frac{dV_1(t)}{dt} &= S(t)\frac{dS(t)}{dt} + \frac{1}{\eta_1}\tilde{f}(t)\frac{d\tilde{f}(t)}{dt} + \frac{1}{\eta_2}\tilde{\beta}(t)\frac{d\tilde{\beta}(t)}{dt} \\
&= S(t)\left[d(t) - \hat{\beta}(t)\text{sgn}(S(t))\right] - \tilde{f}^2(t) - \frac{1}{\eta_2}\dot{\hat{\beta}}(t)\left[\beta - \hat{\beta}(t)\right]
\end{aligned}
\tag{21}
$$

In order to achieve $\frac{dV_1(t)}{dt} \leq 0$, the bound estimation law is designed as:

$$\dot{\hat{\beta}}(t) = \eta_2 S(t)\text{sgn}(S(t)) \tag{22}$$

and (21) can be rewritten as:

$$
\begin{aligned}
\frac{dV_1(t)}{dt} &= S(t)\left[d(t) - \hat{\beta}(t)\text{sgn}(S(t))\right] - \tilde{f}^2(t) - S(t)\text{sgn}(S(t))\left[\beta - \hat{\beta}(t)\right] \\
&= S(t)[d(t) - \beta(t)\text{sgn}(S(t))] - \tilde{f}^2(t) \\
&\leq -(\beta - |d(t)|)|S(t)| - \tilde{f}^2(t) \leq 0
\end{aligned}
\tag{23}
$$

In the adaptation law (23), there is no priori requirement to identify control gain. However, it can be observed from the $\hat{\beta}$-dynamics, that $\dot{\hat{\beta}} = 0$ when S = 0. This condition shows that gain $\hat{\beta}$ is over-estimated (the gain is always increasing), with respect to uncertainties, and induces larger chattering [23,24]. Furthermore, this methodology can be applied only for an ideal SMC, as the objective S = 0 cannot be reached in a real application. In this work, the information provided by the torque estimation (17), which includes uncertainties and perturbations, is used in the adaptation of the gain $\hat{\beta}(t)$ as follows:

$$
\begin{cases}
\dot{\hat{\beta}}(t) = \eta_2 S(t)\text{sgn}(S(t)) & \text{if } |S(t)| > \varepsilon > 0 \\
\hat{\beta}(t) = \bar{\beta}_1|\hat{f}| + \bar{\beta}_2 & \text{if } |S(t)| \leq \varepsilon
\end{cases}
\tag{24}
$$

where, $\bar{\beta}_1 = \hat{\beta}(t^*)$ and $\bar{\beta}_2 > 0$, t^* is the largest time value such that $|S(t^{*-})| > \varepsilon$ and $|S(t^*)| \leq \varepsilon$; ($t^{*-}$ is the time just before t^*).

By supposing that $|S(t)| > \varepsilon$, the adaptive gain control law (24) operates as follows:

- The gain $\hat{\beta}(t)$ increases to reach a value large enough to counteract the bounded uncertainty in (6), until the start of the SMC. The time for this start is t_1.

- When the SMC starts, i.e., $|S(t)| \leq \varepsilon$, from $t = t_1$, $\hat{\beta}(t)$ is carried out using the adaptation (24) with $\bar{\beta}_1 = \hat{\beta}(t_1)$. Using this mechanism, the gain can be decreased and adjusted depending on the actual uncertainties and disturbances.
- However, if the varying uncertainties move the sliding surface $S(t)$ outside the interval $\pm \varepsilon$, then gain adaptation will update in accordance with (24). The gain $\hat{\beta}(t)$ will increase until the sliding mode occurs again at the reaching time instant t_2. As the sliding mode has occurred and $|S(t)| \leq \varepsilon$ from $t = t_2$, $\hat{\beta}(t)$ now follows the gain adaptation law (24) with $\bar{\beta}_1 = \hat{\beta}(t_2)$.

The process will continue operating as mentioned above depending on the condition of $S(t)$.

The proposed SMC scheme with an adaptive mechanism is depicted in Figure 1. The control law i^* is the control law carried out by the SMC speed controller. It will be used as the reference input for the current control loop to carry out control law V^* and operate the generator side DC-DC converter.

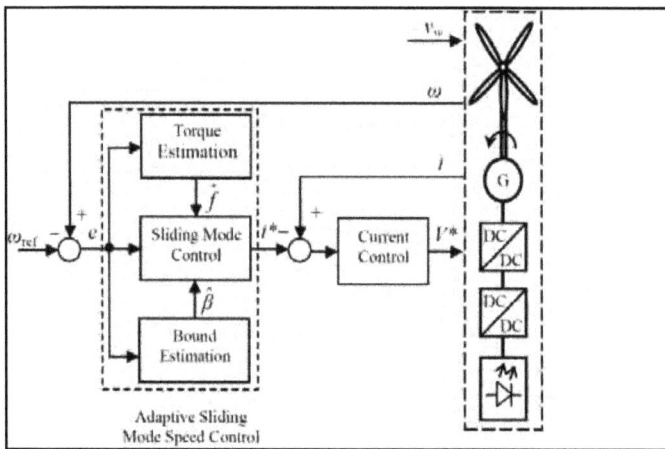

Figure 1. Adaptive sliding mode controller for the DC generator wind turbine.

5. Multivariable Control System of the DC Generator Wind Turbine

5.1. Maximum Power Point Tracking Control

If information on wind turbine characteristics is available, the speed reference ω_{ref} can be computed using optimum tip speed ratio λ_{opt} to maximize power extraction, such that:

$$\omega_{ref}(t) = \frac{\lambda_{opt} v_w(t)}{r} \tag{25}$$

The MPPT control algorithm can be implemented using (25), where wind speed measurements and wind turbine characteristics are required for accurate analysis. Due to measurement inaccuracies and lack of real models, this algorithm is not efficient in tracking the maximum available power. In this work, the MPPT control method, developed in [11], is used for maximum power extraction.

Using variation of the generated power P and the rotor speed ω, the speed reference profile, required in the closed loop control, is carried out by:

$$\frac{d\omega_{ref}(t)}{dt} = \alpha \omega(t) \frac{dP(t)}{dt} \tag{26}$$

where, α is gain. The generated power is carried out by measuring the generator voltage-current as:

$$P(t) = V(t)i(t) \tag{27}$$

In this MPPT algorithm, the choice of the ratio of the change in the generated power and the rotational speed will help improve speed tracking of both high and low wind speeds. Convergence can be achieved quickly by an adequate choice of the positive gain α.

5.2. Generator Current Control

The output of the speed control law (14) is the current reference used in the current control loop, developed from the electrical equation of DC generator (1), and is based on a proportional-integral (PI) regulator and the coupling term:

$$V^*(t) = K_b \omega(t) - \left[k_p(i(t) - i^*(t)) + k_i \int (i(\tau) - i^*(\tau)) d\tau \right] \tag{28}$$

where, i^* is the current reference, V^* is the control law of the current control loop, and k_p and k_i are the proportional and integral gains, respectively.

The wind turbine-generator system is controlled by a cascade control structure; the SMC speed controller is the outer control loop and the current controller is the inner control loop, as shown in Figure 2. The output of the current controller is the firing signal of the MOSFET gate in the DC-DC buck converter.

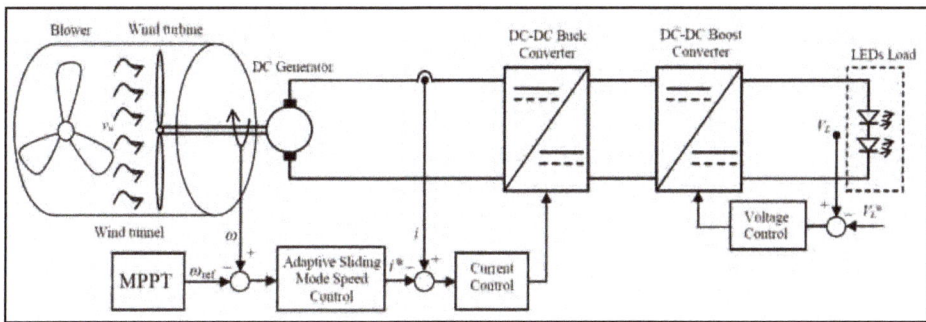

Figure 2. Schematic representation of the adaptive SMC system applied to the wind energy experimental setup.

5.3. Load Voltage Control

The control system on the load side is used to regulate load voltage, in order to be maintained at a level adequate to the LED's specifications. The PI regulator based voltage control is expressed by:

$$U^*(t) = k_p(V_L(t) - V_L^*(t)) + k_i \int (V_L(\tau) - V_L(\tau)) d\tau \tag{29}$$

where, V_L is the load voltage, V_L^* is the voltage reference, U^* is the control law of the load voltage control loop, and k_p and k_i are the proportional and integral gains, respectively.

The control output is the firing signal of the MOSFET gate in the DC-DC boost converter.

6. Wind Turbine Experimental Setup

The wind energy experimental system has a five-blade wind turbine from Quanser Inc. (Markham, ON, Canada), a DC generator, a power electronics interface, and an LED load bank. The speed and load voltage are controlled through the DC-DC converters of the power electronics interface, as shown in Figure 2. Details about the power electronics interface are available in [25].

Control testing under variable load is achieved using five controllable LED banks. Each bank consists of six parallel strings, with two LEDs in a series with resistance in each string, as shown in Figure 3.

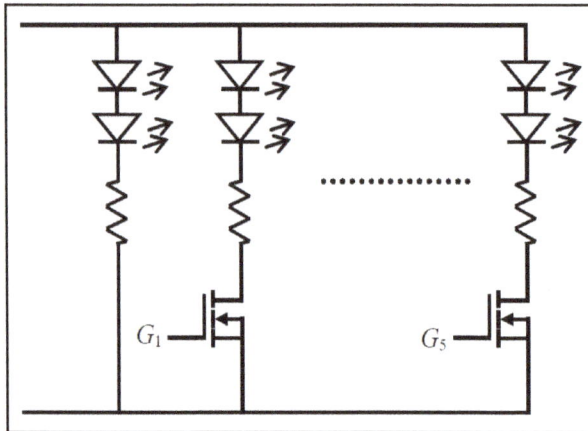

Figure 3. Electronic load.

The experimental wind energy setup is shown in Figure 4. The rotational speed is measured by an encoder. Voltage and current sensors are used for measurement at different points in the system (generator and load), as shown in Figure 4. These measurements are calibrated and received by the data acquisition board Q8-USB and the software package QUARC with MATLAB/Simulink.

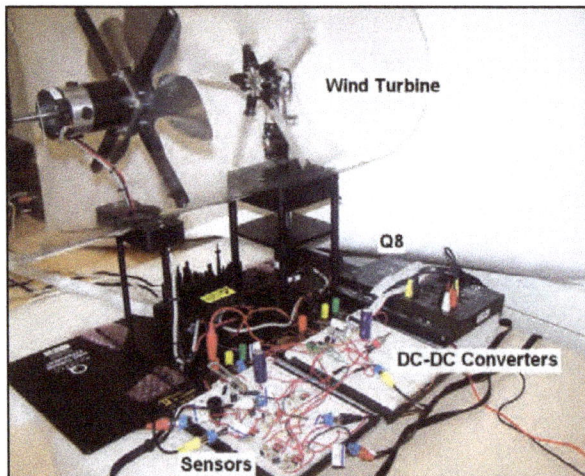

Figure 4. DC generator wind turbine experimental system.

A blower motor is used to generate wind speed and an incremental encoder, mounted on a blower rotor shaft, is used to measure its speed. The proposed control system does not require wind speed to control the wind energy system under maximum power extraction; this control model is a contribution of this work.

This wind turbine experimental setup has been used in other works to investigate different control strategies. In [3], a PI-based pitch control is used to regulate the rotational speed of the wind turbine. In [11], a feedback control strategy with an uncertainties compensator is developed to regulate the speed, whereas, in [12], a predictive control law is developed for the same purpose. In [24], a second-order SMC, based on a super-twisting sliding mode for perturbation and chattering elimination, is applied for speed control on the generator side. Its control gains are fixed and selected by trial and error to achieve good tracking performance. In this work, a different SMC strategy, based on an integral action, is proposed. It offers an adaptive mechanism to update the sliding control gain and an estimation mechanism to compensate the effects of the unknown wind turbine torque and system uncertainties.

7. Experimental Results

The characteristics of the laboratory scale wind energy experimental setup, used to verify the proposed control scheme, are given in Tables 1 and 2. The overall control system is implemented using the MATLAB/Simulink software. The QUARC software is used to communicate to the hardware through the data acquisition board.

Table 1. Characteristics of the wind turbine.

Quantity	Unit	Value
Blade radius (r)	cm	14
Blades number	-	5
Air density (ρ)	kg/m^3	1.14

Table 2. Characteristics of the DC generator.

Quantity	Unit	Value
Resistance (R)	Ω	2.47
Inductance (L)	μH	500
Back-emf constant (K_b)	mV/rpm	7.05
Torque constant (K_i)	mNm/A	67.3
Viscous-friction (B)	-	0.001833
Rotor inertia (J)	g/cm^2	110

In the generator side control, the output signal of the speed controller is used as the reference for the current controller, where its output is transformed to a PWM signal and sent as an analog signal to the MOSFET gate. In the load side converter control, the output signal of the voltage controller is used to create the firing signal for the MOSFET gate of the second converter. The blower is controlled separately to generate a variable wind speed pattern, as shown in Figure 5.

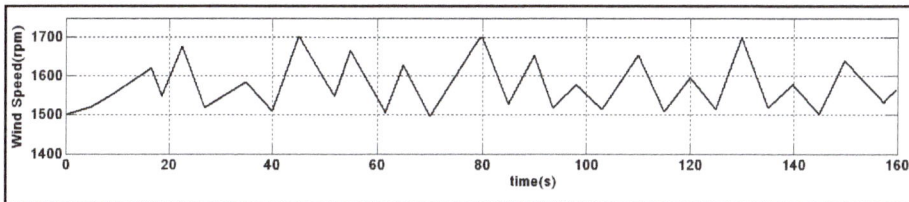

Figure 5. Wind speed profile.

Several experimental tests were conducted to verify the performance of the proposed control scheme to track variable step speed reference and maintain a fixed voltage across the load. The gains of all controllers are shown in Appendix A. The experimental system was run during a time interval of 160 s in order to allow good visualization of the system behavior during the transitions (step changes).

First, a traditional control system, based on PI regulators for speed, current, and voltage control, is carried out. Figure 6 shows that the performance of the speed controller is poor and its improvement depends on the parameter tuning of the three PI controllers. In the majority of cases, tuning is handled by trial and error, which is a time-consuming method as it is difficult to find optimal parameters. In addition, the PI controller is developed for a single input-output linear system, which is not the case in the wind turbine, as the electrical and mechanical equations are coupled. In Figure 7, a sliding mode speed controller, with a constant sliding gain β, is applied, where the same PI regulators as in the first experiment are used for the current and voltage controllers. Speed tracking is improved; however, voltage regulation is affected by the changes in the speed steps. Figure 8 shows speed tracking and voltage regulation using the adaptive SMC speed controller; it is observed that they are enhanced as the sliding gain is online adapted and the effect on the torque turbine is compensated by the estimation.

Friction quantity is considered unknown by the SMC and the tracking performance for speed and voltage is still good, as illustrated in Figure 9, highlighting the robustness of the control system.

In this experiment, the control system is tested under variable loads (as shown in Figure 10), where the switching state of each bank is changed during the operation. It can be observed, from Figure 11, that the performance of the proposed controller is still good and not affected by the load variation.

Finally, the control system is tested using the MPPT and the adaptive SMC system. The speed tracking is shown in Figure 12, where it is observed that the speed reference follows the wind speed profile of Figure 5 and the speed tracking is successfully achieved by the proposed control system.

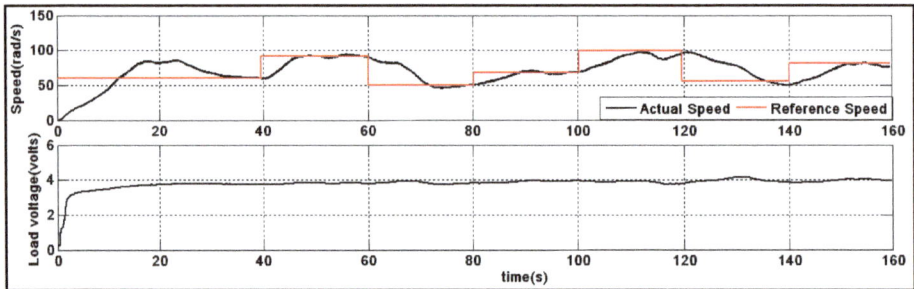

Figure 6. Speed tracking and load voltage under the PI control system.

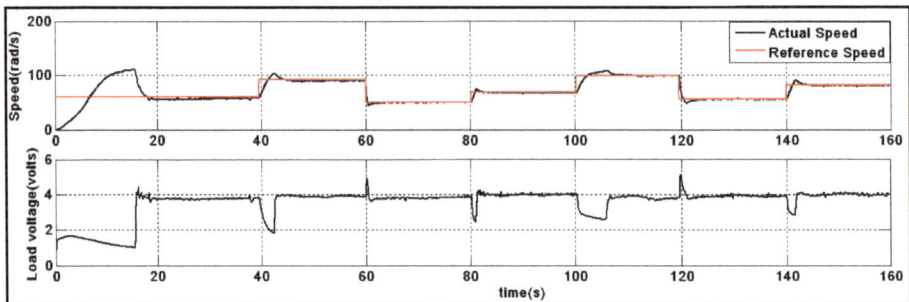

Figure 7. Speed tracking and load voltage under the SMC system.

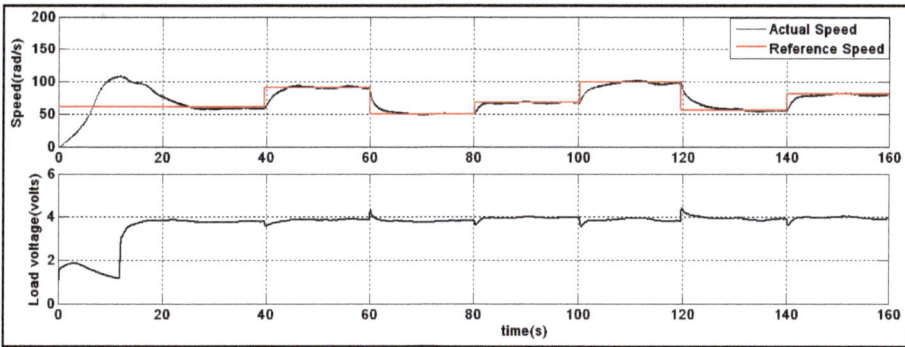

Figure 8. Speed tracking and load voltage under the adaptive SMC system.

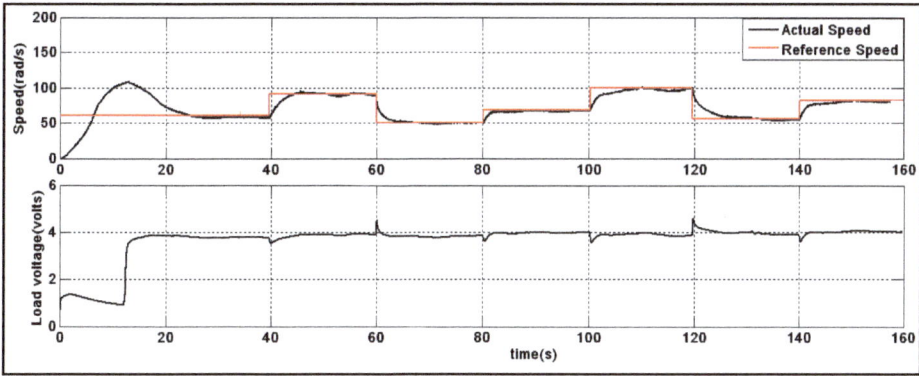

Figure 9. Speed tracking and load voltage under the adaptive SMC system with unmodeled quantity (friction).

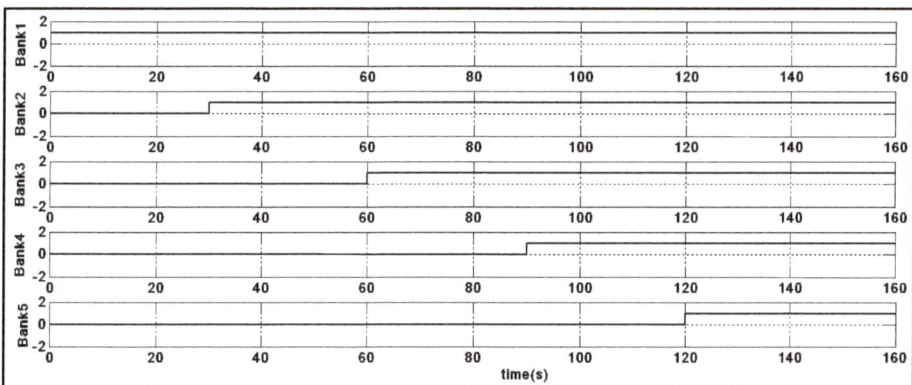

Figure 10. Variable load banks.

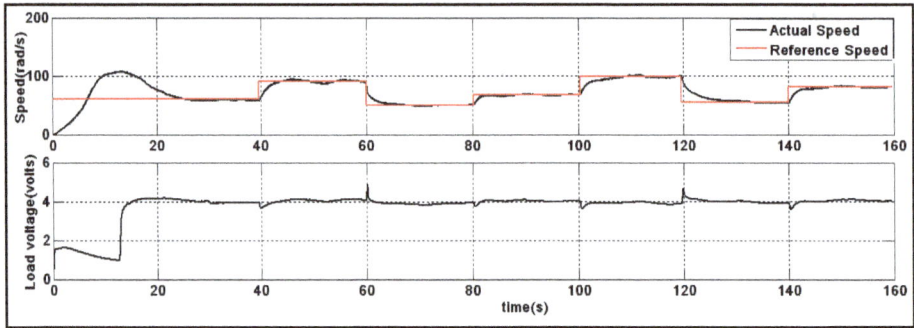

Figure 11. Speed tracking and load voltage under the adaptive SMC system for variable loads.

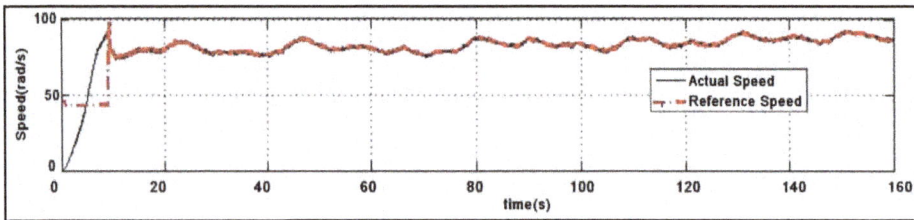

Figure 12. Speed tracking and load voltage under the adaptive SMC system and MPPT control.

8. Conclusions

An adaptive sliding mode controller, based on torque and bound estimation, is presented to deal with speed tracking problems for a DC generator wind turbine experimental system. It is designed to compensate the unknown turbine torque, due to random variation in wind speed and system uncertainties due to parametric variations, unmodeled quantities and external disturbances. The SMC is enhanced by torque estimation to compensate the effect of the turbine, and adaptation of the control gain. The estimation is developed by the Lyapunov theorem to enhance the behavior of the overall control system. Finally, experimental results are presented to demonstrate the effectiveness of the proposed SMC scheme.

Funding: This research received no external funding.

Acknowledgments: The founding sponsors had no role in the design of the study; in the collection, analyses, or interpretation of data; in the writing of the manuscript, and in the decision to publish the results.

Conflicts of Interest: The author declares no conflict of interest.

Appendix A

Table A1. Control gains.

Control	Quantity	Value
Sliding mode	β (fixed)	100
	η_1, η_2	$10^{-4}, 10^4$
Speed PI	k_p, k_i	3, 0.09
Curent PI	k_p, k_i	4, 0.2
Voltage PI	k_p, k_i	20, 0.5

Energies **2018**, *11*, 2238

References

1. Barote, L.; Marinescu, C.; Cirstea, M.N. Control structure for single-phase stand-alone wind-based energy sources. *IEEE Trans. Ind. Electron.* **2013**, *60*, 764–772. [CrossRef]
2. Yamamura, N.; Ishida, M.; Hori, T. A simple wind power generating system with permanent magnet type synchronous generator. In Proceedings of the IEEE 1999 International Conference Power Electronics Drive Systems, Hong Kong, China, 27–29 July 1999; pp. 49–854.
3. Merabet, A.; Rajasekaran, V.; Kerr, J. Modelling and control of a pitch controlled wind turbine experiment workstation. In Proceedings of the 38th Annual Conference IEEE on Industrial Electronics Society (IECON), Montreal, QC, Canada, 25–28 Octorber 2012; pp. 4316–4320.
4. Koutroulis, E.; Kalaitzakis, K. Design of a maximum power tracking system for wind energy conversion Applications. *IEEE Trans. Ind. Electron.* **2006**, *53*, 486–494. [CrossRef]
5. Merabet, A.; Thongam, J.; Gu, J. Torque and pitch angle control for variable speed wind turbines in all operating regimes. In Proceedings of the 10th International Conference on Environment and Electrical Engineering, Roma, Italy, 1–7 May 2011; pp. 1–5.
6. Kazmi, S.M.R.; Goto, H.; Guo, H.-J.; Ichinokura, O. A novel algorithm for fast and efficient speed-sensorless maximum power point tracking in wind energy conversion systems. *IEEE Trans. Ind. Electron.* **2011**, *58*, 29–36. [CrossRef]
7. Arifujjaman, M.; Iqbal, M.T.; Quaicoe, J.E. Maximum power extraction from a small wind turbine emulator using a DC-DC converter controlled by a microcontroller. In Proceedings of the 4th International Conference of Electrical and Computer Engineering, Dhaka, Bangladesh, 19–21 December 2006; pp. 213–216.
8. Lazarov, V.; Roye, D.; Spirov, D.; Zarkov, Z. New control strategy for variable speed wind turbine with DC-DC converters. In Proceedings of the 14th International Power Electronics and Motion Control Conference, Sofia, Bulgaria, 6–8 September 2010; pp. 120–124.
9. Mayo-Maldonado, J.C.; Salas-Cabrera, R.; Cisneros-Villegas, H.; Castillo-Ibarra, R.; Roman-Flores, J.; Hernandez-Colin, M.A. Maximum power point tracking control for a DC-generator/multiplier-converter combination for wind energy applications. In Proceedings of the World Congress on Engineering and Computer Science, San Francisco, USA, 19–21 October 2011; pp. 1–6.
10. Erickson, R.W. DC-DC power converters. In *Wiley Encyclopedia of Electrical and Electronics Engineering*; Webster, J.G., Ed.; John Wiley & Sons: Hoboken, NJ, USA, 2000.
11. Merabet, A.; Islam, M.A.; Beguenane, R.; Trzynadlowski, A.M. Multivariable control algorithm for laboratory experiments in wind energy conversion. *Renew. Energy* **2015**, *83*, 162–170. [CrossRef]
12. Merabet, A.; Islam, M.A.; Beguenane, R. Predictive speed controller for laboratory size wind turbine experiment system. In Proceedings of the Canadian Conference on Electrical and Computer Engineering, Toronto, ON, Canada, 4–7 May 2014; pp. 1–6.
13. Barambones, O. Sliding mode control strategy for wind turbine power maximization. *Energies* **2012**, *5*, 2310–2330. [CrossRef]
14. Evangelista, C.; Puleston, P.; Valenciaga, F. Wind turbine efficiency optimization. Comparative study of controllers based on second order sliding modes. *Int. J. Hydrogen Energy* **2010**, *35*, 5934–5939. [CrossRef]
15. Kairous, D.; Wamkeue, R. DFIG-based fuzzy sliding-mode control of WECS with a flywheel energy storage. *Electr. Power Syst. Res.* **2012**, *93*, 16–23. [CrossRef]
16. Evangelista, C.; Valenciaga, F.; Puleston, P. Multivariable 2-sliding mode control for a wind energy system based on a double fed induction generator. *Int. J. Hydrogen Energy* **2012**, *37*, 10070–10075. [CrossRef]
17. Amimeur, H.; Aouzellag, D.; Abdessemed, R.; Ghedamsi, K. Sliding mode control of a dual-stator induction generator for wind energy conversion systems. *Int. J. Electr. Power* **2012**, *42*, 60–70. [CrossRef]
18. Wai, R.-J.; Lin, C.-M.; Hsu, C.-F. Adaptive fuzzy sliding-mode control for electrical servo drive. *Fuzzy Sets Syst.* **2004**, *143*, 295–310. [CrossRef]
19. Fallahi, M.; Azzadi, S. Fuzzy PID sliding mode controller design for the position control of a DC motor. In Proceedings of the International Conference on Education Technology and Computer, Singapore, 17–20 April 2009; pp. 73–77.
20. Castaneda, C.E.; Loukianov, A.G.; Sanchez, E.N.; Castillo-Toledo, B. Discrete-time neural sliding-mode block control for a DC motor with controlled flux. *IEEE Trans. Ind. Electron.* **2012**, *59*, 1194–1207. [CrossRef]

21. Rhif, A. Stabilizing sliding mode control design and application for a DC motor: Speed control. *Int. J. Instrum. Control Syst.* **2012**, *2*, 39–48. [CrossRef]
22. Beltran, B.; Ahmed-Ali, T.; Benbouzid, M.E. Sliding mode power control of variable-speed wind energy conversion systems. *IEEE Trans. Energy Convers.* **2008**, *23*, 551–558. [CrossRef]
23. Beltran, B.; Ahmed-Ali, T.; Benbouzid, M.E. High Order Sliding Mode Variable-Speed Wind Energy Turbines. *IEEE Trans. Ind. Electron.* **2009**, *56*, 3314–3321. [CrossRef]
24. Merabet, A.; Islam, M.A.; Beguenane, R.; Ibrahim, H. Second-order sliding mode control for variable speed wind turbine experiment system. *Renew. Energy Power Qual. J.* **2014**, *1*, 478–482. [CrossRef]
25. Merabet, A.; Kerr, J.; Rajasekaran, V.; Wight, D. Power electronics circuit for speed control of experimental wind turbine. In Proceedings of the 24th International Conference of Microelectronics, Algiers, Algeria, 17–20 December 2012; pp. 1–4.

energies

MDPI

Article

A Novel Computational Approach for Harmonic Mitigation in PV Systems with Single-Phase Five-Level CHBMI

Rosario Miceli, Giuseppe Schettino and Fabio Viola *

Dipartimento Energia, Ingegneria dell'Informazione e Modelli Matematici (DEIM), University of Palermo, Palermo 90133, Italy; rosario.miceli@unipa.it (R.M.); giuseppe.schettino@unipa.it (G.S.)
* Correspondence: fabio.viola@unipa.it; Tel.: +39-091-238-60253

Received: 30 July 2018; Accepted: 9 August 2018; Published: 13 August 2018

Abstract: In this paper, a novel approach to low order harmonic mitigation in fundamental switching frequency modulation is proposed for high power photovoltaic (PV) applications, without trying to solve the cumbersome non-linear transcendental equations. The proposed method allows for mitigation of the first-five harmonics (third, fifth, seventh, ninth, and eleventh harmonics), to reduce the complexity of the required procedure and to allocate few computational resource in the Field Programmable Gate Array (FPGA) based control board. Therefore, the voltage waveform taken into account is different respect traditional voltage waveform. The same concept, known as "voltage cancelation", used for single-phase cascaded H-bridge inverters, has been applied at a single-phase five-level cascaded H-bridge multilevel inverter (CHBMI). Through a very basic methodology, the polynomial equations that drive the control angles were detected for a single-phase five-level CHBMI. The acquired polynomial equations were implemented in a digital system to real-time operation. The paper presents the preliminary analysis in simulation environment and its experimental validation.

Keywords: photovoltaic systems; multilevel power converter; soft switching; selective harmonic mitigation; phase shifted; voltage cancellation

---

1. Introduction

Nowadays, more than 40% of the carbon dioxide (CO_2) emissions worldwide are caused by the air conditioning, the heating, and electric power systems of buildings. Thus, the optimization of their performances and the reduction of building consumptions can significantly contribute to increasing the sustainability of our planet, trying to fit the International Energy Agency (IEA) requirements, with a perspective of an 80% of reduction by 2050 regarding the global emissions [1].

One of the best practices to reduce CO_2 emissions can be to produce energy locally from photovoltaic (PV) system and use it at the same point [2]. However, at the same time, chasing a reduction in emissions with the use of PV systems can cause a second type of pollution, that of harmonics in the systems connected to electric grid. In such a way, power electronics systems have inherited the task of reducing harmonic pollution.

Multilevel inverters are widely used in different high and medium power industrial application, such as electrical drives, distributed generations, and flexible alternating current transmission system (FACTS). The choice fell on these devices because the advantages of this technology are different; better output voltage waveform, lower harmonic content, lower electromagnetic interferences, less dv/dt stress, reduced necessity of passive filters, lower torque ripple in motor application, and possible fault-tolerant operation [3].

It is well known that the fundamental multilevel topologies include the diode-clamped [4–6], flying capacitor [7], and cascaded H-bridge (CHB) structures [8]. Diode-clamped technology employs diodes to separate direct current (DC) voltage levels from one DC source at its midpoint. The flying capacitor technique substitutes diodes with capacitors. The cascaded H-bridge technique employs different DC sources to create different DC voltage levels.

An interesting review can be found in the literature [9], which also takes into account the inverter's cost, by taking the factors numbers sources, switches, and variety. In the works of [10–12], good reviews are also performed.

Inverters can also be classified according to their applications [10]: renewable energy applications [11–21], automotive [22–27], heavy traction [28,29], power quality [30,31], and industrial drives [32,33]. In the literature [13], the performance in terms of harmonics distortion rates were faced in order to extract maximum power from the PV modules. Again, that authors of [14,15] face the harmonics distortion by implementing a fast switching modulation, but without considering power losses. The authors of [16] recommend a single phase multilevel inverter configuration with three series connected full bridge inverter and a single half bridge inverter. In this case, a reduction of harmonics in terms of total harmonic distortion rate (THD) is evaluated around 9.85%. In another paper [17], a spice model considers not only the THD, but also the storage element. Again, in the literature [18], a simulation of the performance of multilevel inverter is made taking into account the partial shaded condition of PV panels. In the work of [19], a multilevel invert is used to reduce the THD. The simulation generates a 15-level output voltage with 8.12% of total harmonic distortion. The authors of [20] present a three-phase, three-level neutral-point-clamped quasi-Z-source inverter, as a new solution for PV applications. In a similar way, a DC/DC state is in described in the literature [21] to increase the performance of the system.

By considering the renewable applications, for a PV plant, because of the downward trend in the price for the PV modules, the costs of the inverters are progressively standing out while computing the entire cost of the plant [34–37]. Efficiency, size, weight, and reliability have influenced the cost of producing inverters. Now, for a multilevel inverters, efficiency has reached the value of 98% and to achieve the next 1% increase is a very hard challenge, to deal with ever-more advanced modulation techniques.

Multilevel inverter modulation strategies can be classified in high switching frequency pulse width modulation (PWM) and low switching frequency modulation techniques.

The first type of modulation works at higher frequency and the output voltage waveform shows higher order harmonics that can be easily filtered, but high frequencies bring also higher switching losses. In some studies, a comparison among harmonics content in the voltage waveforms, obtained by multicarrier PWM modulation techniques, are reported [38–40].

The second modulation technique allows one to reduce the switching losses to a minimum value and to bound the stress on the power components. Low switching frequencies methods normally perform one or two commutations of the switches during one period of the fundamental, thus creating a staircase waveform. Nevertheless, the output voltage presents waveforms with low order harmonics hard to be filtered.

The efficiency of the system is very important parameter for high power electrical drives applications, which require the reduction of the switching losses and electromagnetic interferences (EMI), so the soft switching modulation techniques, such as selective harmonic elimination (SHE) and selective harmonic mitigation (SHM), were often chosen.

In classical SHE method, the switching angles are obtained by choosing the hth harmonic to be eliminated and by solving the set of non-linear equations. The SHM techniques are used to mitigate concurrently different harmonics by properly choosing the switching angles.

It is well known that the elimination or mitigation technique requires the resolving of non-linear transcendental equations, which requires time and resources, constraints that become cumbersome in the case of limited, but performing hardware such a FPGA-based control board.

Whichever method is used, SHE or SHM, to solve the set of transcendental equations and find the switching angles different approaches are used. The simplest way is based in iterative methods such as Newton–Rhapson. In the literature [41], the Newton–Rhapson iterative method is used to evaluate the switching angles for a seven level inverter. The THD of the output voltage is equal to 11.8%.

In the work of [42], a comparison between various modulation techniques for a five-level cascaded H-bridge multilevel inverter (CHBMI) is carried out. The presented control scheme employed three different pre-defined pattern for the switching angles. The authors achieved a minimum THD of 17.07% for the output voltage waveform.

In the work of [43], an extremely fast optimal solution of harmonic elimination for a five-level multilevel inverter with non-equal dc sources using a novel particle swarm optimization (PSO) algorithm is presented. In this case, the minimum value of THD achieved was 5.44%.

In the work of [44], the set of equations for a seven-level inverter is solved using a Bat algorithm. A BAT algorithm is a recent method for solving numerical global optimization problems, based on the echolocation of microbats.

In the work of [45], an optimal SHM is proposed for a seven-level inverter. The individual harmonic and the THD are mitigated to satisfy three voltage harmonic standard (EN50160 [46], CIGRE JWG C4.07 [47], IEC61000-3-6 [48]).

Again, in another paper [49], a PSO is used for PV sources, the non-linear transcendental equations are solved offline and switching angles are obtained to minimize the THD. DC sources are transformed into identical DC source using the adaptive neuro fuzzy inference system (ANFIS) and constant voltage maximum power point tracker (MPPT) algorithm. The performances obtained in THD were about 3.7%, less than the ones prescribed by IEEE-519 (5%).

Also in another paper [50], an adaptive neuro fuzzy interference system is used for eliminating voltage harmonics. The comparison there proposed show a best performance of ANFIS referred to neuro fuzzy controller (NFC) for a seventh level inverter and active filter. In order to improve the performance of the control, an active filter can also be used [51].

In this work, a novel methodology to selective low order harmonic mitigation is proposed for high power PV applications, without trying to solve the cumbersome non-linear transcendental equations. The purpose of this work consists of defining an approach to mitigate the first-five harmonics (third, fifth, seventh, ninth, and eleventh harmonics), to reduce the complexity of the required algorithm and allocating a few computational resources in the FPGA-based control board. The objective of this paper is to achieve the performances obtained in the literature [16,19,44,45] without using complex structures, but a simple one that can be modified in the future to approximate the novel schemes of the work of [20].

Therefore, the voltage waveform taken into account is different respect traditional voltage waveform. The same concept, known as *"voltage cancelation"* [52], used for single-phase H-bridge inverters, has been applied at a single-phase five-level CHBMI.

This paper is divided in the following sections: Section 2 defines the possible switch states and the output voltage expression of the single phase five-level CHBMI; Section 3 analyses the harmonic content on the voltage waveform taken into account by the Fourier series mathematics formulation; Section 4 defines the proposed method; Section 5 proposed the best polynomial equation to evaluate the control angles; and Section 6 is devoted to the experimental validation.

2. Single-Phase Five-Level CHBMI

The desired output voltage of a multilevel inverter is created by adding different sources of DC voltages. By increasing the number of DC voltage sources, the inverter voltage output waveform assumes an almost sinusoidal waveform, following a fundamental frequency switching scheme.

Figure 1 illustrates the topology circuit of the considered single-phase five-level CHBMI. This type of converters has simple circuital structures, made by two H-bridges connected in series and two series of PV panels, obtaining the DC sources. The converter output voltage, V_{out}, is realized by the states combination of the switches. The switch state can assume only two values, as reported in Equation (1):

$$S_j = \begin{cases} 1 & \text{switch on} \\ 0 & \text{switch off} \end{cases} \tag{1}$$

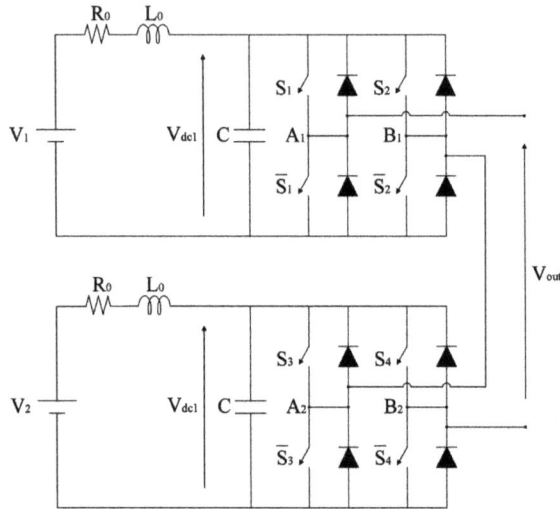

Figure 1. Single-phase five-level cascaded H-bridge multilevel inverter (CHBMI). V_1 and V_2 are the direct current (DC) sources given by photovoltaic (PV) array, with same internal impedance.

With reference to Figure 1, the output voltage V_{out} of the converter can be expressed as follows:

$$V_{out}(t) = V_{dc1}(S_1 - S_2) + V_{dc2}(S_3 - S_4) \tag{2}$$

where V_{dc1} and V_{dc2} are the DC-link voltage of the two H-bridges connected in series, respectively. In this work, V_{dc1} and V_{dc2} of Equation (2) have been considered with the same value, equal to V_{dc}. Thus, Equation (2) can be rewritten in Equation (3).

$$V_{out}(t) = V_{dc}(S_1 - S_2 + S_3 - S_4) \tag{3}$$

The possible combinations of the switch states and the output voltage value $V_{out}(t)$, with reference to Figure 1, are reported in Table 1.

It is interesting to note that there are only two switch state combinations where the output voltage $V_{out}(t)$ is equal to $2V_{dc}$ and $-2V_{dc}$, respectively. Furthermore, there are four switch state combinations to obtain an output voltage equal to V_{dc}, $-V_{dc}$, and zero.

Table 1. States of the switches and output voltage value.

S_1	$\bar{S}_1$	S_2	$\bar{S}_2$	S_3	$\bar{S}_3$	S_4	$\bar{S}_4$	V_{out} (t)
1	0	1	0	1	0	1	0	0
1	0	1	0	1	0	0	1	V_{dc}
1	0	1	0	0	1	1	0	$-V_{dc}$
1	0	1	0	0	1	0	1	0
1	0	0	1	1	0	1	0	V_{dc}
1	0	0	1	1	0	0	1	$2V_{dc}$
1	0	0	1	0	1	1	0	0
1	0	0	1	0	1	0	1	V_{dc}
0	1	1	0	1	0	1	0	$-V_{dc}$
0	1	1	0	1	0	0	1	0
0	1	1	0	0	1	1	0	$-2V_{dc}$
0	1	1	0	0	1	0	1	$-V_{dc}$
0	1	0	1	1	0	1	0	0
0	1	0	1	1	0	0	1	V_{dc}
0	1	0	1	0	1	1	0	$-V_{dc}$
0	1	0	1	0	1	0	1	0

3. Voltage Waveform Analysis

The output voltage waveform of the inverter, as previously mentioned, can be obtained through the arrangements of the states of the switches. Thus, every arrangement synthesizes only one voltage level. By the analysis of Equation (3), the voltage waveforms can be obtained through a separate control of the H-bridge connected in series. Thus, it is possible control the time duration of the voltage level separately.

In order to evaluate the harmonic content of the voltage output of a single-phase five-level CHBMI, a square waveform with amplitude equal to $4V_{dc}$ peak to peak was taken into account. Moreover, it is essential to outline the parameter β that represents the width of angular range, where the amplitude of the output voltage is equal to V_{dc}, as shown in Figure 2. The reference square waveform, with $4V_{dc}$ peak to peak, can be obtained with $\beta = 0$.

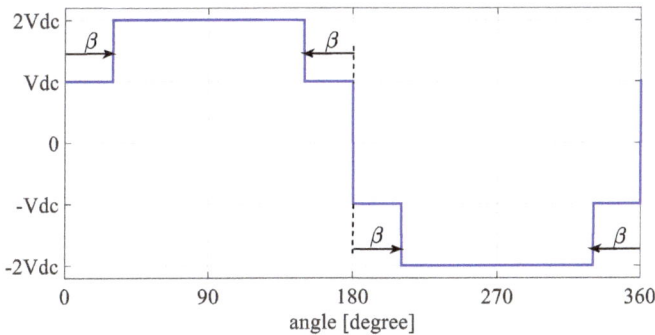

Figure 2. Voltage output waveform with $\beta \neq 0$.

The addition of a stepped voltage level, because $\beta \neq 0$, causes a variation in the harmonic content in the output voltage compared with the reference square waveform. Figure 3 shows the amplitude trend of fundamental harmonic (blue curve), third harmonic (green curve), fifth harmonic (red curve), seventh harmonic (cyan curve), ninth harmonic (purple curve), and eleventh harmonic (yellow curve) versus β expressed in percent respect to the fundamental with $\beta = 0$.

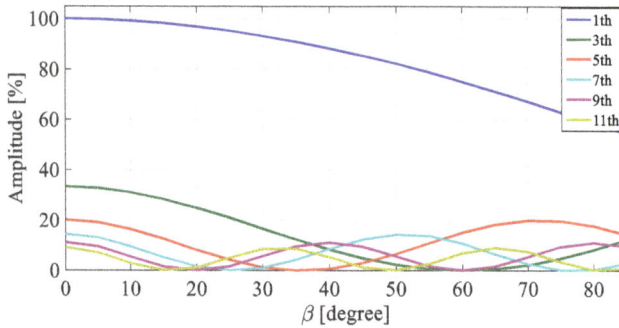

Figure 3. Amplitude of the main harmonic in voltage output versus β. As can be seen, the adjustment of the amplitude of the output waveform varies the harmonic incidence.

As shown in Figure 3, the mutable value of β determines diverse harmonic contents in the output voltage; in particular, there are values of β bringing some harmonic amplitudes to zero. For example, the amplitude of third harmonic is zero for β equal to 60°. Furthermore, it has been noted that the amplitude of fundamental harmonic is reduced with the increase of β and it is not a linear trend.

In literature, modulation techniques for a single-phase H-bridge inverter known as *"voltage cancellation"* are reported [52]. In this work, the same concept as that used for single-level H-bridge inverters has been applied at a single-phase five-level CHBMI, by considering another parameter that can change the harmonic content in output voltage, named α, angular width for which the output voltage is equal to zero, as shown in Figure 4.

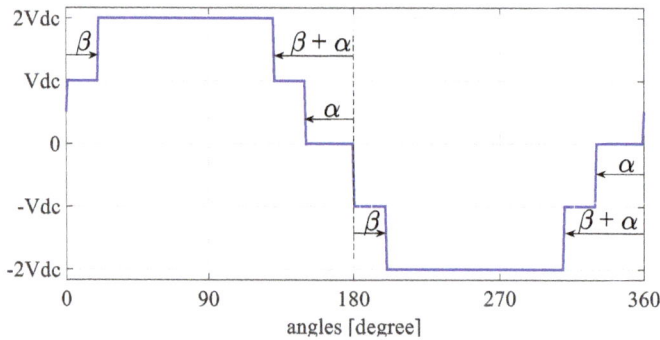

Figure 4. Voltage output waveform with β ≠ 0 and α ≠ 0.

The harmonic content of the voltage waveform shown in Figure 4 was studied by means of the Fourier series. Voltage waveform presents half wave symmetry; thus, amplitude values of even harmonics are zero. In Equation (4), the mathematical formulation of the harmonic amplitude V_h and Fourier coefficients A_h and B_h, where h is the harmonic order, were reported.

$$V_h = \sqrt{A_h^2 + B_h^2}$$
$$A_h = \frac{2V_{dc}}{h\pi}[\sin(h(\pi - \alpha)) - \cos(h\pi)\sin(h(\alpha + \beta)) - \sin(h\beta)] \qquad (4)$$
$$B_h = \frac{2V_{dc}}{h\pi}[1 - \cos(h(\pi - \alpha)) + \cos(h\beta) - \cos(h\pi)\cos(h(\alpha + \beta))]$$

In Equation (5), the voltage waveform in time domain was reported. Obviously, only for odd values of harmonic order.

$$v(t) = \sum_{h=1}^{\infty} [A_h \cos(h\omega t) + B_h \sin(h\omega t)] \tag{5}$$

More in detail, Equation (6) can be used to determine the amplitude of the fundamental harmonic A_1 and the reference harmonics A_{3-11}.

$$
\begin{aligned}
V_1 &= \tfrac{2V_{dc}}{\pi}\sqrt{1 + 2\cos(\alpha) + 4\cos(\beta) + 4\cos(\alpha+\beta) + 2\cos(\alpha+2\beta)} \\
V_3 &= \tfrac{2V_{dc}}{3\pi}\sqrt{1 + 2\cos(3\alpha) + 4\cos(3\beta) + 4\cos(3\alpha+3\beta) + 2\cos(3\alpha+6\beta)} \\
V_5 &= \tfrac{2V_{dc}}{5\pi}\sqrt{1 + 2\cos(5\alpha) + 4\cos(5\beta) + 4\cos(5\alpha+5\beta) + 2\cos(5\alpha+10\beta)} \\
V_7 &= \tfrac{2V_{dc}}{7\pi}\sqrt{1 + 2\cos(7\alpha) + 4\cos(7\beta) + 4\cos(7\alpha+7\beta) + 2\cos(7\alpha+14\beta)} \\
V_9 &= \tfrac{2V_{dc}}{9\pi}\sqrt{1 + 2\cos(9\alpha) + 4\cos(9\beta) + 4\cos(9\alpha+9\beta) + 2\cos(9\alpha+18\beta)} \\
V_{11} &= \tfrac{2V_{dc}}{11\pi}\sqrt{1 + 2\cos(11\alpha) + 4\cos(11\beta) + 4\cos(11\alpha+11\beta) + 2\cos(11\alpha+22\beta)}
\end{aligned}
\tag{6}
$$

The amplitude of fundamental can be varied with α and β values. Figure 5 shows amplitude trend of fundamental versus α and β expressed in percent respect to the fundamental with $\alpha = 0$ and $\beta = 0$. For big values of α and β, there is a change of slope in the trend amplitude because the converter works with only three voltage levels.

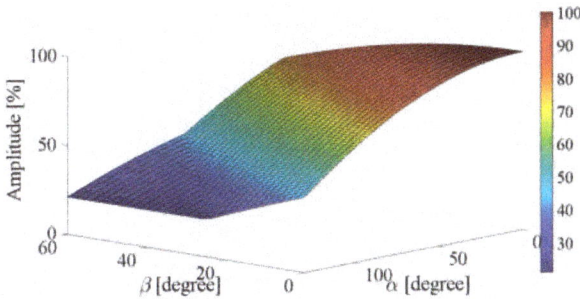

Figure 5. Amplitude trend of fundamental versus α and β. The colour scale is based on the amplitude of the fundamental.

Figures 6–10 show the amplitude trend of third, fifth, seventh, ninth, and eleventh harmonics versus α and β expressed in percent respect to the fundamental with $\alpha = 0$ and $\beta = 0$. It is interesting to note that there are values of α and β that concur together to lowering the considered harmonic amplitude. These areas are colored in darker blue.

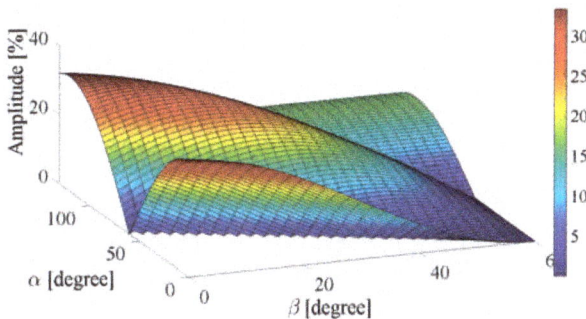

Figure 6. Amplitude trend of third harmonic versus α and β. The colour scale is based on the amplitude of the fundamental.

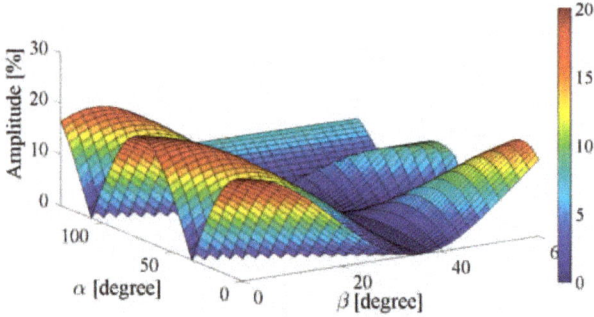

Figure 7. Amplitude trend of fifth harmonic versus α and β. The colour scale is based on the amplitude of the fundamental.

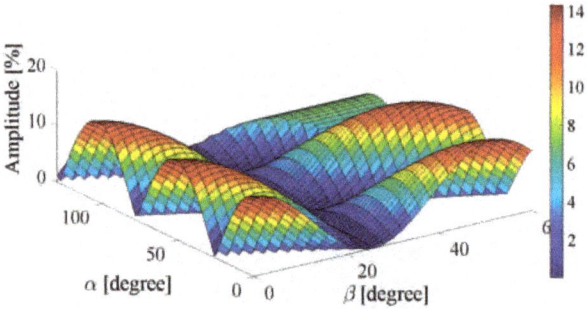

Figure 8. Amplitude trend of seventh harmonic versus α and β. The colour scale is based on the amplitude of the fundamental.

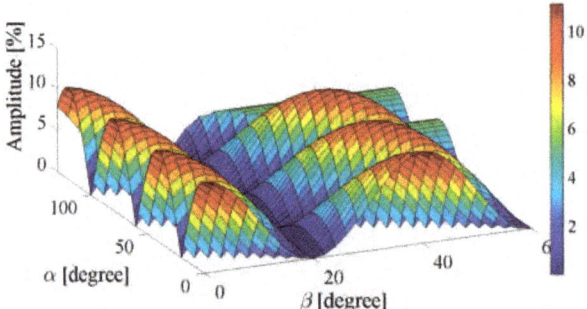

Figure 9. Amplitude trend of ninth harmonic versus α and β. The colour scale is based on the amplitude of the fundamental.

The objective of the next sections, is to propose a new approach for the harmonics mitigation through the individuation of the "working area" (WA), where it is possible to minimize the reference harmonics (third, fifth, seventh, ninth, and eleventh harmonics) without solving non-linear equations.

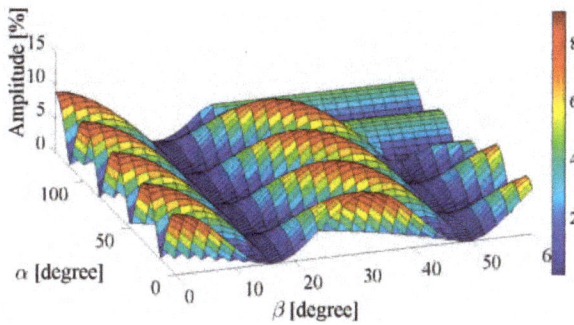

Figure 10. Amplitude trend of eleventh harmonic versus α and β. The colour scale is based on the amplitude of the fundamental.

4. Mitigation Method of Reference Harmonics

As previously described, the main purpose of this work is the mitigation of reference harmonics without solving non-linear equations. In particular, the proposed method is focused on the research of a working area (WA) where the reference harmonics have the minimum values of the amplitude. For the application of this method, it is necessary to calculate the distorting element (DH_{RMS}) of the voltage waveform, which takes into account only the reference harmonics, through Equation (7). In other words, the DH_{RMS} parameter represents the distorting component of the voltage waveform to be mitigated.

$$DH_{RMS} = \sqrt{V_3^2 + V_5^2 + V_7^2 + V_9^2 + V_{11}^2} \tag{7}$$

The amplitude of DH_{RMS} is function of the control angles α and β as can be seen in graphic representation in Figure 11.

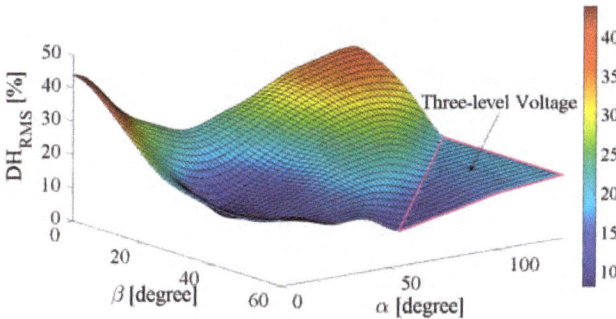

Figure 11. Amplitude trend of the distorting element (DH_{RMS}) versus α and β.

The delimited area with purple lines is the working region where the voltage waveform has only three level voltage. For this reason, this area was neglected for the working points. As can be seen in Figure 11, there is an area with dark blue coloring where the amplitude of the DH_{RMS} is low. Inside this area, it is possible to define the mitigation of reference harmonics. In order to evaluate the WA, a threshold equal to 20% was chosen and through of the level curves the DH_{RMS} and fundamental amplitude were represented. The result obtained is shown in Figure 12.

Figure 12 reports the values of DH_{RMS} with a color map, but also the fundamental amplitude with circumference arcs. By the analysis of Figure 12, it is interesting to note that there is a WA, with dark blue coloring, where the maximum variation of the fundamental amplitude is delimited from approximately 42% to 94%. The clear region has higher values of DH$_{RMS}$ (above 20%) so it can be neglected, and the purple region is the one in which the inverter reduces its levels from five to three, and again can be neglected.

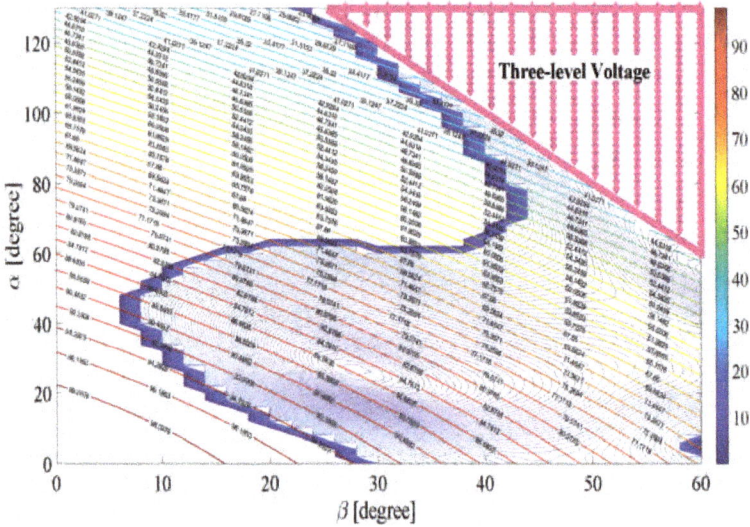

Figure 12. Level curves of fundamental amplitude and DH$_{RMS}$ versus α and β.

The aim of the proposed method is to find the minimum values of the DH$_{RMS}$ for each values of the fundamental amplitude. Thus, the minimum values have to be found in the interceptions of the DH$_{RMS}$ blue lines and circumference arcs of the fundamental amplitude. Therefore, it is possible detected the minimum values of DH$_{RMS}$ and the corresponding control angles, as shown in Figure 13.

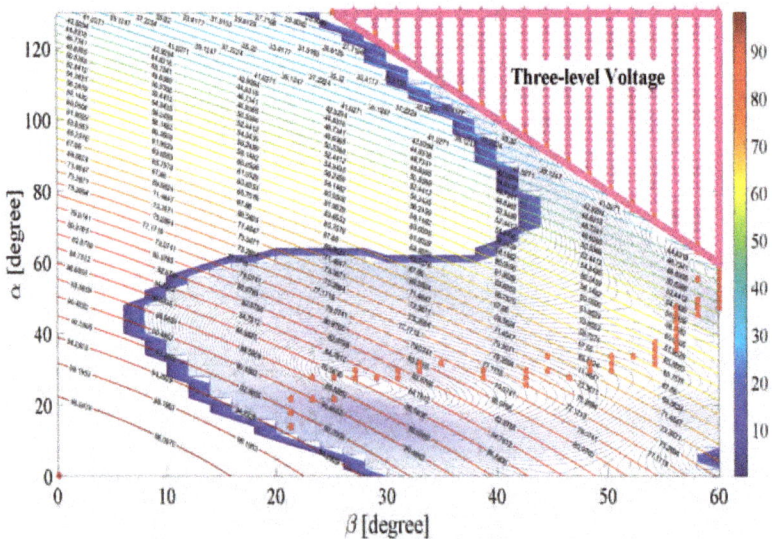

Figure 13. Minimum values of DH$_{RMS}$ versus α and β.

The red dots, shown in Figure 13, represent the trajectory of the minimums values of DH_{RMS}. For each minimum value of DH_{RMS}, there is a pair of value of control angles α and β. In this way, it is possible to obtain mitigation for the chosen harmonic, but the control angles are known without continuity inside the WA. The aim of the next section is to find polynomial equations for the real-time operation of the converter. In this way, all values that reduce the fundamental amplitude from 42% to 94% can be obtained.

5. Polynomial Equations for Real-Time Operation

As stated previously, a WA that minimizes the DH_{RMS} of the considered harmonics was identified. Moreover, for each value of the fundamental amplitude, the minimum values of the DH_{RMS} were detected and are represented as red dots in Figure 13. Thus, the control angles versus the fundamental amplitude in the range from 42% to 94% have been determined, as shown in Figure 14.

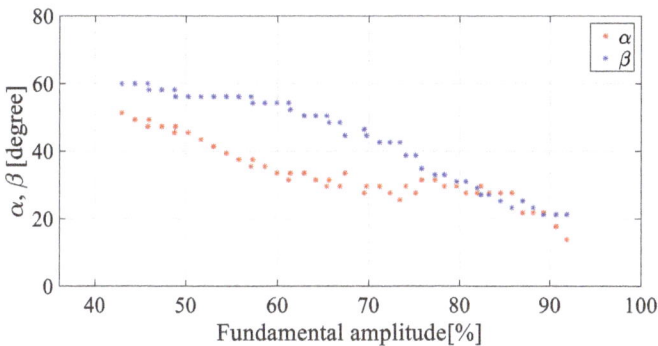

Figure 14. Control angles α and β versus fundamental amplitude.

For real-time operation, it is necessary to know a continuous trend of the control angles α and β inside the WA. Thus, polynomial equations for approximating the trend were evaluated. All equations present a parameter A, indicated in percentage of $100 \cdot (\pi/(8 \cdot V_{dc}))$, which represents the amplitude reference of the voltage waveform. The following five cases, approximating the control angles trends, were taken into account.

In case 1, two second-order polynomial approximations P_α and P_β were used, as shown in Figure 15 and reported in Equation (8):

$$\begin{aligned} P_\alpha &= 0.007890A^2 - 1.631070A + 105.533900 \\ P_\beta &= -0.013164A^2 + 0.881922A + 46.685440 \end{aligned} \tag{8}$$

In case 2, the following two third-order polynomial equations P_α and P_β have been used, and compared with valued that reduce the DH_{RMS}, as shown in Figure 16.

$$\begin{aligned} P_\alpha &= -0.000799A^3 + 0.169602A^2 - 12.233555A + 330.460866 \\ P_\beta &= +0.000456A^3 - 0.105544A^2 + 6.938743A - 81.773030 \end{aligned} \tag{9}$$

In case 3, the range of variation fundamental amplitude, from 42% to 94%, was split into two parts. The first one section involves the variation of the fundamental amplitude from to 42% to 76% and the second section involves the one from 76% to 94%.

For both intervals, two second-order polynomial equations, $P_{\alpha1}$ and $P_{\beta1}$ for the first section and $P_{\alpha2}$ and $P_{\beta2}$ for the second section, were used. Figure 17 shows approximation of case 3.

$$\begin{cases} P_{\alpha1} = +0.020448A^2 - 3.174553A + 151.318608 & \quad 42\% \leq A \leq 76\% \\ P_{\beta1} = -0.020535A^2 + 1.789255A + 19.777618 \\ P_{\alpha2} = -0.080604A^2 + 12.598902A - 462.310004 & \quad 76\% < A \leq 94\% \\ P_{b2} = +0.028942A^2 - 5.789600A + 308.393058 \end{cases} \qquad (10)$$

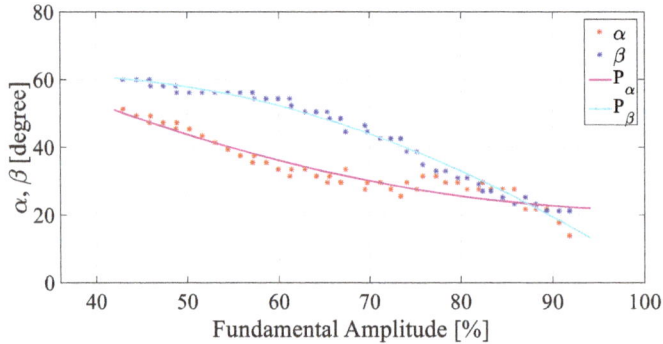

Figure 15. Reconstruction for case 1.

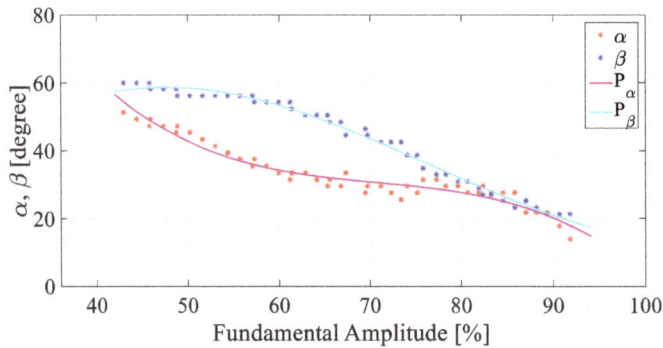

Figure 16. Reconstruction for case 2.

In case 4, the range of variation of the fundamental amplitude, as made for case 3, has been split into two intervals. For the first section, two second-order polynomial equations $P_{\alpha1}$ and $P_{\beta1}$ have been used, while for the second section, two third-order polynomial equations $P_{\alpha2}$ and $P_{\beta2}$ were used. Figure 18 shows the reconstruction of case 4.

$$\begin{cases} P_{\alpha1} = +0.020448A^2 - 3.174553A + 151.318608 & \quad 42\% \leq A \leq 76\% \\ P_{\beta1} = -0.020535A^2 + 1.789255A + 19.777618 \\ P_{\alpha2} = -0.008232A^3 + 2.005310A^2 - 163.278415A + 4472.405127 & \quad 76\% < A \leq 94\% \\ P_{\beta2} = +0.003873A^3 - 0.952589A^2 + 76.969823A - 2013.647026 \end{cases} \qquad (11)$$

Also, for case 5, the range of variation of the fundamental amplitude was split into two sections. For both sections, two third-order polynomial equations, $P_{\alpha 1}$ and $P_{\beta 1}$ for the first section and $P_{\alpha 2}$ and $P_{\beta 2}$ for the second section, have been used. Figure 19 shows the reconstruction for case 5.

$$\begin{cases} P_{\alpha 1} = +0.000799A^3 - 0.122744A^2 + 5.238047A - 10.581937 & 42\% \le A \le 76\% \\ P_{\beta 1} = -0.000674A^3 + 0.100397A^2 - 5.315557A + 156.509770 \\ P_{\alpha 2} = -0.008232A^3 + 2.005310A^2 - 163.278415A + 4472.405127 & 76\% < A \le 94\% \\ P_{\beta 2} = +0.003873A^3 - 0.952589A^2 + 76.969823A - 2013.647026 \end{cases} \tag{12}$$

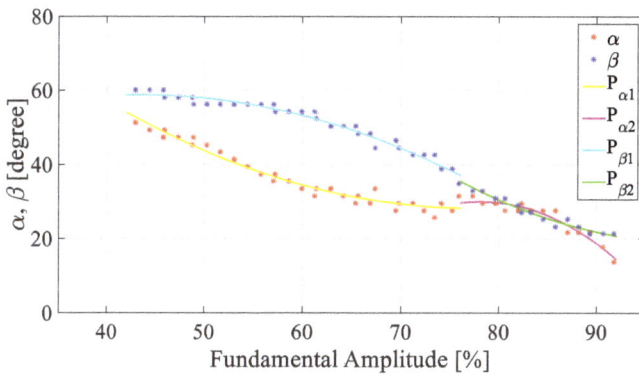

Figure 17. Reconstruction for case 3.

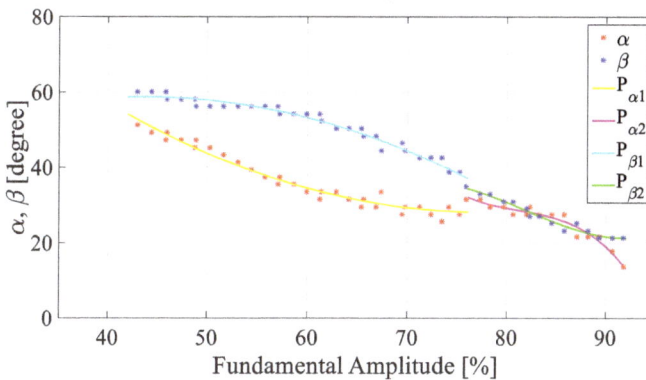

Figure 18. Reconstruction for case 4.

The evaluation of the performance for the five cases was carried out. The aim of the simulations is to reproduce the performances of the converter using the polynomial equations, to evaluate the control angles, as previously described. In Matlab/Simulink® (version 4.1.1, The MathWorks, Inc., Natick, MA, USA) environment mathematic model of the single-phase CHBMI, polynomial equations and logic circuit to generate the gate signals were implemented.

The used tool to compare the performances of the converter, employing different polynomial equations to evaluate the control angles α and β, is the total harmonic distortion rate, THD%, as defined in Equation (13), where V_{rms} is the root mean square value of the voltage waveform and $V_{rms,1}$ is the root mean square of the fundamental amplitude [53].

$$THD\% = \frac{\sqrt{V_{rms}^2 - V_{rms.1}^2}}{V_{rms,1}} \cdot 100 \tag{13}$$

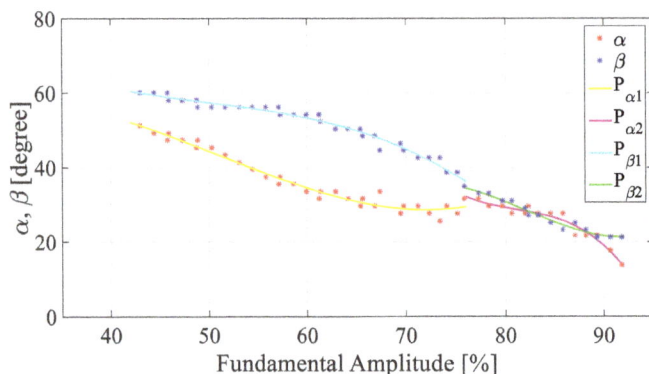

Figure 19. Reconstruction for case 5.

Figure 20 shows comparison among different THD% values obtained for the five different cases. It is interesting to note that in the center of the range, the THD% values are similar for the different cases. Whereas there are different values in the extreme values of the range. In particular, case 1 (red curve) presents the lower values, as illustrated in Figure 20. For this reason, the polynomial equations obtained in case 1 represent the best choice; in real time operation, case 1 requires the lower time execution algorithm respect other cases presented, and it also contributes to better approximating the WA by evaluating the THD%.

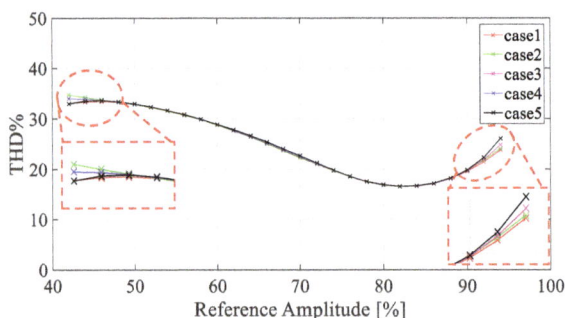

Figure 20. Comparison of total harmonic distortion rate (THD%) values obtained for different cases taken into account.

Figure 21 shows the amplitudes of the obtained reference harmonics. By considering the fundamental amplitude variation from 60% to 88%, the amplitude of the reference harmonics is lower than 10%. The seventh harmonic has the amplitude higher than 10% in the range from 44% to 60%. While, the amplitude of the third harmonic increases after 88% of the reference amplitude.

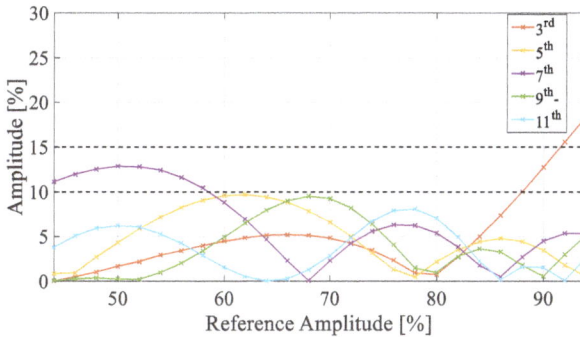

Figure 21. Amplitude of the reference harmonics obtained.

In the next section, an experimental validation of the simulation results was reported. The experimental validation has been carried out by considering only the polynomial equations of case 1.

6. Experimental Validation and Discussion

The aim of this paragraph is to reproduce experimentally the results obtained in the Simulink environment. A single-phase, five level DC/AC converter prototype with a CHBMI circuital structure has been assembled in order to carry out the experimental analysis. The test bench is shown in Figure 22.

Figure 22. A picture of the test bench.

By means of the described test bench, the proposed technique was experimentally tested. The control algorithm was implemented by mean a prototype of FPGA-based (Altera Cyclone III) control board. The use of a FPGA allows the fast execution of control algorithms, parallel elaborations, high numbers of I/O, and high flexibility.

The mathematical operations are managed by a clock signal with frequency equal to 100 MHz and the resolution choice is 32 bit at floating point. The evaluation of the control angles α and β was carried out in 1.01 μs. For the gate signals generation, a simple logic circuit was implemented.

The voltage waveforms have been acquired by the acquisition system with a number of samples equal to 0.5 Ms and in a time interval equal to 20 ms. The acquired experimental samples of the voltage waveforms were processed in Matlab® environment. Figure 23 shows the screenshots of the acquired voltage waveforms and corresponding harmonic spectra for some values of reference amplitude A.

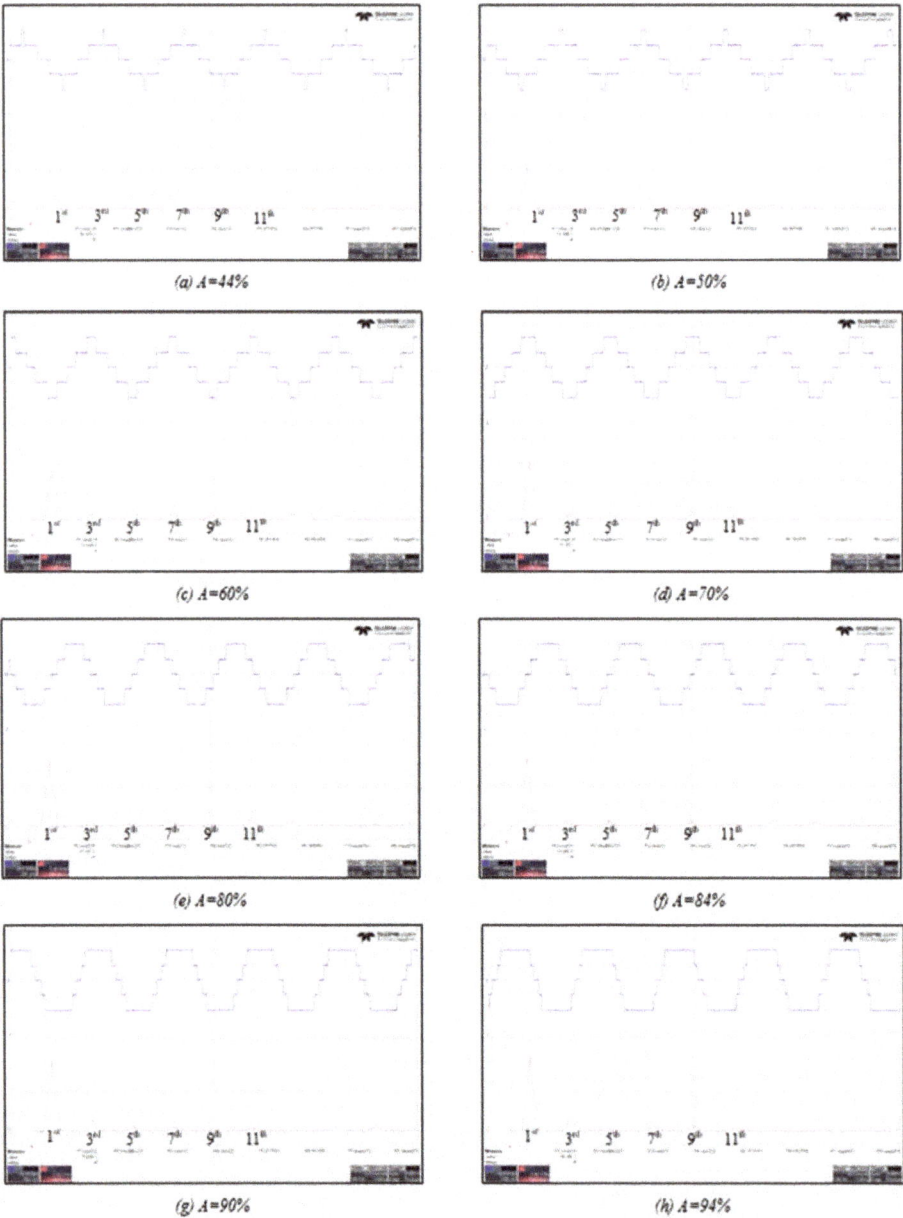

(a) A=44%

(b) A=50%

(c) A=60%

(d) A=70%

(e) A=80%

(f) A=84%

(g) A=90%

(h) A=94%

Figure 23. Screenshot of the voltage waveforms acquired and corresponding harmonic spectra for some values of reference amplitude A.

The used tool to compare the experimental tests with simulations is THD%. Figure 24 shows the comparison between THD% obtained by simulation analysis (blue curve) and THD% obtained by experimental tests (red curve). It is interesting to note that the experimental THD% presents similar values with respect to the simulated ones. In particular, there are only small differences for low values of the reference amplitude.

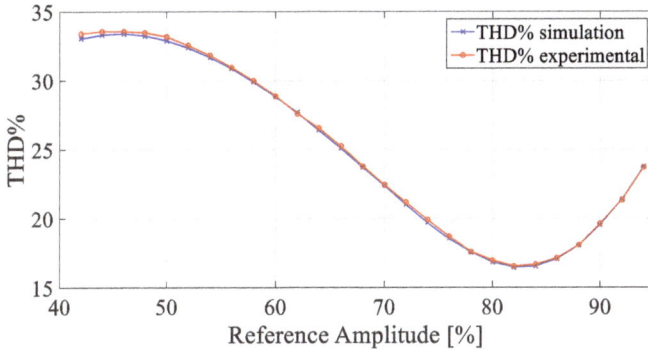

Figure 24. Comparison between THD% obtained by simulation and experimental tests.

For reference amplitude equal to 46%, there is the maximum value of THD% equal to 33.55%. In the range of reference amplitude from 74% to 90%, there are lower values, about 20%. For reference amplitude equal to 82%, there is the minimum value of THD%, which is equal to 16.54%. Such a value is above the proposed ones in the literature [16,19,44,45], but it is obtained for a very simple and more economic structure.

Figure 25 shows the amplitude of fundamental and reference harmonics versus reference amplitude. As observed for simulation results, in the range of reference amplitude from 60% to 88% the amplitude of the reference harmonics is lower than 10%. Moreover, also for the experimental results, there are the same phenomena for the seventh and third harmonics already observed in the simulation results. The fundamental amplitude presents a linear trend inside the working area.

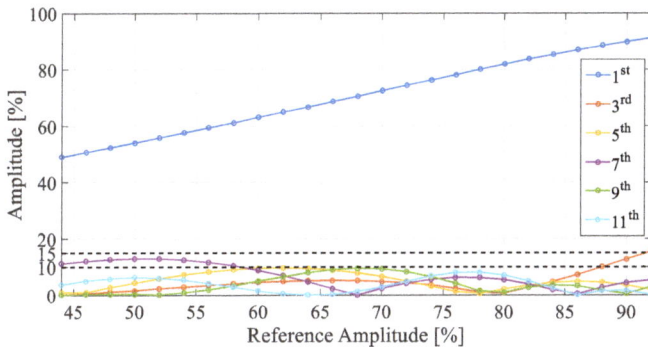

Figure 25. Amplitude of fundamental and main harmonics trend versus reference harmonics. As can be seen, the adjustment of the amplitude of the output of fundamental waveform varies in a linear interval, in which the harmonic incidence can be easily reduced.

7. Conclusions

This paper presents an alternative way to obtain harmonic mitigation for single-phase five-level CHBMI without solving non-linear equation applied in PV systems. Firstly, the so-called "voltage cancellation" technique has been presented, jointly with the analytical expression of the harmonic content on the voltage waveforms.

The proposed method, through the analysis of the distorting component called DH_{RMS}, detects a working area, in which the mitigation of third, fifth seventh, ninth, and eleventh harmonics can be enforced. Thus, through the evaluation of the minimum points of the DH_{RMS}, the corresponding control angles have been obtained. Moreover, the values of the control parameters can be obtained without solving a set of nonlinear transcendental equations, but through a low-time-consuming polynomial equations implementation. In this way, it is possible to enforce a real-time operation with very low computational costs. By experimental validation, the equations obtained allow to drive the converter only in a definite working area, where it is possible change the reference fundamental amplitude from 44% to 94%. Inside this working area, the harmonics mitigation has been limited under 10% in the range from 62% to 88% of the reference amplitude; the seventh harmonic has the amplitude higher than 10% in the range from 44% to 60%, while the amplitude of the third harmonic increases over 10% when the reference amplitude exceeds the 88%.

Author Contributions: The authors have contributed equally to this work. The authors of this manuscript jointly have conceived the theoretical analysis, modeling, simulation and obtained the experimental data.

Funding: This research received no external funding.

Acknowledgments: This work was financially supported by MIUR-Ministero dell'Istruzione dell'Università e della Ricerca (Italian Ministry of Education, University and Research), by SDESLab (Sustainable Development and Energy Saving Laboratory), and LEAP (Laboratory of Electrical Applications) of the University of Palermo.

Conflicts of Interest: The authors declare no conflict of interest.

References

1. Directive 2010/31/EU of the European Parliament and of the Council of 19 May 2010 on the Energy Performance of Buildings. Available online: http://www.buildup.eu/en/practices/publications/directive-201031eu-energy-performance-buildings-recast-19-may-2010 (accessed on 15 January 2018).
2. Miceli, R.; Viola, F. Designing a sustainable university recharge area for electric vehicles: Technical and economic analysis. *Energies* **2017**, *10*, 1604. [CrossRef]
3. Prabaharan, N.; Palanisamy, K. A comprehensive review on reduced switch multilevel inverter topologies, modulation techniques and applications. *Renew. Sustain. Energy Rev.* **2017**, *76*, 1248–1282. [CrossRef]
4. Baker, R.H.; Bannister, L.H. Electric Power Converter. U.S. Patent 3867643, 16 February 1975.
5. Yuan, X.; Barbi, I. Fundamentals of a new diode clamping multilevel inverter. *IEEE Trans. Power Electron.* **2000**, *15*, 711–718. [CrossRef]
6. Rodriguez, J.; Bernet, S.; Steimer, P.K.; Lizama, I.E. A survey on neutral-point-clamped inverters. *IEEE Trans. Ind. Electron.* **2010**, *57*, 2219–2230. [CrossRef]
7. Huang, J.; Corzine, K.A. Extended operation of flying capacitor multilevel inverters. *IEEE Trans. Power Electron.* **2006**, *21*, 140–147. [CrossRef]
8. Kou, X.; Corzine, K.A.; Wielebski, M.W. Overdistention operation of cascaded multilevel inverters. *IEEE Trans. Ind. Appl.* **2006**, *42*, 817–824.
9. Kala, P.; Arora, S. A comprehensive study of classical and hybrid multilevel inverter topologies for renewable energy applications. *Renew. Sustain. Energy Rev.* **2017**, *76*, 905–931. [CrossRef]
10. Venkataramanaiah, J.; Suresh, Y.; Panda, A.K. A review on symmetric, asymmetric, hybrid and single DC sources based multilevel inverter topologies. *Renew. Sustain. Energy Rev.* **2017**, *76*, 788–812. [CrossRef]
11. Jana, J.; Saha, H.; Bhattacharya, K.D. A review of inverter topologies for single-phase grid-connected photovoltaic systems. *Renew. Sustain. Energy Rev.* **2017**, *72*, 1256–1270. [CrossRef]
12. Monica, P.; Kowsalya, M. Control strategies of parallel operated inverters in renewable energy application: A review. *Renew. Sustain. Energy Rev.* **2016**, *65*, 885–901. [CrossRef]
13. Kumar, N.; Saha, T.K.; Dey, J. Modeling, control and analysis of cascaded inverter based grid-connected photovoltaic system. *Int. J. Electr. Power Energy Syst.* **2016**, *78*, 165–173. [CrossRef]
14. Ravi, A.; Manoharan, P.S.; Anand, J.V. Modeling and simulation of three phase multilevel inverter for grid connected photovoltaic systems. *Sol. Energy* **2011**, *85*, 2811–2818. [CrossRef]

15. Rajkumar, M.V.; Manoharan, P.S. FPGA based multilevel cascaded inverters with SVPWM algorithm for photovoltaic system. *Sol. Energy* **2013**, *87*, 229–245. [CrossRef]

16. Prabaharan, N.; Palanisamy, K. Analysis and integration of multilevel inverter configuration with boost converters in a photovoltaic system. *Energy Convers. Manag.* **2016**, *128*, 327–342. [CrossRef]

17. Iero, D.; Carbone, R.; Carotenuto, R.; Felini, C.; Merenda, M.; Pangallo, G.; Della Corte, F.G. SPICE modelling of a complete photovoltaic system including modules, Energy storage elements and a multilevel inverter. *Sol. Energy* **2014**, *107*, 338–350. [CrossRef]

18. Javad, O.; Shayan, E.; Ali, M. Compensation of voltage sag caused by partial shading in grid-connected PV system through the three-level SVM inverter. *Sustain. Energy Technol. Assess.* **2016**, *18*, 107–118.

19. Prabaharan, N.; Palanisamy, K. A Single Phase Grid Connected Hybrid Multilevel Inverter for Interfacing Photo-voltaic System. *Energy Procedia* **2016**, *103*, 250–255. [CrossRef]

20. Husev, O.; Roncero-Clemente, C.; Romero-Cadaval, E.; Vinnikov, D.; Jalakas, T. Three-level three-phase quasi-Z-source neutral-point-clamped inverter with novel modulation technique for photovoltaic application. *Electr. Power Syst. Res.* **2016**, *130*, 10–21. [CrossRef]

21. Lauria, D.; Coppola, M. Design and control of an advanced PV inverter. *Sol. Energy* **2014**, *110*, 533–542. [CrossRef]

22. Hannan, M.A.; Azidin, F.A.; Mohamed, A. Hybrid electric vehicles and their challenges: A review. *Renew. Sustain. Energy Rev.* **2014**, *29*, 135–150. [CrossRef]

23. Zhai, L.; Lee, G.; Gao, X.; Zhang, X.; Gu, Z.; Zou, M. Impact of Electromagnetic Interforance from Power Inverter Drive System on Batteries in Electric Vehicle. *Energy Procedia* **2016**, *88*, 881–888. [CrossRef]

24. Mese, E.; Ayaz, M.; Tezcan, M.M. Design considerations of a multitasked electric machine for automotive applications. *Electr. Power Syst. Res.* **2016**, *131*, 147–158. [CrossRef]

25. Tolbert, L.M.; Peng, F.Z.; Cunnyngham, T.; Chiasson, J.N. Charge balance control schemes for cascade multilevel converter in hybrid electric vehicles. *IEEE Trans. Ind. Electron.* **2002**, *49*, 1058–1064. [CrossRef]

26. Khoucha, F.; Lagoun, S.M.; Marouani, K.; Kheloui, A.; Benbouzid, M.E.H. Hybrid cascaded H-bridge multilevel-inverter induction-motor-drive direct torque control for automotive applications. *IEEE Trans. Ind. Electron.* **2010**, *57*, 892–899. [CrossRef]

27. Ding, X.; Du, M.; Zhou, T.; Guo, H.; Zhang, C. Comprehensive comparison between silicon carbide MOSFETs and silicon IGBTs based traction systems for electric vehicles. *Appl. Energy* **2016**, *194*, 626–634. [CrossRef]

28. Kabalyk, Y. Determination of energy loss in power voltage inverters for power supply of locomotive traction motors. *Procedia Eng.* **2016**, *165*, 1437–1443. [CrossRef]

29. Carpita, M.; Marchesoni, M.; Pellerin, M.; Moser, D. Multilevel converter for traction applications: Small-scale prototype tests results. *IEEE Trans. Ind. Electron.* **2008**, *55*, 2203–2212. [CrossRef]

30. Kumar, Y.P.; Ravikumar, B. A simple modular multilevel inverter topology for the power quality improvement in renewable energy based green building microgrids. *Electr. Power Syst. Res.* **2016**, *140*, 147–161. [CrossRef]

31. Kamel, R.M. New inverter control for balancing standalone micro-grid phase voltages: A review on MG power quality improvement. *Renew. Sustain. Energy Rev.* **2016**, *63*, 520–532. [CrossRef]

32. Mohamed, E.E.; Sayed, M.A. Matrix converters and three-phase inverters fed linear induction motor drives-Performance compare. *Ain Shams Eng. J.* **2016**. Available online: https://doi.org/10.1016/j.asej.2016.02.002 (accessed on 21 March 2016).

33. Trabelsi, M.; Boussak, M.; Benbouzid, M. Multiple criteria for high performance real-time diagnostic of single and multiple open-switch faults in ac-motor drives: Application to IGBT-based voltage source inverter. *Electr. Power Syst. Res.* **2017**, *144*, 136–149. [CrossRef]

34. Prakash, G.; Subramani, C.; Bharatiraja, C.; Shabin, M. A low cost single phase grid connected reduced switch PV inverter based on Time Frame Switching Scheme. *Inter. J. Electr. Power Energy Syst.* **2016**, *2016 77*, 100–111. [CrossRef]

35. Gupta, V.K.; Mahanty, R. Optimized switching scheme of cascaded H-bridge multilevel inverter using PSO. *Inter. J. Electr. Power Energy Syst.* **2015**, *64*, 699–707. [CrossRef]

36. Mesquita, S.J.; Antunes, F.L.M.; Daher, S. A new bidirectional hybrid multilevel inverter with 49-level output voltage using a single dc voltage source and reduced number of on components. *Electr. Power Syst. Res.* **2017**, *143*, 703–714. [CrossRef]

37. Báez-Fernández, H.; Ramírez-Beltrán, N.D.; Méndez-Piñero, M.I. Selection and configuration of inverters and modules for a photovoltaic system to minimize costs. *Renew. Sustain. Energy Rev.* **2016**, *58*, 16–22. [CrossRef]

38. Schettino, G.; Benanti, S.; Buccella, C.; Caruso, M.; Castiglia, V.; Cecati, C.; Di Tommaso, A.O.; Miceli, R.; Romano, P.; Viola, F. Simulation and experimental validation of multicarrier PWM techniques for three-phase five-level cascaded H-bridge with FPGA controller. *Inte. J. Renew. Energy Res.* **2017**, *7*, 1383–1394.

39. Schettino, G.; Buccella, C.; Caruso, M.; Cecati, C.; Castiglia, V.; Miceli, R.; Viola, F. Overview and experimental analysis of MC SPWM techniques for single-phase five level cascaded H-bridge FPGA controller-based. In Proceedings of the IECON 2016-42nd Annual Conference of the IEEE Industrial Electronics Society, Florence, Italy, 23–26 October 2016; pp. 4529–4534.

40. Benanti, S.; Buccella, C.; Caruso, M.; Castiglia, V.; Cecati, C.; Di Tommaso, A.O.; Miceli, R.; Romano, P.; Schettino, G.; Viola, F. Experimental analysis with FPGA controller-based of MC PWM techniques for three-phase five level cascaded H-bridge for PV applications. In Proceedings of the 2016 IEEE International Conference on Renewable Energy Research and Applications (ICRERA), Birmingham, UK, 20–23 November 2016; pp. 1173–1178.

41. Rahim, N.A.; Ping, H.W.; Selvaraj, J. Elimination of harmonics in photovoltaic seven-level inverter with Newton-Raphson optimization. *Procedia Environ. Sci.* **2013**, *2013 17*, 519–528.

42. Kavali, J.; Mittal, A. Analysis of various control schemes for minimal Total Harmonic Distortion in cascaded H-bridge multilevel inverter. *J. Electr. Syst. Inf. Technol.* **2016**, *3*, 428–441. [CrossRef]

43. Al-Othman, A.K.; Abdelhamid, T.H. Elimination of harmonics in multilevel inverters with non-equal dc sources using PSO. In Proceedings of the 2008 13th International Power Electronics and Motion Control Conference (EPE-PEMC 2008), Poznan, Poland, 1–3 September 2008; pp. 756–764.

44. Ganesan, K.; Barathi, K.; Chandrasekar, P.; Balaji, D. Selective harmonic elimination of cascaded multilevel inverter using BAT algorithm. *Procedia Technol.* **2015**, *21*, 651–657. [CrossRef]

45. Marzoughi, A.; Imaneini, H.; Moeini, A. An optimal selective harmonic mitigation technique for high power converters. *Inter. J. Electr. Power Energ. Syst.* **2013**, *49*, 34–39. [CrossRef]

46. Voltage Characteristics of Electricity Supplied by Public Electricity Networks. Available online: http://fs.gongkong.com/files/technicalData/201110/2011100922385600001.pdf (accessed on 15 January 2018).

47. Personen, M.A. Harmonics characteristic parameters methods of study estimates of existing values in the network. *Electra* **1981**, *77*, 35–54.

48. Electromagnetic Compatibility (EMC)–Part 3-7: Limits–Assessment of Emission Limits for the Connection of Fluctuating Installations to MV, HV and EHV Power Systems. Available online: https://ieeexplore.ieee.org/document/6232421/ (accessed on 15 January 2018).

49. Letha, S.S.; Thakur, T.; Kumar, J. Harmonic elimination of a photo-voltaic based cascaded H-bridge multilevel inverter using PSO (particle swarm optimization) for induction motor drive. *Energy* **2016**, *107*, 335–346. [CrossRef]

50. Rao, G.N.; Raju, P.S.; Sekhar, K.C. Harmonic elimination of cascaded H-bridge multilevel inverter based active power filter controlled by intelligent techniques. *Inte. J. Electr. Power Energ. Syst.* **2014**, *61*, 56–63.

51. Panda, A.K.; Patnaik, S.S. Analysis of cascaded multilevel inverters for active harmonic filtering in distribution networks. *Inte. J. Electr. Power Energ. Syst.* **2015**, *66*, 216–226. [CrossRef]

52. Mohan, N.; Undeland, T.M. *Power Electronics: Converters, Applications, and Design*, 3rd ed.; John Wiley & Sons, Inc.: Hoboken, NJ, USA, 2007.

53. Dordevic, O.; Jones, M.; Levi, E. Analytical formulas for phase voltage RMS squared and THD in PWM multiphase systems. *IEEE Trans. Power Electron.* **2015**, *30*, 1645–1656. [CrossRef]

energies

MDPI

Article

An LQR-Based Controller Design for an LCL-Filtered Grid-Connected Inverter in Discrete-Time State-Space under Distorted Grid Environment

Thuy Vi Tran, Seung-Jin Yoon and Kyeong-Hwa Kim *

Department of Electrical and Information Engineering, Seoul National University of Science and Technology, 232 Gongneung-ro, Nowon-gu, Seoul 01811, Korea; tranvithuy@gmail.com (T.V.T.); tmdwls3233@naver.com (S.-J.Y.)
* Correspondence: k2h1@seoultech.ac.kr; Tel.: +82-2-970-6406; Fax: +82-2-978-2754

Received: 18 July 2018; Accepted: 6 August 2018; Published: 8 August 2018

Abstract: In order to alleviate the negative impacts of harmonically distorted grid conditions on inverters, this paper presents a linear quadratic regulator (LQR)-based current control design for an inductive-capacitive-inductive (LCL)-filtered grid-connected inverter. The proposed control scheme is constructed based on the internal model (IM) principle in which a full-state feedback controller is used for the purpose of stabilization and the integral terms as well as resonant terms are augmented into a control structure for the reference tracking and harmonic compensation, respectively. Additionally, the proposed scheme is implemented in the synchronous reference frame (SRF) to take advantage of the simultaneous compensation for both the negative and positive sequence harmonics by one resonant term. Since this leads to the decrease of necessary resonant terms by half, the computation effort of the controller can be reduced. With regard to the full-state feedback control approach for the LCL-filtered grid connected inverter, additional sensing devices are normally required to measure all of the system state variables. However, this causes a complexity in hardware and high implementation cost for measurement devices. To overcome this challenge, this paper presents a discrete-time current full-state observer that uses only the information from the control input, grid-side current sensor, and grid voltage sensor to estimate all of the system state variables with a high precision. Finally, an optimal linear quadratic control approach is introduced for the purpose of choosing optimal feedback gains, systematically, for both the controller and full-state observer. The simulation and experimental results are presented to prove the effectiveness and validity of the proposed control scheme.

Keywords: distorted grid; digital signal processor (DSP) TMS320F28335; grid-connected inverter; internal model; linear quadratic regulator; LCL filter

---

1. Introduction

The increasing interest in grid-connected voltage source inverters (VSI) for renewable energy conversion systems poses a challenge to the current control design of inverter systems. In particular, the current control scheme is responsible for a high quality of injected current to meet the power quality standard of distributed generation such as the IEEE-519 in USA or the IEC 61000-3-2 in Europe [1] even under harmonically distorted grid voltages. Additionally, the filter connected between the utility grid and VSI plays an essential role to attenuate the current in high switching frequency from the pulse width modulated inverter. In general, LCL filters are regarded as being satisfactory for three-phase voltage source grid-connected inverters because they provide a better grid-side current quality with lower costs and a smaller physical size when compared to the conventional L filters. Nevertheless, the disadvantages of using LCL filters include a high-order system and the resonance behavior. As a result, the current control strategies of LCL-filtered inverters are more difficult and complex to stabilize the system.

There are two methods to damp the resonance frequency of the LCL filter: passive damping using additional physical components on LCL circuits, and active damping implemented by modifying the control algorithm. A large number of studies in literature address the controller design in both ways. In [2], a passive resistor is added in series with the filter capacitance with the aim of attenuating the peak of LCL filter resonance. However, the main drawback of this method is that it causes extra losses through heat dissipation and overall reduction of the system efficiency. On the other hand, the active damping approaches are generally preferable and used quite commonly due to the fact that they stabilize the system without increasing the losses. An active damping realized by virtual resistance based on the capacitance current feedback is presented in [3–5]. In particular, Jia. Y et al. [3] presents the capacitance current feedback active damping implemented via a proportional gain of the feedback signal. The stability enhancement and robustness against distorted grid voltages are also discussed in this work. Similarly, the capacitor current feedback loop of the LCL filter is implemented to improve both the damping characteristic and inner-loop stability of a hierarchical control structure [4]. Furthermore, an H-infinity repetitive controller in [5] demonstrates a better performance and efficiency of the inverter by introducing the feedback of capacitor current to damp the resonance. Even though the stabilization can be achieved, those schemes increase the complexity and cost in hardware caused by extra sensors to obtain capacitor currents. As in other approaches, the studies in [6–8] present a state-space control scheme which provides a convenient and straightforward way for resonance damping. In order to avoid extra sensing devices, a full-state observer is also presented in these works, whereupon the number of sensors used in the controller is compatible with the design of the conventional L filter case.

Aside from the resonance of the LCL filter, the issue of grid voltage distortion should be taken into account in a current controller design for a grid-connected inverter. Thus, the adoption of a proportional-resonant (PR) controller in the control strategy was studied widely in both classical and modern control approaches to improve the power quality. Conventionally, the proportional-integral (PI) controllers in rotating frame and the resonant controller in the stationary frame have been studied in detail in [9], which demonstrates that an equivalent control performance can be achieved by these controllers. The research work in [10] uses multiple PI controllers that are implemented in respective reference frames rotating with the fundamental and harmonic frequencies to achieve control objectives such as reference tracking and harmonic compensation. Another approach uses a PR scheme and harmonic compensation control performing at particular frequencies in the stationary frame to restrain the disturbance caused by the distorted grid voltage [3,11–15]. However, since these approaches require two regulators to compensate both the negative and positive sequences, several regulators might be necessary in the control scheme to meet the required total harmonic distortion (THD) performance, which often leads to a significantly heavy computational burden. In order to reduce the complexity of the digital implementation, the PI and resonant (PI-RES) current control scheme constructed in the synchronous reference frame (SRF) has been studied in [16–18] to achieve multiple harmonic compensation with the number of resonant controllers reduced by half. In particular, PI-RES control schemes in the SRF are proposed for active power filters [18] or a three-phase grid converter system [16,17] for the purpose of compensating multiple harmonics.

Aside from the resonant control scheme, an H-infinity repetitive control approach was studied in [6] which presented the robustness against the system parameter variations of the LCL-type grid-connected inverter. In addition, Fu. X et al. investigated a neural network (NN)-based vector control approach for single-phase grid-connected converters to achieve an improved control performance without any damping method for LCL filter [19].

In addition to the typical control structure based on the transfer function design [3,13–15], the internal model (IM) principle proposed by Francis and Woham [20] has been applied to design controllers such as the PI and PR in the state-space. In this regard, several studies considered the IM approach to integrate control terms into the current control structure [17,21–23]. However, such a multivariable design approach also poses a challenge to an appropriate selection of controller gains to stabilize the system as well as to ensure both the desired steady-state and transient-state performances. The current controller design using the direct pole placement method in the state-space has been accomplished in the continuous-time domain [7], as well as in the discrete-time domain [8,21]. Although the controller gains can be chosen based on the open-loop poles and the desired dynamics of the closed-loop system, the pole placement method is not an attractive way in a complex system due to the laborious process to select a large number of state-feedback and controller gains. On the other hand, the studies in [12,17,22] solve the linear matric inequalities derived from the stability condition in the Lyapunov sense to obtain the controller gains systematically. In the same vein, an optimal solution based on the linear quadratic regulator (LQR) has been presented in [23], which optimizes the cost function to calculate the optimal gains of the system.

In regard to the solution to reduce the number of needed sensors while still meeting the requirement on the availability of system state variables for full-state feedback controller, many types of observer have been studied to estimate the system state variables by using only the information from the system input and output signals. In particular, the research works in [7,8,17,21] employ the prediction-type full-state observer, while the study in [24] presents the reduced-order observer. However, there are not many studies regarding the current full-state observer in the discrete-time domain and its performance applied to three-phase LCL-filtered grid-connected inverters, even though it is known to have the advantages that the estimated value is based on the current measurement in comparison with the prediction-type observer and the impact of possible noise from the system output signals can be avoided.

This paper presents a control design methodology for a grid-connected inverter with an LCL filter in the discrete-time state-space, where the current control design is accomplished by a full-state feedback control after incorporating the integral and resonant terms into control structure. In this proposed scheme, the controller is implemented in the SRF in order that the integral control on the DC quantities can ensure zero steady-state current error. Furthermore, four harmonic components in phase currents at the 5th, 7th, 11th and 13th order can be effectively compensated at the same time with only two resonant terms at 6th and 12th order. With an aim of reducing the total number of sensors required for the control of LCL-filtered grid-connected inverters, a current full-state observer is presented in the discrete-time domain with excellent estimation capability. The augmentation of the resonant terms as well as the integral term into an inverter system model causes an increase in the number of feedback gains to be selected. To choose the feedback gains in a systematic way, the optimal linear quadratic control approach is adopted in this paper. By minimizing the cost function to satisfy the stability and robustness requirements of the system, the overall system can be designed in an effective and straightforward way. As a result, both the reference tracking and harmonic compensation capability can be achieved in an LCL-filtered grid-connected inverter with an LQR approach by using only the grid-side current sensor and grid voltage sensor. To demonstrate the effectiveness and validity of the proposed control scheme, the PSIM software-based simulation (9.1, Powersim, Rockville, MD, USA) and experiments have been carried out comprehensively by using a three-phase 2 kVA prototype grid-connected inverter under adverse grid conditions.

2. State-Space Description of a Grid-Connected Inverter with LCL Filter

2.1. Modeling of a Grid-Connected Inverter with LCL Filter

In the SRF, three-phase variables "*abc*" are transformed into two orthogonal DC phasor quantities "*dq*" by means of the Park's transformation as follows:

$$
\begin{bmatrix} f_q \\ f_d \\ f_0 \end{bmatrix} = \frac{2}{3} \begin{bmatrix} \cos(\theta) & \cos(\theta - 2\pi/3) & \cos(\theta + 2\pi/3) \\ \sin(\theta) & \sin(\theta - 2\pi/3) & \sin(\theta + 2\pi/3) \\ 1/2 & 1/2 & 1/2 \end{bmatrix} \begin{bmatrix} f_a \\ f_b \\ f_c \end{bmatrix}
\tag{1}
$$

where f denotes the variable being transformed and θ is the rotating phasor angle.

Figure 1 shows a configuration of a three-phase grid-connected inverter with an LCL filter, in which V_{DC} denotes the DC-link voltage, R_1, R_2, L_1, and L_2 are the filter resistances and filter inductances, respectively, and C is the filter capacitance. In the SRF, the mathematical model of the inverter system can be expressed as follows:

$$\dot{i}_2^q = -\frac{R_2}{L_2}i_2^q - \omega i_2^d + \frac{1}{L_2}v_c^q - \frac{1}{L_2}e^q \tag{2}$$

$$\dot{i}_2^d = -\frac{R_2}{L_2}i_2^d + \omega i_2^q + \frac{1}{L_2}v_c^d - \frac{1}{L_2}e^d \tag{3}$$

$$\dot{i}_1^q = -\frac{R_1}{L_1}i_1^q - \omega i_1^d - \frac{1}{L_1}v_c^q + \frac{1}{L_1}v_i^q \tag{4}$$

$$\dot{i}_1^d = -\frac{R_1}{L_1}i_1^d + \omega i_1^q - \frac{1}{L_1}v_c^d + \frac{1}{L_1}v_i^d \tag{5}$$

$$\dot{v}_c^q = -\omega v_c^d - \frac{1}{C}i_2^q + \frac{1}{C}i_1^q \tag{6}$$

$$\dot{v}_c^d = \omega v_c^q - \frac{1}{C}i_2^d + \frac{1}{C}i_1^d \tag{7}$$

where the superscript "q" and "d" denote the q-axis and d-axis variables, respectively, ω is the angular frequency of the grid voltage, i_1 is the inverter-side current, i_2 is the grid-side current, v_c is the capacitor voltage, e is the grid voltage, and v_i is the inverter output voltage.

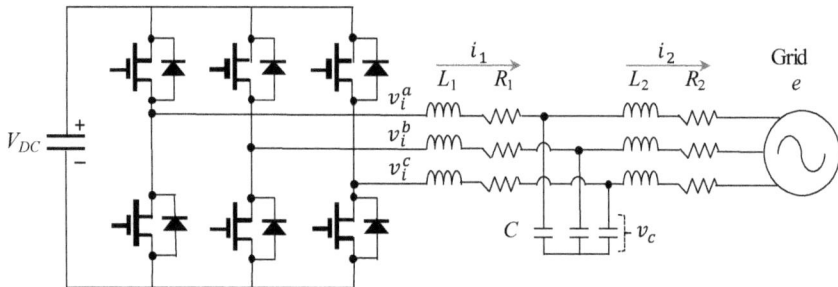

Figure 1. Configuration of a grid-connected inverter with inductive-capacitive-inductive (LCL) filter.

From Equations (2) to (7), the continuous-time representation of inverter system can be expressed in the SRF as:

$$\dot{x}(t) = Ax(t) + Bu(t) + De(t) \tag{8}$$

$$y(t) = Cx(t) \tag{9}$$

where $\mathbf{x} = \begin{bmatrix} i_2^q & i_2^d & i_1^q & i_1^d & v_c^q & v_c^d \end{bmatrix}^T$ is the system state vector, $\mathbf{u} = [v_i^q \ v_i^d]^T$ is the system input vector, $\mathbf{e} = [e^q \ e^d]^T$ is the grid voltage vector, and the system matrices $\mathbf{A}$, $\mathbf{B}$, $\mathbf{C}$, and $\mathbf{D}$ are expressed as:

$$\mathbf{A} = \begin{bmatrix} -R_2/L_2 & -\omega & 0 & 0 & 1/L_2 & 0 \\ \omega & -R_2/L_2 & 0 & 0 & 0 & 1/L_2 \\ 0 & 0 & -R_1/L_1 & -\omega & -1/L_1 & 0 \\ 0 & 0 & \omega & -R_1/L_1 & 0 & -1/L_1 \\ -1/C & 0 & 1/C & 0 & 0 & -\omega \\ 0 & -1/C & 0 & 1/C & \omega & 0 \end{bmatrix} \tag{10}$$

$$\mathbf{B} = \begin{bmatrix} 0 & 0 \\ 0 & 0 \\ 1/L_1 & 0 \\ 0 & 1/L_1 \\ 0 & 0 \\ 0 & 0 \end{bmatrix}, \mathbf{D} = \begin{bmatrix} -1/L_2 & 0 \\ 0 & -1/L_2 \\ 0 & 0 \\ 0 & 0 \\ 0 & 0 \\ 0 & 0 \end{bmatrix}, \mathbf{C} = \begin{bmatrix} 1 & 0 & 0 & 0 & 0 & 0 \\ 0 & 1 & 0 & 0 & 0 & 0 \end{bmatrix} \quad (11)$$

2.2. System Model Discretization

For a digital implementation, the discretized model of inverter system is obtained by using the zero-order hold with the sampling time T_s as [25]:

$$\mathbf{x}(k+1) = \mathbf{A_d}\mathbf{x}(k) + \mathbf{B_d}\mathbf{u}(k) + \mathbf{D_d}\mathbf{e}(k) \quad (12)$$

$$\mathbf{y}(k) = \mathbf{C_d}\mathbf{x}(k) \quad (13)$$

where the matrices $\mathbf{A_d}$, $\mathbf{B_d}$, $\mathbf{C_d}$, and $\mathbf{D_d}$ can be calculated as follows:

$$\mathbf{A_d} = e^{\mathbf{A}T_s} = \mathbf{I} + \frac{\mathbf{A}T_s}{1!} + \frac{\mathbf{A}^2 T_s^2}{2!} + \cdots \quad (14)$$

$$\mathbf{B_d} = \mathbf{A}^{-1}(\mathbf{A_d} - \mathbf{I})\mathbf{B}, \ \mathbf{C_d} = \mathbf{C} \quad (15)$$

$$\mathbf{D_d} = \mathbf{A}^{-1}(\mathbf{A_d} - \mathbf{I})\mathbf{D} \quad (16)$$

3. Proposed Control Scheme

Figure 2 represents the proposed control scheme for a three-phase inverter connected with the utility grid through an LCL filter. The inverter is controlled by the proposed current controller through the space vector pulse width modulation (PWM). Also, the phase-locked loop (PLL) scheme is implemented to generate the phase angle of the grid voltage for the grid synchronization process. The proposed control scheme is constructed by an integral-resonant state feedback controller and a current full-state observer in the discrete-time domain with only the measurements of the grid-side currents and grid voltages. Besides, the current full-state observer is also implemented by using LCL-filter inverter model to estimate the system state variables $\mathbf{x}$ from the control input $\mathbf{u}$ and system outputs $\mathbf{y}$. Those estimated states are used for the full-state feedback controller to stabilize the whole system.

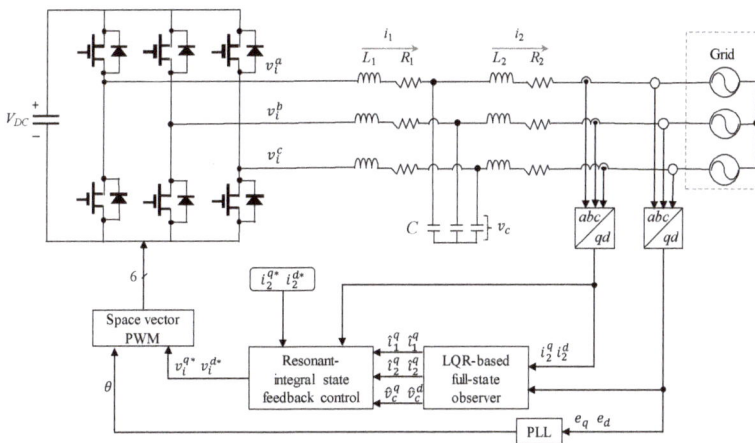

Figure 2. Block diagram of the proposed integral-resonant state feedback current control scheme with observer.

3.1. Internal Model-Based Current Controller

To ensure asymptotic reference tracking as well as disturbance rejection for the harmonics in the orders of 6th and 12th in the SRF, the integral-resonant state feedback control is constructed by augmenting the integral and resonant terms in the discrete-time state-space based on the internal model principle. An integral term in the state-space is expressed as [21,25]:

$$
\begin{bmatrix} \dot{x}_i^q(t) \\ \dot{x}_i^d(t) \end{bmatrix} = \mathbf{A_{Pc}} \begin{bmatrix} x_i^q(t) \\ x_i^d(t) \end{bmatrix} + \mathbf{B_{Pc}} \begin{bmatrix} \varepsilon^q(t) \\ \varepsilon^d(t) \end{bmatrix}
\tag{17}
$$

where $\varepsilon = \begin{bmatrix} \varepsilon^q & \varepsilon^d \end{bmatrix}^T = \mathbf{r} - \mathbf{C_d x}$ is the current error vector, $\mathbf{r} = \begin{bmatrix} i_2^{q*} & i_2^{d*} \end{bmatrix}^T$ is the reference current vector, and $\mathbf{A_{Pc}} = \begin{bmatrix} 0 & 0 \\ 0 & 0 \end{bmatrix}$, $\mathbf{B_{Pc}} = \begin{bmatrix} 1 & 0 \\ 0 & 1 \end{bmatrix}$.

Similarly, resonant terms for the q-axis and d-axis in the state-space are expressed as [17,26]:

$$
\begin{bmatrix} \dot{\delta}_{1i}^q(t) \\ \dot{\delta}_{2i}^q(t) \\ \dot{\delta}_{1i}^d(t) \\ \dot{\delta}_{2i}^d(t) \end{bmatrix} = \mathbf{A_{rci}} \begin{bmatrix} \delta_{1i}^q(t) \\ \delta_{2i}^q(t) \\ \delta_{1i}^d(t) \\ \delta_{2i}^d(t) \end{bmatrix} + \mathbf{B_{rci}} \begin{bmatrix} \varepsilon^q(t) \\ \varepsilon^d(t) \end{bmatrix} \quad \text{for } i = 6,\ 12
\tag{18}
$$

where $\mathbf{A_{rci}} = \begin{bmatrix} 0 & 1 & & \\ -(i\omega)^2 & -2\zeta(i\omega) & & \\ & & 0 & 1 \\ & & -(i\omega)^2 & -2\zeta(i\omega) \end{bmatrix}$, $\mathbf{B_{rci}} = \begin{bmatrix} 0 & 0 \\ 1 & 0 \\ 0 & 0 \\ 0 & 1 \end{bmatrix}$, and ζ is a damping factor.

As the damping ratio ζ is increased, it is well known that the magnitude of the frequency response at the resonant frequency is reduced, and the frequency response is flattened. The purpose of introducing the resonant terms is to effectively compensate the grid-side current harmonics caused by distorted grid voltages with the high gain at selective frequencies. Moreover, the proposed scheme can ensure the tracking performance of grid-side currents by adopting integral terms. Thus, damping ratio ζ is selected as zero in this study, which ensures that the harmonics from distorted voltages can be effectively compensated for by the high gain at selective frequencies.

The system states in Equations (17) and (18) are augmented as:

$$
\dot{\mathbf{z}}_c(t) = \mathbf{A_c z}_c(t) + \mathbf{B_c} \varepsilon(t)
\tag{19}
$$

where $\mathbf{z}_c = \begin{bmatrix} \mathbf{z}_0 & \mathbf{z}_6 & \mathbf{z}_{12} \end{bmatrix}^T$ is the entire state variables for integral and resonant terms with $\mathbf{z}_0 = \begin{bmatrix} x_i^q & x_i^d \end{bmatrix}$, $\mathbf{z}_6 = \begin{bmatrix} \delta_{16}^q & \delta_{26}^q & \delta_{16}^d & \delta_{26}^d \end{bmatrix}$, $\mathbf{z}_{12} = \begin{bmatrix} \delta_{112}^q & \delta_{212}^q & \delta_{112}^d & \delta_{212}^d \end{bmatrix}$:

$$
\mathbf{A_c} = \begin{bmatrix} \mathbf{A_{Pc}} & & \\ & \mathbf{A_{rc6}} & \\ & & \mathbf{A_{rc12}} \end{bmatrix}, \text{ and } \mathbf{B_c} = \begin{bmatrix} \mathbf{B_{Pc}} \\ \mathbf{B_{rc6}} \\ \mathbf{B_{rc12}} \end{bmatrix}
$$

The discrete-time counterparts of $\mathbf{A_c}$ and $\mathbf{B_c}$ can be obtained as:

$$
\mathbf{A_{cd}} = e^{\mathbf{A_c} T_s} = \mathbf{I} + \frac{\mathbf{A_c} T_s}{1!} + \frac{\mathbf{A_c^2} T_s^2}{2!} + \cdots
\tag{20}
$$

$$
\mathbf{B_{cd}} = \mathbf{A_c^{-1}}(\mathbf{A_{cd}} - \mathbf{I})\mathbf{B_c}
\tag{21}
$$

Then, the entire control system can be augmented as follows:

$$\begin{bmatrix} \mathbf{x}(k+1) \\ \mathbf{z_c}(k+1) \end{bmatrix} = \begin{bmatrix} \mathbf{A_d} & \mathbf{0} \\ -\mathbf{B_{cd}C_d} & \mathbf{A_{cd}} \end{bmatrix} \begin{bmatrix} \mathbf{x}(k) \\ \mathbf{z_c}(k) \end{bmatrix} + \begin{bmatrix} \mathbf{B_d} \\ \mathbf{0} \end{bmatrix} \mathbf{u}(k) + \begin{bmatrix} \mathbf{D_d} \\ \mathbf{0} \end{bmatrix} \mathbf{e}(k) + \begin{bmatrix} \mathbf{0} \\ \mathbf{B_{cd}} \end{bmatrix} \mathbf{r}(k) \quad (22)$$

$$\mathbf{y}(k) = \begin{bmatrix} \mathbf{C_d} & \mathbf{0} \end{bmatrix} \begin{bmatrix} \mathbf{x}(k) \\ \mathbf{z_c}(k) \end{bmatrix} \quad (23)$$

Considering the augmented system, the state feedback control is expressed as:

$$\mathbf{u}(k) = -[\mathbf{K_x K_z}] \begin{bmatrix} \mathbf{x}(k) \\ \mathbf{z_c}(k) \end{bmatrix} = \mathbf{u_x}(k) + \mathbf{u_z}(k) \quad (24)$$

where $\mathbf{u_x}(k) = -\mathbf{K_x}\mathbf{x}(k)$ and $\mathbf{u_z}(k) = -\mathbf{K_z}\mathbf{z_c}(k)$.

The augmented system in Equations (22)–(24) can be rewritten in a compact form as:

$$\mathbf{x_e}(k+1) = \mathbf{A_e}\mathbf{x_e}(k) + \mathbf{B_e}\mathbf{u}(k) + \mathbf{D_e}\mathbf{e}(k) + \mathbf{B_{re}}\mathbf{r}(k) \quad (25)$$

$$\mathbf{y}(k) = \mathbf{C_e}\mathbf{x_e}(k) \quad (26)$$

$$\mathbf{u}(k) = -\mathbf{K}\mathbf{x_e}(k) \quad (27)$$

where $\mathbf{K} = [\mathbf{K_x}\ \mathbf{K_z}]$ is a set of feedback gains and $\mathbf{K_z} = [\mathbf{K_{Pz}}\ \mathbf{K_{6z}}\ \mathbf{K_{12z}}]$. The detailed block diagram of the proposed current controller is depicted in Figure 3, where $\mathbf{A_P}$, $\mathbf{A_{r6}}$, $\mathbf{A_{r12}}$, $\mathbf{B_P}$, $\mathbf{B_{r6}}$, and $\mathbf{B_{r12}}$ denote the discrete-time counterparts of $\mathbf{A_{Pc}}$, $\mathbf{A_{rc6}}$, $\mathbf{A_{rc12}}$, $\mathbf{B_{Pc}}$, $\mathbf{B_{rc6}}$, and $\mathbf{B_{rc12}}$, respectively.

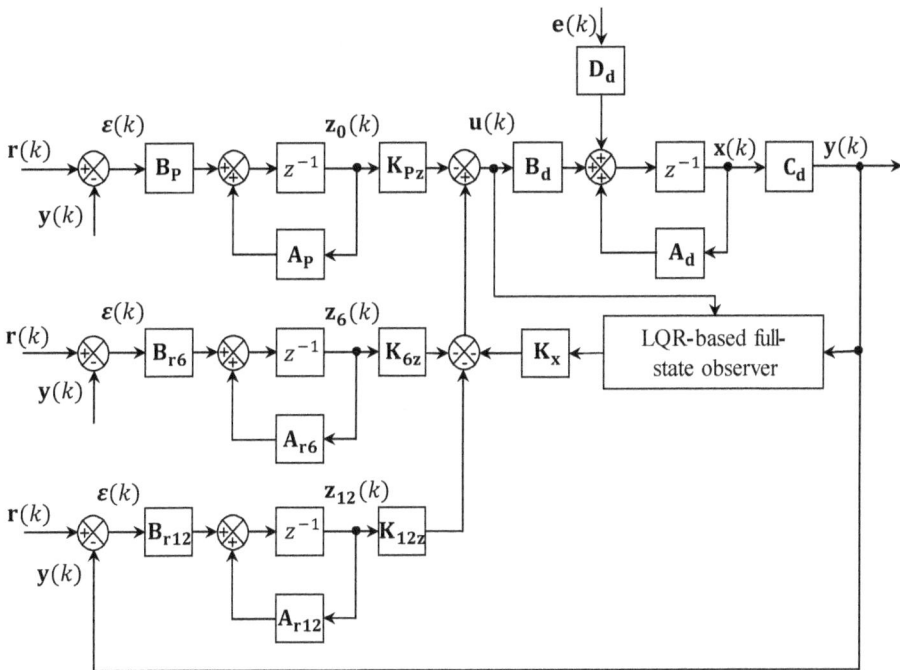

Figure 3. Detailed control block diagram of the proposed current controller.

3.2. Design of an Optimal Feedback Control Using Linear Quadratic Regulator (LQR) Approach

The state feedback control input $\mathbf{u}(k)$ will be an optimal control input to ensure the control performance and system stability if the gain matrix K in system Equations (25)–(27) are evaluated systematically by minimizing the discrete quadratic cost function as follows [27,28]:

$$J = \frac{1}{2} \sum_{k=0}^{\infty} \mathbf{x}_e^T(k)\mathbf{Q}\mathbf{x}_e(k) + \mathbf{u}^T(k)\mathbf{R}\mathbf{u}(k) \tag{28}$$

where $\mathbf{Q}$ is positive semi-definite matrix and $\mathbf{R}$ is positive definite matrix.

To obtain the optimal control input vector $\mathbf{u}(k)$ in closed loop form, an $n \times n$ real symmetric matrix $\mathbf{P}$ with n being the number of state variables should be determined as the solution of the discrete Riccati equation as follows:

$$\mathbf{P} = \mathbf{Q} + A_e^T\mathbf{P}A_e - A_e^T\mathbf{P}B_e\left(\mathbf{R} + B_e^T\mathbf{P}B_e\right)^{-1}B_e^T\mathbf{P}A_e. \tag{29}$$

Then, the gain matrix K can be calculated in terms of $\mathbf{P}$ as follows:

$$K = \mathbf{R}^{-1}B_e^T\left(A_e^T\right)^{-1}(\mathbf{P} - \mathbf{Q}). \tag{30}$$

The optimal control law is obtained by substituting Equations (30) to (27) as:

$$\mathbf{u}(k) = -\mathbf{R}^{-1}B_e^T\left(A_e^T\right)^{-1}(\mathbf{P} - \mathbf{Q})\mathbf{x}_e(k). \tag{31}$$

Then, the whole control system can be re-modeled as:

$$\mathbf{x}_e(k+1) = \left\{A_e - B_e\left[\mathbf{R}^{-1}B_e^T\left(A_e^T\right)^{-1}(\mathbf{P} - \mathbf{Q})\right]\right\}\mathbf{x}_e(k) + D_e\mathbf{e}(k) + B_{re}\mathbf{r}(k). \tag{32}$$

The discrete Riccati Equation (29) can be solved by MATLAB (R2017b, The MathWorks, Inc, Natick, MA, USA) functions "dare" and "dlqr". It is obvious that all the elements in the feedback gain matrix K rely on the choice of the symmetrical weighting matrices $\mathbf{Q}$ and $\mathbf{R}$ which determine the relative importance of the state variable performance and expenditure of energy by control input signals. The larger value of $\mathbf{Q}$ indicates that the system is stabilized with less change in the states, while the smaller $\mathbf{Q}$ implies that the states would be in larger variation. Similarly, the emphasis on $\mathbf{R}$ represents the behavior of the system states inputs. With the larger value of $\mathbf{R}$, the system is stabilized with less control input signals, whereas more energy is used to stabilize the whole system with smaller value of $\mathbf{R}$. In the proposed control scheme, the weighting matrices $\mathbf{Q}$ and $\mathbf{R}$ are selected as:

$$\mathbf{Q} = \begin{bmatrix} 10^{-2} \cdot I^{6 \times 6} & 0^{6 \times 2} & 0^{6 \times 8} \\ 0^{2 \times 6} & 6.3 \times 10^8 \cdot I^{2 \times 2} & 0^{2 \times 8} \\ 0^{8 \times 6} & 0^{8 \times 2} & 6.3 \times 10^8 \cdot I^{8 \times 8} \end{bmatrix}, \quad \mathbf{R} = \begin{bmatrix} 1 & 0 \\ 0 & 1 \end{bmatrix} \tag{33}$$

where $I^{n \times m}$ and $0^{n \times m}$ are the identity and zero matrices with appropriate dimensions, respectively. To improve the transient responses as well as to achieve the control objectives, a large weighting value of 6.3×10^8 is used for the state variables of the IM components z_c, while a quite small value of 10^{-2} is chosen for six system state variables. As a result, a fast reference tracking of state variables and a good suppression capability for the distorted harmonics on grid voltages can be obtained. The simulation and experimental results are presented in next section to demonstrate the performance of the optimal control scheme.

3.3. LQR-Based Current Full-State Observer in Discrete-Time

To realize a full-state feedback controller in the augmented system in Equations (25)–(27), all the system state variables should be available for feedback purpose. However, in a three-phase LCL-filtered grid-connected inverter, the additional sensing devices usually increase the total cost and hardware complexity. Therefore, in the proposed control scheme, an LQR-based discrete-time current full-state observer is employed to produce the estimated signals for the grid-side current $\hat{i}_2$, the inverter-side current $\hat{i}_1$, and the capacitor voltages $\hat{v}_c$.

Regarding to the selection of observer type, there are three alternatives which are the prediction-type observer, the current observer, and the reduced-order observer [20]. In the prediction-type observer, the estimated states $\hat{x}(k)$ are determined based on the past measurement of outputs at $(k-1)T$. This means that the control signal $u_x(k) = -K_x\hat{x}(k)$ does not utilize the most current information on outputs $y(k)$, which leads to the inaccuracy of estimated values and might cause control performance degradation. On the other hand, the reduced-order observer can solve the drawback of the prediction-type observer by using the measured states to estimate remaining unmeasurable states at time kT. However, if the measurement variables are noisy, the imprecise measured states may influence directly the feedback control inputs. For these reasons, the current full-state observer is employed for the three-phase LCL-filtered grid-connected inverter in this paper, which yields a precise estimation capability even under harmonically distorted grid voltage condition.

From the discretized model of the inverter system in Equations (12) and (13), a current full-state observer is given as follows:

$$\bar{x}(k+1) = A_d\hat{x}(k) + B_d u(k) + D_d e(k) \tag{34}$$

$$\hat{x}(k+1) = \bar{x}(k+1) + K_e[y(k+1) - C_d\bar{x}(k+1)] \tag{35}$$

where the symbol "^" denotes the estimated variables, K_e is the observer gain matrix, and $\bar{x}(k+1)$ is the first estimate of the state at time $(k+1)T$. In this type of observer, $\bar{x}(k+1)$ is first calculated from the dynamics of system and input signal at kT, and then this estimation is added with the correction term in Equation (35) when the output signals are measured at time $(k+1)T$. Figure 4 presents a discrete-time current full-state observer with a state-feedback controller, where the estimated state variables are used to construct the state feedback control inputs.

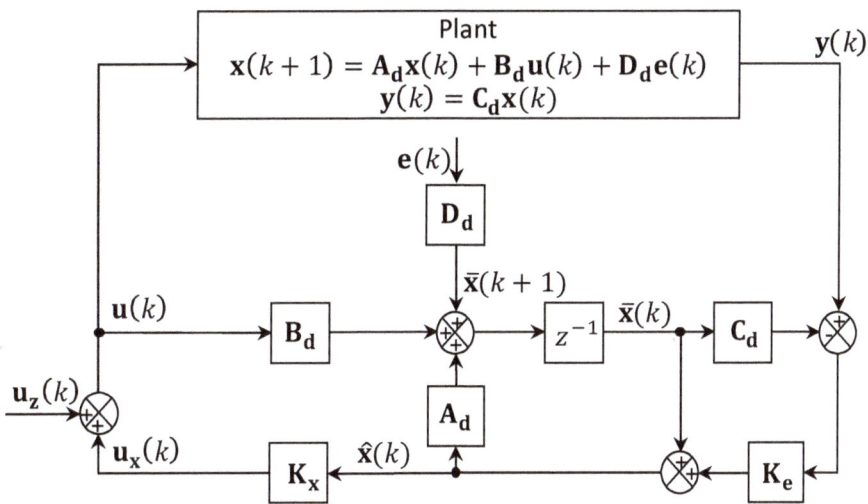

Figure 4. State feedback control using current full-state observer in the discrete-time domain.

In order to determine the observer gain matrix $\mathbf{K_e}$, the estimation error $\tilde{\mathbf{x}}$ is defined as:

$$\tilde{\mathbf{x}}(k) = \mathbf{x}(k) - \hat{\mathbf{x}}(k) \tag{36}$$

Then, the error dynamics of the observer can be obtained by subtracting Equation (35) from Equation (12) as follows:

$$\tilde{\mathbf{x}}(k+1) = (\mathbf{A_d} - \mathbf{K_e}\mathbf{C_d}\mathbf{A_d})\tilde{\mathbf{x}}(k) \tag{37}$$

To ensure that the observer is stable and the estimated states well track the actual ones, the observer gain matrix should be chosen in order that the matrix $(\mathbf{A_d} - \mathbf{K_e}\mathbf{C_d}\mathbf{A_d})$ or $(\mathbf{A_d^T} - (\mathbf{C_d}\mathbf{A_d})^T)\mathbf{K_e^T})$ is stable. By applying a LQR approach similar to the design of a state feedback control, the discrete Riccati equation can be applied for observer design as:

$$\mathbf{P_o} = \mathbf{Q_o} + \mathbf{A_d}\mathbf{P_o}\mathbf{A_d}^T - \mathbf{A_d}\mathbf{P_o}(\mathbf{C_d}\mathbf{A_d})^T(\mathbf{R_o} + (\mathbf{C_d}\mathbf{A_d})\mathbf{P_o}(\mathbf{C_d}\mathbf{A_d})^T)^{-1}(\mathbf{C_d}\mathbf{A_d})\mathbf{P_o}\mathbf{A_d}^T \tag{38}$$

where $\mathbf{P_o}$ is the solution of Riccati equation (38), and $\mathbf{Q_o}$ and $\mathbf{R_o}$ are weighting matrices. Hence, the observer gain $\mathbf{K_e}$ can be calculated in terms of $\mathbf{P_o}$ as follows:

$$\mathbf{K_e} = \mathbf{R_o}^{-1}\mathbf{C_d}\mathbf{A_d}(\mathbf{A_d})^{-1}(\mathbf{P_o} - \mathbf{Q_o}). \tag{39}$$

In this paper, the optimal observer gains can be chosen by utilizing the MATLAB function "dlqr".

4. Simulation Results

In order to verify the feasibility and validity of the proposed current control scheme, simulations were carried out for an LCL-filtered three-phase grid-connected inverter based on the PSIM software. The configuration of the inverter system and the proposed control scheme are depicted in Figure 2. The system parameters are listed in Table 1.

Table 1. System parameters of a grid-connected inverter.

Parameters	Value	Units
DC-link voltage	420	V
Resistance (load bank)	24	Ω
Filter resistance	0.5	Ω
Filter capacitor	4.5	μF
Inverter-side filter inductance	1.7	mH
Grid-side filter inductance	0.9	mH
Grid voltage (line-to line rms)	220	V
Grid frequency	60	Hz

Figure 5 represents three-phase distorted grid voltages used for the simulations. The abnormal grid voltages contain the harmonic components in the order of the 5th, 7th, 11th, and 13th with the magnitude of 5% with respect to the nominal grid voltages, which yields the THD value of 9.99%.

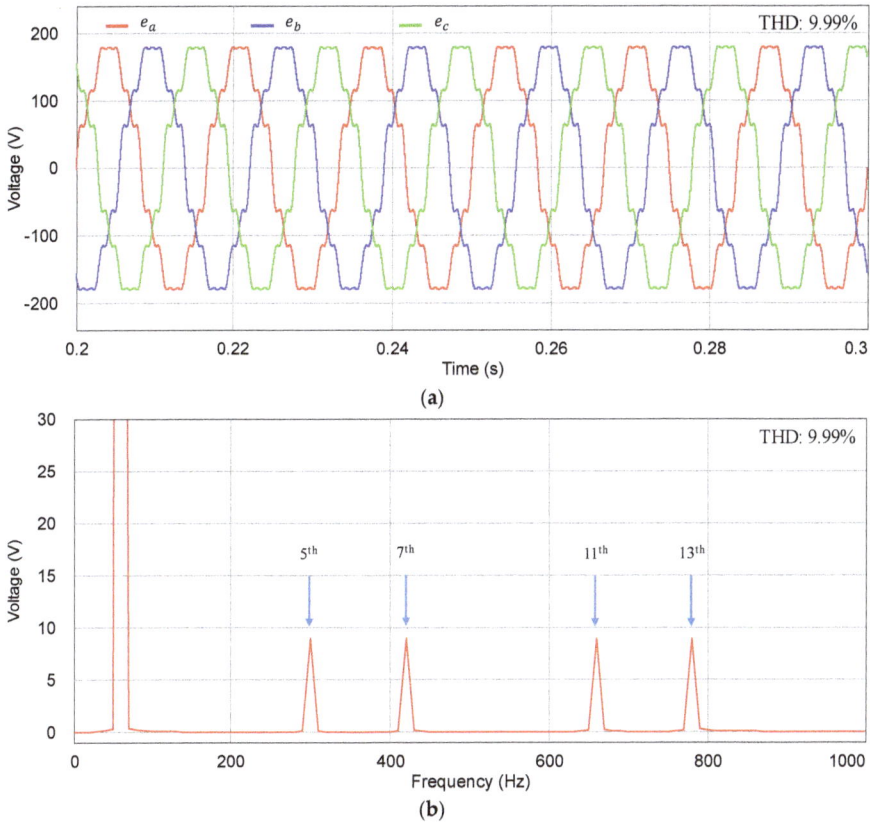

Figure 5. Distorted grid voltages. (**a**) Three-phase distorted grid voltages; (**b**) Fast Fourier transform (FFT) result of *a*-phase voltage.

Figure 6 shows the simulation results of the proposed current control scheme at steady-state under the distorted grid condition as in Figure 5. Figure 6a shows the grid-side current responses at the SRF with the reference currents. As can be observed from Figure 6a, the grid-side currents track the reference values well. Figure 6b,c represent the steady-state responses for the inverter-side current and capacitor voltage, respectively.

To demonstrate the transient performance of the proposed current control scheme, Figure 7 shows the simulation results under the same distorted grid condition when the *q*-axis reference current has a step change from 4 to 7 A at 0.25 s. Similarly, Figure 7a through Figure 7c represents the grid-side current responses, inverter-side current responses, and capacitor voltage responses at the SRF, respectively. As is shown in Figure 7a, the grid-side currents reach the reference very rapidly, which indicates a sufficiently fast transient response of the proposed control scheme. In addition, the fast transient performance of the proposed control scheme can be also inferred from Figure 7b,c.

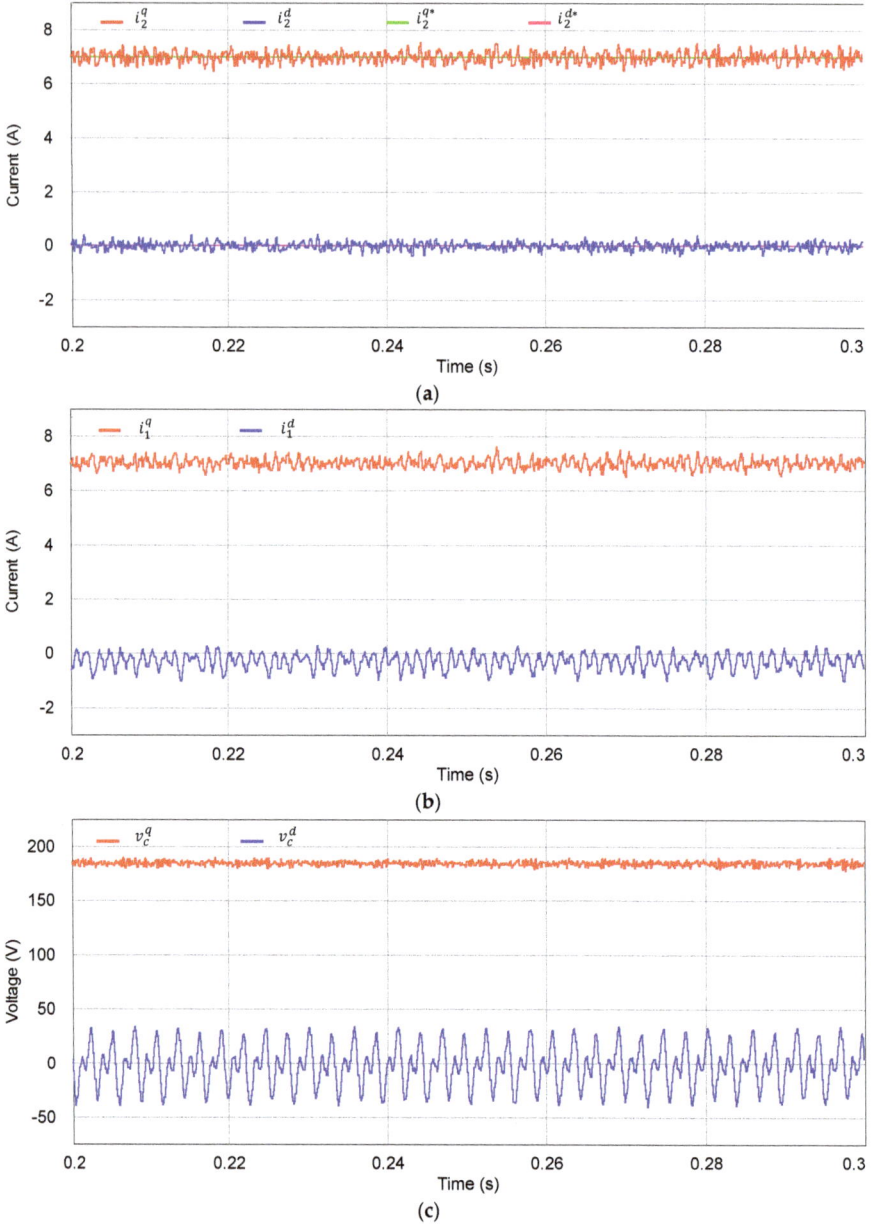

Figure 6. Simulation results for steady-state responses under distorted grid voltage with the proposed controller. (**a**) Grid-side current responses at the synchronous reference frame (SRF); (**b**) Inverter-side current responses at the SRF; (**c**) Capacitor voltage responses at the SRF.

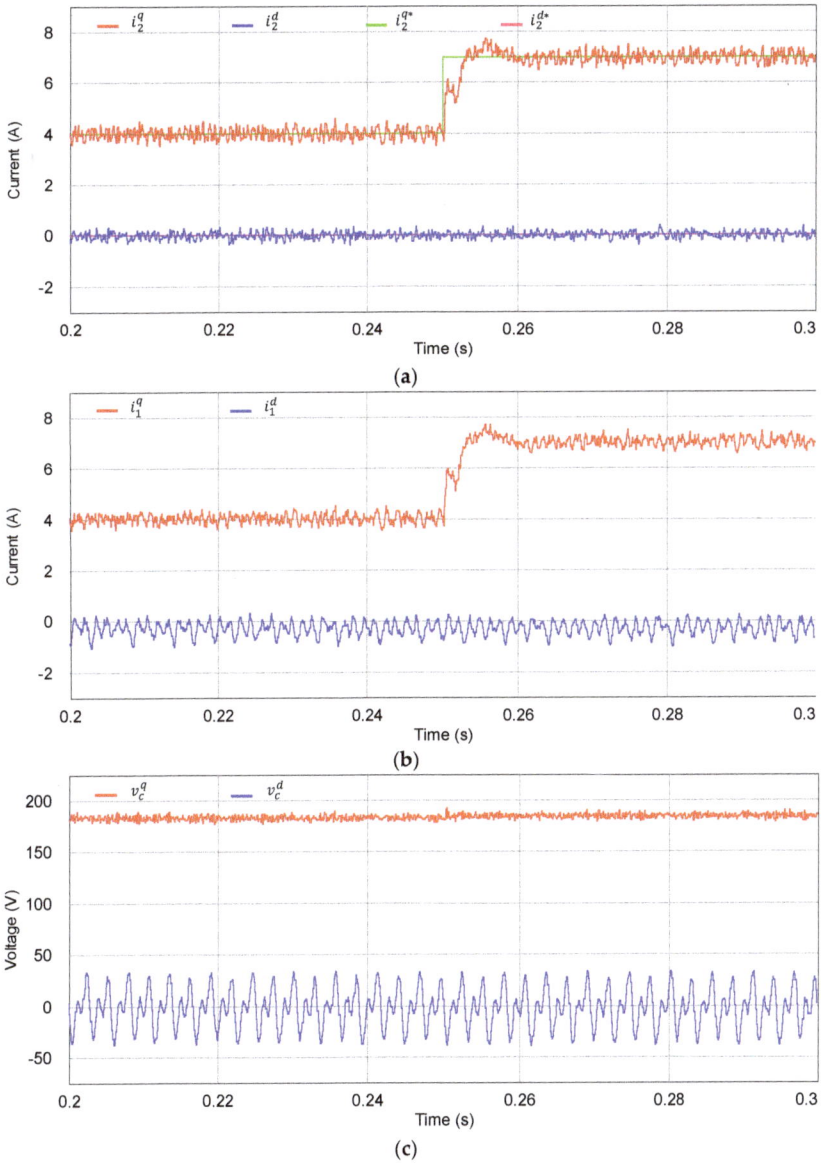

Figure 7. Simulation results for transient responses under distorted grid voltage with the proposed controller. (**a**) Current references and grid-side current responses at the synchronous reference frame (SRF); (**b**) inverter-side current responses at the SRF; (**c**) capacitor voltage responses at the SRF.

Figure 8 shows the simulation results for the inverter states and estimated states using the proposed integral-resonant state feedback control scheme with the discrete-time current full-state observer at the SRF. The optimal observer gains are obtained by using the MATLAB "dlqr" function with given inverter parameters. As can be clearly observed from Figure 8, the estimated states instantly converge to the actual ones even during oscillating transient periods, which confirms a fast and stable operation of the observer.

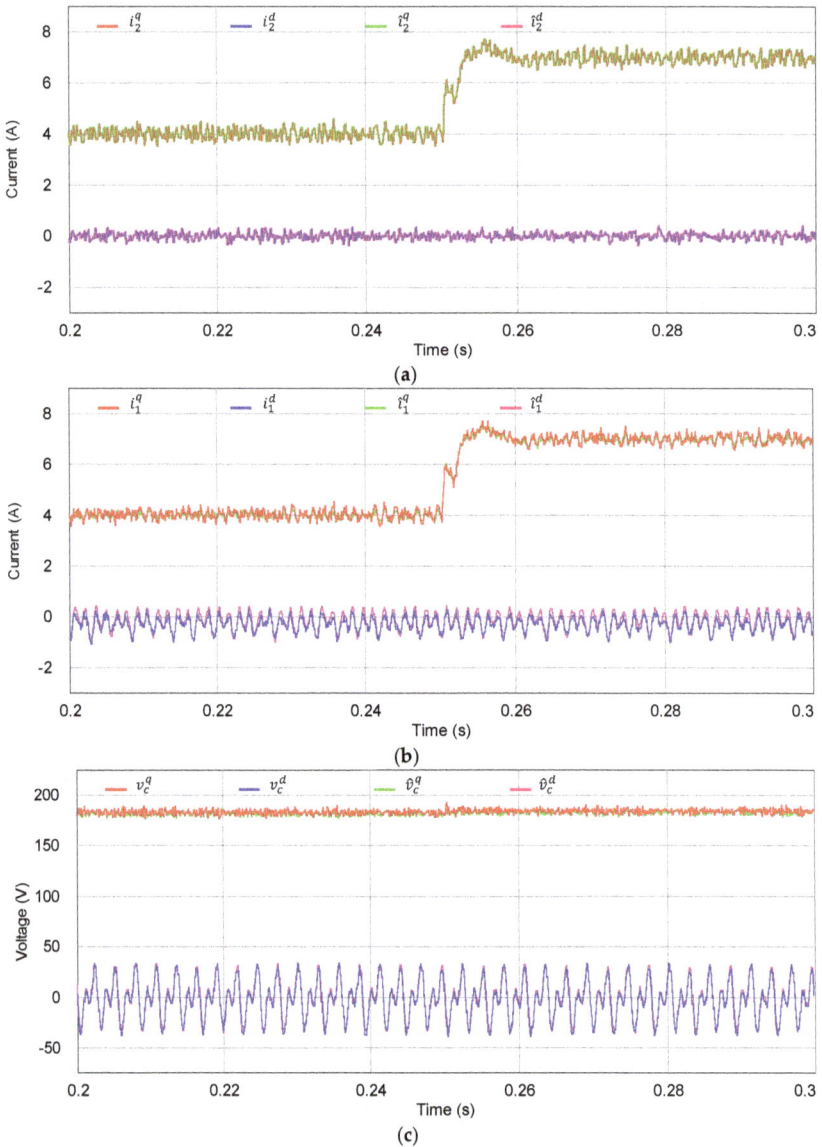

Figure 8. Simulation results for the proposed control scheme with the discrete-time full-state observer under step change in q-axis current reference; states and estimated states. (**a**) Waveforms of grid-side currents and estimated states at the SRF; (**b**) waveforms of inverter-side currents and estimated states at the SRF; (**c**) waveforms of capacitor voltages and estimated states at the SRF.

Figure 9 represents the simulation results for measured three-phase variables using the proposed integral-resonant state feedback control scheme with the discrete-time current full-state observer when the q-axis reference current has a step change. As can be seen from Figure 9a, three-phase grid-side current waveforms remain relatively sinusoidal with a desired transient performance. In fact, the grid-side phase currents have the THD level of 3.57% in this case. Also, Figure 9b,c show actual three-phase inverter-side current waveforms and three-phase capacitor voltage waveforms.

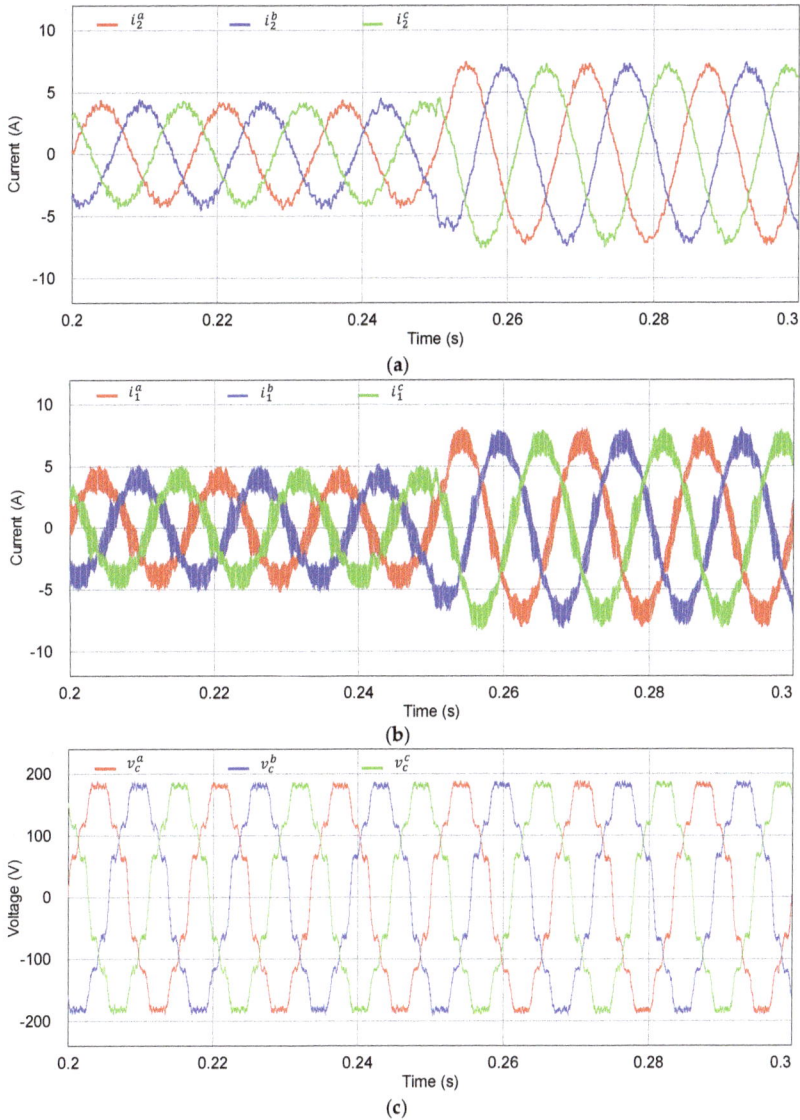

Figure 9. Simulation results for measured three-phase variables with the proposed control scheme under step change in *q*-axis current reference. (**a**) Three-phase grid-side current waveforms; (**b**) three-phase inverter-side current waveforms; (**c**) three-phase capacitor voltage waveforms.

Figure 10 shows the simulation results for the estimated waveforms of three-phase grid-side currents, inverter-side currents, and capacitor voltages under the same condition of Figure 9. In these figures, the estimated three-phase variables are constructed in a DSP by using the estimated states at the SRF to demonstrate the estimating performance of the discrete-time current full-state observer.

Obviously, the estimated three-phase variables are compatible with the actual measured three-phase waveforms in Figure 9.

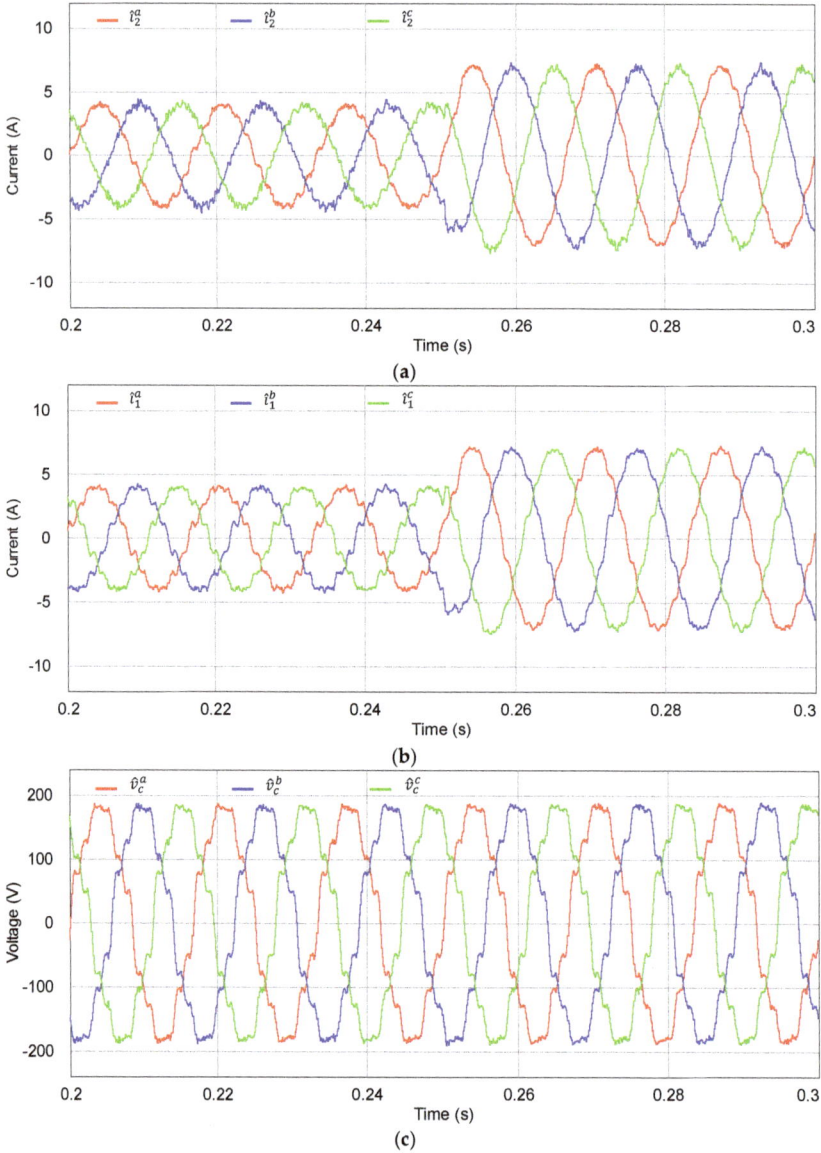

Figure 10. Simulation results for estimated three-phase variables with state observer under step change in *q*-axis current reference. (**a**) Estimated three-phase grid-side currents; (**b**) estimated three-phase inverter-side currents; (**c**) estimated three-phase capacitor voltages.

To verify the quality of injected grid currents for the proposed integral-resonant controller under a distorted grid voltage, Figure 11 shows the FFT result for grid-side *a*-phase current with the harmonic limits specified by the grid interconnection regulation IEEE Std. 1547 [29]. As can be seen clearly, the grid-side phase current yields only small 5th, 7th, 11th, and 13th harmonics components. The resultant THD value is 3.569%, which meets the quality criteria of inverter injected current.

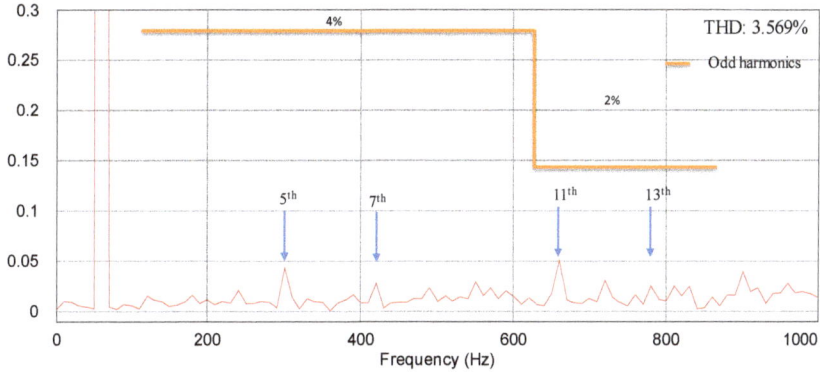

Figure 11. FFT result for grid-side *a*-phase current of the proposed controller under distorted grid voltages.

In order to verify the effectiveness of the proposed current control scheme, the performance of the proposed LQR-based current control is compared to PR plus harmonic compensator (PR + HC) structure presented in [3] under the same parameters as proposed in Table 1 and grid voltage condition presented in Figure 5a. The transfer function of a PI + HC controller is given in the stationary frame as:

$$G = K_P + \frac{K_{r1}s}{s^2 + \omega^2} + \frac{K_{r5}s}{s^2 + (5\omega)^2} + \frac{K_{r7}s}{s^2 + (7\omega)^2} + \frac{K_{r11}s}{s^2 + (11\omega)^2} + \frac{K_{r13}s}{s^2 + (13\omega)^2} \tag{40}$$

where K_P is the proportional gain and K_{ri} is the resonant gain with $i = 1, 5, 7, 11, 13$.

Figure 12 shows the simulation results for the control scheme in [3]. As can be observed from the grid-side current responses in Figure 12a, the PR + HC current control still can compensate effectively the harmonics caused by background voltages. However, the THD value of *a*-phase current is slightly increased to 3.69% in comparison to that obtained from the LCL filter parameters in [3] because the filter inductor values are reduced. Figure 12b presents the simulation results for the reference tracking performance of grid-side currents in the stationary frame.

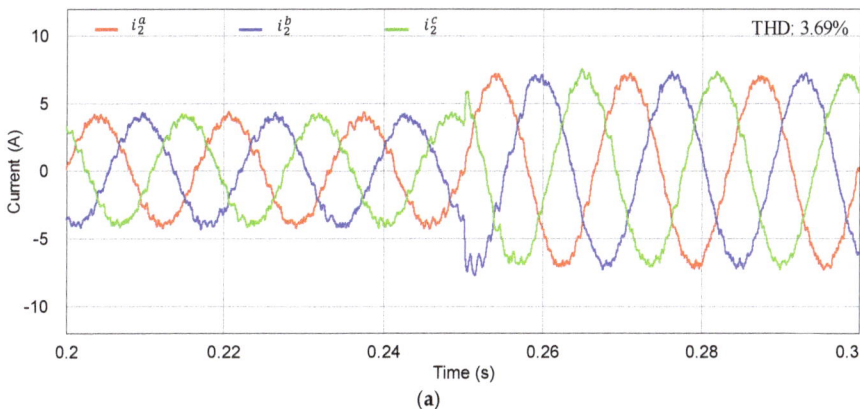

(a)

Figure 12. *Cont.*

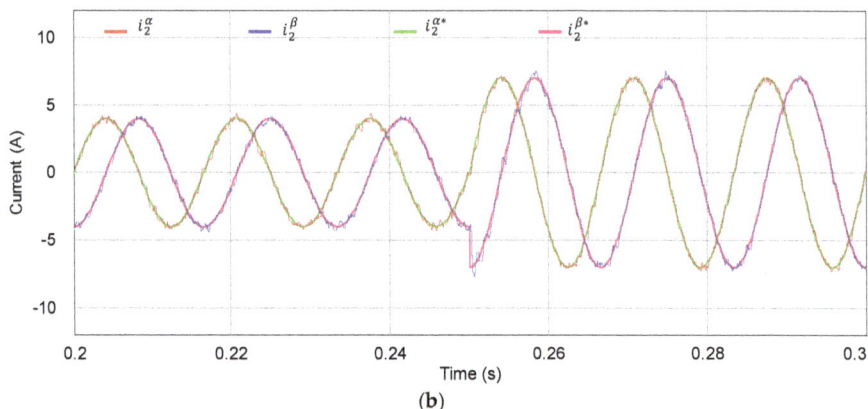

Figure 12. Simulation results for the proportional-resonant plus harmonic compensator (PR + HC) control scheme under step change. (**a**) Three-phase grid-side current waveforms; (**b**) current references and grid-side current responses at the stationary frame.

In spite of the control performance of the study in [3], it is worth mentioning that the main drawback of the work in [3] lies in the requirement for additional current sensing devices. As obviously shown in the control structure, the method in [3] requires the measurement of currents in inverter-side as well as in grid-side to obtain the capacitor current for active damping. Generally, since additional sensing devices cause complexity in the hardware and a high implementation cost, the purpose of this study is to implement a desired control performance by utilizing only the grid-side current sensors and grid voltage sensors.

Furthermore, the control method in [3] requires 5 resonant controllers including the fundamental component for the α-axis and additional 5 resonant controllers for the β-axis to compensate for the harmonic components in the order of the 5th, 7th, 11th, and 13th. On the contrary, since the proposed scheme is designed in the synchronous reference frame, the total of 4 resonant controllers are sufficient for the q- and d- axes to compensate the 6th and 12th order harmonic components. From the viewpoint of digital implementation burden, the proposed control scheme requires only an acceptable level of computation and complexity.

5. Experimental Results

In order to verify the feasibility of the proposed control scheme, the control algorithm is implemented on 32-bit floating-point DSP TMS320F28335 (Texas Instruments, Inc, Dallas, TX, USA) to control a 2 kVA prototype grid-connected inverter [30]. The configuration of the entire system is illustrated in Figure 13a. The sampling period is set to 100 µs, which results in the switching frequency of 10 kHz. Figure 13b depicts the photograph of the experimental test setup. The experimental setup is composed of a three-phase inverter connected to the grid through an LCL filter, a magnetic contactor for grid connecting operations, an AC power source to emulate three-phase grid voltages in the ideal as well as distorted grid conditions, and current and voltage sensors used to measure grid-side currents and grid voltages, respectively.

Figure 14a shows three-phase distorted grid voltages used for the experimental evaluation. Similar to Figure 5 in the simulation, these grid voltages contain the 5th, 7th, 11th, and 13th harmonics with the magnitude of 5% of the fundamental component. Figure 14b presents the FFT results for *a*-phase grid voltage, which shows each harmonic component similar to Figure 5b.

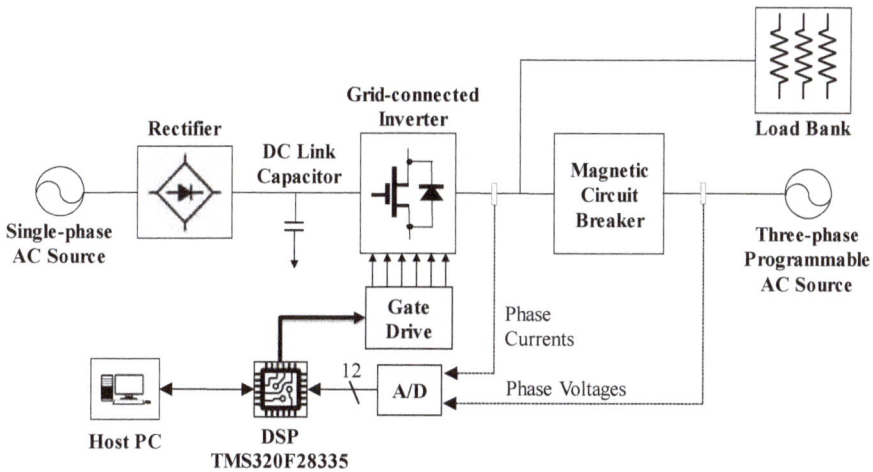

(a)

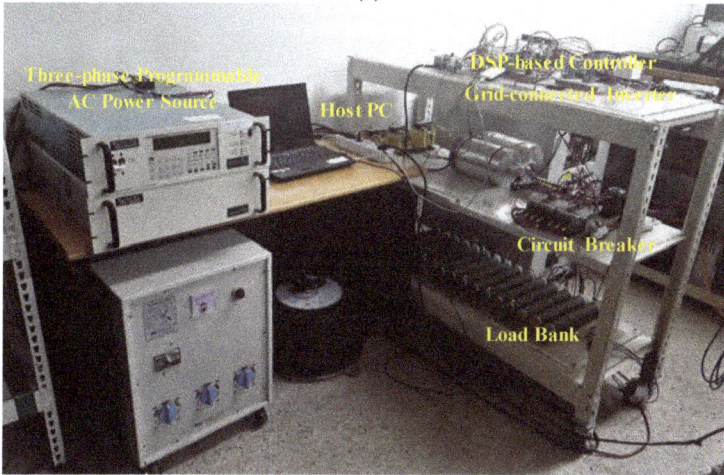

(b)

Figure 13. Configuration of the experimental system. (**a**) Block diagram of the overall system; (**b**) photograph of the experimental test setup.

Figure 15 shows the experimental results for the proposed control scheme under the step change in q-axis current reference from 4 to 6 A. It can be observed from Figure 15a that the inverter output currents can track their references well and instantly reach a new steady-state value, which demonstrates a fast transient response of the proposed control scheme. Figure 15b shows three-phase grid-side current responses. It is confirmed from this figure that the proposed control scheme provides considerable sinusoidal grid-side phase currents, which coincides well with the simulation results in Figure 9a, verifying a stable and reliable operation of the inverter system.

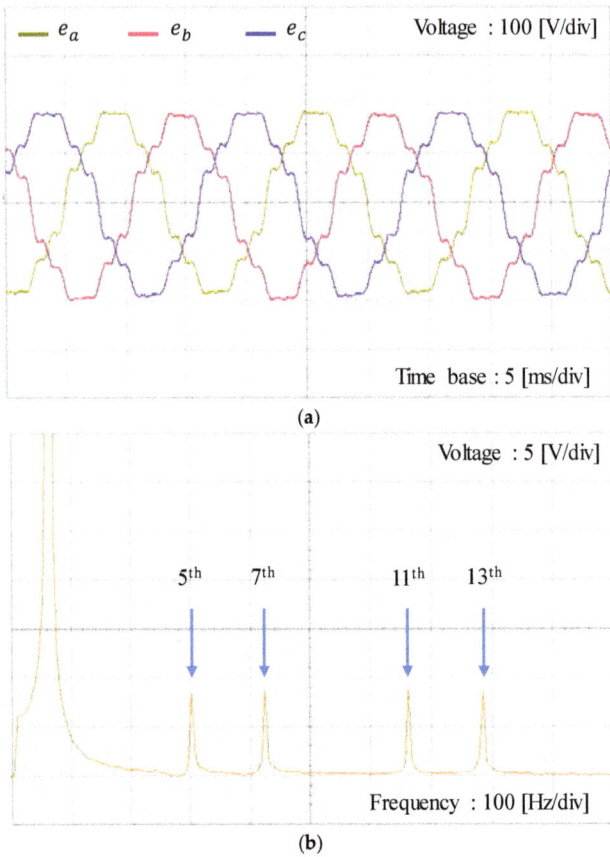

(a)

(b)

Figure 14. Distorted three-phase grid voltages used in the experiments. (a) Three-phase distorted grid voltages; (b) FFT result of *a*-phase grid voltage.

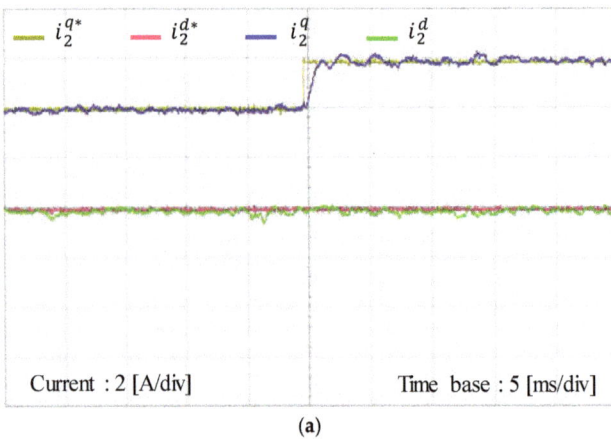

(a)

Figure 15. *Cont.*

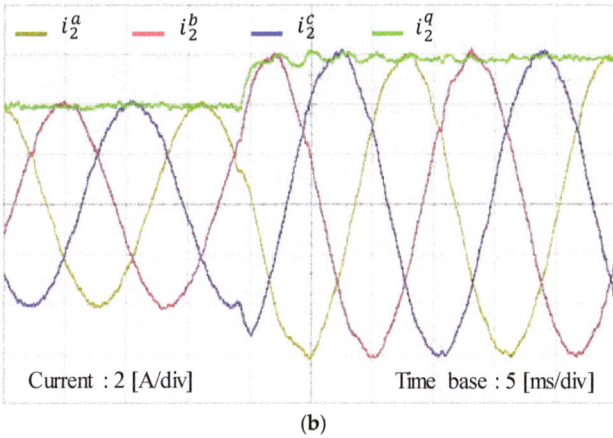

(b)

Figure 15. Experimental results for the proposed control scheme under step change in *q*-axis current reference. (**a**) Grid-side current responses at the SRF; (**b**) three-phase grid-side current responses.

Figure 16 presents the experimental results for the estimated grid-side currents, estimated inverter-side currents, and estimated capacitor voltages by using the discrete-time current full-state observer at the SRF under the step change in *q*-axis current reference. The estimated grid-side currents at the SRF in Figure 16a show similar behavior with actual states in Figure 15a. Also, the experimental estimated waveforms are very similar to the simulation results in Figure 8.

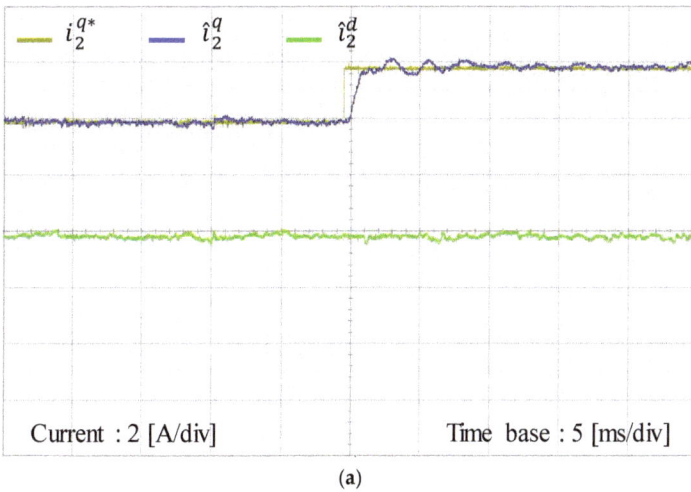

(a)

Figure 16. *Cont.*

(b)

(c)

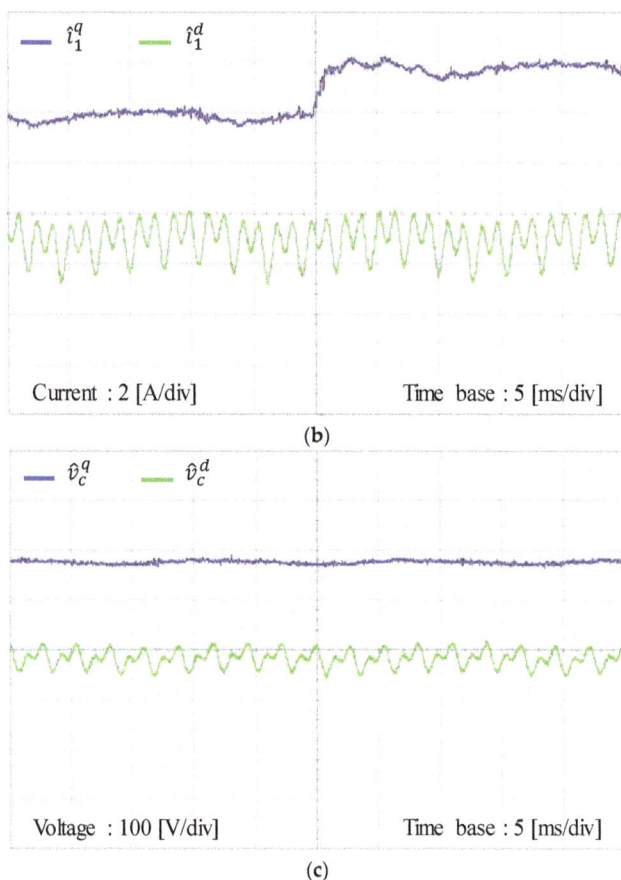

Figure 16. Experimental results for the estimated states with the discrete-time full-state observer under step change in *q*-axis current reference. (**a**) Responses of estimated grid-side currents at the SRF; (**b**) responses of estimated inverter-side currents at the SRF; (**c**) responses of estimated capacitor voltages at the SRF.

Figure 17a shows the experimental results for three-phase grid-side current waveforms at steady-state with the proposed current control scheme under harmonically distorted grid conditions as in Figure 14. Generally, the harmonic distortion on grid voltages directly influence on the grid-side current control performance, reducing the power quality of distributed generation system. However, the three-phase grid-side current waveform of the proposed scheme shows quite sinusoidal phase currents in spite of such a severe harmonic distortion on grid voltages. As is shown in Figure 17b, the FFT result for *a*-phase current shows negligibly small harmonic components in output current, which successfully meets the requirements for the harmonic limits specified by IEEE Standard 519-1992.

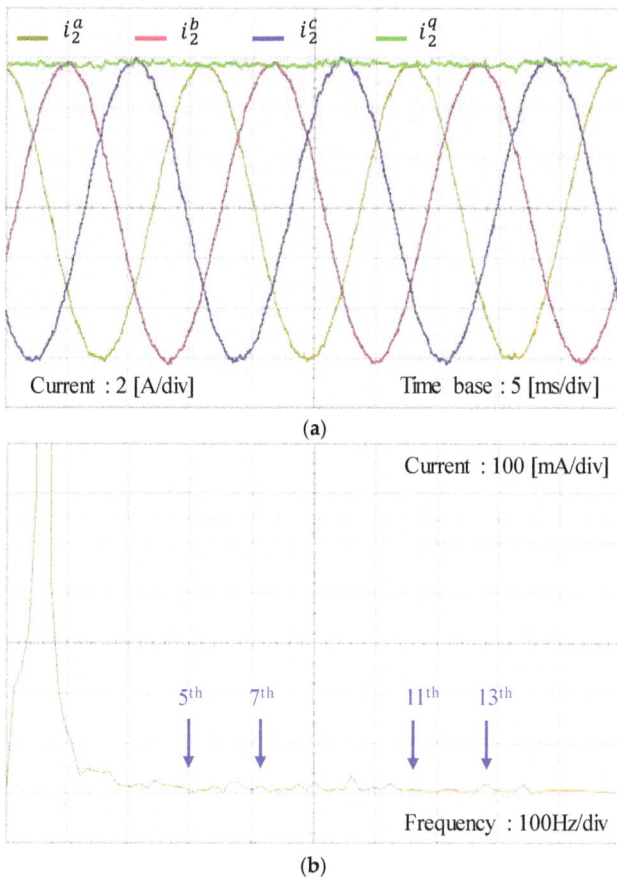

Figure 17. Experimental results for the proposed control scheme at steady-state. (a) Three-phase grid-side current waveforms; (b) FFT result for *a*-phase current.

Figure 18 represents the experimental results for the estimating performance of the discrete-time current full-state observer. To evaluate the estimating performance of the observer by comparison, Figure 18a,b show the estimating performance for grid-side three-phase currents and the comparison of these estimated signals with measured grid-side currents. In these figures, the estimated three-phase variables $\hat{i}_2^a$, $\hat{i}_2^b$, and $\hat{i}_2^c$ are constructed in a DSP by using the estimated states $\hat{i}_2^q$ and $\hat{i}_2^d$ at the SRF. The experimental results are well matched with the simulation results in Figure 10a and validate the stability and reliability of the estimated states by the current full-state observer. Similarly, Figure 18c,d show the estimating performance for inverter-side three-phase currents, and Figure 18e,f for three-phase capacitor voltages, respectively. Also, the estimated three-phase variables $\hat{i}_1^a$, $\hat{i}_1^b$, and $\hat{i}_1^c$ are calculated in a DSP from the estimated currents $\hat{i}_1^q$ and $\hat{i}_1^d$ at the SRF, and $\hat{v}_c^a$, $\hat{v}_c^b$, and $\hat{v}_c^c$ from the estimated capacitor voltages $\hat{v}_c^q$ and $\hat{v}_c^d$ at the SRF, respectively. It can be confirmed from these figures that all the estimated three-phase variables converge to actual measured three-phase variables well.

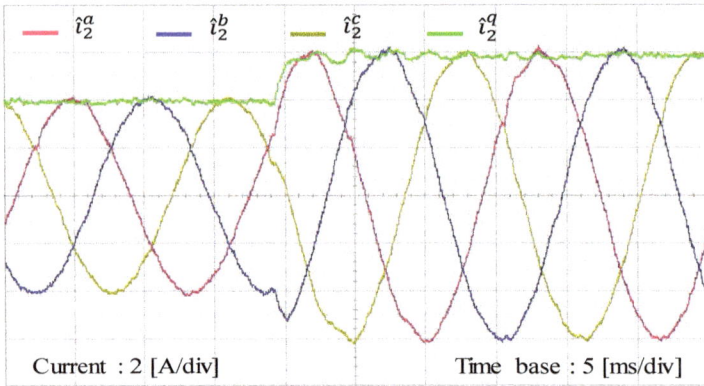

(a)

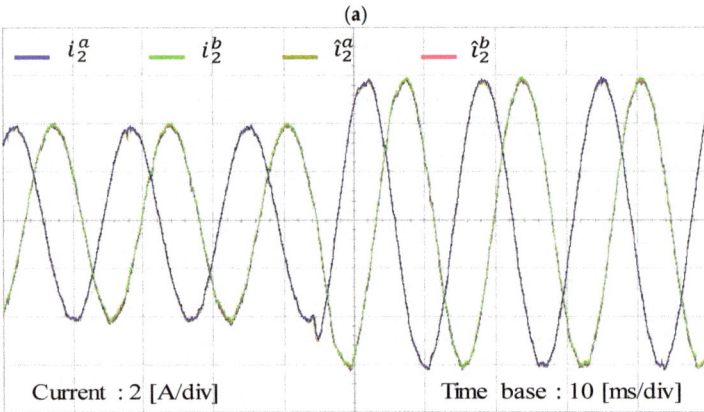

(b)

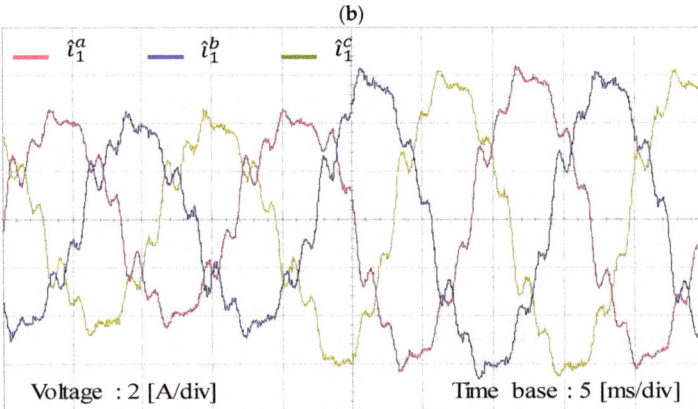

(c)

Figure 18. *Cont.*

(d)

(e)

(f)

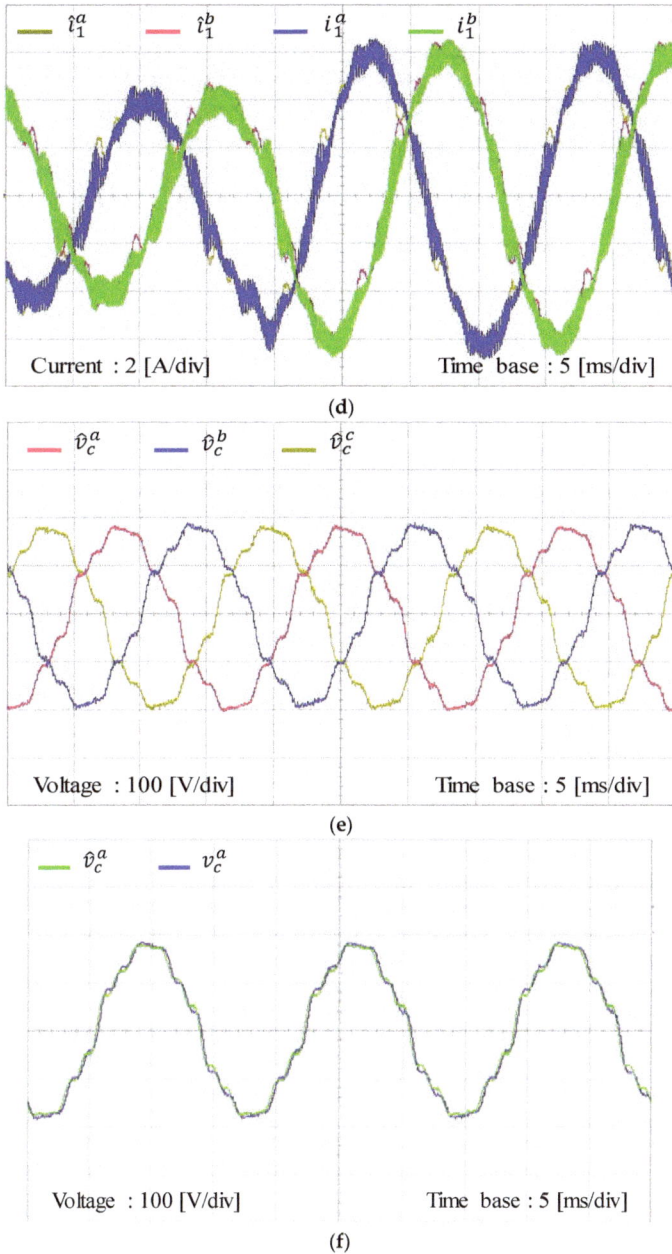

Figure 18. Experimental results for the estimating performance with the discrete-time full-state observer. (a) Estimating performance for grid-side three-phase currents; (b) comparison of the estimated grid-side currents and measured grid-side currents; (c) estimating performance for inverter-side three-phase currents; (d) comparison of the estimated inverter-side currents and measured inverter-side currents; (e) estimating performance for three-phase capacitor voltages; (f) comparison of the estimated capacitor voltage and measured capacitor voltage.

6. Conclusions

This paper has presented an LQR-based current control design for an LCL-filtered grid-connected inverter. The proposed control scheme has been constructed by using the IM principle in the SRF to augment integral and resonant terms into a state feedback control. As a result, the control scheme successfully achieves control objectives such as an asymptotic reference tracking and disturbance rejection, which significantly reduces the impact of grid voltage distortion on the output current. Moreover, since the proposed scheme is implemented in the SRF, both the negative and positive sequence harmonics can be effectively compensated for by only one resonant term, which leads to a decrease in the number of regulators. This feature is usually preferred because the required THD performance can be met with further reduction in computation efforts. Furthermore, with the aim of avoiding the increase of sensing devices in an LCL-filtered grid-connected inverter system in comparison with L-filtered counterpart, a current full-state observer in the discrete-time domain has been discussed in detail to estimate all the state variables. On account of the augmentation of the resonant terms as well as the integral term into the inverter model, an increased number of feedback gains should be selected. To deal with such a limitation, an optimal LQR approach is adopted as a way of choosing the feedback gains systematically, in which the discrete cost function is minimized to satisfy the stability and robustness requirements of the system. As a result, both the system feedback and observer gains can be selected based on the LQR method in an effective and straightforward way. In addition to that, the control objectives such as the reference tracking and harmonic compensation capability can be effectively achieved without increasing the required number of sensing devices.

In order to evaluate the feasibility and validity of the proposed control scheme, the whole control algorithm is implemented on 32-bit floating-point DSP TMS320F28335 to control 2 kVA prototype grid-connected inverter. Comprehensive simulation and experimental results are presented under distorted grid voltage conditions to demonstrate the usefulness of the proposed current control scheme.

Author Contributions: T.V.T., S.-J.Y., and K.-H.K. conceived the main concept of the control structure and developed the entire system. T.V.T. and S.-J.Y. carried out the research and analyzed the numerical data with guidance from K.-H.K. T.V.T., S.-J.Y., and K.-H.K. collaborated in the preparation of the manuscript.

Funding: This work was also supported by the Human Resources Development of the Korea Institute of Energy Technology Evaluation and Planning (KETEP), grant funded by the Korea government Ministry of Trade, Industry and Energy (NO. 20174030201840).

Acknowledgments: This research was supported by Basic Science Research Program through the National Research Foundation of Korea (NRF) funded by the Ministry of Education (NRF-2016R1D1A1B03930975).

Conflicts of Interest: The authors declare no conflict of interest.

References

1. Trinh, Q.N.; Lee, H.H. An advanced current control strategy for three-phase shunt active power filters. *IEEE Trans. Indus. Electr.* **2013**, *60*, 5400–5410. [CrossRef]
2. Peña-Alzola, R.; Liserre, M.; Blaabjerg, F.; Sebastián, R.; Dannehl, J.; Fuchs, F. Analysis of the passive damping losses in LCL-filter-based grid converters. *IEEE Trans. Power Electr.* **2013**, *28*, 2642–2646. [CrossRef]
3. Jia, Y.; Zhao, J.; Fu, X. Direct grid current control of LCL-filtered grid-connected inverter mitigating grid voltage disturbance. *IEEE Trans. Power Electr.* **2014**, *29*, 1532–1541.
4. Han, Y.; Shen, P.; Zhao, X.; Guerrero, J. Control strategies for islanded microgrid using enhanced hierarchical control structure with multiple current-loop damping schemes. *IEEE Trans. Smart Grid* **2017**, *8*, 1139–1153. [CrossRef]
5. Jin, W.; Li, Y.; Sun, G.; Bu, L. H∞ repetitive control based on active damping with reduced computation delay for LCL-type grid-connected inverters. *Energies* **2017**, *10*, 586. [CrossRef]
6. Bolsens, B.; Brabandere, K.D.; Den Keybus, J.V.; Driesen, J.; Belmans, R. Model-based generation of low distortion currents in grid-coupled PWM-inverters using an LCL output filter. *IEEE Trans. Power Electr.* **2006**, *21*, 1032–1040. [CrossRef]

7. Kukkola, J.; Hinkkanen, M. Observer-based state-space current control for a three-phase grid-connected converter equipped with an LCL filter. *IEEE Trans. Indus. Appl.* **2014**, *50*, 2700–2709. [CrossRef]

8. Kukkola, J.; Hinkkanen, M.; Zenger, K. Observer-based state-space current controller for a grid converter equipped with an LCL filter: Analytical method for direct discrete-time design. *IEEE Trans. Indus. Appl.* **2015**, *51*, 4079–4090. [CrossRef]

9. Zmood, D.N.; Holmes, D.G. Stationary frame current regulation of PWM inverters with zero steady-state error. *IEEE Trans. Power Electr.* **2003**, *18*, 814–822. [CrossRef]

10. Lascu, C.; Asiminoaei, L.; Boldea, I.; Blaabjerg, F. High performance current controller for selective harmonic compensation in active power filters. *IEEE Trans. Power Electr.* **2007**, *22*, 1826–1835. [CrossRef]

11. Kulka, A.; Undeland, T.; Vazquez, S.; Franquelo, L.G. Stationary frame voltage harmonic controller for standalone power generation. In Proceedings of the European Conference on Power Electronics Applications, Aalborg, Denmark, 2–5 September 2007; pp. 1–10.

12. Gabe, I.J.; Montagner, V.F.; Pinheiro, H.P. Design and implementation of a robust current controller for VSI connected to the grid through an LCL filter. *IEEE Trans. Power Electr.* **2009**, *24*, 1444–1452. [CrossRef]

13. Gonzatti, R.B.; Ferreira, S.C.; da Silva, C.H.; Pereira, R.R.; da Silva, L.E.B.; Lambert-Torres, G. Using smart impedance to transform high impedance microgrid in a quasi-infinite busbar. *IEEE Trans. Smart Grid* **2017**, *8*, 428–436. [CrossRef]

14. Perez-Estevez, D.; Doval-Gandoy, J.; Yepes, A.G.; Lopez, O.; Baneira, F. Enhanced resonant current controller for grid-connected converters with LCL filter. *IEEE Trans. Power Electr.* **2017**. [CrossRef]

15. Liu, Y.; Wu, W.; He, Y.; Lin, Z.; Blaabjerg, F.; Chung, H.S.H. An efficient and robust hybrid damper for LCL or LLCL-based grid-tied inverter with strong grid-side harmonic voltage effect rejection. *IEEE Trans. Indus. Electr.* **2016**, *63*, 926–936. [CrossRef]

16. Liserre, M.; Teodorescu, R.; Blaabjerg, F. Multiple harmonics control for three-phase grid converter systems with the use of PI-RES current controller in a rotating frame. *IEEE Trans. Power Electr.* **2006**, *21*, 836–841. [CrossRef]

17. Lai, N.B.; Kim, K.H. Robust control scheme for three-phase grid-connected inverters with LCL-filter under unbalanced and distorted grid conditions. *IEEE Trans. Energy Conver.* **2018**, *33*, 506–515. [CrossRef]

18. Bojoi, R.I.; Griva, G.; Bostan, V.; Guerriero, M.; Farina, F.; Profumo, F. Current control strategy for power conditioners using sinusoidal signal integrators in synchronous reference frame. *IEEE Trans. Power Electr.* **2005**, *20*, 1402–1412. [CrossRef]

19. Fu, X.; Li, S. A novel neural network vector control for single-phase grid-connected converters with L, LC and LCL filters. *Energies.* **2017**, *9*, 328. [CrossRef]

20. Francis, B.; Wonham, W. The internal model principle of control theory. *Automatica* **1976**, *12*, 457–465. [CrossRef]

21. Yoon, S.J.; Lai, N.B.; Kim, K.H. A systematic controller design for a grid-connected inverter with LCL filter using a discrete-time integral state feedback control and state observer. *MDPI Energies* **2018**, *11*, 1–20.

22. Lim, J.S.; Park, C.; Han, J.; Lee, Y.I. Robust tracking control of a three-phase DC–AC inverter for UPS applications. *IEEE Trans. Indus. Electr.* **2014**, *61*, 4142–4151. [CrossRef]

23. Hasanzadeh, A.; Edrington, C.S.; Mokhtari, H. A novel LQR based optimal tuning method for IMP-based linear controllers of power electronics/power systems. In Proceedings of the IEEE Conference on Decision and Control, Orlando, FL, USA, 12–15 December 2011; pp. 7711–7716.

24. Perez-Estevez, D.; Doval-Gandoy, J.; Yepes, A.; Lopez, O. Positive- and negative-sequence current controller with direct discrete-time pole placement for grid-tied converters with LCL filter. *IEEE Trans. Power Electr.* **2017**, *32*, 7207–7221. [CrossRef]

25. Franklin, G.; Workman, M.; Powell, J. *Digital Control of Dynamic Systems*; Ellis-Kagle Press: Half Moon Bay, CA, USA, 2006.

26. Maccari, L.; Massing, J.; Schuch, L.; Rech, C.; Pinheiro, H.; Oliveira, R.; Montagner, V. LMI-based control for grid-connected converters with LCL filters under uncertain parameters. *IEEE Trans. Power Electr.* **2014**, *29*, 3776–3785. [CrossRef]

27. Phillips, C.L.; Nagle, H.T. *Digital Control System Analysis and Design*, 3rd ed.; Prentice Hall: Englewood Cliffs, NJ, USA, 1995.

28. Ogata, K. *Discrete Time Control Systems*, 2nd ed.; Prentice-Hall: Englewood Cliffs, NJ, USA, 1995.

29. IEEE. *IEEE Standard for Interconnecting Distributed Resources with Electric Power Systems*; IEEE Std.1547; IEEE: New York, NY, USA, 2003.

30. Texas Instrument. *TMS320F28335 Digital Signal Controller (DSC)—Data Manual*; Texas Instrument: Dallas, TX, USA, 2008.

energies

MDPI

Article

Stability Analysis of Grid-Connected Converters with Different Implementations of Adaptive PR Controllers under Weak Grid Conditions

Xing Li and Hua Lin *

State Key Laboratory of Advanced Electromagnetic Engineering and Technology, School of Electrical and Electronic Engineering, Huazhong University of Science and Technology, Wuhan 430074, China; hust_lx@hust.edu.cn
* Correspondence: lhua@mail.hust.edu.cn; Tel.: +86-27-87543071-315

Received: 10 July 2018; Accepted: 31 July 2018; Published: 1 August 2018

Abstract: Adaptive proportional resonant (PR) controllers, whose resonant frequencies are obtained by the phase-locked loop (PLL), are employed in grid connected voltage source converters (VSCs) to improve the control performance in the case of grid frequency variations. The resonant frequencies can be estimated by either synchronous reference frame PLL (SRF-PLL) or dual second order generalized integrator frequency locked loop (DSOGI-FLL), and there are three different implementations of the PR controllers based on two integrators. Hence, in this paper, system stabilities of the VSC with different implementations of PR controllers and different PLLs under weak grid conditions are analyzed and compared by applying the impedance-based method. First, the $\alpha\beta$-domain admittance matrixes of the VSC are derived using the harmonic linearization method. Then, the admittance matrixes are compared with each other, and the influences of their differences on system stability are revealed. It is demonstrated that if DSOGI-FLL is used, stabilities of the VSC with different implementations of the PR controllers are similar. Moreover, the VSC using a DSOGI-FLL is more stable than that using a SRF-PLL. The simulation and experimental results are conducted to verify the correctness of theoretical analysis.

Keywords: grid-connected converter; adaptive resonant controller; PLL; impedance analysis

1. Introduction

With the increasing energy consumption worldwide, the use of renewable energy sources like wind and solar energies [1,2] in the grid has been growing increasingly. The voltage source converters (VSCs) have many desirable features, such as full controllability, low current harmonics, and high efficiency, thus they are widely used to deliver the power produced by the renewable energies into the grid [3].

Proportional resonant (PR) controllers could control positive-sequence and corresponding negative-sequence grid current at the same time without any additional negative-sequence current controller, and they allow a relatively low computational cost as they are implemented in the stationary frame [3–5]. Hence, PR controllers working on the stationary reference frame are widely used in grid-connected VSCs. Implementations of the PR controllers based on two integrators are widely employed, since no explicit trigonometric functions are needed [5]. Three typical implementations of the resonance term of the current controllers in the prior studies are shown in Figure 1 [5–7]. For future reference, they are called implementation I, II and III, respectively. If the resonant frequency is constant, these implementations are equivalent to each other.

The grid frequency is practically not a constant but within a certain range [8,9], thus the control performance would be inevitable weakened if the center frequency of the resonance controller is set to be constant. If the resonant frequency is set to be the fundamental frequency and the grid frequency deviates from it, there would be a phase shift between the grid current and the corresponding grid voltage, though the aimed power factor is unity [6]. To improve control performance in the case of grid frequency variations, the authors of [6,8,10–13] developed frequency adaptive PR controllers, whose resonant frequencies are not constant values, but are updated online according to the frequency estimated by the phase-locked loop (PLL) system. If the adaptive PR controllers are applied, a unity power factor operation can be achieved [6]. The resonant frequency of the adaptive PR controller is time-variant, thus dynamic properties of the adaptive PR controllers with different implementations might be dramatically different. Consequently, the port characteristics of the converter with different implementations of adaptive PR controllers would not be the same.

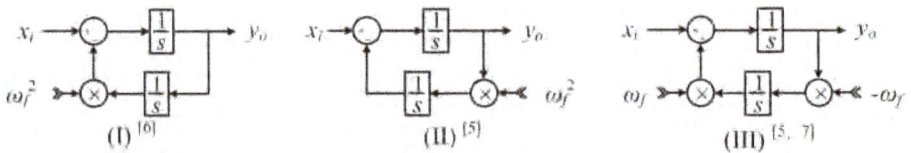

Figure 1. Implementations of the resonant term of frequency adaptive proportional resonant (PR) controllers based on two integrators.

The resonant frequency of the adaptive PR controller can be obtained by grid synchronization techniques based on phase locking approach, e.g., synchronous reference frame-phase locked loop (SRF-PLL) [14] which is widely used for its simplicity and robustness, or frequency locking approach, e.g., dual second-order generalized integrator-frequency locked loop (DSOGI-FLL) [15] which is implemented in the stationary reference frame. As SRF-PLL and DSOGI-FLL use different approaches to obtain the grid frequency, their influences on the port characteristics of the converter with adaptive current controllers might differ be different.

Under weak grid conditions, stability issues introduced by the grid-connected VSCs are of great importance [16–22]. The differences of the port characteristics of the converter introduced by applying different control schemes, including three different implementations of the adaptive PR controllers and the two different PLLs to obtain the resonant frequency, might significantly change the stability of the VSC connected to a weak grid. Hence, it is necessary to compare the robustness of the VSC with different control schemes.

In this paper, the stability issues of the VSC with different implementations of the adaptive PR controllers and different PLLs are studied and analyzed using the impedance-based method [18,23–28]. A suggestion to choose a suitable controllers' implementation and the related PLL for the VSC with adaptive PR controllers is given.

The rest of the paper is organized as follows: in Section 2, the studied system with adaptive current controller is briefly introduced. Section 3 shows how to model the adaptive resonant controller with different implementations of the resonant terms, and the approach to incorporate it into the admittance model of the converter. In Section 4 the effects of the adaptive resonant controllers with different implementations on system stability is analyzed and compared, and the stability of the system using a SRF-PLL for grid synchronization and the system using a DSOGI-FLL for grid synchronization is compared. Section 5 includes experimental verifications of the theoretical analysis. Section 6 concludes this paper.

2. VSC with Adaptive PR Current Controllers

The L-type grid connected converter with grid current regulation working on $\alpha\beta$ reference frame is studied, as depicted in Figure 2. The inductance of the filter is L, and the equivalent series resistance is r. The grid inductance is L_g. The grid voltages, the voltages at the point of common coupling (PCC), the ac-side converter voltages, and the currents delivered to the grid are u_{sk}, u_{gk}, v_k, and i_k ($k = a, b, c$), respectively. The dc input voltage of the converter is V_{dc}.

Grid current references in the stationary reference frame ($i_{\alpha r}$, $i_{\beta r}$) can be obtained by applying an inverse Park transformation to the active, reactive current references (I_{dr}, I_{qr}). In this paper, unit-power factor is considered, i.e., I_{qr} is set to be zero. Subscripts 'α' and 'β' refer to the α-axis and β-axis, respectively, while subscripts 'd' and 'q' refer to the d-axis and q-axis, respectively.

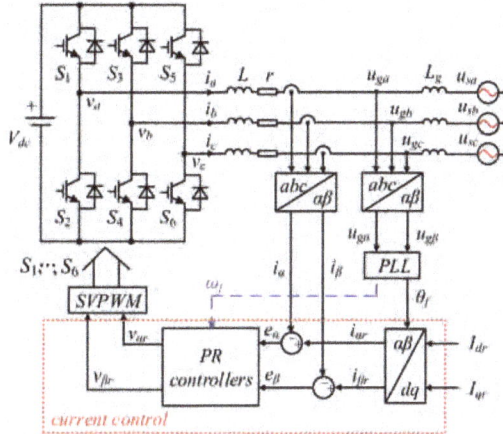

Figure 2. Block diagram of a voltage source converter (VSC) with adaptive PR controllers for grid-connected applications.

2.1. Adaptive PR Controller

The grid current error signals (e_α, e_β) are sent to the PR controllers to generate the reference of the ac-side converter voltages ($v_{\alpha r}$, $v_{\beta r}$). The PR current controllers are defined as follows:

$$H(s) = k_p + k_r \frac{s}{s^2 + \omega_f^2} = k_p + k_r R(s) \tag{1}$$

where ω_f is the resonant frequency, different from the conventional PR controller, the adaptive PR controller uses the frequency estimated by the PLL as the resonant frequency instead of the constant fundamental frequency ω_1; k_p, k_r are the proportional- and resonant-gain of the controller, respectively. The controller gains can be tuned according to: $k_p = \alpha_c L$, $k_i = \alpha_c r$ [18] where α_c is the current control loop bandwidth. $R(s)$ is the resonant term of the controller. $R(s)$ has infinite gain at the resonant frequency and thus it is capable for the grid current to track its reference without steady-state error.

The detailed block diagrams of different implementations of the PR controller based on two integrators are depicted in Figure 3, where $x_{i\alpha}$ and $y_{o\alpha}$ are the input and output of $R(s)$, respectively. The α-axis and β-axis are decoupled, thus only the PR controller implemented in the α-axis is given for simplicity. In time domain, the relationships between $x_{i\alpha}$ and $y_{o\alpha}$ can be derived from Figure 3, as follows:

$$\begin{cases} x_{i\alpha}(t) = \frac{dy_{o\alpha}(t)}{dt} + \omega_f^2(t)\int_0^t y_{o\alpha}(t)dt & \cdots \text{ I} \\ x_{i\alpha}(t) = \frac{dy_{o\alpha}(t)}{dt} + \int_0^t \omega_f^2(t)y_{o\alpha}(t)dt & \cdots \text{ II} \\ x_{i\alpha}(t) = \frac{dy_{o\alpha}(t)}{dt} + \omega_f(t)\int_0^t \omega_f(t)y_{o\alpha}(t)dt & \cdots \text{ III} \end{cases} \tag{2}$$

where Equations (2-I), (2-II) and (2-III) are obtained with implementation I, II, and III of the PR controllers, respectively. If ω_f is constant, the three equations are equivalent to each other. However, for the adaptive PR controllers, ω_f is time varying, the equations are no longer the same. Thus the dynamic properties of the adaptive PR controllers with different implementations might be different, and the transfer function shown in Equation (1) is not enough to describe the dynamic properties of adaptive PR controllers.

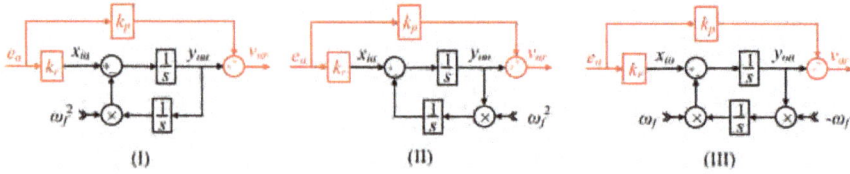

Figure 3. Detailed block diagrams of adaptive PR controllers with different implementations of $R(s)$.

The adaptive PR controllers' outputs are sent to the SVPWM to generate the control signals.

2.2. Grid Synchronization Methods

Two typical grid synchronization methods, i.e., SRF-PLL and DSOGI-FLL as shown in Figure 4 are used to provide the phase angle to calculate the reference currents and the controller's resonant frequency. The block diagram of a typical SRF-PLL is shown in Figure 4a, where $G_{PLL}(s)$ is the PLL controller. Figure 4b shows the block diagram of DSOGI-FLL with frequency locked loop (FLL) gain normalization, where k is the gain of the SOGI, and γ is used to regulate the settling time of the FLL [22]. The DSOGI-FLL consists of four parts: (1) SOGI; (2) positive/negative-sequence calculation block (PNSC) to obtain the positive-sequence voltages ($u_{\alpha+}$, $u_{\beta+}$) and the negative-sequence voltage ($u_{\alpha-}$, $u_{\beta-}$); (3) FLL with FLL gain normalization to estimate the grid frequency; (4) the "atan2" uses the positive-sequence voltages for the phase angle calculation.

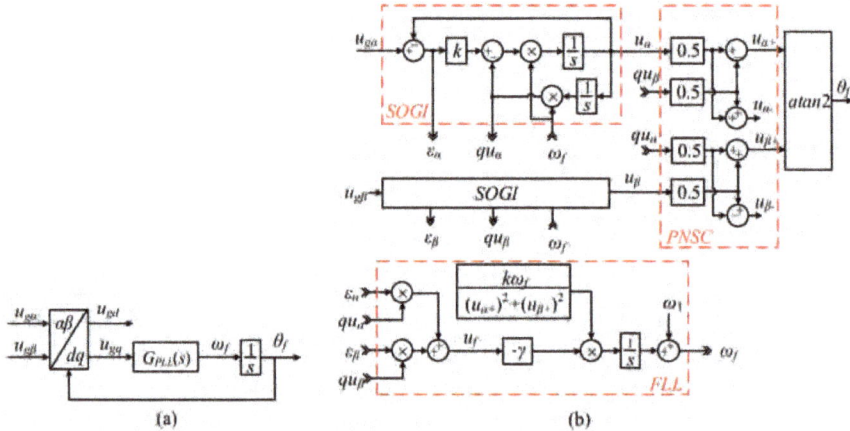

Figure 4. Block diagram of (**a**) a basic synchronous reference frame PLL (SRF-PLL); (**b**) a dual second order generalized integrator frequency locked loop (DSOGI-FLL) based grid synchronization with frequency locked loop (FLL) gain normalization.

With different implementations of the adaptive PR controllers and different PLLs, the stability of the grid-connected converter might be quite different under weak grid conditions.

3. Impedance Modeling

Impedance-based methods were a popular choice in prior studies to analyse the stability of grid-connected converters. Impedance modeling of the converter can be performed in either the *dq*-domain [25] or in the $\alpha\beta$-domain [26,27]. Considering that the grid current regulation working on $\alpha\beta$ reference frame is used, impedance-based method in the $\alpha\beta$-domain is applied in this paper. The $\alpha\beta$-domain converter impedances are derived using harmonic linearization method.

A positive-sequence perturbation at an arbitrary frequency ω_p is injected to the grid to obtain the reflected admittances at the ac terminals. Thus, in three phase variables, there would have positive-sequence components at frequency ω_p and negative-sequence components at frequency $\omega_p - 2\omega_1$ [28]. In this paper, it is defined that $s = j\omega$, where $\omega = \omega_p - \omega_1$.

3.1. Model of the Adaptive PR Controllers

The outputs of the adaptive PR controllers are affected by the error signal and the estimated frequency, as shown in Figure 3, thus the small-signal part of controllers' outputs can be described by Equation (3):

$$\begin{cases} v_{ar}^p(\omega + \omega_1) = H(s + j\omega_1)e_\alpha^p(\omega + \omega_1) + H_p^x(s + j\omega_1)\omega_f(\omega) \\ v_{ar}^n(\omega - \omega_1) = H(s - j\omega_1)e_\alpha^n(\omega - \omega_1) + H_n^x(s - j\omega_1)\omega_f(\omega) \end{cases} \tag{2}$$

where $H(s)$ is defined in Equation (1); $H_p^x(s)$ and $H_n^x(s)$ are used to describe the influence of ω_f on the positive-, negative-sequence components of the current controllers' outputs, respectively. The superscripts of the variables 'p' and 'n' are used to indicate the positive-, negative-sequence of the variables, respectively. The superscripts of the transfer functions 'x' in $H_p^x(s)$ and $H_n^x(s)$ is used to distinguish the implementations of adaptive controllers: $x = i$, ii, and iii indicates that implementation I, II and III is used, respectively.

To obtain the detailed expressions of $H_p^x(s)$ and $H_n^x(s)$, e_α is assumed to be zero, thus from Figure 3, it can be easily obtained that $x_{i\alpha} = 0$, and $y_{o\alpha} = v_{ar}$. If the implementation I of the adaptive PR controller is used, it can be derived from (2-I) by using convolution that:

$$\begin{cases} (s + j\omega_1)v_{ar}^p(\omega + \omega_1) + \left(\omega_1^2 \frac{v_{ar}^p(\omega+\omega_1)}{s+j\omega_1} + 2\omega_1\omega_f(\omega)\frac{V_\alpha[\omega_1]}{j\omega_1}\right) = 0 \\ (s - j\omega_1)v_{ar}^n(\omega - \omega_1) + \left(\omega_1^2 \frac{v_{ar}^n(\omega-\omega_1)}{s-j\omega_1} + 2\omega_1\omega_f(\omega)\frac{V_\alpha[-\omega_1]}{-j\omega_1}\right) = 0 \end{cases} \tag{3}$$

where $V_\alpha[\pm\omega_1] = (V_m + rI_{dr} \pm jL\omega_1 I_{dr})/2$ are the Fourier coefficients of v_{ar} at the positive/negative fundamental frequency, V_m is the magnitude of PCC voltage. For simplicity, set $V_p = 2V_\alpha[\omega_1]$, and $V_n = 2V_\alpha[-\omega_1]$.

The detailed expressions of $H_p^x(s)$ and $H_n^x(s)$ for $x = i$ can be easily obtained from Equations (3) and (4), as shown in Table 1. Following the same procedure, $H_p^x(s)$ and $H_n^x(s)$ for $x = ii$, and iii can also be derived as shown in Table 1.

Table 1. Detailed expressions of $H_p^x(s)$ and $H_n^x(s)$ in Equation (3) with different implementations of the adaptive PR current controllers.

X	i	ii	iii
$H_p^x(s + j\omega_1)$	$\dfrac{jV_p(s+j\omega_1)}{s(s+j2\omega_1)}$	$-\dfrac{V_p\omega_1}{s(s+j2\omega_1)}$	$j\dfrac{V_p}{2s}$
$H_n^x(s - j\omega_1)$	$-\dfrac{jV_n(s-j\omega_1)}{s(s-j2\omega_1)}$	$-\dfrac{V_n\omega_1}{s(s-j2\omega_1)}$	$-j\dfrac{V_n}{2s}$

3.2. Model of the PLL System

The frequency domain forms of ω_f and θ_f are shown in Equation (5). The superscripts 'y' is used to distinguish the PLL that is applied: $y = pll$ means that a SRF-PLL is used; $y = fll$ means that a DSOGI-FLL is used. The expressions of the transfer functions are shown in Table 2 where

$T_{pll}(s) = G_{PLL}(s)/(s + V_m G_{PLL}(s))$ and $D(s) = s^2 + k\omega_1 s + \omega_1{}^2$. The derivations are shown in the Appendix A.

$$\begin{cases} \omega_f(\omega) = H^y_{wp}(s)u^p_{ga}(\omega + \omega_1) + H^y_{wn}(s)u^n_{ga}(\omega - \omega_1) \\ \theta_f(\omega) = H^y_{sp}(s)u^p_{ga}(\omega + \omega_1) + H^y_{sn}(s)u^n_{ga}(\omega - \omega_1) \end{cases} \tag{4}$$

Table 2. Detailed expressions of the transfer functions in Equation (5) with different phase-locked loops (PLLs).

$H^{pll}_{sp}(s) = -jT_{pll}(s)$	$H^{pll}_{wp}(s) = -jsT_{pll}(s)$
$H^{pll}_{sn}(s) = jT_{pll}(s)$	$H^{pll}_{wn}(s) = jsT_{pll}(s)$
$H^{fll}_{sp}(s) = -\dfrac{jk\omega_1(s+j2\omega_1)}{2V_m D(s+j\omega_1)}\left(1 + \dfrac{\lambda\left(\frac{s-j2\omega_1}{D(s-j\omega_1)} + \frac{s+j2\omega_1}{D(s+j\omega_1)}\right)}{1+\lambda\frac{k\omega_1}{s}\left(\frac{s-j\omega_1}{D(s-j\omega_1)} + \frac{s+j\omega_1}{D(s+j\omega_1)}\right)}\right)$	$H^{fll}_{wp}(s) = \dfrac{-j\lambda\frac{k\omega_1}{V_m}\frac{s+j2\omega_1}{D(s+j\omega_1)}}{1+\lambda\frac{k\omega_1}{s}\left(\frac{s-j\omega_1}{D(s-j\omega_1)} + \frac{s+j\omega_1}{D(s+j\omega_1)}\right)}$
$H^{fll}_{sn}(s) = j\dfrac{k\omega_1(s-j2\omega_1)}{2V_m D(s-j\omega_1)}\left(1 + \dfrac{\lambda\left(\frac{s-j2\omega_1}{D(s-j\omega_1)} + \frac{s+j2\omega_1}{D(s+j\omega_1)}\right)}{1+\lambda\frac{k\omega_1}{s}\left(\frac{s-j\omega_1}{D(s-j\omega_1)} + \frac{s+j\omega_1}{D(s+j\omega_1)}\right)}\right)$	$H^{fll}_{wn}(s) = \dfrac{j\lambda\frac{k\omega_1}{V_m}\frac{s-j2\omega_1}{D(s-j\omega_1)}}{1+\lambda\frac{k\omega_1}{s}\left(\frac{s-j\omega_1}{D(s-j\omega_1)} + \frac{s+j\omega_1}{D(s+j\omega_1)}\right)}$

3.3. Model of the Current References

The $\alpha\beta$-domain current references in the frequency domain is shown in Equation (6) [22]:

$$\begin{cases} i^p_{\alpha r}(\omega + \omega_1) = j0.5I_{dr}\theta_f(\omega) \\ i^n_{\alpha r}(\omega - \omega_1) = -j0.5I_{dr}\theta_f(\omega) \end{cases} \tag{5}$$

3.4. Model of the Grid Current Loop

Based on the adaptive PR controller (3), the PLL system (5) and the current references (6), the detailed small-signal model of the grid current with adaptive PR controllers can be obtained, as depicted in Figure 5, where $G_d(s)$ is used to depicts the gain and delays (digital computation delay and the PWM delay) [29] of the converter, as follows:

$$G_d(s) = e^{-1.5sT_s} \tag{6}$$

where T_s is the sampling period.

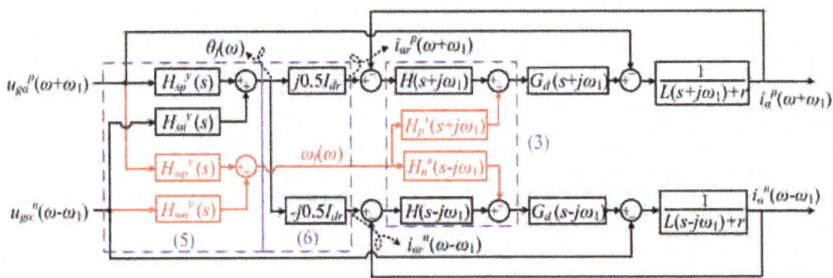

Figure 5. Small signal model of grid current control loop influenced by the PLL system.

Because of the adaptive PR controller, extra branches are introduced in the model of grid current loop, as shown in the red part of Figure 5. Moreover, $H^x_p(s)$, $H^x_n(s)$ (as shown in Table 1) are different for different implementations of adaptive PR controllers, and $H^y_{wp}(s)$, $H^y_{wn}(s)$ (as shown in Table 2) are not the same for different PLLs. Hence, the port characteristics of the grid-connected converters with different implementations of the adaptive PR controllers and different PLLs are different.

The small-signal part of grid current can be obtained from Figure 5, as follows:

$$\begin{cases} i_\alpha^p(\omega + \omega_1) = T_i(s + j\omega_1)i_{\alpha r}^p(\omega + \omega_1) - Y_i(s + j\omega_1)u_{g\alpha}^p(\omega + \omega_1) + T_{wp}(s + j\omega_1)\Delta\omega_f(\omega) \\ i_\alpha^n(\omega - \omega_1) = T_i(s - j\omega_1)i_{\alpha r}^n(\omega - \omega_1) - Y_i(s - j\omega_1)u_{g\alpha}^n(\omega - \omega_1) + T_{wn}(s - j\omega_1)\Delta\omega_f(\omega) \end{cases} \quad (7)$$

where:

$$\begin{cases} Y_i(s) = \frac{1}{H(s)G_d(s)+Ls+r} \\ T_i(s) = H(s)G_d(s)Y_i(s) \\ T_{wp}(s) = H_p^x(s)G_d(s)Y_i(s) \\ T_{wn}(s) = H_n^x(s)G_d(s)Y_i(s) \end{cases} \quad (8)$$

3.5. Admittance Matrix

The admittance matrix of the converter, $Y_{\alpha\beta}(s)$, is defined in Equation (10):

$$\begin{bmatrix} -i_\alpha^p(\omega + \omega_1) \\ -i_\alpha^n(\omega - \omega_1) \end{bmatrix} = \begin{bmatrix} Y_{pp}(s + j\omega_1) & Y_{pn}(s + j\omega_1) \\ Y_{np}(s - j\omega_1) & Y_{nn}(s - j\omega_1) \end{bmatrix} \begin{bmatrix} u_{g\alpha}^p(\omega + \omega_1) \\ u_{g\alpha}^n(\omega - \omega_1) \end{bmatrix} = Y_{\alpha\beta}(s) \begin{bmatrix} u_{g\alpha}^p(\omega + \omega_1) \\ u_{g\alpha}^n(\omega - \omega_1) \end{bmatrix} \quad (9)$$

The elements can be derived from Figure 5, as follows:

$$\begin{cases} Y_{pp}(s + j\omega_1) = Y_i(s + j\omega_1) - \left(j0.5I_{dr}H_{sp}^y(s)T_i(s + j\omega_1) + T_{wp}(s + j\omega_1)H_{wp}^y(s) \right) \\ Y_{pn}(s + j\omega_1) = -j0.5I_{dr}T_i(s + j\omega_1)H_{sn}^y(s) - T_{wp}(s + j\omega_1)H_{wn}^y(s) \\ Y_{np}(s - j\omega_1) = j0.5I_{dr}T_i(s - j\omega_1)H_{sp}^y(s) - T_{wn}(s - j\omega_1)H_{wp}^y(s) \\ Y_{nn}(s - j\omega_1) = Y_i(s - j\omega_1) - \left(-j0.5I_{dr}T_i(s - j\omega_1)H_{sn}^y(s) + T_{wn}(s - j\omega_1)H_{wn}^y(s) \right) \end{cases} \quad (10)$$

4. Impedance-Based Stability Analysis

4.1. Addmittances Analysis and Verifications

Figures 6 and 7 shows the magnitude responses of the admittances of the grid-connected converter with SRF-PLL and DSOGI-FLL, respectively. The solid lines are plotted using the theoretical models in Equation (11), while the circles are the point-by-point numerical simulation results of the admittances for comparison. The parameters used in simulations are as shown in Table 3 with the bandwidth of the SRF-PLL f_{bw_PLL} = 40 Hz (the corresponding parameters of the $G_{PLL}(s)$ is designed based on [30]), and the parameters of the DSOGI-FLL are: k = 1.1, γ = 41. It can be observed from Figures 6 and 7 that for different implementations of adaptive PR controllers and different PLLs, the numerical admittances match the theoretical admittances of the grid-connected converter, which verify the correctness of the proposed admittance model.

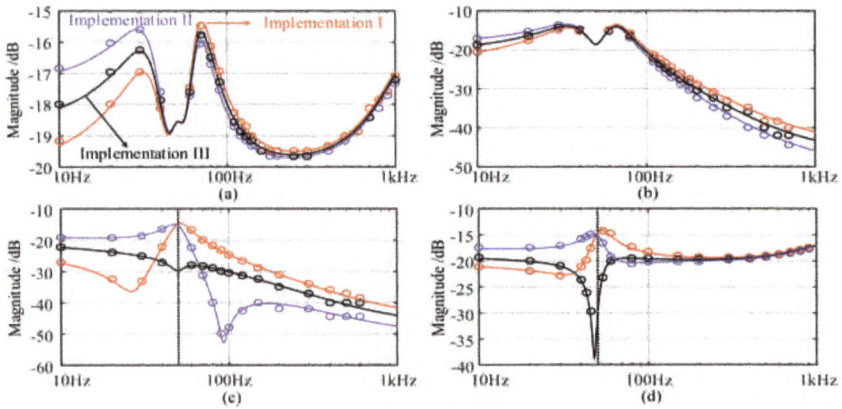

Figure 6. Magnitude responses of admittances of the converter synchronized by a SRF-PLL: (a) $Y_{pp}(s + j\omega_1)$; (b) $Y_{pn}(s + j\omega_1)$; (c) $Y_{np}(s - j\omega_1)$; (d) $Y_{nn}(s - j\omega_1)$. Red line: Implementation I; Green line: implementation II; Black line: implementation III is applied. Circles: numerical simulation results.

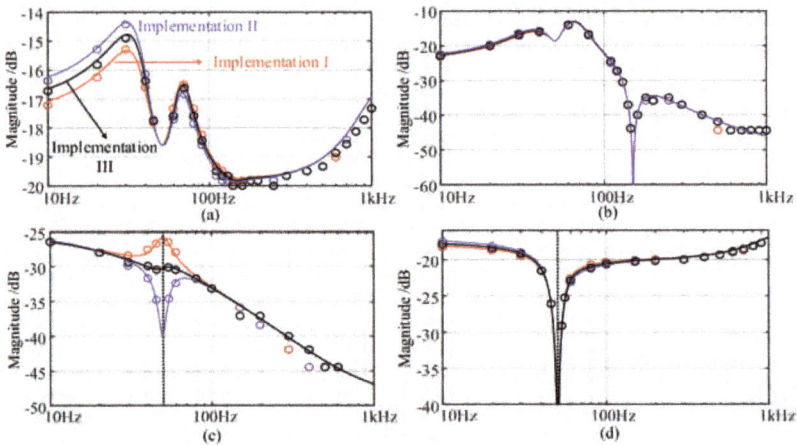

Figure 7. Magnitude responses of admittances of the converter synchronized by a DSOGI-FLL: (a) $Y_{pp}(s + j\omega_1)$; (b) $Y_{pn}(s + j\omega_1)$; (c) $Y_{np}(s - j\omega_1)$; (d) $Y_{nn}(s - j\omega_1)$. Red line: Implementation I; Green line: implementation II; Black line: implementation III is applied. Circles: numerical simulation results.

Table 3. Parameters of grid-tied converter prototype.

Symbol	Description	Value
V_1	Grid phase-neutral peak voltage	$30\sqrt{2}$ V
ω_1	Grid angular frequency	$2\pi \times 50$ rad/s
f_s	Switching frequency	10 kHz
V_{dc}	Dc-link voltage	130 V
L	Inductance of the L-type filter	2 mH
r	Resistance of the filter	$0.2\ \Omega$
α_c	Current control loop bandwidth	$2\pi \times 833$ rad/s
k_i	Proportional gain of ac/dc current controller	10.47
k_r	R parameter of ac/dc current controller	1047
I_{dr}	D channel current reference of VSC	10 A

4.1.1. SRF-PLL is Used for Grid Synchronization

It can be observed from Figure 6 that the $Y_{pp}(s + j\omega_1)$ and $Y_{pn}(s + j\omega_1)$ with different implementations of the adaptive controllers are similar as shown in Figure 6a,b, while $Y_{np}(s - j\omega_1)$ and $Y_{nn}(s - j\omega_1)$ are different, especially at the frequencies around $s - j\omega_1 = j\omega_1$, i.e., $s = j2\omega_1$, as shown in Figure 6c,d. The value of $Y_{np}(s - j\omega_1)$ and $Y_{nn}(s - j\omega_1)$ at $s = j2\omega_1$ can be obtained by Equation (11), as follows:

$$\begin{cases} Y_{nn}(j\omega_1) = -Y_{np}(j\omega_1) = -\left(\frac{I_{dr}}{2} + j\frac{2\omega_1 V_n}{k_r}\right) T_{pll}(j2\omega_1) \cdots \text{ I} \\ Y_{nn}(j\omega_1) = -Y_{np}(j\omega_1) = -\left(\frac{I_{dr}}{2} - j\frac{2\omega_1 V_n}{k_r}\right) T_{pll}(j2\omega_1) \cdots \text{ II} \\ Y_{nn}(j\omega_1) = -Y_{np}(j\omega_1) = -\frac{I_{dr}}{2} T_{pll}(j2\omega_1) \cdots \text{ III} \end{cases} \tag{12}$$

where Equations (12-I), (12-II), and (12-III) are the admittances of the converter with implementation I, II, and III of the adaptive PR controller, respectively.

Obviously, it can be obtained from Equation (12) that for a small k_r as suggested in [16], the magnitudes of $Y_{np}(s - j\omega_1)$ and $Y_{nn}(s - j\omega_1)$ of the converter using implementation III around $s = j2\omega_1$ are much smaller than those using implementation I or II.

4.1.2. DSOGI-FLL is Used for Grid Synchronization

It can be observed from Figure 7 that for different implementations of the adaptive resonant controllers, the converter admittances $Y_{pp}(s + j\omega_1)$, $Y_{pn}(s + j\omega_1)$, $Y_{nn}(s - j\omega_1)$ are similar, while $Y_{np}(s - j\omega_1)$ are different at the frequencies around $s = j2\omega_1$. Due to the effect of the zeros in $H_{wn}^{fll}(s - j\omega_1)$ and $H_{sn}^{fll}(s - j\omega_1)$ (as shown in Table 2) at $s = j2\omega_1$, it can be derived from Equation (11) that:

$$Y_{nn}(j\omega_1) = Y_{pn}(j3\omega_1) = 0 \tag{11}$$

Equation (13) is valid with different implantations of the adaptive PR controllers.

4.2. Stability Analysis

4.2.1. Stability Criterion

System stability is determined by applying general Nyquist stability criterion to the minor loop gain $L_{\alpha\beta}(s) = Z_{g\alpha\beta}(s)Y_{\alpha\beta}(s)$, where $Z_{g\alpha\beta}(s)$ is the impedance matrix of the grid as shown in Equation (14) [28]. The system is stable if the Nyquist curves of the eigenvalues of $L_{\alpha\beta}(s)$, i.e., $\lambda_1(s)$ and $\lambda_2(s)$ as defined in Equation (15), do not encircle the critical point $(-1, j0)$, otherwise, the system is unstable [27,28]:

$$Z_{g\alpha\beta}(s) = \begin{bmatrix} Z_g(s + j\omega_1) & 0 \\ 0 & Z_g(s - j\omega_1) \end{bmatrix} \tag{12}$$

$$\lambda_{1,2}(s) = 0.5\left(Z_g(s + j\omega_1) Y_p(s + j\omega_1) + Z_g(s - j\omega_1)Y_n(s - j\omega_1)\right)\pm$$
$$0.5\sqrt{\begin{array}{l} \left(Z_g(s + j\omega_1) Y_p(s + j\omega_1) + Z_g(s - j\omega_1)Y_n(s - j\omega_1)\right)^2 - \\ 4Z_g(s + j\omega_1) Z_g(s - j\omega_1)\left(Y_{pp}(s + j\omega_1)Y_{nn}(s - j\omega_1) - Y_{pn}(s + j\omega_1)Y_{np}(s - j\omega_1)\right) \end{array}} \tag{13}$$

4.2.2. Stability Analysis with Different PLLs

Figure 8 shows the Nyquist curves of characteristic loci of the grid system using different implementations of the adaptive current controllers and different PLLs. The parameters used in simulations are as shown in Table 3 with $f_{bw\_PLL} = 75$ Hz, the parameters of the DSOGI-FLL are: $k = 1.1$, $\gamma = 41$, and the grid inductance $L_g = 6$ mH.

It can be observed from Figure 8a that if a DSOGI-FLL is used, the stabilities of grid-connected converters with different implementations of the adaptive PR controllers are similar. Since $Y_{pn}(j3\omega_1) = 0$ as derived in Equation (13), $Y_{pn}(s + j\omega_1) \times Y_{np}(s - j\omega_1)$ are very small around

$s = j2\omega_1$. Therefore, based on Equation (15), it can be concluded that although that $Y_{np}(s - j\omega_1)$ are different around $s = j2\omega_1$, the differences would have very limited influence on the $\lambda_1(s)$ and $\lambda_2(s)$.

It can be observed from Figure 8b that if a SRF-PLL is used, the stabilities of grid-connected converters with different implementations of the adaptive PR controllers are quite different. The converter with implementation I is unstable, while the converter with implementation II and III are stable. The stability boundaries of the grid-connected converters with different implementations of the adaptive PR controllers are obtained based on the stability criterion, as depicted in Figure 9. It can be observed from Figure 9 that the system with implementation III has the largest stable region, while the system with implementation I has the smallest stable region. When the short circuit ratio (SCR) is set to be 2.23, the maximum bandwidth to maintain system stability is around 73.3 Hz if implementation I is used, the maximum value extends to about 107.9 Hz if implementation II is applied, and the maximum value can be further raised to about 121.2 Hz if implementation III is applied. Above all, implementation III of the adaptive PR current controllers is suggested to be used for the grid connected converter using SRF-PLL for grid synchronization.

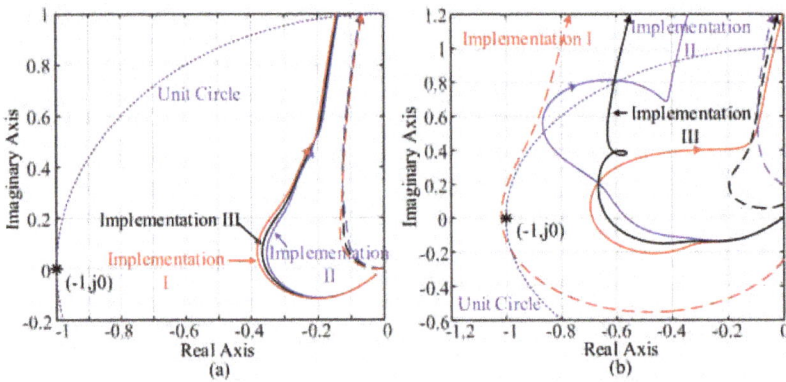

Figure 8. Nyquist curves of characteristic loci of $L_{\alpha\beta}$ with different implementations of the adaptive resonance controllers using (**a**) DSOGI-FLL; (**b**) SRF-PLL for grid synchronization. Solid line: $\lambda_1(s)$; Dash line: $\lambda_2(s)$.

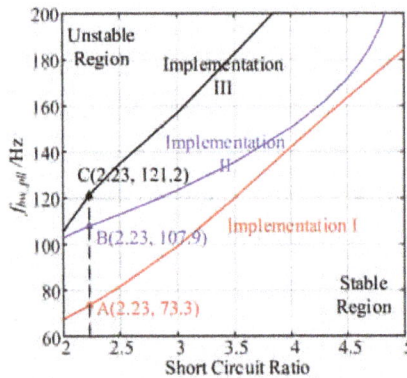

Figure 9. Stability boundaries of the grid-connected converter using different implementations of the adaptive PR controller and SRF-PLL for grid synchronization.

4.2.3. Comparison with Different PLLs

In this subsection, the effects of the SRF-PLL and DSOGI-FLL on stability of the grid connected converter with adaptive resonance current controllers using implementation III are compared.

Based on Equation (6), the dynamic properties of the output angle of a SRF-PLL with $f_{bw\_PLL} = 40$ Hz and a DSOGI-FLL with $k = 1.1$, $\gamma = 41$ are similar, as shown in Figure 10a. Under this circumstance, Figure 10b shows the corresponding dynamic properties of the output frequency for different PLLs. The magnitudes of $H_{wp}^{fll}(s)$ and $H_{wn}^{fll}(s)$ are smaller than that of $H_{wp}^{pll}(s)$ and $H_{wn}^{pll}(s)$, hence it can be concluded that ω_f is more robust if the DSOGI-FLL is used than that if the SRF-PLL is used. The influence of a PLL on system stability depends on the dynamic properties of the output phase angle and the estimated frequency. Hence, the converter using DSOGI-FLL for grid synchronization should be more stable than the converter using SRF-PLL for grid synchronization.

Figure 11 shows the corresponding Nyquist curves of characteristic loci of converters with different synchronization method. The other parameters used in the simulation are as shown in Table 3. It can be concluded from Figure 11 that the converter using DSOGI-FLL for grid synchronization has a larger stability margin than that of the converter using SRF-PLL.

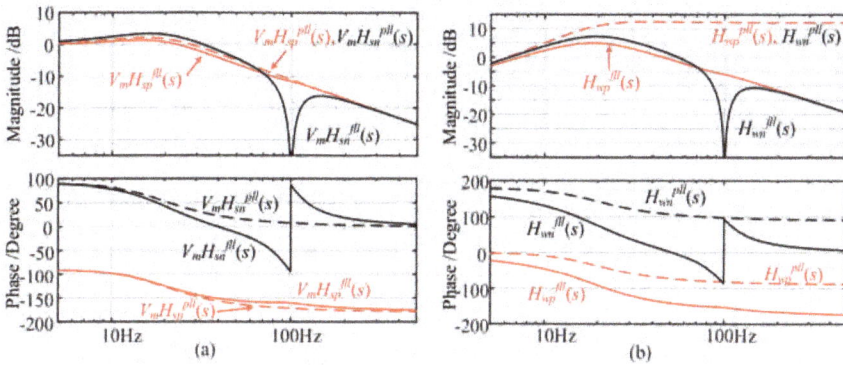

Figure 10. Frequency response of the effects of point of common coupling (PCC) voltages on θ_f in (a) and ω_f in (b). Solid line: DSOGI-FLL is used; Dash line: SRF-PLL is used.

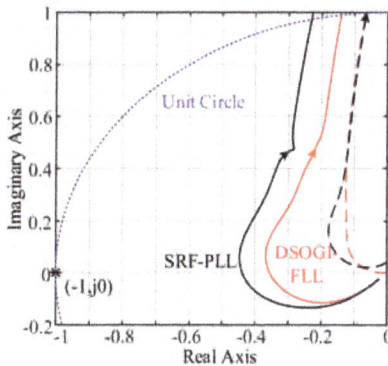

Figure 11. Nyquist curves of characteristic loci of $L_{\alpha\beta}$ using different PLLs. Black line: SRF-PLL; Red line: DSOGI-FLL is used. Solid line: $\lambda_1(s)$; Dash line: $\lambda_2(s)$.

5. Experimental Verifications

A three-phase grid-connected converter has been built and tested to verify proposed analysis. The current controllers, frame transformation and the PLL were implemented in a TMS320F28335 DSP board (Texas Instruments, Inc, Dallas, TX, USA). The grid current is sensed by a TCP0150 current probe (Tektronix, Beaverton, OR, USA) and the bandwidth of the SRF-PLL f_{bw_PLL}, the q-axis component of the grid current i_q, and the estimated frequency ω_f are sent to the D/A in the board as output signals. The output of the D/A cannot be negative, hence, it is programed to have a 15 A offset in i_q and a 200 rad/s offset in ω_f compared to the corresponding actual values. Parameters for this experimental setup are provided in Table 3, which are consistent with the simulation parameters. The parameters of DSOGI-FLL are: $k = 1.1$, $\gamma = 41$, and the short circuit ratio is 2.23 (the corresponding L_g is 6 mH).

Figure 12 are the experimental waveforms of A-phase current of the converter using SRF-PLL for grid synchronization. In Figure 12a, the implementation I of the adaptive resonance controller is applied. At time T_0, the bandwidth of the PLL jumps from 56 Hz to 70 Hz, and after T_0, the grid current diverges. Once the grid currents reach the up-limited value, the system would stop running. In Figure 12b, the implementation II is applied. At time T_0, the bandwidth of the PLL jumps from 102 Hz to 112 Hz, and after T_0, system becomes unstable. In Figure 12c, the implementation III is applied. At time T_0, the bandwidth of the PLL jumps from 114 Hz to 126 Hz, and after T_0, system is no longer stable. The experimental results match the theoretical stability boundaries shown in Figure 9.

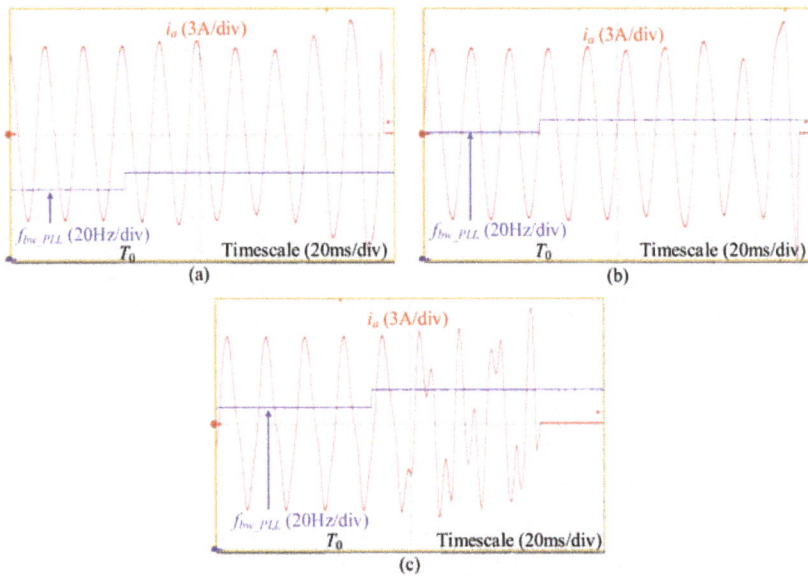

Figure 12. A-phase current waveforms for the converter using (a) implementation I; (b) implementation II; (c) implementation III of the adaptive resonance controllers with f_{bw_pll} changes at time T_0.

Figure 13 shows the experimental waveforms of i_a, ω_f, and i_q of the grid connected converter using DSOGI-FLL for grid synchronization. In Figure 13a–c, implementation I, II and III of the adaptive PR controllers are used, respectively. The active current reference I_{dr} changes from 0 A to 10 A at time T_0, and the dynamic responses of ω_f, and i_q are similar for all implementations of the adaptive PR controllers. The experimental results match the theoretical analysis in Section 4.2.2.

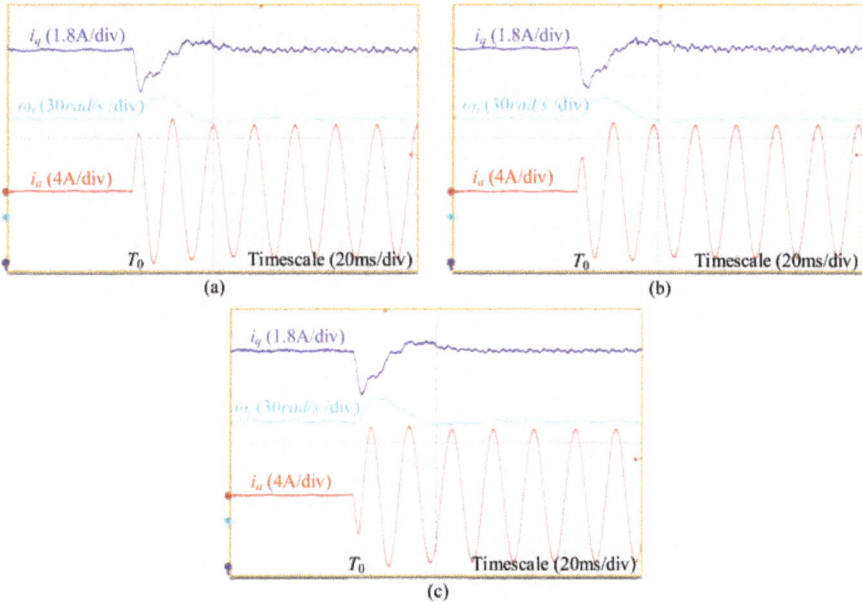

Figure 13. Dynamic responses of the converter using (**a**) implementation I; (**b**) implementation II; (**c**) implementation III of the adaptive resonance controllers with changing active current reference.

Figure 14 shows the experimental waveforms of A-phase current and i_q of the grid connected converter with Implementation III of the adaptive PR controllers. In Figure 14a, the SRF-PLL with 40 Hz bandwidth is applied while in Figure 14b, the DSOGI-FLL is used. The reactive current reference I_{qr} changes from −6 A to 0 A at time T_0. It is can be obtained from Figure 14 that: the percentage overshoot (PO) of i_q in Figure 14a is about 14% (0.84/6) while in Figure 14b the PO is less than 8% (0.48/6). The system with DSOGI-FLL has a stronger damping than that of the system with SRF-PLL. The experimental results verify the effectiveness of the analysis in Section 4.2.3.

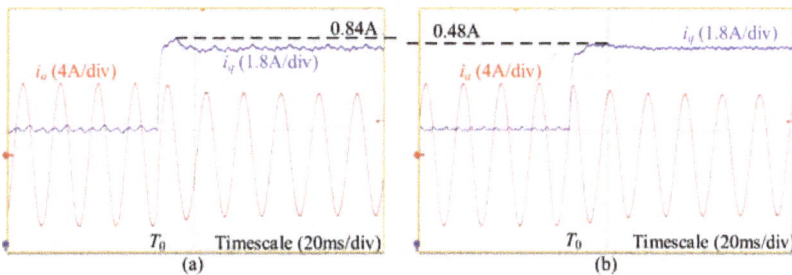

Figure 14. Dynamic response of q-axis component of grid current with its reference changes from −6 A to 0 A at time T_0: (**a**) SRF-PLL; (**b**) DSOGI-FLL is used for grid synchronization.

6. Conclusions

The impedance model and stability of the grid-connected VSCs with adaptive resonance current controllers has been explored in this paper. Based on the proposed small-signal impedance model, some tips should be aware of when implementing adaptive resonant controllers:

(1) If a SRF-PLL is used for grid synchronization, the system using implementation III of the resonant controller has the best stability margin, while the system using implementation I has the worst stability margin under weak grid conditions.

(2) If a DSOGI-FLL is used for grid synchronization, the systems using implementation I, II, and III of the resonant controller have similar stability margins.

(3) The system using a DSOGI-FLL for grid synchronization has a larger stability margin than that of the system using a SRF-PLL if the dynamic property of the output angle of the two synchronization methods are similar.

Experimental results validate the conclusions based on the theoretical analysis. In sum, the implementation of the resonance controller and the grid synchronization method should be carefully chosen for the weak-grid connected converter using adaptive resonance current controllers.

Author Contributions: X.L. proposed the main idea, performed the theoretical analysis, performed the experiments, and wrote the paper; H.L. contributed the experiment materials, and gave some suggestions to the paper writing.

Funding: This research received no external funding.

Conflicts of Interest: The authors declare no conflict of interest.

Appendix A.

Appendix A.1. Modeling of the SRF-PLL

It can be obtained from [22] that, the output angle of the SRF-PLL is:

$$\theta_f(\omega) = -jT_{pll}(s)\left(u_{g\alpha}^p(\omega + \omega_1) - u_{g\alpha}^n(\omega - \omega_1)\right) \tag{A1}$$

where $T_{pll}(s)$ is the close loop gain of the PLL, as follows:

$$T_{pll}(s) = \frac{G_{PLL}(s)/s}{1 + V_m G_{PLL}(s)/s} \tag{A2}$$

The estimated frequency is in fact the differential of the output angle of the PLL, i.e., $\omega_f(\omega) = s\theta_f(\omega)$, hence, from Equation (A1) it can be obtained that:

$$\omega_f(\omega) = -jsT_{pll}(s)\left(u_{g\alpha}^p(\omega + \omega_1) - u_{g\alpha}^n(\omega - \omega_1)\right) \tag{A3}$$

Appendix A.2. Modeling of the DSOGI-FLL

Appendix A.2.1. Modeling of the SOGIs

According to the block diagram shown in Figure 4b, it can be obtained that:

$$\begin{cases} \varepsilon_\alpha^p(\omega + \omega_1) = u_{g\alpha}^p(\omega + \omega_1) - u_\alpha^p(\omega + \omega_1) \\ \varepsilon_\alpha^n(\omega - \omega_1) = u_{g\alpha}^n(\omega - \omega_1) - u_\alpha^n(\omega - \omega_1) \end{cases} \tag{A4}$$

The steady-state value of the u_α is $V_m\cos(\omega_1 t)$, by using convolution, it can be obtained that:

$$\begin{cases} \left(k\varepsilon_\alpha^p(\omega + \omega_1) - qu_\alpha^p(\omega + \omega_1)\right)\frac{\omega_1}{s+j\omega_1} - \frac{V_m}{2j}\frac{1}{s+j\omega_1}\omega_f(\omega) = u_\alpha^p(\omega + \omega_1) \\ \left(k\varepsilon_\alpha^n(\omega - \omega_1) - qu_\alpha^n(\omega - \omega_1)\right)\frac{\omega_1}{s-j\omega_1} + \frac{V_m}{2j}\frac{1}{s-j\omega_1}\omega_f(\omega) = u_\alpha^n(\omega - \omega_1) \end{cases} \tag{A5}$$

$$\begin{cases} qu_\alpha^p(\omega + \omega_1) = \frac{\omega_1}{s+j\omega_1}u_\alpha^p(\omega + \omega_1) + \frac{V_m}{j2\omega_1}\omega_f(\omega) \\ qu_\alpha^n(\omega - \omega_1) = \frac{\omega_1}{s-j\omega_1}u_\alpha^n(\omega - \omega_1) - \frac{V_m}{j2\omega_1}\omega_f(\omega) \end{cases} \tag{A6}$$

With Equations (A4)–(A6), it can be obtained that:

$$\begin{cases} u_\alpha^p(\omega + \omega_1) = \frac{s+j\omega_1}{D(s+j\omega_1)}\left(k\omega_1 u_{g\alpha}^p(\omega + \omega_1) + jV_m\omega_f(\omega)\right) \\ u_\alpha^n(\omega - \omega_1) = \frac{s-j\omega_1}{D(s-j\omega_1)}\left(k\omega_1 u_{g\alpha}^n(\omega - \omega_1) - jV_m\omega_f(\omega)\right) \end{cases} \tag{A7}$$

$$\begin{cases} qu_\alpha^p(\omega + \omega_1) = \frac{k\omega_1^2}{D(s+j\omega_1)}u_{g\alpha}^p(\omega + \omega_1) + \left(\frac{jV_m\omega_1}{D(s+j\omega_1)} + \frac{V_m}{j2\omega_1}\right)\omega_f(\omega) \\ qu_\alpha^n(\omega - \omega_1) = \frac{k\omega_1^2}{D(s-j\omega_1)}u_{g\alpha}^n(\omega - \omega_1) - \left(\frac{jV_m\omega_1}{D(s-j\omega_1)} + \frac{V_m}{j2\omega_1}\right)\omega_f(\omega) \end{cases} \tag{A8}$$

$$\begin{cases} \varepsilon_\alpha^p(\omega + \omega_1) = \frac{s(s+j2\omega_1)}{D(s+j\omega_1)}u_{g\alpha}^p(\omega + \omega_1) - \frac{jV_m(s+j\omega_1)}{D(s+j\omega_1)}\omega_f(\omega) \\ \varepsilon_\alpha^n(\omega - \omega_1) = \frac{s(s-j2\omega_1)}{D(s-j\omega_1)}u_{g\alpha}^n(\omega - \omega_1) + \frac{jV_m(s-j\omega_1)}{D(s-j\omega_1)}\omega_f(\omega) \end{cases} \tag{A9}$$

where $D(s) = s^2 + k\omega_1 s + \omega_1^2$ is used to simplify the expression.

Appendix A.2.2. Modeling of the PNSC

For a balanced positive-, negative-sequence vector, the α-, β-axis components keep the following steady-state relationship on frequency domain:

$$\begin{cases} qu_\alpha^p(\omega + \omega_1) = jqu_\beta^p(\omega + \omega_1) \\ qu_\alpha^n(\omega - \omega_1) = -jqu_\beta^n(\omega - \omega_1) \end{cases} \tag{A10}$$

Thus, the input signals of the "atan2" are:

$$\begin{cases} u_{\alpha+}^p(\omega + \omega_1) = \frac{1}{2}\frac{k\omega_1(s+j2\omega_1)}{D(s+j\omega_1)}u_{g\alpha}^p(\omega + \omega_1) + \frac{V_m}{2}\left(\frac{j(s+j2\omega_1)}{D(s+j\omega_1)} + \frac{1}{2\omega_1}\right)\omega_f(\omega) \\ u_{\alpha+}^n(\omega - \omega_1) = \frac{1}{2}\frac{k\omega_1(s-j2\omega_1)}{D(s-j\omega_1)}u_{g\alpha}^n(\omega - \omega_1) + \frac{V_m}{2}\left(\frac{-j(s-j2\omega_1)}{D(s-j\omega_1)} + \frac{1}{2\omega_1}\right)\omega_f(\omega) \end{cases} \tag{A11}$$

Appendix A.2.3. Modeling of the FLL

The steady state values of ε_α, ε_β are zeros, while qu_α, qu_β are $V_m\sin(\omega_1 t)$ and $-V_m\cos(\omega_1 t)$, respectively. Hence, by using convolution, it can easily be obtained that:

$$u_f(\omega) = -\frac{V_m}{j2}\varepsilon_\alpha^p(\omega + \omega_1) + \frac{V_m}{j2}\varepsilon_\alpha^n(\omega - \omega_1) - \frac{V_m}{2}\varepsilon_\beta^p(\omega + \omega_1) - \frac{V_m}{2}\varepsilon_\beta^n(\omega - \omega_1) \tag{A12}$$

Using the similar steady-state relationship as shown in Equation (A10), Equation (A12) can be simplified as follows:

$$u_f(\omega) = -\frac{V_m}{j}\varepsilon_\alpha^p(\omega + \omega_1) + \frac{V_m}{j}\varepsilon_\alpha^n(\omega - \omega_1) \tag{A13}$$

The steady-state value of u_f is zero, hence the output of the FLL can be written as follows:

$$-\lambda\frac{k\omega_1}{V_m^2}\frac{1}{s}u_f(\omega) = \omega_f(\omega) \tag{A14}$$

With Equations (A9), (A13), and (A14), the output of the FLL can be obtained that:

$$\omega_f(\omega) = \frac{-\lambda\frac{k\omega_1}{V_m}\frac{1}{j}\left(\frac{s-j2\omega_1}{D(s-j\omega_1)}u_{g\alpha}^n(\omega - \omega_1) - \frac{s+j2\omega_1}{D(s+j\omega_1)}u_{g\alpha}^p(\omega + \omega_1)\right)}{1 + \lambda\frac{k\omega_1}{s}\left(\frac{s-j\omega_1}{D(s-j\omega_1)} + \frac{s+j\omega_1}{D(s+j\omega_1)}\right)} \tag{A15}$$

Appendix A.2.4. Modeling of the Output Angle

As demonstrated in [31], the output angle of "atan2" satisfies the following relationship:

$$V_m \theta_f(\omega) = -\frac{j}{2} u^p_{\alpha+}(\omega + \omega_1) + \frac{j}{2} u^n_{\alpha+}(\omega - \omega_1) + \frac{1}{2} u^p_{\beta+}(\omega + \omega_1) + \frac{1}{2} u^n_{\beta+}(\omega - \omega_1) \tag{A16}$$

It is pointed out in [27,28] that the vector at frequency $\omega - \omega_1$ is not positive-sequence, but negative-sequence, hence, Equation (A16) can be simplified to:

$$V_m \theta_f(\omega) = -j u^p_{\alpha+}(\omega + \omega_1) + j u^n_{\alpha+}(\omega - \omega_1) \tag{A17}$$

With Equations (A11), (A15), and (A17), it can be obtained that:

$$\theta_f(\omega) = \frac{j\left(\frac{k\omega_1(s - j2\omega_1)}{D(s - j\omega_1)} u^n_{ga}(\omega - \omega_1) - \frac{k\omega_1(s + j2\omega_1)}{D(s + j\omega_1)} u^p_{ga}(\omega + \omega_1)\right)}{2V_m} \times \left(1 + \frac{\lambda\left(\frac{s - j2\omega_1}{D(s - j\omega_1)} + \frac{s + j2\omega_1}{D(s + j\omega_1)}\right)}{1 + \lambda \frac{k\omega_1}{s}\left(\frac{s - j\omega_1}{D(s - j\omega_1)} + \frac{s + j\omega_1}{D(s + j\omega_1)}\right)}\right) \tag{A18}$$

References

1. Blaabjerg, F.; Chen, Z.; Kjaer, S.B. Power electronics as efficient interface in dispersed power generation systems. *IEEE Trans. Power Electron.* **2004**, *19*, 1184–1194. [CrossRef]
2. Dash, P.P.; Kazerani, M. Dynamic modeling and performance analysis of a grid-connected current-source inverter-based photovoltaic system. *IEEE Trans. Sustain. Energy* **2011**, *2*, 443–450. [CrossRef]
3. Blaabjerg, F.; Teodorescu, R.; Liserre, M.; Timbus, A.V. Overview of control and grid synchronization for distributed power generation systems. *IEEE Trans. Ind. Electron.* **2006**, *53*, 1398–1409. [CrossRef]
4. Zmood, D.N.; Holmes, D.G. Stationary frame current regulation of PWM inverters with zero steady-state error. *IEEE Trans. Power Electron.* **2003**, *18*, 814–822. [CrossRef]
5. Yepes, A.G.; Freijedo, F.D.; Gandoy, J.D.; Lopez, O.; Malvar, J.; Comesana, P.F. Effects of Discretization Methods on the Performance of Resonant Controllers. *IEEE Trans. Power Electron.* **2010**, *25*, 1692–1772. [CrossRef]
6. Yang, Y.; Zhou, K.; Blaabjerg, F. Enhancing the Frequency Adaptability of Periodic Current Controllers with a Fixed Sampling Rate for Grid-Connected Power Converters. *IEEE Trans. Power Electron.* **2016**, *31*, 7232–7285. [CrossRef]
7. Bojoi, R.I.; Griva, G.; Bostan, V.; Guerriero, M.; Farina, F.; Profumo, F. Current Control Strategy for Power Conditioners Using Sinusoidal Signal Integrators in Synchronous Reference Frame. *IEEE Trans. Power Electron.* **2005**, *20*, 1402–1412. [CrossRef]
8. Yang, Y.; Zhou, K.; Wang, H.; Blaabjerg, F.; Wang, D.; Zhang, B. Frequency Adaptive Selective Harmonic Control for Grid-Connected Inverters. *IEEE Trans. Power Electron.* **2015**, *30*, 3912–3924. [CrossRef]
9. Cadaval, E.R.; Spagnuolo, G.; Franquelo, L.G.; Paja, C.A.R.; Suntio, T.; Xiao, W.M. Grid-connected photovoltaic generation plants: Components and operation. *IEEE Ind. Electron. Mag.* **2013**, *7*, 6–20. [CrossRef]
10. Espin, F.G.; Garcera, G.; Patrao, I.; Figueres, E. An adaptive control system for three-phase photovoltaic inverters working in a polluted and variable frequency electric grid. *IEEE Trans. Power Electron.* **2012**, *27*, 4248–4261. [CrossRef]
11. Herran, M.A.; Fischer, J.R.; Gonzalez, S.A.; Judewicz, M.G.; Carugati, I.; Carrica, D.O. Repetitive control with adaptive sampling frequency for wind power generation systems. *IEEE J. Emerg. Sel. Top. Power Electron.* **2014**, *2*, 58–69. [CrossRef]
12. Jorge, S.G.; Busada, C.A.; Solsona, J.A. Frequency-adaptive current controller for three-phase grid-connected converters. *IEEE Trans. Ind. Electron.* **2013**, *60*, 4169–4177. [CrossRef]
13. Timbus, A.V.; Ciobotaru, M.; Teodorescu, R.; Blaabjerg, F. Adaptive Resonant Controller for Grid-Connected Converters in Distributed Power Generation Systems. In Proceedings of the Twenty-First Annual IEEE Applied Power Electronics Conference and Exposition, Dallas, TX, USA, 19–23 March 2006; pp. 1601–1606.

14. Chung, S.K. A phase tracking system for three phase utility interface inverters. *IEEE Trans. Power Electron.* **2000**, *15*, 431–438. [CrossRef]

15. Rodriguez, P.; Luna, A.; Aguilar, R.S.M.; Otadui, I.E.; Teodorescu, R.; Blaabjerg, F. A Stationary Reference Frame Grid Synchronization System for Three-Phase Grid-Connected Power Converters under Adverse Grid Conditions. *IEEE Trans. Power Electron.* **2012**, *27*, 99–112. [CrossRef]

16. Liserre, M.; Teodorescu, R.; Blaabjerg, F. Stability of photovoltaic and wind turbine grid-connected inverters for a large set of grid impedance values. *IEEE Trans. Power Electron.* **2006**, *21*, 263–272. [CrossRef]

17. Wang, X.; Blaabjerg, F.; Wu, W. Modeling and analysis of harmonic stability in an AC power-electronics-based power system. *IEEE Trans. Power Electron.* **2014**, *29*, 6421–6432. [CrossRef]

18. Harnefors, L.; Bongiorno, M.; Lundberg, S. Input-Admittance Calculation and Shaping for Controlled Voltage-Source Converters. *IEEE Trans. Ind. Electron.* **2007**, *54*, 3323–3334. [CrossRef]

19. Alawasa, K.M.; Mohamed, Y.A.R.I.; Xu, W. Active mitigation of subsynchronous interactions between PWM voltage-source converters and power networks. *IEEE Trans. Power Electron.* **2014**, *29*, 121–134. [CrossRef]

20. Zhou, J.Z.; Ding, H.; Fan, S.; Zhang, Y.; Gole, A.M. Impact of short-circuit ratio and phase-locked-loop parameters on the small-signal behavior of a VSC-HVDC converter. *IEEE Trans. Power Deliv.* **2014**, *29*, 2287–2296. [CrossRef]

21. Alvarez, A.E.; Fekriasl, S.; Hassan, F.; Bellmunt, O.G. Advanced Vector Control for Voltage Source Converters Connected to Weak Grids. *IEEE Trans. Power Syst.* **2015**, *30*, 3072–3081. [CrossRef]

22. Li, X.; Lin, H. Multifrequency Small-Signal Model of Voltage Source Converters Connected to a Weak Grid for Stability Analysis. In Proceedings of the 2016 IEEE Applied Power Electronics Conference and Exposition (APEC), Long Beach, CA, USA, 20–24 March 2016; pp. 728–732.

23. Cho, Y.; Lee, C.; Hur, K.; Kang, Y.C.; Muljadi, E. Impedance-Based Stability Analysis in Grid Interconnection Impact Study Owing to the Increased Adoption of Converter-Interfaced Generators. *Energies* **2017**, *10*, 1355. [CrossRef]

24. Sun, J. Impedance-Based Stability Criterion for Grid-Connected Inverters. *IEEE Trans. Power Electron.* **2011**, *26*, 3075–3078. [CrossRef]

25. Wen, B.; Boroyevich, D.; Burgos, R.; Mattavelli, P.; Shen, Z. Analysis of D-Q Small-Signal Impedance of Grid-Tied Inverters. *IEEE Trans. Power Electron.* **2016**, *31*, 675–687. [CrossRef]

26. Cespedes, M.; Sun, J. Impedance Modeling and Analysis of Grid-Connected Voltage-Source Converters. *IEEE Trans. Power Electron.* **2014**, *29*, 1254–1261. [CrossRef]

27. Rygg, A.; Monlinas, M.; Zhang, C.; Cai, X. A modified sequence domain impedance definition and its equivalence to the dq-domain impedance definition for the stability analysis of ac power electronic systems. *IEEE J. Emerg. Sel. Top. Power Electron.* **2016**, *4*, 1383–1396. [CrossRef]

28. Bakhshizadeh, M.K.; Wang, X.; Blaabjerg, F.; Hjerrild, J.; Kocewiak, L.; Bak, C.L.; Hesselbaek, B. Couplings in Phase Domain Impedance Modeling of Grid-Connected Converters. *IEEE Trans. Power Electron.* **2016**, *31*, 6792–6796.

29. Timbus, A.; Liserre, M.; Teodoresce, R.; Rodriguez, P.; Blaabjerg, F. Evaluation of current controllers for distributed power generation systems. *IEEE Trans. Power Electron.* **2009**, *24*, 654–664. [CrossRef]

30. Wang, Y.F.; Li, Y.W. Grid synchronization PLL based on cascaded delayed signal cancellation. *IEEE Trans. Power Electron.* **2001**, *26*, 1987–1997. [CrossRef]

31. Yi, H.; Wang, X.; Blaabjerg, F.; Zhou, F. Impedance Analysis of SOGI-FLL-Based Grid Synchronization. *IEEE Trans. Power Electron.* **2017**, *32*, 7409–7413. [CrossRef]

energies

MDPI

Article

A Novel Two-Stage Photovoltaic Grid-Connected Inverter Voltage-Type Control Method with Failure Zone Characteristics

Xiangwu Yan [1,*], Xueyuan Zhang [1], Bo Zhang [1], Zhonghao Jia [1], Tie Li [2], Ming Wu [3] and Jun Jiang [4]

[1] Key Laboratory of Distributed Energy Storage and Micro-grid of Hebei Province, North China Electric Power University, Baoding 071003, China; 2162213021@ncepu.edu.cn (X.Z.); zhangbo@ncepu.edu.cn (B.Z.); 15032233997@163.com (Z.J.)
[2] State Grid Liaoning Electric Power Research Institute, Shenyang 110006, China; tony_fe@126.com
[3] China Electric Power Research Institute (CEPRI), Beijing 100192, China; wuming@epri.sgcc.com.cn
[4] State Grid Beijing Electric Power Company, Beijing 100031, China; prince_983@163.com
* Correspondence: xiangwuy@ncepu.edu.cn; Tel.: +86-0312-752-2862

Received: 16 May 2018; Accepted: 13 July 2018; Published: 17 July 2018

Abstract: This paper investigates how to develop a two-stage voltage-type grid-connected control method for renewable energy inverters that can make them simulate the characteristics of a synchronous generator governor. Firstly, the causes and necessities of the failure zone are analyzed, and thus the traditional static frequency characteristics are corrected. Then, a novel inverter control scheme with the governor's failure zone characteristics is proposed. An enabling link and a power loop are designed for the inverter to compensate fluctuations and regulate frequency automatically. Outside the failure zone, the inverter participates in the primary frequency regulation by disabling the power loop. In the failure zone, the droop curve is dynamically moved to track the corrected static frequency characteristic by enabling the power loop, resisting the fluctuation of grid frequency. The direct current (DC) bus voltage loop is introduced into the droop control to stabilize the DC bus voltage. Moreover, the designed dispatch instruction interface ensures the schedulability of the renewable energy inverter. Finally, the feasibility and effectiveness of the proposed control method are verified by simulation results from MATLAB (R2016a).

Keywords: failure zone; governor; frequency regulation; inverter; voltage-type control; static frequency characteristics

---

1. Introduction

Energy consumption rises as the development of global industrialization, resulting in more usage of fossil fuels [1]. Carbon dioxide emissions caused by fossil fuels accelerate global warming, leading to serious environmental problems [2]. In order to replace the energy generated by fossil fuels, photovoltaic (PV) power generation systems can be used [3]. For such PV power generation systems, it is important to design a grid-connected inverter that provides reliable AC (alternating current) power to the grid from the PV source's DC (direct current) power [4]. Grid integration of inverters has become increasingly important in distributed generation (DG) systems [5,6]. Many different types of PV inverters have been researched and proposed [7,8]. Generally, when it comes to the topology of photovoltaic grid-connected circuits, there are two types: single-stage inverters and two-stage inverters. The single-stage inverter is simple in structure, but it requires a high input voltage. Many PV modules are used to boost the required high voltage, which have several defects such as the imbalance of hot spots during partial shading, low safety features, and the poor maximum power point tracking

(MPPT) performance [9]. Thus, a DC-DC power-conversion structure which can increase the low PV-source voltage to a high DC-bus voltage was introduced to the single-stage inverter. Inverters with above configuration are named two-stage inverters, which can use the MPPT algorithm more efficiently [10–14]. Therefore, two-stage inverters have the advantage of fewer series-connected PV modules and better MPPT performances in comparison with single-stage inverters.

In the grid-connected mode, the inverters are controlled as current sources [15,16]. In the island mode, there is no grid connection to regulate voltage and frequency profiles, and the inverter is required to determine the voltage and frequency of system, so the inverter generally adopts voltage-type control [17]. Transitions between operation modes can cause deviations in voltage and current, because of the mismatch in frequency, phase of the inverter output voltage and those of the grid voltage [18]. If a voltage-type control structure can still be used in the grid-connected mode, the mode switching process can be avoided and the above problem will be effectively mitigated and eliminated.

Droop control is a voltage control method, which is usually applied to the parallel connection of inverters in distributed uninterruptible power supply systems to provide voltage support for the micro-grid on island mode. Considerable research efforts have been devoted to inverter's voltage-type grid-connected control. De Paiva, et al. [19] added an extra phase loop to traditional droop control, which improved the system's dynamic response and maintained suitable damping performance. Avelar, et al. [20] proposed an improved design of the polynomial model mentioned in [19] and presented a state equation model of an inverter connected to the grid with droop control. However, these methods are suitable for simple single-phase inverters instead of the widely used three-phase inverters. Verma, et al. [21] presented a model of a grid connected inverter operating in grid supporting mode incorporating dynamics of droop control. However, this method does not consider the two-stage application topology combined with PV energy sources. Additionally, none of those control strategies proposed consider fluctuations of the power grid such as fluctuations of grid voltage and frequency.

Recently, the penetration rate of renewable energy generation in the power system has gradually increased. Renewable energy generation is replacing the traditional generation, and it should gradually be equipped with the regulation ability of traditional generators. Therefore, it is a trend for renewable energy sources to share frequency regulation duties on the grid. Zhong [22] proposed the concept of a power electronics-enabled autonomous power system design. This scheme provides a uniform interface mechanism so that renewable energy sources can participate in grid frequency regulation like conventional power supplies. To this end, Yan, et al. [23] combines the droop characteristics and the Virtual Synchronous Generator (VSG) to make the PV storage two-stage inverter have primary frequency regulation characteristics. However, they do not take into account the governor's failure zone. Moreover, their study is based on energy storage, which is uneconomical. In summary, the establishment of a voltage-type two-stage grid-connected photovoltaic system that simulates the characteristics of the failure zone of the synchronous generator governor is of great significance for increasing penetration rate of photovoltaic power generation.

In this paper, the power control scheme of the inverter under the low-voltage line parameters is firstly obtained. Then, starting from the physical structure of the governor, the causes and the necessities of the failure zone and its effect on the frequency regulation are analyzed, and the static frequency characteristics are corrected. Finally, a novel two-stage photovoltaic grid-connected inverter voltage-type control method with the failure zone characteristics is proposed. By enabling the power loop inside the failure zone and disabling the power loop outside the failure zone, the inverter dynamically compensates grid fluctuations and participates in grid frequency regulation. The design of the dispatch interface ensures the schedulability of the inverter. DC voltage loop stabilizes the DC bus voltage, allowing the system to operate without energy storage.

2. Power Transmission Characteristics of Grid-Connected Inverter

The grid-connected equivalent circuit of the voltage-type inverter is shown in Figure 1. As shown in the figure, both the inverter and the grid are simplified to a voltage source. $U\angle\delta$ is the output voltage of the inverter after the filter. $E\angle 0$ is the grid voltage (since the grid capacity is much larger than the inverter capacity, E can be considered as a constant). $Z\angle\theta$ is the impedance between the inverter and power grid.

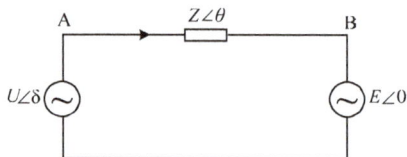

Figure 1. The grid-connected equivalent circuit of voltage-type inverter.

According to the theory of power system analysis, the current flowing through the terminal is:

$$
\begin{aligned}
I &= \frac{U\angle\delta - E\angle 0}{Z\angle\theta} \\
&= \frac{U\cos\delta - E + jU\sin\delta}{Z\angle\theta}
\end{aligned}
\tag{1}
$$

The output active power and reactive power of the inverter is expressed as:

$$
\begin{aligned}
P &= \left(\frac{EU}{Z}\cos\delta - \frac{E^2}{Z}\right)\cos\theta + \frac{EU}{Z}\sin\delta\sin\theta \\
Q &= \left(\frac{EU}{Z}\cos\delta - \frac{E^2}{Z}\right)\sin\theta - \frac{EU}{Z}\sin\delta\cos\theta
\end{aligned}
\tag{2}
$$

where, δ is the power angle.

The studied subject in this paper is a small and medium-sized photovoltaic power generation system, which is usually connected to a low-voltage distribution network. The low-voltage power transmission line is mostly resistive [24]. Therefore, in this scenario, $\theta \approx 0$, so $\sin\theta \approx 0$ and $\cos\theta \approx 1$. For a resistive impedance, $\theta = 90°$. Bring it into Equation (2), Equation (3) can be obtained as:

$$
\begin{aligned}
P &= \frac{EU}{Z}\cos\delta - \frac{E^2}{Z} \\
Q &= -\frac{EU}{Z}\sin\delta
\end{aligned}
\tag{3}
$$

In addition, δ is generally very small, so we have $\sin\delta \approx 0$ and $\cos\delta \approx 1$. Bring the above approximate values into Equation (3), it can be obtained that:

$$
\begin{aligned}
P &\approx \frac{U}{Z}E - \frac{E^2}{Z} \\
Q &\approx -\frac{EU}{Z}\delta
\end{aligned}
\tag{4}
$$

Roughly, Equation (4) can be written as:

$$
\begin{aligned}
P &\sim U \\
Q &\sim -\delta
\end{aligned}
\tag{5}
$$

where ~means in proportion to. Therefore, the conventional droop control strategy can be obtained as:

$$
\begin{aligned}
U &= U_r - k_p(P - P_r) \\
\omega &= \omega_r + k_q(Q - Q_r)
\end{aligned}
\tag{6}
$$

Due to $f = \frac{\omega}{2\pi}$, Equation (6) can be rewritten as:

$$
\begin{aligned}
U &= U_r - k_p(P - P_r) \\
f &= f_r + k_q(Q - Q_r)
\end{aligned}
\tag{7}
$$

where U is the reference amplitude of the inverter's output voltage, f is the frequency reference value of the inverter, U_r is the amplitude of the inverter's rated output voltage, f_r is the rated frequency of the inverter, k_p is the droop coefficient of the inverter's active power, k_q is the droop coefficient of the inverter's reactive power, P is the active power output by the inverter, Q is the reactive power output by the inverter, P_r is the rated active power of the inverter, Q_r is the rated reactive power of the inverter.

Equation (7) indicates that P is approximately linear with U and Q is approximately linear with f. Therefore, by controlling the amplitude and phase of the output voltage of the inverter, decoupled control of the active and reactive power of the inverter can be realized.

3. Failure Zone of Synchronous Generator Governor

3.1. The Cause of Failure Zone

Most steam turbogenerators and hydroturbines in the power system now are equipped with speed governors. The governor's function is to monitor the generator speed and to control the throttle valves that adjust steam flow into the turbine in response to changes in "system speed" or frequency [25]. During the period of actual operation, due to mechanical friction and overlap, the static characteristics of the generator unit differ from the theoretical static characteristics (see Figure 2). In a frequency range around the rated frequency, the governor does not respond to changes in frequency. This frequency range is defined as the failure zone of the governor.

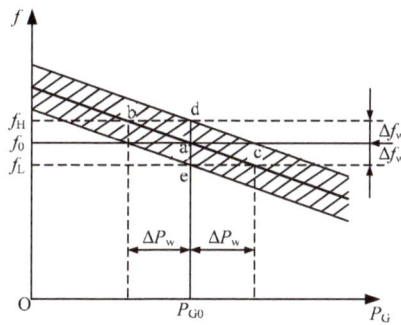

Figure 2. Failure zone of the generator governor.

In Figure 2, the thick solid line is the theoretical static frequency characteristic of the unit, The shaded area is the failure zone, f_0 is the initial frequency of the system, f_H is the upper frequency limit of the failure zone, f_L is the lower frequency limit of the failure zone, Δf_w is the maximum frequency hysteresis of the governor, ΔP_w is the maximum power error of the governor. It should be noted that the failure zone does not exist for a specific frequency or power. On the contrary, the failure zone is ubiquitous, which is determined by the physical characteristics of the synchronous generator governor. From Figure 2, we can see that the failure zone satisfies Equations (8) and (9).

$$\Delta f_w = (f_H - f_L)/2, \tag{8}$$

$$\frac{\Delta f_w}{\Delta P_w} = k_p, \tag{9}$$

where k_p is the droop coefficient of the inverter's active power.

3.2. Static Frequency Characteristics Considering the Failure Zone

This section is a case study of Figure 2, which analyzes the similarities and differences between theoretical static frequency characteristics and actual static frequency characteristics. At the initial state, the system operates at point a, the system frequency is f_0, and the output active power is P_{G0}. For the theoretical static frequency characteristics, the active power generated by the generator will gradually decrease as the grid frequency gradually increases. When the frequency rises to f_H, the active power decreases to $P_{G0} - \Delta P_w$ and the operating point moves to point b. However, there is a failure zone in the actual static frequency characteristics. When the frequency is slightly increased, the power generated by the generator does not change immediately, but remains at P_{G0}. When the frequency rises to f_H, the operating point comes to point d, where the governor completely overcomes friction and passes through overlaps. When the frequency continues to increase, the power output from the inverter gradually decreases from P_{G0} according to the droop coefficient. The reduced process of the grid frequency is similar to the above process, which is not repeated here.

From Equation (7), it can be seen that under low-voltage line conditions, the frequency is no longer approximately linearly related to the active power, but is approximately linearly related to the reactive power. It should be noted that this type of frequency regulation feature does not exist in conventional generators, but it is widely present in renewable energy inverters connected to low-voltage lines. In general, the system operates stably at rated conditions with $f_0 = f_N$. With reference to the failure zone characteristics of the conventional synchronous generator, the static frequency characteristics of the inverter in the low-voltage lines considering the failure zone are obtained (see Figure 3).

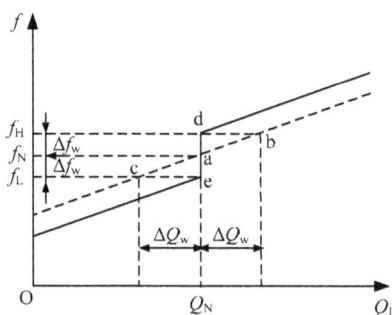

Figure 3. Static frequency characteristics considering failure zone under low-voltage lines.

where Q_I is the reactive power of the inverter and ΔQ_w is the maximum error reactive power of the inverter. The meanings of other variables are consistent with those of Figure 2.

3.3. The Effect of Failure Zone on Frequency Regulation

From the above analysis, it can be seen that the failure zone will shift the theoretical static frequency characteristics, resulting in a "dull" frequency regulation process, which is caused by the physical characteristics of the mechanical hydraulic governor. However, with the advancement of technology, in sensitive governors (such as electro-hydraulic governors), it is necessary to set the failure zone manually. If there is no failure zone or the failure zone is too small, when the frequency of the power system fluctuates, the governor will act unnecessarily, which is not conducive to the healthy operation of synchronous generators/inverters and the frequency stability of the power system. In addition, the failure zone should not be too large. Otherwise, synchronous generators/inverters will lose the capability of frequency regulation, and thus they will fail to provide frequency support for the system actively.

4. Novel Two-Stage Voltage-Type Grid-Connected Photovoltaic Inverter Control Method with Failure Zone Characteristics

4.1. Design Logic of Failure Zone

Synchronous generators in the system must set the failure zone according to the guidelines of the grid company, whose prescribed failure zone range is $|\pm0.033|$ Hz. That is to say, $\Delta f_w = 0.033$ Hz. According to the relevant rules and droop relationship, the inverter control logic with the failure zone characteristics is designed as shown in Equations (10) and (11):

$$\begin{cases} Q = Q_N & |f_g - f_N| < \Delta f_w + \xi \\ Q = Q_N + (f_g - f_H)/k_q & |f_g - f_N| > \Delta f_w + \xi, \ f_g > f_H, \\ Q = Q_N + (f_g - f_L)/k_q & |f_g - f_N| > \Delta f_w + \xi, \ f_g < f_L \end{cases} \tag{10}$$

$$\xi = f(\lambda) \lambda \in [0, 100\%], \ \xi \geq 0, \tag{11}$$

where Δf_w is the maximum frequency hysteresis of the governor, λ is the penetration rate of renewable energy in power system, ξ is the frequency regulation delay of the inverter, indicating the degree to which the renewable energy based inverter lags behind the conventional synchronous generators when participating in the frequency regulation. ξ is a function of λ and there is a negative correlation between them qualitatively. In the initial stage of renewable energy development, λ is very small, and there is no need to consider this issue due to the small capacity of renewable energy. As renewable energy sources gradually increase, the power capacity of the power system connected to the inverter increases. Thus, the inverter should have frequency regulation capability. Otherwise, the frequency regulation capability of the entire power grid will gradually decline. When λ is low, the capacity of the inverter power supply and the frequency regulation capability are small, and the frequency regulation technology is undeveloped. Considering the safety and stability of the power grid, inverters should be less involved in frequency regulation than synchronous generators with large capacity and developed technology. Therefore, the concept of frequency regulation delay of the inverter ξ was introduced so that the inverter involved in frequency regulation behind the synchronous generator. With the increase of λ, the control technology will be more advanced, renewable energy sources will be able to gradually share the frequency regulation tasks of the synchronous generators, so ξ will gradually decrease. It should be noted that the λ and ξ that we introduced in the article are both macroscopic and long-time, because the change of λ is very slow (especially for a huge power grid). For theoretical limit case, when λ reaches 100%, $\xi = 0$, which means that renewable energy inverters have completely replaced the conventional synchronous generators and are qualified for the frequency regulation. Based on the above analysis, the failure zone threshold of the renewable energy inverter $\Delta f_w'$ is set to $\Delta f_w + \xi$. When the frequency deviation is less than $\Delta f_w'$, the inverter outputs constant power. When the frequency deviation increases beyond $\Delta f_w'$, the inverter performs frequency regulation through droop control to limit the fluctuation of the frequency in a wider range. Considering that the inverter should still be able to respond to the dispatch instruction, the operation flowchart shown in Figure 4 is designed.

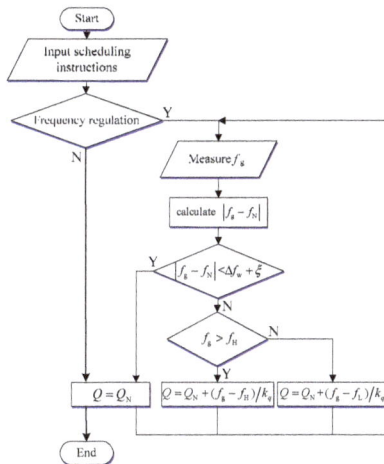

Figure 4. Logic flow chart of the system.

4.2. Novel Two-Stage Photovoltaic Grid-Connected Inverter Voltage-Type Control Method

4.2.1. Active Power-Voltage Control

Considering the economy of operation, the regulation of voltage is generally through balancing locally and near. Therefore, the inverter does not need to adjust the power according to the change of the system voltage, but outputs constant power. An active power loop is formed by adding integral term to the droop control, see Equation (12):

$$U = U_r - \left(k_{pp} + \frac{k_{pi}}{s}\right)(P - P_r),\tag{12}$$

where k_{pp} and k_{pi} are the proportional gain and the integral gain of the active power loop, respectively. For ease of analysis, it is assumed that the irradiance does not change during the analysis of the inverter's active power in this section. For a grid-connected two-stage photovoltaic power generation system, the active power output by the inverter can be reflected by the voltage change on the DC side. In detail, the active power and DC bus voltage satisfy Equation (13):

$$-\Delta P = \frac{1}{2}C_{dc}\Delta U_{dc}{}^2,\tag{13}$$

where ΔP is the amount of changes of the inverter's output active power, C_{dc} is the DC bus capacitance, and ΔU_{dc} is the amount of changes of the DC bus voltage. The rated active power corresponds to the rated DC voltage. If the output active power of inverter increases, the DC bus voltage will drop and vice versa. Therefore, we replace P with $-U_{dc}$, replace P_r with $-U_{dcr}$, and convert the active power loop in Equation (12) into a DC bus voltage loop, as shown in Equation (14):

$$U = U_r - \left(k_{up} + \frac{k_{ui}}{s}\right)(U_{dcr} - U_{dc})\tag{14}$$

The control design of the DC voltage loop allows the system to operate without energy storage and maintain a stable active power output when the grid voltage fluctuates.

4.2.2. Reactive Power-Frequency Control

In this section, the control method outside the failure zone was introduced first. Then, the control method in the failure zone was introduced. In this section, reactive power-frequency control is introduced. The inverter has different characteristics depending on whether the frequency is in the failure zone. In this section, features outside the failure zone are first introduced. Then, the characteristic in failure zone was introduced.

As mentioned earlier, the inverter designed in this paper is a voltage-type inverter. For this kind of inverter, it is common to regulate frequency through droop control. The inverter in this article is in a mainly resistive scenario, so it follows the droop relationship shown in Equation (7). Therefore, outside the governor's failure zone, the inverter can regulate frequency through the relationship of Equation (7).

Conversely, in the failure zone, the output power of the inverter should be kept constant to simulate the characteristics of the synchronous generator governor. The regulation principle in the failure zone is shown in Figure 5

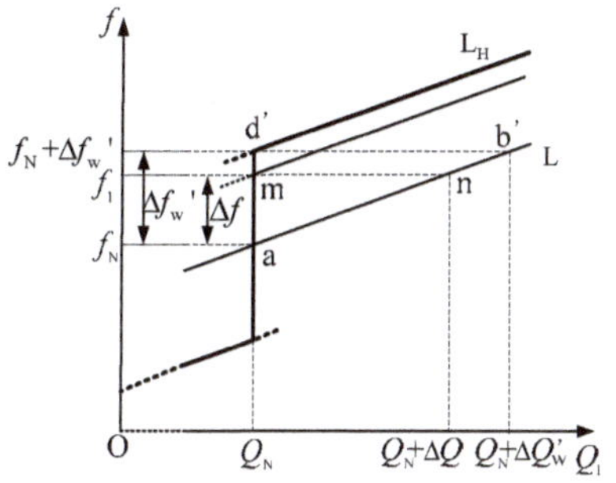

Figure 5. Regulation principle diagram in failure zone.

In Figure 5, the rated frequency of the system is f_N, the rated reactive power of the inverter is Q_N, and the initial operating point is point a. When the grid fluctuates slightly and the frequency of the grid rises by $\Delta f(\Delta f < \Delta f_w')$ to f_1, the frequency of the inverter is also clamped by the grid at f_1. For common inverters, according to the theoretical droop characteristic, the reactive power output from the inverter will increase, and the operation point is moved to point n along the line L. In order to control the inverter's output reactive power as fixed Q_N, the straight line L should be shifted upwards by Δf. During this process, the operating point will be shifted leftwards from point n to point m. Therefore, for any Δf less than $\Delta f_w'$, the operating point can be returned to the straight line $Q = Q_N$ by shifting the droop characteristic line upwards. In this way, the frequent change of inverter output is avoided, and the effect of the conventional synchronous generator governor failure zone is simulated. Specially, when the frequency change Δf is equal to $\Delta f_w'$, the adjusted operating point is point m, and the droop characteristic line is shifted to straight line L_H. If the frequency continues to deviate from $\Delta f_w'$, which indicates a large frequency disturbance, the inverter will participate in the primary frequency regulation according to the droop characteristics shown in L_H, and share the frequency regulation duties with the conventional synchronous generators.

In order to make the renewable energy-based inverter simulate the failure zone characteristics of the synchronous generator governor, the reactive power loop and the enabling link are designed so that the inverter will output constant reactive power within the failure zone threshold to achieve the static frequency characteristics shown in Figure 3. Power loop consists of common droop control and an integral term. If the enabling link is enabled, it guarantees a constant power output. If the enabling link is disabled, the inverter still outputs power according to droop relationship. The principle of reactive power-frequency control is shown in Figure 6.

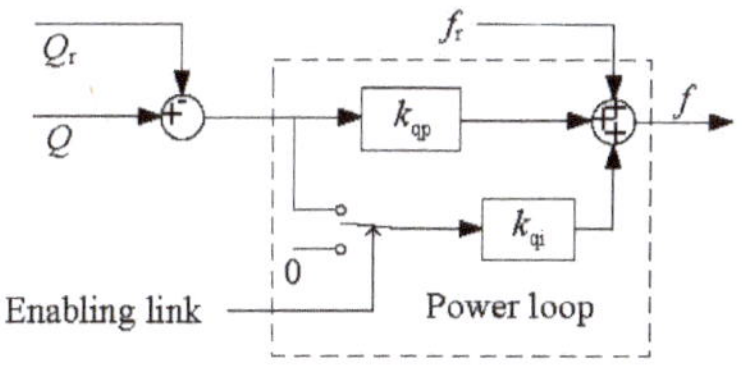

Figure 6. The principle of reactive power-frequency control.

4.2.3. Overall Control Scheme of the Inverter

The above ideas and methods are integrated to obtain a voltage-based control scheme for a two-stage photovoltaic grid-connected inverter with the characteristics of the governor's failure zone, as shown in Figure 7.

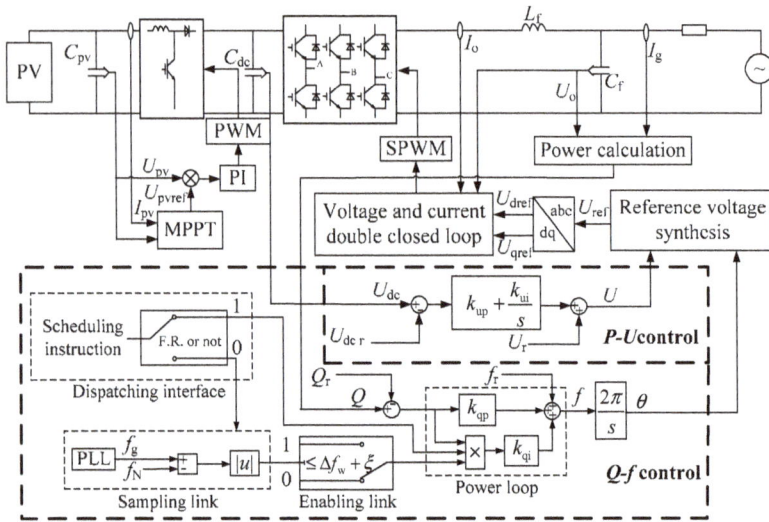

Figure 7. Overall control scheme of inverter. PWM: pulse width modulation; PI: proportional–integral controller; MPPT: maximum power point tracking; SPWM: sinusoidal pulse width modulation; PV: photovoltaic; PLL: phase locked loop.

According to the function and control scheme of the inverter, renewable energy based inverter can be divided in grid-feeding inverter, grid-forming inverter and grid-supporting inverter [26]. The grid-forming inverters can set the voltage amplitude and frequency of the local grid. The grid-feeding power inverters are mainly designed to deliver power to an energized grid. A grid-supporting power converter is in between a grid-feeding and a grid-forming power converter [5]. The inverter in this paper is voltage source based grid-supporting inverter, belonging to the grid-supporting inverter. It should be noted that some of control details are omitted or simplified in consideration of the aesthetic appearance of figure, such as coordinate transformation parts and the voltage-current cascaded control part, etc. The complete control block diagram is shown in Figure A1.

As shown in Figure 7, MPPT is implemented through the control of the boost circuit. The MPPT algorithm generates a photovoltaic voltage reference value U_{pvref}, and the PI regulator implements tracking of U_{pvref}. When the solar irradiance changes, the inverter always outputs the maximum photovoltaic energy to ensure the maximum utilization of renewable energy. The inverter output voltage is controlled by the DC bus voltage loop shown in Section 4.2.1. For renewable energy sources connected to the power grid through inverters, their primary frequency regulation function is a new feature and may be undeveloped. Therefore, this function should be controlled by the dispatch of the power system for the stability and controllability of the power system. Therefore, the inverter should work as follows: when the power dispatch requires the inverter to participate in frequency regulation, the inverter decides whether or not to participate in it through the failed zone; when the power dispatch does not require the inverter to participate in a frequency regulation, the inverter always outputs a constant power even if the frequency is outside the failure zone. Therefore, the frequency control structure includes sampling link, dispatching interface, enabling link and power loop. Dispatch instructions can inputted through the dispatching interface.

If the constant power output of the inverter is required according to the dispatching signal, the dispatching interface outputs 1 to enable the power loop and the power output by the inverter tracks reference value.

If the inverter is required to have frequency regulation capability, the dispatching interface outputs 0 and the signal enters the sampling link. When the upper dispatching provides the inverter with a frequency regulation command, whether or not the failure zone characteristics can be realized depends on the power loop: When the frequency fluctuation is less than $\Delta f_w'$, most of the disturbances are self-recovering. Therefore, the enabling link outputs 1 to simulate the failure zone characteristics of the governor, and the output power remains unchanged. When the frequency fluctuates much larger and it exceeds $\Delta f_w'$, the enabling link outputs 0 to disable the power loop, and the inverter participates in primary frequency regulation.

As can be seen in Figure 7, the inverter reference voltage amplitude can be obtained through *P-U* control. Through the *Q-f* control, the frequency of the inverter reference voltage can be obtained, then the phase of the voltage can be obtained. Using the inverter voltage amplitude and phase, it is possible to synthesize the reference voltage vector of the inverter, which is regarded as a reference value for the voltage and current double closed loop.

It needs to be pointed out that the setting of $\Delta f_w'$ contains the frequency regulation delay of the renewable energy based ξ, and thus $\Delta f_w'$ contains the information on the penetration rate of renewable energy λ. The higher λ is in the system, the more renewable energy inverters are required to undertake frequency regulation tasks, the lower ξ is, the smaller the failure zone threshold $\Delta f_w'$ is, and vice versa. Therefore, this control scheme can achieve the best operating state in an environment with any renewable energy penetration rate λ through flexible parameter settings.

5. Verification

To verify the feasibility of the proposed method, the two-stage grid-connected in. If the enabling link is enabled, generation system was modelled in MATLAB (MathWorks, Inc., Natick, MA, USA) (see Figure 7). The perturbation observation method was used for MPPT [27]. The Boost circuit was used to boost photovoltaic output voltage. The main parameters of the system are shown in Table A1.

5.1. The Dynamic Characteristics of the Source

In order to study the output characteristics of the inverter when the PV output is affected by the environment, the solar irradiance is reduced from 1000 W/m^2 to 800 W/m^2 at the first second and restored to 1000 W/m^2 at the second second. The output of the inverter is shown in Figure 8.

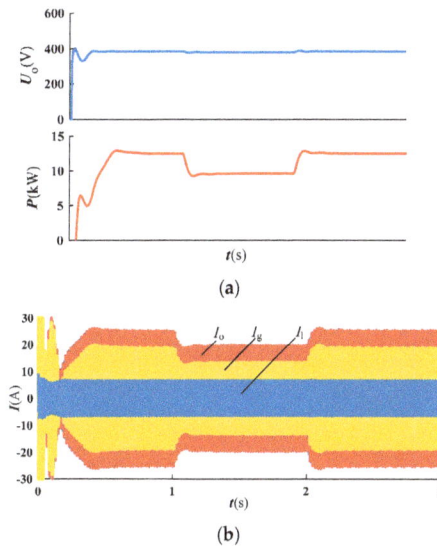

Figure 8. (a) Inverter output phase voltage amplitude and active power; (b) Source-Grid-Load current.

After recovery of solar irradiance, P, I_o, and I_g recover quickly. It can be seen that the proposed control method can maintain the stable operation of the system and ensure the maximum utilization of renewable energy even if the source fluctuates. This means that all the active power generated by the PV can be delivered to the grid to supply the load regardless of the maximum power value.

5.2. Verification of Direct Current (DC) Voltage Loop

In order to verify the effect of the DC voltage loop, the grid voltage amplitude was changed, then the inverter output power and DC voltage are observed. Considering that the laboratory environment is better than the actual operation environment, step tests in the laboratory environment should adopt stricter conditions (larger voltage changes). So we set the voltage step change value to 15%. The grid voltage amplitude is suddenly increased by 57 V ($380 \times 15\%$) at the first second, and the rated value is restored at the second second. The inverter's operating results are shown in Figure 9.

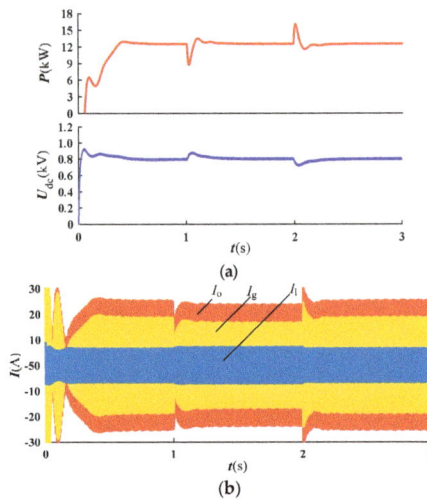

Figure 9. (**a**) DC bus voltage and inverter output active power; (**b**) Source-Grid-Load current.

In Figure 9a, the DC bus voltage and inverter output active power quickly return to their initial values (800 V and 12.5 kW) after the grid fluctuates. During the transient process, the trend of U_{dc} and P is negatively correlated, which is consistent with Equation (14). In Figure 9b, when the grid voltage fluctuates, I_o and I_g change rapidly with P and then quickly restore their rated value. It can be seen that the DC voltage loop not only stabilizes the DC bus voltage but also eliminates the influence of grid voltage disturbance on the output of the inverter, which enhances the stable operation of the system.

5.3. Verification of Failure Zone Characteristics

In this section, the failure zone feature of the system is verified. In order to fully present the role of the failure zone, the verification of this part is divided into two parts. Firstly, the characteristics in the failure zone are observed. Then, the characteristics in the failure zone and outside the failure zone are compared with each other.

In order to observe the characteristics in the failure zone, the frequency variation is set at 0.07 Hz which is lower than $\Delta f_w'$ (0.1 Hz). If the reactive power output from the inverter is still stable at the rated value (0 Var), it means that the inverter has a failure zone characteristic. The grid is set to operate at an initial value of 50 Hz, which is increased to 50.07 Hz at the first second. The inverter operation results are shown in Figure 10, where the reactive power can be reflected by the phase relationship between voltage and current.

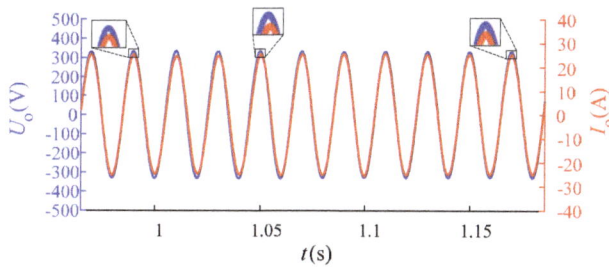

Figure 10. Output voltage and current of inverter inside failure zone.

In initial state, the inverter maintains a unity power factor output according to the rated reactive power Q_r. Therefore, the phase of output voltage and current are the same before 1 the first second. The change in the grid frequency at the first second causes a slight change in the reactive power at the inverter output, which can be indicated by the phase difference between U_o and I_o. The system is in the failure zone for the reason of $0.07 Hz < \Delta f_w'$, thus the reactive power output by the inverter is adjusted to 0 within 0.2 s. The above results show that the failure zone characteristics of the synchronous generator governor can be simulated.

Then, the characteristics inside the failure zone and outside the failure zone were observed together. At the same time, in order to obviously show the differences between the proposed control method, PQ (constant active power and reactive power) control method and droop control method. The results of these three control schemes are presented so that they can be compared with each other. In the following experiment, the grid frequency is increased by Δf (0.05 Hz) from 50 Hz in the first second, the 1.5th second, the 2nd second and the 2.5th second respectively. By comparing with droop control and PQ control, the failure zone characteristics of the proposed control scheme can be observed clearly (see Figure 11).

From the Figure 11, it can be seen that the inverter operates in rated condition in all cases in the initial state. For the droop control in Figure 11a, the grid frequency rose to 50.05 Hz at the first second. The slight frequency fluctuation is lower than $\Delta f_w'$ (0.1 Hz), but the droop controlled inverter still sensitively changes the output power from 0 Var to 500 Var. Note that 500 Var is equal to $\Delta f \cdot k_{qp}$. Then, at the 1.5th second, the 2nd second and the 2.5th second, the inverter outputs 500 Var reactive power for every 0.05 Hz increase in frequency. It is clear that droop control makes the inverter participate in frequency regulation, but the output power also changes when the frequency changes slightly.

For the PQ control in Figure 11b, the grid frequency rose to 50.05 Hz at the first second. The frequency fluctuation is lower than $\Delta f_w'$ (0.1 Hz), so the inverter output reactive power recovers the rated value 0 Var after a brief transient process, which is consistent with expectation. The same situation occurs after the second frequency change. Then, at the 2nd second, the grid frequency increased to 50.15 Hz. At this time, the frequency changes so much that it exceeds the failure zone. However, reactive power output from the inverter still remains unchanged even if the grid frequency continues to rise to 50.2 Hz. As shown in the Figure 11b, PQ control achieves the constant power output of the inverter, but the inverter loses the frequency regulation capability when the frequency deviation is large. In the scenario of low renewable energy penetration, it is acceptable to operate the inverter with a PQ source because there are sufficient synchronous generators that can perform the task of frequency regulation. However, with the popularization of renewable energy, inverters are expected to actively participate in the frequency regulation of the power system in order to share the burden of synchronous generators.

For the proposed novel control in Figure 11c, when the frequency does not change beyond the failure zone, the inverter operates as a PQ source. Once the change of frequency exceeds the failure zone (after 2nd second), the inverter will participate in frequency regulation by adjusting the output power. The proposed novel control scheme enables the system to simulate the failure zone characteristics of a conventional synchronous generator governor. The inverter has a failure zone $|\pm\Delta f_w'|$ Hz, which allows the inverter to make intelligent choices between resisting grid fluctuations and participating in grid frequency regulation based on the actual situations. Note that the failure zone is symmetrical about the rated frequency. The simulation in this paper takes the upper threshold of the failure zone as an example. With the same effect, the lower threshold of the failure zone will not be described in detail herein.

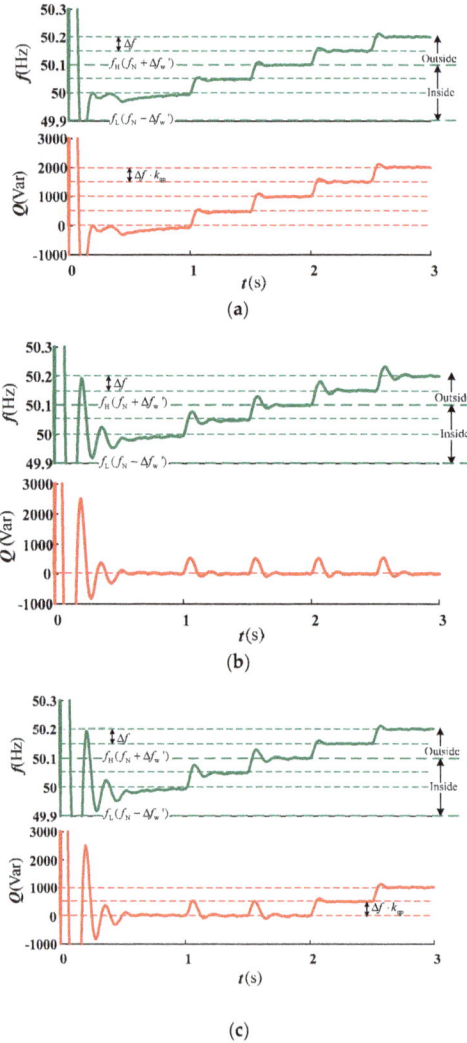

Figure 11. Comparison of droop control, PQ (constant active power and reactive power) control and proposed control (**a**) Droop control; (**b**) PQ control (**c**) Proposed novel control.

5.4. Verification of the Dispatching Interface

Dispatching interfaces should have a higher priority than failure zone, which means that when the dispatch does not require the inverter to participate in a frequency regulation, the inverter always outputs a constant power even if the frequency is outside the failure zone. In order to verify the effectiveness of the dispatching interface, the inverter is initially operated in the rated state. Then a sudden big frequency increase of 0.2 Hz is set at the first second. The frequency regulation instruction dispatched by the dispatching interface is received at the 2nd second. The response of the inverter is shown in Figure 12.

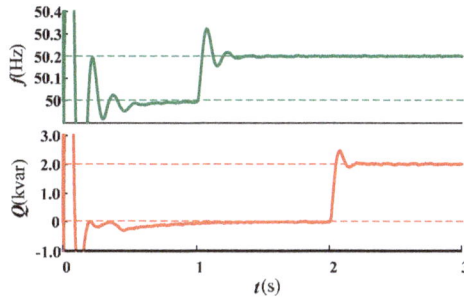

Figure 12. Dispatching response results of inverter.

It can be seen from Figure 12 that the renewable energy based inverter still operates at constant power even though the frequency increment exceeds the failure zone threshold before the 2nd second. At exactly the time when the frequency regulation instruction is sent to the inverter, the inverter immediately participates in primary frequency regulation. The above results verify the effectiveness of the dispatching interface and the schedulability of the renewable energy inverter.

6. Conclusions

With the increase of renewable energy penetration rate, how to make renewable energy based inverters simulate and replace conventional synchronous generator to undertake frequency regulation tasks is a problem that needs to be solved urgently. The conclusions of this article are as follows:

(1) In this paper, based on the characteristics of speed governor system of the conventional generator, the traditional static frequency characteristics are corrected. Then a novel two-stage grid-connected photovoltaic inverter voltage-type control method with the characteristics of the governor's failure zone is proposed. The dynamic balance between resisting fluctuations, participation in frequency regulation and dispatching response is achieved.

(2) Through the improvement of the droop control and the design of the power enabling link, the inverter possesses the failure zone characteristics of the synchronous generator. For small frequency fluctuations inside the failure zone, the inverter maintains a constant output. If the frequency fluctuation exceeds the failure zone, the inverter participates in grid frequency regulation according to the droop relationship.

(3) Whether or not the inverter participates in frequency regulation should be controllable rather than completely autonomous, especially when there are many renewable energy sources. The design of the dispatch interface ensures the schedulability of the inverter.

(4) The frequency regulation delay of renewable energy ζ was introduced to improve inverters' adaptability to renewable energy penetration rate. Therefore, the proposed control scheme can achieve the best operating state in an environment with any renewable energy penetration rate through flexible parameter settings.

(5) A DC voltage loop was designed, which has two roles. On the one hand, it stabilizes the DC bus voltage to achieve operations without energy storage. On the other hand, it ensures that the system is not affected by the grid and delivers the maximum power to the grid stably.

(6) The selection of failure zone thresholds for renewable energy based inverters and coordinated control of multi-inverters can be researched in the future.

Author Contributions: X.Y. and B.Z. conceived the method; X.Z. and B.Z. designed the method; X.Z. and M.W. achieved the method; X.Z., Z.J. and J.J. analyzed the results; X.Z. and T.L. wrote the paper.

Funding: This paper was supported by National Key R&D Program of China (2016YFB0900400), National Science Fundings of Hebei (E2018502134), Scientific Research Program of Hebei University (Z2017132). In addition, this article was funded by 2018 science and technology project of Liaoning Power Grid Corporation "Research on power grid reactive voltage optimization strategy and evaluation index considering the characteristics of source and load fluctuations".

Conflicts of Interest: The authors declare no conflict of interest.

Appendix A

Table A1. Main parameters of the system.

Meaning and Symbols	Value
Maximum power of photovoltaic array P_{pvmax}	12.5 kW
The capacitor on the output side of the photovoltaic array C_{pv}	1000 μF
Capacitor at DC (direct current) bus C_{dc}	2000 μF
Filter inductor L_f	2.2 mH
Filter capacitor C_f	800 μF
The amplitude of the inverter's rated output voltage U_r	$220\sqrt{2}$ V
The reference value of inverter's DC voltage U_{dcr}	800 V
Inverter rated frequency f_r	50 Hz
Inverter rated output reactive power Q_r	0 var
The proportional gain of the reactive power loop k_{qp}	0.0001
The integral gain of the reactive power loop k_{qi}	0.0015
The proportional gain of the DC voltage loop k_{up}	0.5
The integral gain of the DC voltage loop k_{ui}	3
The proportional gain of the voltage loop k_{oup}	1.4
The integral gain of the voltage loop k_{oui}	3.2
The proportional gain of the current loop k_{oip}	1
The integral gain of the current loop k_{oii}	0
The maximum frequency hysteresis of the governor Δf_w	0.033 Hz
The upper limit frequency of failure zone f_H	50.033 Hz
The lower limit frequency of failure zone f_L	49.967 Hz
The frequency regulation delay of the inverter ξ	0.067 Hz
The failure zone threshold of inverter $\Delta f_w'$	0.1 Hz

Appendix B

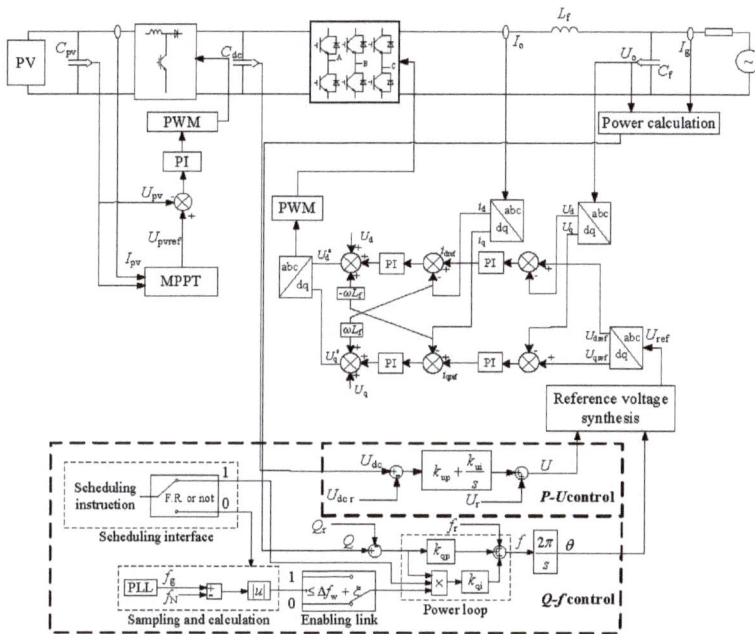

Figure A1. Overall control scheme of inverter (detailed).

References

1. Bose, B.K. Global energy scenario and impact of power electronics in 21st century. *IEEE Trans. Ind. Electron.* **2013**, *60*, 2638–2651. [CrossRef]
2. Zhou, Y.; Gong, D.C.; Huang, B.; Peters, B.A. The impacts of carbon tariff on green supply chain design. *IEEE Trans. Autom. Sci. Eng.* **2017**, *14*, 1542–1555. [CrossRef]
3. Wang, Y.; Lin, X.; Pedram, M. A near-optimal model-based control algorithm for households equipped with residential photovoltaic power generation and energy storage systems. *IEEE Trans. Sustain. Energy* **2016**, *7*, 77–86. [CrossRef]
4. Cho, Y.W.; Cha, W.J.; Kwon, J.M.; Kwon, B.H. Improved single phase transformerless inverter with high power density and high efficiency for grid-connected photovoltaic systems. *IET Renew. Power Gener.* **2016**, *10*, 166–174. [CrossRef]
5. Rocabert, J.; Luna, A.; Blaabjerg, F.; Rodríguez, P. Control of power converters in ac microgrids. *IEEE Trans. Power Electron.* **2012**, *27*, 4734–4749. [CrossRef]
6. Strasser, T.; Andrén, F.; Kathan, J.; Kathan, J.; Kathan, J.; Siano, P.; Siano, P.; Zhabelova, G.; Zhabelova, G.; Vrba, P.; Mařík, V. A review of architectures and concepts for intelligence in future electric energy systems. *IEEE Trans. Ind. Electron.* **2015**, *62*, 2424–2438. [CrossRef]
7. Daher, S.; Schmid, J.; Antunes, F.L.M. Multilevel inverter topologies for stand-alone PV systems. *IEEE Trans. Ind. Electron.* **2008**, *55*, 2703–2712. [CrossRef]
8. Kjaer, S.B.; Pedersen, J.K.; Blaabjerg, F. A review of single phase grid connected inverters for photovoltaic modules. *IEEE Trans. Ind. Appl.* **2005**, *41*, 1292–1306. [CrossRef]
9. Kim, K.A.; Seo, G.S.; Cho, B.H.; Krein, P.T. Photovoltaic hot-spot detection for solar panel substrings using AC parameter characterization. *IEEE Trans. Power Electron.* **2016**, *31*, 1121–1130. [CrossRef]
10. Debnath, D.; Chatterjee, K. Two-stage solar photovoltaic-based standalone scheme having battery as energy storage element for rural deployment. *IEEE Trans. Ind. Electron.* **2015**, *62*, 4148–4157. [CrossRef]

11. Alajmi, B.N.; Ahmed, K.H.; Finney, S.J.; Williams, B.W. A maximum power point tracking technique for partially shaded photovoltaic systems in microgrids. *IEEE Trans. Ind. Electron.* **2013**, *62*, 1596–1606. [CrossRef]

12. Choi, W.Y. Three-level single-ended primary-inductor converter for photovoltaic power conditioning systems. *Sol. Energy* **2016**, *62*, 43–50. [CrossRef]

13. Kanaan, H.; Caron, M.; Al-Haddad, K. Design and implementation of a two-stage grid-connected high efficiency power load emulator. *IEEE Trans. Power Electron.* **2014**, *29*, 3997–4006. [CrossRef]

14. Blaabjerg, F.; Teodorescu, R.; Liserre, M.; Timbus, A.V. Overview of control and grid synchronization for distributed power generation. *IEEE Trans. Ind. Electron.* **2006**, *53*, 1398–1409. [CrossRef]

15. Liu, J.; Cheng, S.; Shen, A. Carrier-overlapping PWM-based hybrid current control strategy applied to two-stage grid-connected PV inverter. *IET Power Electron.* **2018**, *11*, 182–191. [CrossRef]

16. Boukezata, B.; Gaubert, J.P.; Chaoui, A.; Hachemi, M. Predictive current control in multifunctional grid connected inverter interfaced by PV system. *Sol. Energy* **2016**, *391*, 130–141. [CrossRef]

17. Vasquez, V.; Ortega, L.M.; Romero, D.; Ortega, R.; Carranza, O.; Rodriguez, J.J. Comparison of methods for controllers design of single phase inverter operating in island mode in a microgrid: Review. *Renew. Sustain. Energy Rev.* **2017**, *76*, 256–267. [CrossRef]

18. Singh, M.; Khadkikar, V.; Chandra, A.; Varma, R.K. Grid interconnection of renewable energy sources at the distribution level with power quality improvement features. *IEEE Trans. Power Deliv.* **2011**, *26*, 307–315. [CrossRef]

19. De Paiva, E.P.; Vieira, J.B.; De Freitas, L.C.; Farias, V.J.; Coelho, E.A.A. Small signal analysis applied to a single phase inverter connected to stiff AC system using a novel improved power controller. In Proceedings of the Twentieth Annual IEEE Applied Power Electronics Conference and Exposition, Austin, TX, USA, 6–10 March 2005.

20. Avelar, H.J.; Parreira, W.A.; Vieira, J.B.; De Freitas, L.C.G.; Coelho, E.A.A. A State Equation Model of a Single-Phase Grid-Connected Inverter Using a Droop Control Scheme with Extra Phase Shift Control Action. *IEEE Trans. Power Deliv.* **2012**, *59*, 1527–1537. [CrossRef]

21. Verma, V.; Khushalani-Solanki, S.; Solanki, J. Modeling and Criterion for Voltage Stability of Grid Connected Droop Controlled Inverter. In Proceedings of the 2017 North American Power Symposium (NAPS), Morgantown, WV, USA, 17–19 September 2017.

22. Zhong, Q.C. Power electronics-enabled autonomous power systems: Architecture and technical routes. *IEEE Trans. Ind. Electron.* **2017**, *64*, 5907–5918. [CrossRef]

23. Yan, X.W.; Zhang, X.Y.; Zhang, B.; Ma, Y.J.; Wu, M. Research on Distributed PV Storage Virtual Synchronous Generator System and Its Static Frequency Characteristic Analysis. *Appl. Sci.* **2018**, *8*, 532. [CrossRef]

24. Vinayagam, A.; Alqumsan, A.A.; Swarna, K.S.V.; Khoo, S.Y.; Stojcevski, A. Intelligent control strategy in the islanded network of a solar PV microgrid. *Electr. Power Syst. Res.* **2018**, *155*, 93–103. [CrossRef]

25. Grainger, J.J.; Stevenson, W.D.J. *Power System Analysis*; Stephen, E.H., Ed.; McGraw-Hill: Hightstown, NJ, USA, 1994; pp. 562–572.

26. Viinamäki, J.; Kuperman, A.; Suntio, T. Grid-Forming-Mode Operation of Boost-Power-Stage Converter in PV-Generator-Interfacing Applications. *Energies* **2017**, *10*, 1033. [CrossRef]

27. Moradi, M.H.; Reisi, A.R. A hybrid maximum power point tracking method for photovoltaic systems. *Sol. Energy* **2011**, *85*, 2965–2976. [CrossRef]

energies

MDPI

Article

Adaptive-MPPT-Based Control of Improved Photovoltaic Virtual Synchronous Generators

Xiangwu Yan [1,*], Jiajia Li [1], Ling Wang [1], Shuaishuai Zhao [1], Tie Li [2], Zhipeng Lv [3] and Ming Wu [3]

[1]   Key Laboratory of Distributed Energy Storage and Micro-Grid of Hebei Province,
     North China Electric Power University, Baoding 071003, China;
     lijiajia1018@163.com (J.L.); 13697662469@163.com (L.W.); zhao2017shuaishuai@163.com (S.Z.)
[2]   State Grid Liaoning Electric Power Research Institute, Shenyang 110000, China; tony_fe@126.com
[3]   China Electric Power Research Institute (CEPRI), Beijing 100192, China;
     lvzhipeng@epri.sgcc.com.cn (Z.L.); wuming@epri.sgcc.com.cn (M.W.)
*   Correspondence: xiangwuy@ncepu.edu.cn; Tel.: +86-139-0336-5326

Received: 15 June 2018; Accepted: 9 July 2018; Published: 12 July 2018

Abstract: The lack of inertia and damping mechanism of photovoltaic (PV) grid-connected systems controlled by maximum power point tracking (MPPT) poses a challenge for the safety and stability of the grid. Virtual synchronous generator (VSG) technology has attracted wide attention, since it can make PV grid-connected inverter present the external characteristics of a synchronous generator (SG). Nevertheless, traditional PV-VSG is generally equipped with an energy storage device, which leads to many problems, such as increased costs, space occupation, and post-maintenance. Thus, this paper proposes a two-stage improved PV-VSG control method based on an adaptive-MPPT algorithm. When PV power is adequate, the adaptive-MPPT allows the PV to change the operating point within a stable operation area to actualize system supply-demand, matching in accordance to the load or dispatching power demand; when PV power is insufficient, PV achieves traditional MPPT control to reduce power shortage; simultaneously, improved VSG control prevents the DC bus voltage from falling continuously to ensure its stability. The proposed control approach enables the two-stage PV-VSG to supply power to loads or connect to the grid without adding additional energy storage devices, the effectiveness of which in off-grid and grid-connected modes is demonstrated by typical simulation conditions.

Keywords: two-stage photovoltaic power; virtual synchronous generator; adaptive-MPPT (maximum power point tracking); improved-VSG (virtual synchronous generator); power matching

1. Introduction

In recent years, the continued consumption of fossil fuels has caused problems in terms of energy crises and environmental pollution [1]. One effective solution is exploiting and utilizing renewable energy sources, which play a crucial role in the transformation of the energy structure and are guided by the principle of sustainability [2]. Solar energy, due to the characteristics of green, clean, environmental friendliness and sustainability, enables photovoltaic (PV) power generation to be highly valued by countries all over the world [3,4]. Under the policy support of the '12th Five-Year Plan' and '13th Five-Year Plan' [4,5], the PV power market in China has enjoyed rapid development. In 2016, China became the country with the biggest PV power generation capacity in the world, with an accumulated installed capacity of 77.4 GW [3]. Furthermore, the 'Notice on Matters Related to Photovoltaic Power Generation in 2018' [6], issued by the National Development and Reform Commission, Ministry of Finance and National Energy Administration, further promotes the healthy and sustainable development of the PV industry, in which orderly development of distributed PV is sustained, and PV poverty alleviation projects such as roof PVs [2] are supported. The above series of policies indicates that China's PV industry presents great potential for progress.

In the field of PV control technology, because early designers mainly consider economic benefits, PV systems usually adopt a maximum power point tracking (MPPT) control strategy for integration into the grid through a power electronic inverter [7,8]. However, the aforementioned control scheme belongs to passive control, which cannot actively respond to changes of grid frequency. In addition, PV systems do not possess dynamic characteristics consistent with conventional synchronous generators (SG), thanks to the deficiency of inverter inertia and damping [9–11], which is detrimental to improving the stability of frequency and voltage in the case of the grid interference, causing a potential threat to the security and stability of the power grid.

In July 2015, the 'Guidance on Promoting Smart Grid Development' [12] formulated by the National Energy Administration pointed out that "the grid-connected devices with plug-and-play and friendly grid-connected features will be promoted to meet the extensive access requirements of new energy and distributed generation". As a result, with the increasing of PV penetration, its grid-connected equipment needs to gradually alter the idea of "only power generation, regardless of power grid". Under the above background, ensuring high-proportional, large-scale PV power-friendly access from the "source" has become a major issue that needs to be resolved [13]. In view of the above-mentioned factors, some scholars have introduced virtual synchronous generator (VSG) technology [9,14] into the PV system due to its ability to make PV grid-connected inverters behave like SG. Indeed, VSG can offer inertia and damping to suppress rapid variations in voltage and frequency during the transient process; VSG-controlled PV systems autonomously participate in primary regulation to implement responsive interaction with the grid in the steady case.

In numerous studies [15–18], the DC side of the PV inverter is normally assumed to be a constant DC source, and the further investigation of VSG control is based on this foundation. This type of PV-VSG research ignores the potential impact of PV dynamic characteristics on VSG control, hence limiting the development of PV-VSG. For this reason, the cooperative operation mode of PV and energy storage is discussed in [19–22], which is referred to as a PV-storage system for short, in which the PV system adds a bidirectional energy storage device at the DC side to stabilize the DC bus voltage and realizes VSG function through the inverter. In the above PV-storage system, the output power of energy storage compensates for the deficiency of the PV dynamic characteristics and the variations of the load or dispatching requirements. However, in terms of VSG, the power involved in primary regulation derives from the energy storage instead of the PV power source. Moreover, the additional configuration of energy storage increases system cost and maintenance workload, as well as the additional installation space, making it difficult for practical application. Thus, with regard to this issue, a novel control strategy for a stand-alone PV system based on VSG which implements the combination of PV and VSG is proposed in [23], and this PV-VSG control scheme is extended to a grid-connected mode in [24]. The above-mentioned control mechanism eliminates the energy storage configuration of PV-VSG, and takes the PV power source as a virtual prime mover of the virtual synchronous generator, which balances system power according to the actual power demands. Nevertheless, the above PV-VSG control structure is only applied to the centralized PV inverter with single-stage DC/AC topology, which is inappropriate for a string PV inverter with two-stage DC/DC and DC/AC circuits where the networking is flexible and the MPPT voltage adjustment range is wider.

Drawing on the control concept of the single-stage PV-VSG, it is hoped that the two-stage PV-VSG could minimize the energy storage allocation, the realization of which faces two difficulties: First, it is difficult to actualize the combination of PV with VSG, since it has complex dynamic features, operational stability requirements and limited capacity. Next, compared with a single-stage system, the DC bus voltage is no longer the PV voltage, the regulation of which would be more complicated. To deal with these two problems, we can draw lessons from how the wind farm modifies the power tracking curve to achieve the supply-demand matching [25–27]. Under the premise of not adding energy storage, taking the power-voltage output characteristic of PV as a research point, the output power of PV arrays is regulated by a DC/DC converter, which can be directly interfaced with local loads or power grids through post-stage VSG.

On the basis of the analysis above, considering the volatility and finiteness of PV, in this paper, a two-stage improved PV-VSG control approach founded on an adaptive-MPPT algorithm is proposed. When PV power is sufficient, adaptive-MPPT enables the PV system to send power identical to the load or dispatching power requirements, which does not export maximum power all the time; in the case of PV power deficiency, the adaptive-MPPT algorithm turns into traditional MPPT control; meanwhile, improved-VSG heightens the DC bus voltage stability. This control approach obviates the energy storage configuration of the traditional two-stage PV-VSG, thereby reducing investment and maintenance costs. An adaptively regulated two-stage PV-VSG system is established in MATLAB, and the effective combination of the two-stage PV and VSG is verified through simulation in off-grid and grid-connected modes, respectively. In off-grid mode, the novel two-stage PV-VSG sends power in accordance with the load demand and provides voltage and frequency support for load; in grid-connected mode, AC voltage and frequency are sustained by the power grid, and PV-VSG is responsible for delivering power following the dispatching instruction.

2. Overview of Fundamental Problems

The typical PV micro-grid system, which is composed of photovoltaic generation, energy storage device and diesel generator, supplies the local loads or accesses the grid, as illustrated in Figure 1. For ease of reading, the PV system is marked with a red frame in Figure 1.

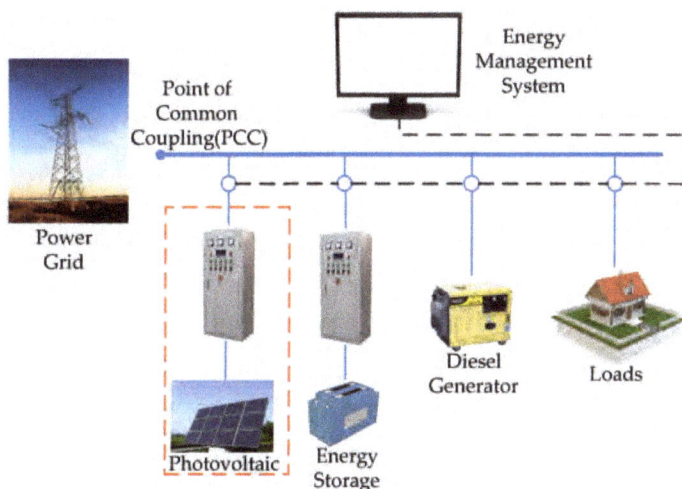

Figure 1. The photovoltaic (PV) micro-grid system.

Picking up the PV system in the above typical PV micro-grid as an important research target, it is desirable for the PV system to achieve friendly access through the application of VSG technology. In a general way, in related works [19,20], a PV-VSG control strategy installed with an energy storage battery on the DC side was adopted, as shown in Figure 2. For traditional PV-VSG control, since PV output power is particularly influenced by environmental factors such as light intensity and temperature [13], the energy storage battery is primarily responsible for balancing power through a DC/DC converter so as to match the required power. However, the above control exhibits two defects: the implementation of the VSG function is overly dependent on the energy storage battery, so the inverter may not continue to work once the energy storage battery failure occurs. Secondly, the energy storage configuration of PV-VSG will greatly increase the expenses of investment, operation and maintenance [23,28].

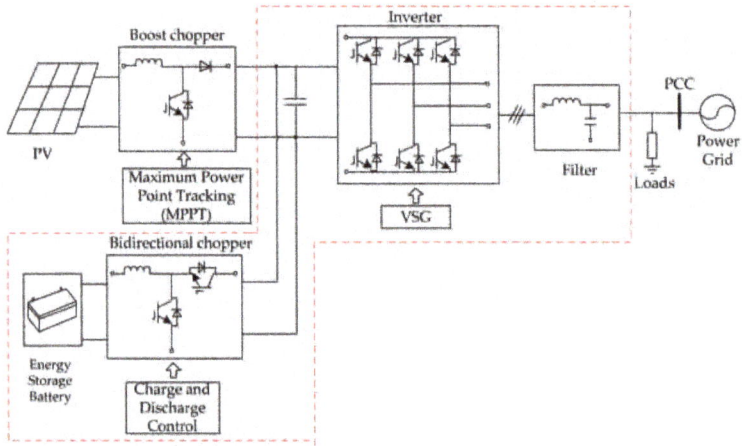

Figure 2. The traditional PV-VSG (photovoltaic-virtual synchronous generator) control oriented for the energy storage battery [19,20].

In order to resolve the inherent problems in Figure 2, a novel two-stage PV-VSG control approach is advanced in Figure 3, which consists of an adaptive-MPPT control and an improved-VSG control. Although the energy storage battery of PV-VSG is removed, the joint control of the adaptive-MPPT and improved-VSG still stabilizes the DC bus voltage and achieves power matching.

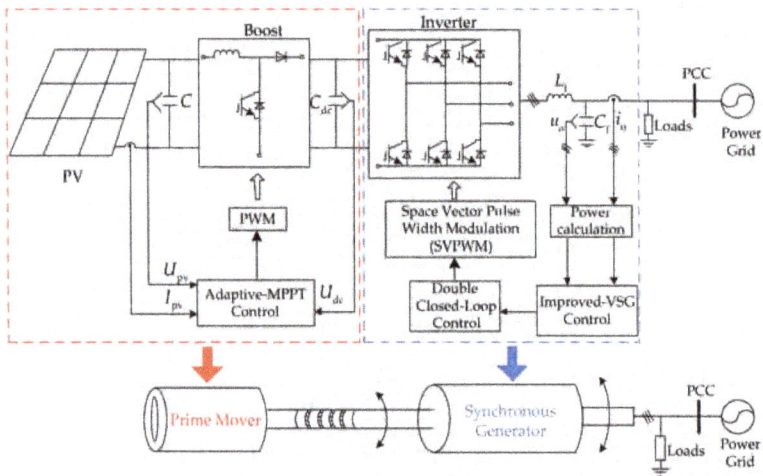

Figure 3. The main circuit topology and control method of the presented two-stage PV-VSG.

As is evident from Figure 3, the pre-stage PV-Boost circuit comprises a prime mover using adaptive-MPPT control, and the post-stage inverter controlled by improved-VSG forms a synchronous generator. In this way, employing the prime mover as a power source drives the synchronous generator to send power to the loads or the power grid. Evidently, unlike the traditional PV-VSG with an energy storage battery as the prime mover in Figure 2, this novel control approach can solve the power balance problem of two-stage PV-VSG from the source and economizes various costs associated with configuring energy storage.

In general, the proposed two-stage PV-VSG possesses two operational scenarios:

(1) PV maximum available power is adequate.

When PV maximum output power is greater than the power required by the load or dispatch instruction, operational power depends on load or dispatching requirements, and the major challenge is guaranteeing system power balance. At this moment, PV controlled by adaptive-MPPT changes the operating point to decrease power output according to load or dispatching power demand, which ensures supply-demand matching.

(2) PV maximum available power is inadequate.

The insufficiency of the PV maximum available power signifies that the maximum output of PV is less than the load or dispatching demand power. Under this scenario, the major challenge is to control PV to provide as much power as possible and to ensure the overall stability of the PV-VSG system. Thus, adaptive-MPPT-controlled PV should operate at the maximum power point (MPP), which exports maximal power to minimize power shortage. Meanwhile, improved-VSG control prevents the DC bus voltage from dropping down continuously, so as to warrant the stability of the DC bus voltage.

Most notably, for the PV system, when the PV maximum available power is inadequate (the operation scenario (2)), if the load or dispatching power demands are still necessary to satisfy, other power supplies, such as energy storage and diesel generators in typical PV micro-grids (as shown in Figure 1), ought to offer power to compensate for the power shortage. This paper mainly aims at the operation scenarios of a single two-stage PV-VSG system and addresses problems of immediately accessing two-stage PV power by way of VSG without supernumerary energy storage devices; consequently, coordination control of the other power supplies and PV systems is no longer necessary to describe.

3. Methods

With regard to the two-stage PV system made up of PV-Boost circuit and inverter, control strategies include pre-stage adaptive-MPPT and post-stage improved-VSG, which are analyzed in detail in the subsequent subsections.

3.1. PV-Boost Control

3.1.1. Overall Control Scheme of Pre-Stage PV-Boost

Figure 4 displays the complete control scheme of the pre-stage PV-Boost. Since the research emphasis of this section is the pre-stage control, the post-stage inverter circuit is omitted, and is discussed in Section 3.2. The PV voltage reference $U_{pv\text{-}ref}$, which is obtained through the adaptive-MPPT control, and the actual value U_{pv} generate the PWM modulated signals D through PI control. The theoretical analysis of adaptive-MPPT algorithm will be elaborated in Section 3.1.2.

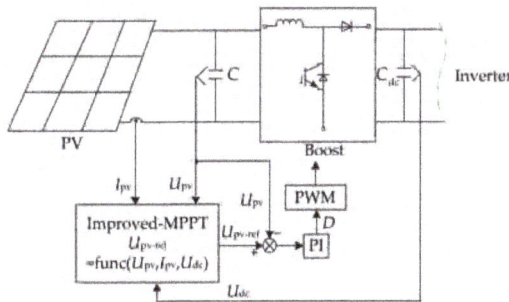

Figure 4. Overview of control scheme for pre-stage PV-Boost.

3.1.2. Adaptive-MPPT Algorithm

Regarding the PV power-voltage (*P-U*) characteristic curve [28–30] (as shown in Figure 5) as a research core point, and taking advantage of regulating function of DC/DC converter on output power of PV cells, the adaptive-MPPT adjusts the working point in the stable operation area to fulfill supply-demand matching on the basis of actual power demands. From Figure 5, abscissa U_{pv} and ordinate P_{pv} represent the PV output voltage and output power, respectively. P_{max} is the PV maximum output power, which corresponds to the voltage U_{mpp}. Additionally, *A*, *B* and *M* are possible PV operating points, where *M* is the maximum power point.

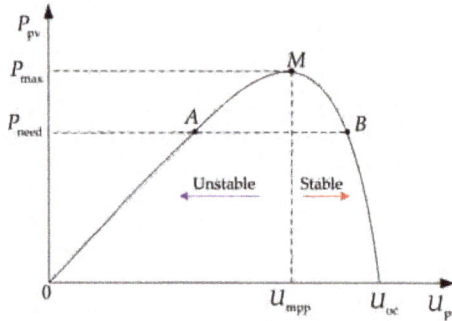

Figure 5. PV Power-voltage (P-U) characteristic curve.

Figure 5 shows that the PV operation area is divided into two regions based on the *M* point. When external power demand P_{need} is less than the PV maximum available power P_{max}, PV exists at two operating points: *A* and *B*. Through quantitative and qualitative verification, Refs. [23,24] indicate that the PV stable operation area is [U_{mpp}, U_{oc}]. Thus, combined with the actual operating conditions, the adaptive-MPPT should be equipped with the dynamic regulation features as follows:

- When $P_{need} < P_{max}$, PV works at *B* point within [U_{mpp}, U_{oc}] to output power equal to P_{need}.
- When $P_{need} \geq P_{max}$, PV works at *M* point to output maximum power P_{max}.

Accordingly, taking into account the DC bus voltage U_{dc}, this paper designs the adaptive-MPPT algorithm based on an improved incremental conductance method in light of the following four control goals, as described in Figure 6.

(1) Ensure that PV operates within the stable operating area [U_{mpp}, U_{oc}].
(2) Whether the DC bus voltage U_{dc} is stable at the set reference value U_{dc-ref} is used as a criterion for judging whether supply and demand match.
(3) When the PV output power is in surplus, adaptive-MPPT causes the DC bus voltage to remain at the set reference value U_{dc-ref} constantly.
(4) When the PV maximum output is insufficient at a given time, $U_{dc} < U_{dc-ref}$, adaptive-MPPT runs MPP to determine maintain maximum output.

From Figure 6, λ is a fixed step, and this adaptive-MPPT algorithm runs as follows:

(1) When the PV system starts for the first time, the slope is $dI_{pv}/dU_{pv} + I_{pv}/U_{pv} > 0$. To prevent PV from operating in unstable areas, the algorithm enables PV run to [U_{mpp}, U_{oc}] with $y = 1$, which ensures accurate tracking in the stable region all the time.
(2) In the stable area, according to difference-value ΔU_{dc} sign of the actual DC bus voltage U_{dc} and set value U_{dc-ref}, PV regulates output power to meet supply-demand matching, i.e., $P_{pv} = P_{need}$. There are three main situations.

- When $\Delta U_{dc}(k) > 0$, in this case, $P_{pv}(k) > P_{need}(k)$, the PV output power should be reduced, so the voltage judgement sign is $x = -1$.
- When $\Delta U_{dc}(k) < 0$, in this case, $P_{pv}(k) < P_{need}(k)$, the PV output power ought to increase, so the voltage judgement sign is $x = 1$.
- When $\Delta U_{dc}(k) = 0$, in this case, $P_{pv}(k) = P_{need}(k)$, the voltage judgement sign is $x = 0$.

Nevertheless, since the actual adjustment direction is opposite to the voltage judgment sign in the stable area, the PV regulates with $y = -x$. Most notably, If the PV maximum output power is less than the load or the dispatch demand invariably, ΔU_{dc} is always less than zero, so $y = -x = -1$. Thus, this algorithm jumps out of the ΔU_{dc} judgment step and turns into the traditional MPPT control based on the conductance increment method.

(3) The actual step size $y \times \lambda$ is obtained, thereby refreshing the PV voltage value U_{pv}.

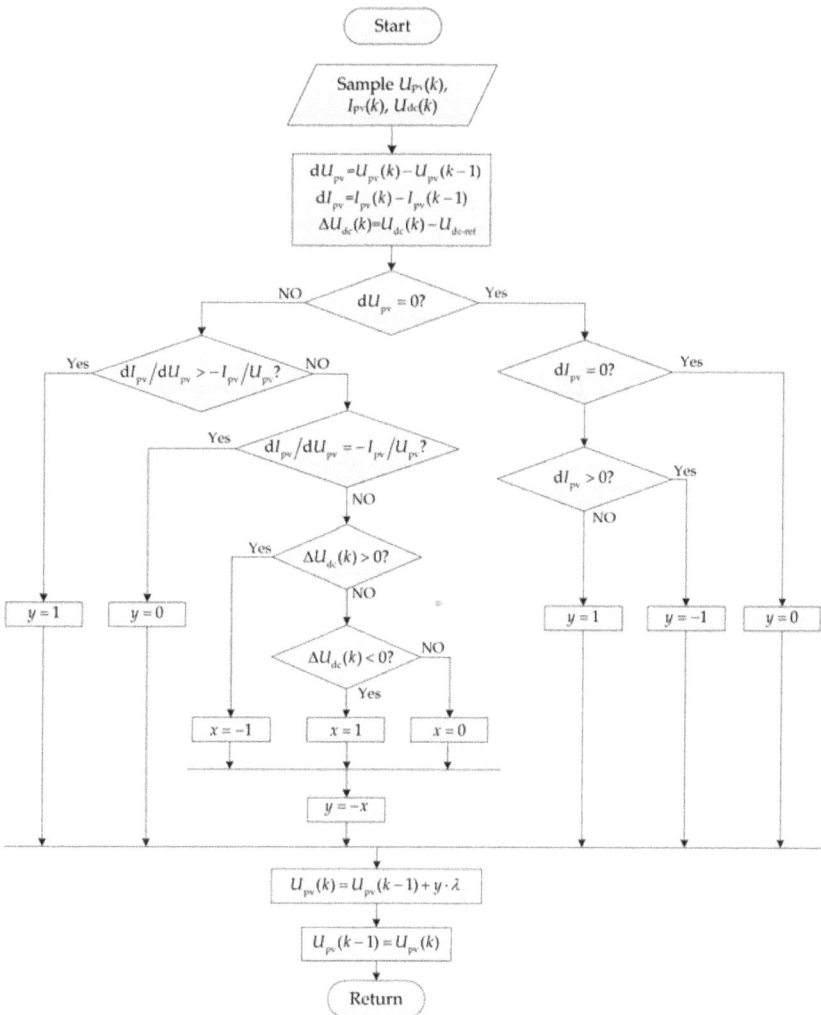

Figure 6. Adaptive-MPPT algorithm.

In conclusion, compared to traditional MPPT control, which always outputs maximum power [29,30], the proposed adaptive-MPPT algorithm, which is applied to the pre-stage PV-DC/DC circuit, possesses two contributions:

- In the case of sufficient PV power, adaptive-MPPT enables two-stage PV to transmit power in accordance with the load or dispatching requirements, while guaranteeing DC bus voltage $U_{dc} = U_{dc-ref}$.
- Under conditions where PV maximum output power is inadequate, adaptive-MPPT automatically switches to traditional MPPT control, always outputting maximum power to decrease the power shortage. At this moment, the DC bus voltage is no longer controlled.

3.2. Inverter Control

3.2.1. VSG Basic Modeling

The classical second-order model of synchronous generator (SG) is introduced into the VSG control to emulate the inertia and damping characteristics [31,32], including the rotor motion equation and the stator voltage equation, as expressed in Equations (1) and (2). In addition, droop control is used for mimicking the frequency regulator and excitation controller of SG, as shown in Equation (3).

$$J\frac{d\omega}{dt} = \frac{P_m - P_e}{\omega} - D(\omega - \omega_g) \tag{1}$$

$$e_{abc} = u_{abc} + Ri_{abc} + L\frac{di_{abc}}{dt} \tag{2}$$

where P_m is the mechanical power; P_e is the electromagnetic power; J and D represent the inertia and damping, respectively; ω is the rotor angular frequency; ω_g is the actual angular frequency of the grid; e_{abc}, u_{abc} and i_{abc} are the excitation voltage, terminal voltage and stator current of SG respectively, which correspond to input voltage, output voltage and output current in the inverter; R is the armature resistance and L is the synchronous reactance.

$$\begin{cases} U = U_N + D_p(P_N - P) \\ \omega_g = \omega_N + D_q(Q_N - Q) \end{cases} \tag{3}$$

where P_N and Q_N are the rated active power and reactive power, respectively; P and Q are the active power and reactive power of VSG; D_p is the P-ω droop coefficient and D_q is the Q-U droop coefficient; U_N is the rated voltage amplitude; ω_N is the rated angular frequency.

The combination of the classical second-order model and droop control constitutes the VSG basic model, whose control structure is depicted in Figure 7. VSG output power simulates the electromagnetic power, i.e., $P = P_e$.

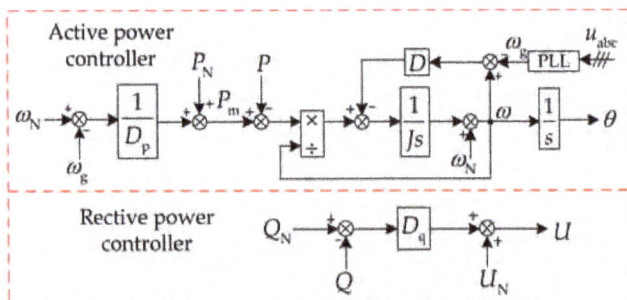

Figure 7. Basic control for VSG.

3.2.2. Improved-VSG Control

When the load or grid demands exceed the PV maximum available power, in order to diminish the power shortage, the DC bus capacitor C_{dc} will discharge. If this power shortage is larger, the DC bus voltage may constantly drop or even collapse, which endangers the stability of the system. To settle this problem, Figure 8 puts forward the improved-VSG control method which comprises of VSG basic control and additional control.

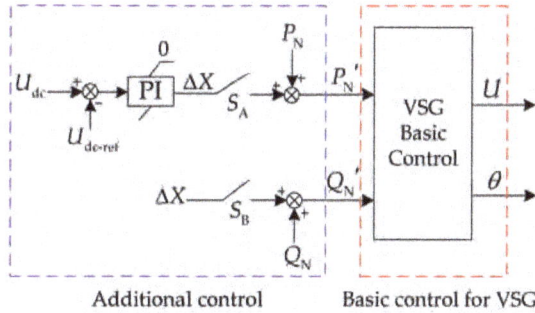

Figure 8. Improved-VSG control.

From Figure 8, the VSG basic control enables the PV inverter to take on the same inertia, damping and primary regulation characteristics as the SG, and the additional control is in charge of avoiding the DC bus voltage from falling continuously to ensure stability. For additional control, the difference between the measured DC bus voltage U_{dc} and the set-point $U_{dc\text{-ref}}$ is regulated by PI control to obtain an additional control variable ΔX; moreover, $\Delta X \leq 0$. The working mode of the additional control is as below:

- In off-grid mode, switch S_B is closed, while switch S_A is open, so ΔX is introduced into the reactive power loop to revise the reactive power reference Q_N', i.e., $Q_N' = Q_N + \Delta X \leq Q_N$.
- In grid-connected mode, switch S_A is closed. Since voltage is sustained by the bulk power system, switch S_B is open. Active power reference P_N' is modified by ΔX, i.e., $P_N' = P_N + \Delta X \leq P_N$.

The basic principles of additional control are given by the following two aspects:

(1) When the PV output power is surplus, pre-stage adaptive-MPPT changes the working point to achieve $P_{pv} = P_{need}$, so that the DC bus voltage can stabilize at the reference value $U_{dc\text{-ref}}$. Thus, $\Delta X = 0$, the additional control is inoperative, in this case, the adaptive-MPPT cooperates with VSG basic control.

(2) When the PV maximum output is insufficient, i.e., $P_{max} < P_{need}$, although MPPT keeps PV outputting maximum power at all times, it still cannot meet the load or dispatch power requirements. Consequently, the DC bus capacitance will discharge, with the result that U_{dc} failed to keep at $U_{dc\text{-ref}}$. If the power difference-value is higher, U_{dc} will continue to fall until it collapses. In such situations, additional control takes effect.

  - In off-grid mode, load power is related to voltage. ΔX acts on the reactive power loop to indirectly decrease the voltage amplitude, so as to reduce the inverter output power, which lowers the decline degree of U_{dc} to improve the PV steadiness.
  - In grid-connected mode, the dispatching power is greater than the PV maximum output, resulting in insufficient power. ΔX is led into the active power loop to lessen dispatching power reference value, which prevents U_{dc} from falling ceaselessly.

Even if the DC bus voltage U_{dc} is jointly controlled by the pre-stage adaptive-MPPT and the post-stage improved-VSG, the two stages of control do not affect each other. When $P_{pv} > P_{need}$, the pre-stage U_{dc} plays a role, but the post-stage U_{dc} does not perform due to $\Delta X = 0$; when $P_{pv} < P_{need}$, the adaptive-MPPT changes into the traditional MPPT, in which U_{dc} is not governed. At this point, the post-stage U_{dc} is under control on account of $\Delta X < 0$.

To prove the effectiveness, we compared improved-VSG control with VSG basic control through simulation in off-grid and grid-connected mode. Before 1 s, $P_{max} > P_{need}$, and after 1 s, $P_{max} < P_{need}$. The simulation results are shown in Figure 9, in which the blue line represents the improved-VSG control and the red line is the VSG basic control.

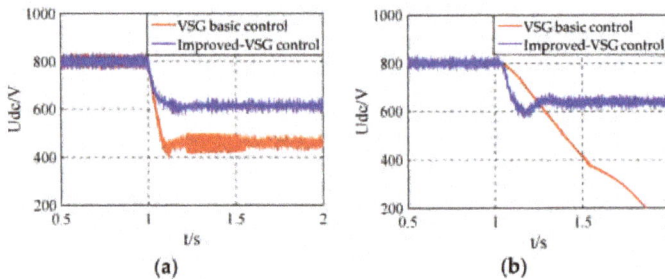

Figure 9. Comparison of DC bus voltage U_{dc} (**a**) Off-grid mode; (**b**) Grid-connected mode.

Apparently, as can be observed from the red line in Figure 9, when $P_{max} < P_{need}$, U_{dc} with VSG, basic control drops considerably whether in off-grid mode or grid-connected mode. Especially in the grid-connected mode, this case is more serious. However, when adopting the improved-VSG control, although PV maximum output cannot meet the power demand, it is manifestly known from the blue line in Figure 9 that the improved-VSG control can forestall incessant falling of U_{dc} and drastically reduce the U_{dc} drop degree to guarantee PV system stability. Therefore, later simulation validation will make use of the proposed improved-VSG control.

In summary, the improved-VSG that governs the post-stage inverter is provided with following properties:

- In situations of adequate PV power, the adaptive-MPPT-controlled pre-stage DC/DC converter accomplishes stability of the DC bus voltage, which can be considered a constant DC source. At this point, post-stage improved-VSG control mainly causes the inverter to present inertia, damping and primary regulation characteristics of the SG, that is, achieving VSG basic function.
- When the PV power is inadequate, the DC bus voltage is not managed by pre-stage DC/DC circuit anymore, since adaptive-MPPT changes into a traditional MPPT. Under this condition, additional control of the improved-VSG is effective to prevent the continuous drop of the DC bus voltage, thus enhancing the stability of the PV system.

3.2.3. Complete Control Scheme of Post-Stage Inverter

For the studied case, an overview of the applied control scheme for the post-stage inverter is displayed in Figure 10. The DC-side capacitor C_{dc} should be attached to the PV-Boost circuit, whose control method is discussed in Figure 4.

The control structure of the inverter has two layers of cascaded controllers, including the power controller, which is comprised of improved-VSG control, and a double closed-loop controller, which is formed of voltage outer-loop control and current inner-loop control. The outer-layer power controller basically emulates the main characteristic of the SG, and ensures the stability of the DC bus voltage when the PV power is inadequate. The inner-layer double closed-loop controller availably enhances the wave quality of the inverter output. Afterwards, the inner-loop current control provides the modulation index for SVPWM which generates the pulse signals of the inverter switching transistor.

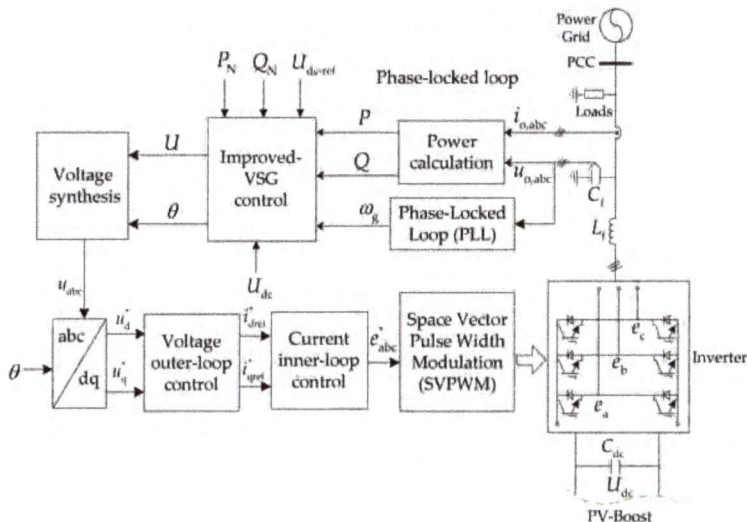

Figure 10. Complete control scheme of post-stage inverter.

4. Results

4.1. Simulation System and Simulation Parameters

The integrated system model of a single two-stage PV-VSG connected to the loads or power grid is built on the MATLAB/Simulink, as shown in Figure 11. Under the off-grid/grid-connected mode, the performance of the proposed control is validated.

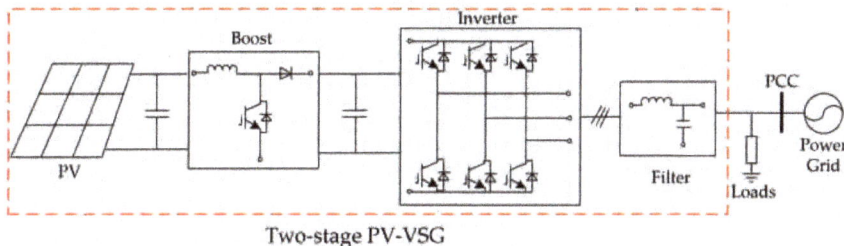

Two-stage PV-VSG

Figure 11. Schematic diagram of a single two-stage PV-VSG connects to the loads/power grid.

The photovoltaic cell type is First Solar FS-380. The DC/DC converter is a Boost circuit, which is controlled by the adaptive-MPPT in Section 3.1. The DC/AC inverter adapts the improved-VSG control presented in Section 3.2.

The main component parameters of the investigated system are reported in Table 1.

Table 1. Simulation parameters.

	Parameters	Values
Boost circuit parameters	PV-side capacitance, C	30 μF
	Inductance, L	1 mH
	DC side capacitance, C_{dc}	5000 μF
Filter parameters	The series inductance of the filter, L_f	10 mH
	The parallel capacitance of the filter, C_f	350 μF
System parameters	Reference value of DC voltage, $U_{dc\text{-}ref}$	800 V
	Rated frequency	50 Hz
	The rated phase voltage of power system	220 V
	Inverter switching frequency	5 kHz
Control parameters	The $P\text{-}\omega$ droop coefficient, D_p	0.0003
	The $Q\text{-}U$ droop coefficient, D_q	0.003
	The virtual inertia of VSG, J	0.1
	The virtual damping of VSG, D	20
	The proportionality factor of additional control in improved-VSG control, P_{Udc}	50
	The integration factor of additional control in improved-VSG control, I_{Udc}	0.01

4.2. Verification Process

For the typical PV microgrid system shown in Figure 1, there are two operating ways: off-grid mode and grid-connected mode. Focusing on the PV system in the above microgrid, we attest the effectiveness that pre-stage PV power implements direct access through post-stage VSG in the absence of allocating energy storage device. Thus, taking the single two-stage PV system in Figure 11 as an example, the variations of the load or scheduling demand power and PV maximum output are respectively set up to verify the proposed control method.

4.2.1. Off-Grid Mode

Under the off-grid mode, since this paper only simulates a single PV system, the inverter output voltage will decrease due to the power shortage when PV maximum power is not enough. In this case, the power shortage should be supplemented by other power sources of the PV microgrid in Figure 1, such as an energy storage battery or diesel generator, to ensure power quality, which is not the research emphasis in this paper, so the coordination control of the other power sources and PV system is omitted. This part is also outlined in Section 2 (2).

When the system works in off-grid mode, the two-stage PV-VSG which supplies to the loads provides the frequency and voltage support for the loads.

(1) Variation of Load Demand

The external environment remains constant. PV maximum available power is P_{max} = 15 kW and the corresponding output voltage is U_{mpp} = 370 V. Before 1 s, the power demand of the load is 12 kW, which is reduced to 10 kW at 1 s but increased to 18 kW at 1.5 s. Figure 12 displays the system dynamic response waveforms in the case of varying load requirements.

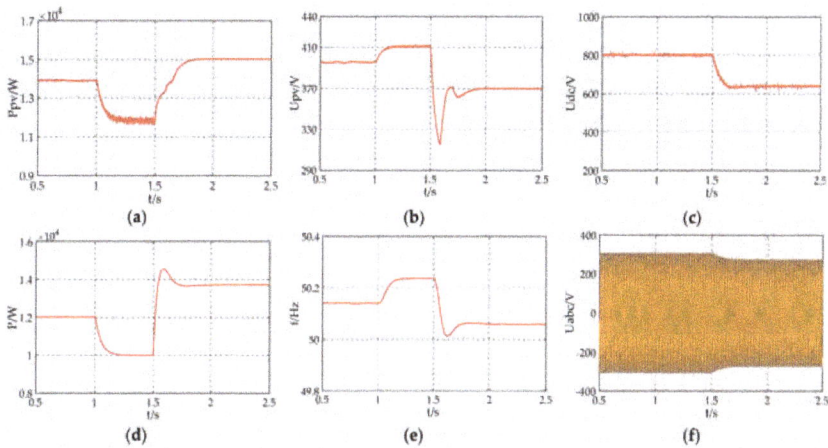

Figure 12. Waveforms in the off-grid mode when load demand changes (**a**) PV output power; (**b**) PV output voltage; (**c**) DC bus voltage; (**d**) Inverter output power; (**e**) VSG frequency; (**f**) Inverter output voltage.

Before 1 s and between 1~1.5 s, the load demand power is 12 kW and 10 kW, respectively, which is less than PV maximum output power 15 kW, i.e., $P_{max} > P_{need}$, so the PV power is surplus. Thus, it can be seen in Figure 12a,d that PV power controlled by adaptive-MPPT changes the operating point to make inverter output power match the load demand. Additionally, before 1.5 s, Figure 12b shows that PV output voltage is greater than $U_{mpp} = 370$ V, i.e., $U_{pv} > U_{mpp}$, which means that PV works in the stable area $[U_{mpp}, U_{oc}]$. Furthermore, DC bus voltage stabilizes at the set value 800 V owing to system power balance in Figure 12c, and inverter output voltage is identical to the AC rated voltage in Figure 12f. After 1.5 s, the load demand power increases to 18 kW, which is greater than the PV maximum output power of 15 kW, i.e., $P_{max} < P_{need}$; as a result, the PV power is inadequate. In this case, Figure 12a,b indicates that PV works at MPP point (370 V, 15 kW) to output maximum power. In addition, due to power shortage, the DC bus voltage reduces to 625 V in Figure 12c and the inverter output voltage falls below the rated AC voltage in Figure 12f. During the above power regulation process, Figure 12e shows that VSG frequency is involved in regulating according to the active droop coefficient.

(2)   Variation of PV Maximum Output Power

The load demand power remains at 10 kW continuously. The variation of the PV maximum output power is emulated by changing light intensity, whose parameter settings are shown in Table 2. Figure 13 shows the simulation waveforms when changing PV maximum output power.

Table 2. Light intensity parameters.

Time (s)	Light Intensity (W/m^2)	P_{max} (kW)	U_{mpp} (V)
Before 1 s	1000	15	370
1~1.5 s	1200	18.2	380
After 1.5 s	700	9.5	342

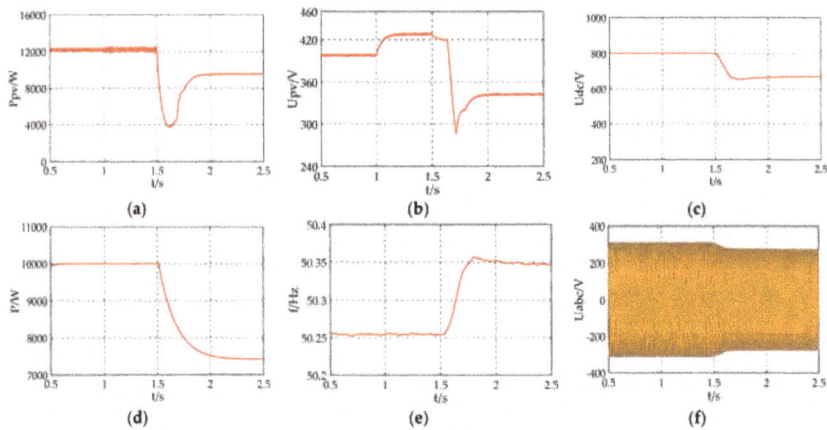

Figure 13. Waveforms in the off-grid mode when PV maximum output power changes (**a**) PV output power; (**b**) PV output voltage; (**c**) DC bus voltage; (**d**) Inverter output power; (**e**) VSG frequency; (**f**) Inverter output voltage.

Before 1 s and between 1~1.5 s, Table 2 illustrates that the PV maximum power is 15 kW and 18.2 kW, respectively, which are greater than the load demand power 10 kW, i.e., $P_{max} > P_{need}$. Figure 13a,d demonstrates that the PV power decreases the output power to enable inverter output of 10 kW power, which is equal to the load requirement. Despite the increased light intensity, the PV output remains unchanged due to the constant load demand in Figure 13a. We can see from Figure 13b that PV output voltage is 398 V before 1 s and 428 V between 1~1.5 s, which are more than 370 V and 380 V, respectively, so the PV power based on the adaptive-MPPT algorithm adjusts the working point in the stable region $[U_{mpp}, U_{oc}]$. The VSG frequency in Figure 13e also stays constant on account of power demand invariability. Due to supply-demand matching before 1.5 s, the DC bus voltage can hold at 800 V in Figure 13c and the amplitude of the inverter output voltage is 311.13 V. After 1.5 s, $P_{max} < P_{need}$, it can be seen from Figure 13a,b that PV power runs at the MPP point (9.5 kW, 342 V), which proves that adaptive-MPPT can be transformed into a traditional MPPT control when PV maximum is inadequate. VSG frequency regulation in Figure 13e depends on the droop characteristic. Figure 13c shows that the improved-VSG let the DC bus voltage stabilize at 670 V. Due to the existence of power shortage, the inverter output voltage is below the rated amplitude 311.13 V.

4.2.2. Grid-Connected Mode

In grid-connected mode, the AC voltage is backed by the bulk power gird, and the VSG is mainly aimed at contenting the dispatching power demand.

(1)   Variation of Dispatching Power Demand

During the simulation process, the external environment is configured on be fixed, that is, the PV maximum available power is P_{max} = 15 kW, which corresponds to output voltage U_{mpp} = 370 V. Before 1 s, the dispatching power instruction was 11 kW, which decreased to 9 kW at 1 s and increased to 16 kW at 1.5 s. In this case, the simulation results are expressed in Figure 14.

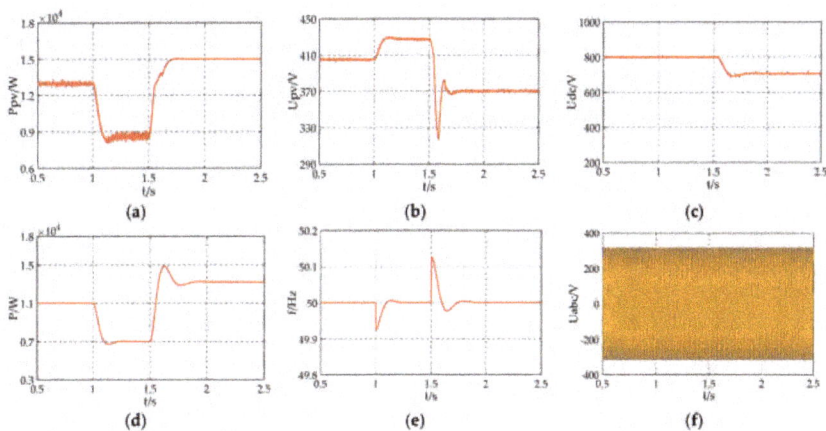

Figure 14. Waveforms in the grid-connected mode when dispatching power demand changes (**a**) PV output power; (**b**) PV output voltage; (**c**) DC bus voltage; (**d**) Inverter output power; (**e**)VSG frequency; (**f**) Inverter output voltage.

Before 1 s and between 1~1.5 s, the PV maximum available power 15 kW exceeded the dispatching demand power of 11 kW and 9 kW, i.e., $P_{max} > P_{need}$. Thus, can be observed in Figure 14a,b,d that PV adaptively regulates in the stable interval ($U_{pv} > 370$ V), so that the inverter outputs power of 11 kW and 7 kW, which are identical to the dispatch orders. On account of system power balance, the DC bus voltage maintains at 800 V in Figure 14c. After 1.5 s, the dispatching requirements surpass the PV maximum power, i.e., $P_{max} = 15$ kW $< P_{need} = 16$ kW. At this time, PV power operates at the MPP point (15 kW, 370 V) to achieve full output. Moreover, improved-VSG causes the DC bus voltage to settle at 710 V, so as to ensure its stability. Throughout the adjustment process, because of the rigid support offered by the bulk power grid, the VSG frequency in Figure 14e maintains at the rated value of 50 Hz after slight fluctuations, and the inverter output voltage in Figure 14f is the same as the AC voltage of the bulk power grid.

(2)   Variation of PV Maximum Output Power

The dispatch power demand is consistently 12 kW. The simulation conditions are in agreement with the variation of PV maximum output power under the off-grid model. The light intensity is 1000 W/m^2 before 1 s, but it rises to 1200 W/m^2 at 1 s and then weakens to 700 W/m^2 at 1.5 s. Other relevant PV parameters of the PV power and voltage are indicated in Table 2. Figure 15 gives the simulation waveforms in grid-connected mode when the PV maximum output power changes.

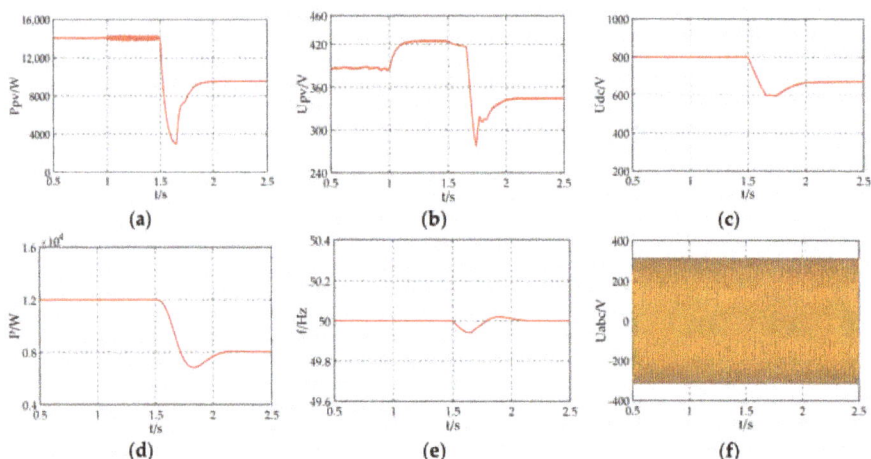

Figure 15. Waveforms in the grid-connected mode when PV maximum output power changes (a) PV output power; (b) PV output voltage; (c) DC bus voltage; (d) Inverter output power; (e) VSG frequency; (f) Inverter output voltage.

Before 1 s, when the light intensity is 1000 W/m^2, the PV maximum output power is 15 kW, which is more than the dispatching demand 12 kW. Accordingly, from Figure 15a,b,d, the PV managed by adaptive-MPPT alters the working point in the stable voltage range ($U_{pv} > 370$ V) to enable inverter power output to meet dispatching power need of 12 kW. At 1 s, the light intensity is amplified to 1200 W/m^2 so that MPP point is (18.2 kW, 380 V). Even if the light intensity strengthens, PV still operates in the stable region ($U_{pv} > 380$ V) to send invariant power owing to the constant dispatching power demand. Moreover, since this above regulation ensures supply-demand matching, Figure 15c shows that the DC bus voltage remains at the set value of 800 V. After 1.5 s, the light intensity weakens to 700 W/m^2, at this moment, PV maximum output power fails to satisfy the dispatching power requirement, i.e., P_{max} = 9.5 kW < P_{need} = 12 kW, so it can be seen from Figure 15a,b that PV power transforms into traditional MPPT operation (9.5 kW, 342 V) to lower power shortage. Additionally, improved-VSG control takes effect to stabilize DC bus voltage at 655 V. Similar to the variation of the dispatching power demand in grid-connected mode, in the steady state, the VSG frequency in Figure 15e and inverter output voltage in Figure 15f are identical to the frequency and voltage of the bulk power grid.

To summarize, based on the simulation results and simulation analysis of the Figures 12–15, the effectiveness of the proposed two-stage PV-VSG approach based on adaptive-MPPT has been demonstrated. Whether in off-grid mode or grid-connected mode, no matter when the power demand changes or the light intensity changes, PV power can operate in a stable area to adjust adaptively, and PV-VSG can provide the most effective output power matching under different load/dispatching power requirements, guaranteeing system stability in conditions of insufficient PV power. Accordingly, compared with traditional two-stage PV-VSG equipped with additional energy storage on the DC bus [19–22], the proposed control method reduces the cost of investment and operation on account of not configuring energy storage.

5. Conclusions

In order to address the increasing energy crisis and environmental pollution, photovoltaic power generation is highly regarded because of its sustainable nature, which can develop a circular economy. In terms of PV control, PV systems built on economic benefits mostly work in MPPT mode, which is not in a position to satisfy the operational demands of the future power system owing to the inability to provide inertia and damping support for the grid. VSG technology is emerging as an attractive solution for the above problem, but traditional PV-VSG ordinarily assembles energy storage devices, which cause many limitations in terms of various costs. Therefore, following consideration of dynamic characteristics of PV output, adaptive-MPPT-based control of an improved two-stage PV-VSG is proposed. The suggested strategy permits PV-VSG to inject power to the loads or grid in light of the power requirements under the circumstances of adequate PV power; in the case of inadequate PV power, the PV system can implement full output due to the conversion of adaptive-MPPT to traditional MPPT, and improved-VSG enhances the stability of PV-VSG. The accuracy of the proposed model has been proven through MATLAB/Simulink. The main contribution of this paper is that a two-stage PV-VSG can be interfaced with loads or the power grid without requiring energy storage allocation. The presented approach can be applied in high-permeability PV regions, future grids with access to a large number of PVs, and in some areas hoping to reduce costs, space occupation, and post-maintenance, and takes on better scalability.

Author Contributions: X.Y. proposed the research direction, adaptive-MPPT method and improved-VSG strategy. J.L. completed the adaptive-MPPT algorithm and improved-VSG control approach. X.Y., J.L., L.W., S.Z., T.L., Z.L. and M.W. performed the verification and analyzed the results. J.L. wrote the paper.

Acknowledgments: This paper was supported by the National Key R&D Program of China (2016YFB0900400); Natural Science Foundation of Hebei Province (E2018502134); State Grid Corporation of Science and Technology Project (PD71-17-008); Liaoning Power Grid Corporation's 2018 Science and Technology Project "Research on the Reactive Power and Voltage Optimization Strategy and Evaluation Index Considering Source and Load Fluctuation Characteristics".

Conflicts of Interest: The authors declare no conflict of interest.

References

1. Yang, X.; Song, Y.; Wang, G.; Wang, W. A comprehensive review on the development of sustainable energy strategy and implementation in China. *IEEE Trans. Sustain. Energy* **2010**, *1*, 57–65. [CrossRef]
2. Cucchiella, F.; D'Adamo, I.; Gastaldi, M. Economic Analysis of a Photovoltaic System: A Resource for Residential Households. *Energies* **2017**, *10*, 814. [CrossRef]
3. Zsiborács, H.; Hegedűsné Baranyai, N.; Vincze, A.; Háber, I.; Pintér, G. Economic and Technical Aspects of Flexible Storage Photovoltaic Systems in Europe. *Energies* **2018**, *11*, 1445. [CrossRef]
4. Liu, J.; Long, Y.; Song, X. A Study on the Conduction Mechanism and Evaluation of the Comprehensive Efficiency of Photovoltaic Power Generation in China. *Energies* **2017**, *10*, 723.
5. Han, P.; Lin, Z.; Wang, L.; Fan, G.; Zhang, X. A Survey on Equivalence Modeling for Large-Scale Photovoltaic Power Plants. *Energies* **2018**, *11*, 1463. [CrossRef]
6. National Development and Reform Commission, Ministry of Finance, National Energy Administration. Notice on Matters Related to Photovoltaic Power Generation in 2018. Available online: http://www.ndrc.gov.cn/gzdt/201806/t20180601_888639.html (accessed on 31 May 2018).
7. Su, M.; Luo, C.; Hou, X.; Yuan, W.; Liu, Z.; Han, H.; Guerrero, J.M. A communication-free decentralized control for grid-connected cascaded PV inverters. *Energies* **2018**, *11*, 1375. [CrossRef]
8. Chen, D.; Jiang, J.; Qiu, Y.; Zhang, J.; Huang, F. Single-stage three-phase current-source photovoltaic grid-connected inverter high voltage transmission ratio. *IEEE Trans. Power Electron.* **2017**, *32*, 7591–7601. [CrossRef]
9. Zhong, Q.C.; Weiss, G. Synchronverters: Inverters that mimic synchronous generators. *IEEE Trans. Ind. Electron.* **2011**, *58*, 1259–1267. [CrossRef]
10. Chen, D.; Xu, Y.; Huang, A.Q. Integration of dc microgrids as virtual synchronous machines into the ac grid. *IEEE Trans. Ind. Electron.* **2017**, *99*, 7455–7466. [CrossRef]
11. Zhong, Q.C. Virtual synchronous machines: A unified interface for grid integration. *IEEE Power Electron. Mag.* **2016**, *3*, 18–27. [CrossRef]
12. National Energy Administration. Guidance on Promoting Smart Grid Development. Available online: http://www.ndrc.gov.cn/gzdt/201507/t20150706_736625.html (accessed on 6 June 2018).
13. Zheng, T.; Chen, L.; Liu, W.; Guo, Y.; Mei, S. Multi-mode operation control for photovoltaic virtual synchronous generator considering the dynamic characteristics of primary source. *Proc. CSEE* **2017**, *37*, 454–464. (In Chinese)
14. Shintai, T.; Miura, Y.; Ise, T. Oscillation damping of a distributed generator using a virtual synchronous generator. *IEEE Trans. Power Del.* **2014**, *29*, 668–676. [CrossRef]

15. Shi, K.; Zhou, G.; Xu, P.; Ye, H.; Tan, F. The integrated switching control strategy for grid-connected and islanding operation of micro-grid inverters based on a virtual synchronous generator. *Energies* **2018**, *11*, 1544. [CrossRef]

16. Yao, G.; Lu, Z.; Wang, Y.; Benbouzid, M.; Moreau, L. A virtual synchronous generator based hierarchical control scheme of distributed generation systems. *Energies* **2017**, *10*, 2049. [CrossRef]

17. Zheng, T.; Chen, L.; Guo, Y.; Mei, S. Comprehensive control strategy of virtual synchronous generator under unbalanced voltage conditions. *IET Gener. Transm. Dis.* **2018**, *12*, 1621–1630. [CrossRef]

18. Wu, H.; Ruan, X.; Yang, D.; Chen, X.; Zhao, W.; Lv, Z.; Zhong, Q.C. Small-signal modeling and parameters design for virtual synchronous generators. *IEEE Trans. Ind. Electron.* **2016**, *63*, 4292–4303. [CrossRef]

19. Yan, X.; Zhang, X.; Zhang, B.; Ma, Y.; Wu, M. Research on distributed PV storage virtual synchronous generator system and its static frequency characteristic analysis. *Appl. Sci.* **2018**, *8*, 532. [CrossRef]

20. Mao, M.; Qian, C.; Ding, Y. Decentralized coordination power control for islanding microgrid based on PV/BES-VSG. *CPSS TPEA* **2018**, *3*, 14–24. [CrossRef]

21. Liu, J.; Miura, Y.; Ise, T. Comparison of dynamic characteristics between virtual synchronous generator and droop control in inverter-based distributed generators. *IEEE Trans. Power Electron.* **2016**, *31*, 3600–3611. [CrossRef]

22. Gao, B.; Xia, C.; Chen, N.; Cheema, K.; Yang, L.; Li, C. Virtual synchronous generator based auxiliary damping control design for the power system with renewable generation. *Energies* **2017**, *10*, 1146. [CrossRef]

23. Guo, Y.; Chen, L.; Li, K.; Zheng, T.; Mei, S. A novel control strategy for stand-alone photovoltaic system based on virtual synchronous generator. In Proceedings of the 2016 IEEE Power and Energy Society General Meeting (PESGM), Boston, MA, USA, 17–21 July 2016; pp. 1–5.

24. Mei, S.; Zheng, T.; Chen, L.; Li, C.; Si, Y.; Guo, Y. A comprehensive consensus-based distributed control strategy for grid-connected PV-VSG. In Proceedings of the 2016 35th Chinese Control Conference (CCC), Chengdu, China, 27–29 July 2016; pp. 10029–10034.

25. Fu, Y.; Wang, Y.; Zhang, X. Integrated wind turbine controller with virtual inertia and primary frequency responses for grid dynamic frequency support. *IET Renew. Power Gen.* **2017**, *11*, 1129–1137. [CrossRef]

26. Zhang, Z.S.; Sun, Y.Z.; Lin, J.; Li, G.J. Coordinated frequency regulation by doubly fed induction generator-based wind power plants. *IET Renew. Power Gen.* **2012**, *6*, 38–47. [CrossRef]

27. Wang, Y.; Meng, J.; Zhang, X.; Xu, L. Control of PMSG-based wind turbines for system inertial response and power oscillation damping. *IEEE Trans. Sustain. Energy* **2015**, *6*, 565–574. [CrossRef]

28. Hua, T.; Yan, X.; Fan, W. Research on power point tracking algorithm considered spinning reserve capacity in gird-connected photovoltaic system based on VSG control strategy. In Proceedings of the 2017 IEEE 3rd International Future Energy Electronics Conference and ECCE Asia (IFEEC 2017—ECCE Asia), Kaohsiung, Taiwan, 3–7 June 2017; pp. 2059–2063.

29. Tang, L.; Xu, W.; Mu, C. Analysis for step-size optimisation on MPPT algorithm for photovoltaic systems. *IET Power Electron.* **2017**, *10*, 1647–1654. [CrossRef]

30. Killi, M.; Samanta, S. Modified perturb and observe MPPT algorithm for drift avoidance in photovoltaic systems. *IEEE Trans. Ind. Electron.* **2015**, *62*, 5549–5559. [CrossRef]

31. Zhang, B.; Yan, X.; Li, D.; Zhang, X.; Han, J.; Xiao, X. Stable operation and small-signal analysis of multiple parallel DG inverters based on a virtual synchronous generator scheme. *Energies* **2018**, *11*, 203. [CrossRef]

32. D'Arco, S.; Suul, J.A. Equivalence of virtual synchronous machines and frequency-droops for converter-based microgrids. *IEEE Trans. Smart Grid* **2014**, *5*, 394–395. [CrossRef]

![energies logo] *energies*

MDPI

Article

A New Maximum Power Point Tracking (MPPT) Algorithm for Thermoelectric Generators with Reduced Voltage Sensors Count Control †

Zakariya M. Dalala [1,*], Osama Saadeh [1], Mathhar Bdour [1] and Zaka Ullah Zahid [2]

[1] Energy Engineering Department, German Jordanian University, Amman 11180, Jordan; osama.saadeh@gju.edu.jo (O.S.); madher.bdour@gju.edu.jo (M.B.)

[2] Department of Electrical Engineering, University of Engineering and Technology, Lahore 54890, Pakistan; zuzahid@uetpeshawar.edu.pk

* Correspondence: zakariya.dalalah@gju.edu.jo; Tel.: +962-795-788-085

† This paper is an extended version of our paper published in: Zakariya M. Dalala, Zaka Ullah Zahid. "New MPPT algorithm based on indirect open circuit voltage and short circuit current detection for thermoelectric generators", 2015 IEEE Energy Conversion Congress and Exposition (ECCE), 2015, also presented at 2016 IEEE International Energy Conference (ENERGYCON), 2016.

Received: 15 June 2018; Accepted: 10 July 2018; Published: 12 July 2018

Abstract: This paper proposes a new maximum power point tracking (MPPT) algorithm for thermoelectric generators (TEG). The new-presented method is based on implementing an indirect open circuit voltage detection and short circuit current estimation methods, which will be used to directly control the TEG interface power converter, resulting in reaching the maximum power point (MPP) in minimal number of steps. Two modes of operation are used in the proposed algorithm, namely the perturb and observe (P&O) method for fine-tuning and the transient mode for coarse tracking of the MPP during fast changes that occur to the temperature gradient across the structure. A novel voltage sensing technique as well is proposed in this work, to reduce the number of voltage sensors used to control and monitor the power converter. The proposed strategy employs a novel approach to sense two different voltages using the same voltage sensor. The input and output voltage information is collected from an intermediate point in the converter. The reconstructed voltages are used in the control loops as well as for monitoring the battery output or load voltages. Simulation and experimental results are provided to validate the effectiveness of the proposed algorithm and the sensing technique.

Keywords: maximum power point tracking; open circuit voltage; perturb and observe; thermoelectric generator

1. Introduction

Due to the rising costs and the scarce availability and depletion of traditional fossil fuels, efforts have intensified in looking for new sources that are preferably renewable in nature. The most abundant among renewable sources are solar and wind. Increasing the overall system's efficiency is another major concern, which has tremendous overall system effect, and has increased interest in other small resources as well. Thermoelectric generators (TEG) perfectly fit into this category [1]. TEGs are solid-state devices, which generate electricity directly from heat, in what is known as the Seebeck effect [2]. The temperature gradient across the structure of the device is maintained to generate lucrative amounts of electrical energy that may be used to charge batteries and to add load support. The conversion efficiency of TEG material is characterized by the figure of merit (ZT), which is defined as $ZT = (S^2\sigma/k)T$, where S, σ, T, and k are Seebeck coefficient, electrical conductivity, operating temperature, and the thermal conductivity of the material, respectively [3]. *ZT* for most TEG materials is around 1, however, feasible designs currently adapt materials with *ZT* around 2~3.

Historically, TEGs have low efficiencies in the range of 5%, which has kept their use limited to specialized medical, remote sensing, space, and military applications [4]. However, the recent trend for harvesting energy at all scales, coupled with the technological development of new power processing devices and circuits, has placed the TEG on the list of viable energy sources that may be exploited and improved for commercial use. The recent cost reduction of TEG manufacturing, in addition to its silent operation, maintenance free service, and long life-time (~25–30 years) has increased attention to this vital energy production tool [5,6]. Recent development in the thermoelectric material manufacturing and the introduction of nanotechnology in the process has boosted new generations of TEGs with higher efficiency ranges that are predicted to reach 15–20% [3].

The TEG can be used as an energy-harvesting device that recovers waste heat from sources such as industrial processes, which will result in an overall increase in the efficiency of these processes. They may also be utilized to harness power form some renewable sources such as solar thermal and geothermal energy sources. Tailored applications for TEG systems include low power autonomous sensors and wearable human diagnosis devices [7,8]. TEGs can now be successfully employed to scavenge energy from the rejected heat by many processes such as automotive [9], stove tops [10], geothermal [11] and power plants [12]. Heat rejected from exhaust gas systems are most likely to be the most attractive area of application for TEG energy-harvesting systems, and will be the real motivation for development.

Adaptation of power electronic converter interfacing has improved the capability of controlling non-conventional power sources as well as maximizing energy-harvesting due to the implementation of maximum power point tracking (MPPT) algorithms. To increase the generated power from the TEG module, a MPPT algorithm should be employed, to push the TEG's operating point to its optimum location. A DC voltage source in series with an internal resistance is used to model the TEG characteristics and can be used to verify MPPT algorithms under different dynamics [13]. The MPPT algorithm adaptively changes the load to match the internal resistance of the TEG module to secure maximum power transfer.

Many MPPT algorithms have been used for TEG systems, most of which were originally developed for photovoltaic (PV) systems [14–17]. The perturb and observe (P&O) method is the most widely used, due to its simple implementation and reliability. The most attractive feature of the P&O algorithm is its independence of actual circuit parameters and the variation of source/load conditions. However, this method does have some drawbacks, which are discussed in [18]. Thus, modified MPPT algorithms are of recent and current research focus to overcome the flaws presented by conventional algorithms.

A PV module's I-V characteristics are not linear, while in the case of the TEG, linearity governs the relationship between the voltage and current. The optimum operating point is related to the characteristics of the TEG module, as follows:

$$V_{MPP} = \frac{V_{OC}}{2} \text{ and } I_{MPP} = \frac{I_{sc}}{2} \tag{1}$$

where V_{MPP} is the MPP voltage and I_{MPP} is the MPP current. V_{OC} is the open circuit voltage and I_{SC} is the short circuit current of the TEG. V_{OC} or I_{SC} is measured for proper operation of some of the main MPPT algorithms, [19–21], in which it requires disconnecting the converter from the TEG module during the measurement action. Therefore, there is no energy flow during measurement, which can be at high frequency, especially in fast-dynamics systems where the temperature gradient is rapidly changing. In [22], an open circuit voltage measurement technique that does not require disconnecting the TEG form the converter has been proposed. But, it can only be applied to converters with discontinuous input current, such as the buck and the buck-boost converters. The proposed method does not apply to continuous input current topologies, such as the boost converter.

In this paper, an expansion to the work in [23] is presented where an innovative method for measuring both V_{OC} and I_{SC} without requiring disconnecting the converter form the TEG is proposed. The proposed technique is universal, and may be applied to all different types of converters and it is independent of the nature of the converter employed. The TEG module's linear characteristics are exploited to derive the measurement approach. The measurement output will then be employed in the MPPT algorithm that is proposed in this paper. There are two modes of operation for the proposed MPPT algorithm, the P&O mode, which is used for fine tracking of the MPP and it is deployed when the temperature gradient across the structure is held nearly constant or it shows a slow rate of change with time. The other mode of operation is the transient mode, and it takes over when there is a fast or sudden change in the open circuit voltage due to a change in the temperature gradient. During this operating mode, the proposed V_{OC} measurement method will be applied, to rapidly adjust for the new MPP location.

To maintain the MPPT algorithm controllability over a wide range of operating conditions, modeling and control of the power converter is essential for stable operation. To ensure that control theory is applicable in such circuits, feedback from various points in the system might be needed, such as the voltages and currents at both the input and output ports of the converters. Voltage and current sensing might not be the most trivial part of the design, as the placement of the sensors and their cost and accuracy, heavily affect the final product. Several researchers have attempted to minimize the number of sensors that were used for the control loops' implementation [24–30]. Some work has utilized estimation algorithms for some variables depending on the system models. Estimation techniques suffer from a stability point of view, as the dynamics of the power converter might shift the estimation accuracy beyond acceptable levels in addition to their complex implementation task. Reconstruction techniques are better, in the sense that one measured variable is decomposed into two or more variables using direct physical relations, which gives these methods the real time virtue over estimation techniques.

In this work, a novel voltage sensing approach is proposed in the DC/DC converters as well. The boost converter is selected as an example, but with the full capability to replicate the proposed method to other switching-type power converters. The voltage sensor is placed between two terminals of the circuit in which one is the switching terminal. Voltage equations during the 'on' and 'off' times of the converter's main switch are constructed and voltage relations to the input and output terminals are derived. Simulation results and experimental verification show the effectiveness of the proposed MPPT algorithm and the voltage sensing technique.

2. System Configuration

2.1. TEG Characteristics

The principle of operation of the TEG module is based on the Seebeck effect, which states that an electromotive force appears between two semiconductors of different doping when a temperature gradient across the structure exists [17,31]. The developed voltage across the terminals of the TEG module is expressed by:

$$V_L = V_{OC} - R_{int}I_L \tag{2}$$

where V_L is the load voltage and I_L is the load current. R_{int} is the internal resistance of the TEG module. V_{OC} is related to the TEG material and temperature difference, as shown in (3).

$$V_{OC} = \alpha \Delta T \tag{3}$$

where α is the Seebeck coefficient of the material and ΔT is the temperature difference between the hot and cold sides of the module. Higher voltage and current levels may be achieved with series and parallel connection of several modules.

The TEG can be electrically modeled under a constant temperature gradient as a voltage source with a series connected internal resistance [13]. Figure 1 shows the typical TEG electrical characteristics, where it shows the current versus the voltage and power for the TEG module at different temperature gradients. It is clear that the optimum selection of the current will guarantee MPP operation, which is achieved by matching the load of the converter to the TEG internal resistance. Dynamic matching of the load for max power transfer is done by utilizing dc-dc converters. The boost converter is the most utilized power conditioning circuit for TEG modules, due to the high voltage gain ratio, simple to construct, and easy to control. Figure 2 shows the boost converter that will be used in this paper, along with the TEG electrical model.

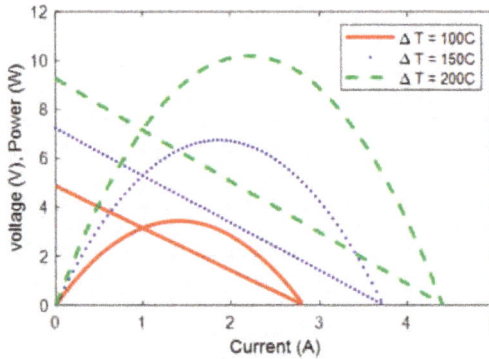

Figure 1. Typical electrical characteristics of thermoelectric generators (TEG).

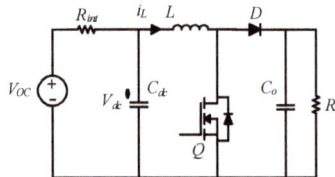

Figure 2. TEG model connected with boost converter.

The input capacitor C_{dc} is used as a buffer for the output voltage of the TEG module and to support any load transients, but not too large as not to mask the true dynamics of the input source. As the inductor current is the output current of the TEG, it should be regulated to the optimum value to insure MPP operation. Another option to ensure MPP operation is to regulate the input voltage to its optimum value as well. Voltage control is usually slower than current control, but that highly depends on the input side capacitance. Dual loop control with current as the internal loop is also possible. This has the advantage of controlling both the transient voltages across the capacitor, and currents through the inductor, to guarantee that they are within the components' ratings. In this paper, current control will be utilized to operate the converter at MPP and voltage control will be used to meet the battery charging requirements in the case of battery charging applications. In the case of a boost converter supplying a dynamic load, the voltage control loop is needed to regulate the output dc bus that interfaces with that load.

2.2. Open Circuit Voltage and Short Circuit Current Detection

From (2) and Figure 1, it can clearly be seen that the MPP occurs when both the load current and the load voltage are regulated, such that the input impedance of the converter matches the internal resistance of the TEG module, and in this case, (1) is derived.

In essence, the MPPT point lies at the mid-point of the I-V characteristics line. Several methods have been proposed for measuring the open circuit voltage, which can then be used as a control input for the converter to guide the MPPT algorithm toward the optimum point. However, these methods need to physically disconnect the converter from the TEG to measure the port voltage and then use it for converter control. Measuring the voltage must happen at a frequency that is much higher than the dynamic speed of the system temperature variation to ensure that it captures all variations. This introduces system losses as no energy flows to the load during measurement. In this paper, an innovative approach to measure the voltage without the need to disconnect the converter from the TEG is proposed, as following:

First, the boost converter is assumed to be current controlled and that the P&O MPPT algorithm is deployed under normal operating conditions of the TEG. Normal conditions are defined as when the temperature gradient variation across the structure is slow. In this case, the current reference variation from the P&O algorithm will not induce large power increments at the output of the TEG, nor will it cause the output voltage of the TEG to vary sharply, since the step size of the reference current applied from the P&O algorithm is small enough to precisely track the MPP of the TEG.

During sharp transients in the open circuit voltage of the TEG due to sudden temperature variation across the structure, first: the reference current to the boost converter is kept constant and it is not changed by the MPPT algorithm. This is represented by points 1–4 in Figure 3. Point 1 is the optimum current command point for the TEG converter at $\Delta T = 100\,^{\circ}$C. When there is a sharp transient in the temperature gradient, the operating point over the power curve moves directly from point 2 to point 4, where the current is still fixed at point 1, which is the same as point 3. The circuit waits until it settles to the new voltage level at point 3. This operating point is recorded as (I_1, V_1). A shift in the current command with suitable step size is then executed (e.g., $I_2 = I_1 + \Delta I$) to push the operating point to point 5 in the I–V characteristics line. This step size depends on the system's size and the TEG characteristics. The converter responds by regulating to the new current level at point 5, which will change the voltage at the output of the TEG to a new value (V_2). This is another point that is recorded. The linear characteristics of the TEG enable the construction of a straight line equation using two points (3 and 5) along its path, which are $(I_1, V_1), (I_2, V_2)$. From these two points, the straight-line equation is formed, and the middle point (point 6) is determined by setting either the current or the voltage in the equation to zero. In this case, two points of operation is sufficient to measure both V_{OC} or I_{SC}. Ultimately, point 7 can be reached from point 3 in only two steps.

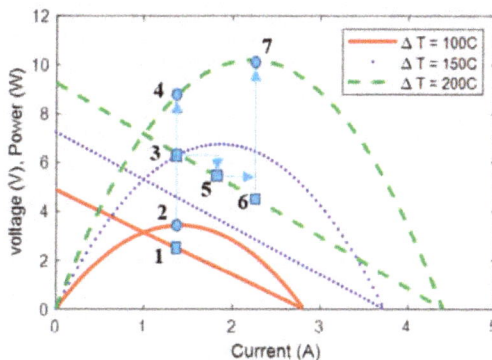

Figure 3. Proposed maximum power point tracking (MPPT) algorithm principle.

It should be noted that the shift in the current command (between point 3 and point 5 for example) is in the direction of the power increase or decrease during transients. During the measurement routine, the converter still delivers power to the output, hence, collecting more energy as compared to traditional methods. The measurement is independent of any offset that might exist in the voltage sensors, which can produce less accurate readings using traditional measurement techniques. Moreover, the measurement is fast as it takes only two iterations of the MPPT algorithm.

3. Proposed MPPT Control Algorithm

A flow chart for the proposed MPPT algorithm is shown in Figure 4. First, the variables are measured and ΔI and ΔP are calculated. If ΔP is less than a certain threshold, then it is concluded that the system is to run in the P&O mode, where incremental changes in power levels are small due to fine steps in the current command. If there is a sudden change in the open circuit voltage, due to sharp transients in the temperature gradient, large power increment is detected, and thus, the algorithm is operated in the transient mode. In this operating mode, the voltage and current readings of the last P&O step are recorded (i.e., before the algorithm moves to the transient mode). Then, the current is incremented by a sufficient magnitude to obtain distinct readings for the voltage and current to establish the other point needed to construct the line characteristics ($y = mx + b$) for the TEG. The estimated relation is then used to find V_{OC} and I_{SC}. If the converter is current controlled, then the optimum current command would be $\frac{I_{SC}}{2}$, which guarantees moving the operating point directly to the new MPP. Next, the system moves back to the normal P&O mode of operation.

The proposed MPPT algorithm uses the P&O mode for fine-tuning to the MPP, and the transient mode for the direct relocation of the operating point to the new MPP when there is a sudden change in the temperature gradient across the structure. The step size for the P&O mode is designed to be small in order to capture the closest point to the MPP and can be adaptive to be a scaled value of the measured power increment. The step size for the transient mode ΔI, is chosen such that the readings of the two points on the characteristics line are distinctive. It can be a design parameter that can be tuned for best performance.

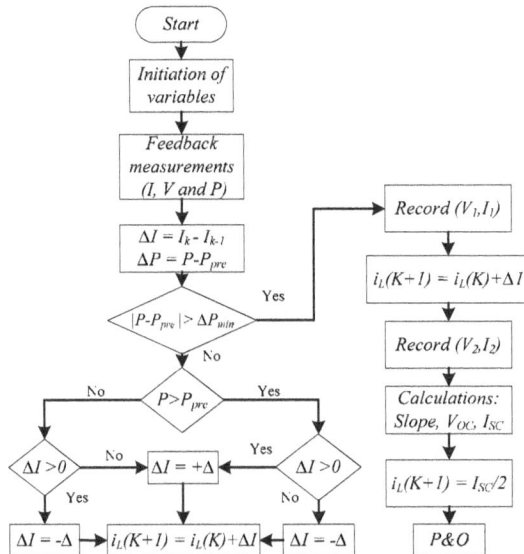

Figure 4. Proposed MPPT algorithm flow chart.

4. Proposed Voltage Sensing Technique

Figure 5 shows the schematic diagram of the boost converter under consideration. The output load in this case is a battery. The MPPT algorithm requires the collection of the power production information through the measurements of the input current and voltage. Moreover, the control objective of the MPPT algorithm is to tune the input voltage or current to its optimum point through dual control loops. While the input voltage is being measured and used in the control loops and MPPT algorithm, continuous monitoring of the output voltage is required to shift the controller from MPPT control to load matching (voltage mode control) in the case that the battery approaches its full charge (light load conditions). In the case where the converter is supplying a dynamic load instead of the battery, then control of the dc bus is needed, and in all cases, the output voltage information is needed to complete the cycle of the controller. The collection of the output voltage information is not intended in this paper to be used for the previously proposed MPPT algorithm, rather it is can be used to further control and monitor the load side, irrespective of load type.

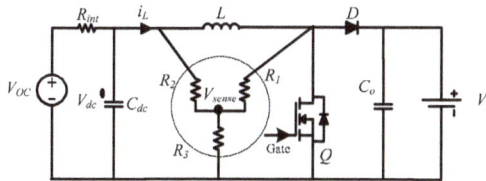

Figure 5. DC/DC boost converter with Thermoelectric Generator input.

For both control modes, MPPT and voltage mode control, the input and output voltages are needed. Usually two voltage sensors are placed at the input and output ports of the converter to acquire the voltage information in a real time manner. It should be noted that it is critical to have real time and direct measurements of the voltage for sensitive control implementation, especially if the battery is sensitive and excess charge cannot be permitted. Thus, utilizing estimation technique for either voltage might not be allowed.

In Figure 5, the proposed voltage sensing circuitry is connected as a Y-connected resistive network. The voltage is sensed at the neutral point of the network as shown. The resistors R_1, R_2, and R_3 are design variables and they are large enough not to cause additional resistive losses in the circuit, common values ranges between 10 kΩ and 500 kΩ. The sensed voltage (V_{sense}) is used to reconstruct the input and output voltage information at the same time. The idea of the connection of the Y-network is to attach one terminal to a reference point through R_3, and the second terminal to be constantly attached to a steady voltage, in this case V_{in} through R_2. The third terminal is connected to a switching voltage pole. In this case, to the switch voltage through R_1.

To derive the relation between V_{sense} and the input and output voltages' information, the converter is analyzed during the ON and OFF times of the switch, as shown in Figure 6.

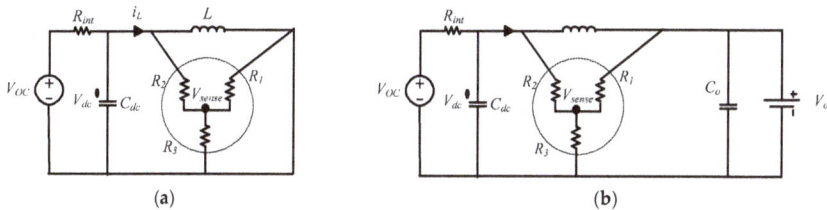

(a)

(b)

Figure 6. Equivalent circuit of the converter during (**a**) ON time (**b**) OFF time.

While the converter is operating under mode 1 (ON-time), the sensed voltage V_{sense} in Figure 6a can be described, as follows:

$$V_{sense} = \frac{R_1 \parallel R_3}{R_1 \parallel R_3 + R_2} V_{in} \tag{4}$$

where V_{in} is the input voltage to the boost converter and it is across the capacitor C_{dc} as V_{dc} in Figure 6. Equation (4) is reduced to (5) below.

$$V_{sense} = \frac{R_1 R_3}{R_1 R_3 + R_1 R_2 + R_2 R_3} V_{in} = \frac{R_1 R_3}{R_{eq}} V_{in} \tag{5}$$

where $R_{eq} = R_1 R_3 + R_1 R_2 + R_2 R_3$.

During the off-time, which is mode 2 of the boost converter operation, Figure 6b holds. In this case, the input and output voltages are considered as sources and the superposition principle is applied to find the V_{sense}. First, V_o is killed and V_{in} is kept, which yields Equation (6).

$$V'_{sense} = \frac{R_1 \parallel R_3}{R_1 \parallel R_3 + R_2} V_{in} = \frac{R_1 R_3}{R_{eq}} V_{in} \tag{6}$$

Then, V_{in} is killed and V_o is kept, yielding the following:

$$V''_{sense} = \frac{R_2 \parallel R_3}{R_2 \parallel R_3 + R_1} V_o = \frac{R_2 R_3}{R_{eq}} V_o \tag{7}$$

Then, the sensed voltage during mode 2 is shown in (8).

$$V_{sense} = V'_{sense} + V''_{sense} = \frac{R_1 R_3 V_{in} + R_2 R_3 V_o}{R_{eq}} \tag{8}$$

From (5) and (8), it is apparent that V_{sense} is closely related to both V_{in} and V_o. To examine the proportionality, the averaged voltage over one switching period with duty cycle D can be found, as follows:

$$\overline{V}_{sense-avg} = \frac{DR_1 R_3}{R_{eq}} V_{in} + \frac{(1-D)}{R_{eq}} [R_1 R_3 V_{in} + R_2 R_3 V_o]$$
$$\overline{V}_{sense-avg} = \frac{R_1 R_3}{R_{eq}} V_{in} + \frac{(1-D) R_2 R_3}{R_{eq}} V_o \tag{9}$$

Examining (9), the average sensed voltage carries the information of the input and output average voltages. Which means that the input and output voltages can be reconstructed by utilizing the instantaneous measurements of the voltage V_{sense}, as follows:

To reconstruct V_{in}, Equation (5) is utilized, where it shows that during the turn ON time the sensed voltage V_{sense} is related to the input voltage V_{in} through the factor $\frac{R_1 R_3}{R_{eq}}$. Taking samples of V_{sense} during the ON times results in discrete samples of V_{in}, which are fed in to establish continuous measurement for V_{in}.

For V_o reconstruction, and during off time, Equation (8) holds. However, V_{sense} does not include V_o information only, but V_{in} as well. During one switching period, the input voltage is not assumed to greatly vary, especially with the existence of the input capacitor C_{dc}, which holds the variation in the input voltage under slow dynamics compared to a time of one switching period. Thus, V_{in} information during the ON time are utilized to reconstruct V_o in (8).

In a digital signal processing environment, one sample per switching period is needed for each circuit variable to carry-on the control and monitoring tasks. Usually at the end of the PWM cycle. However, in the case of the proposed technique, two samples per switching period are needed for the V_{sense} to be able to reconstruct V_{in} and V_o. The first sample should be taken at the middle of the turn-on time to establish V_{in}, and the other sample should be taken at the middle of the turn-off time to retrieve V_o. Two samples per period can be easily carried out using simple and low cost processors. Analogue recovery of the voltage signals using simple and low cost analogue electronics is possible too.

5. Simulation Results and Discussion

5.1. System Implementation

A simple model of the TEG is selected for simulation to verify the proposed MPPT concept. The TEG model has an open circuit voltage of 10 V and an internal resistance of 2 Ω. The simulation design parameters are shown in Table 1.

Table 1. Design parameters of the TEG module and Boost converter.

TEG	Value [Unit]
Nominal Voltage	10 [V]
R_{int}	2 [Ω]
Boost Converter	**Value [Unit]**
L	100 [μH]
C_{dc}	10 [μF]
C_o	470 [μF]
R_L	2 [Ω]
f_s	50 [kHz]

The schematic diagram of the designed energy-harvester controller is shown in Figure 7, along with a photo for the test setup developed in the lab. In Figure 7a, the voltage V_{sense} is fed to the DSP, where the sample during the ON time leads to V_{in} through Equation (5). The determined V_{in} sample along with the voltage V_{sense} are utilized to reconstruct the voltage V_o through Equation (8). The output voltage can be used to implement control/monitoring for the output port of the converter. The sensed current and input voltage are fed along with the input power (P_{dc}) to the MPPT algorithm that is detailed in the flow chart in Figure 4. The MPPT controller takes the voltage, current, and power readings as inputs and generates an optimum current command to drive the converter to the optimum operating point, and hence, obtaining the MPP. The generated current command (i_L^{ref}) is directly related to the mode of operation, whether it is P&O mode or transient mode, as explained in Section 3. It should be highlighted here that both current control and voltage control are possible in this circuit, as both the open circuit voltage and short circuit current are available by utilizing the same procedure explained in Section 2.2. The duty-to-current transfer function is derived, as follows:

$$\frac{i_{L(s)}}{d(s)} = \frac{C_o V_o s + 2(1-D)I_L}{LC_o s^2 + \frac{L}{R_o}s + (1-D)^2} \tag{10}$$

The transfer function of the boost converter in (10) is operating point dependent. To properly design a robust controller, the worst case scenario must be considered. A practical realistic upper limit on the conversion ratio of the boost converter is five times. This leads to an upper limit of the duty cycle of 0.8. A conservative value of $D = 0.9$ will be selected for the controller design. Using the parameters shown in Table 1, and considering the nominal value for I_L of 2.5 A, which represents the optimum current value at the nominal TEG voltage of 10 V and optimum input resistance of the boost converter of 2 Ω, the plant transfer function becomes:

$$\frac{i_{L(s)}}{d(s)} = \frac{0.01128s + 0.5}{4.7 \times 10^{-8}s^2 + 5 \times 10^{-6}s + 0.01} \tag{11}$$

The compensator designed is proportional-integral (PI) type with the following parameters:

$$G_{ci}(s) = 0.075 + \frac{75}{s} \tag{12}$$

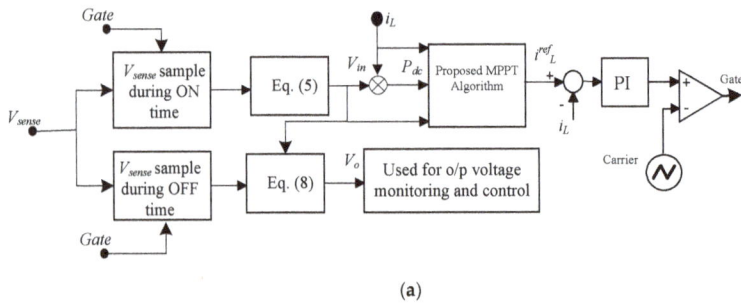

(a)

(b)

Figure 7. (a) Schematic diagram of the TEG energy-harvester control diagram. **(b)** Hardware photo of the test setup.

5.2. Proposed MPPT Algorithm Simulation

The PSIM simulator is used to simulate the boost convert controlled using the proposed MPPT algorithm. The P&O algorithm step size is tuned to 0.1 A after experimenting different values for optimum performance for the TEG size given. The time frame over which the power readings are taken and averaged is 0.6 ms, which means that the voltage and current samples are averaged every 0.6 ms, and then the power sample is generated for use in the MPPT algorithm. This averaging routine is important because it is needed to handle any transient overshoot or undershoot that may occur when the current reference is changed due to the tracking process [32].

A sudden change in the open circuit voltage is simulated in Figure 8, and the MPPT algorithm is set to track the MPP. In the figure, initially, $V_{OC} = 10$ V and the MPP is tracked using a current controlled boost converter. The current command is around 2.5 A, as can be seen in the figure. A transition from 10 V to 14 V is imposed. The algorithm detects large jump in the power measured. This condition results in the algorithm deciding to shift to the transition mode. At the start of this mode, the current command is set as the previous value used in the P&O mode. Averaging of voltage and current measurements are done and recorded. The current is then increased or decreased in the same direction of the power transition by a constant step that is known to the designer and optimized depending on the system size and TEG characteristics. In this case, the jump is 1 A in the direction of the power increment after the transition. The new voltage and current values are recorded. Using these two points, the TEG characteristics line is constructed and the short circuit current and open circuit voltage are estimated. In this case, I_{SC} is used to define the new current level. In the simulation, the current command starts at around 2.7 A immediately after the transition, then moved to 3.7 A where a new voltage level is recorded. The reconstructed relation defines the new current level to be 3.5 A, which matches the new MPP location. The transition to 14 V means that the new current command for the new MPP is 3.5 A, which is $\frac{I_{SC}}{2}$. The system then returns to the P&O mode where fine tracking for the MPP takes place.

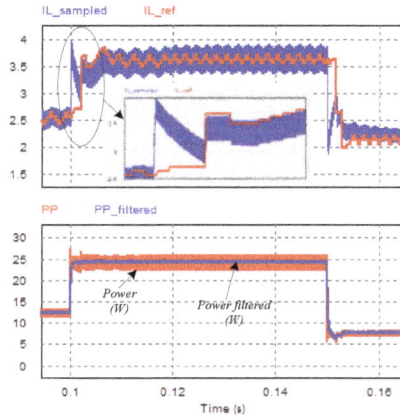

Figure 8. Simulation results of the proposed MPPT algorithm.

A transition down by six volts is later commanded, and the same scenario occurs once again by utilizing the transition mode. A new I_{SC} value is detected and used to move the operating point directly to the new MPP. As can be seen from the simulation results, the proposed technique of MPP tracking does help in identifying anomalies in the power signal feedback, and it is able to excite the indirect detection method of the voltage and current to "rapidly" adapt the current command to the new operating condition.

The advantages of the proposed algorithm includes fast tracking for the MPP under a dynamically changing environment, where in only three steps, it moves the operating point to the vicinity of the new MPP. The current I_{SC} and voltage V_{OC} detection is independent of the operating point or conditions. These figures (V_{oc} and I_{sc}), may change due to temperature variation, humidity, or even loading conditions. However, the proposed detection and measurement method is adaptive to such conditions and it is very simple to implement.

5.3. Proposed Voltage Sensing Technique Simulation

The circuit shown in Figure 5 is simulated in PSIM software. The resistors' values of the voltage sensor are chosen to be 1 KΩ each. Figure 9 shows the simulation results of the algorithm. The input source takes step changes and the MPPT algorithm follows the changes. The actual V_{in} and V_o are shown in blue and the reconstructed voltages through only V_{sense} are shown in red. The reconstruction technique clearly shows excellent effectiveness and it can capture even acute dynamics. In fact the reconstructed voltages are the ones that are used to close the control loop of the power converter and the MPPT algorithm still runs with full stability. Figure 10 shows the sampled intervals of V_{sense}, along with the actual V_{in}. It is clearly shown that the sampled intervals of the sensed voltage follow the contour of the actual input voltage, which in turn enables the reconstruction of V_{in}. Similar figures can be generated for V_o.

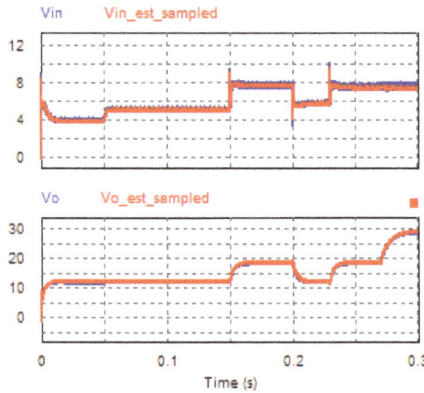

Figure 9. Input and output voltages. Actual (blue) and Reconstructed (red).

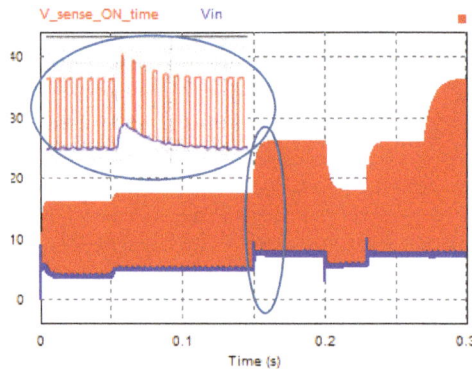

Figure 10. Simulated input actual voltage and sampled V_{sense} intervals during ON-time.

6. Experimental Results

An in-lab hardware prototype was built and tested to verify the proposed algorithms. A boost converter built with components' parameters shown in Table 1. The sensing resistors, as shown in Figure 7, were selected to be 1 KΩ each. A TMS320F28335 DSP processor is utilized to build and execute the proposed algorithms. The input source is a DC voltage source in series with a resistor to simulate the behavior of a TEG module. The power level of the input source can be tuned to represent a single TEG module or a complete set of TEG modules.

Figure 11 shows the proper operation of the boost converter where the current, input voltage to the boost, and gate signal are shown. Smooth controlled operation is verified first and then the proposed MPPT algorithm was executed and tested against conventional P&O MPPT algorithm, as shown in Figure 12. At time t_1, the input power is increased by increasing the input source voltage in a step change behavior from 10 V to 20 V, and at t_2 falls back to 10 V before repeating the same scenario starting at t_3. In the time interval $[t_1 - t_2]$, the proposed MPPT algorithm is operating. As can be seen in the current waveform, the algorithm detects sudden change in input power, and accordingly, activates the transient mode. In this mode, two successive steps in the current command are introduced by the algorithm and proper voltage and current measurements are taken to estimate the open circuit voltage, and hence the position of the MPP. The algorithm steers the converter to operate at the estimated MPP and at the same time folds back to the regular P&O mode for fine tuning and to detect slow dynamics changes in input power. The exact same scenario happens exactly when the input power drops suddenly at time instant t_2.

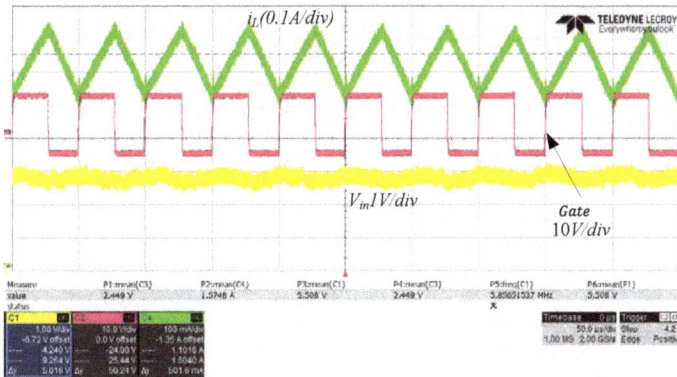

Figure 11. Experimental waveforms for the current, PWM signal, and input voltage of the operating boost converter.

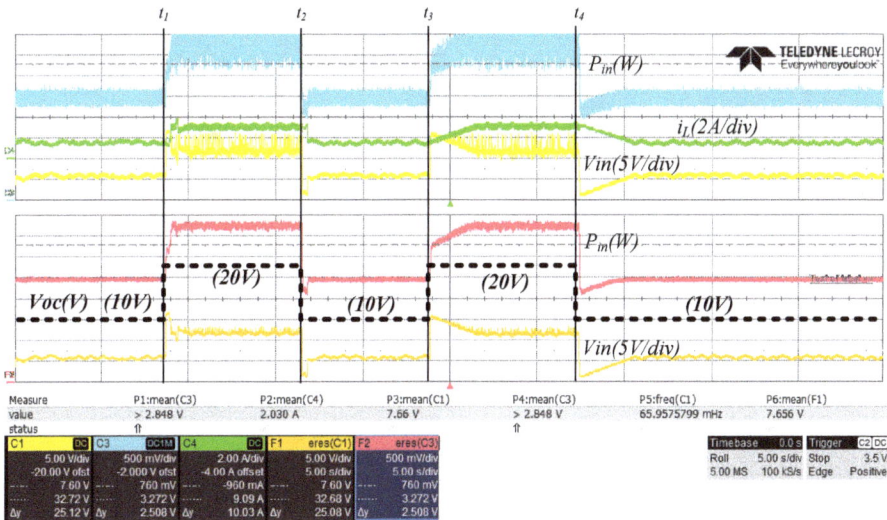

Figure 12. Experimental comparison waveforms for the current, voltage, and input power under the conventional perturb and observe (P&O) and the proposed MPPT algorithms. (Lower part is filtered version of the upper part).

The whole experimental procedure is repeated for time instants t_3 and t_4 where only P&O mode is operating. The delay in tracking the new MPP is clearly seen due to the natural behavior of P&O algorithm of applying small steps towards the new set point. Of course, some tuning can facilitate the conventional P&O algorithms but with flaws, as discussed in [18]. The enhancement of the proposed MPPT algorithm over conventional ones can be clearly seen from the figure.

The proposed voltage sensing technique was implemented and tested using the exact same setup used for the MPPT algorithm verification. The sensed voltage (V_{sense}) and the reconstructed input (V_{in}) and output (V_o) voltages are shown in Figure 13. The sensed voltage is factorized using (5) and

(8), as shown in Figure 7, to reconstruct the input and output voltages. Excellent matching to the derivation and simulation can be seen. In Figure 13, the traces are configured to better see the close behavior of the sensed voltage as compared to the input and output voltages. As shown in the figure, V_{sense} is a switched waveform that resembles Equations (5) and (8). Noting that V_{sense} does not coincide with either V_{in} or V_o, but the traces in the figure are configured to show the interrelation between the variables.

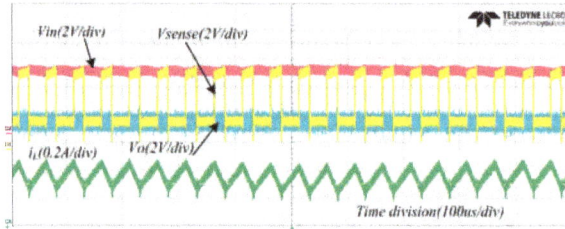

Figure 13. Experimental waveforms for the current, estimated input and output voltages, and the sensed voltage, as highlighted in Figure 5.

In Figure 14, voltage sensing technique was tested under transient conditions, where it clearly shows how the reconstructed voltages follow the input and output voltages smoothly and justify using it as the only voltage to be sensed in order to completely close the control loop of the converter and to apply continuous monitoring over the output voltage in the case of battery charging application.

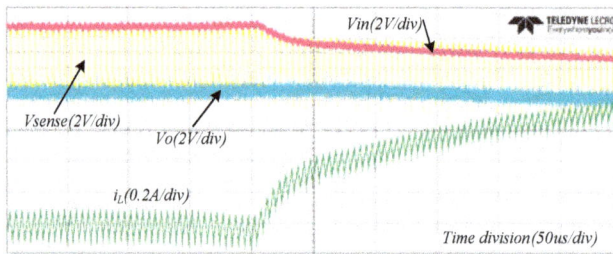

Figure 14. Experimental waveforms for the current, estimated input and output voltages and the sensed voltage as highlighted in Figure 5 during transient dynamics.

7. Conclusions

In this manuscript, the TEGs conversion system is considered, where a new MPPT control method is proposed. This new method is based on a novel strategy to measure the open circuit voltage and short circuit current of the TEG without using any special physical connection or disconnection form the source. The proposed detection method utilizes the inherent linear characteristics of the TEG I-V curve, and it is independent of operating or environmental conditions. During normal conditions, the P&O algorithm is used for fine tracking, where, as during fast changing dynamics, the transient mode is adapted. In the transient mode, the open circuit voltage is estimated, and conveniently, the operating point is moved towards $\frac{V_{oc}}{2}$ where the MPP operation is maintained. Even during the normal operating conditions where the P&O is activated, verification of the proper MPP tracking might be needed and that can be achieved by utilizing the estimated open circuit voltage.

Moreover, for low computational overhead systems, where a minimum number of computations is desired, the proposed algorithm can serve very well without engaging the P&O mode. Spaced samples are enough to establish the measurements for the open circuit voltage and the operating point, which can be steered towards $\frac{V_{oc}}{2}$ where MPPT is guaranteed.

The proposed algorithm utilizes the input voltage for tracking and the output voltage for controlling the load demand. Usually, the TEG power electronic converter interfaces with a battery or DC link capacitor, where the dc voltage at the output is a control variable and needs to be continuously measured and controlled. Thus, a novel voltage sensing technique is developed to reduce the used sensors count. Through a single point voltage measurements, the input and output voltage information were reconstructed in a real time manner with excellent accuracy. Simulation and hardware results show the effectiveness and verify the operation of the proposed methods.

Author Contributions: Conceptualization, Z.D.; Methodology, Z.D.; Software, Z.Z.; Validation, Z.D. and O.S.; Formal Analysis, M.B.; Investigation, Z.D.; Resources, Z.D.; Data Curation, M.B.; Writing-Original Draft Preparation, Z.D.; Writing-Review & Editing, O.S.; Visualization, Z.Z.; Supervision, Z.D.; Project Administration, O.S.; Funding Acquisition, Z.D.

Funding: This research was funded by 'Support to Research, Technological Development & Innovation in Jordan' (SRTD–II), an EU funded project managed by the Higher Council for Science & Technology under grant number AR-164.

Conflicts of Interest: The authors declare no conflict of interest.

References

1. Rowe, D. Thermoelectric waste heat recovery as a renewable energy source. *Int. J. Innov. Energy Syst. Power* **2006**, *1*, 13–23.
2. Riffat, S.B.; Ma, X. Thermoelectrics: A review of present and potential applications. *Appl. Therm. Eng.* **2003**, *23*, 913–935. [CrossRef]
3. Zhang, X.; Zhao, L.-D. Thermoelectric materials: Energy conversion between heat and electricity. *J. Materiomics* **2015**, *1*, 92–105. [CrossRef]
4. Rowe, D. Thermoelectrics, an environmentally-friendly source of electrical power. *Renew. Energy* **1999**, *16*, 1251–1256. [CrossRef]
5. Mehta, R.J.; Zhang, Y.; Karthik, C.; Singh, B.; Siegel, R.W.; Borca-Tasciuc, T.; Ramanath, G. A new class of doped nanobulk high-figure-of-merit thermoelectrics by scalable bottom-up assembly. *Nat. Mater.* **2012**, *11*, 233–240. [CrossRef] [PubMed]
6. Biswas, K.; He, J.; Blum, I.D.; Wu, C.-I.; Hogan, T.P.; Seidman, D.N.; Dravid, V.P.; Kanatzidis, M.G. High-performance bulk thermoelectrics with all-scale hierarchical architectures. *Nature* **2012**, *489*, 414–418. [CrossRef] [PubMed]
7. Dalola, S.; Ferrari, M.; Ferrari, V.; Guizzetti, M.; Marioli, D.; Taroni, A. Characterization of Thermoelectric Modules for Powering Autonomous Sensors. *IEEE Trans. Instrum. Meas.* **2009**, *58*, 99–107. [CrossRef]
8. Deng, F.; Qiu, H.; Chen, J.; Wang, L.; Wang, B. Wearable Thermoelectric Power Generators Combined with Flexible Super Capacitor for Low-power Human Diagnosis Devices. *IEEE Trans. Ind. Electron.* **2016**, *64*, 1477–1485. [CrossRef]
9. Risse, S.; Zellbeck, H. Close-coupled exhaust gas energy recovery in a gasoline engine. *Res. Therm. Manag.* **2013**, *74*, 54–61. [CrossRef]
10. Champier, D.; Bédécarrats, J.P.; Kousksou, T.; Rivaletto, M.; Strub, F.; Pignolet, P. Study of a TE (thermoelectric) generator incorporated in a multifunction wood stove. *Energy* **2011**, *36*, 1518–1526. [CrossRef]
11. Sutera, C.; Jovanovica, Z.R.; Steinfeld, A. A 1 kW$_e$ thermoelectric stack for geothermal power generation—Modeling and geometrical optimization. *Appl. Energy* **2012**, *99*, 379–385. [CrossRef]
12. Siviter, J.; Knox, A.; Buckle, J.; Montecucco, A.; McCulloch, E. Megawatt scale energy recovery in the Rankine cycle. In Proceedings of the 2012 IEEE Energy Conversion Congress and Exposition (ECCE), Raleigh, NC, USA, 15–20 September 2012; pp. 1374–1379.
13. Xu, Y.; Yuan, Y.; Fu, J. Modeling and design for a thermoelectric charger. In Proceedings of the 2012 IEEE International Symposium on Industrial Electronics (ISIE), Hangzhou, China, 28–31 May 2012; pp. 383–386.
14. Kasa, N.; Iida, T.; Liang, C. Flyback Inverter Controlled by Sensorless Current MPPT for Photovoltaic Power System. *IEEE Trans. Ind. Electron.* **2005**, *52*, 1145–1152. [CrossRef]

15. Koizumi, H.; Mizuno, T.; Kaito, T.; Noda, Y.; Goshima, N.; Kawasaki, M.; Nagasaka, K.; Kurokawa, K. A Novel Microcontroller for Grid-Connected Photovoltaic Systems. *IEEE Trans. Ind. Electron.* **2006**, *53*, 1889–1897. [CrossRef]

16. Kuo, Y.-C.; Liang, T.-J.; Chen, J.-F. Novel maximum-power-point-tracking controller for photovoltaic energy conversion system. *IEEE Trans. Ind. Electron.* **2001**, *48*, 594–601.

17. Kim, R.-Y.; Lai, J.-S. A Seamless Mode Transfer Maximum Power Point Tracking Controller For Thermoelectric Generator Applications. *IEEE Trans. Power Electron.* **2008**, *23*, 2310–2318.

18. Raza Kazmi, S.M.; Goto, H.; Hai-Jiao, G.; Ichinokura, O. Review and critical analysis of the research papers published till date on maximum power point tracking in wind energy conversion system. In Proceedings of the 2010 IEEE Energy Conversion Congress and Exposition (ECCE), Atlanta, GA, USA, 12–16 September 2010; pp. 4075–4082.

19. Jungmoon, K.; Minseob, S.; Junwon, J.; Heejun, K.; Chulwoo, K. A DC-DC boost converter with variation tolerant MPPT technique and efficient ZCS circuit for thermoelectric energy harvesting applications. In Proceedings of the 19th Asia and South Pacific Design Automation Conference ASP-DAC 2014, Singapore, 20–23 January 2014; pp. 35–36.

20. Schwartz, D.E. A maximum-power-point-tracking control system for thermoelectric generators. In Proceedings of the 2012 3rd IEEE International Symposium on Power Electronics for Distributed Generation Systems (PEDG), Aalborg, Denmark, 25–28 June 2012; pp. 78–81.

21. Laird, I.; Lu, D.D.C. High Step-Up DC/DC Topology and MPPT Algorithm for Use with a Thermoelectric Generator. *IEEE Trans. Power Electron.* **2013**, *28*, 3147–3157. [CrossRef]

22. Montecucco, A.; Knox, A.R. Maximum Power Point Tracking Converter Based on the Open-Circuit Voltage Method for Thermoelectric Generators. *IEEE Trans. Power Electron.* **2015**, *30*, 828–839. [CrossRef]

23. Dalala, Z.M.; Zahid, Z.U. New MPPT algorithm based on indirect open circuit voltage and short circuit current detection for thermoelectric generators. In Proceedings of the 2015 IEEE Energy Conversion Congress and Exposition (ECCE), Montreal, QC, Canada, 20–24 September 2015; pp. 1062–1067.

24. Tolani, S.; Joshi, S.; Sensarma, P. Dual loop digital control of UPS inverter with reduced sensor count. In Proceedings of the 2016 IEEE International Conference on Power Electronics, Drives and Energy Systems (PEDES), Trivandrum, India, 14–17 December 2016; pp. 1–6.

25. Uno, M.; Kukita, A. Current sensorless single-switch voltage equalizer using multi-stacked buck-boost converters for photovoltaic modules under partial shading. In Proceedings of the 2015 9th International Conference on Power Electronics and ECCE Asia (ICPE-ECCE Asia), Seoul, Korea, 1–5 June 2015; pp. 645–651.

26. Dallago, E.; Finarelli, D.G.; Gianazza, U.P.; Barnabei, A.L.; Liberale, A. Theoretical and Experimental Analysis of an MPP Detection Algorithm Employing a Single-Voltage Sensor Only and a Noisy Signal. *IEEE Trans. Power Electron.* **2013**, *28*, 5088–5097. [CrossRef]

27. dos Santos, W.M.; Martins, D.C. Digital MPPT technique for PV panels with a single voltage sensor. In Proceedings of the Intelec 2012, Scottsdale, AZ, USA, 30 September–4 October 2012; pp. 1–8.

28. Ciani, L.; Catelani, M.; Mancini, M.; Simoni, E. A novel technique for power inverter control based on a single voltage sensor. In Proceedings of the 2009 IEEE Instrumentation and Measurement Technology Conference, Singapore, 5–7 May 2009; pp. 1167–1170.

29. Mallik, A.; Khaligh, A. Control of a Three-Phase Boost PFC Converter Using a Single DC-Link Voltage Sensor. *IEEE Trans. Power Electron.* **2017**, *32*, 6481–6492. [CrossRef]

30. Mukherjee, S.; Shamsi, P.; Ferdowsi, M. Control of a Single-Phase Standalone Inverter without an Output Voltage Sensor. *IEEE Trans. Power Electron.* **2017**, *32*, 5601–5612. [CrossRef]

31. Lineykin, S.; Ben-Yaakov, S. Modeling and analysis of thermoelectric modules. In Proceedings of the Twentieth Annual IEEE Applied Power Electronics Conference and Exposition, APEC 2005, Austin, TX, USA, 6–10 March 2005; Volume 2013, pp. 2019–2023.

32. Dalala, Z.M.; Zahid, Z.U.; Wensong, Y.; Younghoon, C.; Jih-Sheng, L. Design and Analysis of an MPPT Technique for Small-Scale Wind Energy Conversion Systems. *IEEE Trans. Energy Convers.* **2013**, *28*, 756–767. [CrossRef]

energies

MDPI

Article

Energy Management for Smart Multi-Energy Complementary Micro-Grid in the Presence of Demand Response

Yongli Wang, Yujing Huang *, Yudong Wang, Haiyang Yu, Ruiwen Li and Shanshan Song

School of Economics and Management, North China Electric Power University, Changping District,
Beijing 102206, China; wyl_2001_ren@163.com (Y.W.); yudongwang@ncepu.edu.cn (Y.W.);
HaiyangYU@ncepu.edu.cn (H.Y.); liruiwen@ncepu.edu.cn (R.L.); licheelily@ncepu.edu.cn (S.S.)
* Correspondence: yujinghuang@ncepu.edu.cn

Received: 25 March 2018; Accepted: 11 April 2018; Published: 18 April 2018

Abstract: With the application and the rapid advancement of smart grid technology, the practical application and operation status of multi-energy complementary microgrids have been widely investigated. In the paper presented, the optimal operation of a solar unit, a storage battery and combined cooling, heating and power is studied via an economic optimization model implemented in General Algebraic Modeling Systems (GAMS). The model represents an optimization strategy for the economic operation of a microgrid considering demand response programs in different scenarios, and it is intended for the targets of minimizing the operating cost of the microgrid and maximizing the efficiency of renewable energy utilization. In addition, a multi-time electricity price response model based on user behavior and satisfaction is established, and the core value of the model is to describe the mechanism and effect of participation in electricity price demand response. In order to verify the accuracy of the model proposed, we design the dispatch strategy of a microgrid under different states considering demand response, and use genetic algorithm to solve the optimization problems. On the other hand, the application of methodology to a real case study in Suzhou demonstrates the effectiveness of this model to solve the economic dispatch of the microgrid's renewable energy park.

Keywords: multi-energy complementary; microgrid; demand response; operation optimization; electricity price

---

1. Introduction

Energy is the basis of human existence and development. Human beings are facing increasingly serious energy shortage and environmental damage, and the development of clean energy is an inevitable trend of social progress. At present, most of the renewable energy is generated by distributed generation (DG), with the widespread application of distributed generation technology, the grid-connected generation of distributed generation and other issues are gradually prominent. As microgrid technology provides a new technical approach for the large-scale application of renewable energy grid-connected power generation, the optimal operation of a microgrid with demand response (DR) has attracted more and more attention [1].

In recent years, many institutions have studied the microgrid, these studies include the access and control technology of distributed power supply, energy efficiency management and economic optimization operation of microgrid and so on. Guo et al. built a model to supply electricity for residents living in the remote and less developed areas, which includes a power grid extension mode and a microgrid mode [2]. Niu et al. proposed a multi-objective optimal energy management framework for the integrated electrical and natural gas network (IEGN) with combined cooling, heat, and power (CCHP) plants [3]. Hossain E et al. introduce the concept and progress of microgrid technology, and used a storage-based load side compensation technique to enhance stability of microgrids [4,5]. Emmanuel et al. identified various ways national green computing campaigns can be carried out in Africa's sociocultural context, and applied a metaheuristic algorithm to the stochastic optimization problem to search for the best-known green computing awareness creation solution [6]. Zhang et al. proposed the composition, tasks and flow of the energy efficiency management system, and described the mathematical modeling method of economic dispatch and optimal operation [7]. Zhao et al. introduced the general structure of a home network energy management system based on smart grid, and proposes an effective home power dispatching method [8]. Jin et al. establishes a multi-objective optimization with principal constraints through a large-scale MG model with flexible loading, which leads to the derivation of a strategy containing uncertainty, and takes a real project to evaluate the uncertainty and demand response potential [9]. Brearley et al. attempt to reexamine the basic concepts of a microgrid and to study the issues and various protection strategies a microgrid faces in protecting the environment [10]. Meghwani et al. presented a noncell protection scheme that uses a locally measured DC microgrid (DCMG) and discusses a threshold calculation method for protection schemes, and it is validated on ring DCMG architectures with different conditions [11].

The optimal dispatching model for a stand-alone microgrid is of great importance to its operation reliability and economy. Wang et al. aimed at addressing the difficulties in improving the operational economy and maintaining the power balance under uncertain load demand and renewable generation, and propose a new two-time scale multi-objective optimization model to optimize the operational cost of the microgrid based on an efficient microgrid energy market [12]. Okoye C O et al. have carried on the thorough research in the microgrid modeling aspect, and established the operation model of microgrid, which has been widely used [13–15]. Saffari M et al. focus on the study of microgrid operation optimization strategy, established the operation optimization model and different constraints; their research results have a significant role in promoting the research in this field [16,17].

The development of smart grid provides powerful technical support for demand response, it is an important technical means of demand side management (DSM). Under the demand response mechanism, users can respond to the price or incentive signal, and change the normal power consumption mode to optimize power consumption and Increase the efficiency of the use of system resources. As studied by Xiao L et al., an improved method of wind energy utilization based on demand response is proposed, and an opportunity constraint decision model of wind power utilization is established to obtain the optimal solution of demand response resource scheduling [18,19]. The results show that demand response can promote photovoltaic power generation (PV) consumption and realize the economic operation of microgrid. Yu et al. propose a two-stage, robust optimization-based model for coordinated investment of DG and DRF, aiming at accommodating the uncertainties of renewables and load demand [20]. Pan et al. propose an electric vehicle (EV) operation strategy based on electricity price incentive policy and establishes the economic dispatch model to realize stable operation of EV when joining in the power grid [21]. Li Yuan et al. present a method of setting gear based on demand response time-sharing ladder price, and obtains the optimal gear of right price to analyze the response effect and energy-saving benefit [22]. Fan et al. propose an optimized operation model for the microgrid on the user side, which includes the zoning strategy of real-time electricity price and the control strategy of controllable load, and the results show that the feasibility and effectiveness of the optimization model to provide theoretical support for low-cost operation on the user side [23]. Gao et al. analysis the benefits, drivers, and barriers of DR, and the status of international DRS was discussed through a broad review of existing projects in different regions [24]. Lu Xiaonan et al. proposes an energy sharing model based on demand response price for the microgrids of PV manufacturers [25]. Abdelaziz et al. conduct a comprehensive assessment of DR barriers and "sociotechnical-economic" in the context of smart grid, and discusses the contributing factors and the complexity of the energy system as well as the features [26]. Besides, a model of residential energy system operation optimization based on price response is proposed, and a new load clustering method is adopted to disperse the response capacity of residential areas [27].

In summary, a demand response program is one of the key technologies in demand side management; it can promote the development of renewable energy power generation technology, and it is also conducive to the economic operation of a microgrid. In the process of a microgrid operation, the operational cost and the better electricity services have always been important issues of concern to users, and electricity price policies and incentive policies also have received extensive attention. Therefore, a demand response program and grid operation characteristics need to be considered together in the process of microgrid operation optimization research. Compared with the existing research on demand response and microgrid operation optimization, the existing research has made some contributions in these two aspects, however, few articles combine these two aspects for operation optimization research, and existing articles have not studied the overall system operation and the system internal load response characteristics under the premise of considering the demand response mechanism and user response behavior. To fill this gap, this paper presents a multi-energy complementary operation model of a microgrid with PV, electric energy storage (EES) and CCHP considering the multi-period electricity price response strategy. In the work presented, a demand response mechanism of price is considered to establish the optimal operating model of a microgrid, the optimization model can achieve the goal of minimizing the operating cost of a microgrid, and the optimal coordination among various loads, renewable energy power generation, electric energy storage system and user comfort are considered in the modeling process. In addition, a microgrid demonstration project in Suzhou is used to verify the role of the coordinated optimization model in responding to grid scheduling and controlling user power costs in this paper.

In the result of this paper, firstly the structure and composition of a microgrid and a multi-period electricity price DR model are analyzed. Secondly, the modeling process of microgrid operation optimization is described in detail, and the operation strategy of a microgrid under different scenarios is introduced in Section 3. After these, simulation process and a result analysis of the model are described in Section 4. Lastly, Section 5 summarizes the important conclusions drawn in this study.

2. Smart Microgrid DR Model

Microgrid is a small regional power generation, which contains clean energy power, electric energy storage device, electronic device, load and automatic demand response system [28]. Microgrid system can realize self-control, protection and management, and it can run not only in the grid-connected operation grid, but also under the stand-alone operation. Microgrid is a concept of the traditional large power grid, which is a network composed of multiple distributed power sources and loads, and it is connected to the conventional power grid through static switches. With the development of intelligent power technology, the development of a microgrid can promote the large-scale access of DG and renewable energy, and realize high-reliability supply of various energy forms of load. In addition, it is an effective means to realize the popularization of smart grid, and lead the transition from traditional power grid to smart grid.

In this paper, a microgrid including distributed photovoltaic power generation (PV), combined cooling heating and power (CCHP) and electric energy storage (EES) is constructed. And the demand response mechanism is fully considered in the modeling process of the microgrid, which is embodied in Figure 1. In a microgrid system, power generation unit and user unit are two important modules, and the user's demand directly determines the operating condition of microgrid system. When users receive the stimulation of information such as electricity price, policy and the like of the outside world, they can change the power consumption behavior reasonably. At the same time, the operating conditions of each power generation unit and energy storage in a microgrid will change accordingly, which will ultimately lead to the change of system operating costs.

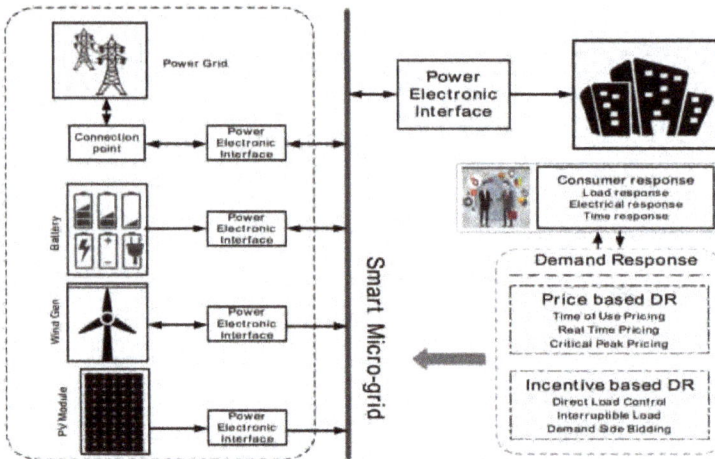

Figure 1. Typical structure of a multi-energy microgrid.

As analyzed above, the implementation of time-of-use (TOU) price has become an effective demand response method in China. Through the response of DR prices, it can not only achieve peak load shifting, but also improve the benefits of a microgrid and the stability of the system [29]. Customers are one of the key elements in demand response. Response behavior and satisfaction of customers directly affect the implementation effect of demand response. Therefore, this paper establishes a demand response model based on multi-period electricity price, considering both response behavior and satisfaction of users.

2.1. Customer Response Behavior Model

In the price-based demand response mechanism, the price factor is sensitive to user behavior. Users respond to changes of prices by adjusting their own electricity consumption behavior, which is embodied in the reduction of electricity consumption. As a special commodity, the change of the price of electric energy will affect the purchase behavior of consumers. At the equilibrium point of the electric power market curve, the electric price k and the electric quantity E are approximately linearly related [30], and the expression is:

$$E = -\omega k + \psi \tag{1}$$

where ω, ψ are the coefficients of the relationship between electricity and electricity prices respectively.

Under the time-of-use price (TOU) of demand response, users respond to change of price by adjusting the electricity structure. Changes in electricity price not only affect the load of its own period, but also affect other periods of load, and it is a multi-period response [31]. This effect can be described by the elasticity of electricity price elasticity:

$$\theta = \frac{\Delta E}{E}\left(\frac{\Delta k}{k}\right)^{-1} \tag{2}$$

where θ is the elastic coefficient of the electricity price; $E, \Delta E$ are the amount of electricity and the amount of electricity; $k, \Delta k$ are electricity price and its variation. According to (2), the change value of the electric quantity after the users responds to the change of the electric price can be obtained. The equation is as follows:

$$\Delta E = \theta \cdot E \cdot \frac{\Delta k}{k} \tag{3}$$

The customer response behavior is the real embodiment of the demand response effect, which can be described comprehensively by establishing the electricity price demand balance relationship and the elasticity matrix of multi-period electricity price.

$$
I = \begin{bmatrix} \theta_{ff} & \theta_{fp} & \theta_{fg} \\ \theta_{pf} & \theta_{pp} & \theta_{pg} \\ \theta_{gf} & \theta_{gp} & \theta_{gg} \end{bmatrix}
\tag{4}
$$

where I is the elasticity coefficient matrix; $\theta_{ff}, \theta_{pp}, \theta_{gg}$ are TOU price elastic coefficient of peak section, flat section and valley section; the rest is the coefficient of cross period.

By combining Equations (1)–(4), we can get a mathematical model of the response price. The demand response model based on electricity price established in this paper is as follows:

$$
E_{TOU} = E_0 + \begin{bmatrix} E_{0,f} & 0 & 0 \\ 0 & E_{0,p} & 0 \\ 0 & 0 & E_{0,g} \end{bmatrix} \cdot \begin{bmatrix} \theta_{ff} & \theta_{fp} & \theta_{fg} \\ \theta_{pf} & \theta_{pp} & \theta_{pg} \\ \theta_{gf} & \theta_{gp} & \theta_{gg} \end{bmatrix} \cdot \begin{bmatrix} \Delta k_f / k_0 \\ \Delta k_p / k_0 \\ \Delta k_g / k_0 \end{bmatrix}
\tag{5}
$$

where E_0 is the amount of power before the Demand Responds; E_{TOU} is the amount of power after the Demand Responds; k_f, k_p, k_g, k_0 are peak section, flat section, valley section of the demand response price and fixed price respectively, $E_{TOU} = \begin{bmatrix} E_f & E_p & E_g \end{bmatrix}^T$ and $E_0 = \begin{bmatrix} E_{0,f} & E_{0,p} & E_{0,g} \end{bmatrix}^T$.

2.2. Customer Satisfaction Model

Customer satisfaction is an important factor to measure the effect of demand response, and it is also a prerequisite for customers to participate in intelligent power program. In the demand response scheduling of a microgrid system, if the satisfaction of customers is not considered, the optimization result may have a bad effect on the attitude of the customers, and it is difficult to achieve the purpose of coordinating the power consumption of customers to achieve the purpose of economic operation of smart grid [32]. Therefore, this paper takes the electricity consumption mode and the change of the electricity expenditure as the evaluation index of customer satisfaction. At the same time, it is used as a constraint in the optimal scheduling model. Customer satisfaction modeling is shown below:

(1) Electricity Satisfaction M_e

$$
M_e = 1 - \frac{\sum\limits_{t=0}^{T} |\Delta E_t|}{\sum\limits_{t=0}^{T} E_t}
\tag{6}
$$

where $\sum\limits_{t=0}^{T} |\Delta E_t|$ is the sum of unsigned values of the amount of power change during the optimization period before and after optimization; $\sum\limits_{t=0}^{T} E_t$ is the total power consumption before optimization.

(2) Electricity expenditure satisfaction M_c

$$
M_c = 1 - \frac{\sum\limits_{t=0}^{T} \Delta C_t}{\sum\limits_{t=0}^{T} C_t}
\tag{7}
$$

where $\sum\limits_{t=0}^{T} |\Delta E_t|$ is the sum of the unsigned values of the amount of change in electricity cost before and after the optimization. $\sum\limits_{t=0}^{T} C_t$ is the total electricity cost before optimization.

3. Microgrid Operation Optimization Model

3.1. Objective Function Optimization

The operation cost of a microgrid is mainly composed of generation cost and environmental cost. The power generation cost is mainly the fuel cost and maintenance cost of the generator set. And the environmental cost mainly includes the emission cost and punishment cost of pollutants in the power generation process. During the operation of a microgrid, due to the stimulation of electricity price and other factors, the output of a nonclean power supply, such as CCHP, in the system will increase and the increase of output will bring electricity sales revenue to the system. At the same time, it will produce certain pollutant emissions. Considering the problems of economic operation and environmental protection, a multi-objective function is established this paper. The target of the objective function is to minimize the cost of power generation and achieve the goal of minimizing the environmental cost. On the basis of the above analysis, an optimization model for the operation of a microgrid is established in this section. The model is mainly oriented to commercial building microgrid with CCHP, PV and EES. In this study, the electrical load of the T period is predicted at each T-1 period, because the turbine engine response time is very fast than the steam turbine, the optimization period is set to 5 min, and the output of each unit for the next 5 min is adjusted during the T-1 period. The model mentioned can achieve the target of minimizing the cost of a microgrid system under the common constraints of operating conditions and environmental factors, and the objective function is as follows:

Objective function I:

$$F_1 = \min C_{op} = \int_0^T \left(C_{pv} P_{pv} + \frac{P_{fuel} V}{60} + C_{bat,dep} + C_{B-grid} \right) \tag{8}$$

$$C_{B-grid} = \begin{cases} k_B P_{grid} & , P_{grid} > 0 \\ 0 & , P_{grid} = 0 \\ k_S P_{grid} & , P_{grid} < 0 \end{cases} \tag{9}$$

where C_{op} is the operation cost of microgrid system, yuan; C_{pv} is the power generation cost of the PV system, yuan/kW; P_{fuel} is the price of natural gas in the system, yuan/m^3; V is the volume of natural gas consumed, Nm^3/h; $C_{bat,dep}$ is the operation cost of energy storage, yuan/kWh; K_b is the purchase price, yuan/kWh; K_s is the selling price of microgrid in scheduling time, determined by the tariff policy, yuan/kWh; P_{grid} is the energy exchange power, kW. If $P_{grid} > 0$, purchase power from the grid; If $P_{grid} < 0$, sell power to the grid.

Objective function II:

$$F_2 = \min C_E = \int_0^T \left(\sum_{j=1}^m (\delta_{E,j} Q_j + \zeta_j) \right) \tag{10}$$

where C_E is the environmental costs; $\delta_{E,j}$ is cost of per kilogram pollutant j, yuan/kg; Q_j is the emissions of pollutants, kg; ζ_j is the emission tax on pollutants.

3.2. Constraints

3.2.1. PV Output Constraint

The output power of the photovoltaic power generation system is constrained by the safety of the system, and the PV power constraint ranges are as follows [13]:

$$P_{pv}^{\min}(t) < P_{pv}(t) < P_{pv}^{\max}(t) \tag{11}$$

where $P_{PV}(t)$ is the output power of PV system at time t, kW; $P_{pv}^{max}(t)$ and $P_{pv}^{min}(t)$ are the maximum and minimum values of PV power, kW.

3.2.2. CCHP Output Constraint

Safety and stability are important factors during the operation of CCHP system. In order to ensure that CCHP operates in a safe and stable environment, the output power must remain within the maximum output power, and the power constraints are as follows [14]:

$$P_e^{min}(t) < P_e(t) < P_e^{max}(t) \tag{12}$$

where $P_e(t)$ is the output power of CCHP at time t, kW; $P_e^{max}(t)$ and $P_e^{min}(t)$ are are the maximum and minimum values of CCHP, kW.

3.2.3. Power Balance

In the process of microgrid operation, power is the key factor of system operation. In a microgrid system, power demand mainly comes from the core load and electric equipment, and power supply mainly includes the DG and power. The energy supply and demand in that system should be balanced. The system power balance constraints are as follows [15].

$$\sum_{k=1}^{N} P_{k-i}(t) + P_{gird}(t) + P_{EES}(t) - P_{loss} - P_{core} - P_{load} = 0 \tag{13}$$

$$P_{load} = \sum_{i=1}^{M} P_i(t)x_i(t) \tag{14}$$

where $P_i(t)$ is the power of the adjustable load in the microgrid, kW; P_{core} is the core load power in the system, kW; P_{load} is the total load involved in demand response scheduling in the system, kW; P_{loss} is the load lost in the distribution network, kW; $P_{gird}(t)$ is the total load power value bought by the microgrid system from the grid, kW; $P_{EES}(t)$ is the power of EES at time t, kW.

3.2.4. EES Constraints

EES must ensure its safe and economic operation, and it also stores as much clean power as possible. During the operation of the energy storage system, the state of charge (SOC) and battery life are the key factors that affect the operation of the energy storage system [16].

$$\begin{cases} P_{storage}(t) \leq P_{dis-max} & P_{storage}(t) > 0 \\ |P_{storage}(t)| \leq P_{ch-max} & P_{storage}(t) < 0 \end{cases} \tag{15}$$

$$SOC_{min} < SOC(t) < SOC_{max} \tag{16}$$

where $P_{dis-max}$ is maximum value of discharge power at time t, kW; P_{ch-max} is the maximum value of charging power at time t, kW; SOC_{max} and SOC_{min} are the upper and lower limits of the remaining capacity.

3.2.5. Power Purchase Cost Constraint

Energy cost is one of the important factors that different users must consider when participating in demand response program. Demand response encourages users to participate in load reduction and load transfer plan, under the premise of meeting the user comfort requirements. At the same time, it is necessary to ensure that the system can use distributed power generation as much as possible to reduce the operation cost of microgrid system [17].

$$\sum_{t \in (T_{0,f,p,g})} (k_f + k_p + k_g - k_0)P_d(t)\Delta t \leq 0 \tag{17}$$

where $P_d(t)$ is the total power load of the system; $T_{0,f,p,g}$ is the time period corresponding to the fixed electricity price and the peak-to-valley electricity price; k_f, k_p, k_g, k_0 are peak section, flat section, valley section of the demand response price and fixed price respectively.

3.2.6. Customer Satisfaction Constraints

Customer satisfaction mainly includes two aspects: user satisfaction with electricity consumption and electricity expenditure after participating in the demand response.

$$\begin{cases} M_e > M_{e,min} \\ M_e > M_{e,min} \end{cases} \tag{18}$$

where $M_{e,min}, M_{c,min}$ is the minimum value of satisfaction with electricity mode and the minimum value of satisfaction with electricity expenditure.

3.3. Model Solution Method and Operating Strategy

3.3.1. Operating Strategy

The power sources for the load are DG, grid power supply and energy storage. In order to achieve the purpose of cost savings, DG is taken as the first priority of power supply energy, followed by energy storage batteries, and the last is the power grid power supply. Based on the above three types of microgrid resources, this paper considers the demand response mechanism to coordinate and optimize the distributed, energy storage and response load. A flow chart of the proposed microgrid optimization model is shown in Figure 2.

The main considerations are the power cost of the microgrid (the amount of power supplied by the grid) and the charging and discharging capacity of the energy storage, and the management between the three energy sources is carried out according to the basic strategy as described below:

If $P_{grid} > 0$ DG and EES capacity is much smaller than the load demand of system, the system must purchase some power from the grid to achieve its energy balance. In this situation, the SOC of the energy storage system is considered by the dispatching center firstly. If the remaining power is less than the total load of the system, then the output state of distributed power supply should be included in the scope of scheduling. According to the operating state of the system and market conditions and other factors, the quantity and price of electricity purchased are determined, and the dispatching center issues dispatching instructions to each energy unit in the system.

Load state: The amount of electricity generated by DG and the amount of electricity in the Energy Storage is greater than the load demand.

$$P_{load} - P_{PV}(t) - P_{storage}(t) < 0 \tag{19}$$

Operating Strategy: The system first inputs excess power into the energy storage, then sell electricity to the grid.

$$P_{grid}(t) = P_{PV}(t) - P_{load} - \Delta P_{storage}(t) \tag{20}$$

If $P_{grid} < 0$, DG and EES capacity is much larger than the total load, the system can Sell excess clean power to the grid or other microgrid systems. In view of the above situation, the operation state of energy storage system is first scheduled by the system. Then, the SOC of EES and the charging requirement are judged. Finally, the amount of power that the system sells to the grid is determined and dispatch instructions are issued to the system.

Load state: The amount of electricity generated by the DG and the amount of electricity in the Energy Storage are not enough to meet the load demand.

$$P_{load} - P_{PV}(t) - P_{storage}(t) > 0 \tag{21}$$

Operating Strategy: Consumers buy electricity from the grid.

$$P_{grid}(t) = P_{load} - P_{PV}(t) - P_{storage}(t) \tag{22}$$

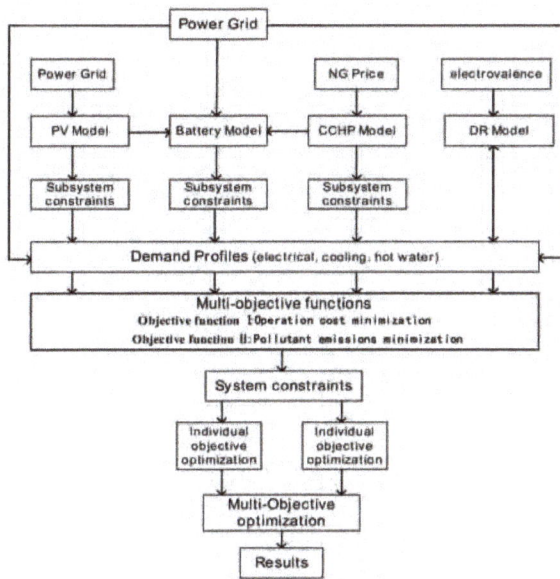

Figure 2. Flow chart of microgrid optimization model.

3.3.2. Model Solution Method

In general, there is a mutual restriction between the objective functions in the multi-objective constraint problem. In the process of multi-objective optimization, the realization of one objective may have a bad effect on other objectives. Therefore, it is difficult to evaluate the advantages and disadvantages of the multi-objective problem solution. Compared with a general optimization problem, the solving method of multi-objective optimization problem is not unique, but exists in the optimal solution set composed of multiple solutions. The set mentioned above is called Pareto, and the element in the set are called Pareto optimal or noninferior optimal [33].

The multi-objective optimization questions can be described as follows:

$$\begin{cases} \min F(z) = \min([f_1(z), f_2(z), f_3(z), \cdots f_n(z),]^T), & s.t. \ z \in \Omega \\ J(z) \leq 0 \\ H(z) = 0 \end{cases} \tag{23}$$

where z is the optimization variable; $f_n(z)$ is the optimization target; Ω is a collection of all possible solutions; $J(z)$ and $H(z)$ are the set of constraints of inequality and equality.

Considering the output of distributed generation and load level, the constraint value for each variable is set. Based on the scheduling strategy of demand response mechanism, the model presented is solved by Non-domimated Sorting Genetic Algorithm II (NSGA-II). In order to verify the performance of NSGA-II, this paper applied it to different scenarios of microgrid scheduling. The population size (NP) is set to 500, and the value of the maximum variation generations is 100, and the constraint processing method Ref. [34]. The flow chart of the NSGA-II proposed is shown in Figure 3.

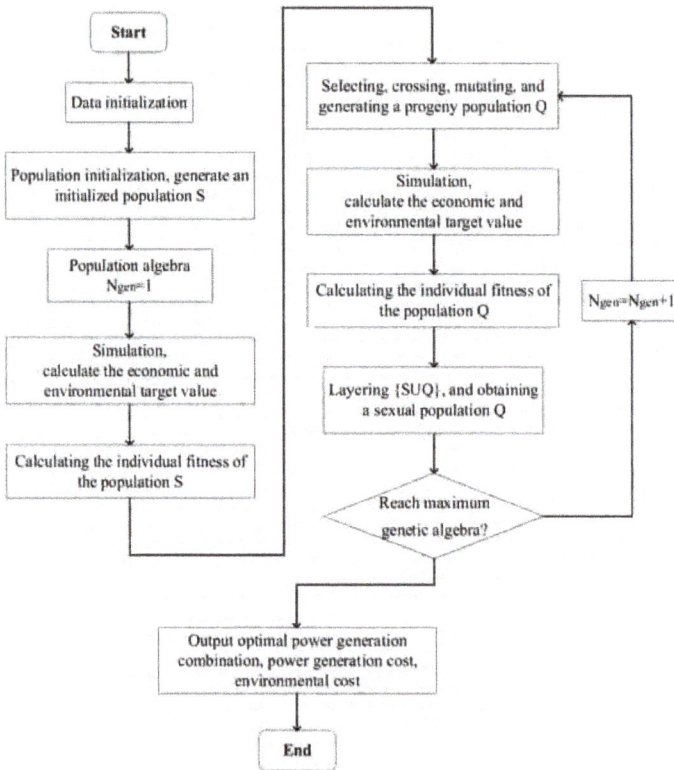

Figure 3. The flow chart of the NSGA-II based on demand response.

4. Case Study

4.1. Parameters

In order to verify the accuracy of the optimization model established in this paper, a typical microgrid including CCHP, PV and EES system is optimized and analyzed in this study. Typical load data for 1440 min of the day is selected for analysis, and the power predicted of CCHP, PV, and loads is shown in the Figure 4. In this microgrid, the capacity of CCHP is 1.5 MW, PV is 400 kW, electric load is 667 kW. And the EES capacity is 300 kWh, charge and discharge efficiency of battery is 90%, the actual cumulative throughput is 1.99×10^7 Ah. The TOU price is shown in the Table 1.

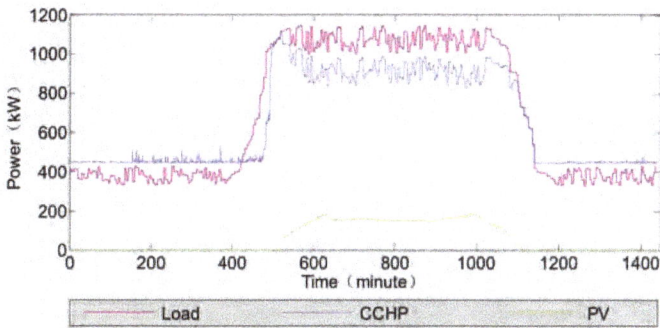

Figure 4. Daily variation curves of micropower and load.

Table 1. Sectional electricity price of demand response.

Time Windows	Price (Yuan/kWh)		
	Low Price	**Medium Price**	**High Price**
	0:00~6:00 18:00~24:00	6:00~10:00 15:00~18:00	10:00~15:00
Purchase	0.5522	0.8185	1.2035
Sell	0.65	0.65	0.65

4.2. DR Study Results

According to the electricity price of the DR policy, the fixed electricity price is 0.9 yuan/kWh, the purchase price and the sale price are TOU price, as shown in Table 1 above. This section sets the parameters of the electricity and electricity price curve based on the elasticity coefficient of electricity and electricity price in Ref. [35]. The parameter is shown in Table 2.

Table 2. Parameters of relation curve between electricity and electricity price.

Time	Low Price	Medium Price	High Price
(ω, ψ)	(6.5,65)	(5.0,60)	(4.0,58)

According to Equations (4) and (5), an electricity and electricity price elasticity matrix M is obtain.

$$I = \begin{bmatrix} -1.0235 & 0.1123 & 0.1058 \\ 0.1206 & -1.0089 & 0.0986 \\ 0.1351 & 0.1209 & -1.0165 \end{bmatrix} \tag{24}$$

According to the demand response model, the load characteristics before and after demand response are obtained (shown in Table 3), and the system load curve is also obtained (shown in Figure 4). Peak valley difference of the system decrease after the demand response is shown in Figure 5.

Table 3. Load characteristics before and after DR.

Situations	Peak of Load/kW	Valley of Load/kW	Peak-Valley Difference of Load/kW	Me	Mc
Before DR	1136.93	330.31	806.62	1	0
After DR	1100.19	326.15	774.04	0.97	0.75

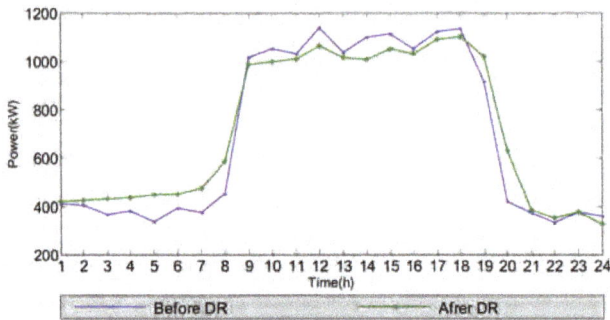

Figure 5. Load curve before and after DR.

According to the calculation results in Table 3, the peak valley difference of the system before the demand response is 806.62 kW, and after the demand response, it is 774.04 kW, the effect of Peak Shaving has been achieved. In addition, after DR, the satisfaction degree of electricity consumption satisfaction M_e is 0.97, and the satisfaction degree of electricity expenditure M_c is 0.75.

Under the demand response mechanism, electricity consumers can respond to the market price signal based on their load demand and preference, and change the power consumption mode and load demand independently. The load mainly conditions of the electric vehicle and the lighting before and after demand response is shown in Figures 6 and 7 respectively.

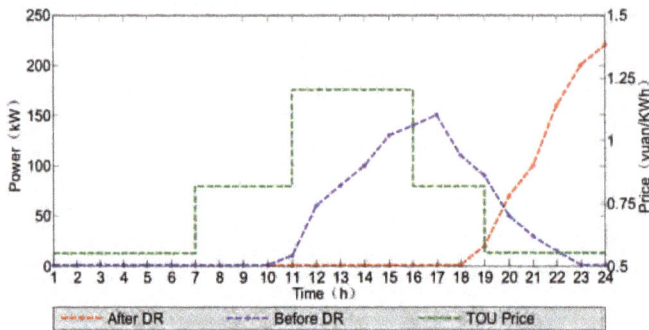

Figure 6. Electric vehicles load before and after DR.

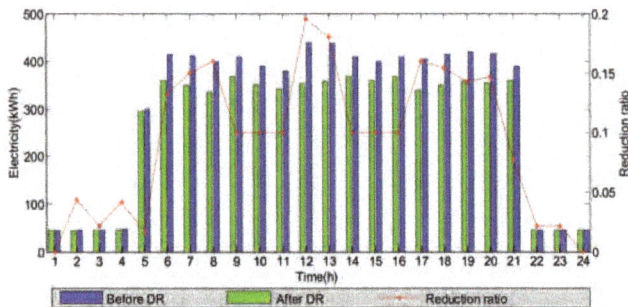

Figure 7. Lighting Load before and after DR.

In Figure 6, the charging time of the electric vehicle is mainly concentrated at 10:00 to 18:00 and 20:00 to 24:00. According to TOU price, users are willing to charge in the low night electricity price period, which can not only meet the needs of vehicles, but also save charging costs [36].

In Figure 7, the load of lighting system is not adjusted before 6:00 and after 21:00 due to the user's habits. Between 6:00 and 21:00, the load reduction of the lighting system reached more than 30%. In addition, the maximum load reduction ratio can reach 10~20% after the implementation of the demand response program during the peak power consumption period.

4.3. Operation Optimization

4.3.1. Scene Analysis

In this section, three scenarios are set up to study the mechanism of demand response project on microgrid economic operation.

Scene 1: The system operates in off-grid microgrid. The power in the system comes from PV, CCHP and EES system, and the DG of the system can meet the power demand of consumers.

Scene 2: The system operates in grid-connected microgrid. Without the introduction of demand response, the major power suppliers in the system are distributed PV system, CCHP system, EES system and grid, and the electricity price is fixed. It is assume that the user preferentially use the power from DG.

Scene 3: The system operates in grid-connected microgrid. With the introduction of demand response, the major power suppliers in the system are distributed PV system, CCHP system, EES system and grid, and the electricity price is TOU price. It is assume that the user preferentially use the power from DG.

4.3.2. Optimization Results

According to the three cases presented above, this section randomly selected typical-day load in the microgrid for operation optimization, the operation results are shown in Figures 8–10.

Figure 8 shows the output of the different unit in Scene1.In this scene, the system does not exchange energy with the grid, and the power demand of the system is provided by combined energy supply system composed of PV, CCHP and EES. The operation cost of the system is mainly DG cost and charging and discharging cost of EES.

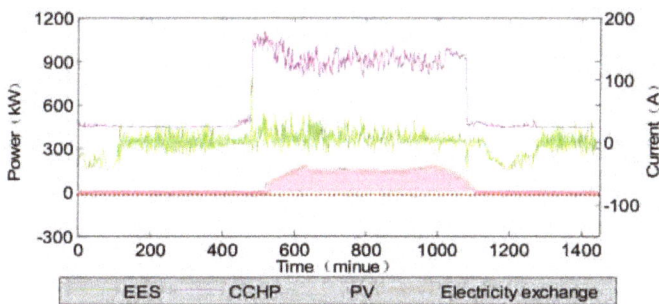

Figure 8. The output of micropower in Scene 1.

Figure 9 shows the output of the units in Scene 2. In this state, the consumers first use DG power and energy storage. The surplus electricity is delivered to grid, the insufficient electricity is purchased from grid, and the electricity purchase price is the fixed price.

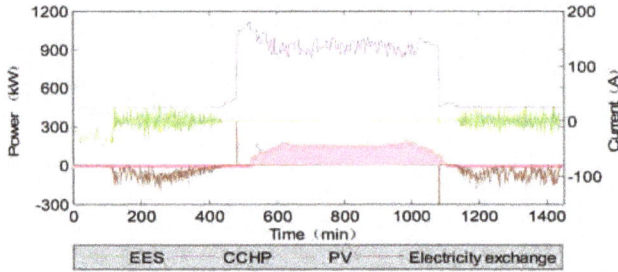

Figure 9. The output of micropower in Scene 2.

Figure 10 shows the output of different unit in Scene 3. In this state, because of the interval distribution of electrovalence, the daytime price is higher, the power of CCHP and the output of ES and discharge in the system are improved compared with scenario 2 in order to reduce the cost of electricity. And the output of each unit of the system is more sensitive to electricity prices, and the output of each power supply and the change of electric quantity exchange are more obvious.

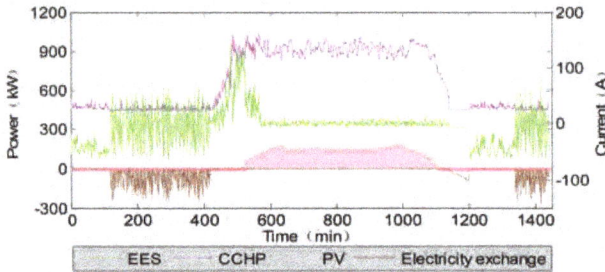

Figure 10. The output of micropower in Scene 3.

4.3.3. Optimization Process Analysis

In the microgrid system, the PV power is a clean power. However, the CCHP system drives the gas turbine to turn through the consumption of gas to generate electricity and heat, which produces a large amount of pollutants during the combustion process. In addition, electricity is mainly derived from thermal power generation in China. In the context of environmental issues, the operation of microgrid should not only consider the economy, but also consider whether it is environmentally friendly. Therefore, there is Pareto with both economic and environmental protection.

In addition to the above discussion, microgrid economic/environmental multi-objective scheduling problem is often faced with complex situations such as operation mode switching, electricity price change and so on, so the algorithm should have better adaptability to the problem. In this section, three scenarios are simulated respectively: TOU price mode, fixed price mode and isolated grid mode. And the extreme solution optimal target value, average target value and algorithm operation time in Pareto solution set are compared. The optimization results of three schemes in wardrobe optimization period (5 min) are shown in Table 4 and Figure 11.

Table 4. Extreme solutions and the average operation time under different operation modes of the microgrid.

Scenarios	Optimization Result/Yuan		NSGA-II
Scene 1: TOU price	Generation cost	optimal value	46.980
		Average value	47.206
	Environmental cost	optimal value	1.246
		Average value	1.255
Scene 2: fixed price	Generation cost	optimal value	45.450
		Average value	47.268
	Environmental cost	optimal value	1.538
		Average value	1.541
Scene 3: isolated grid mode	Generation cost	optimal value	50.4
		Average value	51.685
	Environmental cost	optimal value	1.325
		Average value	1.336
Optimized time records/s		Average value	19.675

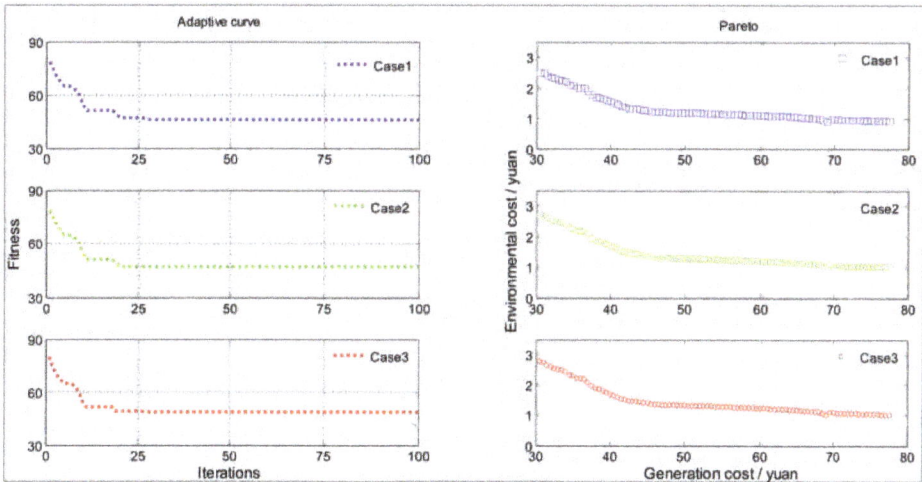

Figure 11. Adaptive curve and Pareto optimization frontier of NASG-II.

With initial data and parameters, the proposed NSGA-II algorithm is used to solve the operation optimization problem with considering a variety of constrains. For proving the effectiveness of the improvement of NSGA-II algorithm, the algorithm is used in the operation optimization of three different microgrid operating states, and the fitness value and Pareto solvable set are also recorded in detail. It is clear that when the algorithm is run in scenario 1, the energy cost decreases until the algorithm converges after about 34 iterations, and the total running time of the program is about 18.957 s within the interval T. Similar trends can be revealed in the other two scenarios. The minimum value of each population variation generation is recorded, and the fitness curve in different situation is generated iteratively, as shown in Figure 11.

The points in Figure 11 represent the solutions distributed within the target space as the algorithm evolves to the last generation. As can be seen from Table 4 and Figure 11, the optimization of NSGA-II under different operation modes is ideal and can better reflect the preferences of decision makers. At the same time, the sub-solution set corresponding to each weight has certain diversity.

4.3.4. Optimization Result Analysis

According to the objective function presented in this paper and the three scenarios set by the work presented, we can get the cost composition of the system under different operating conditions. For the multi-energy complementary microgrid system established in this paper, the operating cost of the system is mainly composed of power generation cost and environmental cost. The cost of power generation mainly includes the operation and maintenance cost of the generating unit and the cost of fuel. The environmental cost is mainly the environmental cost due to the discharge of the pollutants. Table 5 shows system cost structure under different operating conditions.

Table 5. System cost structure under different operating conditions.

Cost/Yuan	Generation Cost/Yuan					Environmental Cost/Yuan	Operation Cost/Yuan
	CCHP	PV	EES	Electricity Exchange			
				Purchase	Sell		
Scene 1	12,600	1120	800	0	0	381.3	14,901.3
Scene 2	12,260	1120	770	250	−870	359.9	13,889.9
Scene 3	12,510	1120	820	240	−1600	443.7	13,533.7

According to Table 4 and optimization results, when the system operates in off-grid, the output of PV system can't meet the requirements of the system, and the CCHP and the EES are required to supply power to the system, so that the operation cost is high. When the system operates in grid-connected, the system sells excess power in response to changes in electricity prices and to obtain economic benefits, so that the operation cost is lower than that in the off-grid operation. After the DR program, the power from the grid has little change, but the power delivered to grid increases obviously. The amount of electricity sold per day is up to 583.051~803.302 kWh. In the microgrid with PV, ES and CCHP, the operation cost mainly comes from the operation and maintenance cost and fuel cost of the system. The composition of the system cost in different scenarios is shown in Figure 12.

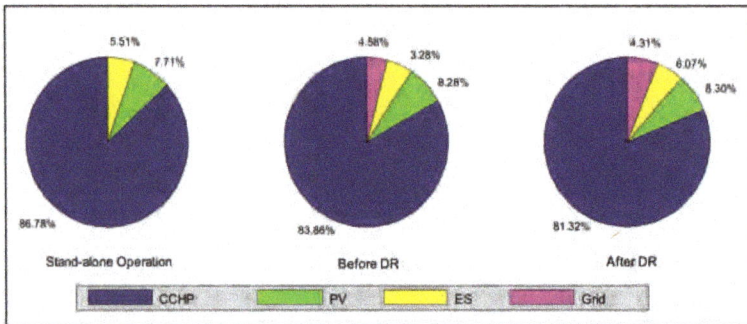

Figure 12. Composition of the system cost in different scenarios.

The cost components of the system show significant differences under three operating conditions. As far as the overall trend is concerned, the operating cost of the CCHP system accounts for more than 80% of the total cost of the system, which directly determines the operating effect of the system. And the direct reason for this result is that the price of natural gas fluctuates greatly with the market and the price of industrial gas is higher in China. In addition, the proportion of system cost has also changed significantly with the implementation of demand response strategy.

5. Conclusions

In the work presented, the operation optimization of a multi-energy complementary microgrid containing DG and EES is conducted. Besides these, a demand response model and the optimization model of a microgrid based on TOU price are established, and some research results are obtained. On the one hand, the coordinated optimization strategy of a multi-energy complementary microgrid is designed in this paper, which realizes the optimization of load distribution and system cost minimization in space domain. On the other hand, this paper establishes an integral optimization objective function considering the demand response and user satisfaction constraints, which has a great promotion effect on the economical and efficient operation of the system with the demand response strategy. In addition, on the basis of the operation optimization model presented, this paper makes a more reasonable scheduling for microgrid operation, the building of a microgrid operation, and the optimization results that the optimization strategy can reduce the power consumption of building.

Acknowledgments: This work was supported by "the Fundamental Research Funds for the Central Universities" (2018ZD13).

Author Contributions: All the authors gave equal contributions in writing the paper.

Conflicts of Interest: The authors declare no conflict of interest.

Nomenclature

The parameters in this article are described below.

C_{op}	the operation cost of system
P_{fuel}	System gas price, yuan/m^3
I	PV current
R_{sh}	the equivalent parallel impedance
J	curve fitting parameter
I_{ph}, I_d	the current generated by the light and the reflected missing current
D_A	the actual depth of discharge
C_A	actual capacity
ω, ψ	the coefficients of the relationship between electricity and electricity prices
$E, \Delta E$	the amount of power and its change
M_e	Electricity satisfaction
$P_e(t)$	the output power of the CCHP, kW
C_{B-grid}	the electricity cost of the microgrid during the Scheduling period, yuan/kWh
$\eta_e(t)$	the power generation efficiency of the CCHP
η_l	the heat loss coefficient of the CCHP
P_{grid}	the energy exchange power, kW
P_{core}	the core load power in the system/kW
P_{loss}	the load lost in the distribution network/kW
SOC_{max}	the upper limit of the remaining capacity/kW
$P_{ch,max}$	the maximum value of charging power at time t, kW
$P_{dis,max}$	maximum value of discharge power at time t, kW
$M_{c,min}$	the minimum value of satisfaction with electricity expenditure
$P_{DG,max}$	the maximum power of DG
C_i	the Generation cost, yuan/kW
R_s	the equivalent series impedance
q	the electronic charge
k	boltzmann constant
L_R	the cycle number of stored energy under rated discharge depth and rated discharge current
$d_{act,t}$	the actual discharge current ampere hours in the unit of time
C_{bat}	the initial investment cost of energy storage

θ	the elastic coefficient of the electricity price
$k, \Delta k$	the price and its variation
M_c	Electricity expenses satisfaction
P_i	the output of DG during the Scheduling period, kw
$C_{bat,dep}$	the operation cost of energy storage, yuan/kWh
K_b	the purchase price, yuan/kWh
K_s	is the selling price of microgrid in scheduling time, yuan/kWh
$P_{PV}(t)$	the output power of PV system/kW
$P_i(t)$	the power of the adjustable load in the microgrid/kW
P_{load}	the total load involved in demand response scheduling in the system/kW
$P_{EES}(t)$	the power of EES at time t/kW
$P_d(t)$	the total power load of the system
$T_{0,f,p,g}$	the time period corresponding to the fixed electricity price and the peak-to-valley electricity price
$M_{e,min}$	the minimum value of satisfaction with electricity mode

References

1. Zhu, X. *Research and Application of Key Technology Based on PCS Intelligent Micro-Grid System*; Anhui University: Hefei, China, 2015.
2. Guo, S.; Zhao, H.; Zhao, H. The Most Economical Mode of Power Supply for Remote and Less Developed Areas in China: Power Grid Extension or Micro-Grid? *Sustainability* **2017**, *9*, 910. [CrossRef]
3. Ming, N.; Wei, H.; Guo, J.; Su, L. Research on Economic Operation of Grid-Connected Micro-grid. *Power Syst. Technol.* **2010**, *34*, 38–42.
4. Hossain, E.; Perez, R.; Padmanaban, S.; Mihet-Popa, L.; Blaabjerg, F.; Ramachandaramurthy, V.K. Sliding Mode Controller and Lyapunov Redesign Controller to Improve Microgrid Stability: A Comparative Analysis with CPL Power Variation. *Energies* **2017**, *10*, 1959. [CrossRef]
5. Yuan, Y.; Li, Z.; Feng, Y.; Zuo, W. Development purposes, orientations and prospects of Micro-grid in China. *Auto. Electr. Power Syst.* **2010**, *34*, 59–63. (In Chinese)
6. Okewu, E.; Misra, S.; Maskeliūnas, R.; Damaševičius, R.; Fernandez-Sanz, L. Optimizing Green Computing Awareness for Environmental Sustainability and Economic Security as a Stochastic Optimization Problem. *Sustainability* **2017**, *9*, 1857. [CrossRef]
7. Zhang, J.; Su, L.; Chen, Y.; Jing, S.U.; Wang, L. Energ management of Micro-grid and its control strategy. *Power Syst. Technol.* **2011**, *35*, 24–28. (In Chinese)
8. Zhao, Z.; Lee, W.C.; Shin, Y.; Song, K.B. An Optimal Power Scheduling Method Applied in Home Energy Management System Based on Demand Response. *ETRI J.* **2013**, *4*, 1391–1400. [CrossRef]
9. Jin, M.; Feng, W.; Liu, P.; Marnay, C.; Spanos, C. MOD-DR: Micro-grid optimal dispatch with demand response. *Appl. Energy* **2017**, *187*, 758–776. [CrossRef]
10. Brearley, B.J.; Prabu, R.R. A review on issues and approaches for Micro-grid protection. *Renew. Sustain. Energy Rev.* **2017**, *67*, 988–997. [CrossRef]
11. Meghwani, A.; Srivastava, S.; Chakrabarti, S. A Non-Unit Protection Scheme for DC Micro-grid Based on Local Measurements. *IEEE Trans. Power Deliv.* **2017**, *32*, 172–181. [CrossRef]
12. Wang, F.; Zhou, L.; Ren, H.; Liu, X. Search Improvement Process-Chaotic Optimization-Particle Swarm Optimization-Elite Retention Strategy and Improved Combined Cooling-Heating-Power Strategy Based Two-Time Scale Multi-Objective Optimization Model for Stand-Alone Microgrid Operation. *Energies* **2017**, *10*, 1936. [CrossRef]
13. Okoye, C.O.; Oranekwu-Okoye, B.C. Economic feasibility of solar PV system for rural electrification in Sub-Sahara Africa. *Renew. Sustain. Energy Rev.* **2018**, *82*, 2537–2547. [CrossRef]
14. Igualada, L.; Corchero, C.; Cruz-Zambrano, M.; Heredia, F.J. Optimal Energy Management for a Residential Microgrid Including a Vehicle-to-Grid System. Available online: https://upcommons.upc.edu/handle/2117/20642 (accessed on 25 March 2018).

15. Li, W.T.; Yuen, C.; Hassan, N.U.; Tushar, W.; Wen, C.K.; Wood, K.L.; Hu, K.; Liu, X. Demand Response Management for Residential Smart Grid: From Theory to Practice. Available online: http://xueshu.baidu.com/s?wd=paperuri%3A%28c31a841b8e868b2ef596e1524f3b5937%29&filter= sc_long_sign&tn=SE_xueshusource_2kduw22v&sc_vurl=http%3A%2F%2Fieeexplore.ieee.org% 2Fdocument%2F7336481%2F&ie=utf-8&sc_us=15685181391879326455 (accessed on 25 March 2018).

16. Saffari, M.; Gracia, A.D.; Fernández, C.; Belusko, M.; Boer, D.; Cabeza, L.F. Optimized demand side management (DSM) of peak electricity demand by coupling low temperature thermal energy storage (TES) and solar PV. *Appl. Energy* **2018**, *211*, 604–616. [CrossRef]

17. Wang, P.; Wang, W.; Meng, N.; Liu, H.; Feng, H.; Xu, D. Optimal sizing of DC micro-grids based on comprehensive evaluation of operating modes and operating targets. *Power Syst. Technol.* **2016**, *40*, 741–748.

18. Xin, A.I.; Xiao, L. Chance constrained model for wind power usage based on demand response. *J. North China Electr. Power Univ.* **2011**, *3*, 4.

19. Zhang, D.; Tong, Y.; Jin, X.; Liang, J. Optimal energy storage configuration of Micro-grid considering demand response. *Power Electr.* **2016**, *50*, 107–109.

20. Yu, Y.; Wen, X.; Zhao, J.; Xu, Z.; Li, J. Co-Planning of Demand Response and Distributed Generators in an Active Distribution Network. *Energies* **2018**, *11*, 354. [CrossRef]

21. Pan, Z.; Gao, C. Economic dispatch of electric vehicles based on demand response. *Electr. Power Constr.* **2015**, *36*, 139–145. (In Chinese)

22. Li, Y.; Luo, Q.; Song, Y.Q.; Xu, J.; Cai, L.; Gu, J. Study on tiered level determination of TOU & tiered pricing for residential electricity based on demand response. *Power Syst. Prot. Control* **2012**, *40*, 65–74. (In Chinese)

23. Fan, W.; Zhou, N.; Liu, N.; Lin, X.; Zhang, J.; Lei, J. Multi-objective optimization operation method of user -side Micro-grid based on demand response. *Power Syst. Clean Energy* **2016**, *32*, 17–23. [CrossRef]

24. Gao, C.; Li, Q.; Li, H.; Zhai, H.; Zhang, L. Integration method and operation mechanism of demand response resources based on load aggregator business. *Auto. Electr. Power Syst.* **2013**, *37*, 78–86.

25. Lu, X.; Sun, K.; Huang, L.; Xiao, X.; Guerrero, J.M. The improved droop control method of distributed energy storage system in islanding AC Micro-grid. *Electr. Power Syst. Auto.* **2013**, *37*, 80–185. [CrossRef]

26. Abdelaziz, M.M.A.; Farag, H.E.; El-Saadany, E.F. Optimum Reconfiguration of Droop-Controlled Islanded Micro-grids. *IEEE Trans. Power Syst.* **2016**, *31*, 2144–2153. [CrossRef]

27. Paterakis, N.G.; Erdinç, O.; Catalão, J.P.S. An overview of Demand Response: Key-elements and international experience. *Renew. Sustain. Energy Rev.* **2017**, *69*, 871–891. [CrossRef]

28. Mengelkamp, E.; Gärttner, J.; Rock, K.; Kessler, S.; Orsini, L.; Weinhardt, C. Designing micro-grid energy markets. *Appl. Energy* **2018**, *210*, 870–880. [CrossRef]

29. Zhou, N.; Fan, W.; Liu, N.; Zhang, J.; Lei, J. Battery storage multi-objective optimization for capacity configuration of PV-based Micro-grid considering demand response. *Power Syst. Technol.* **2016**, *40*, 1709–1716.

30. Alamaniotis, M.; Tsoukalas, L.H. *Integration of Price Anticipation and Self-Elasticity for Purchase Decisions and Planning in Price Directed Electricity Markets*; IET: Stevenage, UK, 2016.

31. Chen, C.; Hu, B.; Xie, K.; Wan, L.; Xiang, B. A peak-valley TOU price model considering power system reliability and power purchase risk. *Power Syst. Technol.* **2014**, *38*, 2141–2148.

32. Bie, Z.; Hu, G.; Xie, H.; Gengfeng, L. Optimal dispatch for wind power integrated systems considering demand response. *Autom. Electr. Power Syst.* **2014**, *38*, 115–120. [CrossRef]

33. Khierkhah, A.S.; Nobari, A.; Hajipour, V. A Pareto-based approach to optimise aggregate production planning problem considering reliable supplier selection. *Int. J. Serv. Oper. Manag.* **2018**, *29*, 59. [CrossRef]

34. Sun, Y.; Lin, F.; Xu, H. Multi-objective Optimization of Resource Scheduling in Fog Computing Using an Improved NSGA-II. *Wirel. Pers. Commun.* **2018**, *1*, 1–17. [CrossRef]

35. Liu, W.; Niu, S.; Xu, H. Optimal Planning of Battery Energy Storage Considering Reliability Benefit and Operation Strategy in Active Distribution System. Available online: https://www.researchgate.net/publication/301664508_Optimal_planning_of_battery_energy_storage_considering_reliability_benefit_and_operation_strategy_in_active_distribution_system (accessed on 25 March 2018).

36. Chen, X.; Yin, J.; Wang, W.; Wu, L.; Tang, F. Approaches to diminish large unsprung mass negative effects of wheel side drive electric vehicles. *J. Adv. Mech. Des. Syst. Manuf.* **2016**, *10*, JAMDSM0064. [CrossRef]

![energies logo] *energies*

MDPI

Article

State-of-Charge Balancing Control of a Modular Multilevel Converter with an Integrated Battery Energy Storage

Hui Liang [1,2], Long Guo [3], Junhong Song [1,2,*], Yong Yang [1,2], Weige Zhang [1,2] and Hongfeng Qi [4]

[1]   National Active Distribution Network Technology Research Center (NANTEC), Beijing Jiaotong University, Beijing 100044, China; hliang@bjtu.edu.cn (H.L.); 16126062@bjtu.edu.cn (Y.Y.); wgzhang@bjtu.edu.cn (W.Z.)
[2]   Collaborative Innovation Center of Electric Vehicles in Beijing, Beijing Jiaotong University, Beijing 100044, China
[3]   Huawei Technologies Co., Ltd., Shenzhen 518000, China; guolong5@huawei.com
[4]   CRRC Industrial Institute Co., Ltd., Beijing 100071, China; qihongfeng@crrcgc.cc
*   Correspondence: 15121466@bjtu.edu.cn; Tel.: +86-188-1070-0721

Received: 5 March 2018; Accepted: 4 April 2018; Published: 9 April 2018

Abstract: With the fast development of the electric vehicle industry, the reuse of second-life batteries in vehicles are becoming more attractive, however, both the state-of-charge (SOC) inconsistency and the capacity inconsistency of second-life batteries have limits in their utilization. This paper focuses on the second-life batteries applied battery energy storage system (BESS) based on modular multilevel converter (MMC). By analyzing the power flow characteristics among all sources within the MMC-BESS, a three-level SOC equilibrium control strategy aiming to battery capacity inconsistency is proposed to balance the energy of batteries, which includes SOC balance among three-phase legs, SOC balance between the upper and lower arms of each phase, and SOC balance of submodules within each arm. In battery charging and discharging control, by introducing power regulations based on battery capacity proportion of three-phase legs, capacity deviation between the upper and lower's arm, and the capacity coefficient of the submodule into the SOC feedback control loop, SOC balance of all battery modules is accomplished, thus effectively improving the energy utilization of second-life battery energy storage system. Finally, the effectiveness and feasibility of the proposed methods are verified by results obtained from simulations and the experimental platform.

Keywords: modular multilevel converter; battery energy storage system; state-of-charge balancing; second-life battery

---

1. Introduction

With the fast-growing commercial application of electric vehicles, there will be a substantial increase of the batteries retired from these vehicles, leading to a great waste of resources if the batteries are directly thrown out. By expanding the useful life of these retired batteries for second use, the total battery life cycle cost can be easily reduced and the utilization of the battery could also be greatly improved [1], which is of great significance to promote the replaceable developments of the electric vehicle industry. The most economical way of reusing second-life batteries is the battery energy storage system (BESS). In conventional battery energy storage systems, a large number of batteries are connected in series or in parallel in a battery pack, which requires a higher battery consistency in practical applications. However, due to the high capacity inconsistency and high cost of module reconstruction in second-life batteries, the large number of series/parallel applications and "short-board" effect will reduce the total capacity utilization of the energy storage system, which will affect the energy and capacity utilization efficiency.

Flexible group technology is an effective method to solve the problem of the high battery inconsistency [2,3]. Different from the conventional battery group composed of a large number of single batteries directly connected in series and in parallel, the flexible group energy storage system is consist of cascaded submodules combining the low-voltage battery pack with converters. Charging or discharging the current of each battery module is controlled independently based on the state parameters, effectively reducing the requirement of battery capacity consistency and the cost of regrouping. Thus, the capacity utilization efficiency and cycle life of batteries can be improved while meeting the requirements of energy storage systems. Consequently, the efficient utilization of the retired power batteries is realized.

Various topologies can be used in flexible group energy storage system [4,5]. In the application where power flows among the AC grid, the DC bus and battery, the MMC-BESS has its superior advantages of overcoming the "short-board" effect. By dispersedly connecting the low-voltage battery pack to the DC side of each submodule, this topology combines the merits of both MMC and BESS, which is suitable for hybrid AC/DC micro-grids and high-voltage direct current (HVDC) power system. Meanwhile, advanced modeling [6–8], control systems, and modulation [9,10] has developed MMC greatly. The battery capacity utilization of the whole MMC-BESS is limited by the submodule with the highest or lowest state of charge (SOC), therefore the SOC equilibrium control becomes the essential part of improving battery capacity utilization. Since SOC is directly related to battery capacity, the capacity inconsistency can easily result in divergent real-time SOC. When second-life batteries are widely used in battery packs in MMC-BESS, in addition to capacity inconsistency of batteries in the same arm, the total battery capacity between the upper and lower arm as well as the total capacity among different phases are also inconsistent, leading to greater SOC inconsistency at each level of battery modules. Thus, the conventional SOC equilibrium control strategy has limited applications and new control methods are in great urge.

SOC balancing control and fault-tolerant control are essential for the MMC-BESS to improve the efficiency and reliability of capacity utilization. In Reference [11], the zero-sequence voltage injection method is able to balance the SOC among different phases, however, the calculation of zero-sequence voltage injection involves complex mathematical calculation, leading to higher requirement of control hardware. By sorting the SOCs of all submodules, SOC balancing control can be realized using the carrier-based disposition pulse width modulation (PWM) method [12]. However, the complexity increases dramatically with the increase of the number of submodules. Reference [13] proposed a simple and easy closed-loop method to achieve SOC balancing among submodules within an arm and phase legs, while the SOC balancing problem between the upper and lower arms is not under full consideration. Some literature focuses on the MMC-BESS applied in vehicles, in which using AC-circulating current to balance SOC between the lower and upper arms; the current only contain positive sequence and negative sequence to protect the current from flowing to the DC source [14]. As SOH can also be used to improve utilization of battery, the author adopted dc and ac circulating current as well as modulation index of each submodule to achieve the tracking of SOC, thus effectively improving the cycle time of battery system [15]. In [16], the capacity energy in both upper and lower arm can be controlled by adjusting the circulating current after bypassing the fault submodule, resulting in the SOC rebalancing even under fault operation. Reference [17] focuses on the hybrid MMC energy storage system consisting of half-bridge and full-bridge topologies, which highly integrating different voltage and current injection methods for both interphase and intra-phase SOC equalization. Although various SOC equilibrium control methods were proposed in the previous literature, the impact of capacity inconsistency has not been fully considered. In the condition where the inconsistency index goes higher, the control error may turn greater, resulting in lower battery capacity utilization.

To overcome the shortage of conventional SOC equalization methods under the operation of battery capacity inconsistency, after the analysis of the power transfer relationship of MMC-BESS, this paper proposes a three-level SOC balancing control strategy. The SOC closed-loop control strategy is implemented to adjust the power command from phase level to each submodule, and then regular both the DC circulating current and AC current. By respectively adjusting the phase power and arm power, the power of submodules can be reconfigured. To solve the battery capacity inconsistency problem, this paper proposed a novel control method based on power regulations and SOC equalization control to synchronously converge the SOC equilibrium among different battery packs with various battery capacity, and eventually achieves the goal of the same SOC of all battery modules of MMC-BESS, effectively improving the utilization of second-life batteries. Both the simulation model and experimental platform of a three-phase 24-module energy storage system have been established to verify the effectiveness of the proposed control strategy.

2. Operating Principles

2.1. Topology and Modulation Strategy

The schematic diagram of the MMC-BESS is shown in Figure 1. Three-phase legs are connected in parallel to a common DC grid and the midpoint is connected to an AC grid through the grid inductor Lg. Each leg consists of upper and lower arms, the arm inductor La and equivalent series impedance Ra. There are N cascaded submodules in each arm, where low-voltage battery packs and half bridge are embedded. The bypass switch of each submodule will be closed once a failure in this submodule happens. The power devices T_1 and T_2 are in complementary operation, which means the submodule cannot output negative voltages. Some researchers add DC-DC converter between battery and half bridge to reduce the current ripple of battery [18].

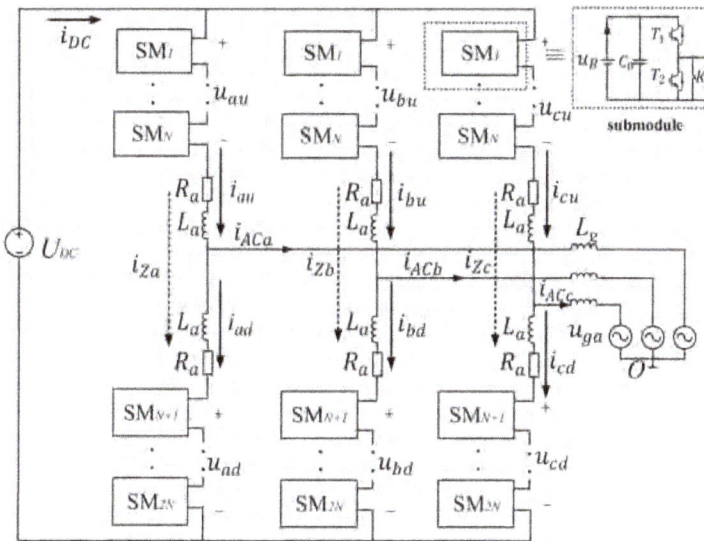

Figure 1. The configuration of the MMC-BESS and its submodule.

The driving signals of the power devices are generated by the carrier phase shift modulation in this paper. A schematic diagram of the carrier phase-shift modulation strategy is shown in Figure 2. N carrier signals for the submodules within the same arm have a $2\pi/N$ phase shift. The reference voltage for the upper and lower arms are opposite in phase. In this way, the output voltage of the converter has $2N + 1$ levels and its dominant harmonic component is $2N \cdot fSW$. Therefore, the harmonic performance is acceptable even at a low switching frequency.

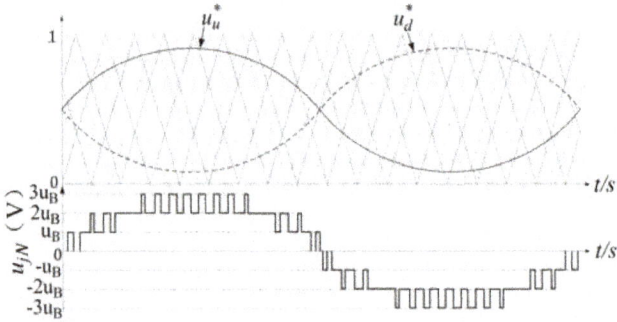

Figure 2. A diagram of the carrier phase-shift modulation (7 levels).

2.2. Power Flow Analysis

MMC-BESS is a three-port power converter system connected to the AC grid, the DC link, and the batteries. The power flow analysis is the basis of the control strategy design. The high-frequency component of the voltage and current is neglected in this paper for better analysis. In the following discussion, $j \in \{a, b, c\}$ represents different phases, $k \in \{u, d\}$ refers to the upper and lower arms of the same phase leg, and $i \in \{1, 2, \ldots, N\}$ represents the number of submodules located in one arm.

As described in Reference [14], the output voltage of submodule u_{jki} consists of three parts: the AC grid component u_{ACjki}, the DC circulating component u_{DCjki}, and the AC circulating component u_{Xjki}. Since most of the second harmonic current flows through the batteries, the second harmonic circulating current can be neglected in this system [19]. The AC components of the submodules output voltage (including u_{ACjki} and u_{Xjki}) in the same arm have the same phase angle. To keep the symmetrical system, the total upper arm voltage and the total lower arm voltage should be presented as follows:

$$
\begin{aligned}
u_{ju} &= \sum\nolimits_{i=1}^{N} u_{jui} = u_{DCj} + u_{Xj} - u_{ACj} \\
u_{jd} &= \sum\nolimits_{i=1}^{N} u_{jdi} = u_{DCj} + u_{Xj} + u_{ACj}
\end{aligned}
\tag{1}
$$

where u_{ACj} is the drive voltage of the AC current i_{ACj}. u_{DCj} is the DC drive voltage of the DC circulating current i_{DCj} and $u_{DCj} \approx U_{DC}/2$ when R_a is small enough. u_{Xj} is the drive voltage corresponding to the AC circulating current i_{Xj}. The arm currents i_{jk} are composed of these three components accordingly:

$$
\begin{aligned}
i_{ju} &= i_{DCj} + i_{Xj} + \frac{i_{ACj}}{2} \\
i_{jd} &= i_{DCj} + i_{Xj} - \frac{i_{ACj}}{2}
\end{aligned}
\tag{2}
$$

i_{DCj} and i_{Xj} compose the circulating current i_{Zj}, which is a common component in both the upper and lower arms:

$$
i_{Zj} = \frac{i_{ju} + i_{jd}}{2} = i_{DCj} + i_{Xj}
\tag{3}
$$

After applying the KVL method to the system in Figure 1, the following relationships are found:

$$
\begin{aligned}
L_a \frac{di_{Xj}}{dt} + i_{Xj} R_a &= -u_{Xj} \\
2 i_{DCj} R_a &= U_{DC} - 2 u_{DCj} \\
\left(\frac{L_a}{2} + L_g \right) \frac{di_{ACj}}{dt} + \frac{R_a}{2} i_{ACj} &= u_{gj} - u_{ACj}
\end{aligned}
\tag{4}
$$

Assuming that all submodules in the same arm can be regard as a single module, then the total output active power is equal to the total arm battery power. By multiplying the voltage components in Equation (1) and current components in Equation (2) one by one, all of the instantaneous active power and reactive powers can be found. Only the average active power is studied in this paper and the total arm battery power of the upper and lower arms, P_{Bju} and P_{Bjd}, result in the following equations:

$$p_{Bju} = -\frac{i_{DCj}U_{DC}}{2} - \frac{\hat{i}_{gj}\hat{u}_{ACj}}{2}\cos\theta + \left(\frac{\hat{i}_{Xj}\hat{u}_{ACj}}{2}\cos\phi_1 + \frac{\hat{i}_{gj}\hat{u}_{Xj}}{2}\cos\phi_2\right)$$

$$p_{Bjd} = -\frac{i_{DCj}U_{DC}}{2} - \frac{\hat{i}_{gj}\hat{u}_{ACj}}{2}\cos\theta - \left(\frac{\hat{i}_{Xj}\hat{u}_{ACj}}{2}\cos\phi_1 + \frac{\hat{i}_{gj}\hat{u}_{Xj}}{2}\cos\phi_2\right)$$

(5)

where θ and φ_2 are the phase angles of I_{ACj}, corresponding to U_{ACj} and U_{Xj}, respectively. φ_1 is the phase angle of I_{Xj} and U_{ACj}. The third term in Equation (5) is the AC circulating power yielded by the AC circulating current I_{Xj} and voltage U_{Xj}. Though I_{Xj} is usually considered the current generating power loss, it is employed to shift power between arms in the same phase.

The expressions of the battery pack power of each submodule in the upper and lower arms, P_{Bjui} and P_{Bjdi}, are similar to Equation (5). This results in the following relationships:

$$P_{Bjui} = -I_{DCj}U_{DCjui} - \frac{\hat{i}_{gj}\hat{u}_{ACjui}}{2}\cos\theta + \left(\frac{\hat{i}_{Xj}\hat{u}_{ACjui}}{2}\cos\phi_1 + \frac{\hat{i}_{gj}\hat{u}_{Xjui}}{2}\cos\phi_2\right)$$

$$P_{Bjdi} = -I_{DCj}U_{DCjdi} - \frac{\hat{i}_{gj}\hat{u}_{ACjdi}}{2}\cos\theta - \left(\frac{\hat{i}_{Xj}\hat{u}_{ACjdi}}{2}\cos\phi_1 + \frac{\hat{i}_{gj}\hat{u}_{Xjdi}}{2}\cos\phi_2\right)$$

(6)

Comparing Equation (5) with Equation (6), it is clear that when the magnitude of the three voltage components of U_{jki} is proportional to the corresponding components of the total arm output voltage U_{jk} with the factor k_i, the battery power P_{Bjki} is also proportional to the total arm battery power P_{Bjk} with k_i. Thus, the battery power of each submodule in an arm can be distributed by adjusting the output voltage ratio k_i.

Based on Equation (5), the total leg battery power is derived from the following:

$$p_{Bj} = p_{Bju} + p_{Bjd} = -p_{DCj} - p_{gj} = -i_{DCj}U_{DC} - \hat{i}_{gj}\hat{u}_{ACj}\cos\theta \tag{7}$$

To keep the grid currents balanced, each grid power P_{ACj} is made equal. Therefore, P_{Bj} can be changed by managing the DC circulating current I_{DCj}. The power transfer between the upper and lower arm batteries can be controlled by modifying the AC circulating current I_{Xj} and the individual battery power in the same arm can be controlled by adjusting the output voltage ratio k_i. In this way, the individual power control of each battery pack can be achieved.

3. SOC Balancing Control Strategy

During the operation of the BESS, the SOCs of the battery packs will gradually become unequal, which will decrease the capacity utilization efficiency of the batteries. Thus, SOC balancing control is essential. The definition of the SOC is given by the following equation:

$$f_{SOC} = \frac{\text{Storaged charges}}{\text{Nominal capacity}} \times 100\% \tag{8}$$

The SOC of each cell is estimated by the following equation:

$$f_{SOC(t)} = f_{SOC(t_0)} + \frac{1}{3600E_B}\int_{t_0}^{t} p_B(t)dt \tag{9}$$

where E_B is the battery nominal energy and $p_B(t)$ is the instantaneous battery power. E_B can be gained by multiplying battery voltage u_B and its capacity.

As shown in Equation (9), dynamic SOC is a first-order process with integral behavior and the changing rate is directly related to the battery power $p_B(t)$. Combining the previous power flow analysis, this paper proposes a three-level SOC balancing control strategy, including the phase legs SOC balancing control, upper and lower arms SOC balancing control, and individual submodules SOC balancing control. Define $\overline{SOC}_{jk}$ as the mean SOC value of all the battery submodules in the same arm; $\overline{SOC}_j$ as the average value of all battery packs' SOC in the same phase leg and $\overline{SOC}_{abc}$ as the average SOC for each phase leg. Besides, since the battery capacity inconsistency goes higher, adjusting power based on the battery capacity proportion of three-phase legs, capacity deviation between upper and lower's arm and capacity coefficient of submodule will directly balance the energy of all battery, thus improving the utilization of second-life batteries.

3.1. Phase-Leg Balancing

Influenced by different operation modes, the total battery power demand P_B^* is determined by the DC-link power P_{DC}^* and the AC grid power P_{AC}^*. The relationship between these three power demands is as follows:

$$P_B^* = -P_{AC}^* - P_{DC}^* \tag{10}$$

P_B^* is distributed to all battery packs. The basic power demand of each battery pack P_{Bav} is given by the following equation:

$$P_{Bav} = \frac{P_B^*}{n_{SM}} \tag{11}$$

where n_{SM} is the number of the submodules in the system (6N in the normal operation).

The total power reference of the phase leg P_{Bj}^*, as shown in Figure 3, is obtained by combining the difference regulated by proportional controller of the average of phase leg $\overline{SOC}_{abc}$ and the average of all battery packs' SOC in the same phase leg $\overline{SOC}_j$ with the adjustment power based on capacity.

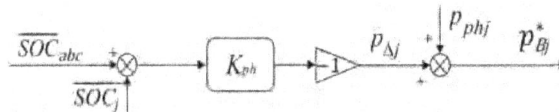

Figure 3. The phase-leg SOC balancing controller.

$$p_{Bj}^* = p_{phj} + p_{\Delta j} \tag{12}$$

where p_{phj} is the power regulations based on battery capacity proportion of three-phase legs

$$p_{phj} = \frac{\sum\limits_{i=1}^{N}(C_{jui} + C_{jdi})}{\frac{1}{3}\sum\limits_{i=1}^{N}(C_{Aui} + C_{Bui} + C_{Cui} + C_{Adi} + C_{Bdi} + C_{Cdi})} - 2NP_{Bav} \tag{13}$$

$p_{\Delta j}$ is generated via the P controller by

$$p_{\Delta j} = K_{ph}\left(\overline{SOC}_{abc} - \overline{SOC}_j\right) \tag{14}$$

According to reference power of each phase, DC circulating can be deduced,

$$i^*_{DCj} = \frac{\left(-p^*_{Bj} - P^*_{AC}/3\right)}{U_{DC}} \tag{15}$$

The circulating of each phase leg i_{Zj} can be obtained by adding upper arm current and lower arm within the same phase leg, as shown in Figure 4. The DC circulating current i_{DCj} of each leg is obtained through a low-pass filter whose cut-off frequency is less than 50 Hz, and a PI controller is employed to track i_{DCj}, and achieving SOC balance of the phase leg.

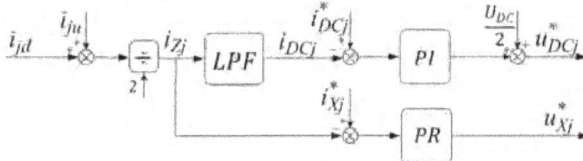

Figure 4. Block diagram for circulating current control.

3.2. Upper and Lower Arm Balancing

As shown in Equation (5), the AC circulating current I_{Xj} can convert the power $p_{\Delta armjud}$ between the upper and lower arms. The deviation in the power reference is obtained through proportional control for the SOC difference of the upper and lower arms and the different of power reference based on arm capacity, as follows:

$$p^*_{\Delta armjud} = K_{arm}\left(\overline{SOC}_{ju} - \overline{SOC}_{jd}\right) + \frac{1}{2}\left(p_{armju} - p_{armjd}\right) = p_{\Delta jud} + \frac{1}{2}\left(p_{armju} - p_{armjd}\right) \tag{16}$$

The deviation between p_{armju} and p_{armjd} is power reference based on battery capacity deviation between upper and lower's arm, the power transfer from upper and lower arm based on capacity is calculated as follow,

$$p_{armju} - p_{armjd} = \left(\frac{\sum\limits_{n=1}^{N} C_{jui}}{\frac{1}{2}\sum\limits_{n=1}^{N}\left(C_{jui} + C_{jdi}\right)} - \frac{\sum\limits_{n=1}^{N} C_{jdi}}{\frac{1}{2}\sum\limits_{n=1}^{N}\left(C_{jui} + C_{jdi}\right)}\right)Np_{Bav} \tag{17}$$

To prevent from DC-grid current distortion caused by SOC balancing control, the three-phase AC circulating currents should only be composed of positive and negative sequence components, as shown in Figure 5. The calculating method is described in detail in [20], and the magnitude and phase angle of the positive and negative sequence currents are derived from the given power to be shifted between the upper and lower arms. As shown in the lower part in Figure 4, a proportional resonant (PR) controller is employed to adjust I_{Xj}.

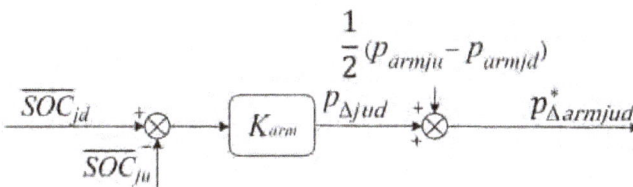

Figure 5. The upper and lower arms SOC balancing controller.

Equations (14) and (16) show that the SOC balancing rate is determined by the proportional coefficients K_{ph} and K_{arm}. However, the value of these coefficients has to be limited in case of over-modulation.

3.3. Submodule Balancing

The objective of the former two SOC balancing control methods is to generate equal average SOC of each arm. The SOC balancing of the submodules within an arm is implemented by adjusting the given power of each submodule p^*_{Bjki} is shown in Figure 6, as follows:

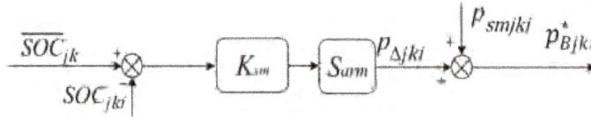

Figure 6. The individual submodule SOC balancing controller.

The variable S_{arm} is the power regulating direction. Since battery packs within an arm have the same current direction, the value of S_{arm} is -1 when the total arm battery power p^*_{Bjki} is less than zero. Namely, the batteries are in charge and the value of S_{arm} turns to 1 when p^*_{Bjki} is greater than zero.

$$p^*_{Bjki} = P_{smjki} + p_{\Delta jki} \tag{18}$$

where $p_{\Delta jki}$ is generated by a proportional controller as follows:

$$p_{\Delta jki} = S_{arm} K_{sm} \left(\overline{SOC}_{jk} - \overline{SOC}_{jki} \right) \tag{19}$$

p_{smjkii} is power regulations based on battery capacity coefficient of submodule within the same arm.

$$p_{smji} = \frac{C_{jui}}{\frac{1}{N} \sum\limits_{n=1}^{N} C_{jui}} N p_{Bav} \tag{20}$$

By multiplying k_i with the arm voltage reference u^*_{jk}, the total arm active power of each submodule can be obtained.

The power ratio factor k_i is calculated as

$$k_i = \frac{p^*_{Bjki}}{\sum_{i=1}^{N} p^*_{Bjki}} \tag{21}$$

Similarly, the coefficients need to be limited for avoiding over-modulation. Ignoring the voltage drop in the arm inductors and grid inductors, the limitation of k_i can be written as

$$K_{sm} \leq \left(\frac{U_{Bnor}}{m U_{Bmin}} - \frac{Cmax}{C_{avr}} \right) \frac{P_{Bav}}{\Delta SOC_{max}} \tag{22}$$

where ΔSOC_{max} is the maximal SOC difference among the SOCs of the battery packs and the corresponding average arm SOC, u_{Bmin} is the minimum battery pack voltage. m is rated modulation ratio, C_{avr} and C_{max} is average capacity and maximum capacity within an arm.

In this paper, the direct current control based on the dq axis is employed to control the AC grid's current. The general control structure of the MMC-BESS is shown in Figure 7.

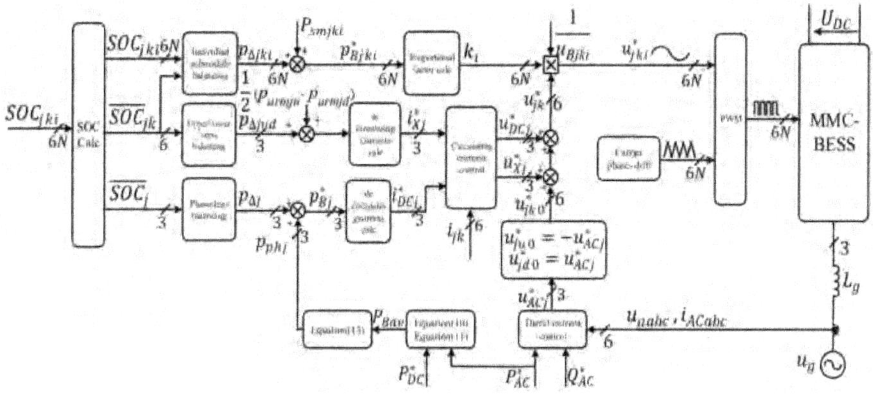

Figure 7. The general control structure of the MMC-BESS.

4. Simulation Results

To demonstrate the feasibility of the proposed SOC balancing control strategy under both normal operation and fault-tolerant operation, a simulation model based on the topology shown in Figure 1 was built in MATLAB/Simulink. Table 1 summarizes the parameters of the simulation model. The initial SOC values of the 24 battery modules are randomly set from 80.0% to 83.0%, and various capacity is preset as listed in Table 2.

Table 1. The parameters of the simulation system.

Item		Value
Output voltage (phase, peak)	$\hat{u}_g$	311 V
DC-link voltage	u_{DC}	750 V
Arm inductance	L	1 mH
Submodules in one arm	N	4
Equivalent series resistance	R	0.1 Ω
Grid inductance	L_g	1.5 mH
Submodule capacitance	C_0	2200 μF
Nominal battery voltage	u_B	250 V
Nominal battery capacity	C_B	10 Ah
Switching Frequency	f_w	5000 Hz

Table 2. The initial SOCs and capacity of the 24 battery modules.

Phase	Arm	Submodules SOC and Capacity(Ah)				Mean SOC/Capacity(Ah)	Mean SOC/Capacity(Ah)
		1	2	3	4		
a	u	80.7%/8	80.3%/8	82.9%/8	81.4%/8	81.325%/8	80.975%/8
	d	80.1%/8	81.2%/8	80.7%/8	80.5%/8	80.625%/8	
b	u	82.7%/10	80.0%/11	80.2%/13	81.6%/14	81.125%/12	81.388%/10
	d	80.8%/8	81.2%/8	82.6%/8	82.0%/8	81.65%/8	
c	u	81.8%/10	83.0%/10	82.2%/10	81.9%/10	82.225%/10	81.763%/12
	d	81.8%/14	81.1%/14	80.8%/14	81.5%/14	81.3%/14	

First of all, only three-level SOC balance is implemented, the DC link absorbs energy from the system and its reference power is kept at −37.5 kW. The AC grid conveys 93.3 kW to the system for 240 s. So, the battery is charged during this time. Then, adding the power regulation based on capacity proportion in the second time simulation, to verify the method proposed in this paper. The power configuration is the same as the first time. Finally, changing the AC power from 93.3 kW to −64.65 kW to test the strategy when the batteries are discharged.

Figures 8 and 9 reveal that the control strategy proposed in the paper have less influence on the AC output current and DC-link current. The global and local zoomed-in waveforms of the circulating currents are illustrated in Figure 10. It shows that the SOC balancing control generates a large circulating current at the begin of simulation for both 3-level SOC balance control and power adjustment based on capacity is working, then the circulating decrease gradually, finally it become stable. At the end of simulation, power adjustment based on capacity play great role in the circulating and adjustment from 3-level SOC balance control is little.

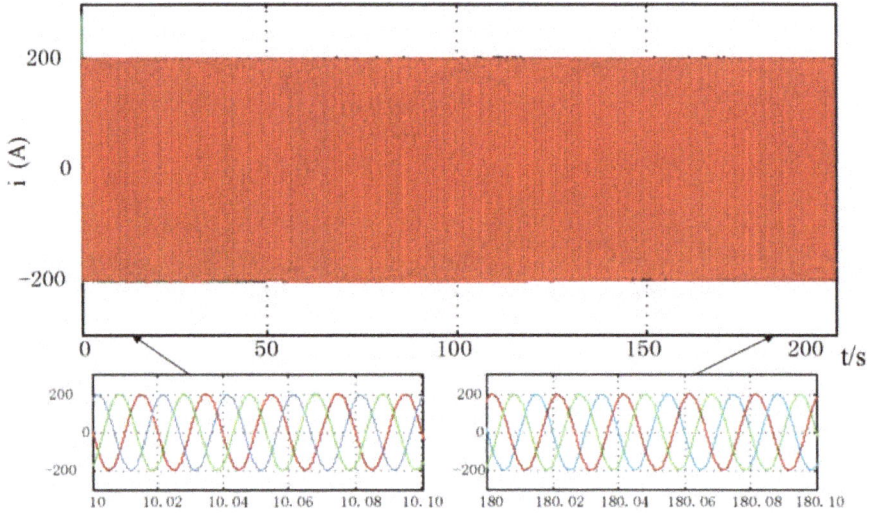

Figure 8. The AC grid output currents.

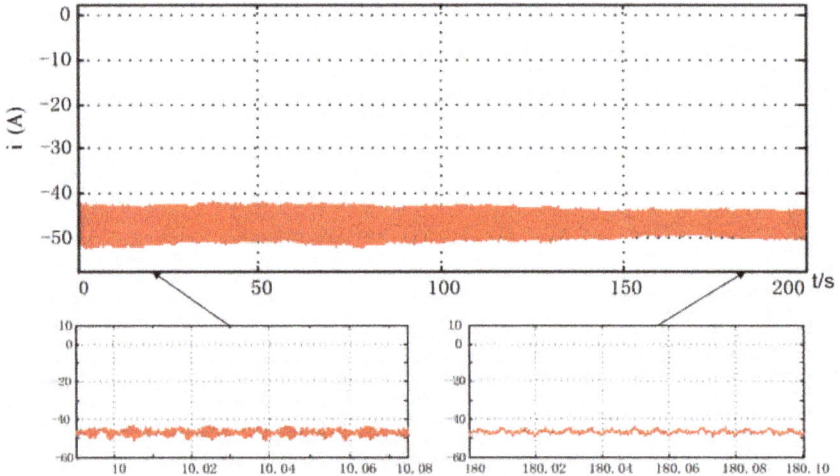

Figure 9. The DC-link current.

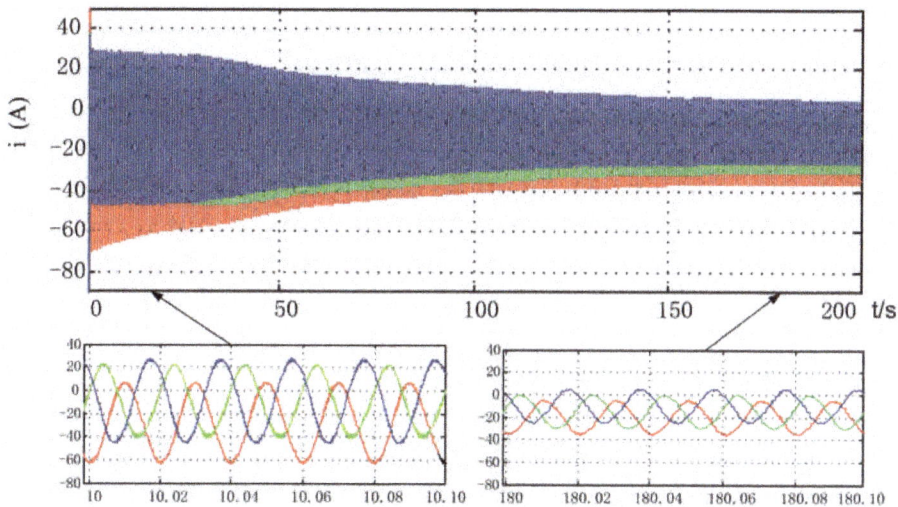

Figure 10. The circulating currents of the three-phase legs.

Figure 11 illustrates the simulation results of the 3-level SOC balancing control without power adjustment in charge mode. Figure 12 illustrates the simulation results of the 3-level SOC balancing control combined with power adjustment based on capacity in charge mode. Figure 13 reveals the simulation results of the 3-level SOC balancing control combined with power adjustment based on capacity in discharge mode.

In Figure 11, only the three-level SOC balance control strategy is implemented. The Figure 11a shows that the SOC of all battery almost converge at last, however, the convergence is poor for the batteries with various capacity. The maximum SOC difference of all battery reduce to 0.6%. Besides, in Figure 11b, the SOC difference of three-phase leg reduced from 0.45% to 0.2%. It is more obvious that the capacity has influence the SOC balance. In Figure 11c,d, the SOC difference of upper and lower arm is less than 0.001% in phase A with same capacity, but it is 0.4% in phase B with various capacity.

In Figure 12, power adjustment based on capacity is added to the simulation and the convergence of SOC gets better contrast with Figure 11. The maximum different SOC of all battery becomes 0.1% at last, and the different SOC of the three-phase leg is reduced to less than 0.01% in Figure 12b. The deviation of upper and lower arm has decreased to 0.05% in Figure 12d. The maximum SOC difference in the upper arm of phase B has also reduced, which is 0.05% less than 0.18% showed in Figure 11f. The obvious contrast of Figures 11 and 12 reveal that the three-level SOC balance control combine with power regulation related to capacity can balance the batteries with different capacity. Finally, the three-level SOC balancing and power adjustment based on capacity in discharge mode is simulated in Figure 13.

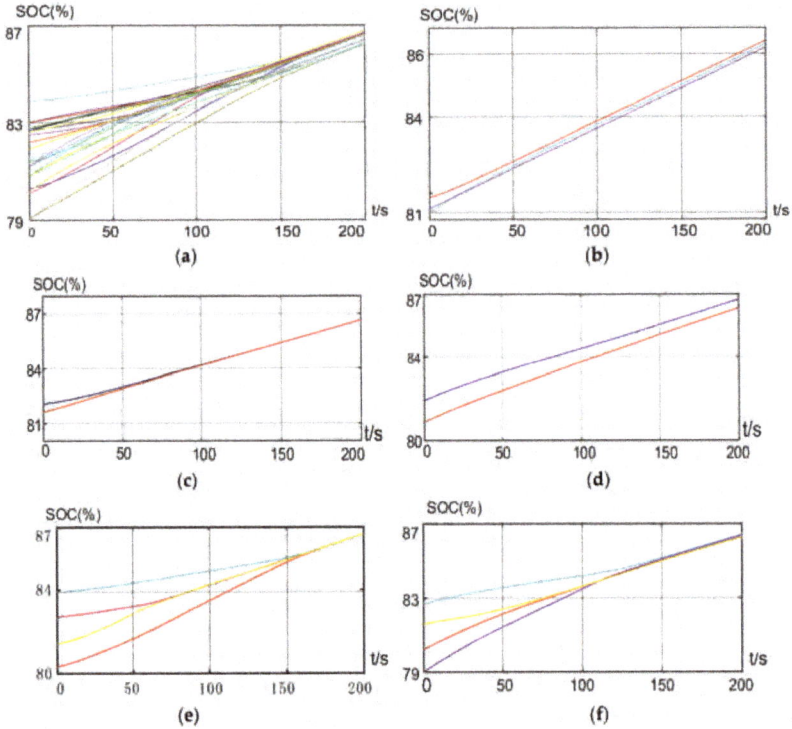

Figure 11. Three-level SOC balancing without power adjustment in charge mode. (**a**) The SOC of all 24 battery modules. (**b**) The SOC balancing among the three phase legs. (**c**) The SOC of upper and lower arm within phase A. (**d**) The SOC of upper and lower arm within phase B. (**e**) Submodule SOC of upper arm in phase A. (**f**) Submodule SOC of upper arm in phase B.

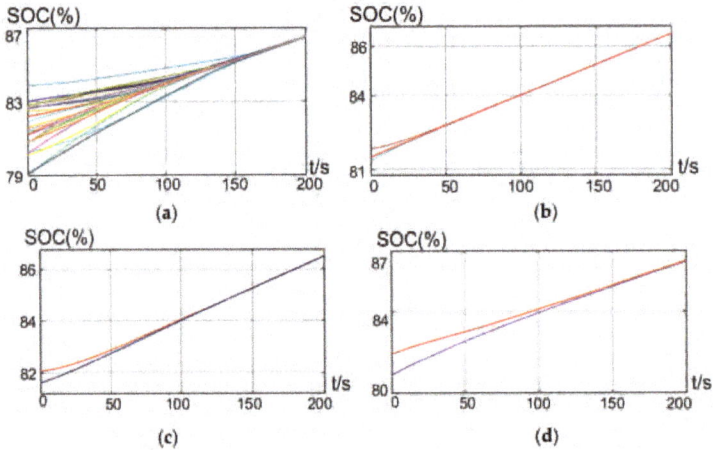

Figure 12. *Cont.*

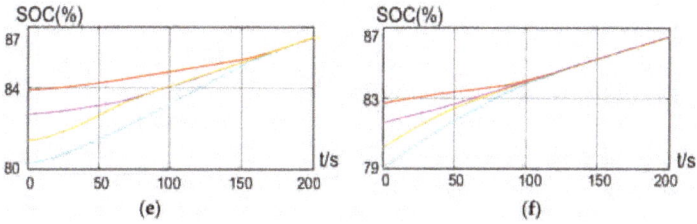

Figure 12. Three-level SOC balancing with power adjustment based on capacity in charge mode. (**a**) The SOC of all 24 battery modules. (**b**) The SOC balancing among the three phase legs. (**c**) The SOC of upper and lower arm in phase A. (**d**) The SOC of upper and lower arm in phase B. (**e**) Submodule SOC of upper arm in phase A. (**f**) Submodule SOC of upper arm in phase B.

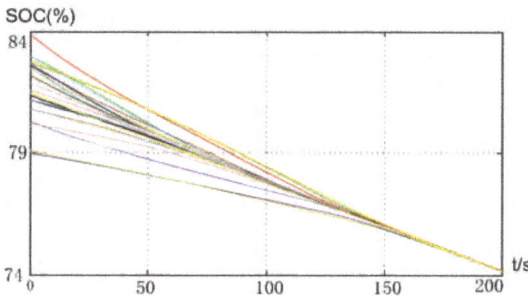

Figure 13. Three-level SOC balancing and power adjustment based on capacity in discharge mode.

5. Experimental Results

To verify the effectiveness of the proposed control strategy, a prototype was built in the lab as shown in Figure 14. The parameters of the experimental system are shown in Table 3. Due to the large number of submodules, a digital signal processor (DSP) and a field-programmable gate array (FPGA) are employed in this prototype. Since the foundation of the SOC balancing control strategy is the individual battery power control, this paper firstly validates the feasibility of the internal power flow control, then verifies the three-level SOC balance control strategy of MMC-BESS.

Figure 14. The prototype system.

Table 3. The parameters of the experimental system.

Item		Value
Output voltage (phase, peak)	$\hat{u}_g$	65 V
DC-link voltage	u_{DC}	180 V
Arm inductance	L	1 mH
Grid inductance	L_g	0.6 mH
Submodule capacitance	C_0	2200 μF
Nominal battery voltage	u_B	45 V
Switching Frequency	f_w	5000 Hz
Battery Pack Capacity		6 AH
Battery Type		ternary lithium battery
Battery Pack Grouping Method		12 series 2 parallel
Rated Battery Pack Voltage		44 V

Figures 15 and 16 show the waveforms of the grid current and the DC-link voltage and current. Figure 17 shows the output voltages of the converter, which has nine levels. Figure 18 illustrates the currents of the 9 battery submodules (au, ad, bu, bd, cu1, cu2, cu3, cu4, and cd). The average value of the battery current is analyzed by the scope and marked on the image. It can be seen that the average current of the battery submodules in phase a, b, and c decrease in turn and that the average current of the upper arm is greater than the lower arm in phase b. Meanwhile, the battery currents of the submodules within the upper arm of phase c decrease with an equal difference.

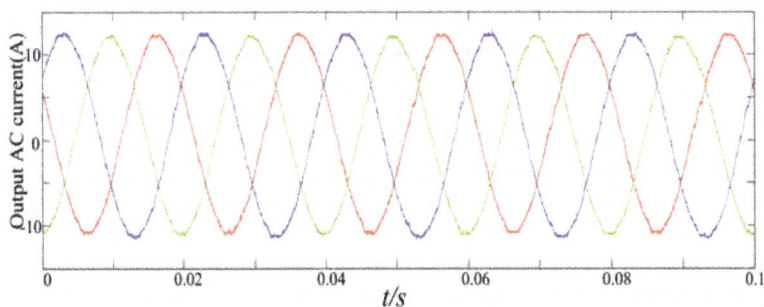

Figure 15. The experiment result of the AC output current.

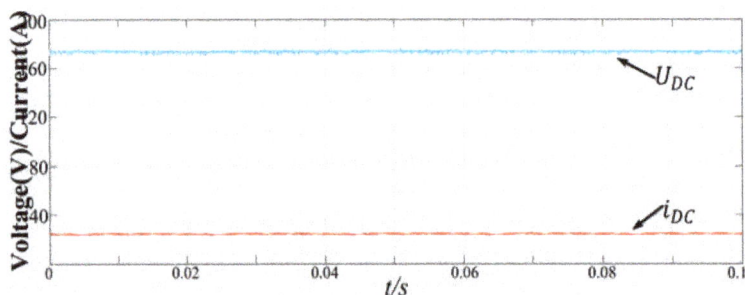

Figure 16. The waveforms of the DC-link voltage and current.

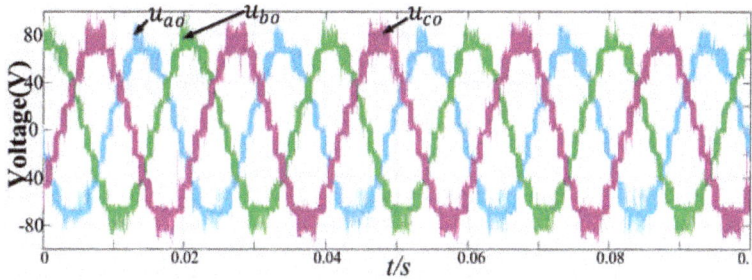

Figure 17. The waveforms of the converter output voltages.

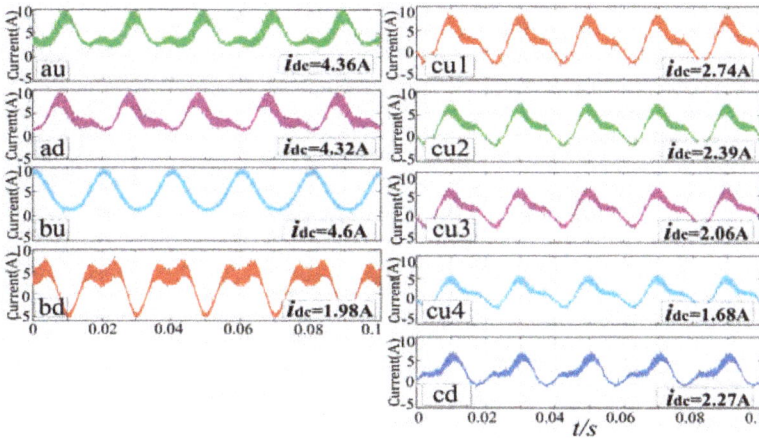

Figure 18. The currents of the battery modules.

Figure 19 illustrates the waveforms of three-phase circulating currents, which contain the DC and AC circulating components. These waveforms are measured at the same time in Figure 17. By comparing the phase angle of the converter output voltages and circulating currents of each phase, 90° and −90° is discovered in phase A and phase C, respectively. In this case, the AC circulating currents do not transfer power between the arms within the same phase leg. However, the circulating current in phase B has the opposite phase angle compared with the converter output voltage and thus, the circulating current transfers substantial active power from the lower arm to the upper arm of phase B. Since the total battery power reference of each phase leg has decreased from phase A to phase C, the DC components of the circulating currents decrease correspondingly.

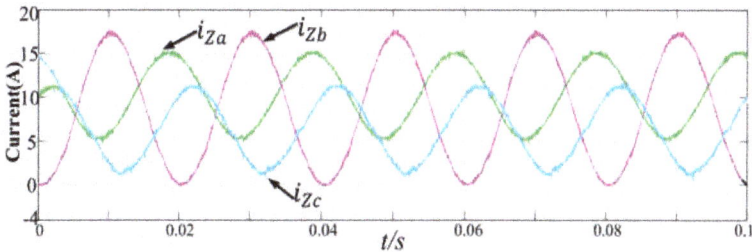

Figure 19. The waveforms of the three-phase circulating currents.

If the power of the DC link is greater than the AC side, the batteries will be charged, and vice versa. Since each battery module has an individual reference power, the total battery reference power of each phase and each arm will be different.

Figure 20 shows the experimental results of the SOC balancing control of the three-level MMC-BESS. Owing to the limitation of experimental conditions, the battery module capacity is basically the same, the results under various batteries capacities has been verified in the previous simulation. Figure 20a shows the SOC of all the battery modules in the system. The SOC difference of the battery decreases from 15.64% (at the beginning state) to 1.67% (after 40 min), which verifies the effectiveness of the proposed balancing strategy. Figure 20b,c represent the trend of the three-phase SOC and the bridge arm SOC, respectively. In a certain period, the three-phase and bridge arm SOC also tend to be consistent. The Coulomb integral method was used in this strategy for SOC estimation due to some inevitable error existing in the current sampling. However, there is still a little deviation in the SOC estimation, which will affect the SOC convergence results at the end of the equilibrium process. However, this experiment generally conforms to the theoretical expectation and verifies the correctness of the theory.

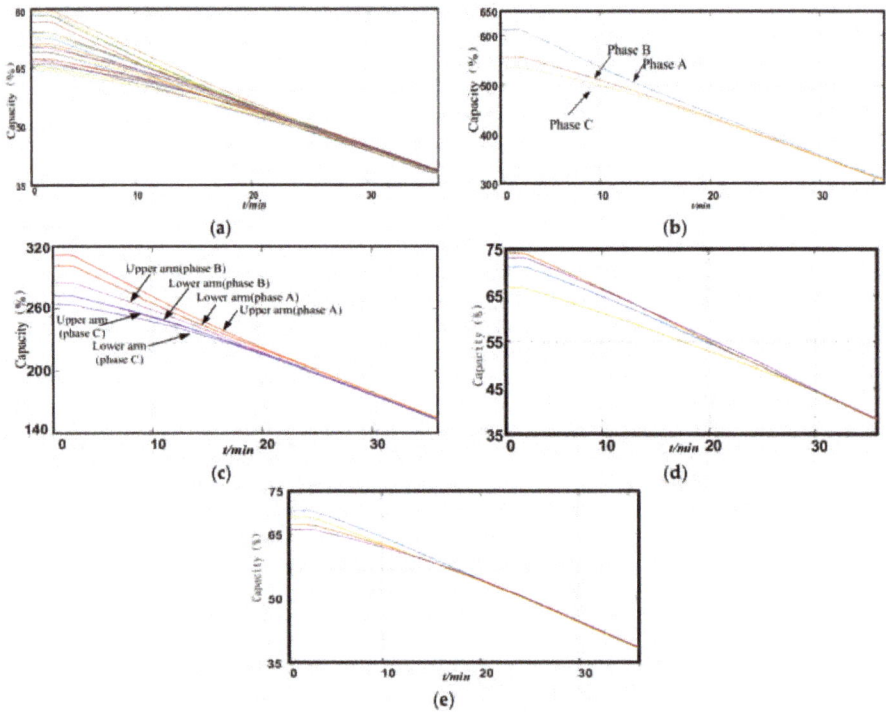

Figure 20. The experimental results of the three-level SOC balancing. (**a**) The SOC balancing of each module. (**b**) The interphase SOC balancing. (**c**) The intra-phase SOC balancing. (**d**) The SOC balancing of the upper arm of phase B. (**e**) The SOC balancing of the lower arm of phase B.

6. Conclusions

This paper focuses on the second-life battery used in MMC-BESS, and presents the shortcomings of both SOC and capacity inconsistency. The internal power flow in the AC grid, battery and DC link is analyzed. The results show that the fundamental component of both DC and AC circulating current can be used to adjust the total battery power of phase legs and arms respectively, and the power of submodules can be changed by adjusting the submodule output voltage. Based on the above-mentioned results, a three-level SOC balance control strategy is proposed: adjusting the power related to the capacity ratio of three-phase leg; considering the difference of the upper and lower leg's capacity, and the proportional capacity of the submodule collaborating on closed-loop control of the SOC to balance the SOC of MMC-BESS. Eventually, the batteries' SOC balance in MMC-BESS is achieved. Finally, the effectiveness and feasibility of the proposed methods are verified by results obtained from simulations and the experimental platform.

Acknowledgments: This work is financially supported by National Natural Science Foundation of China under Key Program 61633015.

Author Contributions: In this paper, author contributions are as follows: Hui Liang, Long Guo and Junhong Song conceived and designed the experiments; Hui Liang, Long Guo and Yong Yang performed the experiments; Hui Liang analyzed the data; Junhong Song, Hongfeng Qi and Weige Zhang contributed reagents/materials/analysis tools; Hui Liang, Junhong Song and Long Guo wrote the paper.

Conflicts of Interest: The authors declare no conflicts of interest.

Abbreviations

The following abbreviations are used in this manuscript:

MMC	Modular Multilevel Converter
SOC	State of Charge
BESS	Battery Energy Storage System
HVDC	High-voltage Direct Current
PWM	Pulse Width Modulation

References

1. Neubauer, J.; Pesaran, A. The ability of battery second use strategies to impact plug-in electric vehicle prices and serve utility energy storage applications. *J. Power Sources* **2011**, *196*, 10351–10358. [CrossRef]
2. Diao, W.; Jiang, J. Flexible Grouping for Enhanced Energy Utilization Efficiency in Battery Energy Storage Systems. *Energies* **2016**, *9*, 498. [CrossRef]
3. Diao, W.; Xue, N. Active battery cell equalization based on residual available energy maximization. *Applied Energy* **2018**, *15*, 690–698. [CrossRef]
4. Li, Q.; Liang, H. Modular Battery Energy Storage System Based on One Integrated Primary multi-Secondaries Transformer and its Independent Control Strategy. In Proceedings of the IEEE Transportation Electrification Conference and Expo, (ITEC Asia-Pacific), Harbin, China, 7–10 August 2017.
5. Song, J.; Zhang, W. Fault-Tolerant Control for a Flexible Group Battery Energy Storage System Based on Cascaded Multilevel Converters. *Energies* **2018**, *11*, 171. [CrossRef]
6. Maharjan, L.; Yamagishi, T. Active-Power Control of Individual Converter Cells for a Battery Energy Storage System Based on a Multilevel Cascade PWM Converter. *IEEE Trans. Power Electron.* **2012**, *27*, 1099–1107. [CrossRef]
7. Mehrasa, M.; Pouresmaeil, E.; Zabihi, S.; Vechiu, I.; Catalão, J.P. A multi-loop control technique for the stable operation of modular multilevel converters in HVDC transmission systems. *Int. J. Electr. Power Energy Syst.* **2018**, *96*, 194–207. [CrossRef]
8. Mehrasa, M.; Pouresmaeil, E.; Taheri, S.; Vechiu, I.; Catalão, J.P. Novel Control Strategy for Modular Multilevel Converters Based on Differential Flatness Theory. *IEEE J. Emerg. Sel. Top. Power Electron.* **2017**, *99*, 1–11. [CrossRef]
9. Mei, J.; Xiao, B.; Shen, K.; Tolbert, L.M.; Zheng, J.Y. Modular Multilevel Inverter with New Modulation Method and Its Application to Photovoltaic Grid-Connected Generator. *IEEE Trans. Power Electron.* **2013**, *28*, 5063–5073. [CrossRef]
10. Meshram, P.M.; Borghate, V.B. A Simplified Nearest Level Control (NLC) (NLC) Voltage Balancing Method for Modular Multilevel Converter (MMC). *IEEE Trans. Power Electron.* **2015**, *30*, 450–462. [CrossRef]
11. Vasiladiotis, M.; Rufer, A. Analysis and Control of Modular Multilevel Converters With Integrated Battery Energy Storage. *IEEE Trans. Power Electron.* **2015**, *30*, 163–175. [CrossRef]

12. Quraan, M.; Yeo, T. Design and Control of Modular Multilevel Converters for Battery Electric Vehicles. *IEEE Trans. Power Electron.* **2015**, *31*, 507–517. [CrossRef]

13. Zhang, L.; Gao, F. Interlinking modular multilevel converter of hybrid AC-DC distribution system with integrated battery energy storage. In Proceedings of the Energy Conversion Congress and Exposition, Montreal, QC, Canada, 20–24 September 2015; pp. 70–77.

14. Quraan, M.; Tricoli, P.; D'Arco, S.; Piegari, L. Efficiency Assessment of Modular Multilevel Converters for Battery Electric Vehicles. *IEEE Trans. Power Electron.* **2016**, *32*, 2041–2051. [CrossRef]

15. Diao, W.; Jiang, J.; Zhang, C.; Liang, H. Energy state of health estimation for battery packs based on the degradation and inconsistency. *Energy Procedia* **2017**, *142*, 6581–6591. [CrossRef]

16. Hillers, A.; Biela, J. Fault-tolerant operation of the modular multilevel converter in an energy storage system based on split batteries. In Proceedings of the European Conference on Power Electronics and Applications, Lappeenranta, Finland, 26–28 August 2014; pp. 1–8.

17. Chen, Q.; Li, R. Analysis and Fault Control of Hybrid Modular Multilevel Converter with Integrated Battery Energy Storage System. *IEEE J. Emrg. Sel. Top. Power Electron.* **2017**, *5*, 64–79. [CrossRef]

18. Ma, Y.J.; Lin, H.; Wang, Z.; Wang, T. Capacitor voltage balancing control of modular multilevel converters with energy storage system by using carrier phase-shifted modulation. In Proceedings of the IEEE Applied Power Electronics Conference and Exposition, Tampa, FL, USA, 26–30 March 2017; pp. 1821–1828.

19. Tu, Q.; Xu, Z. Mechanism analysis on the circulating current in modular multilevel converter based HVDC. *High Volt. Eng.* **2010**, *36*, 547–552. (In Chinese)

20. Soong, T.; Lehn, P. Internal Power Flow of a Modular Multilevel Converter with Distributed Energy Resources. *Emrg. Sel. Top. Power Electron. IEEE J.* **2014**, *2*, 1127–1138. [CrossRef]

MDPI

St. Alban-Anlage 66

4052 Basel

Switzerland

Tel. +41 61 683 77 34

Fax +41 61 302 89 18

www.mdpi.com

Energies Editorial Office

E-mail: energies@mdpi.com

www.mdpi.com/journal/energies

www.ingramcontent.com/pod-product-compliance
Lightning Source LLC
Chambersburg PA
CBHW051700210326
41597CB00032B/5316